CAMBRIDGE LIBRARY COLLECTION

Books of enduring scholarly value

Mathematical Sciences

From its pre-historic roots in simple counting to the algorithms powering modern desktop computers, from the genius of Archimedes to the genius of Einstein, advances in mathematical understanding and numerical techniques have been directly responsible for creating the modern world as we know it. This series will provide a library of the most influential publications and writers on mathematics in its broadest sense. As such, it will show not only the deep roots from which modern science and technology have grown, but also the astonishing breadth of application of mathematical techniques in the humanities and social sciences, and in everyday life.

The Collected Mathematical Papers

Arthur Cayley (1821-1895) was a key figure in the creation of modern algebra. He studied mathematics at Cambridge and published three papers while still an undergraduate. He then qualified as a lawyer and published about 250 mathematical papers during his fourteen years at the Bar. In 1863 he took a significant salary cut to become the first Sadleirian Professor of Pure Mathematics at Cambridge, where he continued to publish at a phenomenal rate on nearly every aspect of the subject, his most important work being in matrices, geometry and abstract groups. In 1882 he spent five months at Johns Hopkins University, and in 1883 became president of the British Association for the Advancement of Science. Publication of his Collected Papers - 967 papers in 13 volumes plus an index volume - began in 1889 and was completed after his death under the editorship of his successor in the Sadleirian Chair. This volume contains 89 papers mostly published between 1883 and 1889.

The Collected Mathematical Papers

VOLUME 12

ARTHUR CAYLEY

CAMBRIDGE
UNIVERSITY PRESS

CAMBRIDGE UNIVERSITY PRESS

Cambridge New York Melbourne Madrid Cape Town Singapore São Paolo Delhi

Published in the United States of America by Cambridge University Press, New York

www.cambridge.org
Information on this title: www.cambridge.org/9781108005043

This edition first published 1897
This digitally printed version 2009

ISBN 978-1-108-00504-3

MATHEMATICAL PAPERS.

London: C. J. CLAY and SONS,
CAMBRIDGE UNIVERSITY PRESS WAREHOUSE,
AVE MARIA LANE.
Glasgow: 263, ARGYLE STREET.

Leipzig: F. A. BROCKHAUS.
New York: THE MACMILLAN COMPANY.

THE COLLECTED

MATHEMATICAL PAPERS

OF

ARTHUR CAYLEY, Sc.D., F.R.S.,

LATE SADLERIAN PROFESSOR OF PURE MATHEMATICS IN THE UNIVERSITY OF CAMBRIDGE.

VOL. XII.

CAMBRIDGE:
AT THE UNIVERSITY PRESS.
1897.

CAMBRIDGE:
PRINTED BY J. AND C. F. CLAY,
AT THE UNIVERSITY PRESS.

ADVERTISEMENT.

THE present volume contains 89 papers, numbered 799 to 887, published for the most part in the years 1883 to 1889.

The Table for the twelve volumes is

Vol.	I.	Numbers	1	to	100,
,,	II.	,,	101	,,	158,
,,	III.	,,	159	,,	222,
,,	IV.	,,	223	,,	299,
,,	V.	,,	300	,,	383,
,,	VI.	,,	384	,,	416,
,,	VII.	,,	417	,,	485,
,,	VIII.	,,	486	,,	555,
,,	IX.	,,	556	,,	629,
,,	X.	,,	630	,,	705,
,,	XI.	,,	706	,,	798,
,,	XII.	,,	799	,,	887.

A. R. FORSYTH.

19 *May*, 1897.

CONTENTS.

[An Asterisk means that the paper is not printed in full.]

CLASSIFICATION.

799.

ON CURVILINEAR COORDINATES.

[From the *Quarterly Journal of Pure and Applied Mathematics*, vol. XIX. (1883), pp. 1—22.]

THE present memoir is based upon Mr Warren's "Exercises in Curvilinear and Normal Coordinates," *Camb. Phil. Trans.* t. XII. (1877), pp. 455—502, and has for a principal object the establishment of the six differential equations of the second order corresponding to his six equations for normal coordinates: but the notation is different; the results are more general, inasmuch as I use throughout general curvilinear coordinates instead of his normal coordinates; and as regards my six equations for general curvilinear coordinates, the terms containing differential coefficients of the first order are presented under a different form.

If the position of a point in space is determined by the rectangular coordinates x, y, z; then p, q, r being each of them a given function of x, y, z, we have conversely x, y, z, each of them a given function of p, q, r, which are thus in effect coordinates serving to determine the position of the point, and are called curvilinear coordinates.

But it is not in the first instance necessary to regard x, y, z as rectangular coordinates, or even as Cartesian coordinates at all; we are simply concerned with the two sets of variables x, y, z and p, q, r, each variable of the one set being a given function of the variables of the other set; and, in particular, the x, y, z are regarded as being each of them a given function of the p, q, r.

Except as regards the symbols ξ, η, ζ presently mentioned, the suffixes 1, 2, 3 refer to the variables p, q, r respectively, and are used to denote differentiations in regard to these variables, viz.

$$x_1,\ x_2,\ x_3,\ x_{11},\ x_{12},\ x_{22},\ \ldots$$

are written to denote

$$\frac{dx}{dp},\ \frac{dx}{dq},\ \frac{dx}{dr},\ \frac{d^2x}{dp^2},\ \frac{d^2x}{dp\,dq},\ \frac{d^2x}{dq^2},\ \&c.,$$

and so in other cases; in particular,

$$\begin{array}{lll} x_1,\ x_2,\ x_3 & \text{denote} & \frac{dx}{dp},\ \frac{dx}{dq},\ \frac{dx}{dr}, \\ y_1,\ y_2,\ y_3 & \text{,,} & \frac{dy}{dp},\ \frac{dy}{dq},\ \frac{dy}{dr}, \\ z_1,\ z_2,\ z_3 & \text{,,} & \frac{dz}{dp},\ \frac{dz}{dq},\ \frac{dz}{dr}. \end{array}$$

The minors formed with these differential coefficients are denoted by suffixed letters ξ, η, ζ, thus

$$\begin{array}{lllll} \xi_1,\ \xi_2,\ \xi_3 & \text{denote} & y_2z_3 - y_3z_2, & y_3z_1 - y_1z_3, & y_1z_2 - y_2z_1, \\ \eta_1,\ \eta_2,\ \eta_3 & \text{,,} & z_2x_3 - z_3x_2, & z_3x_1 - z_1x_3, & z_1x_2 - z_2x_1, \\ \zeta_1,\ \zeta_2,\ \zeta_3 & \text{,,} & x_2y_3 - x_3y_2, & x_3y_1 - x_1y_3, & x_1y_2 - x_2y_1, \end{array}$$

so that, as regards these letters ξ, η, ζ, the suffixes do *not* denote differentiations.

The determinant $\begin{vmatrix} x_1, & y_1, & z_1 \\ x_2, & y_2, & z_2 \\ x_3, & y_3, & z_3 \end{vmatrix}$ is put $= L$:

and the symbols (a, b, c, f, g, h), (A, B, C, F, G, H) are defined as follows:

$$\begin{array}{ll} \mathrm{a} = x_1^2 + y_1^2 + z_1^2, & A = \xi_1^2 + \eta_1^2 + \zeta_1^2, \\ \mathrm{b} = x_2^2 + y_2^2 + z_2^2, & B = \xi_2^2 + \eta_2^2 + \zeta_2^2, \\ \mathrm{c} = x_3^2 + y_3^2 + z_3^2, & C = \xi_3^2 + \eta_3^2 + \zeta_3^2, \\ \mathrm{f} = x_2x_3 + y_2y_3 + z_2z_3, & F = \xi_2\xi_3 + \eta_2\eta_3 + \zeta_2\zeta_3, \\ \mathrm{g} = x_3x_1 + y_3y_1 + z_3z_1, & G = \xi_3\xi_1 + \eta_3\eta_1 + \zeta_3\zeta_1, \\ \mathrm{h} = x_1x_2 + y_1y_2 + z_1z_2, & H = \xi_1\xi_2 + \eta_1\eta_2 + \zeta_1\zeta_2. \end{array}$$

We have then, further,

$$\begin{vmatrix} x_1, & y_1, & z_1 \\ x_2, & y_2, & z_2 \\ x_3, & y_3, & z_3 \end{vmatrix} = L, \qquad \begin{vmatrix} \mathrm{a}, & \mathrm{h}, & \mathrm{g} \\ \mathrm{h}, & \mathrm{b}, & \mathrm{f} \\ \mathrm{g}, & \mathrm{f}, & \mathrm{c} \end{vmatrix} = L^2,$$

$$\begin{vmatrix} \xi_1, & \eta_1, & \zeta_1 \\ \xi_2, & \eta_2, & \zeta_2 \\ \xi_3, & \eta_3, & \zeta_3 \end{vmatrix} = L^2, \qquad \begin{vmatrix} A, & H, & G \\ H, & B, & F \\ G, & F, & C \end{vmatrix} = \frac{1}{L^2},$$

$$(A, B, C, F, G, H) = (\mathrm{bc} - \mathrm{f}^2,\ \mathrm{ca} - \mathrm{g}^2,\ \mathrm{ab} - \mathrm{h}^2,\ \mathrm{gh} - \mathrm{af},\ \mathrm{hf} - \mathrm{bg},\ \mathrm{fg} - \mathrm{ch}),$$

$$L^2(\mathrm{a}, \mathrm{b}, \mathrm{c}, \mathrm{f}, \mathrm{g}, \mathrm{h}) = (BC - F^2,\ CA - G^2,\ AB - H^2,\ GH - AF,\ HF - BG,\ FG - CH),$$

which equations are at once proved, and are fundamental ones in the theory.

It is convenient to add that we have

$$\left(\begin{array}{ccc} \frac{dp}{dx}, & \frac{dp}{dy}, & \frac{dp}{dz} \\ \frac{dq}{dx}, & \frac{dq}{dy}, & \frac{dq}{dz} \\ \frac{dr}{dx}, & \frac{dr}{dy}, & \frac{dr}{dz} \end{array}\right) = \frac{1}{L}\left(\begin{array}{ccc} \xi_1, & \eta_1, & \zeta_1 \\ \xi_2, & \eta_2, & \zeta_2 \\ \xi_3, & \eta_3, & \zeta_3 \end{array}\right),$$

that is, $\frac{dp}{dx} = \frac{1}{L}\xi_1$, &c.; and, further,

$$dx^2 + dy^2 + dz^2 = \quad (\mathrm{a},\ \mathrm{b},\ \mathrm{c},\ \mathrm{f},\ \mathrm{g},\ \mathrm{h}\between dp,\ dq,\ dr)^2,$$

$$dp^2 + dq^2 + dr^2 = \frac{1}{L^2}(A,\ B,\ C,\ F,\ G,\ H\between dx,\ dy,\ dz)^2.$$

Differentiating the values of a, b, c, f, g, h, we have $\tfrac{1}{2}\mathrm{a}_1 = x_1x_{11} + y_1y_{11} + z_1z_{11}$, which may be written in the abbreviated form $\tfrac{1}{2}\mathrm{a}_1 = 1.11$; similarly

$$\mathrm{f}_1 = x_2x_{13} + y_2y_{13} + z_2z_{13} + x_3x_{12} + y_3y_{12} + z_3z_{12},$$

which in like manner may be written $\mathrm{f}_1 = 2.13 + 3.12$, and so in other cases; observe that, in the duad part of any symbol, the order of the numbers is immaterial, $2.13 = 2.31$. The whole system of equations is

$$\begin{array}{lll} \tfrac{1}{2}\mathrm{a}_1 = 1.11 & ,\ \tfrac{1}{2}\mathrm{a}_2 = 1.12 & ,\ \tfrac{1}{2}\mathrm{a}_3 = 1.13 \quad , \\ \tfrac{1}{2}\mathrm{b}_1 = 2.12 & ,\ \tfrac{1}{2}\mathrm{b}_2 = 2.22 & ,\ \tfrac{1}{2}\mathrm{b}_3 = 2.23 \quad , \\ \tfrac{1}{2}\mathrm{c}_1 = 3.13 & ,\ \tfrac{1}{2}\mathrm{c}_2 = 3.23 & ,\ \tfrac{1}{2}\mathrm{c}_3 = 3.33 \quad , \end{array}$$

$$\begin{array}{lll} \mathrm{f}_1 = 2.13 + 3.12, & \mathrm{f}_2 = 2.23 + 3.22, & \mathrm{f}_3 = 2.33 + 3.23, \\ \mathrm{g}_1 = 3.11 + 1.13, & \mathrm{g}_2 = 3.12 + 1.23, & \mathrm{g}_3 = 3.13 + 1.33, \\ \mathrm{h}_1 = 1.12 + 2.11, & \mathrm{h}_2 = 1.22 + 2.12, & \mathrm{h}_3 = 1.23 + 2.13. \end{array}$$

These may also be written

$$\begin{array}{lll} 1.11 = \tfrac{1}{2}\mathrm{a}_1 & ,\ 1.22 = \mathrm{h}_2 - \tfrac{1}{2}\mathrm{b}_1 & ,\ 1.33 = \mathrm{g}_3 - \tfrac{1}{2}\mathrm{c}_1 \quad , \\ 2.11 = \mathrm{h}_1 - \tfrac{1}{2}\mathrm{a}_2 & ,\ 2.22 = \tfrac{1}{2}\mathrm{b}_2 & ,\ 2.33 = \mathrm{f}_3 - \tfrac{1}{2}\mathrm{c}_2 \quad , \\ 3.11 = \mathrm{g}_1 - \tfrac{1}{2}\mathrm{a}_3 & ,\ 3.22 = \mathrm{f}_2 - \tfrac{1}{2}\mathrm{b}_3 & ,\ 3.33 = \tfrac{1}{2}\mathrm{c}_3 \quad , \\ 1.23 = \tfrac{1}{2}(-\mathrm{f}_1 + \mathrm{g}_2 + \mathrm{h}_3), & 1.31 = \tfrac{1}{2}\mathrm{a}_3 & ,\ 1.12 = \tfrac{1}{2}\mathrm{a}_2 \quad , \\ 2.23 = \tfrac{1}{2}\mathrm{b}_3 & ,\ 2.31 = \tfrac{1}{2}(\mathrm{f}_1 - \mathrm{g}_2 + \mathrm{h}_3), & 2.12 = \tfrac{1}{2}\mathrm{b}_1 \quad , \\ 3.23 = \tfrac{1}{2}\mathrm{c}_2 & ,\ 3.31 = \tfrac{1}{2}\mathrm{c}_1 & ,\ 3.12 = \tfrac{1}{2}(\mathrm{f}_1 + \mathrm{g}_2 - \mathrm{h}_3). \end{array}$$

It is to be observed that we can, from each system of three equations, express a set of second differential coefficients of the x, y, z in terms of the first differential coefficients of the a, b, c, f, g, h; thus the three equations containing 11, written at length, are

$$\begin{aligned} x_1 \,.\, x_{11} + y_1 \,.\, y_{11} + z_1 \,.\, z_{11} &= \tfrac{1}{2}\mathrm{a}_1, \\ x_2 \;\text{,,}\; + y_2 \;\text{,,}\; + z_2 \;\text{,,}\; &= \mathrm{h}_1 - \tfrac{1}{2}\mathrm{a}_2, \\ x_3 \;\text{,,}\; + y_3 \;\text{,,}\; + z_3 \;\text{,,}\; &= \mathrm{g}_1 - \tfrac{1}{2}\mathrm{a}_3, \end{aligned}$$

three linear equations for the determination of x_{11}, y_{11}, z_{11}; hence for x_{11} we have

$$x_{11} \begin{vmatrix} x_1, & y_1, & z_1 \\ x_2, & y_2, & z_2 \\ x_3, & y_3, & z_3 \end{vmatrix} = \tfrac{1}{2}\mathrm{a}_1 (y_2 z_3 - y_3 z_2) + (\mathrm{h}_1 - \tfrac{1}{2}\mathrm{a}_2)(y_3 z_1 - y_1 z_3) + (\mathrm{g}_1 - \tfrac{1}{2}\mathrm{a}_3)(y_1 z_2 - y_2 z_1),$$

or, what is the same thing,

$$L x_{11} = \tfrac{1}{2}\mathrm{a}_1 \,.\, \xi_1 + (\mathrm{h}_1 - \tfrac{1}{2}\mathrm{a}_2)\,\xi_2 + (\mathrm{g}_1 - \tfrac{1}{2}\mathrm{a}_3)\,\xi_3,$$

and so for y_{11}, z_{11}; or if (as in the sequel) we desire the value of a linear function $\alpha x_{11} + \beta y_{11} + \gamma z_{11}$, calling this for a moment $\square$, we join to the foregoing a new equation

$$\alpha x_{11} + \beta y_{11} + \gamma z_{11} = \square,$$

and then, eliminating the three quantities, we have

$$\begin{vmatrix} \alpha, & \beta, & \gamma, & \square \\ x_1, & y_1, & z_1, & \tfrac{1}{2}\mathrm{a}_1 \\ x_2, & y_2, & z_2, & \mathrm{h}_1 - \tfrac{1}{2}\mathrm{a}_2 \\ x_3, & y_3, & z_3, & \mathrm{g}_1 - \tfrac{1}{2}\mathrm{a}_3 \end{vmatrix} = 0,$$

giving $L\square$ as a linear function of $\tfrac{1}{2}\mathrm{a}_1$, $\mathrm{h}_1 - \tfrac{1}{2}\mathrm{a}_2$, $\mathrm{g}_1 - \tfrac{1}{2}\mathrm{a}_3$.

We can in like manner form the expressions for the second differential coefficients of the a, b, c, f, g, h: these will of course contain third differential coefficients of the x, y, z.

Writing down only what is wanted, we have

$$\begin{aligned} \tfrac{1}{2}\mathrm{a}_{22} &= 12 \,.\, 12 + 1 \,.\, 122, \\ \tfrac{1}{2}\mathrm{b}_{11} &= 12 \,.\, 12 + 2 \,.\, 112, \\ \mathrm{h}_{12} &= 12 \,.\, 12 + 11 \,.\, 22 + 1 \,.\, 122 + 2 \,.\, 112, \end{aligned}$$

where of course $12 \,.\, 12$ denotes $x^2_{12} + y^2_{12} + z^2_{12}$, $1 \,.\, 122$ denotes $x_1 \,.\, x_{122} + y_1 \,.\, y_{122} + z_1 \,.\, z_{122}$, and so in other cases: it follows that

$$\tfrac{1}{2}(\mathrm{a}_{22} + \mathrm{b}_{11} - 2\mathrm{h}_{12}) = 12 \,.\, 12 - 11 \,.\, 22,$$

so that the third differential coefficients of x, y, z, which enter into the expression of $\mathrm{a}_{22} + \mathrm{b}_{11} - 2\mathrm{h}_{12}$ destroy each other, and this combination contains really only second differential coefficients of x, y, z.

Similarly

$$\begin{aligned} f_{31} &= 31.23+33.12+\quad 3.123 \qquad\qquad +2.133,\\ g_{23} &= 31.23+33.12+\quad 3.123+1.233\\ c_{12} &= 2.31.23 \qquad\quad +2.3.123\\ h_{33} &= 2.31.23 \qquad\qquad\qquad +1.233+2.133, \end{aligned}$$

and thence

$$f_{31}+g_{23}-c_{12}-h_{33}=-2\,(31.23-33.12),$$

so that here again we have a combination containing only second differential coefficients of x, y, z.

There are thus, in all, the six combinations

$$\begin{aligned} &b_{33}+c_{22}-2f_{23} \ldots\ldots\ldots\ldots (\mathfrak{A}),\\ &c_{11}+a_{33}-2g_{31} \ldots\ldots\ldots\ldots (\mathfrak{B}),\\ &a_{22}+b_{11}-2h_{12} \ldots\ldots\ldots\ldots (\mathfrak{C}),\\ &g_{12}+h_{31}-a_{23}-f_{11} \ldots\ldots\ldots\ldots (\mathfrak{F}),\\ &h_{23}+f_{12}-b_{31}-g_{22} \ldots\ldots\ldots\ldots (\mathfrak{G}),\\ &f_{31}+g_{23}-c_{12}-h_{33} \ldots\ldots\ldots\ldots (\mathfrak{H}), \end{aligned}$$

each really containing only second differential coefficients of x, y, z; and we thus understand how each of these combinations may be expressible in terms of the first differential coefficients of (a, b, c, f, g, h). We have, in fact, for thus expressing these combinations the six equations called ($\mathfrak{A}$), ($\mathfrak{B}$), ($\mathfrak{C}$), ($\mathfrak{F}$), ($\mathfrak{G}$), ($\mathfrak{H}$) about to be obtained, and which are the generalisations of Warren's six equations for normal coordinates.

I consider the several determinants of the form

$$\begin{vmatrix} x_{11}, & y_{11}, & z_{11} \\ x_1, & y_1, & z_1 \\ x_2, & y_2, & z_2 \end{vmatrix},$$

in all 18, since the suffixes for the top row may be 11, 22, 33, 23, 31, 12, and those for the second and third rows 2, 3; 3, 1; or 1, 2. Each determinant is a linear function of second differential coefficients of x, y, z (thus the determinant written down is $=\xi_3x_{11}+\eta_3y_{11}+\zeta_3z_{11}$), and as such it can, by what precedes, be expressed by means of the first differential coefficients of the a, b, c, f, g, h. Thus, if the determinant above written down be called $\square$, writing ξ_3, η_3, ζ_3 for α, β, γ, we have

$$\begin{vmatrix} \xi_3, & \eta_3, & \zeta_3, & \square \\ x_1, & y_1, & z_1, & \tfrac{1}{2}a_1 \\ x_2, & y_2, & z_2, & h_1-\tfrac{1}{2}a_2 \\ x_3, & y_3, & z_3, & g_1-\tfrac{1}{2}a_3 \end{vmatrix}=0,$$

that is,

$$L.\square=-\begin{vmatrix} & \xi_3, & \eta_3, & \zeta_3 \\ \tfrac{1}{2}a_1, & x_1, & y_1, & z_1 \\ h_1-\tfrac{1}{2}a_2, & x_2, & y_2, & z_2 \\ g_1-\tfrac{1}{2}a_3, & x_3, & y_3, & z_3 \end{vmatrix},$$

where the right-hand side is

$$\begin{aligned} = \quad & \tfrac{1}{2}\mathrm{a}_1 \,.\, \xi_1\xi_3 + \eta_1\eta_3 + \zeta_1\zeta_3 \\ & -(\mathrm{h}_1 - \tfrac{1}{2}\mathrm{a}_2)\,.-(\xi_2\xi_3 + \eta_2\eta_3 + \zeta_2\zeta_3) \\ & +(\mathrm{g}_1 - \tfrac{1}{2}\mathrm{a}_3)\,.\,\xi_3^2 + \eta_3^2 + \zeta_3^2, \end{aligned}$$

which is

$$= G\,.\,\tfrac{1}{2}\mathrm{a}_1 + F(\mathrm{h}_1 - \tfrac{1}{2}\mathrm{a}_2) + C(\mathrm{g}_1 - \tfrac{1}{2}\mathrm{a}_3),$$

or, as this might be written,

$$= (G,\ F,\ C \between \mathrm{a}_1,\ \mathrm{h}_1,\ \mathrm{g}_1) - \tfrac{1}{2}(G,\ F,\ C \between \mathrm{a}_1,\ \mathrm{a}_2,\ \mathrm{a}_3).$$

Retaining the former form, the result is

$$L\,.\,\square = G\,.\,\tfrac{1}{2}\mathrm{a}_1 + F(\mathrm{h}_1 - \tfrac{1}{2}\mathrm{a}_2) + C(\mathrm{g}_1 - \tfrac{1}{2}\mathrm{a}_3),$$

and it is now very easy to write down the complete system of the 18 equations; viz. if the determinant above written down be called 11.1.2, and so in other cases, then we have

	2.3	A	H	G
	3.1	H	B	F
	1.2	G	F	C
L.11	„ =	$\tfrac{1}{2}\mathrm{a}_1$	$\mathrm{h}_1 - \tfrac{1}{2}\mathrm{a}_2$	$\mathrm{g}_1 - \tfrac{1}{2}\mathrm{a}_3$
„ 22	„ =	$\mathrm{h}_2 - \tfrac{1}{2}\mathrm{b}_1$	$\tfrac{1}{2}\mathrm{b}_2$	$\mathrm{f}_2 - \tfrac{1}{2}\mathrm{b}_3$
„ 33	„ =	$\mathrm{g}_3 - \tfrac{1}{2}\mathrm{c}_1$	$\mathrm{f}_3 - \tfrac{1}{2}\mathrm{c}_2$	$\tfrac{1}{2}\mathrm{c}_3$
„ 23	„ =	$\tfrac{1}{2}(-\mathrm{f}_1 + \mathrm{g}_2 + \mathrm{h}_3)$	$\tfrac{1}{2}\mathrm{b}_3$	$\tfrac{1}{2}\mathrm{c}_2$
„ 31	„ =	$\tfrac{1}{2}\mathrm{a}_3$	$\tfrac{1}{2}(\mathrm{f}_1 - \mathrm{g}_2 + \mathrm{h}_3)$	$\tfrac{1}{2}\mathrm{c}_1$
„ 12	„ =	$\tfrac{1}{2}\mathrm{a}_2$	$\tfrac{1}{2}\mathrm{b}_1$	$\tfrac{1}{2}(\mathrm{f}_1 + \mathrm{g}_2 - \mathrm{h}_3)$:

read for instance

$$L\,.\,11\,.\,1\,.\,2 = G\,.\,\tfrac{1}{2}\mathrm{a}_1 + F(\mathrm{h}_1 - \tfrac{1}{2}\mathrm{a}_2) + C(\mathrm{g}_1 - \tfrac{1}{2}\mathrm{a}_3),$$

the equation obtained above.

There are eighteen functions as shown by the following diagram:

		2.3	3.1	1.2
(22) (33) − (23)^2	2.3			
(33) (11) − (31)^2	3.1			
(11) (22) − (12)^2	1.2			
(31) (12) − (11) (23)	2.3			
(12) (23) − (22) (31)	3.1			
(23) (31) − (33) (12)	1.2,			

viz. in any line of the diagram the bracketed duads may belong to any one at pleasure, but all to the same, of the three pairs 2.3, 3.1, 1.2; thus the first line might be

$$(22.3.1)(33.3.1)-(23.3.1)^2,$$

or, instead of the 3.1, we might have 2.3 or 1.2. But of the 18 functions I distinguish 6, viz. those in which for the six lines respectively the pairs are 2.3, 3.1, 1.2, 2.3, 3.1, 1.2, as shown in the diagram. Each of these six functions can be obtained under two different forms, and by equating these we have the equations ($\mathfrak{A}$), ($\mathfrak{B}$), ($\mathfrak{C}$), ($\mathfrak{F}$), ($\mathfrak{G}$), ($\mathfrak{H}$), before referred to; thus ($\mathfrak{C}$) is obtained by equating two different forms of the function $(11.1.2)(22.1.2)-(12.1.2)^2$; and ($\mathfrak{H}$) is obtained by equating two different forms of the function $(23.1.2)(31.1.2)-(33.1.2)(12.1.2)$.

The determinants 11.1.2, &c., may be denoted by accented letters a, b, c, f, g, h, as follows:

$$\left|\begin{array}{cccccc|l} x_{11}, & x_{22}, & x_{33}, & x_{23}, & x_{31}, & x_{12} & x_2,\ x_3 \\ y_{11}, & y_{22}, & y_{33}, & y_{23}, & y_{31}, & y_{12} & y_2,\ y_3 \\ z_{11}, & z_{22}, & z_{33}, & z_{23}, & z_{31}, & z_{12} & z_2,\ z_3 \end{array}\right. = \mathrm{a}',\ \mathrm{b}',\ \mathrm{c}',\ \mathrm{f}',\ \mathrm{g}',\ \mathrm{h}',$$

$$\left|\begin{array}{cc|l} ,, & ,, & x_3,\ x_1 \\ & & y_3,\ y_1 \\ & & z_3,\ z_1 \end{array}\right. = \mathrm{a}'',\ \mathrm{b}'',\ \mathrm{c}'',\ \mathrm{f}'',\ \mathrm{g}'',\ \mathrm{h}'',$$

$$\left|\begin{array}{cc|l} ,, & ,, & x_1,\ x_2 \\ & & y_1,\ y_2 \\ & & z_1,\ z_2 \end{array}\right. = \mathrm{a}''',\ \mathrm{b}''',\ \mathrm{c}''',\ \mathrm{f}''',\ \mathrm{g}''',\ \mathrm{h}''';$$

viz.

$$\left|\begin{array}{ccc} x_{11}, & x_2, & x_3 \\ y_{11}, & y_2, & y_3 \\ z_{11}, & z_2, & z_3 \end{array}\right| = \mathrm{a}',\ \&\mathrm{c}.$$

In this notation, ($\mathfrak{C}$) is obtained by equating two values of $\mathrm{a}'''\mathrm{b}'''-(\mathrm{h}''')^2$, and ($\mathfrak{H}$) by equating two values of $\mathrm{f}'''\mathrm{g}'''-\mathrm{c}'''\mathrm{h}'''$.

The forms which I call the second are those given by the immediate substitution of the foregoing values of (11.1.2), &c.; to obtain the first forms, I proceed to calculate those of ($\mathfrak{C}$) and ($\mathfrak{H}$).

Forming by the ordinary rule the product of the determinants (11.1.2) and (22.1.2), which are

$$\left|\begin{array}{ccc} x_{11}, & y_{11}, & z_{11} \\ x_1, & y_1, & z_1 \\ x_2, & y_2, & z_2 \end{array}\right| \text{ and } \left|\begin{array}{ccc} x_{22}, & y_{22}, & z_{22} \\ x_1, & y_1, & z_1 \\ x_2, & y_2, & z_2 \end{array}\right|,$$

this is
$$\begin{vmatrix} 11.22, & 1.11, & 2.11 \\ 1.22, & 1.1\,, & 1.2 \\ 2.22, & 1.2\,, & 2.2 \end{vmatrix},$$

where 11 . 22 denotes $x_{11}x_{22}+y_{11}y_{22}+z_{11}z_{22}$, and the like for the other symbols. In like manner, the square of the determinant (12 . 1 . 2), that is, of
$$\begin{vmatrix} x_{12}, & y_{12}, & z_{12} \\ x_1\,, & y_1\,, & z_1 \\ x_2\,, & y_2\,, & z_2 \end{vmatrix},$$
is
$$\begin{vmatrix} 12.12, & 1.12, & 2.12 \\ 1.12, & 1.1\,, & 1.2 \\ 2.12, & 2.1\,, & 2.2 \end{vmatrix},$$

or, observing that in the two resulting determinants the terms 11 . 22 and 12 . 12 are multiplied by the same factor, the expression for the difference gives
$$(11.1.2)(22.1.2)-(12.1.2)^2$$
$$=(11.22-12.12)\begin{vmatrix} 1.1, & 1.2 \\ 1.2, & 2.2 \end{vmatrix}+\begin{vmatrix} 0\,, & 1.11, & 2.11 \\ 1.22, & 1.1\,, & 1.2 \\ 2.22, & 1.2\,, & 2.2 \end{vmatrix}-\begin{vmatrix} 0\,, & 1.12, & 2.12 \\ 1.12, & 1.1\,, & 1.2 \\ 2.12, & 1.2\,, & 2.2 \end{vmatrix},$$

containing 11 . 22 − 12 . 12, which, by what precedes, is
$$=-\tfrac{1}{2}(a_{22}+b_{11}-2h_{12});$$

the other terms are also known, viz. the whole value is
$$=-\tfrac{1}{2}(a_{22}+b_{11}-2h_{12})(ab-h^2)+\begin{vmatrix} 0\quad, & \tfrac{1}{2}a, & h_1-\tfrac{1}{2}a_2 \\ h_2-\tfrac{1}{2}b_1, & a, & h \\ \tfrac{1}{2}b_2, & h, & b \end{vmatrix}-\begin{vmatrix} 0\,, & \tfrac{1}{2}a_2, & \tfrac{1}{2}b_1 \\ \tfrac{1}{2}a_2, & a\,, & h \\ \tfrac{1}{2}b_1, & h\,, & b \end{vmatrix},$$

which is
$$\begin{aligned}=&-\tfrac{1}{2}(a_{22}+b_{11}-2h_{12})(ab-h^2)\\ &+a\,(\tfrac{1}{4}b_1^2-\tfrac{1}{2}b_2h_1+\tfrac{1}{4}a_2b_2)\\ &+b\,(\tfrac{1}{4}a_2^2-\tfrac{1}{2}a_1h_2+\tfrac{1}{4}a_1b_1)\\ &+h\,(\tfrac{1}{4}a_1b_2-\tfrac{1}{4}a_2b_1+h_1h_2-\tfrac{1}{2}a_2h_2-\tfrac{1}{2}b_1h_1),\end{aligned}$$
$=(ab-h^2)\,k$, where k is the measure of curvature.

In the same way, we have
$$(23.1.2)(31.1.2)-(33.1.2)(12.1.2)$$
$$=\begin{vmatrix} 23.31, & 1.31, & 2.31 \\ 1.23, & 1.1\,, & 1.2 \\ 2.23, & 1.2\,, & 2.2 \end{vmatrix}-\begin{vmatrix} 12.33, & 1.33, & 2.33 \\ 1.12, & 1.1\,, & 1.2 \\ 2.12, & 1.2\,, & 2.2 \end{vmatrix},$$

which is

$$= (23.31 - 12.33) \begin{vmatrix} 1.1, & 1.2 \\ 1.2, & 2.2 \end{vmatrix} + \begin{vmatrix} 0\ , & 1.13, & 2.13 \\ 1.23, & 1.1\ , & 1.2 \\ 2.23, & 1.2\ , & 2.2 \end{vmatrix} - \begin{vmatrix} 0\ , & 1.33, & 2.33 \\ 1.12, & 1.1\ , & 1.2 \\ 2.12, & 1.2\ , & 2.2 \end{vmatrix},$$

containing $23.31 - 12.33$, which by what precedes is

$$= -\tfrac{1}{2}(g_{23} + f_{31} - c_{12} - h_{33});$$

the other terms are also known, and the value is

$$= -\tfrac{1}{2}(g_{23} + f_{31} - c_{12} - h_{33})(ab - h^2)$$

$$+ \begin{vmatrix} 0 & , & \tfrac{1}{2}a_3, & \tfrac{1}{2}(f_1 - g_2 + h_3) \\ \tfrac{1}{2}(-f_1 + g_2 + h_3), & & a\ , & h \\ \tfrac{1}{2}b_3 & , & h\ , & b \end{vmatrix} - \begin{vmatrix} 0\ , & g_3 - \tfrac{1}{2}c_1, & f_3 - \tfrac{1}{2}c_2 \\ \tfrac{1}{2}a_2, & a\ , & h \\ \tfrac{1}{2}b_1, & h\ , & b \end{vmatrix},$$

or finally this is

$$\begin{aligned} = & -\tfrac{1}{2}(g_{23} + f_{31} - c_{12} - h_{33})(ab - h^2) \\ & + a(\tfrac{1}{2}b_1 f_3 - \tfrac{1}{4}b_3 f_1 - \tfrac{1}{4}b_1 c_2 + \tfrac{1}{4}b_3 g_2 - \tfrac{1}{4}b_3 h_3) \\ & + b(\tfrac{1}{2}a_2 g_3 - \tfrac{1}{4}a_3 g_2 - \tfrac{1}{4}a_2 c_1 + \tfrac{1}{4}a_3 f_1 - \tfrac{1}{4}a_3 h_3) \\ & + h\{\tfrac{1}{4}a_3 b_3 + \tfrac{1}{4}b_1 c_1 + \tfrac{1}{4}a_2 c_2 - \tfrac{1}{2}b_1 g_3 - \tfrac{1}{2}a_2 f_3 + \tfrac{1}{4}h_3^2 - \tfrac{1}{4}(f_1 - g_2)^2\}. \end{aligned}$$

The remaining four of the six functions can of course be obtained from the two just found by a cyclical interchange of the letters and suffix-numbers 1, 2, 3, and it is not worth while to write down the values.

The two values of $(11.1.2)(22.1.2) - (12.1.2)^2$ are

$$\begin{aligned} -(a_{22} + b_{11} - 2h_{12})(ab - h^2) & + a(\tfrac{1}{4}b_1^2 - \tfrac{1}{2}b_2 h_1 + \tfrac{1}{4}a_2 b_2) \\ & + b(\tfrac{1}{4}a_2^2 - \tfrac{1}{2}a_1 h_2 + \tfrac{1}{4}a_1 b_1) \\ & + h\{\tfrac{1}{4}(a_1 b_2 - a_2 b_1) + h_1 h_2 - \tfrac{1}{2}b_1 h_1 - \tfrac{1}{2}a_2 h_2), \end{aligned}$$

and

$$\begin{aligned} & \{G.\tfrac{1}{2}a_1 + F(h_1 - \tfrac{1}{2}a_2) + C(g_1 - \tfrac{1}{2}a_3)\}.\{G(h_2 - \tfrac{1}{2}b_1) + F.\tfrac{1}{2}b_2 + C(f_2 - \tfrac{1}{2}b_3)\} \\ & \quad - \{G.\tfrac{1}{2}a_2 + F.\tfrac{1}{2}b_1 + C.\tfrac{1}{2}(f_1 + g_2 - h_3)\}^2, \end{aligned}$$

where, in the first value, for a, b, h, $ab - h^2$ we must write $L^{-2}(BC - F^2)$, $L^{-2}(CA - G^2)$, $L^{-2}(FG - CH)$, and L^2C; making this change, multiplying by 4 and equating, we obtain

$$\begin{aligned} & -2L^2C(a_{22} + b_{11} - 2h_{12}) \\ & + (BC - F^2)(b_1^2 - 2b_2 h_1 + a_2 b_2) \\ & + (CA - G^2)(a_2^2 - 2a_1 h_2 + a_1 b_1) \\ & + (FG - CH)(a_1 b_2 - a_2 b_1 + 4h_1 h_2 - 2b_1 h_1 - 2a_2 h_2) \\ & - \{Ga_1 + F(2h_1 - a_2) + C(2g_1 - a_3)\}\{G(2h_2 - b_1) + Fb_2 + C(2f_2 - b_3)\} \\ & + \{Ga_2 + Fb_1 + C(f_1 + g_2 - h_3)\}^2 \qquad = 0. \end{aligned}$$

Developing the fourth and fifth lines, it appears that in this expression the coefficients of F^2, G^2 and FG each of them vanish; the whole equation is thus divisible by C, and omitting this factor throughout, the equation becomes

$$\begin{aligned}
0 = &- 2L^2 (\mathrm{a}_{22} + \mathrm{b}_{11} - 2\mathrm{h}_{12}) \\
&+ A (\mathrm{a}_2^2 - 2\mathrm{a}_1\mathrm{h}_2 + \mathrm{a}_1\mathrm{b}_1) \\
&+ B (\mathrm{b}_1^2 - 2\mathrm{b}_2\mathrm{h}_1 + \mathrm{a}_2\mathrm{b}_2) \\
&+ C \{- \mathrm{a}_3\mathrm{b}_3 + 2\mathrm{a}_3\mathrm{f}_2 + 2\mathrm{b}_3\mathrm{g}_1 - 4\mathrm{f}_2\mathrm{g}_1 + (\mathrm{f}_1 + \mathrm{g}_2 - \mathrm{h}_3)^2\} \\
&+ F \{(\mathrm{a}_3\mathrm{b}_2 - \mathrm{a}_2\mathrm{b}_3) + 2 (\mathrm{b}_1\mathrm{g}_2 - \mathrm{b}_2\mathrm{g}_1) + 2 (\mathrm{b}_3\mathrm{h}_1 - \mathrm{b}_1\mathrm{h}_3) + 2 (\mathrm{a}_2\mathrm{f}_2 + \mathrm{b}_1\mathrm{f}_1 - 2\mathrm{h}_1\mathrm{f}_2)\} \\
&+ G \{\mathrm{a}_1\mathrm{b}_3 - \mathrm{a}_3\mathrm{b}_1 + 2 (\mathrm{a}_2\mathrm{f}_1 - \mathrm{a}_1\mathrm{f}_2) + 2 (\mathrm{a}_3\mathrm{h}_2 - \mathrm{a}_2\mathrm{h}_3) + 2 (\mathrm{a}_2\mathrm{g}_2 + \mathrm{b}_1\mathrm{g}_1 - 2\mathrm{g}_1\mathrm{h}_2)\} \\
&+ H \{ \qquad\qquad 2 (\mathrm{a}_2\mathrm{h}_2 + \mathrm{b}_1\mathrm{h}_1 - 2\mathrm{h}_1\mathrm{h}_2)\};
\end{aligned}$$

this is substantially the required equation ($\mathfrak{C}$), but the form of it may be greatly simplified.

Forming the identity,

$$\begin{aligned}
0 = 2L [(\mathrm{a}_2 - \mathrm{h}_1) L_2 + (\mathrm{b}_1 - \mathrm{h}_2) L_1] = &- A (\mathrm{a}_1\mathrm{b}_1 - \mathrm{a}_1\mathrm{h}_2 - \mathrm{a}_2\mathrm{h}_1 + \mathrm{a}_2^2) \\
&- B (\mathrm{b}_1^2 - \mathrm{b}_1\mathrm{h}_2 - \mathrm{b}_2\mathrm{h}_1 + \mathrm{a}_2\mathrm{b}_2) \\
&- C (\mathrm{b}_1\mathrm{c}_1 + \mathrm{a}_2\mathrm{c}_2 - \mathrm{c}_1\mathrm{h}_2 - \mathrm{c}_2\mathrm{h}_1) \\
&- 2F (\mathrm{b}_1\mathrm{f}_1 + \mathrm{a}_2\mathrm{f}_2 - \mathrm{f}_1\mathrm{h}_2 - \mathrm{f}_2\mathrm{h}_1) \\
&- 2G (\mathrm{b}_1\mathrm{g}_1 + \mathrm{a}_2\mathrm{g}_2 - \mathrm{g}_1\mathrm{h}_2 - \mathrm{g}_2\mathrm{h}_1) \\
&- 2H (\mathrm{a}_2\mathrm{h}_2 + \mathrm{b}_1\mathrm{h}_1 - 2\mathrm{h}_1\mathrm{h}_2),
\end{aligned}$$

we add hereto the last preceding equation; the coefficients of A, B, F, G, H thus assume new and simple forms, but the coefficient of C requires a further transformation.

Assume

$$\begin{aligned}
\Omega = &(\mathrm{b}_1\mathrm{c}_1 - \mathrm{f}_1^2) + (\mathrm{c}_2\mathrm{a}_2 - \mathrm{g}_2^2) + (\mathrm{a}_3\mathrm{b}_3 - \mathrm{h}_3^2) \\
&+ (\mathrm{g}_2\mathrm{h}_3 + \mathrm{g}_3\mathrm{h}_2 - \mathrm{a}_2\mathrm{f}_3 - \mathrm{a}_3\mathrm{f}_2) + (\mathrm{h}_3\mathrm{f}_1 + \mathrm{h}_1\mathrm{f}_3 - \mathrm{b}_1\mathrm{g}_3 - \mathrm{b}_3\mathrm{g}_1) + (\mathrm{f}_1\mathrm{g}_2 + \mathrm{f}_2\mathrm{g}_1 - \mathrm{c}_1\mathrm{h}_2 - \mathrm{c}_2\mathrm{h}_1);
\end{aligned}$$

then, if we add to the equation C multiplied by this value and subtract $C\Omega$, the coefficient of C takes its proper form, and the equation is

$$\begin{aligned}
&- 2L^2 (\mathrm{a}_{22} + \mathrm{b}_{11} - 2\mathrm{h}_{12}) \\
&+ 2L [(\mathrm{b}_1 - \mathrm{h}_2) L_1 + (\mathrm{a}_2 - \mathrm{h}_1) L_2] - C\Omega \\
&+ A \{- (\mathrm{a}\mathrm{h}_2 - \mathrm{a}_2\mathrm{h})\} \\
&+ B \{ \ (\mathrm{b}_1\mathrm{h}_2 - \mathrm{b}_2\mathrm{h}_1)\} \\
&+ C \{- (\mathrm{a}_2\mathrm{f}_3 - \mathrm{a}_3\mathrm{f}_2) + (\mathrm{b}_3\mathrm{g}_1 - \mathrm{b}_1\mathrm{g}_3) - (\mathrm{g}_2\mathrm{h}_3 - \mathrm{g}_3\mathrm{h}_2) - (\mathrm{h}_3\mathrm{f}_1 - \mathrm{h}_1\mathrm{f}_3) + 3 (\mathrm{f}_1\mathrm{g}_2 - \mathrm{f}_2\mathrm{g}_1)\} \\
&+ F \{- (\mathrm{a}_2\mathrm{b}_3 - \mathrm{a}_3\mathrm{b}_2) + 2 (\mathrm{b}_1\mathrm{g}_2 - \mathrm{b}_2\mathrm{g}_1) + 2 (\mathrm{b}_3\mathrm{h}_1 - \mathrm{b}_1\mathrm{h}_3) - 2 (\mathrm{h}_1\mathrm{f}_2 - \mathrm{h}_2\mathrm{f}_1)\} \\
&+ G \{- (\mathrm{a}_3\mathrm{b}_1 - \mathrm{a}_1\mathrm{b}_3) - 2 (\mathrm{a}_1\mathrm{f}_2 - \mathrm{a}_2\mathrm{f}_1) - 2 (\mathrm{a}_2\mathrm{h}_3 - \mathrm{a}_3\mathrm{h}_2) - 2 (\mathrm{g}_1\mathrm{h}_2 - \mathrm{g}_2\mathrm{h}_1)\} \\
&+ H \{- (\mathrm{a}_1\mathrm{b}_2 - \mathrm{a}_2\mathrm{b}_1)\} \\
&= 0;
\end{aligned}$$

where, as a verification, observe that the letters (a, b), (f, g) may be interchanged if at the same time we interchange the suffixes 1 and 2.

The two values of (23 . 1 . 2)(31 . 1 . 2) − (33 . 1 . 2)(12 . 1 . 2) are

$$\begin{aligned}
&-\tfrac{1}{2}(g_{23}+f_{31}-c_{12}-h_{33})(ab-h^2)\\
&+a\,(\tfrac{1}{2}b_1f_3-\tfrac{1}{4}b_3f_1-\tfrac{1}{4}b_1c_2+\tfrac{1}{4}b_3g_2-\tfrac{1}{4}b_3h_3)\\
&+b\,(\tfrac{1}{2}a_2g_3-\tfrac{1}{4}a_3g_2-\tfrac{1}{4}a_2c_1+\tfrac{1}{4}a_3f_1-\tfrac{1}{4}a_3h_3)\\
&+h\,\{\tfrac{1}{4}a_3b_3+\tfrac{1}{4}b_1c_1+\tfrac{1}{4}a_2c_2-\tfrac{1}{2}b_1g_3-\tfrac{1}{2}a_2f_3+\tfrac{1}{4}h_3^2-\tfrac{1}{4}(f_1-g_2)^2\},
\end{aligned}$$

and

$$\begin{aligned}
&\{G\,.\,\tfrac{1}{2}(-f_1+g_2+h_3)+F\,.\,\tfrac{1}{2}b_3+C\,.\,\tfrac{1}{2}c_2\}\,\{G\,.\,\tfrac{1}{2}a_3+F\,.\,\tfrac{1}{2}(f_1-g_2+h_3)+C\,.\,\tfrac{1}{2}c_1\}\\
&-\{G\;(g_3-\tfrac{1}{2}c_1)+F\,(f_3-\tfrac{1}{2}c_2)+C\,.\,\tfrac{1}{2}c_3\}\,\{G\,.\,\tfrac{1}{2}a_2+F\,.\,\tfrac{1}{2}b_1+C\,.\,\tfrac{1}{2}(f_1+g_2-h_3)\},
\end{aligned}$$

and here for a, b, h, $ab-h^2$, substituting their values, multiplying by 4, and equating, we have

$$\begin{aligned}
&-2L^2C\,(g_{23}+f_{31}-c_{12}-h_{33})\\
&+(BC-F^2)(2b_1f_3-b_3f_1-b_1c_2+b_3g_2-b_3h_3)\\
&+(CA-G^2)(2a_2g_3-a_3g_2-a_2c_1+a_3f_1-a_3h_3)\\
&+(FG-CH)(a_3b_3+b_1c_1+a_2c_2-2b_1g_3-2a_2f_3+h_3^2-f_1^2+2f_1g_2-g_2^2)\\
&-\{G\,(-f_1+g_2+h_3)+Fb_3+Cc_2\}\,\{Ga_3+F(f_1-g_2+h_3)+Cc_1\}\\
&+\{G\,(2g_3-c_1)+F\,(2f_3-c_2)+Cc_3\}\,\{Ga_2+Fb_1+C\,(f_1+g_2-h_3)\}=0.
\end{aligned}$$

Here again the terms in F^2, FG, G^2 all vanish, hence the whole equation divides by C; throwing this factor out, we have

$$\begin{aligned}
&-2L^2\,(g_{23}+f_{31}-c_{12}-h_{33})\\
&+A\,(\;2a_2g_3-a_3g_2\;-\;a_2c_1+\;a_3f_1\;-\;a_3h_3)\\
&+B\,(\;2b_1f_3-b_3f_1\;-\;b_1c_2+\;b_3g_2-\;b_3h_3)\\
&+C\,(-\;c_1c_2+c_3f_1\;+\;c_3g_2-\;c_3h_3)\\
&+F\,\{-(b_3c_1-b_1c_3)-2c_2f_1+2f_1f_3+2g_2f_3-2f_3h_3\}\\
&+G\,(\;a_2c_3-a_3c_2\;-2c_1g_2+2f_1g_3+2g_2g_3-2g_3h_3)\\
&+H\,(-\;a_3b_3-b_1c_1\;-\;a_2c_2+2b_1g_3+2a_2f_3\;-h_3^2+f_1^2-2f_1g_2+g_2^2)=0,
\end{aligned}$$

which is substantially the required equation ($\mathfrak{H}$); but the form has to be altered. First, multiply by 2, then forming the expression

$$\begin{aligned}
2L\,[(f_3-c_2)\,L_1+(g_3-c_1)\,L_2+(f_1+g_2-2h_3)\,L_3]&\\
=\quad(f_3-\;c_2)\,(Aa_1+Bb_1+Cc_1+2Ff_1+2Gg_1+2Hh_1)&\\
+\,(g_3-\;c_1)\,(Aa_2+Bb_2+Cc_2+2Ff_2+2Gg_2+2Hh_2)&\\
+\,(f_1+\;g_2-2h_3)\,(Aa_3+Bb_3+Cc_3+2Ff_3+2Gg_3+2Hh_3),&
\end{aligned}$$

this is

$$\begin{aligned}
= \;\; & A\,(-\,a_1c_2 - a_2c_1 + a_1f_3 + a_3f_1 + a_2g_3 + a_3g_2 - 2a_3h_3)\\
+ \;\; & B\,(-\,b_1c_2 - b_2c_1 + b_1f_3 + b_3f_1 + b_2g_3 + b_3g_2 - 2b_3h_3)\\
+ \;\; & C\,(-2c_1c_2 + c_1f_3 + c_3f_1 + c_2g_3 + c_3g_2 - 2c_3h_3)\\
+ \;\; & 2F\,(-\,c_1f_2 - c_2f_1 + 2f_1f_3 + f_2g_3 + f_3g_2 - 2f_3h_3)\\
+ \;\; & 2G\,(-\,c_1g_2 - c_2g_1 + f_1g_3 + f_3g_1 + 2g_2g_3 - 2g_3h_3)\\
+ \;\; & 2H\,(-\,c_1h_2 - c_2h_1 + g_2h_3 + g_3h_2 + h_3f_1 + h_1f_3 - 2h_3^2);
\end{aligned}$$

then adding this expression

$$(f_3 - c_2)\,L_1 + (g_3 - c_1)\,L_2 + (f_1 + g_2 - 2h_3)\,L_3$$

and subtracting its foregoing value, we have an equation with new coefficients of A, B, C, F, G, H, all of which, except that of H, are in the proper form, and for the coefficient of H, recurring to the foregoing value of Ω, we must add to the equation $2H$ multiplied by the value of Ω and subtract $2H\Omega$. The final result is

$$\begin{aligned}
& -4L^2\,(g_{23} + f_{31} - c_{12} - h_{33})\\
& + 2L\,[(f_3 - c_2)\,L_1 + (g_3 - c_1)\,L_2 + (f_1 + g_2 - 2h_3)\,L_3] - 2H\Omega\\
& \quad + A\,\{\;3\,(a_2g_3 - a_3g_2) - (c_1a_2 - c_2a_1) + (a_3f_1 - a_1f_3)\}\\
& \quad + B\,\{-3\,(b_3f_1 - b_1f_3) - (b_1c_2 - b_2c_1) - (b_2g_3 - b_3g_2)\}\\
& \quad + C\,\{\;(c_3f_1 - c_1f_3) - (c_2g_3 - c_3g_2)\}\\
& \quad + 2F\,\{-\,(b_3c_1 - b_1c_3) + (c_1f_2 - c_2f_1) - (f_2g_3 - f_3g_2)\}\\
& \quad + 2G\,\{-\,(c_2a_3 - c_3a_2) - (c_1g_2 - c_2g_1) - (f_3g_1 - f_1g_3)\}\\
& \quad + 2H\,\{-\,(b_3g_1 - b_1g_3) + (a_2f_3 - a_3f_2)\} = 0,
\end{aligned}$$

which, it will be observed, remains unaltered by the interchange of a and b, f and g A and B, F and G, the suffix numbers 1 and 2 being at the same time interchanged.

We have thus the required six equations, in which

$$L^2 = abc - af^2 - bg^2 - ch^2 + 2fgh,$$

$$\Omega = b_1c_1 - f_1^2 + c_2a_2 - g_2^2 + a_3b_3 - h_3^2$$

$$+ (g_2h_3 + g_3h_2 - a_2f_3 - a_3f_2) + (h_3f_1 + h_1f_3 - b_3g_1 - b_1g_3) + (f_1g_2 + f_2g_1 - a_1f_2 - a_2f_1),$$

and where I write also, for shortness, ab12 for $a_1b_2 - a_2b_1$, &c.; viz. the equations are

$$\begin{aligned}
0 = & -2L^2\,(b_{33} + c_{22} - 2f_{23}) \qquad\qquad (\mathfrak{A}).\\
& + 2L\,[\;.\quad (c_2 - f_3)\,L_2 + (b_3 - f_2)\,L_3] \quad - A\Omega\\
& + A\,\{\;.\; - \text{bg}31 + \text{ch}12 + 3\,.\,\text{gh}23 - \text{hf}31 - \text{fg}12\}\\
& + B\,\{-\,\text{bf}23\}\\
& + C\,\{+\,\text{cf}23\}\\
& + F\,\{-\,\text{bc}23\}\\
& + G\,\{-\,\text{bc}31 + 2\,.\,\text{cf}12 \quad . \quad + 2\,.\,\text{ch}23 \quad . \quad - 2\,.\,\text{fg}23\}\\
& + H\,\{-\,\text{bc}12 - 2\,.\,\text{bf}31 - 2\,.\,\text{bg}23 \quad . \quad . \quad - 2\,.\,\text{hf}23\}
\end{aligned}$$

$$
\begin{array}{rl}
0 = & -\, 2L^2(\mathrm{c}_{11} + \mathrm{a}_{33} - 2\mathrm{g}_{31}) \\
& +\, 2L\,[(\mathrm{c}_1 - \mathrm{g}_3)\,L_1 \quad . \quad + (\mathrm{a}_3 - \mathrm{g}_1)\,L_3] \quad - B\Omega \\
& +\; A\,\{ \quad . \quad \mathrm{ag}31 \quad \} \\
& +\; B\,\{ \quad \mathrm{af}23 \quad . \quad -\; \mathrm{ch}12 - \mathrm{gh}23 + 3\,.\,\mathrm{hf}31 - \mathrm{fg}31\} \\
& +\; C\,\{ \quad -\; \mathrm{cg}31 \quad \} \\
& +\; F\,\{-\,\mathrm{ca}23 \quad . \quad -\,2\,.\,\mathrm{cg}12 - 2\,.\,\mathrm{ch}31 - 2\,.\,\mathrm{fg}31\} \\
& +\; G\,\{-\,\mathrm{ca}31 \quad \} \\
& +\; H\,\{-\,\mathrm{ca}12 + 2\,.\,\mathrm{af}31 + 2\,.\,\mathrm{ag}23 \quad 2\,.\,\mathrm{gh}31\}
\end{array}
\qquad (\mathfrak{B}).
$$

$$
\begin{array}{rl}
0 = & -\, 2L^2(\mathrm{a}_{22} + \mathrm{b}_{11} - 2\mathrm{h}_{12}) \\
& +\, 2L\,[(\mathrm{b}_1 - \mathrm{h}_2)\,L_1 + (\mathrm{a}_2 - \mathrm{h}_1)\,L_2 \quad . \quad] \quad - C\Omega \\
& +\; A\,\{ \quad . \quad . \quad -\; \mathrm{ah}12 \quad \} \\
& +\; B\,\{ \quad . \quad . \quad +\; \mathrm{bh}12 \quad \} \\
& +\; C\,\{-\,\mathrm{af}23 + \mathrm{bg}31 \quad . \quad -\; \mathrm{gh}23 - \mathrm{hf}31 + 3\,.\,\mathrm{fg}12\} \\
& +\; F\,\{-\,\mathrm{ab}23 \quad . \quad . \quad + 2\,.\,\mathrm{bg}12 + 2\,.\,\mathrm{bh}31 - 2\,.\,\mathrm{hf}12\} \\
& +\; G\,\{-\,\mathrm{ab}31 \quad . \quad -\,2\,.\,\mathrm{af}12 \quad . \quad + 2\,.\,\mathrm{ah}23 - 2\,.\,\mathrm{gh}12\} \\
& +\; H\,\{-\,\mathrm{ab}12 \quad \}
\end{array}
\qquad (\mathfrak{C}).
$$

$$
\begin{array}{rl}
0 = & -\, 4L^2(\mathrm{g}_{12} + \mathrm{h}_{31} - \mathrm{a}_{23} - \mathrm{f}_{11}) \\
& +\, 2L\,[(\mathrm{g}_2 + \mathrm{h}_3 - 2\mathrm{f}_1)\,L_1 + (\mathrm{g}_1 - \mathrm{a}_3)\,L_2 + (\mathrm{h}_1 - \mathrm{a}_2)\,L_3] \quad - 2F\Omega \\
& +\; A\,\{ \quad . \quad \mathrm{ag}12 - \mathrm{ah}31 \quad \} \\
& +\; B\,\{-\,\mathrm{ab}23 + \mathrm{bg}12 + 3\,.\,\mathrm{bh}31 \quad \} \\
& +\; C\,\{-\,\mathrm{ca}23 - 3\,.\,\mathrm{cg}12 - \mathrm{ch}31 \quad \} \\
& +\, 2F\,\{ \quad . \quad + \mathrm{bg}31 - \mathrm{ch}12 \quad \} \\
& +\, 2G\,\{-\,\mathrm{ca}12 + \mathrm{ag}23 \quad . \quad -\,\mathrm{gh}31\} \\
& +\, 2H\,\{-\,\mathrm{ab}31 \quad . \quad -\; \mathrm{ah}23 - \mathrm{gh}12\}
\end{array}
\qquad (\mathfrak{F}).
$$

$$
\begin{array}{rl}
0 = & -\, 4L^2(\mathrm{h}_{23} + \mathrm{f}_{12} - \mathrm{b}_{31} - \mathrm{g}_{22}) \\
& +\, 2L\,[(\mathrm{f}_2 - \mathrm{b}_3)\,L_1 + (\mathrm{h}_3 + \mathrm{f}_1 - 2\mathrm{g}_2)\,L_2 + (\mathrm{h}_2 - \mathrm{b}_1)\,L_3] \quad - 2G\Omega \\
& +\; A\,\{-\,\mathrm{ab}31 - 3\,.\,\mathrm{ah}23 - \mathrm{af}12 \quad \} \\
& +\; B\,\{ \quad . \quad \mathrm{bh}23 - \mathrm{bf}12 \quad \} \\
& +\; C\,\{-\,\mathrm{bc}31 + \mathrm{ch}23 + 3\,.\,\mathrm{cf}12 \quad \} \\
& +\, 2F\,\{-\,\mathrm{bc}12 \quad . \quad -\; \mathrm{bf}31 - \mathrm{hf}23\} \\
& +\, 2G\,\{ \quad . \quad + \mathrm{ch}12 - \mathrm{af}23 \quad . \quad \} \\
& +\, 2H\,\{-\,\mathrm{ab}23 + \mathrm{bh}31 \quad . \quad -\,\mathrm{hf}12\}
\end{array}
\qquad (\mathfrak{G}).
$$

$$
\begin{aligned}
0 = & - 4L^2 (f_{31} + g_{23} - c_{12} - h_{33}) \qquad (\mathfrak{H}). \\
& + 2L\,[(f_3 - c_2)\,L_1 + (g_3 - c_1)\,L_2 + (f_1 + g_2 - 2h_3)\,L_3] \quad - 2H\Omega \\
& + \ A\,\{- \text{ca}12 + \quad \text{af}31 + 3\,.\,\text{ag}23 \qquad \} \\
& + \ B\,\{- \text{bc}12 - 3\,.\,\text{bf}31 - \quad \text{bg}23 \qquad \} \\
& + \ C\,\{ \quad . \qquad \text{cf}31 - \quad \text{cg}23 \qquad \} \\
& + \ F\,\{- \text{bc}31 + \quad \text{cf}12 \qquad . \quad - \text{fg}23\} \\
& + \ G\,\{- \text{ca}23 \qquad . \qquad - \quad \text{cg}12 - \text{fg}31\} \\
& + \ H\,\{ \quad . \quad + \quad \text{af}23 - \quad \text{bg}31 \quad . \quad \}
\end{aligned}
$$

It would be possible in these equations to introduce the symbols *AB*12, &c., in place of ab12, &c., and then writing $A = B = C = 1$, all these symbols other than those where the letters are *GH*, *HF* or *FG* would vanish, and we should obtain Mr Warren's six equations for normal coordinates. But in the general case it would seem that there is not any advantage in the introduction of the new symbols *AB*12, &c., and I retain by preference the equations in the form in which I have given them.

To the foregoing may be joined a symmetrical equation obtained (as by Mr Warren) by multiplying the several equations by a, b, c, f, g, h respectively, and adding; the result is in the first instance obtained in the form

$$- 2L^2\Theta + 2L\Psi - 3L^2\Omega + \square = 0 \ldots\ldots\ldots\ldots\ldots\ldots\ldots\ldots (\mathfrak{M}),$$

where

$$
\begin{aligned}
\Theta = & \quad \text{a}\,(\text{b}_{33} + \text{c}_{22} - 2\text{f}_{23}) \\
& + \text{b}\,(\text{c}_{11} + \text{a}_{33} - 2\text{g}_{31}) \\
& + \text{c}\,(\text{a}_{22} + \text{b}_{11} - 2\text{h}_{12}) \\
& + 2\text{f}\,(\text{g}_{12} + \text{h}_{31} - \text{a}_{23} - \text{f}_{11}) \\
& + 2\text{g}\,(\text{h}_{23} + \text{f}_{12} - \text{b}_{31} - \text{g}_{22}) \\
& + 2\text{h}\,(\text{f}_{31} + \text{g}_{23} - \text{c}_{12} - \text{h}_{33}).
\end{aligned}
$$

For Ψ, collecting the terms which contain L_1, L_2, L_3 respectively, and attending to the values of (A, B, C, F, G, H),

$$= (\text{bc} - \text{f}^2,\ \text{ca} - \text{g}^2,\ \text{ab} - \text{h}^2,\ \text{gh} - \text{af},\ \text{hf} - \text{bg},\ \text{fg} - \text{ch}),$$

this is easily reduced to the form

$$\Psi = (A_1 + H_2 + G_3)\,L_1 + (H_1 + B_2 + F_3)\,L_2 + (G_1 + F_2 + C_3)\,L_3.$$

The term in Ω is

$$
\begin{aligned}
& = - (A\text{a} + B\text{b} + C\text{c} + 2F\text{f} + 2G\text{g} + 2H\text{h})\,\Omega, \\
& = - 3L^2\Omega, \text{ as above.}
\end{aligned}
$$

For the calculation of $\square$, collecting the terms, we have $\square =$

23	A	B	C	F	G	H	31	A	B	C	F	G	H	12	A	B	C	F	G	H
bc				− a			bc			− g	− 2h	− a		bc		− h		− 2g		− a
ca			− f	− b	− 2h		ca					− b		ca	− h				− 2f	− b
ab		− f		− c		− 2g	ab	− g				− c	− 2f	ab						− c
af		b	− c		− 2g	2h	af	h					2b	af	− g				− 2c	
bf		− a					bf		− 3h		− 2g		− 2a	bf		− g				
cf			a				cf			h				cf			3g	2h	2a	
ag	3h				2f	2b	ag	b						ag	f					
bg		− h				− 2a	bg	− a		+ c	2f		− 2h	bg		f		2c		
cg			− h				cg			− b				cg			− 3f	− 2b	− 2h	
ah	− 3g				− 2c	− 2f	ah	− f						ah	− c					
bh		g					bh		3f		2c		2g	bh		c				
ch			g		2a		ch			− f	− 2b			ch	a	− b		− 2f	+ 2g	
gh	− 3a	− b	− c				gh					− 2f	− 2b	gh					− 2c	− 2f
hf				− 2g		− 2a	hf	− a	3b	− c				hf				− 2c		− 2g
fg				− 2h	− 2a		fg				− 2b	− 2h		fg	− a	− b	+ 3c			

viz. this is

$$\square = \text{bc23} \,.\, -\text{a}F + \text{bc31}\,(-\text{g}C - 2\text{h}F - \text{a}G) + \text{bc12}\,(-\text{h}B - 2\text{g}F - \text{a}H) + \&\text{c}.$$

Some however of the coefficients require reduction; for instance, that of bc31 is $= -(\text{a, h, g} \between G, F, C) - \text{h}F, = -\text{h}F$; and so $-\text{h}B - 2\text{g}F - \text{a}H$ is $= -(\text{a, h, g} \between H, B, F) - \text{g}F, = -\text{g}F$. After these reductions, the value is found to be $\square =$

bc23 . − aF	+ bc31 . − hF	+ bc12 . − gF
+ ca23 . − hG	+ ca31 . − bG	+ ca12 . − fG
+ ab23 . − gH	+ ab31 . − fH	+ ab12 . − cH
+ af23 (− gG + hH)	+ af31 (bH − fG)	+ af12 (fH − cG)
+ bf23 . − aB	+ bf31 . − hB	+ bf12 . − gB
+ cf23 . aC	+ cf31 . hC	+ cf12 . gC
+ ag23 . hA	+ ag31 . bA	+ ag12 . fA
+ bg23 (gF − aH)	+ bg31 (− hH + fF)	+ bg12 (cF − gH)
+ cg23 . − hC	+ cg31 . − bC	+ cg12 . − fC
+ ah23 . − gA	+ ah31 . − fA	+ ah12 . − cA
+ bh23 . gB	+ bh31 . fB	+ bh12 . cB
+ ch23 (aG − hF)	+ ch31 (hG − bF)	+ ch12 (− fF + gG)
+ gh23 (3aA − bB − cC)	+ gh31 . 2hA	+ gh12 . 2gA
+ hf23 . 2hB	+ hf31 (− aA + 3bB − cC)	+ hf12 . 2fB
+ fg23 . 2gC	+ fg31 . 2fC	+ fg12 (− aA − bB + 3cC);

or, arranging the terms in a different order, this may be written

$$
\begin{array}{lllll}
\square = & & & & \\
A\,\{ & \mathrm{a}\,(3\,.\,\mathrm{gh}23-\mathrm{hf}31-\mathrm{fg}12)+\mathrm{b}\,.\,\mathrm{ag}31-\mathrm{c}\,.\,\mathrm{ah}12 & & & -\mathrm{c}\,.\,\mathrm{ah}12 \\
 & & +\mathrm{f}\,(\mathrm{ag}12-\mathrm{ah}31) & +\mathrm{g}\,(2\,.\,\mathrm{gh}12-\mathrm{ah}23) & +\mathrm{h}\,(2\,.\,\mathrm{gh}31+\mathrm{ag}23)\} \\
+B\,\{ & -\mathrm{a}\,.\,\mathrm{bf}23 & +\mathrm{b}\,(-\mathrm{gh}23+3\,.\,\mathrm{hf}31-\mathrm{fg}12) & +\mathrm{c}\,.\,\mathrm{bh}12 & \\
 & & +\mathrm{f}\,(2\,.\,\mathrm{hf}12+\mathrm{bh}31) & +\mathrm{g}\,(\mathrm{bh}23-\mathrm{bf}12) & +\mathrm{h}\,(2\,.\,\mathrm{hf}23-\mathrm{bf}31)\} \\
+C\,\{ & \mathrm{a}\,.\,\mathrm{cf}23 & -\mathrm{b}\,.\,\mathrm{cg}31 & & +\mathrm{c}\,(-\mathrm{gh}23-\mathrm{hf}31+3\,.\,\mathrm{fg}12) \\
 & & +\mathrm{f}\,(2\,.\,\mathrm{fg}31-\mathrm{cg}12) & +\mathrm{g}\,(2\,.\,\mathrm{fg}23+\mathrm{cf}12) & +\mathrm{h}\,(\mathrm{cf}31-\mathrm{cg}23)\} \\
+F\,\{ & -\mathrm{a}\,.\,\mathrm{bc}23 & -\mathrm{b}\,.\,\mathrm{ch}31 & & +\mathrm{c}\,.\,\mathrm{bg}12 \\
 & & +\mathrm{f}\,(\mathrm{bg}31-\mathrm{ch}12) & +\mathrm{g}\,(\mathrm{bg}23-\mathrm{bc}12) & -\mathrm{h}\,(\mathrm{ch}23+\mathrm{bc}31)\} \\
+G\,\{ & +\mathrm{a}\,.\,\mathrm{ch}23 & -\mathrm{b}\,.\,\mathrm{ca}31 & & -\mathrm{c}\,.\,\mathrm{af}12 \\
 & & -\mathrm{f}\,(\mathrm{af}31+\mathrm{ca}12) & +\mathrm{g}\,(\mathrm{ch}12-\mathrm{af}23) & +\mathrm{h}\,(\mathrm{ch}31-\mathrm{ca}23)\} \\
+H\,\{ & -\mathrm{a}\,.\,\mathrm{bg}23 & +\mathrm{b}\,.\,\mathrm{af}31 & & -\mathrm{c}\,.\,\mathrm{ab}12 \\
 & & +\mathrm{f}\,(\mathrm{af}12-\mathrm{ab}31) & -\mathrm{g}\,(\mathrm{bg}12+\mathrm{ab}23) & +\mathrm{h}\,(\mathrm{af}23-\mathrm{bg}31)\}.
\end{array}
$$

Attending to the values of A, B, C, F, G, H, we have

$$
\begin{array}{lll}
\multicolumn{3}{l}{A_{11}+B_{22}+C_{33}+2F_{23}+2G_{31}+2H_{12}} \\
= & \mathrm{bc}_{11}+\mathrm{cb}_{11}-2\mathrm{ff}_{11} & +2\,(\mathrm{b}_1\mathrm{c}_1-\mathrm{f}_1{}^2) \\
+ & \mathrm{ca}_{22}+\mathrm{ac}_{22}-2\mathrm{gg}_{22} & +2\,(\mathrm{c}_2\mathrm{a}_2-\mathrm{g}_2{}^2) \\
+ & \mathrm{ab}_{33}+\mathrm{ba}_{33}-2\mathrm{hh}_{33} & +2\,(\mathrm{a}_3\mathrm{b}_3-\mathrm{h}_3{}^2) \\
+2\,\{ & \mathrm{gh}_{23}+\mathrm{hg}_{23}-\mathrm{af}_{23}-\mathrm{fa}_{23}+ & \mathrm{g}_2\mathrm{h}_3+\mathrm{g}_3\mathrm{h}_2-\mathrm{a}_2\mathrm{f}_3-\mathrm{a}_3\mathrm{f}_2\,\} \\
+2\,\{ & \mathrm{hf}_{31}+\mathrm{fh}_{31}-\mathrm{bg}_{31}-\mathrm{gb}_{31}+ & \mathrm{h}_3\mathrm{f}_1+\mathrm{h}_1\mathrm{f}_3-\mathrm{b}_3\mathrm{g}_1-\mathrm{b}_1\mathrm{g}_3\} \\
+2\,\{ & \mathrm{fg}_{12}+\mathrm{gf}_{12}-\mathrm{ch}_{12}-\mathrm{hc}_{12}+ & \mathrm{f}_1\mathrm{g}_2+\mathrm{f}_2\mathrm{g}_1-\mathrm{c}_1\mathrm{h}_2-\mathrm{c}_2\mathrm{h}_1\},
\end{array}
$$

which is, in fact, $=2\Theta+2\Omega$.

Hence the foregoing equation ($\mathfrak{M}$) may also be written

$$-L^2\,(A_{11}+B_{22}+C_{33}+2F_{23}+2G_{31}+2H_{12})+2L\Psi-L^2\Omega+\square=0,$$

where Ψ, Ω, $\square$ have their before-mentioned values.

In the particular case where $\mathrm{f}=0$, $\mathrm{g}=0$, $\mathrm{h}=0$, we have

$$A,\ B,\ C,\ F,\ G,\ H=\mathrm{bc},\ \mathrm{ca},\ \mathrm{ab},\ 0,\ 0,\ 0\,;\ \ L^2=\mathrm{abc}\,;$$

$$\Omega=\mathrm{b}_1\mathrm{c}_1+\mathrm{c}_2\mathrm{a}_2+\mathrm{a}_3\mathrm{b}_3\,;$$

the equation ($\mathfrak{A}$) becomes

$$-2\mathrm{abc}\,(\mathrm{b}_{33}+\mathrm{c}_{22})+[\mathrm{c}_2\,(\mathrm{a}_2\mathrm{bc}+\mathrm{b}_2\mathrm{ca}+\mathrm{c}_2\mathrm{ab})+\mathrm{b}_3\,(\mathrm{a}_3\mathrm{bc}+\mathrm{b}_3\mathrm{ca}+\mathrm{c}_3\mathrm{ab})]-\mathrm{bc}\,(\mathrm{b}_1\mathrm{c}_1+\mathrm{c}_2\mathrm{a}_2+\mathrm{a}_3\mathrm{b}_3)=0,$$

that is,

$$-2\mathrm{abc}\,(\mathrm{b}_{33}+\mathrm{c}_{22})-\mathrm{bcb}_1\mathrm{c}_1+\mathrm{ca}\,(\mathrm{b}_2\mathrm{c}_2+\mathrm{b}_3{}^2)+\mathrm{ab}\,(\mathrm{b}_3\mathrm{c}_3+\mathrm{c}_2{}^2)=0,$$

and the equation ($\mathfrak{F}$) becomes

$$4abc\,.\,a_{23} - [a_3(a_2bc + b_2ca + c_2ab) + a_2(a_3bc + b_3ca + c_3ab)]$$
$$- ca(a_2b_3 - a_3b_2) - ab(c_2a_3 - c_3a_2) = 0,$$

that is,

$$4abc\,.\,a_{23} - 2bc\,.\,a_2a_3 - 2ca\,.\,a_2b_3 - 2ab\,.\,a_3b_2 = 0.$$

Dividing by $-2abc$ and $4abc$ respectively, and completing the system, the six equations are

$$b_{33} + c_{22} - \frac{b_3^2}{2b} - \frac{c_2^2}{2c} + \frac{b_1c_1}{2a} - \frac{b_2c_2}{2b} - \frac{b_3c_3}{2c} = 0 \quad \ldots\ldots\ldots\ldots\ldots\ldots(\mathfrak{A}),$$

$$c_{11} + a_{33} - \frac{c_1^2}{2c} - \frac{a_3^2}{2a} - \frac{c_1a_1}{2a} + \frac{c_2a_2}{2b} - \frac{c_3a_3}{2c} = 0 \quad \ldots\ldots\ldots\ldots\ldots\ldots(\mathfrak{B}),$$

$$a_{22} + b_{11} - \frac{a_2^2}{2a} - \frac{b_1^2}{2b} - \frac{a_1b_1}{2a} - \frac{a_2b_2}{2b} + \frac{a_3b_3}{2c} = 0 \quad \ldots\ldots\ldots\ldots\ldots\ldots(\mathfrak{C}),$$

$$a_{23} - \frac{a_2a_3}{2a} - \frac{a_2b_3}{2b} - \frac{a_3c_2}{2c} = 0 \quad \ldots\ldots\ldots\ldots\ldots\ldots(\mathfrak{F}),$$

$$b_{31} - \frac{b_1a_3}{2a} - \frac{b_3b_1}{2b} - \frac{b_3c_1}{2c} = 0 \quad \ldots\ldots\ldots\ldots\ldots\ldots(\mathfrak{G}),$$

$$c_{12} - \frac{c_1a_2}{2a} - \frac{c_2b_1}{2b} - \frac{c_1c_2}{2c} = 0 \quad \ldots\ldots\ldots\ldots\ldots\ldots(\mathfrak{H}).$$

These are in fact Lamé's equations, *Leçons sur les coordonnées curvilignes, etc.*, Paris (1859), pp. 76, 78, viz. the first of the equations (8), p. 76, is

$$\frac{d^2H}{d\rho_1 d\rho_2} = \frac{1}{H_1}\frac{dH}{d\rho_1}\frac{dH_1}{d\rho_2} + \frac{1}{H_2}\frac{dH}{d\rho_2}\frac{dH_2}{d\rho_1},$$

which, in the notation of the present paper, is

$$(\sqrt{a})_{23} = \frac{1}{\sqrt{b}}(\sqrt{a})_2(\sqrt{b})_3 + \frac{1}{\sqrt{c}}(\sqrt{a})_3(\sqrt{c})_2.$$

Here

$$(\sqrt{a})_2 = \tfrac{1}{2}\frac{a_2}{\sqrt{a}},$$

$$(\sqrt{a})_{23} = \tfrac{1}{2}\left(\frac{a_2}{\sqrt{a}}\right)_3 = \tfrac{1}{2}\left\{\frac{a_{23}}{\sqrt{a}} - \frac{\tfrac{1}{2}a_2a_3}{a\sqrt{a}}\right\};$$

and the equation therefore is

$$\frac{a_{23}}{\sqrt{a}} - \frac{\tfrac{1}{2}a_2a_3}{a\sqrt{a}} = \tfrac{1}{2}\frac{1}{\sqrt{b}}\frac{a_2}{\sqrt{a}}\frac{b_3}{\sqrt{b}} + \tfrac{1}{2}\frac{1}{\sqrt{c}}\frac{a_3}{\sqrt{a}}\frac{c_2}{\sqrt{c}},$$

which, multiplying by $\sqrt{a}$, gives the foregoing equation

$$a_{23} - \frac{a_2a_3}{a} - \frac{a_2b_3}{b} - \frac{a_3c_2}{c} = 0.$$

The first of the equations (9) p. 78 is

$$\frac{d}{d\rho_1}\left(\frac{1}{H_1}\frac{dH}{d\rho_1}\right)+\frac{d}{d\rho}\left(\frac{1}{H}\frac{dH_1}{d\rho}\right)+\frac{1}{H_2^2}\frac{dH}{d\rho_2}\frac{dH_1}{d\rho_2}=0,$$

which, in the notation of the present paper, is

$$\left\{\frac{(\sqrt{a})_2}{\sqrt{b}}\right\}_2+\left\{\frac{(\sqrt{b})_1}{\sqrt{a}}\right\}_1+\frac{1}{c}(\sqrt{a})_3(\sqrt{b})_3=0.$$

This gives first

$$\tfrac{1}{2}\left(\frac{a_2}{\sqrt{a}\sqrt{b}}\right)_2+\tfrac{1}{2}\left(\frac{b_1}{\sqrt{a}\sqrt{b}}\right)_1+\tfrac{1}{4}\frac{a_3b_3}{c\sqrt{a}\sqrt{b}}=0,$$

and then

$$\begin{aligned}&\tfrac{1}{2}\frac{a_{22}}{\sqrt{a}\sqrt{b}}-\tfrac{1}{4}\frac{a_2^2}{a\sqrt{a}\sqrt{b}}-\tfrac{1}{4}\frac{a_2b_2}{b\sqrt{a}\sqrt{b}}\\&+\tfrac{1}{2}\frac{b_{11}}{\sqrt{a}\sqrt{b}}-\tfrac{1}{4}\frac{b_1^2}{b\sqrt{a}\sqrt{b}}-\tfrac{1}{4}\frac{a_1b_1}{a\sqrt{a}\sqrt{b}}\end{aligned}+\tfrac{1}{4}\frac{a_3b_3}{c\sqrt{a}\sqrt{b}}=0,$$

which, on multiplying by $\sqrt{a}\sqrt{b}$ and reducing, gives the foregoing equation

$$a_{22}+b_{11}-\frac{a_2^2}{a}-\frac{b_1^2}{b}-\frac{a_1b_1}{a}-\frac{a_2b_2}{b}+\frac{a_3b_3}{c}=0.$$

800.

NOTE ON THE STANDARD SOLUTIONS OF A SYSTEM OF LINEAR EQUATIONS.

[From the *Quarterly Journal of Pure and Applied Mathematics*, vol. XIX. (1883), pp. 38—40.]

To fix the ideas, the equations are assumed to be without constant terms. Supposing the system to be insufficient for the determination of the ratios of the unknown quantities, then regarding the unknown quantities as having a definite order of arrangement, there are certain solutions which may be regarded as standard solutions. Take the unknown quantities to be A, B, C, D, E, F, G, &c.; then assuming $A=0$, or else A, B, each $=0$, or else A, B, C, each $=0$, as many equations as possible, the system as thus modified will have a definite solution; for instance, the assumed equations A, B, C, D, E, each $=0$, may give a definite solution in which F is not $=0$, and it may then for convenience be put $=1$. We have thus a solution beginning with $F=1$. This being so, there will be a solution or solutions with $F=0$; we cannot then have A, B, C, D, E, each $=0$, but we again assume $A=0$, or else A, B, each $=0$, as many equations as possible; suppose A, B, C, each $=0$ give a definite solution, with D not $=0$; and then taking it for convenience to be $=1$, we have a solution beginning $D=1$, and for which $F=0$. Going on in this manner we obtain, it may be, a solution beginning $B=1$, and for which $D=0$, $F=0$; and so on, the process stopping, if not sooner, with a solution beginning $A=1$, and with the initial letters of the preceding solutions, each $=0$. We have in this manner the system of standard solutions, of a form such as

	A	B	C	D	E	F	G ...
=	0	0	0	0	0	1	*
=	0	0	0	1	*	0	*
=	0	1	0	0	*	0	*
=	1	0	*	0	*	0	*

where the * denotes a value which is not in general $=0$, but which may in any particular case happen to be so.

For instance, let it be required for the binary quartic $(a, b, c, d, e\between x, y)^4$, to find the asyzygetic seminvariants of the degree 4 and weight 6. Assuming for the seminvariant the value in the left-hand column of the diagram, the unknown coefficients being A, B, C, D, E, F, G, this must be reduced to zero by the operation

$$a\partial_b + 2b\partial_c + 3c\partial_d + 4d\partial_e;$$

and we thus obtain as many equations as there are terms of the degree 4 and weight 5, as appearing by the second column

			$a\partial_b + 2b\partial_c + 3c\partial_d + 4d\partial_e$	
A	a^2ce	a^2be	$2C + 2A$	$=0,$
B	„ d^2	„ cd	$D + 6B + 4A$	$=0,$
C	ab^2e	ab^2d	$3F + 2D + 4C$	$=0,$
D	„ bcd	„ bc^2	$2G + 6E + 3D$	$=0,$
E	„ c^3	a^0b^3c	$4G + 3F$	$=0;$
F	a^0b^3d			
G	„ b^2c^2			

viz. the equations are

A	B	C	D	E	F	G	
2		+ 2					$=0,$
4	+ 6		+ 1				$=0,$
		4	+ 2		+ 3		$=0,$
			3	+ 6		+ 2	$=0,$
					3	+ 4	$=0.$

We have first a solution beginning $B=1$, and secondly a solution beginning $A=1$, with $B=0$: the resulting two seminvariants, say P and Q, are

		$P=$	$Q=$	I	II
A	a^2ce	0	1	1	1
B	„ d^2	1	0	− 1	0
C	ab^2e	0	− 1	− 1	− 1
D	„ bcd	− 6	− 4	+ 2	− 4
E	„ c^3	+ 4	3	− 1	+ 3
F	a^0b^3d	+ 4	4		+ 4
G	„ b^2c^2	− 3	− 3		+ 3 .

As is known, there is no irreducible solution, but only the composite forms

$$\text{I} \;= a\,(ace - ad^2 - b^2e + 2bcd - c^3),$$
$$\text{II} = (ac - b^2)\,(ae - 4bd + 3c^2),$$

the developed values of which are given above: II (as a form beginning $A = 1$ and with $B = 0$) can be nothing else than, and is in fact $= Q$: and so I (as a form beginning with $A = 1$, $B = -1$) can be nothing else than, and is in fact $= Q - P$; that is, we have

$$P = -\text{I} + \text{II}, \quad \text{or} \quad \text{I} = -P + Q,$$
$$Q = \quad \text{II}, \qquad \text{II} = \quad Q;$$

and so, in general, we have a standard set of values for the asyzygetic seminvariants of a given degree and weight; or, what is the same thing, for the covariants of a given deg-order.

801.

ON SEMINVARIANTS.

[From the *Quarterly Journal of Pure and Applied Mathematics*, vol. XIX. (1883), pp. 131—138.]

THE present paper is a somewhat fragmentary one, but it contains some results which seem to me to be worth putting on record.

I consider here not any binary quantic in particular, but the whole series $(a, b, c \between x, y)^2$, $(a, b, c, d \between x, y)^3$, &c.; or in a somewhat different point of view, I consider the indefinite series of coefficients $(a, b, c, d, e, \ldots)$; here, instead of covariants and invariants, we have only seminvariants; viz. a seminvariant is a function reduced to zero by the operator

$$\Delta = a\partial_b + 2b\partial_c + 3c\partial_d + \ldots;$$

for instance, seminvariants are

$$a,\quad ac - b^2,\quad a^2d - 3abc + 2b^3,\quad a^2d^2 + 4ac^3 + 4b^3d - 6abcd - 3b^2c^2,$$
$$ae - 4bd + 3c^2,\quad ace - ad^2 - b^2e + 2bcd - c^3,\ \&c.$$

A seminvariant is of a certain degree θ in the coefficients, and of a certain weight w (viz. the coefficients $a, b, c, d, \ldots$ are reckoned as being of the weights $0, 1, 2, 3, \ldots$ respectively); it is, moreover, of a certain rank ρ; viz. according as the highest letter therein is $a, c, d, e, \ldots$ (it is never b), the rank is taken to be $0, 2, 3, 4, \ldots$, and we have $w =$ or $< \tfrac{1}{2}\rho\theta$. The seminvariant may be regarded as belonging to a quantic $(a, \ldots \between x, y)^n$, the order of which, n, is equal to or greater than ρ; viz. in regard to such quantic the seminvariant, say A, is the leading coefficient of a covariant

$$(A, B, \ldots, K \between x, y)^{\mu},$$

where the weights of the successive coefficients are $w, w+1, \ldots$ up to $n\theta - w$; hence number of terms less unity, that is, μ, is $= n\theta - 2w$; the least value of μ is thus $= \rho\theta - 2w$, which is either zero, or positive; in the former case, $w = \tfrac{1}{2}\rho\theta$, the seminvariant is an invariant of the quantic $(a, \ldots \between x, y)^{\rho}$, the order of which is equal to the rank of the seminvariant; but if $w < \tfrac{1}{2}\rho\theta$, then it is the leading coefficient of a covariant $(A, B, \ldots, K \between x, y)^{\rho\theta - 2w}$ of the same quantic $(a, \ldots \between x, y)^{\rho}$; and in every case, taking $n > \rho$, the seminvariant is the leading coefficient A of a covariant

$$(A, B, \ldots, K \between x, y)^{n\theta - 2w}$$

of a quantic $(a, \ldots \between x, y)^n$.

Take A as belonging to the quantic $(a, \ldots \between x, y)^n$; corresponding to such quantic, we have an operator Λ of the same rank n, viz.

$$\begin{aligned}\Lambda &= 2b\partial_a + c\partial_b && \text{for } n=2,\\ &= 3b\partial_a + 2c\partial_b + d\partial_c && \quad ,, \quad 3,\\ &= 4b\partial_a + 3c\partial_b + 2d\partial_c + e\partial_d && \quad ,, \quad 4,\\ &\vdots && \qquad \vdots\end{aligned}$$

Operating with Λ on A, we have a series of terms

$$A,\ \Lambda A,\ \Lambda^2 A,\ \ldots,\ \Lambda^{n\theta-2w} A,$$

but the next term $\Lambda^{n\theta-2w+1}A$, and of course every succeeding term, is $=0$, and this being so, the coefficients of the covariant $(A, B, \ldots, K \between x, y)^{n\theta-2w}$ are

$$(1, \tfrac{1}{1}\Lambda, \tfrac{1}{1\,.\,2}\Lambda^2, \ldots) A,$$

or what is the same thing, each coefficient is obtained from the next preceding one by the formulæ

$$B = \tfrac{1}{1}\Lambda A, \quad C = \tfrac{1}{2}\Lambda B, \quad D = \tfrac{1}{3}\Lambda C, \ldots.$$

The coefficients A and K, B and $J, \ldots$ are derived one from the other by reversal of the order of the coefficients of $(a, b, \ldots \between x, y)^n$, with or without a change of sign, and thus it is only necessary to calculate up to the middle coefficient, or pair of coefficients; and we obtain, moreover, a verification.

Calculating in this manner the covariant

$$(A, B, \ldots, K)^{\rho\theta-2w},$$

which belongs to the quantic $(a, \ldots \between x, y)^\rho$, if we herein change $a, b, c, \ldots$ into $ax+by$, $bx+cy$, $cx+dy, \ldots$ we obtain the covariant belonging to the quantic $(a, \ldots \between x, y)^{\rho+1}$; and in this covariant making the like change, or what is the same thing, in the first-mentioned covariant changing $a, b, c, \ldots$ into $(a, b, c \between x, y)^2$, $(b, c, d \between x, y)^2$, $(c, d, e \between x, y)^2, \ldots$ we have the covariant belonging to $(a, \ldots \between x, y)^{\rho+2}$; and in like manner we obtain the covariant belonging to the quantic $(a, \ldots \between x, y)^n$ of any given order n.

In particular, if $w = \tfrac{1}{2}\rho\theta$, that is, if the given seminvariant be an invariant of $(a, \ldots \between x, y)^\rho$, then we obtain the series of covariants directly from A by therein changing $a, b, c, \ldots$ into $ax+by$, $bx+cy$, $cx+dy, \ldots$ and in the result making the like change; or what is the same thing, in A changing $a, b, c, \ldots$ into $(a, b, c \between x, y)^2$, $(b, c, d \between x, y)^2$, $(c, d, e \between x, y)^2, \ldots$: and so on until we obtain the covariant for the quantic $(a, \ldots \between x, y)^n$ of the given order n.

A seminvariant which cannot be expressed as a rational and integral function of lower seminvariants is said to be irreducible. The theory is distinct from that of the irreducible covariants of a quantic of a given order; for instance, as regards the cubic $(a, b, c, d \between x, y)^3$, we have the irreducible covariant (invariant)

$$a^2d^2 + 4ac^3 + 4b^3d - 6abcd - 3b^2c^2,$$

but this is not an irreducible seminvariant; it is

$$\begin{aligned}&= (ac-b^2)(ae-4bd+3c^2)\\ &\quad - a\,.\,(ace-ad^2-b^2e-c^3+2bcd),\end{aligned}$$

or, what is the same thing, there is not for the quartic $(a, b, c, d, e \between x, y)^4$, or for the higher quantics, any *irreducible* covariant having this for the leading coefficient.

We may consider the question to determine the number of asyzygetic seminvariants of a given degree and weight. For instance, taking the weights up to 12, so that the series of letters extends as far as m, then for the degrees 1, 2, 3 we have as follows:

$W=$	0	1	2	3	4	5	6	7	8	9	10	11	12
Deg. 1	a	b	c	d	e	f	g	h	i	j	k	l	m
Nos.	1	1	1	1	1	1	1	1	1	1	1	1	1
Diff.	0	0	0	0	0	0	0	0	0	0	0	0	0
Deg. 2	a^2	ab	ac	ad	ae	af	ag	ah	ai	aj	ak	al	am
			b^2	bc	bd	be	bf	bg	bh	bi	bj	bk	bl
					c^2	cd	ce	cf	cg	ch	ci	cj	ck
							d^2	de	df	dg	dh	di	dj
									e^2	ef	eg	eh	ei
											f^2	fg	fh
													g^2
Nos.	1	1	2	2	3	3	4	4	5	5	6	6	7
Diff.	1	0	1	0	1	0	1	0	1	0	1	0	1
Deg. 3	a^3	a^2b	a^2c	a^2d	a^2e	a^2f	a^2g	a^2h	a^2i	a^2j	a^2k	a^2l	a^2m
			ab^2	abc	abd	abe	abf	abg	abh	abi	abj	abk	abl
					ac^2	acd	ace	acf	acg	ach	aci	acj	ack
							ad^2	ade	adf	adg	adh	adi	adj
									ae^2	aef	aeg	aeh	aei
											af^2	afg	afh
													ag^2
				b^3	b^2c	b^2d	b^2e	b^2f	b^2g	b^2h	b^2i	b^2j	b^2k
						bc^2	bcd	bce	bcf	bcg	bch	bci	bcj
								bd^2	bde	bdf	bdg	bdh	bdi
										be^2	bef	beg	beh
												bf^2	bfg
							c^3	c^2d	c^2e	c^2f	c^2g	c^2h	c^2i
									cd^2	cde	cdf	cdg	cdh
											ce^2	cef	ceg
													cf^2
										d^3	d^2e	d^2f	d^2g
												de^2	def
													e^3
Nos.	1	1	2	3	4	5	7	8	10	12	14	16	19
Diff.	1	0	1	1	1	2	1	2	2	2	2	2	3

For the degree 1, the line of differences shows that the only seminvariant is ($W=0$), the seminvariant a.

For the degree 2, the line of differences 1, 0, 1, 0, ..., shows that the number of seminvariants is $=1$ for each even degree, $=0$ for each odd degree; thus for the weight 0 there is a seminvariant $=a^2$, which of course is not irreducible; while for each of the other even weights we have a single irreducible seminvariant; as is well known, the forms are

$W=$ 2	4	6	8	10	12
$ac+1$	$ae+1$	$ag+1$	$ai+1$	$ak+1$	$am+1$
b^2-1	$bd-4$	$bf-6$	$bh-8$	$bj-10$	$bl-12$
	c^2+3	$ce+15$	$cg+28$	$ci+45$	$ck+66$
		d^2-10	$df-56$	$dh-120$	$dj-220$
			e^2+35	$eg+210$	$ei+495$
				f^2-126	$fh-792$
					g^2+462

For degree 3, the line of differences shows that for

W	$=0$	1	2	3	4	5	6	7	8	9	10	11	12
Nos. are	$=1$	0	1	1	1	1	2	1	2	2	2	2	3;

but inasmuch as for each even weight there is a quadric seminvariant, which multiplied by a gives a cubic seminvariant, to obtain the number of irreducible cubic seminvariants we subtract

1	0	1	0	1	0	1	0	1	0	1	0	1
0	0	0	1	0	1	1	1	1	2	1	2	2,

or the numbers of irreducible cubic seminvariants are as in the line last written down.

There is a convenience however in giving, for each even weight, as well the rejected reducible covariant; and the entire series of results is found to be

$W=$ 0	2	3	4	5	6		7	8	
a^3+1	a^2c+1	a^2d+1	a^2e+1	a^2f+1	a^2g+1		a^2h+1	a^2i+1	
	ab^2-1	$abc-3$	$abd-4$	$abe-5$	$abf-6$		$abg-7$	$abh-8$	
		b^3+2	ac^2+3	$acd+2$	$ace+15$	$+1$	$acf+9$	$acg+28$	$+1$
				b^2d+8	ad^2-10	-1	$ade-5$	$adf-56$	-3
				bc^2-6	b^2e	-1	b^2f+12	ae^2+35	$+2$
					bcd	$+2$	$bce-30$	b^2g	-1
					c^3	-1	bd^2+20	bcf	$+3$
							c^2d	bde	-1
								c^2e	-3
								cd^2	$+2$

9		10		11		12		
a^2j+1		a^2k+1		a^2l+1		a^2m	$+1$	
$abi-9$		$abj-10$		$abk-11$		abl	-12	
ach	$+2$	$aci+45$	$+1$	$acj+35$	$+2$	ack	$+66$	$+3$
$adg+42$	-7	$adh-120$	-4	$adi-75$	-9	adj	-220	-15
$aef-36$	$+5$	$aeg+210$	$+8$	$aeh+90$	$+14$	aei	$+495$	$+40$	$+1$
b^2h+36	-2	af^2-126	-5	$afg-42$	-7	afh	-792	-70	-4
$bcg-126$	$+7$	b^2i	-1	b^2j+20	-2	ag^2	$+462$	$+42$	$+3$
$bdf-108$	$+22$	bch	$+4$	$bci-90$	$+9$	b^2k		-3	
be^2+180	-25	bdg	-4	$bdh+240$	$+16$	bcj		$+15$	
c^2f+270	-27	bef	$+2$	$beg-420$	-63	bdi		-25	-4
$cde-450$	$+45$	c^2g	-4	bf^2+252	$+42$	beh		$+30$	$+12$
d^3+200	-20	cdf	$+8$	c^2h	-30	bfg		-14	-8
		ce^2	-5	cdg	$+70$	c^2i		-15	$+3$
		d^2e		cef	-21	cdh		$+40$	-8
				d^2f	-56	ceg		-70	-22
				de^2	$+35$	cf^2		$+42$	$+24$
						d^2g			$+24$
						def			-36
						e^3			$+15$

The canonical form given for the quintic in my Tenth Memoir on Quantics [693] belongs to a series, viz. writing now the small roman letters (instead of the italic letters) for the series of coefficients, and using the italic letters $a, c, d, e, f, \ldots$ to denote seminvariants, they are as follows:

0 $a = \mathrm{a}$,
2 $c = \mathrm{ac - b^2}$,
3 $d = \mathrm{a^2d - 3abc + 2b^3}$ (= f in the tenth memoir),
4 $e = \mathrm{ae - 4bd + 3c^2}$ (= b in the tenth memoir),
5 $f = \mathrm{a^2f - 5abe + 2acd + 8b^2d - 6bc^2}$,
6 $g = \mathrm{ag - 6bf + 15ce - 10d^2}$,
7 $h = \mathrm{a^2h - 7abg + 9acf - 5ade + 12b^2f - 30bce + 20bd^2}$,
8 $i = \mathrm{ai - 8bh + 28cg - 56df + 35e^2}$,
&c.

Writing also (instead of d in the tenth memoir)

$$\epsilon = \mathrm{ace - ad^2 - b^2e + 2bcd - c^3},$$

so that the equation $a^3d - a^2bc + 4c^3 - f^2 = 0$ of the tenth memoir, is in the present notation $a^3\epsilon - a^2ec + 4c^3 - d^2 = 0$, then the series of canonical forms is

Quadric $(1,\ 0,\ c \between x, y)^2$,
Cubic $(1,\ 0,\ c,\ d \between x, y)^3$,
Quartic $(1,\ 0,\ c,\ d,\ a^2e - 3c^2 \between x, y)^4$,
Quintic $(1,\ 0,\ c,\ d,\ a^2e - 3c^2,\ a^2f - 2cd \between x, y)^5$,
&c.

the series of coefficients being

1,	0,	c,	d,	a^2e	$+1$,	a^2f	$+1$,	a^4g	$+\ 1$,	a^4h	$+1$,	a^6i	$+\ 1$,
				c^2	-3	cd	-2	a^2ce	$-\ 15$	a^2cf	-9	a^4cg	$-\ 28$
								c^3	$+45$	a^2de	$+5$	a^4e^2	$-\ 35$
								d^2	$+10$	c^2d	$+3$	a^2c^2e	$+\ 630$
												a^2df	$+\ 56$
												c^4	-1575
												cd^2	$-\ 392$

these values being, in fact, the expressions in terms of the seminvariants a, c, d, &c. of

1,	0,	ac,	$\mathrm{a^2d}$,	$\mathrm{a^3e}$,	$\mathrm{a^4f}$,	$\mathrm{a^5g}$,	$\mathrm{a^6h}$,	$\mathrm{a^7i}$
		$-\mathrm{b^2}$,	$-\ 3\mathrm{abc}$,	$-\ 4\mathrm{a^2bd}$,	$-\ 5\mathrm{a^3be}$,	$-\ 6\mathrm{a^4bf}$,	$-\ 7\mathrm{a^5bg}$,	$-\ 8\mathrm{a^6bh}$
			$+\ 2\ \mathrm{b^3}$,	$+\ 6\mathrm{ab^2c}$,	$+\ 10\mathrm{a^2b^2d}$,	$+\ 15\mathrm{a^3b^2e}$,	$+\ 21\mathrm{a^4b^2f}$,	$+\ 28\mathrm{a^5b^2g}$
				$-\ 3\ \mathrm{b^4}$,	$-\ 10\mathrm{ab^3c}$,	$-\ 20\mathrm{a^2b^3d}$,	$-\ 35\mathrm{a^3b^3e}$,	$-\ 56\mathrm{a^4b^3f}$
					$+\ 4\ \mathrm{b^5}$,	$+\ 15\mathrm{ab^4c}$,	$+\ 35\mathrm{a^2b^4d}$,	$+\ 70\mathrm{a^3b^4e}$
						$-\ 5\ \mathrm{b^6}$,	$-\ 21\mathrm{ab^5c}$,	$-\ 56\mathrm{a^2b^5d}$
							$+\ 6\ \mathrm{b^7}$,	$+\ 28\mathrm{ab^6c}$
								$-\ 7\ \mathrm{b^8}$

I annex verifications of the foregoing values:

	$a^2e =$	$-3c^2$	
a^3e	1		+1
a^2bd	−4		−4
a^2c^2	+3	−3	0
ab^2c		+6	+6
b^4		−3	−3

	$a^2f =$	$-2cd$	
a^4f	+1		+1
a^3be	−5		−5
a^3cd	+2	−2	0
a^2b^2d	+8	+2	+10
a^2bc^2	−6	+6	0
ab^3c		−10	−10
b^5		+4	+4

	$a^4g =$	$-15a^2ce$	$+45c^3$	$+10d^2$	
a^5g	+1				1
a^4bf	−6				−6
a^4ce	+15	−15			0
a^4d^2	−10			+10	0
a^3b^2e		+15			+15
a^3bcd		+60		−60	0
a^3c^3		−45	+45		0
a^2b^3d		−60		+40	−20
$a^2b^2c^2$		+45	−135	+90	0
ab^4c			+135	−120	+15
b^6			−45	+40	−5

	$a^4h =$	$-9a^2cf$	$+5a^2de$	$+3c^2d$	
a^6h	+1				+1
a^5bg	−7				−7
a^5cf	+9	−9			0
a^5de	−5		+5		0
a^4b^2f	+12	+9			+21
a^4bce	−30	+45	−15		0
a^4bd^2	+20		−20		0
a^4c^2d		−18	+15	+3	0
a^3b^3e		−45	+10		−35
a^3b^2cd		−54	+60	−6	0
a^3bc^3		+54	−45	−9	0
a^2b^4d		+72	−40	+3	+35
$a^2b^3c^2$		−54	+30	+24	0
ab^5c				−21	−21
b^7				+6	+6

	$a^6i =$	$-28a^4cg$	$-35a^4e^2$	$+56a^2df$	$+630a^2c^2e^2$	$-1575c^4$	$-392c^2d$	
a^7i	+ 1							+ 1
a^6bh	− 8							− 8
cg	+ 28	− 28						0
df	− 56				+ 56			0
e^2	+ 35		− 35					0
a^5b^2g		+ 28						+ 28
bcf		+ 168			− 168			0
bde			+ 280		− 280			0
c^2e		− 420	− 210	+ 630				0
cd^2		+ 280			+ 112		− 392	0
a^4b^3f		− 168			+ 112			− 56
b^2ce		+ 420		− 1260	+ 840			0
b^2d^2		− 280	− 560		+ 448		+ 392	0
bc^2d			+ 840	− 2520	− 672		+ 2352	0
c^4			− 315	+ 1890		− 1575		0
a^3b^4e				+ 630	− 560			+ 70
b^3cd				+ 5040	− 1120		− 3920	0
b^2c^3				− 3780	+ 1008	+ 6300	− 3528	0
a^2b^5d				− 2520	+ 896		+ 1568	− 56
b^4c^2				+ 1890	− 672	− 9450	+ 8232	0
$a\,b^6d$						+ 6300	− 6272	+ 28
a^0b^8						− 1575	+ 1568	− 7

It would be interesting to obtain the general law for the expressions of the canonical coefficients in terms of the seminvariants.

802.

NOTE ON CAPTAIN MACMAHON'S PAPER, "ON THE DIFFERENTIAL EQUATION $X^{-\frac{2}{3}}dx + Y^{-\frac{2}{3}}dy + Z^{-\frac{2}{3}}dz = 0$."

[From the *Quarterly Journal of Pure and Applied Mathematics*, vol. XIX. (1883), pp. 182—184.]

IN general, if $f, = (x, y, 1)^3, = 0$ be the equation of a cubic curve, and if

$$d\omega = \frac{dx}{\frac{df}{dy}}, = \frac{-dy}{\frac{df}{dx}},$$

then if 1, 2, 3 are the intersections of the curve by an arbitrary right line, the coordinates of these points being (x_1, y_1), (x_2, y_2), (x_3, y_3) respectively, we have, by Abel's theorem,

$$d\omega_1 + d\omega_2 + d\omega_3 = 0,$$

viz. this is the differential relation corresponding to the integral relation which expresses that the three points are the intersections of the cubic curve by a right line, or say to the integral equation

$$\nabla, = \begin{vmatrix} y_1, & y_2, & y_3 \\ x_1, & x_2, & x_3 \\ 1, & 1, & 1 \end{vmatrix}, = 0,$$

in which equation y_1, y_2, y_3 are regarded as functions of x_1, x_2, x_3 respectively, given by means of the equations $f_1 = 0$, $f_2 = 0$, $f_3 = 0$ which express that the points are on the cubic curve. See my "Memoir on the Abelian and Theta Functions," [819].

In particular, if the equation of the curve is

$$f, = \tfrac{1}{3}\{y^3 - (A + 3Bx + 3Cx^2 + Dx^3)\}, = \tfrac{1}{3}(y^3 - X), = 0,$$

then

$$d\omega = \frac{dx}{y^2}, \quad = \frac{dx}{X^{\frac{2}{3}}}:$$

and corresponding to the differential relation

$$X_1^{-\frac{2}{3}}\, dx_1 + X_2^{-\frac{2}{3}}\, dx_2 + X_3^{-\frac{2}{3}}\, dx_3 = 0,$$

we have the integral relation

$$\nabla, \quad = \begin{vmatrix} X_1^{\frac{1}{3}}, & X_2^{\frac{1}{3}}, & X_3^{\frac{1}{3}} \\ x_1\;, & x_2\;, & x_3 \\ 1\;, & 1\;, & 1 \end{vmatrix} = 0,$$

viz. this last equation, as containing no arbitrary constant, is a particular integral of the differential equation.

If instead of x_1, x_2, x_3 we write x, y, z, then we have

$$\nabla = \begin{vmatrix} X^{\frac{1}{3}}, & Y^{\frac{1}{3}}, & Z^{\frac{1}{3}} \\ x\;, & y\;, & z \\ 1\;, & 1\;, & 1 \end{vmatrix}, \quad = 0,$$

as a particular integral of the differential equation

$$X^{-\frac{2}{3}}\, dx + Y^{-\frac{2}{3}}\, dy + Z^{-\frac{2}{3}}\, dz = 0.$$

To rationalize the integral equation, write

$$\alpha,\ \beta,\ \gamma = y - z,\ z - x,\ x - y \text{ (so that } \alpha + \beta + \gamma = 0),$$

the equation is

$$\alpha X^{\frac{1}{3}} + \beta Y^{\frac{1}{3}} + \gamma Z^{\frac{1}{3}} = 0;$$

and we thence have

$$\alpha^3 X + \beta^3 Y + \gamma^3 Z = 3\alpha\beta\gamma X^{\frac{1}{3}} Y^{\frac{1}{3}} Z^{\frac{1}{3}}.$$

The left-hand side is

$$\begin{aligned} &\alpha^3\,(A + 3Bx + 3Cx^2 + Dx^3) \\ &+ \beta^3\,(A + 3By + 3Cy^2 + Dy^3) \\ &+ \gamma^3\,(A + 3Bz + 3Cz^2 + Dz^3); \end{aligned}$$

or assuming

$$\alpha',\ \beta',\ \gamma' = x\,(y - z),\ y\,(z - x),\ z\,(x - y) \text{ (so that } \alpha' + \beta' + \gamma' = 0),$$

this is

$$= A\,(\alpha^3 + \beta^3 + \gamma^3) + 3B\,(\alpha^2\alpha' + \beta^2\beta' + \gamma^2\gamma') + 3C\,(\alpha\alpha'^2 + \beta\beta'^2 + \gamma\gamma'^2) + D\,(\alpha'^3 + \beta'^3 + \gamma'^3).$$

But taking λ arbitrary, and

$$a,\ b,\ c = \alpha + \lambda\alpha',\quad \beta + \lambda\beta',\quad \gamma + \lambda\gamma',$$

then

$$a + b + c = 0,$$

whence

$$a^3 + b^3 + c^3 = 3abc;$$

or substituting for a, b, c their values, and comparing the coefficients of the several powers of λ,

$$\begin{aligned}
\alpha^3 + \beta^3 + \gamma^3 &= 3\alpha\beta\gamma,\\
\alpha^2\alpha' + \beta^2\beta' + \gamma^2\gamma' &= \alpha'\beta\gamma + \beta'\gamma\alpha + \gamma'\alpha\beta = \alpha\beta\gamma\,(x + y + z),\\
\alpha\alpha'^2 + \beta\beta'^2 + \gamma\gamma'^2 &= \alpha\beta'\gamma' + \beta\gamma'\alpha' + \gamma\alpha'\beta' = \alpha\beta\gamma\,(yz + zx + xy),\\
\alpha'^3 + \beta'^3 + \gamma'^3 &= 3\alpha'\beta'\gamma' = 3\alpha\beta\gamma\,.\,xyz.
\end{aligned}$$

Hence we have

$$\alpha^3X + \beta^3Y + \gamma^3Z = 3\alpha\beta\gamma\,\{A + B(x+y+z) + C(yz + zx + xy) + Dxyz\},$$

or the integral equation is

$$\{A + B(x+y+z) + C(yz + zx + xy) + Dxyz\} = X^{\frac{1}{3}}Y^{\frac{1}{3}}Z^{\frac{1}{3}},$$

that is,

$$\{A + B(x+y+z) + C(yz + zx + xy) + Dxyz\}^3 = XYZ,$$

the elegant result given by Capt. MacMahon at the beginning of his paper.

The author in a letter to me, dated Jan. 13, 1883, remarks that the particular integral of the equation in question

$$X^{-\frac{2}{3}}\,dx + Y^{-\frac{2}{3}}\,dy + Z^{-\frac{2}{3}}\,dz = 0,$$

is expressible as a determinant in a rational form as follows. Writing it $XYZ = P^3$, where

$$P = A + B(x+y+z) + C(yz + zx + xy) + Dxyz,$$

then the form is

$$\nabla, = \begin{vmatrix} 1, & \left(\frac{1}{3}P\frac{dX}{dx} - X\frac{dP}{dx}\right)^3, & X \\ 1, & \left(\frac{1}{3}P\frac{dY}{dy} - Y\frac{dP}{dy}\right)^3, & Y \\ 1, & \left(\frac{1}{3}P\frac{dZ}{dz} - Z\frac{dP}{dz}\right)^3, & Z \end{vmatrix} = 0,$$

for, as shown by Captain MacMahon in his paper, each of the three terms such as

$$Z\left(\tfrac{1}{3}P\frac{dY}{dy} - Y\frac{dP}{dy}\right)^3 - Y\left(\tfrac{1}{3}P\frac{dZ}{dz} - Z\frac{dP}{dz}\right)^3,$$

which compose the determinant, is divisible by $XYZ - P^3$.

It may be added that we have identically

$$\tfrac{1}{3}P\frac{dX}{dx} - X\frac{dP}{dx} = (AC - B^2)(2x - y - z) + (AD - BC)(x^2 - yz) + (BD - C^2)(x^2y + x^2z - 2xyz),$$

and of course like values for the other two expressions in the determinant.

803.

ON MR ANGLIN'S FORMULA FOR THE SUCCESSIVE POWERS OF THE ROOT OF AN ALGEBRAICAL EQUATION.

[From the *Quarterly Journal of Pure and Applied Mathematics*, vol. XIX. (1883), pp. 223, 224.]

SUPPOSE $x^m - px^{m-1} + qx^{m-2} - \ldots = 0$, then the successive powers x^m, x^{m+1}, x^{m+2}, &c. of x can be expressed in the form $Px^{m-1} - Qx^{m-2} + Rx^{m-3} - \&c.$ Mr Anglin has obtained for this purpose a very elegant formula, with a demonstration which (it occurred to me) might be presented under a somewhat simplified form; and he has permitted me to draw up the present Note.

Take, for greater convenience, the equation to be

$$x^4 - px^3 + qx^2 - rx + s = 0,$$

and let $h_1, h_2, h_3, \ldots$ be the sums of the homogeneous products of the roots, of the orders 1, 2, 3, &c. respectively; then, writing also $h_0 = 1$, we have

$$\begin{aligned}
h_1 &= h_0 p,\\
h_2 &= h_1 p - h_0 q,\\
h_3 &= h_2 p - h_1 q + h_0 r,\\
h_4 &= h_3 p - h_2 q + h_1 r - h_0 s,\\
h_5 &= h_4 p - h_3 q + h_2 r - h_1 s,\\
&\vdots
\end{aligned}$$

And this being so, starting from the equation

$$x^4 = px^3 - \quad qx^2 + \quad rx - s,$$

that is,

$$= h_1 x^3 - h_0 qx^2 + h_0 rx - h_0 s,$$

we obtain successively

$$\begin{aligned}
x^5 = {} & h_1(px^3 - qx^2 + rx - s) \\
& - h_0qx^3 + h_0rx^2 - h_0sx \\
= {} & h_2x^3 - (h_1q - h_0r)\,x^2 + (h_1r - h_0s)\,x - h_1s, \\
x^6 = {} & h_2(px^3 - qx^2 + rx - s) \\
& - (h_1q - h_0r)\,x^3 + (h_1r - h_0s)\,x^2 - h_1sx \\
= {} & h_3x^3 - (h_2q - h_1r + h_0s)\,x^2 + (h_2r - h_1s)\,x - h_2s, \\
x^7 = {} & h_3(px^3 - qx^2 + rx - s) \\
& - (h_2q - h_1r + h_0s)\,x^3 + (h_2r - h_1s)\,x^2 - h_2sx \\
= {} & h_4x^3 - (h_3q - h_2r + h_1s)\,x^2 + (h_3r - h_2s)\,x - h_3s,
\end{aligned}$$

and so on, the characteristic feature being that by the introduction of the symbols h, the coefficient of x^3 presents itself at each step as a monomial, and the coefficients of the lower powers require no reduction. It is obvious that the process is a perfectly general one, and that for the equation

$$x^m - p_1x^{m-1} + p_2x^{m-2} - \ldots + (-)^m p_m = 0,$$

the formula is

$$\begin{aligned}
x^{m+\theta} = {} & \qquad h_{\theta+1}\,x^{m-1} \\
& - \quad (h_\theta p_2 - h_{\theta-1}p_3 + \ldots\ )\,x^{m-2} \\
& + \quad (h_\theta p_3 - h_{\theta-1}p_4 + \ldots\ )\,x^{m-3} \\
& \qquad \vdots \\
& + (-)^{s-1}\,(h_\theta p_s - h_{\theta-1}p_{s+1} + \ldots)\,x^{m-s} \\
& \qquad \vdots \\
& + (-)^{m-1}\,.\,h_\theta p_m \qquad x^0,
\end{aligned}$$

where, as regards each power of x, the series forming the coefficient thereof is continued as far as possible, that is, up to the term which contains p_m or h_0 as the case may be.

804.

ON THE ELLIPTIC-FUNCTION SOLUTION OF THE EQUATION $x^3+y^3-1=0$.

[From the *Proceedings of the Cambridge Philosophical Society*, vol. IV. (1883), pp. 106—109.]

I HAD occasion to find elliptic-function expressions for the coordinates (x, y) of a point on the cubic curve $x^3+y^3=1$. These are derivable from the formulæ given, Legendre, *Fonctions Elliptiques*, t. I. pp. 185, 186, for the reduction to elliptic integrals of the integral $R=\int \frac{dr}{(1-z^3)^{\frac{2}{3}}}$. Legendre, writing

$$z=\frac{\sqrt{4y^3-1}-\sqrt{3}}{\sqrt{4y^3-1}+\sqrt{3}},$$

and then

$$m^3=2 \text{ and } m^2y=1+x^2,$$

finds first

$$R=m\sqrt{3}\int\frac{dx}{\sqrt{x^4+3x^2+3}};$$

and then writing $r=\sqrt[4]{3}$, $x=\tan\frac{1}{2}\phi$, and $c^2=\frac{1}{4}(2-r^2)$, finds

$$R=\tfrac{1}{2}mr\int\frac{d\phi}{\sqrt{1-c^2\sin^2\phi}};$$

we have therefore only to write $\sin\phi=\operatorname{sn} u$, to modulus

$$c, =\tfrac{1}{2}\sqrt{2-\sqrt{3}},$$

and we thence obtain an expression for z in terms of the elliptic functions $\operatorname{sn} u$, $\operatorname{cn} u$, $\operatorname{dn} u$.

Writing x instead of z, and k for c, then

$$m=\sqrt[3]{2},\quad r=\sqrt[4]{3};\quad k=\tfrac{1}{2}\sqrt{2-r^2},\quad k'=\tfrac{1}{2}\sqrt{2+r^2}.$$

Working out the substitutions, the resulting formulæ are

$$x = \frac{2r \operatorname{sn} u \operatorname{dn} u - (1 + \operatorname{cn} u)^2}{2r \operatorname{sn} u \operatorname{dn} u + (1 + \operatorname{cn} u)^2},$$

$$y = \frac{m(1 + \operatorname{cn} u)\{1 + r^2 + (1 + r^2) \operatorname{cn} u\}}{2r \operatorname{sn} u \operatorname{dn} u + (1 + \operatorname{cn} u)^2},$$

where the modulus is k as above; and these values give

$$x^3 + y^3 = 1,$$

$$\frac{dx}{(1 - x^3)^{\frac{2}{3}}}, \quad = \frac{-dy}{(1 - y^3)^{\frac{2}{3}}} = \tfrac{1}{2} mr\, du.$$

The verification is interesting enough; starting from the expression for x, and for shortness representing it by

$$x = \frac{A - B}{A + B},$$

we have

$$1 - x^3 = \frac{2B(3A^2 + B^2)}{(A + B)^3}, \quad = \frac{m^3 (1 + \operatorname{cn} u)^2 (3A^2 + B^2)}{\{2r \operatorname{sn} u \operatorname{dn} u + (1 + \operatorname{cn} u)^2\}^3}.$$

We find

$$3A^2 + B^2 = 12r^2 \operatorname{cn}^2 u \operatorname{dn}^2 u + (1 + \operatorname{cn} u)^4,$$

$$= (1 + \operatorname{cn} u)\{12r^2 (1 - \operatorname{cn} u)(k'^2 + k^2 \operatorname{cn}^2 u) + (1 + \operatorname{cn} u)^3\},$$

where the term in { } is a perfect cube

$$= [1 + \operatorname{cn} u + r^2 (1 - \operatorname{cn} u)]^3.$$

The last-mentioned expression is, in fact,

$$= (1 + \operatorname{cn} u)^3 + r^2 (1 - \operatorname{cn} u)[3(1 + \operatorname{cn} u)^2 + 3r^2 (1 + \operatorname{cn} u)(1 - \operatorname{cn} u) + r^4 (1 - \operatorname{cn} u)^2],$$

where the second term is

$$= 12r^2 (1 - \operatorname{cn} u)[\tfrac{1}{2}(1 + \operatorname{cn}^2 u) + \tfrac{1}{4} r^2 (1 - \operatorname{cn}^2 u)],$$

that is, it is

$$= 12r^2 (1 - \operatorname{cn} u)(k'^2 + k^2 \operatorname{cn}^2 u)).$$

We have consequently

$$1 - x^3 = \frac{m^3 (1 + \operatorname{cn} u)^3 \{1 + r^2 + (1 - r^2) \operatorname{cn} u\}^3}{\{2r \operatorname{sn} u \operatorname{dn} u + (1 + \operatorname{cn} u)^2\}^3},$$

or extracting the cube root y, $= \sqrt{1 - x^3}$, has its foregoing value: and the differential expressions are then verified.

Suppose $y = 1$, we have

$$(m - 1)(1 + \operatorname{cn} u)^2 + mr^2 (1 - \operatorname{cn}^2 u) = 2r \operatorname{sn} u \operatorname{dn} u,$$

that is,

$$(m-1)^2(1+\text{cn}\,u)^3+2m(m-1)r^2(1+\text{cn}\,u)^2(1-\text{cn}\,u)$$
$$+3m^2(1+\text{cn}\,u)(1-\text{cn}\,u)^2=r^2(1-\text{cn}\,u)\{4-4k^2(1-\text{cn}^2\,u)\},$$

or observing that the right-hand side is

$$=r^2(1-\text{cn}\,u)\{(1+\text{cn}\,u)^2+(1-\text{cn}\,u)^2+r^2(1+\text{cn}\,u)(1-\text{cn}\,u)\},$$

and multiplying by $\frac{1}{3}r^2$, the equation becomes

$$0=\tfrac{1}{3}(m-1)^2r^2(1+\text{cn}\,u)^3+(2m^2-2m+1)(1+\text{cn}\,u)^2(1-\text{cn}\,u)$$
$$+(m^2-1)r^2(1+\text{cn}\,u)(1-\text{cn}\,u)^2-(1-\text{cn}\,u)^3;$$

viz. this is

$$0=\{\tfrac{1}{3}r^2(m^2-1)(1+\text{cn}\,u)-(1-\text{cn}\,u)\}^3,$$

as is immediately verified: hence writing $\frac{1}{3}r^2=\dfrac{1}{r^2}$, we have for the value in question, $y=1$,

$$(m^2-1)(1+\text{cn}\,u)-r^2(1-\text{cn}\,u)=0,$$

or say

$$m^2(1+\text{cn}\,u)=(1+\text{cn}\,u)+r^2(1-\text{cn}\,u),$$

that is,

$$\text{cn}\,u=\frac{r^2+1-m^2}{r^2-1+m^2},$$

which is one of the values of $\text{cn}\,u$ derived from the equation $x=0$; but this equation $x=0$ gives, not the foregoing equation, but

$$m^6(1+\text{cn}\,u)^3=\{(1+\text{cn}\,u)+r^2(1-\text{cn}\,u)\}^3,$$

viz. the three values of $\text{cn}\,u$ are the foregoing value and the two values obtained therefrom by changing m into ωm and $\omega^2 m$ respectively, ω being an imaginary cube root of unity. In fact, the curve $x^3+y^3=1$ has at the point $x=0$, $y=1$ an inflexion, the tangent being $y=1$, so that this line meets the curve in the point counting three times; but the line $x=0$ meets the curve in the point, and besides in two imaginary points.

805.

NOTE ON ABEL'S THEOREM.

[From the *Proceedings of the Cambridge Philosophical Society*, vol. IV. (1883), pp. 119—122.]

CONSIDERING Abel's theorem in so far as it relates to the first kind of integrals, and as a differential instead of an integral theorem, the theorem may be stated as follows:

We have a fixed curve $f(x, y, 1) = 0$ of the order m; this implies a relation $f'(x)\,dx + f'(y)\,dy = 0$, between the differentials dx, dy of the coordinates of a point on the curve; and we may therefore write

$$d\omega = \frac{dx}{f'(y)} = -\frac{dy}{f'(x)},$$

and, instead of dx or dy, use $d\omega$ to denote the displacement of a point (x, y) on the curve.

Taking for greater simplicity the fixed curve to be a curve without nodes or cusps, and therefore of the deficiency $\frac{1}{2}(m-1)(m-2)$, we consider its mn intersections by a variable curve $\phi(x, y, 1) = 0$ of the order n. And then, if $(x, y, 1)^{m-3}$ denote an arbitrary rational and integral function of (x, y) of the order $m-3$, the theorem is that we have between the displacements $d\omega_1, d\omega_2, \ldots, d\omega_{mn}$ of the mn points of intersection, the relation

$$\Sigma (x, y, 1)^{m-3}\, d\omega = 0,$$

where the left-hand side is the sum of the values of $(x, y, 1)^{m-3}\, d\omega$, belonging to the mn points of intersection respectively.

For the proof, observe that, varying in any manner the curve ϕ, we obtain

$$\frac{d\phi}{dx}\,dx + \frac{d\phi}{dy}\,dy + \delta\phi = 0,$$

where $\delta\phi$ is that part which depends on the variation of the coefficients, of the whole variation of ϕ; viz. if $\phi = ax^n + bx^{n-1}y + \ldots$, then $\delta\phi = x^n da + x^{n-1}y\,db + \ldots$; $\delta\phi$ is thus, in regard to the coordinates (x, y), a rational and integral function of the order n. Writing in this equation

$$dx,\ dy = \frac{df}{dy}\,d\omega,\quad -\frac{df}{dx}\,d\omega,$$

the equation becomes

$$\left(\frac{d\phi}{dx}\frac{df}{dy} - \frac{d\phi}{dy}\frac{df}{dx}\right)d\omega + \delta\phi = 0,$$

or say

$$-J(f, \phi)\,d\omega + \delta\phi = 0,$$

that is,

$$d\omega = \frac{\delta\phi}{J(f, \phi)};$$

and then multiplying each side by the arbitrary function $(x, y, 1)^{m-3}$, we have

$$\Sigma\,(x, y, 1)^{m-3}\,d\omega = \Sigma\,\frac{(x, y, 1)^{m-3}}{J(f, \phi)}\,\delta\phi,$$

where $\delta\phi$ being of the order n in the variables, the numerator is a rational and integral function of (x, y) of the order $m+n-3$: hence by a theorem contained in Jacobi's paper "Theoremata nova algebraica circa systema duarum æquationum inter duas variabiles," *Crelle*, t. XIV. (1835), pp. 281—288, [*Ges. Werke*, t. III., pp. 285—294], the sum on the right-hand side is $=0$: hence the required result $\Sigma\,(x, y, 1)^{m-3}\,d\omega = 0$.

Observing that $(x, y, 1)^{m-3}$ is an arbitrary function, the equation just obtained breaks up into the equations

$$\Sigma\,d\omega = 0,\quad \Sigma x\,d\omega = 0,\quad \Sigma y\,d\omega = 0, \ldots,\quad \Sigma x^{m-3}\,d\omega = 0, \ldots,\quad \Sigma y^{m-3}\,d\omega = 0,$$

viz. the number of equations is

$$1 + 2 + \ldots + (m-2),\quad = \tfrac{1}{2}(m-1)(m-2),$$

which is $=p$, the deficiency of the curve.

Suppose the fixed curve $f(x, y, 1) = 0$ is a cubic, $m=3$, and we have the single relation $\Sigma\,d\omega = 0$, where the summation refers to the $3n$ points of intersection of the cubic and of the variable curve of the order n, $\phi(x, y, 1) = 0$.

In particular, if this curve be a line, $n=1$, and the equation is $d\omega_1 + d\omega_2 + d\omega_3 = 0$; here the two points (x_1, y_1), (x_2, y_2) taken at pleasure on the cubic, determine the line, and they consequently determine uniquely the third point of intersection (x_3, y_3); there should thus be a single equation giving the displacement $d\omega_3$ in terms of the displacements $d\omega_1$, $d\omega_2$; viz. this is the equation just found

$$d\omega_1 + d\omega_2 + d\omega_3 = 0.$$

So if the variable curve be a conic, $n=2$; and we have between the displacements of the six points the relation

$$d\omega_1 + d\omega_2 + \ldots + d\omega_6 = 0:$$

here five of the points determine the conic, and they therefore determine uniquely the sixth point; and there should be between the displacements a single relation as just found.

If the variable curve be a cubic, $n=3$, and we have between the displacements of the nine points the relation

$$d\omega_1 + d\omega_2 + \ldots + d\omega_9 = 0:$$

here eight of the points do *not* determine the cubic ϕ, but they nevertheless determine the ninth point, viz. (reproducing the reasoning which establishes this well-known and fundamental theorem as to cubic curves) if $\phi_0=0$ be a particular cubic through the 8 points, then the general cubic is $\phi_0+kf=0$, and the intersections with $f=0$ are given by the equations $\phi_0=0$, $f=0$; whence the ninth point is independent of k, and is determined uniquely by the 8 points. There should thus be a single relation between the displacements, viz. this is the relation just found.

And so if the variable curve be a quartic, or curve of any higher order, it appears in like manner that there should be a single relation between the displacements; this relation being in fact the foregoing relation $\Sigma d\omega = 0$.

But take the fixed curve to be a quartic, $m=4$: then we have between the displacements $d\omega$ the relation

$$\Sigma (x,\ y,\ 1)\, d\omega = 0,$$

that is, the three equations

$$\Sigma x\, d\omega = 0,\quad \Sigma y\, d\omega = 0,\quad \Sigma\, d\omega = 0.$$

If the variable curve is a conic, $n=2$, then there are 8 points of intersection; 5 of these taken at pleasure determine the conic, and they consequently determine the remaining 3 points of intersection: hence there should be 3 equations. And so if the variable curve be a curve of any higher order, then by considerations similar to those made use of in the case where the first curve is a cubic it appears that the number of equations between the displacements $d\omega$ should always be $=3$.

But if the variable curve be a line, $n=1$, then the number of the points of intersection is $=4$: 2 of these taken at pleasure determine the line, and they consequently determine the remaining 2 points of intersection; and the number of equations between the displacements $d\omega$ should thus be $=2$. But by what precedes, we have the 3 equations

$$\begin{aligned} d\omega_1 + d\omega_2 + d\omega_3 + d\omega_4 &= 0, \\ x_1 d\omega_1 + x_2 d\omega_2 + x_3 d\omega_3 + x_4 d\omega_4 &= 0, \\ y_1 d\omega_1 + y_2 d\omega_2 + y_3 d\omega_3 + y_4 d\omega_4 &= 0; \end{aligned}$$

here the 4 points of intersection are on a line $y = ax + b$; we have therefore $y_1 = ax_1 + b, \ldots, y_4 = ax_4 + b$; the equations between the $d\omega$'s give

$$(y_1 - ax_1 - b)\, d\omega_1 + \ldots + (y_4 - ax_4 - b)\, d\omega_4 = 0,$$

that is, is a single relation $0 = 0$; or the 3 equations thus reduce themselves to 2 independent equations.

Again, if the fixed curve be a quintic, $m = 5$, there are here between the displacements the 6 equations

$$\Sigma x^2\, d\omega = 0,\quad \Sigma xy\, d\omega = 0,\quad \Sigma y^2\, d\omega = 0,$$
$$\Sigma x\, d\omega = 0,\quad \Sigma y\, d\omega = 0,\quad \Sigma\, d\omega = 0;$$

the two cases in which the number of independent equations is less than 6 are (i) when the variable curve is a line, and (ii) when the variable curve is a conic. For the line $n = 1$, and the number should be $= 3$. We have the above 6 equations; but the equation of the line is $ax + by + c = 0$, that is, we have $ax_1 + by_1 + c = 0$, &c.; we deduce the 3 identical equations

$$\Sigma x\, (ax + by + c) = 0,\quad \Sigma y\, (ax + by + c) = 0,\quad \Sigma\, (ax + by + c) = 0,$$

and the number of independent equations is thus $6 - 3, = 3$ as it should be.

So when the variable curve is a conic, $n = 2$; the number of independent equations should be $= 5$. The points of intersection lie on a conic $(a, b, c, f, g, h \between x, y, 1)^2 = 0$; we have therefore the several equations $(a, b, c, f, g, h \between x_1, y_1, 1)^2 = 0$, &c.: we have therefore the single identical equation

$$\Sigma\, (a, b, c, f, g, h \between x, y, 1)^2\, d\omega = 0,$$

and the number of independent equations is $6 - 1, = 5$ as it should be.

Obviously the like considerations apply to the case where the fixed curve is a curve of any given order whatever.

806.

DETERMINATION OF THE ORDER OF A SURFACE.

[From the *Messenger of Mathematics*, vol. XII. (1883), pp. 29—32.]

[ON p. lxx of the Prolegomena to C. Taylor's *Introduction to the Geometry of Conics* (1881) occurs the following passage:

"*Proof and extension of Newton's Descriptio Organica.*

"Let two angles AOB and $A\omega B$ of given magnitudes turn about O and ω respectively, and let the intersection A trace a curve of the nth order. For a given position of the arm OB there are n positions of A and therefore n of B. When OB is in the position $O\omega$ the n B's coincide with ω, which is therefore an n-fold point on the locus of B, as is also the point O; and since any line through O (or ω) meets the locus of B in n other points, the locus is of the order $2n$. Its order is the same when $A\omega B$ is a zero-angle or straight line.

"Let a given trihedral angle $O(ABC)$—or a plane OBC and a line OA rigidly attached to it—turn about O, and let a variable plane through a fixed point ω meet OA in A and the plane OBC in BC; then if the line BC describes a ruled surface of the order n the point A describes a surface of the order $4n$."

And in a foot-note it is stated that the author is indebted to Professor Cayley for the determination of the order of this surface. The following paper contains Professor Cayley's determination, which was communicated by him to Mr Taylor.]*

LEMMA. Take (a, b, c, f, g, h) the six coordinates of a line; these are connected by the equation

$$\text{af} + \text{bg} + \text{ch} = 0.$$

[* *l.c.*, p. 29.]

In order that the line may belong to a ruled surface, we must have between the coordinates three more equations, say these are

$$F\,(\mathrm{a, b, c, f, g, h}) = 0,$$
$$G\,(\mathrm{a} \quad " \quad) = 0,$$
$$H\,(\mathrm{a} \quad " \quad) = 0,$$

of the orders p, q, r respectively; then the scroll is of the order $n = 2pqr$.

For expressing that the line meets an arbitrary line (a′, b′, c′, f′, g′, h′), we have the linear relation

$$\mathrm{f'a + g'b + h'c + a'f + b'g + c'h} = 0;$$

the five relations determine the ratios a : b : c : f : g : h, and the number of systems of values is = product of orders $= 2 . p . q . r . 1$, $= 2pqr$, viz. this is the number of lines meeting the arbitrary line; or, what is the same thing, it is = order of ruled surface.

Consider now a trihedral angle $OABC$ rotating about a fixed point O which may be taken for the origin, and consider a fixed point ω. Let the lines OB, OC each meet a line L, and the plane ωL intersect OA in a point P; then it is to be shown that, if L is a line of a ruled surface of the order n, the locus of P is a surface of the order $4n$.

Observe that, for a given position of the line L, the position of the lines OB, OC is not determinate, but that the angle BOC has any position at pleasure in the plane OL, or say it rotates round the line ON which is the normal at O to the plane OL; the line OA therefore also rotates about ON, being always inclined to it at a determinate angle θ, or the locus of OA is a right cone axis ON and semi-aperture $= \theta$. The points P are the intersections of the several lines of the cone with the fixed plane ωL, or say that (for the given position of L) the locus of P is the conic C which is the intersection of the cone in question by the plane ωL. And then varying the position of L, the required surface is the locus of the corresponding conics C.

Take now

$$Ax + By + Cz + D = 0, \quad A'x + B'y + C'z + D' = 0,$$

for the equations of a particular line L; then writing

$$AD' - A'D,\ BD' - B'D,\ CD' - C'D,\ BC' - B'C,\ CA' - C'A,\ AB' - A'B = (\mathrm{a, b, c, f, g, h})$$

respectively, these are the six coordinates of the line L.

The equation of the plane OL is

$$(AD' - A'D)\,x + (BD' - B'D)\,y + (CD' - C'D)\,z = 0,$$

that is,

$$\mathrm{a}x + \mathrm{b}y + \mathrm{c}z = 0;$$

and the equations of the normal ON are therefore

$$\frac{x}{\mathrm{a}} = \frac{y}{\mathrm{b}} = \frac{z}{\mathrm{c}};$$

we have therefore for the cone

$$\frac{\mathrm{a}x + \mathrm{b}y + \mathrm{c}z}{\sqrt{(x^2 + y^2 + z^2)}\,\sqrt{(\mathrm{a}^2 + \mathrm{b}^2 + \mathrm{c}^2)}} = \cos\theta;$$

or if, for convenience, $\cos^2\theta = k$, then the equation of the cone is

$$(\mathrm{a}x + \mathrm{b}y + \mathrm{c}z)^2 - k(\mathrm{a}^2 + \mathrm{b}^2 + \mathrm{c}^2)(x^2 + y^2 + z^2) = 0;$$

say this is

$$\{\mathrm{a}^2 - k(\mathrm{a}^2 + \mathrm{b}^2 + \mathrm{c}^2),\ \mathrm{b}^2 - k(\mathrm{a}^2 + \mathrm{b}^2 + \mathrm{c}^2),\ \mathrm{c}^2 - k(\mathrm{a}^2 + \mathrm{b}^2 + \mathrm{c}^2),\ \mathrm{bc},\ \mathrm{ca},\ \mathrm{ab}\}(x, y, z)^2 = 0 \ldots(1).$$

Taking then (x_0, y_0, z_0) for the coordinates of the point ω, the equation of the plane ωL is

$$\frac{Ax + By + Cz + D}{Ax_0 + By_0 + Cz_0 + D} - \frac{A'x + B'y + C'z + D'}{A'x_0 + B'y_0 + C'z_0 + D'} = 0,$$

viz. it is

$$(BC' - B'C)(yz_0 - y_0z) + \ldots + (AD' - A'D)(x - x_0) + \ldots = 0,$$

that is,

$$\mathrm{f}(yz_0 - y_0z) + \mathrm{g}(zx_0 - z_0x) + \mathrm{h}(xy_0 - x_0y) + \mathrm{a}(x - x_0) + \mathrm{b}(y - y_0) + \mathrm{c}(z - z_0) = 0,$$

or say

$$x(\mathrm{h}y_0 - \mathrm{g}z_0 + \mathrm{a}) + y(-\mathrm{h}x_0 + \mathrm{f}z_0 + \mathrm{b}) + z(\mathrm{g}x_0 - \mathrm{f}y_0 + \mathrm{c}) + (-\mathrm{a}x_0 - \mathrm{b}y_0 - \mathrm{c}z_0) = 0 \ldots(2),$$

viz. (1) and (2) are the equations of the conic C, and the coordinates (a, b, c, f, g, h) satisfy of course the equation

$$\mathrm{af} + \mathrm{bg} + \mathrm{ch} = 0 \ldots\ldots(3).$$

Considering now the line L as belonging to a ruled surface, the coordinates (a, ...) satisfy as before three equations

$$F(\mathrm{a}, \mathrm{b}, \mathrm{c}, \mathrm{f}, \mathrm{g}, \mathrm{h}) = 0 \ldots\ldots(4),$$

$$G(\quad\text{,,}\quad) = 0 \ldots\ldots(5),$$

$$H(\quad\text{,,}\quad) = 0 \ldots\ldots(6),$$

of the orders p, q, r respectively, and we can from the six equations eliminate a, b, c, f, g, h. The resulting equation $\nabla = 0$ contains the coefficients of (1) in the order $1.2.p.q.r = 2pqr$ (which is the product of the orders of the other 5 equations), and the coefficients of (2) in the order $2.2.p.q.r = 4pqr$, (which is the product of the orders of the other 5 equations). But the coefficients of (1) being quadric functions of (x, y, z), and those of (2) being linear functions of (x, y, z), the aggregate order in (x, y, z) is

$$2.2pqr + 1.4pqr, = 8pqr;$$

or, since the order of the ruled surface is $n, = 2pqr$, the order of the locus is $= 4n$; which is the above-stated theorem.

807.

A PROOF OF WILSON'S THEOREM.

[From the *Messenger of Mathematics*, vol. XII. (1883), p. 41.]

LET n be a prime number; and imagine n points, the vertices of a regular polygon; any polygon which can be formed with these n points as vertices is either regular or else it is one of a set of n equal and similar polygons. For instance, $n=5$, the polygon as shown in the figure is one of a set of 5 equal and similar

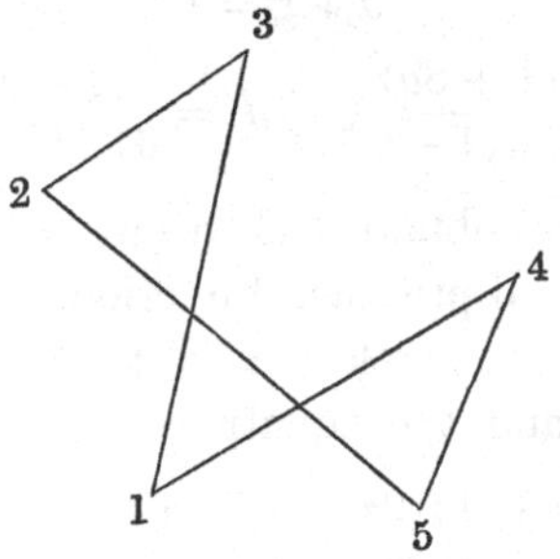

polygons; in fact, if the points taken in their cyclical order, but beginning at pleasure with any one of the 5 points are called 1, 2, 3, 4, 5, then we have 5 such polygons 13254; and so in general. The whole number of polygons is $\frac{1}{2}.1.2.3\ldots(n-1)$; and the number of the regular polygons is $\frac{1}{2}(n-1)$; hence the number of the remaining polygons is $=\frac{1}{2}(n-1)\{1.2\ldots(n-2)-1\}$; and this number must therefore be divisible by n; that is, $1.2\ldots(n-1)-n+1$ is divisible by n; or, what is the same thing, $1.2\ldots(n-1)+1$ is divisible by n, which is the theorem in question.

808.

NOTE ON A FORM OF THE MODULAR EQUATION IN THE TRANSFORMATION OF THE THIRD ORDER.

[From the *Messenger of Mathematics*, vol. XII. (1883), pp. 173, 174.]

IN my *Treatise on Elliptic Functions*, pp. 214—216, writing only $\frac{1}{J}$, $\frac{1}{J'}$ instead of Ω, Ω', and α, β instead of α', β', I have shown as follows: viz. if k, λ denote as usual the original modulus, and the transformed modulus, and if

$$J=\frac{(k^4+14k^2+1)^3}{108k^2(1-k^2)^4},\quad J'=\frac{(\lambda^4+14\lambda^2+1)^3}{108\lambda^2(1-\lambda^2)^4},$$

then the relation between J and J' can be found by the elimination of α, β from the equations

$$\alpha+\beta=1,$$

$$J=\frac{(1+8\alpha)^3}{64\alpha(1-\alpha)^3},\quad J'=\frac{(1+8\beta)^3}{64\beta(1-\beta)^3}.$$

By a very slight change we obtain the result given by Prof. Klein in his paper, "Ueber die Transformation der elliptischen Functionen, &c.," *Math. Ann.* t. XIV. (1879), pp. 111—172; viz., see p. 143, the relation is to be obtained by the elimination of τ, τ' from the equation $\tau\tau'=1$, and the equations

$$J:J-1:1=(\tau-1)(9\tau-1)^3:(27\tau^2-18\tau-1)^2:-64\tau;$$
$$J':J'-1:1=(\tau'-1)(9\tau'-1)^3:(27\tau'^2-18\tau'-1)^2:-64\tau';$$

these last equations being equivalent to two equations only in virtue of the identity

$$(\tau-1)(9\tau-1)^3+64\tau=(27\tau^2-18\tau-1)^2,$$

and the like identity in τ'.

In fact, writing $\alpha=\frac{\tau}{\tau-1}$, $\beta=\frac{\tau'}{\tau'-1}$, the equation $\alpha+\beta=1$ becomes $\tau\tau'=1$; and then for α, β substituting their values, we have

$$J=\frac{(9\tau-1)^3(\tau-1)}{-64\tau},\quad J'=\frac{(9\tau'-1)^3(\tau'-1)}{-64\tau'},$$

which are the formulæ in question.

809.

SCHRÖTER'S CONSTRUCTION OF THE REGULAR PENTAGON.

[From the *Messenger of Mathematics*, vol. XII. (1883), p. 177.]

THE following construction of the regular pentagon, analogous to the more complicated one for the polygon of 17 sides, is given in the paper, H. Schröter, Zur *v. Staudt*schen "Construction des regulären Siebenzehnecks," *Crelle*, t. LXXV. (1873), pp. 13—24.

Take in a circle AB, CD diameters at right angles to each other, and at C, D draw in the directions A to B and B to A respectively, the lines Cc, $=4$ radii, and Dd, $=1$ radius; draw cd meeting the circle in the points E and F; draw CE and CF cutting AB in the points e and f respectively, and through these at right angles to AB draw the chords 34 and 25 respectively, then we have $A2345$ a regular pentagon.

810.

NOTE ON A SYSTEM OF EQUATIONS.

[From the *Messenger of Mathematics*, vol. XII. (1883), pp. 191, 192.]

THE equations are

$$x^2 = ax + by, \quad xy = cx + dy, \quad y^2 = ex + fy,$$

where

$$\frac{b}{d} = \frac{a-d}{c-f} = \frac{c}{e};$$

or, what is the same thing, if a, b, c, d are given, then

$$e = \frac{cd}{b}, \quad f = c - \frac{d(a-d)}{b};$$

and this being so, the equations are equivalent to two independent equations; viz. starting from the first and the second equations, we have

$$dx^2 \;\; - bxy = (ad - bc)\, x,$$

that is,

$$dx \;\; - by \;\; = (ad - bc):$$

and thence

$$dxy - by^2 \;\; = (ad - bc)\, y,$$

or

$$d(cx + dy) - by^2 = (ad - bc)\, y;$$

which, attending to the values of e and f, is the third equation

$$y^2 = ex + fy.$$

We have

$$\frac{x}{y} = \frac{ax + by}{cx + dy}, \quad \frac{y}{x} = \frac{ex + fy}{cx + dy};$$

that is,

$$cx^2 - (a-d)\,xy - by^2 = 0,$$
$$ex^2 - (c-f)\,xy - dy^2 = 0,$$

and eliminating the (x, y) from these equations we have an equation $\Omega = 0$ as the condition that the original three equations may have a single common root; the before-mentioned equations $\frac{b}{d} = \frac{a-d}{c-f} = \frac{c}{e}$, are the conditions in order that the three equations may have two common roots, that is, that there may be *two* systems of (x, y) satisfying the three equations.

We have, moreover, $y(x-d) = cx$, $x(y-c) = dy$, and substituting these values, say $y = \frac{cx}{x-d}$ and $x = \frac{dy}{y-c}$, in the first and third equations respectively, they become

$$x - a = \frac{bc}{x-d},\quad y - f = \frac{de}{y-c},$$

that is,

$$x^2 - (a+d)\,x + ad - bc = 0,$$
$$y^2 - (c+f)\,y + cf - de = 0,$$

which are quadric equations for x and y respectively; it is easy to express the second equation (like the first) in terms of (a, b, c, d), and the first equation (like the second) in terms of (c, d, e, f), but the forms are less simple.

Suppose $(a, b, c, d) = (-1, -1, 1, 0)$, then we have $(e, f) = (0, 1)$, the two equations in $x : y$ become $x^2 + xy + y^2 = 0$, and $0 = 0$ respectively; those in x and y become $x^2 + x + 1 = 0$, $y^2 - 2y + 1 = 0$ respectively; this is right, for the three equations are

$$x^2 = -x - y,\quad xy = x,\quad y^2 = y;$$

viz. from the third equation we have $y = 1$, a value satisfying the second equation, and then the first equation becomes $x^2 + x + 1 = 0$; or, if we please, $x^2 + xy + y^2 = 0$, the values in fact being $x = \omega$, an imaginary cube root of unity, and $y = 1$.

In the general case, the values (x, y) may be regarded as units in a complex numerical theory, viz. if (a, b, c, d, e, f) are integers, and p, q, p', q', P, Q are also integers, then the product of the two complex integers $px + qy$ and $p'x + q'y$ will be a complex integer $Px + Qy$.

811.

ON THE LINEAR TRANSFORMATION OF THE THETA FUNCTIONS.

[From the *Messenger of Mathematics*, vol. XIII. (1884), pp. 54—60.]

THE functions referred to are the single Theta Functions; these may be defined as doubly infinite products, as was in fact done in my "Mémoire sur les fonctions doublement périodiques," *Liouv.* t. x. (1845), pp. 385—420, [25]; and it is interesting to consider from this point of view the theory of their linear transformation: this I propose to do in the present paper, adopting throughout the notation of Smith's* "Memoir on the Theta and Omega Functions."

The periods K, iK' are, in general, imaginary quantities

$$K = A + Bi,$$
$$iK' = C + Di,$$

where $AD - BC$ is positive; writing then $\omega = \dfrac{iK'}{K}$, and $q = e^{i\pi\omega}$, also for shortness

$$(q1) = 2q^{\frac{1}{4}}\Pi_1^\infty (1 - q^{2n})^3,$$

where $q^{\frac{1}{4}}$ denotes $e^{\frac{1}{4}i\pi\omega}$, the expression of the odd theta-function $\vartheta_1(x, \omega)$ as a doubly infinite product is

$$\vartheta_1(x,\ \omega) = (q1)\, x\Pi\Pi\left(1 + \frac{x}{m\pi + n\omega\pi}\right),\quad \left(\frac{\mu}{\nu} = \infty\right),$$

where (m, n) have any positive or negative integer values (the combination $m = 0$, $n = 0$ excluded) from $m = -\mu$ to μ, and $n = -\nu$ to ν, μ and ν being each ultimately infinite but so that μ is infinite in comparison with ν; this condition in regard to the limits is indicated by $\mu/\nu = \infty$; and similarly $\nu/\mu = \infty$ would indicate that ν was infinite in comparison with μ.

[* Smith's *Collected Mathematical Papers*, vol. II., pp. 415—621.]

The condition as to the limits might be that (m, n) have any positive or negative values (excluding as before) such that the modulus of $m+n\omega$ does not exceed a positive value T, which is ultimately taken to be infinite; this condition may be indicated by mod $=\infty$.

The values of the double product corresponding to the different conditions as to the limits are not equal, but they differ only by an exponential factor, the exponent being a multiple of x^2; we thus have

$$x\Pi\Pi\left(1+\frac{x}{mK+niK'}\right)\left(\frac{\mu}{\nu}=\infty\right)=\exp(\nabla x^2)\, x\Pi\Pi\left(1+\frac{x}{mK+niK'}\right)(\text{mod}=\infty),$$

where ∇ is a determinate value, depending on K and K'; and similarly

$$x\Pi\Pi\left(1+\frac{x}{m\Lambda+ni\Lambda'}\right)\left(\frac{\mu}{\nu}=\infty\right)=\exp(\Box x^2)\, x\Pi\Pi\left(1+\frac{x}{m\Lambda+ni\Lambda'}\right)(\text{mod}=\infty),$$

where $\Box$ is a determinate value depending in like manner on Λ, Λ'.

We have, then, as above

$$\begin{aligned}\vartheta_1(x,\omega)&=(q1)\, x\Pi\Pi\left(1+\frac{x}{m\pi+n\omega\pi}\right)\left(\frac{\mu}{\nu}=\infty\right)\\ &=(q1)\frac{\pi}{K}\frac{Kx}{\pi}\Pi\Pi\left(1+\frac{\frac{Kx}{\pi}}{mK+niK'}\right)\\ &=(q1)\frac{\pi}{K}\exp\left(\nabla\frac{K^2x^2}{\pi^2}\right)\frac{Kx}{\pi}\Pi\Pi\left(1+\frac{\frac{Kx}{\pi}}{mK+niK'}\right)(\text{mod}=\infty),\end{aligned}$$

viz. we have thus defined $\vartheta_1(x,\omega)$ as a doubly infinite product with the limiting condition (mod $=\infty$); if for x we write $\frac{\pi x}{h}$, h arbitrary, we have

$$\vartheta_1\left(\frac{\pi x}{h},\ \omega\right)=(q1)\frac{\pi}{K}\exp\left(\nabla\frac{K^2x^2}{h^2}\right)\frac{Kx}{h}\Pi\Pi\left(1+\frac{\frac{Kx}{h}}{mK+niK'}\right)(\text{mod}=\infty),$$

and similarly, if $\Omega=\frac{i\Lambda'}{\Lambda}$, $Q=e^{i\pi\Omega}$, then

$$\begin{aligned}\vartheta_1\left\{(a+b\Omega)\frac{\pi x}{h},\ \Omega\right\}&=(Q1)\frac{\pi}{\Lambda}\exp\left\{(a+b\Omega)^2\,\Box\,\frac{\Lambda^2x^2}{h^2}\right\}\\ &\quad\times(a+b\Omega)\frac{\Lambda x}{h}\Pi\Pi\left(1+\frac{(a+b\Omega)\frac{\Lambda x}{h}}{m\Lambda+ni\Lambda'}\right)(\text{mod}=\infty).\end{aligned}$$

In the case of a linear transformation, we have

$$\omega=\left|\begin{matrix}a, & b\\ c, & d\end{matrix}\right|\times\Omega,\ \text{ that is, }\ \omega=\frac{c+d\Omega}{a+b\Omega},$$

where a, b, c, d are positive or negative integers such that $ad-bc=+1$; it is to be shown that the two infinite products are in this case identical; this being so, we have

$$\frac{\vartheta_1\left\{(a+b\Omega)\frac{\pi x}{h},\ \Omega\right\}}{\vartheta_1\left\{\frac{\pi x}{h},\ \omega\right\}}=\frac{(Q1)}{(q1)}\exp\left\{\{(a+b\Omega)^2\,\Box\Lambda^2-\nabla K^2\}\frac{x^2}{h^2}\right\},$$

viz. the two functions differ only by a constant factor and by an exponential factor, the exponent being a multiple of x^2; after all reductions, this factor is found to be

$$=\exp\left(-i\pi b\,(a+b\Omega)\frac{x^2}{h^2}\right).$$

We have

$$\omega=\frac{c+d\Omega}{a+b\Omega},$$

or since

$$\omega=\frac{iK'}{K},\quad \Omega=\frac{i\Lambda'}{\Lambda},$$

this is

$$\frac{iK'}{K}=\frac{c\Lambda+di\Lambda'}{a\Lambda+bi\Lambda'},$$

or say

$$\frac{1}{M}K=a\Lambda+bi\Lambda',$$

$$\frac{1}{M}iK'=c\Lambda+di\Lambda',$$

either of which equations may be taken as a definition of the multiplier M. We have

$$\frac{K}{M\Lambda}=a+b\Omega,$$

$$\begin{aligned}\frac{1}{M}(mK+niK')&=(am+cn)\,\Lambda+(bm+dn)\,i\Lambda'\\&=m'\Lambda+n'i\Lambda',\end{aligned}$$

if

$$m'=am+cn,$$

$$n'=bm+dn.$$

Here to any integer values of $(m,\ n)$ there correspond integer values of m', n'; and conversely, in virtue of the equation $ad-bc=1$, to any integer values of m', n' there correspond integer values of m, n. The two products are

$$\Pi\Pi\left(1+\frac{\frac{Kx}{Mh}}{m'\Lambda+n'i\Lambda'}\right),\ (\text{mod}=\infty),$$

$$\Pi\Pi\left(1+\frac{\frac{(a+b\Omega)\,\Lambda x}{h}}{m\Lambda+ni\Lambda'}\right),\ (\text{mod}=\infty).$$

But, as above, we have $\dfrac{K}{M} = (a + b\Omega)\Lambda$: and then, observing that in the first of the two products we may for m', n' write m, n, it at once appears that the two products are identical.

The exponential factor, writing therein $(a + b\Omega)\Lambda = \dfrac{K}{M}$, becomes

$$\exp\left\{\left(\frac{\square}{M^2} - \nabla\right)\frac{K^2x^2}{h^2}\right\}.$$

The values of ∇, $\square$ are at once obtained by means of a formula* given in my Memoir, viz. we have

$$\nabla = -\tfrac{1}{2}(B + \beta),$$

where

$$B = \frac{\pi(\omega\upsilon + \omega'\upsilon')}{\Omega\Upsilon \bmod (\omega\upsilon' - \omega'\upsilon)},$$

$$\beta = \frac{\pi i}{\Omega\Upsilon}\frac{(\omega\upsilon' - \omega'\upsilon)}{\bmod (\omega\upsilon' - \omega'\upsilon)}.$$

Comparing with the present notation

$$\Omega = \omega + \omega' i, \quad = A + Bi = K,$$

$$\Upsilon = \upsilon + \upsilon' i, \quad = C + Di = K'i,$$

so that Ω, Υ denote K, $K'i$, and ω, ω', υ, υ' denote A, B, C, D respectively: $\omega\upsilon' - \omega'\upsilon$ is thus $= AD - BC$, which has been assumed to be positive; hence also $\bmod(\omega\upsilon' - \omega'\upsilon) = AD - BC$, and the formula becomes

$$\nabla = -\tfrac{1}{2}\pi\left\{\frac{AC + BD}{i(AD - BC)} + 1\right\}\frac{1}{KK'}.$$

Now writing

$$\Lambda = A_1 + B_1 i,$$

$$i\Lambda' = C_1 + D_1 i,$$

then we have

$$\frac{1}{M}(A + Bi) = a\Lambda + bi\Lambda' = a(A_1 + B_1 i) + b(C_1 + D_1 i),$$

$$\frac{1}{M}(C + Di) = c\Lambda + di\Lambda' = c(A_1 + B_1 i) + d(C_1 + D_1 i);$$

consequently, if

$$M = \rho(\cos\theta + i\sin\theta),$$

[* *Collected Mathematical Papers*, t. I., p. 164. The denominator factor $\Omega\Upsilon$ has been omitted (p. 165) by mistake.]

we have

$$\frac{1}{\rho}(\ \ A\cos\theta + B\sin\theta) = aA_1 + bC_1,$$

$$\frac{1}{\rho}(-A\sin\theta + B\cos\theta) = aB_1 + bD_1,$$

$$\frac{1}{\rho}(\ \ C\cos\theta + D\sin\theta) = cA_1 + dC_1,$$

$$\frac{1}{\rho}(-C\sin\theta + D\cos\theta) = cB_1 + dD_1,$$

and thence

$$\frac{1}{\rho^2}(AD - BC) = (ad - bc)(A_1D_1 - B_1C_1), \quad = (A_1D_1 - B_1C_1).$$

Hence $A_1D_1 - B_1C_1$ is positive, and we have

$$\square = -\tfrac{1}{2}\pi\left\{\frac{A_1C_1 + B_1D_1}{i(A_1D_1 - B_1C_1)} + 1\right\}\frac{1}{\Lambda\Lambda'}.$$

Take K_1 the conjugate of K, Λ_1 the conjugate of Λ, then

$$K_1 = A - Bi, \quad \Lambda_1 = A_1 - B_1 i,$$

$$iK' = C + Di, \quad i\Lambda' = C_1 + D_1 i.$$

We have

$$iK_1K' = AC + BD + i(AD - BC),$$

and therefore

$$\frac{K_1K'}{AD - BC} = \frac{AC + BD}{i(AD - BC)} + 1, \quad \nabla = \frac{-\tfrac{1}{2}\pi}{AD - BC}\,\frac{K_1}{K},$$

and similarly

$$\square = \frac{-\tfrac{1}{2}\pi}{A_1D_1 - B_1C_1}\,\frac{\Lambda_1}{\Lambda}.$$

The exponential is

$$\left(\frac{\square}{M^2} - \nabla\right)\frac{K^2x^2}{h^2};$$

and we have

$$\frac{\square}{M^2} - \nabla = \frac{-\tfrac{1}{2}\pi}{M^2(A_1D_1 - B_1C_1)}\,\frac{\Lambda_1}{\Lambda} + \frac{\tfrac{1}{2}\pi}{AD - BC}\,\frac{K_1}{K},$$

which is

$$= \frac{-\tfrac{1}{2}\pi}{M^2(A_1D_1 - B_1C_1)}\,\frac{\Lambda_1}{\Lambda} + \frac{\tfrac{1}{2}\pi}{\rho^2(A_1D_1 - B_1C_1)}\,\frac{K_1}{K},$$

$$= \frac{-\tfrac{1}{2}\pi}{A_1D_1 - B_1C_1}\left(\frac{1}{M^2}\,\frac{\Lambda_1}{\Lambda} - \frac{1}{\rho^2}\,\frac{K_1}{K}\right).$$

But $\rho/M = \cos\theta - i\sin\theta$, or calling this for a moment P, then $1/M^2 = P^2/\rho^2$, and the formula may be written

$$\frac{\square}{M^2} - \nabla = \frac{-\tfrac{1}{2}\pi P}{\rho^2(A_1D_1 - B_1C_1)}\left(P\frac{\Lambda_1}{\Lambda} - P^{-1}\frac{K_1}{K}\right)$$

$$= \frac{-\tfrac{1}{2}\pi P}{\rho^2(A_1D_1 - B_1C_1)}\{(\cos\theta - i\sin\theta)\,\Lambda_1 K - (\cos\theta + i\sin\theta)\,\Lambda K_1\}\frac{1}{K\Lambda}.$$

The term in { } is

$$(\cos\theta - i\sin\theta)(A + Bi)(A_1 - B_1 i) - (\cos\theta + i\sin\theta)(A - Bi)(A_1 + B_1 i),$$

$$= \quad 2\cos\theta\,[-(AB_1 - A_1B)\,i] - 2i\sin\theta\,(AA_1 + BB_1),$$

$$= -2i\,\{(AB_1 - A_1B)\cos\theta + (AA_1 + BB_1)\sin\theta\},$$

$$= -2i\,\{B_1(A\cos\theta + B\sin\theta) - A_1(-A\sin\theta + B\cos\theta)\},$$

$$= -2i\rho\,\{B_1(aA_1 + bC_1) - A_1(aB_1 + bD_1)\},$$

$$= +2i\rho b\,(A_1D_1 - B_1C_1).$$

Hence

$$\frac{\square}{M^2} - \nabla = \frac{-\tfrac{1}{2}\pi P}{\rho^2(A_1D_1 - B_1C_1)}\,2ib\rho\,(A_1D_1 - B_1C_1)\frac{1}{K\Lambda}$$

$$= -\frac{i\pi bP}{\rho}\frac{1}{K\Lambda} = -\frac{i\pi b}{M}\frac{1}{K\Lambda},$$

and the exponential thus is

$$= \exp\left(-\frac{i\pi b}{M}\frac{1}{K\Lambda}\frac{K^2x^2}{h^2}\right), \quad = \exp\left(-i\pi b\frac{K}{M\Lambda}\frac{x^2}{h^2}\right);$$

or, since $\frac{K}{M\Lambda} = (a + b\Omega)$, this is

$$= \exp\left(-i\pi b\,(a + b\Omega)\frac{x^2}{h^2}\right);$$

and we have thus the required formula

$$\frac{\vartheta_1\left\{(a + b\Omega)\frac{\pi x}{h},\ \Omega\right\}}{\vartheta_1\left\{\qquad\frac{\pi x}{h},\ \omega\right\}} = \frac{(Q1)}{(q1)}(a + b\Omega)\exp\left(-i\pi b\,(a + b\Omega)\frac{x^2}{h^2}\right).$$

812.

ON ARCHIMEDES' THEOREM FOR THE SURFACE OF A CYLINDER.

[From the *Messenger of Mathematics*, vol. XIII. (1884), pp. 107, 108.]

THE measure of the surface of a cylinder was first obtained by Archimedes in his Treatise on the Sphere and Cylinder (Book I., Prop. XIV.), *Œuvres d'Archimède*, par F. Peyrard, 4° Paris, 1807, pp. 26—31; viz. Archimedes showed that the surface of the cylinder was equal to the area of a circle, having its radius a mean proportional between the height and the diameter of the circular base $[S = 2\pi ah, = \pi \{\sqrt{(2a \,.\, h)}\}^2]$.

The following is *in effect* his demonstration:

He considers regular polygons (with the same number of sides) inscribed in and circumscribed about a circle; and, as regards the cylinder, the prisms standing on these polygons.

Say for the circular base of the cylinder we have

$S^\times$ surface of circumscribed prism,
S „ „ cylinder,
S° „ „ inscribed prism;

and for the circle, having its radius a mean proportional between the height and the diameter of the circular base,

$B^\times$ area of circumscribed polygon,
B „ „ circle,
B° „ „ inscribed polygon,

where the four polygons referred to by $S^\times$, S°, $B^\times$, B° have all of them the same number of sides.

It is in the preceding propositions (by means of an axiom as to curve lines) shown that

$$S^\times > S > S^\circ, \quad B^\times > B > B^\circ;$$

and it is further shown that

$$S^\times = B^\times, \quad S^\circ = B^\circ.$$

It is moreover shown that, by taking the number of sides sufficiently large, the ratio $B^\times : B^\circ$, or say the fraction $B^\times/B^\circ$ (which is greater than 1) may be made less than any given quantity $1+\epsilon$.

It is then to be shown that $S = B$.

If not, then

either

$$B < S.$$

This being so, it is possible to make

$$B^\times/B^\circ < S/B,$$

that is,

$$S^\times/B^\circ < S/B,$$

or

$$S^\times/S < B^\circ/B,$$

which is absurd, since

$$S^\times/S > 1; \quad B^\circ/B < 1;$$

or else

$$B > S.$$

This being so, it is possible to make

$$B^\times/B^\circ < B/S,$$

that is,

$$B^\times/S^\circ < B/S,$$

or

$$B^\times/B < S^\circ/S,$$

which is absurd, since

$$B^\times/B > 1; \quad S^\circ/S < 1;$$

and consequently $S = B$, the theorem in question.

I take the opportunity of referring to two theorems by Archimedes, Lemmas, Prop. V. and VI., Peyrard, pp. 429—435, which relate to the contacts of circles. We have in each of them the figure which he calls the Arbelon, viz. if A, C, B are points in this order on the same straight line, then the figure consists of the three semicircles on the diameters AC, CB, and AB respectively, and the Arbelon is the space included between the three semi-circumferences.

In Prop. V., we have also the common tangent at C to the two semicircles AC, CB; this divides the Arbelon into two mixtilinear triangles (each bounded by the common tangent, one of the smaller semicircles, and a portion of the larger semicircle), and inscribing each of these a circle, the theorem is that the two inscribed circles are of equal magnitude.

In Prop. VI., the theorem is that the radii of the smaller semicircles being as $3 : 2$, then the radius of the circle inscribed in the Arbelon is to the diameter of the larger semicircle as 6 to 19. But it is noticed that the demonstration would apply to any other value of the ratio; and, in fact, if the radii of the two smaller circles are as $a : b$, then the radius of the inscribed circle is to the diameter of the larger semicircle as ab to $a^2 + ab + b^2$, which is the general form of the theorem.

813.

[NOTE ON MR GRIFFITHS' PAPER "ON A DEDUCTION FROM THE ELLIPTIC-INTEGRAL FORMULA $y = \sin(A+B+C+\ldots)$".]

[From the *Proceedings of the London Mathematical Society*, vol. xv. (1884), p. 81.]

CONSIDER, for instance,

the cubic transformation

$$y = \frac{x[1+2\alpha' - (1+\alpha')^2 x^2]}{1-\alpha^2 x^2},$$

where $\alpha^2 + \alpha'^2 = 1$.

This implies

$$\sqrt{1-y^2} = \frac{\sqrt{1-x^2}\,[1-(1+\alpha')^2 x^2]}{1-\alpha^2 x^2},$$

viz., $\sqrt{1-y^2}$ = a rational multiple of $\sqrt{1-x^2}$.

Also the quadric transformation

$$z = \frac{1-(1+\beta'^2)\,x^2}{1-\beta^2 x^2};$$

where $\beta^2 + \beta'^2 = 1$.

This implies

$$\sqrt{1-z^2} = \frac{\sqrt{1-x^2}\,.\,2\beta' x}{1-\beta^2 x^2},$$

viz., $\sqrt{1-z^2}$ = a rational multiple of $\sqrt{1-x^2}$.

Hence, assuming

$$u = yz - \sqrt{1-y^2}\,\sqrt{1-z^2},$$

which is a rational function

$$= \frac{x(a_0 - a_2 x^2 + a_4 x^4)}{1-\alpha^2 x^2\,.\,1-\beta^2 y^2},$$

we have

$$\sqrt{1-u^2} = y\sqrt{1-z^2} + z\sqrt{1-y^2},$$

which is $= \sqrt{1-x^2}$ multiplied by a like rational function.

That is, in defining the a_0, a_2, a_4, functions of the two arbitrary coefficients α, β, as above, we have in effect so determined them that $\sqrt{1-u^2}$ shall be $=\sqrt{1-x^2}$ multiplied by a rational function of x.

We can then further determine a_0, a_2, a_4 in such wise that the change of x into $\frac{1}{kx}$ shall change u into $\frac{1}{\lambda u}$; and, this being so, making the change in $\sqrt{1-u^2}$, we obtain $\sqrt{1-\lambda^2u^2}$ in the form, $\sqrt{1-k^2x^2}$ multiplied by a rational function of x; viz. u is a function of x such that

$$\frac{du}{\sqrt{1-u^2 \,.\, 1-\lambda^2u^2}} = \frac{Mdu}{\sqrt{1-x^2 \,.\, 1-k^2x^2}}.$$

The theory is thus in effect Jacobi's—with the *novelty* of combining two lower transformations in such wise that the assumed expression for u as a rational function of x shall give

$\sqrt{1-u^2}=\sqrt{1-x^2}$ multiplied by a rational function of x.

It is not necessary that the equations

$y =$ rational function of x and $z =$ rational function of x

should be elliptic-function transformations. All that is required is that they should be such as to give $\sqrt{1-y^2}$ and $\sqrt{1-z^2}$ each $=\sqrt{1-x^2}$ multiplied by a rational function of x.

814.

ON DOUBLE ALGEBRA.

[From the *Proceedings of the London Mathematical Society*, vol. xv. (1884), pp. 185—197. Read April 3, 1884.]

1. I CONSIDER the Double Algebra formed with the extraordinary symbols, or "extraordinaries" x, y, which are such that

$$\begin{aligned} x^2 &= ax + by, \\ xy &= cx + dy, \\ yx &= ex + fy, \\ y^2 &= gx + hy, \end{aligned}$$

or, as these equations may also be written,

	x	y
x	$(a,\ b)$	$(c,\ d)$
y	$(e,\ f)$	$(g,\ h)$

where a, b, c, d, e, f, g, h are ordinary symbols, or say coefficients; all coefficients being commutative and associative *inter se* and with the extraordinaries x, y.

The system depends in the first instance on the eight parameters a, b, c, d, e, f, g, h; but we may, instead of the extraordinaries x, y, consider the new extraordinaries connected therewith by the linear relations $\xi = \alpha x + \beta y$, $\eta = \gamma x + \delta y$, where the coefficients α, β, γ, δ may be determined so as to establish between the eight parameters any four relations at pleasure (or, what is the same thing, α, β, γ, δ are what I call "apoclastic" constants): and the number of parameters is thus properly $8-4, =4$.

2. The extraordinaries here considered are not in general associative; differing herein from the imaginaries of Peirce's Memoir, "Linear Associative Algebra" (1870),

reprinted in the *American Mathematical Journal*, t. IV. (1881), pp. 97—227, which, as appears by the title, refers only to associative imaginaries. I recall some definitions and results. The symbol x is said to be *idempotent* if $x^2 = x$, *nilpotent* if $x^2 = 0$; and the systems of associative symbols are expressed as much as may be by means of such idempotent and nilpotent symbols: thus the linear systems are $(a_1)\ x^2 = x$, $(b_1)\ x^2 = 0$. A double system composed of independent symbols, that is, symbols x, y each belonging to its own linear system, and moreover such that $xy = yx = 0$, is said to be "mixed"; thus the mixed double systems are

	x	y
x	x	0
y	0	y

,

	x	y
x	x	0
y	0	0

,

	x	y
x	0	0
y	0	0

.

But Peirce excludes these from consideration, attending only to the pure systems, which he finds to be

(a_2)

	x	y
x	x	y
y	y	0

, (b_2)

	x	y
x	x	y
y	0	0

, (c_2)

	x	y
x	y	0
y	0	0

.

To these, however, should be added the system

(d_2)

	x	y
x	x	0
y	y	0

;

see *post*, No. 19.

3. In the general theory, where the symbols are not in the first instance taken to be associative, we may of course establish between the coefficients such relations as will make the symbols associative; and the question presents itself to show how in this case the system reduces itself to one of Peirce's systems. This I considered in my note "On Associative Imaginaries," *Johns Hopkins University Circular*, No. 15 (1882), p. 211 [822]; I there obtained, as the general form of the commutative and associative system,

$$x^2 = ax + by,$$

$$xy = yx = cx + dy,$$

$$y^2 = \frac{cd}{b}x + \frac{d^2 + bc - ad}{b}y,$$

the relation of which to Peirce's system was, as I there remarked, pointed out to me by Mr C. S. Peirce: this will be considered in the sequel, Nos. 13 to 19.

4. Starting now with the general equations

$$x^2 = ax + by,$$
$$xy = cx + dy,$$
$$yx = ex + fy,$$
$$y^2 = gx + hy,$$

we may attempt to find an extraordinary ξ, $= \alpha x + \beta y$ (α, β coefficients), such that $\xi^2 = K\xi$ (K a coefficient). In general, K is not $=0$, and, when it is not $=0$, it may without loss of generality be taken to be $=1$; we have then $\xi^2 = \xi$, ξ an *idempotent* symbol. But K may be $=0$; and then $\xi^2 = 0$, ξ a *nilpotent* symbol. To include the two cases, I retain K, it being understood that, when K is not $=0$, it may be taken to be $=1$. We have

$$\xi^2 = \alpha^2 (ax + by) + \alpha\beta\,(\overline{c + e\,x + d + f y}) + \beta^2 (gx + hy)$$
$$= \{a\alpha^2 + (c+e)\,\alpha\beta + g\beta^2\}\,x + \{b\alpha^2 + (d+f)\,\alpha\beta + h\beta^2\}\,y.$$

Hence, when this is $= K\xi$, that is,

$$= K(\alpha x + \beta y),$$

we have

$$\frac{\alpha}{\beta} = \frac{a\alpha^2 + (c+e)\,\alpha\beta + g\beta^2}{b\alpha^2 + (d+f)\,\alpha\beta + h\beta^2},$$

a cubic equation for the determination of the ratio $\alpha : \beta$; and, for any particular value of the ratio, we can in general determine the absolute magnitudes, so that

$$K, = \frac{1}{\alpha}\{a\alpha^2 + (c+e)\,\alpha\beta + g\beta^2\}, \quad = \frac{1}{\beta}\{b\alpha^2 + (d+f)\,\alpha\beta + h\beta^2\},$$

shall be $=1$. If, however, for the given value of the ratio we have

$$a\alpha^2 + (c+e)\,\alpha\beta + g\beta^2 = 0, \quad b\alpha^2 + (d+f)\,\alpha\beta + h\beta^2 = 0,$$

one of these equations, of course, implying the other, then the value of K is $=0$.

5. It follows that there are in general three idempotent symbols ξ, η, ζ, that is, extraordinaries such that $\xi^2 = \xi$, $\eta^2 = \eta$, $\zeta^2 = \zeta$. The cubic equation may, however, have two equal roots, or three equal roots, or it may vanish identically; in this last case, any linear function $\alpha x + \beta y$ is in general idempotent. But (as will be considered in detail further on) we may, instead of an idempotent symbol or symbols, have a nilpotent symbol or symbols. It might be convenient to use the term Potency for a symbol which is in general idempotent, but which may be nilpotent. Writing $\dfrac{\alpha}{\beta} = \dfrac{-y}{x}$, we obtain a cubic equation $\Omega = (x, y)^3 = 0$, where obviously the linear factors of Ω are the just-mentioned functions ξ, η, ζ; that is, we have ξ, η, ζ as the linear factors of the cubic function

$$\Omega, = gx^3 + (h - c - e)\,x^2y + (a - d - f)\,xy^2 + by^3;$$

each such factor, except in the case where it is nilpotent, being determined so that it shall be idempotent. The cubic function of course vanishes identically if

$$g = 0,\quad h - c - e = 0,\quad a - d - f = 0,\quad b = 0.$$

6. Two extraordinaries $\xi, = \alpha x + \beta y$; $\eta, = \gamma x + \delta y$ (α, β, γ, δ coefficients); may be such that $\xi\eta = 0$: this, of course, does not imply $\eta\xi = 0$; we have, in fact,

$$\begin{aligned}\xi\eta &= (\alpha x + \beta y)(\gamma x + \delta y)\\ &= \alpha\gamma x^2 + \alpha\delta xy + \beta\gamma yx + \beta\delta y^2\\ &= \alpha\gamma(ax + by) + \alpha\delta(cx + dy) + \beta\gamma(ex + fy) + \beta\delta(gx + hy)\\ &= (a\alpha\gamma + c\alpha\delta + e\beta\gamma + g\beta\delta)x + (b\alpha\gamma + d\alpha\delta + f\beta\gamma + h\beta\delta)y;\end{aligned}$$

and the required condition is satisfied if

$$a\alpha\gamma + c\alpha\delta + e\beta\gamma + g\beta\delta = 0,$$
$$b\alpha\gamma + d\alpha\delta + f\beta\gamma + h\beta\delta = 0.$$

Writing these equations first under the form

$$\gamma(a\alpha + e\beta) + \delta(c\alpha + g\beta) = 0,$$
$$\gamma(b\alpha + f\beta) + \delta(d\alpha + h\beta) = 0,$$

and then under the form

$$\alpha(a\gamma + c\delta) + \beta(e\gamma + g\delta) = 0,$$
$$\alpha(b\gamma + d\delta) + \beta(f\gamma + h\delta) = 0,$$

we have

$$(a\alpha + e\beta)(d\alpha + h\beta) - (b\alpha + f\beta)(c\alpha + g\beta) = 0,$$

a quadric equation for the determination of $\alpha : \beta$; and then

$$(a\gamma + c\delta)(f\gamma + h\delta) - (b\gamma + d\delta)(e\gamma + g\delta) = 0,$$

a quadric equation for the determination of $\gamma : \delta$; that is, there are two values of the left-hand factor ξ, and two values of the right-hand factor η. But, of course, these correspond each to each, viz. either factor being given, the other factor is determined uniquely.

Writing successively $\frac{\alpha}{\beta} = \frac{-y}{x}$, and $\frac{\gamma}{\delta} = \frac{-y}{x}$, we have the quadric functions $(x, y)^2$,

$$\Phi = (eh - fg)x^2 + (-ah - de + bg + cf)xy + (ad - bc)y^2,$$
$$\Phi_1 = (ch - dg)x^2 + (-ah - cf + bg + de)xy + (af - be)y^2,$$

where the linear factors of Φ are the two values ξ_1, ξ_2 of the left-hand factor ξ, and the linear factors of Φ_1 are the two values η_1, η_2 of the right-hand factor η.

7. In the commutative case, $c = e$, and $d = f$, we have

$$\Phi = \Phi_1 = (ch - dg)x^2 + (-ah + bg)xy + (ad - bc)y^2;$$

here $\xi\eta = 0$, $\eta\xi = 0$, and the values may be taken to be $(\xi_1, \eta_1) = (\xi, \eta)$, $(\xi_2, \eta_2) = (\eta, \xi)$, so that $\xi_1\xi_2 = \xi\eta, = \eta_1\eta_2$; that is, $\Phi = \Phi_1$, as above. The value of Ω is

$$\Omega = gx^3 + (h - 2c)\, x^2y + (a - 2d)\, xy^2 + by^3.$$

8. In the commutative and associative case, taking ξ, η, ζ to be the three idempotent symbols, $\xi^2 = \xi$, $\eta^2 = \eta$, $\zeta^2 = \zeta$, we have

$$\xi(\eta - \xi\eta) = \xi\eta - \xi^2\eta = \xi\eta - \xi\eta = 0;$$

and in this manner we have the six equations

$$\xi(\eta - \xi\eta) = 0,\quad \xi(\zeta - \xi\zeta) = 0;\quad \eta(\xi - \eta\xi) = 0,\quad \eta(\zeta - \eta\zeta) = 0;\quad \zeta(\xi - \zeta\xi) = 0,\quad \zeta(\eta - \zeta\eta) = 0;$$

viz. regarding the right-hand factor as being in each case expressed as a linear function of x, y, we have apparently six products of two linear factors, each $= 0$. There is only one such product $\Phi = 0$; hence, disregarding coefficients, each of the six products must be $= \Phi$, or it must be identically $= 0$, viz. this will be the case if the second factor be $= 0$. We hence conclude that two of the symbols ξ, η, ζ, suppose ξ and η, must be factors of Φ, viz. Φ must be $= \xi\eta$. We have $\Omega = \xi\eta\zeta$; consequently $\Omega = \zeta\Phi$, that is, two of the three linear factors of Ω are the symbols ξ, η, which are such that $\xi\eta\,(= \eta\xi) = 0$. To complete the theory, observe that ζ must be a linear function of ξ, η, $= a\xi + b\eta$ suppose (a, b coefficients, neither of them $= 0$); we thence have

$$\zeta^2,\ = a^2\xi^2 + b^2\eta^2,\ = a^2\xi + b^2\eta,\ = \zeta,\ = a\xi + b\eta;$$

that is, $(a^2 - a)\,\xi + (b^2 - b)\,\eta = 0$; whence $a = 1$, $b = 1$, and therefore $\zeta = \xi + \eta$; hence also $\xi\zeta = \xi$ and $\eta\zeta = \eta$; $\zeta - \xi\zeta = \eta$, $\zeta - \eta\zeta = \xi$. The six products consequently are $\xi\eta$, $\xi\eta$, $\eta\xi$, $\eta\xi$, $\zeta 0$, $\zeta 0$, each $= \Phi$ or identically $= 0$.

9. In verification of the theorem that for the commutative and associative system the cubic function Ω contains the quadric function Φ as a factor, we may write, as above,

$$g = \frac{cd}{b},\quad h = \frac{d^2 + bc - ad}{b},$$

values which give

$$\begin{aligned} b\Phi &= (bc^2 - acd)\, x^2 + (-ad^2 - abc + a^2d + bcd)\, xy + (abd - b^2c)\, y^2 \\ &= -(ad - bc)\,\{cx^2 + (d - a)\, xy - by^2\}, \end{aligned}$$

$$\begin{aligned} b\Omega &= cdx^3 + (d^2 - bc - ad)\, x^2y + (ab - 2bd)\, xy^2 + b^2y^3, \\ &= (dx - by)\,\{cx^2 + (d - a)\, xy - by^2\}, \end{aligned}$$

which gives the theorem in question. And observe further that

$$\begin{aligned} (dx - by)^2 &= d^2(ax + by) - 2bd\,(cx + dy) + b^2\left(\frac{cd}{b}\, x + \frac{d^2 + bc - ad}{b}\, y\right), \\ &= (ad - bc)\,(dx - by); \end{aligned}$$

that is, disregarding coefficients, the two idempotent symbols ξ, η are the linear factors of $cx^2 + (d - a)\, xy - by^2$, and the third idempotent symbol ζ is $= dx - by$.

10. Introducing coefficients in order to make the symbols ξ, η, ζ idempotent, and writing accordingly

$$\xi = \frac{1}{K}\{cx + \tfrac{1}{2}(d - a + \sqrt{\nabla})\, y\}, \quad \nabla = (d-a)^2 + 4bc,$$

so that

$$\eta = \frac{1}{L}\{cx + \tfrac{1}{2}(d - a - \sqrt{\nabla})\, y\}, \quad \xi\eta = \frac{c}{KL}\{cx^2 + (d-a)\, xy - by^2\},$$

$$\zeta = \frac{1}{P}(dx - by),$$

we have to verify that it is possible to determine K, L, P so that $\xi^2 = \xi$, $\eta^2 = \eta$, $\zeta^2 = \zeta$, $\zeta = \xi + \eta$. The last equation gives

$$\frac{d}{P} = \frac{c}{K} + \frac{c}{L},$$

$$-\frac{2b}{P} = \frac{d-a+\sqrt{\nabla}}{K} + \frac{d-a-\sqrt{\nabla}}{L},$$

and we thence have

$$\frac{d(d-a) + 2bc - d\sqrt{\nabla}}{P} = -\frac{2c\sqrt{\nabla}}{K},$$

$$\frac{d(d-a) + 2bc + d\sqrt{\nabla}}{P} = \frac{2c\sqrt{\nabla}}{L},$$

and we can from the equation $\zeta^2 = \zeta$ find P; viz. comparing the coefficients of x, we have

$$\frac{d}{P} = \frac{1}{P^2}\left(ad^2 - 2b\, dc + b^2\frac{cd}{b}\right), \quad = \frac{d}{P^2}(ad - bc), \text{ that is, } P = (ad - bc),$$

or the values of K and L are

$$\frac{\{d(d-a) + 2bc - d\sqrt{\nabla}\}}{ad - bc} = -\frac{2c\sqrt{\nabla}}{K},$$

$$\frac{\{d(d-a) + 2bc + d\sqrt{\nabla}\}}{ad - bc} = +\frac{2c\sqrt{\nabla}}{L},$$

which should agree with the values of K and L found from the equations $\xi^2 = \xi$, $\eta^2 = \eta$, respectively. Comparing the coefficients of x, the first of these equations gives

$$\frac{c}{K} = \frac{1}{K^2}\left\{c^2a + c(d - a + \sqrt{\nabla})\, c + \tfrac{1}{4}(d - a + \sqrt{\nabla})^2\frac{cd}{b}\right\}$$

$$= \frac{c}{4bK^2}\left\{4abc + 4bc(d-a) + d\{(d-a)^2 + \nabla\} + 2\{2bc + d(d-a)\}\sqrt{\nabla}\right\}$$

$$= \frac{c}{2bK^2}\left\{d\nabla + \{d(d-a) + 2bc\}\sqrt{\nabla}\right\},$$

that is,

$$K = \frac{1}{2b}\sqrt{\nabla}\,\{d(d - a + 2bc) + d\sqrt{\nabla}\},$$

and the equation for K becomes

$$\frac{d(d-a)+2bc-d\sqrt{\nabla}}{ad-bc}=\frac{-4bc}{d(d-a)+2bc+d\sqrt{\nabla}};$$

that is,

$$\{d(d-a)+2bc\}^2-d^2\{(d-a)^2+4bc\}=-4bc(ad-bc),$$

which is right; and similarly the equation for L leads to this same equation.

11. We may now establish, on the principles appearing in No. 5, the different forms of the system. Using *idem* and *nil* as abbreviations for idempotent and nilpotent respectively, there are in all 11 cases.

(i) 3 idems. Taking two of these to be x and y, the system is

$$x^2=x, \quad xy=cx+dy, \quad yx=ex+fy, \quad y^2=y.$$

Hence $\Omega=(1-c-e)x^2y+(1-d-f)xy^2$, so that the third factor is

$$(1-c-e)x+(1-d-f)y.$$

This must not reduce itself to x or y, for, if so, there would be a twofold idem; viz. as negative conditions we must have $c+e\neq 1$, $d+f\neq 1$.

And we have

$$\{(1-c-e)x+(1-d-f)y\}^2=\{1-(c+e)(d+f)\}[(1-c-e)x+(1-d-f)y],$$

which must be an idem: viz. we have the further negative condition $(c+e)(d+f)\neq 1$.

(ii) 2 idems and 1 nil. This arises from (i) by assuming therein

$$(c+e)(d+f)=1, \text{ say } d+f=\frac{1}{c+e}:$$

for then, writing $z=-(c+e)x+y$, we have

$$\begin{aligned} z^2&=(c+e)^2x-(c+e)\{(c+e)x+(d+f)y\}+y,\\ &=\{1-(c+e)(d+f)\}y, =0; \end{aligned}$$

viz. z is a nil. And, if in the equations instead of the idem y we introduce the nil z, then the equations assume the form

$$x^2=x, \quad xz=[(c+e)d-e]x+dz, \quad zx=[(c+e)f-c]x+fz, \quad z^2=0;$$

with the idem $y=(c+e)x+z$: hence the negative conditions $c+e\neq 1$ or 0, implying $d+f\neq 1$.

But the equations are obtained in a more simple form by taking x for the idem and y for the nil; viz. we then have $x^2=x$, $xy=cx+dy$, $yx=ex+fy$, $y^2=0$: we must then have $z, =-(c+e)x+(1-d-f)y$, for an idem; this gives $z^2=-(c+e)(d+f)z$, and we have the negative conditions $c+e\neq 0$, $d+f\neq 0$ or 1.

(iii) 1 idem and 2 nils. This may be deduced from (ii) by writing therein $d+f=0$; for then z, $=-(c+e)x+y$, is a nil. The equations are $x^2=x$, $xy=cx+dy$, $yx=ex-dy$, $y^2=0$: and if, instead of x, we introduce therein z by the equation $z=-(c+e)x+y$, the equations become $y^2=0$, $yz=[-e+d(c+e)]y-ez$, $zy=[-c-d(c+e)]y+cz$, $z^2=0$, with the negative condition $c+e\neq 0$.

But it is more simple to take x, y as the nils: the equations then are $x^2=0$, $xy=cx+dy$, $yx=ex+fy$, $y^2=0$. We must have z, $=(c+e)x+(d+f)y$, an idem: this gives $z^2=(c+e)(d+f)z$; and we have the negative conditions $c+e\neq 0$, $d+f\neq 0$.

(*) We cannot have three nils. For in (iii), to make z a nil, we must have $c+e=0$ or $d+f=0$, and in the two cases respectively z, $=(c+e)x+(d+f)y$, becomes $=x$ and $=y$; so that x or y is a twofold nil. Or, what comes to the same thing, we have

$$\Omega=-(c+e)x^2y-(d+f)xy^2,$$

and Ω has a twofold factor if $c+e=0$ or $d+f=0$.

(iv) A twofold idem and a onefold idem. Taking x for the twofold idem and y for the onefold idem, Ω must reduce itself to $(1-c-e)x^2y$, viz. we must have $d+f=1$, or say $f=1-d$. The equations are $x^2=x$, $xy=cx+dy$, $yx=ex+(1-d)y$, $y^2=y$; and we have the negative condition $c+e\neq 1$, for otherwise Ω would vanish identically.

(v) A twofold idem and a onefold nil. Taking x for the twofold idem and y for the onefold nil, then the equations are $x^2=x$, $xy=cx+dy$, $yx=ex+(1-d)y$, $y^2=0$; and we have the negative condition $c+e\neq 0$.

(vi) A twofold nil and a onefold idem. Taking these to be x and y, then $d+f=0$, and the equations are $x^2=0$, $xy=cx+dy$, $yx=ex-dy$, $y^2=y$; and we have the negative condition $c+e\neq 1$.

(vii) A twofold nil and a onefold nil. Taking these to be x and y, we have $d+f=0$, and the equations are $x^2=0$, $xy=cx+dy$, $yx=ex-dy$, $y^2=0$; with the negative condition $c+e\neq 0$.

(viii) A threefold idem. Taking this to be x, then Ω must reduce itself to gx^3, viz. we must have $h=c+e$, $1=d+f$; and the equations are $x^2=x$, $xy=cx+dy$, $yx=ex+(1-d)y$, $y^2=gx+(c+e)y$; we have the negative condition $g\neq 0$, for otherwise Ω would vanish identically.

(ix) A threefold nil. Taking this to be x, then we must have $h=c+e$, $0=d+f$; the equations are $x^2=0$, $xy=cx+dy$, $yx=ex-dy$, $y^2=gx+(c+e)y$; and there is again the negative condition $g\neq 0$.

(x) $\Omega=0$ identically: infinity of idems, 1 nil. Ω will vanish identically if $g=0$, $h=c+e$, $a=d+f$, $b=0$. If there is 1 idem, there will be an infinity of idems, and 1 nil. For, assume an idem x, $x^2=x$; and, if possible, let there be no other idem; then there will be a nil y, $y^2=0$. We have $c+e=0$, $d+f=1$; and the equations are $x^2=x$, $xy=cx+dy$, $yx=-cx+(1-d)\,y$, $y^2=0$; whence $xy+yx=y$. Taking α, β arbitrary coefficients, we have

$$(\alpha x+\beta y)^2=\alpha^2 x+\alpha\beta y,\ =\alpha\,(\alpha x+\beta y)\,;$$

hence $\alpha x+\beta y$ is an idem, except in the case $\alpha=0$, when it is the original nil y.

If besides the idem x we have an idem y, then the conditions are $c+e=1$, $d+f=1$: the equations are

$$x^2=x,\quad xy=cx+dy,\quad yx=(1-c)\,x+(1-d)\,y,\quad y^2=y\,;$$

whence $xy+yx=x+y$. Considering the combination $\alpha x+\beta y$, we have

$$(\alpha x+\beta y)^2=\alpha^2 x+\alpha\beta\,(x+y)+\beta^2 y,\ =(\alpha+\beta)\,(\alpha x+\beta y).$$

This is an idem, except in the case $\alpha+\beta=0$, when it is a nil; or, say we have the single nil $x-y$. We have thus again an infinity of idems, 1 nil.

(xi) $\Omega=0$ identically; an infinity of nils. Taking the two nils x and y, the conditions are $c+e=0$, $d+f=0$; the equations are $x^2=0$, $xy=cx+dy$, $yx=-cx-dy$, $y^2=0$; whence $xy+yx=0$. Considering the arbitrary combination $\alpha x+\beta y$, we have

$$(\alpha x+\beta y)^2=\alpha\beta\,(xy+yx),\ =0,$$

viz. $\alpha x+\beta y$ is a nil; or there are an infinity of nils.

12. The different cases may be grouped together as follows:—

A. 2 idems, (i), (ii), (iv), (x).

Equations $x^2=x$, $xy=cx+dy$, $yx=ex+fy$, $y^2=y$.

B. 1 idem and 1 nil, (ii), (iii), (v), (vi), (x).

Equations $x^2=x$, $xy=cx+dy$, $yx=ex+fy$, $y^2=0$.

C. 2 nils, (iii), (vii), (xi).

Equations $x^2=0$, $xy=cx+dy$, $yx=ex+fy$, $y^2=0$.

D. Threefold idem, (viii).

Equations $x^2=x$, $xy=cx+dy$, $yx=ex-(1-d)\,y$, $y^2=gx+(c+e)\,y$.

E. Threefold nil, (ix).

Equations $x^2=0$, $xy=cx+dy$, $yx=ex-dy$, $y^2=gx+(c+e)\,y$.

The several cases of A, B, C respectively are distinguished by negative conditions which need not be here repeated.

13. I consider, as in my Note before referred to, the conditions in order that the system may be associative. We have the 8 products, x^3, x^2y, xyx, xy^2, yx^2, y^2x, yxy, y^3, giving rise to equations $x\,.\,x^2=x^2\,.\,x$, $x\,.\,xy=x^2\,.\,y$, ..., $y\,.\,y^2=y^2\,.\,y$, which, on putting therein for x^2, xy, yx, y^2 their values, must be satisfied identically. We thus obtain in the first instance 16 relations, but some of these are repeated, and we have actually only 12 relations; viz. the relations are

$$\begin{array}{ll}
\text{(twice)} & b\,(c-e)=0, \\
 & b\,(f-d)=0, \\
 & g\,(c-e)=0, \\
\text{(twice)} & g\,(f-d)=0, \\
\text{(twice)} & bg-cd=0, \\
\text{(twice)} & bg-ef=0, \\
 & c\,(c-h)+g\,(d-a)=0, \\
 & d\,(d-a)+b\,(c-h)=0, \\
 & e\,(e-h)+g\,(f-a)=0, \\
 & b\,(e-h)+f\,(f-a)=0, \\
 & a\,(c-e)-cf+de=0, \\
 & h\,(f-d)-cf+de=0.
\end{array}$$

14. From the first four equations it appears that either $b=0$ and $g=0$, or else $c=e$ and $d=f$. I attend first to the latter case, viz. we have here the commutative system

$$x^2=ax+by,\quad xy=yx=cx+dy,\quad y^2=gx+hy.$$

In order that this may be associative, we must still have the relations

$$bg-cd=0,$$
$$c\,(c-h)+g\,(d-a)=0,$$
$$d\,(d-a)+b\,(c-h)=0,$$

or, as they may be written,

$$\left|\begin{array}{ccc} b, & -c, & d-a \\ -d, & g, & c-h \end{array}\right|=0.$$

These are satisfied by $g=\dfrac{cd}{b}$, $h=\dfrac{d^2+bc-ad}{b}$, and we have thus the commutative and associative system of the Note.

Every system is of the form A, B, C, D, or E; and it can be shown that the commutative and associative system is not of the form D. For, if D were commutative, we should have $e=c$, $d=\frac{1}{2}$, viz. the equations will be $x^2=x$, $xy=yx=cx+\frac{1}{2}y$, $y^2=gx+2cy$, that is,

$$a,\ b,\ c,\ d,\ g,\ h=1,\ 0,\ c,\ \tfrac{1}{2},\ g,\ 2c;$$

and the last of the three relations, viz. $d\,(d-a)+b\,(c-h)=0$, would thus be $\frac{1}{2}(\frac{1}{2}-1)=0$, which is not satisfied. Hence the commutative and associative system can only be of one of the forms A, B, C, and E.

15. First, if the form be A, B, or C; there will be the two idem-or-nil symbols x and y, that is, we may assume $b=0$, $g=0$; and the associative conditions then become $cd=0$, $c(c-h)=0$, $d(d-a)=0$, viz. for the forms A, B, C,

$$\text{A.}\quad x^2=x,\quad y^2=y,$$
$$\text{B.}\quad x^2=x,\quad y^2=0,$$
$$\text{C.}\quad x^2=0,\quad y^2=0,$$

these are $cd=0$, $c(c-1)=0$, $d(d-1)=0$; $c=0$ or 1, $d=0$ or 1,

" $cd=0$, $c^2=0$, $d(d-1)=0$; $c=0$, $d=0$ or 1,

" $cd=0$, $c^2=0$, $d^2=0$; $c=0$, $d=0$.

But for the form A, if $c=0$, $d=1$, that is, $xy=yx=y$, then, writing $z=x-y$, we have $z^2=z$, $yz=zy=0$, $y^2=y$. And similarly, if $c=1$, $d=0$, that is, $xy=yx=x$, then, writing $z=-x+y$, we have $z^2=z$, $zx=xz=0$, $x^2=x$. That is, each of these is reduced to the first case $c=0$, $d=0$; that is, $x^2=x$, $xy=yx=0$, $y^2=y$.

For the form B, if $c=0$, $d=1$, then the system is $x^2=x$, $xy=yx=y$, $y^2=0$; and this cannot be reduced to the first case $x^2=x$, $xy=yx=0$, $y^2=0$.

For the form C, there is only one case, as above.

For the form E, we have $a=0$, $b=0$, ($c=e$, $d=0$, in order that the system may be commutative), $h=2c$, viz. the equations must be $x^2=0$, $xy=yx=cx$, $y^2=gx+2cy$. The associative conditions then give $c=0$; or, the system is $x^2=0$, $xy=yx=0$, $y^2=gx$. Writing $\frac{x}{g}$ instead of x, and for convenience interchanging x and y, the equations are $x^2=y$, $xy=yx=0$, $y^2=0$.

16. The commutative associative system is thus seen to be reducible as follows:—

A. system is $x^2=x$, $xy=yx=0$, $y^2=y$, first mixed system, see No. 2.

B. " $x^2=x$, $xy=yx=y$, $y^2=0$, Peirce's system (a_2),

or else

B. " $x^2=x$, $xy=yx=0$, $y^2=0$, second mixed system.

C. " $x^2=0$, $xy=yx=0$, $y^2=0$, third mixed system.

E. " $x^2=y$, $xy=yx=0$, $y^2=0$, Peirce's system (c_2).

I said, at the end of my Note before referred to, that it had been pointed out to me "that my system [the commutative associative system], in the general case $ad-bc$ not $=0$, is expressible as a mixture of two algebras of the form (a_1), see *American Journal of Mathematics*, vol. IV., p. 120; whereas, if $ad-bc=0$, it is reducible to the form (c_2), see p. 122 (*l.c.*)." The accurate conclusion is as above, that the commutative associative system is either a mixed system of one of the three forms, or else a system (a_2), or (c_2).

17. Considering next the non-commutative associative systems, we have here, *ante*, No. 14, $b=0$, $g=0$; and the relations which remain to be satisfied then are

$$cd=0,\ ef=0,\ c(c-h)=0,\ d(d-a)=0,\ e(e-h)=0,\ f(f-a)=0,$$
$$a(c-e)-cf+de=0,\ h(f-d)-cf+de=0.$$

The first equation gives $cd=0$, that is, $c=0$ or $d=0$; but we may attend exclusively to the case $c=0$, for the case $d=0$ may be deduced from this by the interchange of x, y. We have then $ef=0$; and it will be convenient to separate the cases

I. $c=0,\ e=0,\ f=0$, giving $d(d-a)=0,\quad dh=0$,

II. $c=0,\ e=0,\ f\neq 0$, „ $d(d-a)=0,\ f-a=0,\ h(f-d)=0$,

III. $c=0,\ e\neq 0,\ f=0$, „ $d(d-a)=0,\ e-h=0,\quad (d-a)=0,\ d(e-h)=0$,

that is, $d-a=0,\ e-h=0$.

18. We have thus five cases;

I. (a) $d=0$: $x^2=ax$, $xy=yx=0$, $y^2=hy$: commutative, and so included in what precedes.

I. (b) $d=a$, $h=0$: $x^2=ax$, $xy=ay$, $yx=0$, $y^2=0$: or, writing as we may do $a=1$, this is $x^2=x$, $xy=y$, $yx=0$, $y^2=0$; which is Peirce's system (b_2).

II. (c) $d=f=a$: $x^2=ax$, $xy=yx=ay$, $y^2=hy$: commutative, and so included in what precedes.

II. (d) $d=0$, $f=a$, $h=0$: $x^2=ax$, $xy=0$, $yx=ay$, $y^2=0$; or, writing as we may do $a=1$, this is $x^2=x$, $xy=0$, $yx=y$, $y^2=0$; which is the system (d_2).

III. (e) $d=a$, $e=h$: $x^2=ax$, $xy=ay$, $yx=hx$, $y^2=hy$; or, writing as we may do $a=1$, $h=1$, this is $x^2=x$, $xy=y$; $yx=x$; $y^2=y$. Introducing here the new symbol z, $=x-y$, we have $z^2=0$, $xz=z$, $zx=0$, $yz=z$, $zy=0$. Thus x, z form the system $x^2=x$, $xz=z$, $zx=0$, $z^2=0$ (or, what is the same thing, y, z form a system $y^2=y$, $yz=z$, $zy=0$, $z^2=0$); each of these is Peirce's system (b_2).

The conclusion is that every non-commutative associative system is either Peirce's system (b_2), or else the omitted system (d_2). Hence, disregarding the mixed systems, every associative system is either (a_2), (b_2), (c_2), or (d_2).

19. It may be proper to show that the systems (b_2), $x^2=x$, $xy=y$, $yx=0$, $y^2=0$, and (d_2), $x^2=x$, $xy=0$, $yx=y$, $y^2=0$, or say

	a	b	c	d	e	f	g	h
(b_2)	1	0	0	1	0	0	0	0
(d_2)	1	0	0	0	0	1	0	0,

are really distinct from each other. Observe that they each belong to the case (x), $\Omega=0$, an infinity of idems and 1 nil; viz. in each of them writing $z=x+\beta y$, β an arbitrary coefficient, we have $z^2=x^2+\beta(xy+yx)+\beta^2y^2$, $=x+\beta y$, $=z$, we have z an idem, and y is the only nil. And, this being so, we have in the first system $zy=y$, $yz=0$, viz. the system is $z^2=z$, $zy=y$, $yz=0$, $y^2=0$, retaining, when we write z for x, its original form. And similarly, in the second system, $zy=0$, $yz=y$; viz. the system is $z^2=z$, $zy=0$, $yz=y$, $y^2=0$, retaining, when we write therein z for x, its original form. The two are thus distinct systems, in no wise transformable the one into the other.

815.

THE BINOMIAL EQUATION $x^p - 1 = 0$; QUINQUISECTION. SECOND PART.

[From the *Proceedings of the London Mathematical Society*, vol. XVI. (1885), pp. 61—63.]

IN the paper, "The binomial equation $x^p - 1 = 0$; quinquisection," *Proc. Lond. Math. Soc.*, t. XII. (1881), pp. 15, 16, [764], I considered for an exponent $p = 5n + 1$, the five periods X, Y, Z, W, T connected by the equations

$$\begin{array}{l|lllll} & X, & Y, & Z, & W, & T \\ \hline X^2 = & a, & b, & c, & d, & e \\ XY = & f, & g, & h, & i, & j \\ XZ = & k, & l, & m, & n, & o, \end{array}$$

and the equations deduced from these by cyclical permutations of the periods and of the coefficients of each set; but I did not obtain completely even the linear relations connecting the coefficients. I since found, by induction from the examples given in the Table 1, that the coefficients could be expressed linearly in terms of the linearly independent integer numbers α, β, f, k as follows: viz. introducing for convenience the new number θ, such that

$$\alpha + \beta + \theta = \tfrac{1}{5}(p - 1),$$

then the expressions in question are

$$\begin{array}{lllll} a,\ b,\ c\ ,\ d,\ e = -1 - 2\theta + \alpha + \beta, & -\theta - \alpha - \beta + f, & -\theta - \alpha - \beta + k, & -\alpha - 2\beta - k, & -2\alpha - \beta - f, \\ f,\ g,\ h\ ,\ i\ ,\ j = \quad f, & \theta - \alpha - f, & \alpha, & \beta, & \alpha, \\ k,\ l,\ m,\ n,\ o = \quad k, & \alpha, & \theta - \beta - k, & \beta, & \beta, \end{array}$$

and I found further that, substituting these values of the coefficients in the 20 quadric relations referred to in the former paper, the 20 relations reduced themselves to two equations only, viz. these were

$$\theta(-2\alpha + \beta + k) + 3\alpha^2 - \beta^2 \quad + \quad \alpha(f - k - 1) - \beta f + f^2 - 2fk = 0,$$

$$-\theta^2 + \theta(3\beta + 2k + f) \ + \alpha^2 - \alpha\beta - 3\beta^2 - \alpha k + \beta(1 - f - k) - k^2 - 2fk = 0.$$

The final result thus is that the coefficients are expressed as functions of the five numbers α, β, f, k, θ, connected by the linear equation $\alpha + \beta + \theta = \frac{1}{5}(p-1)$, and the two quadric equations. I remark that formulæ equivalent to these were obtained and proved by Mr F. S. Carey in his Trinity Fellowship Dissertation, 1884; viz. writing $n = \frac{1}{5}(p-1)$, his formulæ were

$$\begin{array}{llllll} a,\ b,\ c,\ d,\ e = & \alpha - n, & \beta - n, & \gamma - n, & \delta - n, & \epsilon - n, \\ f,\ g,\ h,\ i,\ j = & \beta, & \epsilon, & \rho, & \sigma, & \rho, \\ k,\ l,\ m,\ n,\ o = & \gamma, & \rho, & \delta, & \sigma, & \sigma, \end{array}$$

with the three linear relations

$$\begin{aligned} \alpha + \beta + \gamma + \delta + \epsilon &= n - 1, \\ \beta + \epsilon + 2\rho + \sigma &= n, \\ \gamma + \delta + \rho + 2\sigma &= n, \end{aligned}$$

and the two quadric relations

$$\begin{aligned} \delta^2 + \gamma^2 + 2\sigma\alpha + (\rho - \sigma)(\delta + \gamma) - 2\rho(\rho + \sigma) &= (\delta - \gamma)(\beta - \epsilon), \\ \beta^2 + \epsilon^2 + 2\rho\alpha + (\sigma - \rho)(\beta + \epsilon) - 2\sigma(\rho + \sigma) &= (\gamma - \delta)(\beta - \epsilon), \end{aligned}$$

the coefficients being thus expressed in terms of the seven numbers α, β, γ, δ, ϵ, ρ, σ connected by five equations. The equivalence of the two sets of formulæ may be shown without difficulty.

To the Table 2 of the Quintic Equations, given in the paper, may be added the following result from Legendre's *Théorie des Nombres*, Ed. 3, t. II., p. 213,

p	η^5	η^4	η^3	η^2	η	1	
641	1	+ 1	− 256	− 564	+ 5238	− 5120	= 0,

calculated by him for the isolated case $p = 641$.

816.

ON THE BITANGENTS OF A PLANE QUARTIC.

[From *Crelle's Journal der Mathem.*, t. XCIV. (1883), pp. 93—115; *Camb. Phil. Soc. Proc.*, t. IV. (1883), p. 321.]

RIEMANN in the paper "Zur Theorie der *Abel*schen Functionen für den Fall $p=3$," *Werke*, pp. 456—479, has given a remarkably elegant solution of the problem of the bitangents of a plane quartic. But his formulæ may be improved by a slight change; viz. we may in his first equation $x+y+z+\xi+\eta+\zeta=0$ introduce coefficients so as to bring this into the same form as the other three equations. It thus appears that, instead of his $3+3$ equations of the forms

$$\frac{x}{1-\beta\gamma}+\frac{y}{1-\gamma\alpha}+\frac{z}{1-\alpha\beta}=0 \text{ and } \frac{\xi}{\alpha(\gamma-\beta)}+\frac{\eta}{\beta(\gamma-\alpha)}+\frac{\zeta}{\gamma(\beta-\alpha)}=0$$

respectively, we have $6+6$ equations of like forms; but these two systems each of 6 equations are equivalent to each other, so that instead of $6+16+3+3=28$, we have $6+16+6(=6)=28$ equations for the 28 bitangents*. I make another slight change of notation by introducing the single letters f, g, h to denote the reciprocals $\frac{1}{a}$, $\frac{1}{b}$, $\frac{1}{c}$; and I consider the whole question as follows. The theory is based on the equations

$$\begin{aligned}
a\,x+b\,y+c\,z+f\,\xi+g\,\eta+h\,\zeta&=0,\\
a_1x+b_1y+c_1z+f_1\xi+g_1\eta+h_1\zeta&=0,\\
a_2x+b_2y+c_2z+f_2\xi+g_2\eta+h_2\zeta&=0,\\
a_3x+b_3y+c_3z+f_3\xi+g_3\eta+h_3\zeta&=0,
\end{aligned}$$

where $af=bg=ch=a_1f_1=\&c.=c_3h_3=1$: and the coefficients a, b, c, a_1, b_1, c_1, a_2, b_2, c_2 are arbitrary. The equations $x=0$, $y=0$, $z=0$ represent any three given lines: and

* It will appear further on that the equation of each of the last-mentioned 6 bitangents can be expressed in 8 different forms.

considering ξ, η, ζ to be determined as linear functions of x, y, z by means of the first three equations, then (the coefficients a, b, c, a_1, b_1, c_1, a_2, b_2, c_2 being determined accordingly) the equations $\xi=0$, $\eta=0$, $\zeta=0$ will also represent any three given lines. But observe that, if the equation of the first of these lines is $x+my+nz=0$, ξ is not = an arbitrary multiple $\theta(x+my+nz)$ of the linear function $x+my+nz$, but the constant factor θ has a completely determinate value: and the like as regards η and ζ respectively.

The coefficients a_3, b_3, c_3 of the fourth equation are such that this fourth equation is a mere consequence of the other three: viz. we must have

$$\begin{aligned} a_3 &= \lambda a + \lambda_1 a_1 + \lambda_2 a_2,\\ b_3 &= \lambda b + \lambda_1 b_1 + \lambda_2 b_2,\\ c_3 &= \lambda c + \lambda_1 c_1 + \lambda_2 c_2,\\ f_3 &= \lambda f + \lambda_1 f_1 + \lambda_2 f_2,\\ g_3 &= \lambda g + \lambda_1 g_1 + \lambda_2 g_2,\\ h_3 &= \lambda h + \lambda_1 h_1 + \lambda_2 h_2, \end{aligned}$$

or, what is the same thing, we must have

$$\left\| \begin{matrix} a, & b, & c, & f, & g, & h \\ a_1, & b_1, & c_1, & f_1, & g_1, & h_1 \\ a_2, & b_2, & c_2, & f_2, & g_2, & h_2 \\ a_3, & b_3, & c_3, & f_3, & g_3, & h_3 \end{matrix} \right\| = 0,$$

viz. each of the determinants formed with four columns out of this matrix is equal to 0.

Using the equations in λ, λ_1, λ_2, and writing for shortness

$$\begin{aligned} &a_1 f_2 + a_2 f_1, \quad a_2 f + a f_2, \quad a f_1 + a_1 f = F,\ F_1,\ F_2,\\ &b_1 g_2 + b_2 g_1, \quad b_2 g + b g_2, \quad b g_1 + b_1 g = G,\ G_1,\ G_2,\\ &c_1 h_2 + c_2 h_1, \quad c_2 h + c h_2, \quad c h_1 + c_1 h = H,\ H_1,\ H_2; \end{aligned}$$

then, forming the products $a_3 f_3$, $b_3 g_3$, $c_3 h_3$, we find

$$\begin{aligned} 1-\lambda^2-\lambda_1^2-\lambda_2^2 &= F\lambda_1\lambda_2 + F_1\lambda_2\lambda + F_2\lambda\lambda_1,\\ \text{,,} \qquad &= G\lambda_1\lambda_2 + G_1\lambda_2\lambda + G_2\lambda\lambda_1,\\ \text{,,} \qquad &= H\lambda_1\lambda_2 + H_1\lambda_2\lambda + H_2\lambda\lambda_1, \end{aligned}$$

or, as these equations may be written,

$$1-\lambda^2-\lambda_1^2-\lambda_2^2 : \lambda_1\lambda_2 : \lambda_2\lambda : \lambda\lambda_1$$
$$= \begin{vmatrix} F, & F_1, & F_2 \\ G, & G_1, & G_2 \\ H, & H_1, & H_2 \end{vmatrix} : \begin{vmatrix} 1, & F_1, & F_2 \\ 1, & G_1, & G_2 \\ 1, & H_1, & H_2 \end{vmatrix} : \begin{vmatrix} 1, & F_2, & F \\ 1, & G_2, & G \\ 1, & H_2, & H \end{vmatrix} : \begin{vmatrix} 1, & F, & F_1 \\ 1, & G, & G_1 \\ 1, & H, & H_1 \end{vmatrix},$$

say for shortness

$$=\Pi : \Lambda : \Lambda_1 : \Lambda_2.$$

We have thus

$$\frac{\lambda^2}{1-\lambda^2-\lambda_1^2-\lambda_2^2}=\frac{\Lambda_1\Lambda_2}{\Pi\Lambda},$$

$$\frac{\lambda_1^2}{1-\lambda^2-\lambda_1^2-\lambda_2^2}=\frac{\Lambda_2\Lambda}{\Pi\Lambda_1},$$

$$\frac{\lambda_2^2}{1-\lambda^2-\lambda_1^2-\lambda_2^2}=\frac{\Lambda\Lambda_1}{\Pi\Lambda_2},$$

and thence

$$\frac{1}{1-\lambda^2-\lambda_1^2-\lambda_2^2}=\frac{1}{\Pi}\left(\frac{\Lambda_1\Lambda_2}{\Lambda}+\frac{\Lambda_2\Lambda}{\Lambda_1}+\frac{\Lambda\Lambda_1}{\Lambda_2}\right),$$

equations which give λ, λ_1, λ_2, and consequently a_3, b_3, c_3, f_3, g_3, h_3 in terms of known quantities, the several expressions depending on the single radical

$$\sqrt{\frac{1}{\Pi}\left(\frac{\Lambda_1\Lambda_2}{\Lambda}+\frac{\Lambda_2\Lambda}{\Lambda_1}+\frac{\Lambda\Lambda_1}{\Lambda_2}\right)}.$$

Observe that we have rationally

$$\lambda : \lambda_1 : \lambda_2 = \frac{1}{\Lambda} : \frac{1}{\Lambda_1} : \frac{1}{\Lambda_2}.$$

I consider now the quartic curve

$$\sqrt{x\xi}+\sqrt{y\eta}+\sqrt{z\zeta}=0,$$

and I write down the equations of the 28 bitangents, each with its triple-theta characteristic as given by Riemann, and also with its duad symbol, derived from Hesse's method. I assume that these characteristics and duad symbols are properly attached to the several lines: and I insert also the current nos. 1 to 28. The equations are:

Current No.	Duad Symbol	Characteristic	
1	18	111 111	$x=0,$
2	28	001 011	$y=0,$
3	38	011 001	$z=0,$
4	23	010 010	$\xi=0,$
5	13	100 110	$\eta=0,$
6	12	110 100	$\zeta=0,$

Current No.	Duad Symbol	Characteristic	
7	48	101 100	$ax+by+cz=0,\quad f\xi+g\eta+h\zeta=0,$
8	14	010 011	$f\xi+by+cz=0,\quad ax+g\eta+h\zeta=0,$
9	24	100 111	$ax+g\eta+cz=0,\quad f\xi+by+h\zeta=0,$
10	34	110 101	$ax+by+h\zeta=0,\quad f\xi+g\eta+cz=0,$
11	58	100 101	$a_1x+b_1y+c_1z=0,$
12	15	011 010	$f_1\xi+b_1y+c_1z=0,$
13	25	101 110	$a_1x+g_1\eta+c_1z=0,$
14	35	111 100	$a_1x+b_1y+h_1\zeta=0,$
15	68	110 010	$a_2x+b_2y+c_2z=0,$
16	16	001 101	$f_2\xi+b_2y+c_2z=0,$
17	26	111 001	$a_2x+g_2\eta+c_2z=0,$
18	36	101 011	$a_2x+b_2y+h_2\zeta=0,$
19	78	010 110	$a_3x+b_3y+c_3z=0,$
20	17	101 001	$f_3\xi+b_3y+c_3z=0,$
21	27	011 101	$a_3x+g_3\eta+c_3z=0,$
22	37	000 111	$a_3x+b_3y+h_3\zeta=0,$
23	67	100 100	$\frac{x}{bc-b_1c_1}+\frac{y}{ca-c_1a_1}+\frac{z}{ab-a_1b_1}=0,\quad \frac{\xi}{g_2h_2-g_3h_3}+\frac{\eta}{h_2f_2-h_3f_3}+\frac{\zeta}{f_2g_2-f_3g_3}=0,$ $\frac{x}{bc-b_1c_1}+\frac{\eta}{bb_1(ca_1-c_1a)}-\frac{\zeta}{cc_1(ab_1-a_1b)}=0,\quad \frac{\xi}{g_2h_2-g_3h_3}+\frac{y}{g_2g_3(h_2f_3-h_3f_2)}-\frac{z}{h_2h_3(f_2g_3-f_3g_2)}=0,$ $\frac{-\xi}{aa_1(bc_1-b_1c)}+\frac{y}{ca-c_1a_1}+\frac{\zeta}{cc_1(ab_1-a_1b)}=0,\quad \frac{-x}{f_2f_3(g_2h_3-g_3h_2)}+\frac{\eta}{h_2f_2-h_3f_3}+\frac{z}{h_2h_3(f_2g_3-f_3g_2)}=0,$ $\frac{\xi}{aa_1(bc_1-b_1c)}-\frac{\eta}{bb_1(ca_1-c_1a)}+\frac{z}{ab-a_1b_1}=0,\quad \frac{x}{f_2f_3(g_2h_3-g_3h_2)}-\frac{y}{g_2g_3(h_2f_3-h_3f_2)}+\frac{\zeta}{f_2g_2-f_3g_3}=0,$
24	57	110 011	$\frac{x}{bc-b_2c_2}+\frac{y}{ca-c_2a_2}+\frac{z}{ab-a_2b_2}=0,$
25	56	010 111	$\frac{x}{bc-b_3c_3}+\frac{y}{ca-c_3a_3}+\frac{z}{ab-a_3b_3}=0,$
26	45	001 001	$\frac{x}{b_2c_2-b_3c_3}+\frac{y}{c_2a_2-c_3a_3}+\frac{z}{a_2b_2-a_3b_3}=0,$
27	46	011 110	$\frac{x}{b_3c_3-b_1c_1}+\frac{y}{c_3a_3-c_1a_1}+\frac{z}{a_3b_3-a_1b_1}=0,$
28	47	111 010	$\frac{x}{b_1c_1-b_2c_2}+\frac{y}{c_1a_1-c_2a_2}+\frac{z}{a_1b_1-a_2b_2}=0.$

As regards each of the bitangents 7 to 22, the equivalence of the two forms of equation is at once seen by means of the fundamental equations $ax+by+cz+f\xi+g\eta+h\zeta=0$: as regards the remaining bitangents 23 to 28, the equation of each of these is expressible in eight different forms, which are written down in full for the bitangent 23; the equivalence of the eight forms of equation must of course be proved.

I proceed to show that each of the 28 lines is, in fact, a bitangent to the quartic curve

$$\sqrt{x\xi}+\sqrt{y\eta}+\sqrt{z\zeta}=0,$$

or, what is the same thing, that writing this equation in the form

$$\Omega=x^2\xi^2+y^2\eta^2+z^2\zeta^2-2yz\eta\zeta-2zx\zeta\xi-2xy\xi\eta=0,$$

then that by means of any one of these equations Ω becomes a perfect square.

This is obviously the case for each of the equations 1, 2, 3, 4, 5, 6.

For the equations 7, 8, 9, 10, observe that the equation of the quartic may be written:

$$\sqrt{axf\xi}+\sqrt{byg\eta}+\sqrt{czh\zeta}=0,$$

or, what is the same thing,

$$\Omega=a^2x^2\,.\,f^2\xi^2+\ldots=0.$$

Hence writing x, y, z, ξ, η, ζ in place of ax, by, cz, $f\xi$, $g\eta$, $h\zeta$, so that now

$$x+y+z+\xi+\eta+\zeta=0,$$

the equation of the line 7 is expressed in the two equivalent forms $x+y+z=0$, $\xi+\eta+\zeta=0$. We have for Ω its original value

$$\begin{aligned}\Omega&=x^2\xi^2+y^2\eta^2+z^2\zeta^2-2yz\eta\zeta-2zx\zeta\xi-2xy\xi\eta,\\&=(z\zeta-x\xi-y\eta)^2-4xy\xi\eta;\end{aligned}$$

and writing herein $z=-x-y$, $\zeta=-\xi-\eta$, we have $z\zeta-x\xi-y\eta=x\eta+y\xi$ and consequently $\Omega=(x\eta+y\xi)^2-4xy\xi\eta$, $=(x\eta-y\xi)^2$, a perfect square. The like proof applies to the equations 8, 9, 10: and then clearly the result applies to the equations 11, 12, 13, 14; 15, 16, 17, 18; 19, 20, 21, 22.

We come now to the equation of the line 23, say this is in the first instance taken to be

$$\frac{x}{bc-b_1c_1}+\frac{y}{ca-c_1a_1}+\frac{z}{ab-a_1b_1}=0;$$

if from this equation, by means of the two fundamental equations

$$\begin{aligned}ax+by+cz+f\xi+g\eta+h\zeta&=0,\\a_1x+b_1y+c_1z+f_1\xi+g_1\eta+h_1\zeta&=0,\end{aligned}$$

we eliminate first (y, z), secondly (z, x), and thirdly (x, y), it is found that ξ, η, ζ will also disappear in the three cases respectively, and that the resulting equations are

$$\frac{x}{bc - b_1c_1} + \frac{\eta}{bb_1(ca_1 - c_1a)} - \frac{\zeta}{cc_1(ab_1 - a_1b)} = 0,$$

$$-\frac{\xi}{aa_1(bc_1 - b_1c)} + \frac{y}{ca - c_1a_1} + \frac{\zeta}{cc_1(ab_1 - a_1b)} = 0,$$

$$\frac{\xi}{aa_1(bc_1 - b_1c)} - \frac{\eta}{bb_1(ca_1 - c_1a)} + \frac{z}{ab - a_1b_1} = 0,$$

so that each of these is, in fact, equivalent to the original equation in (x, y, z): and observe that, adding together the three equations, we reproduce the original equation in (x, y, z).

Write for shortness

$$\begin{aligned} P &= bc - b_1c_1, \quad \alpha = bc_1 - b_1c, \\ Q &= ca - c_1a_1, \quad \beta = ca_1 - c_1a, \\ R &= ab - a_1b_1, \quad \gamma = ab_1 - a_1b, \end{aligned}$$

then the equation in (x, y, z) is

$$QRx + RPy + PQz = 0,$$

so that, eliminating y and z, we find

$$\begin{vmatrix} b\ , & c\ , & ax + f\xi + g\eta + h\zeta \\ b_1\ , & c_1\ , & a_1x + f_1\xi + g_1\eta + h_1\zeta \\ RP, & PQ, & QRx \end{vmatrix} = 0.$$

In this equation, the coefficient of x is

$$\begin{aligned} &(bc_1 - b_1c)\,QR + (ca_1 - c_1a)\,RP + (ab_1 - a_1b)\,PQ, \\ = \ &(bc_1 - b_1c)\,\{a^2bc + a_1^2b_1c_1 - aa_1(bc_1 + b_1c)\} \\ &+ (ca_1 - c_1a)\,\{ab^2c + a_1b_1^2c_1 - bb_1(ca_1 + c_1a)\} \\ &+ (ab_1 - a_1b)\,\{abc^2 + a_1b_1c_1^2 - cc_1(ab_1 + a_1b)\}, \\ = &- aa_1(b^2c_1^2 - b_1^2c^2) - bb_1(c^2a_1^2 - c_1^2a^2) - cc_1(a^2b_1^2 - a_1^2b^2), \\ = &- (bc_1 - b_1c)(ca_1 - c_1a)(ab_1 - a_1b), \\ = &- \alpha\beta\gamma. \end{aligned}$$

The coefficient of ξ is

$$= RP(cf_1 - c_1f) - PQ(bf_1 - b_1f),$$

which $\left(\text{observing that } cf_1 - c_1f, = \frac{c}{a_1} - \frac{c_1}{a}, \text{ is } = \frac{1}{aa_1}Q, \text{ and } bf_1 - b_1f, = \frac{b}{a_1} - \frac{b_1}{a}, \text{ is } = \frac{1}{aa_1}R\right)$ is $= 0$.

The coefficient of η is

$$RP(cg_1 - c_1g) - PQ(bg_1 - b_1g),$$

which is

$$= RP\frac{1}{bb_1}(bc - b_1c_1) - PQ\frac{1}{bb_1}(b^2 - b_1^2),$$

$$= \frac{P}{bb_1}\{(bc - b_1c_1)(ab - a_1b_1) - (b^2 - b_1^2)(ca - c_1a_1)\},$$

$$= \frac{P}{bb_1}\{b_1^2ca + b^2c_1a_1 - bb_1(ac_1 + a_1c)\}$$

$$= -\frac{P}{bb_1}\gamma\alpha;$$

and similarly the coefficient of ζ is

$$= \frac{P}{cc_1}\alpha\beta.$$

We have thus in x, η, ζ the equation

$$-\alpha\beta\gamma x - \frac{P}{bb_1}\gamma\alpha\eta + \frac{P}{cc_1}\alpha\beta\zeta = 0,$$

or, what is the same thing,

$$\frac{x}{P} + \frac{\eta}{bb_1\beta} - \frac{\zeta}{cc_1\gamma} = 0:$$

and in like manner, by the elimination of (z, x) and (x, y) respectively,

$$-\frac{\xi}{aa_1\alpha} + \frac{y}{Q} + \frac{\zeta}{cc_1\gamma} = 0,$$

$$\frac{\xi}{aa_1\alpha} - \frac{\eta}{bb_1\beta} + \frac{z}{R} = 0,$$

which are the required three equations.

We have next to show that the line is a bitangent—write for a moment $aa_1\alpha$, $bb_1\beta$, $cc_1\gamma = \lambda$, μ, ν, then the three equations give

$$x,\ y,\ z = P\left(-\frac{\eta}{\mu} + \frac{\zeta}{\nu}\right),\quad Q\left(-\frac{\zeta}{\nu} + \frac{\xi}{\lambda}\right),\quad R\left(-\frac{\xi}{\lambda} + \frac{\eta}{\mu}\right),$$

and we thence have

$$\begin{aligned} \Omega = {} & P^2\left(-\frac{\xi\eta}{\mu} + \frac{\zeta\xi}{\nu}\right)^2 \\ & + Q^2\left(-\frac{\eta\zeta}{\nu} + \frac{\xi\eta}{\lambda}\right)^2 \\ & + R^2\left(-\frac{\zeta\xi}{\lambda} + \frac{\eta\zeta}{\mu}\right)^2 \\ & - 2QR\left(-\frac{\eta\zeta}{\nu} + \frac{\xi\eta}{\lambda}\right)\left(-\frac{\zeta\xi}{\lambda} + \frac{\eta\zeta}{\mu}\right) \\ & - 2RP\left(-\frac{\zeta\xi}{\lambda} + \frac{\eta\zeta}{\mu}\right)\left(-\frac{\xi\eta}{\mu} + \frac{\zeta\xi}{\nu}\right) \\ & - 2PQ\left(-\frac{\xi\eta}{\mu} + \frac{\zeta\xi}{\nu}\right)\left(-\frac{\eta\zeta}{\nu} + \frac{\xi\eta}{\lambda}\right); \end{aligned}$$

which, putting further $P\lambda$, $Q\mu$, $R\nu = \text{a, b, c}$, becomes

$$\begin{aligned}\Omega = & \frac{\eta^2\zeta^2}{\mu^2\nu^2}(\text{b}+\text{c})^2\\ &+\frac{\zeta^2\xi^2}{\nu^2\lambda^2}(\text{c}+\text{a})^2\\ &+\frac{\xi^2\eta^2}{\lambda^2\mu^2}(\text{a}+\text{b})^2\\ &+\frac{2\xi^2\eta\zeta}{\lambda^2\mu\nu}\{-\text{a}(\text{a}+\text{b}+\text{c})+\text{bc}\}\\ &+\frac{2\xi\eta^2\zeta}{\lambda\mu^2\nu}\{-\text{b}(\text{a}+\text{b}+\text{c})+\text{ca}\}\\ &+\frac{2\xi\eta\zeta^2}{\lambda\mu\nu^2}\{-\text{c}(\text{a}+\text{b}+\text{c})+\text{ab}\}.\end{aligned}$$

But here

$$\begin{aligned}\text{a} &= aa_1(bc_1-b_1c)(bc-b_1c_1) = aa_1\{(b^2+b_1^2)cc_1-(c^2+c_1^2)bb_1\},\\ \text{b} &= bb_1(ca_1-c_1a)(ca-c_1a_1) = bb_1\{(c^2+c_1^2)aa_1-(a^2+a_1^2)cc_1\},\\ \text{c} &= cc_1(ab_1-a_1b)(ab-a_1b_1) = cc_1\{(a^2+a_1^2)bb_1-(b^2+b_1^2)aa_1\}.\end{aligned}$$

Hence

$$\text{a}+\text{b}+\text{c}=0,$$

and we obtain

$$\begin{aligned}\Omega &= \left(\frac{\text{a}\eta\zeta}{\mu\nu}+\frac{\text{b}\zeta\xi}{\nu\lambda}+\frac{\text{c}\xi\eta}{\lambda\mu}\right)^2,\\ &= \left\{\frac{aa_1\alpha P}{bb_1cc_1\beta\gamma}\eta\zeta+\frac{bb_1\beta Q}{cc_1aa_1\gamma\alpha}\zeta\xi+\frac{cc_1\gamma R}{aa_1bb_1\alpha\beta}\xi\eta\right\}^2,\\ &= \frac{1}{(aa_1bb_1cc_1\alpha\beta\gamma)^2}(a^2a_1^2\alpha^2P\eta\zeta+b^2b_1^2\beta^2Q\zeta\xi+c^2c_1^2\gamma^2R\xi\eta)^2,\end{aligned}$$

a perfect square, and the line 23 is thus a bitangent.

We have next to prove the equivalence of the two equations

$$\frac{x}{bc-b_1c_1}+\frac{y}{ca-c_1a_1}+\frac{z}{ab-a_1b_1}=0,$$

and

$$\frac{\xi}{g_2h_2-g_3h_3}+\frac{\eta}{h_2f_2-h_3f_3}+\frac{\zeta}{f_2g_2-f_3g_3}=0.$$

Starting from the former equation written in the form

$$-QRx-RPy-PQz=0,$$

and combining herewith the three fundamental equations

$$\begin{aligned}a\,x+b\,y+c\,z+f\,\xi+g\,\eta+h\,\zeta&=0,\\ a_1x+b_1y+c_1z+f_1\xi+g_1\eta+h_1\zeta&=0,\\ a_2x+b_2y+c_2z+f_2\xi+g_2\eta+h_2\zeta&=0,\end{aligned}$$

if we eliminate from these the (x, y, z), we obtain the following equation in (ξ, η, ζ)

$$\begin{vmatrix} a, & b, & c, & f\xi+g\eta+h\zeta \\ a_1, & b_1, & c_1, & f_1\xi+g_1\eta+h_1\zeta \\ a_2, & b_2, & c_2, & f_2\xi+g_2\eta+h_2\zeta \\ -QR, & -RP, & -PQ & \end{vmatrix}=0,$$

which, observing that the values of $-QR$, $-RP$, $-PQ$ are

$$-a^2bc-a_1^2b_1c_1+aa_1(bc_1+b_1c),\quad -ab^2c-a_1b_1^2c_1+bb_1(ca_1+c_1a),\quad -abc^2-a_1b_1c_1^2+cc_1(ab_1+a_1b),$$

is immediately transformed into

$$\begin{vmatrix} a, & b, & c, & f\xi+g\eta+h\zeta \\ a_1, & b_1, & c_1, & f_1\xi+g_1\eta+h_1\zeta \\ a_2, & b_2, & c_2, & f_2\xi+g_2\eta+h_2\zeta \\ aa_1(bc_1+b_1c), & bb_1(ca_1+c_1a), & cc_1(ab_1+a_1b), & \begin{cases} abc\,(f\xi+g\eta+h\zeta) \\ +a_1b_1c_1(f_1\xi+g_1\eta+h_1\zeta) \end{cases} \end{vmatrix}=0$$

Here the coefficient of ξ is

$$\begin{vmatrix} a, & b, & c, & f \\ a_1, & b_1, & c_1, & f_1 \\ a_2, & b_2, & c_2, & f_2 \\ aa_1(bc_1+b_1c), & bb_1(ca_1+c_1a), & cc_1(ab_1+a_1b), & abcf+a_1b_1c_1f_1 \end{vmatrix}$$

which is of the form $lf+l_1f_1+l_2f_2$; and we find without difficulty

$$\begin{aligned} l = {} & a\,a_2(bc-b_1c_1)(bc_1-b_1c), & = {} & a\,a_2P\alpha \\ & +b\,b_2\,(ca-c_1a_1)(ca_1-c_1a) & & +b\,b_2Q\beta \\ & +c\,c_2\,(ab-a_1b_1)(ab_1-a_1b) & & +c\,c_2R\gamma; \\ l_1 = {} & -a_1a_2(bc-b_1c_1)(bc_1-b_1c), & = {} & -a_1a_2P\alpha \\ & -b_1b_2(ca-c_1a_1)(ca_1-c_1a) & & -b_1b_2Q\beta \\ & -c_1c_2(ab-a_1b_1)(ab_1-a_1b) & & -c_1c_2R\gamma; \\ l_2 = {} & -(bc_1-b_1c)(ca_1-c_1a)(ab_1-a_1b) = -\alpha\beta\gamma. \end{aligned}$$

Hence

$$lf+l_1f_1+l_2f_2=(af-a_1f_1)\,a_2P\alpha+(bf-b_1f_1)\,b_2Q\beta+(cf-c_1f_1)\,c_2R\gamma-\alpha\beta\gamma f_2,$$

which, observing that

$$af-a_1f_1=0,\quad bf-b_1f_1=-\frac{\gamma}{aa_1},\quad cf-c_1f_1=\frac{\beta}{aa_1},$$

becomes

$$=-b_2Q\frac{\beta\gamma}{aa_1}+c_2R\frac{\beta\gamma}{aa_1}-\alpha\beta\gamma f_2,$$

or, in the last term writing $f_2 = \frac{1}{a_2}$, this is

$$= \frac{\beta\gamma}{aa_1a_2}\{-a_2b_2Q + a_2c_2R - aa_1\alpha\},$$

$$= \frac{\beta\gamma}{aa_1a_2}\{-a_2b_2(ca - c_1a_1) + a_2c_2(ab - a_1b_1) - aa_1(bc_1 - b_1c)\},$$

$$= \frac{\beta\gamma}{aa_1a_2}\begin{vmatrix} ab\ , & ac\ , & 1 \\ a_1b_1, & a_1c_1, & 1 \\ a_2b_2, & a_2c_2, & 1 \end{vmatrix}, \ = \beta\gamma \begin{vmatrix} b\ , & c\ , & f \\ b_1, & c_1, & f_1 \\ b_2, & c_2, & f_2 \end{vmatrix};$$

viz. representing for shortness the last-mentioned determinant by (bcf), the coefficient of ξ is $= \beta\gamma(bcf)$. Similarly the coefficients of η, ζ are equal $\gamma\alpha(cag)$, and $\alpha\beta(abh)$, respectively. Hence, for convenience dividing by $\alpha\beta\gamma$, the original equation

$$\frac{x}{bc - b_1c_1} + \frac{y}{ca - c_1a_1} + \frac{z}{ab - a_1b_1} = 0,$$

expressed in terms of (ξ, η, ζ), becomes

$$\frac{1}{\alpha}(bcf)\xi + \frac{1}{\beta}(cag)\eta + \frac{1}{\gamma}(abh)\zeta = 0;$$

and it is to be shown that this is, in fact, equivalent to

$$\frac{\xi}{g_2h_2 - g_3h_3} + \frac{\eta}{h_2f_2 - h_3f_3} + \frac{\zeta}{f_2g_2 - f_3g_3} = 0.$$

We ought therefore to have

$$(g_2h_2 - g_3h_3)\frac{1}{\alpha}(bcf) = (h_2f_2 - h_3f_3)\frac{1}{\beta}(cag) = (f_2g_2 - f_3g_3)\frac{1}{\gamma}(abh),$$

and it will be sufficient to prove the first of these equalities, say

$$(g_2h_2 - g_3h_3)\beta(bcf) = (h_2f_2 - h_3f_3)\alpha(cag).$$

Multiplying by c_2c_3, this becomes

$$\begin{aligned} &c_3[g_2\beta(bcf) - f_2\alpha(cag)] \\ &- c_2g_3\beta(bcf) + c_2f_3\alpha(cag) = 0, \end{aligned}$$

viz. this is

$$\begin{aligned} &(\lambda c + \lambda_1c_1 + \lambda_2c_2)[g_2\beta(bcf) - f_2\alpha(cag)] \\ &- (\lambda g + \lambda_1g_1 + \lambda_2g_2)c_2\beta(bcf) \\ &+ (\lambda f + \lambda_1f_1 + \lambda_2f_2)c_2\alpha(cag) = 0. \end{aligned}$$

The term in λ_2 disappears, and the equation is

$$\lambda\ [(c\,g_2 - c_2 g)\,\beta\,(bcf) - (c\,f_2 - c_2 f)\,\alpha\,(cag)]$$
$$+\,\lambda_1\,[(c_1 g_2 - c_2 g_1)\,\beta\,(bcf) - (c_1 f_2 - c_2 f_1)\,\alpha\,(cag)] = 0.$$

But we have $\lambda : \lambda_1 = \frac{1}{\Lambda} : \frac{1}{\Lambda_1}$, that is, $\Lambda\lambda - \Lambda_1\lambda_1 = 0$, or if we write for Λ and Λ_1 their values $(1F_1F_2)$, $(=-(1F_2F_1))$, and $(1F_2F)$ respectively, then the equation connecting λ, λ_1 is

$$\lambda\,(1F_2F_1) + \lambda_1\,(1F_2F) = 0,$$

where $(1F_2F_1)$ and $(1F_2F)$ denote the determinants

$$\begin{vmatrix} 1, & af_1 + a_1 f, & af_2 + a_2 f \\ 1, & bg_1 + b_1 g, & bg_2 + b_2 g \\ 1, & ch_1 + c_1 h, & ch_2 + c_2 h \end{vmatrix},\quad \begin{vmatrix} 1, & af_1 + a_1 f, & a_1 f_2 + a_2 f_1 \\ 1, & bg_1 + b_1 g, & b_1 g_2 + b_2 g_1 \\ 1, & ch_1 + c_1 h, & c_1 h_2 + c_2 h_1 \end{vmatrix}.$$

Multiplying by cc_1c_2, the equation may be written

$$\lambda \begin{vmatrix} af, & af_1 + a_1 f, & af_2 + a_2 f \\ bg, & bg_1 + b_1 g, & bg_2 + b_2 g \\ cc_1c_2, & c_2(c^2 + c_1^2), & c_1(c^2 + c_2^2) \end{vmatrix} + \lambda_1 \begin{vmatrix} a_1 f_1, & af_1 + a_1 f, & a_1 f_2 + a_2 f_1 \\ b_1 g_1, & bg_1 + b_1 g, & b_1 g_2 + b_2 g_1 \\ cc_1c_2, & c_2(c^2 + c_1^2), & c_1(c^2 + c_2^2) \end{vmatrix} = 0.$$

This should agree with the equation in (λ, λ_1) obtained above: and it can, in fact, be shown that the coefficients

$$(c\,g_2 - c_2 g)\,\beta\,(bcf) - (c\,f_2 - c_2 f)\,\alpha\,(cag),$$

and

$$(c_1 g_2 - c_2 g_1)\,\beta\,(bcf) - (c_1 f_2 - c_2 f_1)\,\alpha\,(cag),$$

are equal to the two determinants respectively; it will be sufficient to prove one of the two relations, say

$$-(cg_2 - c_2 g)(ca_1 - c_1 a)(bcf) + (cf_2 - c_2 f)(bc_1 - b_1 c)(cag)$$
$$+ \begin{vmatrix} af, & af_1 + a_1 f, & af_2 + a_2 f \\ bg, & bg_1 + b_1 g, & bg_2 + b_2 g \\ cc_1c_2, & c_2(c^2 + c_1^2), & c_1(c^2 + c_2^2) \end{vmatrix} = 0;$$

where it will be recollected that (bcf) and (cag) stand for the determinants

$$\begin{vmatrix} b, & c, & f \\ b_1, & c_1, & f_1 \\ b_2, & c_2, & f_2 \end{vmatrix},\quad \begin{vmatrix} c, & a, & g \\ c_1, & a_1, & g_1 \\ c_2, & a_2, & g_2 \end{vmatrix},$$

respectively: each term is thus of the seventh order in all the letters conjointly, but is quadric in f, f_1, f_2, g, g_1, g_2. Instead of α, β, γ used hitherto to denote $bc_1 - b_1 c$,

$ca_1 - c_1 a$, $ab_1 - a_1 b$, it will be convenient to call these $\alpha_2, \beta_2, \gamma_2$, and to consider the complete system of coefficients

$$\alpha,\ \alpha_1,\ \alpha_2 = b_1c_2 - b_2c_1,\ b_2c - bc_2,\ bc_1 - b_1c;\quad \beta,\ \beta_1,\ \beta_2 = c_1a_2 - c_2a_1,\ c_2a - ca_2,\ ca_1 - c_1a;$$
$$\gamma,\ \gamma_1,\ \gamma_2 = a_1b_2 - a_2b_1,\ a_2b - ab_2,\ ab_1 - a_1b;$$

and to denote by ∇ the determinant formed with $\alpha, \beta, \gamma;\ \alpha_1, \beta_1, \gamma_1;\ \alpha_2, \beta_2, \gamma_2$. If we then expand in terms of the products $(f, f_1, f_2)(g, g_1, g_2)$, we find first

$$\begin{aligned}
&-(cg_2 - c_2g)(ca_1 - c_1a)(bcf) + (cf_2 - c_2f)(bc_1 - b_1c)(cag)\\
&= -(cg_2 - c_2g)\beta_2(af + a_1f_1 + a_2f_2) + (cf_2 - c_2f)\alpha_2(\beta g + \beta_1g_1 + \beta_2g_2),\\
&= -\alpha_2\beta + \alpha\beta_2\,(= -c_1\nabla)\,c_2fg\\
&\quad -\alpha_2\beta_1c_2fg_1\\
&\quad +\alpha_1\beta_2c_2f_1g\\
&\quad -\beta_2(\ \alpha c + \alpha_2c_2 = -\alpha_1c_1)fg_2\\
&\quad -\alpha_2(-\beta c - \beta_2c_2 = \ \beta_1c_1)f_2g\\
&\quad -\alpha_1\beta_2cf_1g_2\\
&\quad +\alpha_2\beta_1cf_2g_1;
\end{aligned}$$

and next

$$\begin{vmatrix} af, & af_1 + a_1f, & af_2 + a_2f \\ bg, & bg_1 + b_1g, & bg_2 + b_2g \\ cc_1c_2, & c_2(c^2 + c_1^2), & c_1(c^2 + c_2^2) \end{vmatrix}$$

$$\begin{aligned}
&= cc_1c_2\gamma + c_2(c^2 + c_1^2)\gamma_1 + c_1(c^2 + c_2^2)\gamma_2\,(= c_1c_2\nabla + c^2(c_1\gamma_2 + c_2\gamma_1))fg\\
&\quad + c_1(\beta_1c_2b + c^2ab)fg_1\\
&\quad + c_1(\alpha_1c_2a - c^2ab)f_1g\\
&\quad + c_2(\beta_2c_1b - c^2ab)fg_2\\
&\quad + c_2(\alpha_2c_1a + c^2ab)f_2g\\
&\quad + c_1c_2abcf_1g_2\\
&\quad - c_1c_2abcf_2g_1.
\end{aligned}$$

Uniting the two parts, we ought to have

$$\begin{aligned}
0 &= c^2(c_2\gamma_1 + c_1\gamma_2)fg\\
&\quad + [\beta_1c_2(c_1b - \alpha_2) + c_1c^2ab]fg_1\\
&\quad + [\alpha_1c_2(c_1a + \beta_2) - c_1c^2ab]f_1g\\
&\quad + [c_1\beta_2(c_2b + \alpha_1) - c_2c^2ab]fg_2\\
&\quad + [c_1\alpha_2(c_2a - \beta_1) + c_2c^2ab]f_2g\\
&\quad + c(c_1c_2ab - \alpha_1\beta_2)f_1g_2\\
&\quad - c(c_1c_2ab - \alpha_2\beta_1)f_2g_1.
\end{aligned}$$

Substituting for $\alpha_1, \beta_1, \gamma_1, \alpha_2, \beta_2, \gamma_2$, then after a slight reduction it is found that the whole equation divides by c, and omitting this factor, we have

$$\begin{aligned}
0 = c\,&[(c_2a_2 - c_1a_1)\,b - (c_2b_2 - c_1b_1)\,a]\,fg \\
&+ [\;\; cc_1ab + b_1c_2(c_2a - ca_2) \;\;]\,fg_1 \\
&+ [- cc_1ab + a_1c_2(b_2c - bc_2) \;\;]\,f_1g \\
&+ [- cc_2ab + c_1b_2(ca_1 - c_1a) \;\;]\,fg_2 \\
&+ [\;\; cc_2ab + c_1a_2(bc_1 - b_1c) \;\;]\,f_2g \\
&+ [- c^2a_1b_2 + ca\,b_2c_1 + bc\,a_1c_2 \;\;]\,f_1g_2 \\
&+ [\;\; c^2a_2b_1 - bc\,a_2c_1 - ca\,b_1c_2 \;\;]\,f_2g_1.
\end{aligned}$$

Writing here $af = bg = \ldots = 1$, the equation becomes

$$\begin{aligned}
0 = c\,(c_2a_2 - c_1a_1)&f - c\,(c_2b_2 - c_1b_1)\,g \\
&+ cc_1bg_1 + c_2(c_2a - ca_2)\,f \\
&- cc_1af_1 + c_2(b_2c - bc_2)\,g \\
&- cc_2bg_2 + c_1(ca_1 - c_1a)\,f \\
&+ cc_2af_2 + c_1(bc_1 - b_1c)\,g \\
&- c^2 + acc_1f_1 + bcc_2g_2 \\
&+ c^2 - bcc_1g_1 - cac_2f_2,
\end{aligned}$$

where all the terms in f_1, g_1 destroy each other; the equation thus becomes

$$\begin{aligned}
0 = &\;(\;\; cc_2a_2 - cc_1a_1 + c_2^2a - cc_2a_2 + cc_1a_1 - c_1^2a)\,f \\
&+ (- cc_2b_2 + cc_1b_1 + cc_2b_2 - c_2^2b + c_1^2b - cc_1b_1)\,g
\end{aligned}$$

that is,

$$0 = (c_2^2 - c_1^2)\,af + (-c_2^2 + c_1^2)\,bg.$$

Or again writing $af = bg = 1$, we have the identity $0 = 0$. This completes the proof of the equivalence of the two equations

$$\frac{x}{bc - b_1c_1} + \frac{y}{ca - c_1a_1} + \frac{z}{ab - a_1b_1} = 0, \quad \frac{\xi}{g_2h_2 - g_3h_3} + \frac{\eta}{h_2f_2 - h_3f_3} + \frac{\zeta}{f_2g_2 - f_3g_3} = 0.$$

We have obviously three new forms derived from the equation in (ξ, η, ζ) in like manner as we had previously three new forms derived from the equation in (x, y, z); the equivalence of the 8 forms to each other is thus shown, and the proof is now complete for each of the 28 bitangents.

Cambridge, 15 *Dec.* 1882.

In what follows, instead of the four equations of the form

$$ax + by + cz + f\xi + g\eta + h\zeta = 0,$$

I take the first equation in Riemann's form with the coefficients unity. The equations thus are

$$\begin{aligned} x + y + z + \xi + \eta + \zeta &= 0, \\ a\,x + b\,y + c\,z + f\,\xi + g\,\eta + h\,\zeta &= 0, \\ a_1x + b_1y + c_1z + f_1\xi + g_1\eta + h_1\zeta &= 0, \\ a_2x + b_2y + c_2z + f_2\xi + g_2\eta + h_2\zeta &= 0, \end{aligned}$$

and if we assume

$$a_2 = \lambda + \lambda_1 a + \lambda_2 a_1, \text{ \&c.}$$

then we have

$$\begin{aligned} 1 - \lambda^2 - \lambda_1^2 - \lambda_2^2 &= \lambda_1\lambda_2(af_1 + a_1f) + \lambda_2\lambda(a_1 + f_1) + \lambda\lambda_1(a + f), \\ \text{,,} \quad &= \lambda_1\lambda_2(bg_1 + b_1g) + \lambda_2\lambda(b_1 + g_1) + \lambda\lambda_1(b + g), \\ \text{,,} \quad &= \lambda_1\lambda_2(ch_1 + c_1h) + \lambda_2\lambda(c_1 + h_1) + \lambda\lambda_1(c + h), \end{aligned}$$

and consequently

$$\lambda_1\lambda_2 : \lambda_2\lambda : \lambda\lambda_1 = P : Q : R,$$

or say

$$\lambda : \lambda_1 : \lambda_2 = QR : RP : PQ,$$

where

$$P,\ Q,\ R = \begin{vmatrix} 1, & a_1 + f_1, & a + f \\ 1, & b_1 + g_1, & b + g \\ 1, & c_1 + h_1, & c + h \end{vmatrix},\quad \begin{vmatrix} 1, & a + f, & af_1 + a_1f \\ 1, & b + g, & bg_1 + b_1g \\ 1, & c + h, & ch_1 + c_1h \end{vmatrix},\quad \begin{vmatrix} 1, & af_1 + a_1f, & a_1 + f_1 \\ 1, & bg_1 + b_1g, & b_1 + g_1 \\ 1, & ch_1 + c_1h, & c_1 + h_1 \end{vmatrix}.$$

Suppose that one of these determinants, to fix the ideas, say Q, is $= 0$; we have $\lambda : \lambda_1 : \lambda_2 = 0 : RP : 0$; that is, λ and λ_2 each $= 0$; and consequently $\lambda_1 = 1$: that is, we have $a_2, b_2, c_2, f_2, g_2, h_2 = a, b, c, f, g, h$, the fourth equation identical with the second, or say the second equation,

$$ax + by + cz + f\xi + g\eta + h\zeta = 0,$$

is a double equation: as just seen, the condition for this is

$$Q = \begin{vmatrix} 1, & a + f, & af_1 + a_1f \\ 1, & b + g, & bg_1 + b_1g \\ 1, & c + h, & ch_1 + c_1h \end{vmatrix} = 0,$$

say this is

$$L(af_1 + a_1f) + M(bg_1 + b_1g) + N(ch_1 + c_1h) = 0,$$

if for shortness

$$L,\ M,\ N = b + g - c - h,\quad c + h - a - f,\quad a + f - b - g.$$

A special solution is if $a_1, b_1, c_1 = a^2, b^2, c^2$; for then

$$a_1, b_1, c_1, f_1, g_1, h_1 = a^2, b^2, c^2, f^2, g^2, h^2;$$

consequently $af_1 + a_1f,\ bg_1 + b_1g,\ ch_1 + c_1h = a + f,\ b + g,\ c + h$: and the condition in question $Q = 0$ is thus satisfied.

It is to be shown that, in the general case of the equation $Q = 0$, the curve is a nodal quartic having the node at the point

$$x : y : z : \xi : \eta : \zeta = fL : gM : hN : aL : bM : cN,$$

and that in the special case $a_1, b_1, c_1 = a^2, b^2, c^2$ the node is a fleflecnode, viz. a node which is a point of inflexion on each of its two branches. In the former case, the six tangents from the node are

$$\begin{aligned} ax + by + cz &= 0 \quad \text{or} \quad f\xi + g\eta + h\zeta = 0,\\ f\xi + by + cz &= 0 \quad\ ,, \quad\ ax + g\eta + h\zeta = 0,\\ ax + g\eta + cz &= 0 \quad\ ,, \quad\ f\xi + by + h\zeta = 0,\\ ax + by + h\zeta &= 0 \quad\ ,, \quad\ f\xi + g\eta + cz = 0,\end{aligned}$$

$$\frac{x}{1-bc} + \frac{y}{1-ca} + \frac{z}{1-ab} = 0, \quad \text{or} \quad \frac{\xi}{gh - g_1h_1} + \frac{\eta}{hf - h_1f_1} + \frac{\zeta}{fg - f_1g_1} = 0,$$

$$\frac{x}{bc - b_1c_1} + \frac{y}{ca - c_1a_1} + \frac{z}{ab - a_1b_1} = 0, \quad ,, \quad \frac{\xi}{1-gh} + \frac{\eta}{1-hf} + \frac{\zeta}{1-fg} = 0.$$

In the latter case, the same equations represent the four tangents from the node and the two tangents at the node respectively.

For the proof, we assume that the four lines

$$\begin{aligned} ax + by + cz &= 0,\\ f\xi + by + cz &= 0,\\ ax + g\eta + cz &= 0,\\ ax + by + h\zeta &= 0,\end{aligned}$$

have a common intersection: this gives $ax = f\xi$, $by = g\eta$, $cz = h\zeta$; or say

$$x,\ y,\ z,\ \xi,\ \eta,\ \zeta = fL,\ gM,\ hN,\ aL,\ bM,\ cN,$$

where for the moment L, M, N are indeterminate quantities: but substituting these values in the first, second and third equations, we find

$$\begin{aligned} (a+f)\,L + {} & (b+g)\,M + {} & (c+h)\,N &= 0,\\ L + {} & M + {} & N &= 0,\\ (af_1 + a_1f)\,L + {} & (bg_1 + b_1g)\,M + {} & (ch_1 + c_1h)\,N &= 0,\end{aligned}$$

giving for L, M, N the values $b + g - c - h$, $c + h - a - f$, $a + f - b - g$ already attributed to these letters.

And we see further that the point in question

$$x,\ y,\ z,\ \xi,\ \eta,\ \zeta, = fL,\ gM,\ hN,\ aL,\ bM,\ cN,$$

is a point on each of the lines

$$\frac{x}{1-bc} + \frac{y}{1-ca} + \frac{z}{1-ab} = 0,$$

$$\frac{x}{bc-b_1c_1} + \frac{y}{ca-c_1a_1} + \frac{z}{ab-a_1b_1} = 0.$$

In fact, for the first of these lines, we have only to verify the equation

$$fL(1-ca)(1-ab) + gM(1-ab)(1-bc) + hN(1-bc)(1-ca) = 0,$$

that is,

$$L(f-b-c+abc) + M(g-c-a+abc) + N(h-a-b+abc) = 0,$$

viz. this is

$$L(a+f) + M(b+g) + N(c+h) + (-a-b-c+abc)(L+M+N) = 0,$$

which is right. And similarly, for the second line, we have to verify that

$$fL(ca-c_1a_1)(ab-a_1b_1) + gM(ab-a_1b_1)(bc-b_1c_1) + hN(bc-b_1c_1)(ca-c_1a_1) = 0,$$

that is,

$$L(abc - bc_1a_1 - ca_1b_1 + fa_1{}^2b_1c_1) + \&c. = 0,$$

or, multiplying by $f_1g_1h_1$, this is

$$L(abcf_1g_1h_1 - bg_1 - ch_1 + a_1f) + M(abcf_1g_1h_1 - ch_1 - af_1 + b_1g) + N(abcf_1g_1h_1 - af_1 - bg_1 + c_1h) = 0,$$

viz. this is

$$(abcf_1g_1h_1 - af_1 - bg_1 - ch_1)(L+M+N) + (af_1 + a_1f)L + (bg_1 + b_1g)M + (ch_1 + c_1h)N = 0.$$

We have thus six of the double tangents meeting at the point

$$x,\ y,\ z,\ \xi,\ \eta,\ \zeta, = fL,\ gM,\ hN,\ aL,\ bM,\ cN,$$

which implies that this point is a node on the curve: and we at once see that the values satisfy the equation $\sqrt{x\xi} + \sqrt{y\eta} + \sqrt{z\zeta} = 0$ of the curve; viz. the equation for the values in question becomes $\sqrt{L^2} + \sqrt{M^2} + \sqrt{N^2} = 0$, and the rationalised equation is satisfied as containing the factor $L+M+N$ which is $=0$.

But to show *à posteriori* with greater distinctness that the point is, in fact, a node, observe that, the rationalised equation being

$$\Omega = x^2\xi^2 + y^2\eta^2 + z^2\zeta^2 - 2yz\eta\zeta - 2zx\zeta\xi - 2xy\xi\eta,$$

the differential $d\Omega$ is

$$= (xd\xi + \xi dx)(x\xi - y\eta - z\zeta) + (yd\eta + \eta dy)(-x\xi + y\eta - z\zeta) + (zd\zeta + \zeta dz)(-x\xi - y\eta + z\zeta),$$

which for the values in question becomes

$$= L\,(f d\xi + a dx)(L^2 - M^2 - N^2) + M\,(g d\eta + b dy)(-L^2 + M^2 - N^2) + N\,(h d\zeta + c dz)(-L^2 - M^2 + N^2);$$

and this, in virtue of $L + M + N = 0$, and therefore $L^2 - M^2 - N^2 = 2MN$, &c., is

$$= 2LMN\,(a\,dx + b\,dy + c\,dz + f\,d\xi + g\,d\eta + h\,d\zeta),$$

which is $= 0$ in virtue of the second equation.

Consider now in the special case $a_1, b_1, c_1 = a^2, b^2, c^2$, the line

$$\frac{x}{1 - bc} + \frac{y}{1 - ca} + \frac{z}{1 - ab} = 0;$$

combining this with the first and second equations

$$x + y + z + \xi + \eta + \zeta = 0,$$
$$ax + by + cz + f\xi + g\eta + h\zeta = 0,$$

we deduce

$$x = -F\left(\frac{\eta}{b\beta} - \frac{\zeta}{c\gamma}\right),$$

$$y = -G\left(\frac{\zeta}{c\gamma} - \frac{\xi}{a\alpha}\right),$$

$$z = -H\left(\frac{\xi}{a\alpha} - \frac{\eta}{b\beta}\right),$$

if for shortness $F, G, H, \alpha, \beta, \gamma = bc - 1,\ ca - 1,\ ab - 1,\ b - c,\ c - a,\ a - b$. Observe that we have $L = b + g - c - h = (b - c)\left(1 - \frac{1}{bc}\right)$, that is, $bcL = \alpha F$; and similarly $caM = \beta G$, and $abN = \gamma H$. And if for a moment we write further

$$\xi, \eta, \zeta, I, J, K, = a\alpha X,\ b\beta Y,\ c\gamma Z,\ a\alpha F,\ b\beta G,\ c\gamma H,$$

values which give

$$I + J + K = (abc - a)(b - c) + (abc - b)(c - a) + (abc - c)(a - b), \quad = 0,$$

then the equation $\sqrt{x\xi} + \sqrt{y\eta} + \sqrt{z\zeta} = 0$ of the curve becomes

$$\sqrt{IX(Y - Z)} + \sqrt{JY(Z - X)} + \sqrt{KZ(X - Y)} = 0,$$

which in the rationalised form is

$$I^2X^2(Y - Z)^2 + J^2Y^2(Z - X)^2 + K^2Z^2(X - Y)^2$$
$$- 2JKYZ(Z - X)(X - Y) - 2KIZX(X - Y)(Y - Z) - 2IJXY(Y - Z)(Z - X) = 0,$$

viz. this is

$$Y^2Z^2(J^2 + 2JK + K^2) + \ldots + 2YZ(JK - I^2 - IJ - IK) + \ldots = 0,$$

which, in virtue of $I + J + K = 0$, becomes

$$(IYZ + JZX + KXY)^2 = 0,$$

that is, the quartic is met by the line in question at the intersections (each taken twice) of the line with the conic

$$IYZ + JZX + KXY = 0,$$

that is,

$$Fa^2\alpha^2 . \eta\zeta + Gb^2\beta^2 . \zeta\xi + Hc^2\gamma^2 . \xi\eta = 0.$$

But the equation of the line in terms of ξ, η, ζ is

$$\frac{\xi}{gh - g_1h_1} + \frac{\eta}{hf - h_1f_1} + \frac{\zeta}{fg - f_1g_1} = 0,$$

that is,

$$\frac{\xi}{gh - g^2h^2} + \frac{\eta}{hf - h^2f^2} + \frac{\zeta}{fg - f^2g^2} = 0,$$

or, what is the same thing,

$$\frac{b^2c^2\xi}{F} + \frac{c^2a^2\eta}{G} + \frac{a^2b^2\zeta}{H} = 0,$$

and this is clearly a tangent to the conic; viz. the rationalised form of

$$\sqrt{Fa^2\alpha^2 . \frac{b^2c^2}{F}} + \sqrt{Gb^2\beta^2 . \frac{c^2a^2}{G}} + \sqrt{Hc^2\gamma^2 . \frac{a^2b^2}{H}} = 0,$$

that is, $\sqrt{\alpha^2} + \sqrt{\beta^2} + \sqrt{\gamma^2} = 0$ is satisfied in virtue of $\alpha + \beta + \gamma = 0$. Hence the four intersections coincide together, or the line has with the curve a quadruple intersection at the node, that is, it is a tangent at the node to a branch having an inflexion. And similarly the other line

$$\frac{x}{bc - b^2c^2} + \frac{y}{ca - c^2a^2} + \frac{z}{ab - a^2b^2} = 0, \text{ that is, } \frac{\xi}{1 - gh} + \frac{\eta}{1 - hf} + \frac{\zeta}{1 - fg} = 0,$$

is a tangent at the node to the other branch having also an inflexion; thus the node is a fleflecnode, and the lines are the tangents to the two branches.

I continue the investigation of the special case $a_1, b_1, c_1, = a^2, b^2, c^2$. The three equations give

$$\begin{aligned}
-(h - f)(f - g)\,\xi + (a - g)(a - h)\,x + (b - g)(b - h)\,y + (c - g)(c - h)\,z = 0,\\
-(f - g)(g - h)\,\eta + (a - h)(a - f)\,x + (b - h)(b - f)\,y + (c - h)(c - f)\,z = 0,\\
-(g - h)(h - f)\,\zeta + (a - f)(a - g)\,x + (b - f)(b - g)\,y + (c - f)(c - g)\,z = 0;
\end{aligned}$$

writing

$A, B, C, F, G, H;\ \alpha, \beta, \gamma, = a^2 - 1, b^2 - 1, c^2 - 1, bc - 1, ca - 1, ab - 1;\ b - c, c - a, a - b,$

and also

$$u = \frac{1}{\alpha\beta\gamma}(Ax + By + Cz),$$

then we find

$$\begin{aligned}
\xi &= a^2(\alpha Fu - x),\\
\eta &= b^2(\beta Gu - y),\\
\zeta &= c^2(\gamma Hu - z),
\end{aligned}$$

say for shortness these values are

$$\xi,\ \eta,\ \zeta = pu - a^2x,\quad qu - b^2y,\quad ru - c^2z,$$

where

$$p,\ q,\ r = a^2\alpha F,\ b^2\beta G,\ c^2\gamma H.$$

We have moreover $f^2\beta\gamma\,.\,\xi = \beta\gamma(\alpha Fu - x)$, $= F(Ax + By + Cz) - \beta\gamma x$; or observing that $GH - AF = -\beta\gamma$, &c., we have

$$f^2\beta\gamma\xi = GHx + BFy + CFz,$$
$$g^2\gamma\alpha\eta = AGx + HFy + CGz,$$
$$h^2\alpha\beta\zeta = AHx + BHy + FGz.$$

It is to be shown that the tangents from the fleflecnode

$$ax + by + cz = 0,$$
$$f\xi + by + cz = 0,$$
$$ax + g\eta + cz = 0,$$
$$ax + by + h\zeta = 0,$$

touch the quartic at its intersections with the line $u = 0$.

To prove this, for the first line we have $ax + by + cz = 0$, $f\xi + g\eta + h\zeta = 0$, and consequently $x\xi - y\eta - z\zeta = bhy\zeta + cgz\eta$. But the equation of the curve is

$$\Omega = (x\xi - y\eta - z\zeta)^2 - 4yz\eta\zeta = 0;$$

and this becomes thus

$$\Omega = (bhy\zeta + cgz\eta)^2 - 4bcghyz\eta\zeta = (bhy\zeta - cgz\eta)^2 = 0;$$

and we thus find that, for the four lines respectively, the equations for the points of contact are

$$ax + by + cz = 0,\quad bhy\zeta - cgz\eta = 0,$$
$$f\xi + by + cz = 0,\quad\quad ''$$
$$ax + g\eta + cz = 0,\quad gh\eta\zeta - bcyz = 0,$$
$$ax + by + h\zeta = 0,\quad\quad ''\quad .$$

But for the first line we have

$$\begin{aligned} bhy\zeta - cgz\eta &= bhyc^2(\gamma Hu - z) - cgzb^2(\beta Gu - y),\\ &= bcy(\gamma Hu - z) - bcz(\beta Gu - y),\\ &= bcu(\gamma Hy - \beta Gz), \end{aligned}$$

so that the intersections with the conic are given by the equations $\gamma Hy - \beta Gz = 0$, and $u = 0$ respectively: the first of these corresponds to the fleflecnode (in fact, we have $\gamma H\,.\,gM - \beta G\,.\,hN = 0$, viz. multiplying by abc, this is

$$\gamma H\,.\,caM - \beta G\,.\,abN = 0,$$

which is right), hence the second corresponds to the point of contact, that is, the point of contact lies on the line $u = 0$. And similarly, for each of the other three tangents, the point of contact lies on the same line $u = 0$.

It is clear that writing $T_1=0$, $T_2=0$, $T_3=0$, $T_4=0$ for the four tangents from the fleflecnode and $T_5=0$, $T_6=0$ for the tangents at this point, the equation of the curve must be expressible in the form

$$\lambda u^2\, T_5 T_6 + \mu T_1 T_2 T_3 T_4 = 0.$$

The reduction to this form is, I think, effected by means of the formulæ which give ξ, η, ζ in terms of x, y, z, more readily than by means of simpler formulæ for ξ, η, ζ in terms of x, y, z and u. We have

$$\Omega = x^2\xi^2 + y^2\eta^2 + z^2\zeta^2 - 2yz\eta\zeta - 2zx\zeta\xi - 2xy\xi\eta = 0,$$

and hence

$$\begin{aligned}\Omega' = f^4g^4h^4\alpha^2\beta^2\gamma^2\Omega = \;& g^4h^4\alpha^2x^2\,(GHx + BFy + CFz)^2\\ &+ h^4f^4\beta^2y^2(AGx + HFy + CGz)^2\\ &+ f^4g^4\gamma^2z^2\,(AHx + BHy + FGz)^2\\ &- 2f^4g^2h^2\beta\gamma yz\,(AGx + HFy + CGz)(AHx + BHy + FGz)\\ &- 2g^4h^2f^2\gamma\alpha zx\,(AHx + BHy + FGz)(GHx + BFy + CFz)\\ &- 2h^4f^2g^2\alpha\beta xy(GHx + BFy + CFz)(AGx + HFy + CGz) = 0.\end{aligned}$$

The result after some reductions, and putting for shortness

$$\begin{aligned} h^2\beta CG - g^2\gamma BH &= \mathfrak{A},\\ f^2\gamma AH - h^2\alpha CF &= \mathfrak{B},\\ g^2\alpha BF - f^2\beta AG &= \mathfrak{C},\end{aligned}$$

is

$$\begin{aligned}\Omega' = \;& g^4h^4G^2H^2\alpha^2x^4 + h^4f^4H^2F^2\beta^2y^4 + f^4g^4F^2G^2\gamma^2z^4\\ &+ 2h^2f^4HF\beta\,.\,\mathfrak{A}y^3z + 2g^2f^4FG\gamma\,.\,\mathfrak{A}yz^3\\ &+ 2f^2g^4FG\gamma\,.\,\mathfrak{B}z^3x + 2h^2g^4GH\alpha\,.\,\mathfrak{B}zx^3\\ &+ 2g^2h^4GH\alpha\,.\,\mathfrak{C}x^3y + 2f^2h^4HF\beta\,.\,\mathfrak{C}xy^3\\ &+ f^4y^2z^2\,(\mathfrak{A}^2 - 2g^2h^2\beta\gamma F^2GH)\\ &+ g^4z^2x^2\,(\mathfrak{B}^2 - 2h^2f^2\gamma\alpha FG^2H)\\ &+ h^4x^2y^2\,(\mathfrak{C}^2 - 2f^2g^2\alpha\beta FGH^2)\\ &+ 2g^2h^2x^2yz\,\{-\mathfrak{B}\mathfrak{C} - f^2GH\,(2f^2\beta\gamma A^2 + g^2\gamma\alpha BH + h^2\alpha\beta CG)\}\\ &+ 2h^2f^2xy^2z\,\{-\mathfrak{C}\mathfrak{A} - g^2HF\,(f^2\beta\gamma AH + 2g^2\gamma\alpha B^2 + h^2\alpha\beta CF)\}\\ &+ 2f^2g^2xyz^2\,\{-\mathfrak{A}\mathfrak{B} - h^2FG\,(f^2\beta\gamma AG + g^2\gamma\alpha BF + 2h^2\alpha\beta C^2)\} = 0.\end{aligned}$$

We have

$$u = \frac{1}{\alpha\beta\gamma}(Ax + By + Cz),$$

$$T_5 = \frac{x}{1-bc} + \frac{y}{1-ca} + \frac{z}{1-ab},$$

$$T_6 = \frac{x}{bc-b^2c^2} + \frac{y}{ca-c^2a^2} + \frac{z}{ab-a^2b^2},$$

or, as these equations may be written,

$$\begin{aligned} \alpha\beta\gamma u &= Ax + By + Cz, \\ -FGHT_5 &= GHx + HFy + FGz, \\ -abcFGHT_6 &= aGHx + bHFy + cFGz; \end{aligned}$$

whence

$$abcF^2G^2H^2 . \alpha\beta\gamma . uT_5T_6 = (Ax + By + Cz)^2 (GHx + HFy + FGz)(aGHx + bHFy + cFGz),$$

where the coefficient of x^4 is $= aA^2G^2H^2$.

Again $T_1 = ax + by + cz$. For T_2 we have

$$\begin{aligned} bc\beta\gamma T_2 &= abcf\beta\gamma (f\xi + by + cz) \\ &= abc(GHx + BFy + CFz) + bc\beta\gamma(by + cz), \end{aligned}$$

where the coefficient of x is $abcGH$. The coefficient of y is

$$\begin{aligned} abc(b^2 - 1)(bc - 1) + b^2c(c - a)(a - b), \quad &= bc(ab^3c - b^2c - a^2b + a), \\ &= bc(ab - 1)(b^2c - a), \\ &= bcH(b^2c - a), \end{aligned}$$

and similarly the coefficient of z is $= bcG(bc^2 - a)$. We have thus the expression for T_2; and forming in like manner the expressions for T_3 and T_4, we may write

$$\begin{aligned} T_1 &= ax + by + cz, \\ bc\beta\gamma T_2 &= abcGHx + bcH(b^2c - a)y + bcG(bc^2 - a)z, \\ ca\gamma\alpha T_3 &= caH(ca^2 - b)x + abcHFy + caF(c^2a - b)z, \\ ab\alpha\beta T_4 &= abG(a^2b - c)x + abF(ab^2 - c)y + abcFGz, \end{aligned}$$

whence in $a^2b^2c^2\alpha^2\beta^2\gamma^2T_1T_2T_3T_4$ the coefficient of x^4 is

$$= a^4b^2c^2 . G^2H^2 . (ca^2 - b)(a^2b - c) = a^4b^2c^2 . G^2H^2(bcA^2 - a^2\alpha^2).$$

We hence obtain the required identity: viz. this is

$$\begin{aligned} a^6b^6c^6\Omega' = a^3b^3c^3 . abcF^2G^2H^2\alpha\beta\gamma . uT_5T_6 \\ -1 . \quad a^2b^2c^2\alpha^2\beta^2\gamma^2 . T_1T_2T_3T_4; \end{aligned}$$

for here comparing the terms in x^4, the factor G^2H^2 divides out, and omitting it, we have

$$a^6b^2c^2\alpha^2 = a^3b^3c^3 . aA^2 - a^4b^2c^2 . (bcA^2 - a^2\alpha^2),$$

which is right. If for Ω' we substitute its value, $= f^4g^4h^4\alpha^2\beta^2\gamma^2\Omega$, then the identity divides by $a^2b^2c^2\alpha\beta\gamma$, and omitting this factor, we obtain the equation of the curve in the form

$$\alpha\beta\gamma\Omega = a^2b^2c^2FGH . uT_5T_6 - \alpha\beta\gamma . T_1T_2T_3T_4 = 0;$$

the required form, putting in evidence the two tangents at the fleflecnode, and the four tangents from this point.

Cambridge, 3rd January, 1883.

817.

ON THE SIXTEEN-NODAL QUARTIC SURFACE.

[From *Crelle's Journal der Mathem.*, t. XCIV. (1883), pp. 270—272.]

RIEMANN'S theory of the bitangents of a plane quartic leads at once to a very simple form of the equation of the sixteen-nodal surface: viz. if ξ, η, ζ denote linear functions of the coordinates (x, y, z, w) such that identically

$$x + y + z + \xi + \eta + \zeta = 0,$$
$$ax + by + cz + f\xi + g\eta + h\zeta = 0,$$

(where $af = bg = ch = 1$), then the quartic surface

$$\sqrt{x\xi} + \sqrt{y\eta} + \sqrt{z\zeta} = 0$$

has the sixteen singular tangent planes (each touching it along a conic)

$$x = 0,\quad y = 0,\quad z = 0,\quad \xi = 0,\quad \eta = 0,\quad \zeta = 0,$$
$$x + y + z = 0,\quad ax + by + cz = 0,$$
$$\xi + y + z = 0,\quad f\xi + by + cz = 0,$$
$$x + \eta + z = 0,\quad ax + g\eta + cz = 0,$$
$$x + y + \zeta = 0,\quad ax + by + h\zeta = 0,$$
$$\frac{x}{1 - bc} + \frac{y}{1 - ca} + \frac{z}{1 - ab} = 0,\quad \frac{\xi}{1 - gh} + \frac{\eta}{1 - hf} + \frac{\zeta}{1 - fg} = 0:$$

and it is thus a sixteen-nodal surface.

I have formerly given the equation of this surface under the form

$$\sqrt{x(X - w)} + \sqrt{y(Y - w)} + \sqrt{z(Z - w)} = 0,$$

where

$$\begin{array}{ll}
\alpha + \beta + \gamma = 0, & X = \alpha(\gamma'\gamma''y - \beta'\beta''z), \\
\alpha' + \beta' + \gamma' = 0, & Y = \beta(\alpha'\alpha''z - \gamma'\gamma''x), \\
\alpha'' + \beta'' + \gamma'' = 0, & Z = \gamma(\beta'\beta''x - \alpha'\alpha''y), \\
P = \frac{x}{\alpha} + \frac{y}{\beta} + \frac{z}{\gamma}, & X' = \alpha'(\gamma''\gamma y - \beta''\beta z), \\
P' = \frac{x}{\alpha'} + \frac{y}{\beta'} + \frac{z}{\gamma'}, & Y' = \beta'(\alpha''\alpha z - \gamma''\gamma x), \\
P'' = \frac{x}{\alpha''} + \frac{y}{\beta''} + \frac{z}{\gamma''}, & Z' = \gamma'(\beta''\beta x - \alpha''\alpha y), \\
 & X'' = \alpha''(\gamma\gamma'y - \beta\beta'z), \\
 & Y'' = \beta''(\alpha\alpha'z - \gamma\gamma'x), \\
 & Z'' = \gamma''(\beta\beta'x - \alpha\alpha'y),
\end{array}$$

and where the equations of the sixteen singular tangent planes are

$$\begin{array}{llll}
x = 0, & y = 0, & z = 0, & w = 0, \\
X - w = 0, & Y - w = 0, & Z - w = 0, & P = 0, \\
X' - w = 0, & Y' - w = 0, & Z' - w = 0, & P' = 0, \\
X'' - w = 0, & Y'' - w = 0, & Z'' - w = 0, & P'' = 0;
\end{array}$$

see *Crelle's Journal*, vol. LXXIII. (1871), pp. 292, 293, [442], and also *Proc. Lond. Math. Soc.*, vol. III. (1871), p. 251*, [454].

To identify the two forms, using $x', y', z', \xi', \eta', \zeta'$ for the new form, I assume

$$x', y', z', \xi', \eta', \zeta' = lx, my, nz, p(X-w), q(Y-w), r(Z-w),$$

where $lp = mq = nr = 1$; and so convert the equation

$$\sqrt{x(X-w)} + \sqrt{y(Y-w)} + \sqrt{z(Z-w)} = 0$$

into

$$\sqrt{x'\xi'} + \sqrt{y'\eta'} + \sqrt{z'\zeta'} = 0.$$

The constants (l, m, n, p, q, r) and (a, b, c, f, g, h), where $af = bg = ch = 1$, are then to be determined so that we may have identically

$$\begin{aligned}
x' + y' + z' + \xi' + \eta' + \zeta' &= 0, \\
ax' + by' + cz' + f\xi' + g\eta' + h\zeta' &= 0,
\end{aligned}$$

and we thus obtain 8 new equations to be satisfied by the 12 constants, viz. these are

$$\begin{array}{llll}
l & + r \,.\, \gamma\beta'\beta'' & - q \,.\, \beta\gamma'\gamma'' & = 0, \\
m & + p \,.\, \alpha\gamma'\gamma'' & - r \,.\, \gamma\alpha'\alpha'' & = 0, \\
n & + q \,.\, \beta\alpha'\alpha'' & - p \,.\, \alpha\beta'\beta'' & = 0, \\
p & + q & + r & = 0, \\
al & + hr \,.\, \gamma\beta'\beta'' & - gq \,.\, \beta\gamma'\gamma'' & = 0, \\
bm & + fp \,.\, \alpha\gamma'\gamma'' & - hr \,.\, \gamma\alpha'\alpha'' & = 0, \\
cn & + gq \,.\, \beta\alpha'\alpha'' & - fp \,.\, \alpha\beta'\beta'' & = 0, \\
fp & + gq & + hr & = 0.
\end{array}$$

[* This Collection, vol. VII., p. 282.]

But substituting for a, b, c, l, m, n their values $\frac{1}{f}, \frac{1}{g}, \frac{1}{h}, \frac{1}{p}, \frac{1}{q}, \frac{1}{r}$, we have in all 8 equations for the determination of qr, rp, pq, gh, hf, fg; viz. if for greater convenience we introduce the new symbols $\mathfrak{A}$, $\mathfrak{B}$, $\mathfrak{C} = qr\alpha'\alpha''$, $rp\beta'\beta''$, $pq\gamma'\gamma''$, then the 8 equations are

$$\frac{1}{\beta\gamma} + \frac{\mathfrak{B}}{\beta} - \frac{\mathfrak{C}}{\gamma} = 0,$$

$$\frac{1}{\gamma\alpha} + \frac{\mathfrak{C}}{\gamma} - \frac{\mathfrak{A}}{\alpha} = 0,$$

$$\frac{1}{\alpha\beta} + \frac{\mathfrak{A}}{\alpha} - \frac{\mathfrak{B}}{\beta} = 0,$$

$$\frac{\alpha'\alpha''}{\mathfrak{A}} + \frac{\beta'\beta''}{\mathfrak{B}} + \frac{\gamma'\gamma''}{\mathfrak{C}} = 0,$$

$$\frac{1}{\beta\gamma} + hf.\frac{\mathfrak{B}}{\beta} - fg.\frac{\mathfrak{C}}{\gamma} = 0,$$

$$\frac{1}{\gamma\alpha} + fg.\frac{\mathfrak{C}}{\gamma} - gh.\frac{\mathfrak{A}}{\alpha} = 0,$$

$$\frac{1}{\alpha\beta} + gh.\frac{\mathfrak{A}}{\alpha} - hf.\frac{\mathfrak{B}}{\beta} = 0,$$

$$\frac{\alpha'\alpha''}{\mathfrak{A}gh} + \frac{\beta'\beta''}{\mathfrak{B}hf} + \frac{\gamma'\gamma''}{\mathfrak{C}fg} = 0.$$

But in virtue of the equation $\alpha + \beta + \gamma = 0$ the first four equations are equivalent to three equations only, and they determine $\mathfrak{A}$, $\mathfrak{B}$, $\mathfrak{C}$, that is, p, q, r, which give at once l, m, n; and similarly the second four equations are equivalent to three equations only, and $\mathfrak{A}$, $\mathfrak{B}$, $\mathfrak{C}$ being known they determine gh, hf, fg, that is, f, g, h, which give at once a, b, c: the identification of the two forms is thus completed.

Cambridge, 11th January, 1883.

818.

NOTE IN CONNEXION WITH THE HYPERELLIPTIC INTEGRALS OF THE FIRST ORDER.

[From *Crelle's Journal der Mathem.*, t. XCVIII. (1885), pp. 95, 96.]

In the early paper by Mr Weierstrass "Zur Theorie der *Abel*schen Functionen," *Crelle's Journal*, t. XLVII. (1854), pp. 289—306, we have pp. 302, 303, certain equations (43), and (stated to be deduced from them) an equation (49). Taking for greater simplicity $n = 2$, the equations (43) written at full length are

$$(43)\quad \begin{cases} K_{11}J_{12} - K_{12}J_{11} + K_{21}J_{22} - K_{22}J_{21} = 0, & K'_{11}J'_{12} - K'_{12}J'_{11} + K'_{21}J'_{22} - K'_{22}J'_{21} = 0, \\ K_{11}J'_{12} - K'_{12}J_{11} + K_{21}J'_{22} - K'_{22}J_{21} = 0, & K_{12}J'_{11} - K'_{11}J_{12} + K_{22}J'_{21} - K'_{21}J_{22} = 0, \\ K_{11}J'_{11} - K'_{11}J_{11} + K_{21}J'_{21} - K'_{21}J_{21} = \tfrac{1}{2}\pi, & K_{12}J'_{12} - K'_{12}J_{12} + K_{22}J'_{22} - K'_{22}J_{22} = \tfrac{1}{2}\pi; \end{cases}$$

viz. in the theory of the hyperelliptic functions depending on the radical

$$\sqrt{x - a_0 \,.\, x - a_1 \,.\, x - a_2 \,.\, x - a_3 \,.\, x - a_4},$$

these are relations between the eight integrals K of the first kind, and the eight integrals J of the second kind. Each equation contains both K's and J's, and there is not in the paper any express mention of a relation between the K's only, which occurs in Rosenhain's Memoir, and is a leading equation in the theory. But taking as before $n = 2$, and for the G's which occur in (49) substituting their values as obtained from the preceding equations (46) and (47), the equation becomes

$$(49)\quad K_{11}K'_{21} - K_{21}K'_{11} + K_{12}K'_{22} - K_{22}K'_{12} = 0,$$

which is the equation in question: it is the equation $\omega_0 v_3 - \omega_3 v_0 + \omega_1 v_2 - \omega_2 v_1 = 0$ of Hermite's Memoir "Sur la théorie de la transformation des fonctions Abéliennes," *Comptes Rendus*, t. XL. (1855).

It is interesting to see how the equation (49) is derived from the equations (43). I write for greater convenience

$$\begin{matrix} K_{11}, & K_{12}, & K_{21}, & K_{22}, & K'_{11}, & K'_{12}, & K'_{21}, & K'_{22}, & J_{11}, & J_{12}, & J_{21}, & J_{22}, & J'_{11}, & J'_{12}, & J'_{21}, & J'_{22} \\ = A, & B, & C, & D, & A', & B', & C', & D', & \alpha, & \beta, & \gamma, & \delta, & \alpha', & \beta', & \gamma', & \delta'. \end{matrix}$$

The given equations then are

$$(43)\quad \begin{cases} A\beta\ - B\alpha\ + C\delta\ - D\gamma\ = 0, & A'\beta' - B'\alpha'\ + C'\delta' - D'\gamma' = 0, \\ A\beta' - B'\alpha + C\delta' - D'\gamma = 0, & A'\beta\ - B\alpha'\ + C'\delta\ - D\gamma'\ = 0, \\ A\alpha'\ - A'\alpha + C\gamma' - C'\gamma = \tfrac{1}{2}\pi, & B\beta'\ - B'\beta\ + D\delta'\ - D'\delta\ = \tfrac{1}{2}\pi; \end{cases}$$

and it is required to show that these lead to the relation

$$(49)\quad AC' - A'C + BD' - B'D = 0.$$

From the first and fourth equations, and from the second and third equations of (43), we deduce

$$(AC' - A'C)\,\beta\ + (C\alpha' - C'\alpha)\,B\ + (C\gamma' - C'\gamma)\,D\ = 0,$$

$$(AC' - A'C)\,\beta' + (C\alpha' - C'\alpha)\,B' + (C\gamma' - C'\gamma)\,D' = 0;$$

and again from the first and third equations, and from the second and fourth equations of (43), we deduce

$$(BD' - B'D)\,\alpha\ + (D\beta' - D'\beta)\,A\ + (D\delta' - D'\delta)\,C\ = 0,$$

$$(BD' - B'D)\,\alpha' + (D\beta' - D'\beta)\,A' + (D\delta' - D'\delta)\,C' = 0.$$

These pairs of equations give respectively

$$AC' - A'C : C\alpha' - C'\alpha : C\gamma' - C'\gamma = BD' - B'D : D\beta' - D'\beta : -(B\beta' - B'\beta),$$

and

$$AC' - A'C : C\alpha' - C'\alpha : -(A\alpha' - A'\alpha) = BD' - B'D : D\beta' - D'\beta : D\delta' - D'\delta;$$

whence putting for shortness $A\alpha' - A'\alpha$, $B\beta' - B'\beta$, $C\gamma' - C'\gamma$, $D\delta' - D'\delta = \mathrm{a, b, c, d}$, we have

$$\frac{AC' - A'C}{BD' - B'D} = -\frac{\mathrm{c}}{\mathrm{b}} = -\frac{\mathrm{a}}{\mathrm{d}};\ \text{whence } \mathrm{ab} = \mathrm{cd}.$$

But the last two of the equations (43) are

$$\mathrm{a} + \mathrm{c} = \tfrac{1}{2}\pi,\quad \mathrm{b} + \mathrm{d} = \tfrac{1}{2}\pi;$$

we have thus $\mathrm{a} + \mathrm{c} = \mathrm{b} + \mathrm{d}$, $= \mathrm{b} + \dfrac{\mathrm{ab}}{\mathrm{c}}$, $= \dfrac{\mathrm{b}}{\mathrm{c}}(\mathrm{a} + \mathrm{c})$; or, since $\mathrm{a} + \mathrm{c}$, $= \tfrac{1}{2}\pi$, is not $= 0$, this gives $\mathrm{b} = \mathrm{c}$, whence also $\mathrm{a} = \mathrm{d}$, and we have

$$\frac{AC' - A'C}{BD' - B'D} = -1,$$

that is,

$$AC' - A'C + BD' - B'D = 0,$$

the required equation.

Cambridge, 10th September, 1884.

819.

ON TWO CASES OF THE QUADRIC TRANSFORMATION BETWEEN TWO PLANES.

[From the *Johns Hopkins University Circulars*, No. 13 (1882), pp. 178, 179.]

SEEKING for the coordinates x_3, y_3, z_3 of the third point of intersection of the cubic curve $x^3+y^3+z^3+6lxyz=0$ by the line through any two points (x_1, y_1, z_1), (x_2, y_2, z_2) on the curve, the expressions present themselves in the form

$$x_3 : y_3 : z_3 = P+2lA : Q+2lB : R+2lC,$$

where

$$P = x_1y_1y_2^2 + z_1x_1z_2^2 - y_1^2x_2y_2 - z_1^2z_2x_2, \quad A = x_1^2y_2z_2 - y_1z_1x_2^2,$$
$$Q = y_1z_1z_2^2 + x_1y_1x_2^2 - z_1^2y_2z_2 - x_1^2x_2y_2, \quad B = y_1^2z_2x_2 - z_1x_1y_2^2,$$
$$R = z_1x_1x_2^2 + y_1z_1y_2^2 - x_1^2z_2x_2 - y_1^2y_2z_2, \quad C = z_1^2x_2y_2 - x_1y_1z_2^2;$$

but it is known that, in virtue of

$$U_1 = x_1^3+y_1^3+z_1^3+6lx_1y_1z_1=0, \quad U_2 = x_2^3+y_2^3+z_2^3+6lx_2y_2z_2=0,$$

which connect the coordinates (x_1, y_1, z_1) and (x_2, y_2, z_2), we have $P:Q:R=A:B:C$*, so that the coordinates (x_3, y_3, z_3) of the third point of intersection may be expressed indifferently in the two forms

$$x_3 : y_3 : z_3 = P : Q : R, \quad \text{and} \quad x_3 : y_3 : z_3 = A : B : C.$$

But these considered irrespectively of the equations $U_1=0$, $U_2=0$, are distinct formulæ, each of them separately establishing a correspondence between the three points (x_1, y_1, z_1), (x_2, y_2, z_2), (x_3, y_3, z_3), or if we regard one of these points as a fixed point, then a correspondence between the remaining two points, or if we consider these as belonging each to its own plane, then a correspondence between two planes.

* See Sylvester on Rational Derivation of Points on Cubic Curves, *Amer. Jour. of Math.* vol. III. p. 62.

Writing for convenience (a, b, c) for the coordinates of the fixed point, and (x_1, y_1, z_1), (x_2, y_2, z_2) for those of the other two points, the formulæ with A, B, C give thus the correspondence

$$x_2 : y_2 : z_2 = bcx_1^2 - a^2y_1z_1 : cay_1^2 - b^2z_1x_1 : abz_1^2 - c^2x_1y_1,$$

which is the first of the two cases in question. These equations give reciprocally

$$x_1 : y_1 : z_1 = bcx_2^2 - a^2y_2z_2 : cay_2^2 - b^2z_2x_2 : abz_2^2 - c^2x_2y_2,$$

or the correspondence is a (1, 1) quadric correspondence.

The formulæ with P, Q, R give in like manner

$$x_2 : y_2 : z_2 = a(ax_1^2 + by_1^2 + cz_1^2) - x_1(a^2x_1 + b^2y_1 + c^2z_1), \text{ \&c.},$$

or if for shortness

$$\Omega_1 = ax_1^2 + by_1^2 + cz_1^2, \quad \Theta_1 = a^2x_1 + b^2y_1 + c^2z_1,$$

then

$$x_2 : y_2 : z_2 = a\Omega_1 - x_1\Theta_1 : b\Omega_1 - y_1\Theta_1 : c\Omega_1 - z_1\Theta_1,$$

which is the second of the two cases. We have reciprocally

$$x_1 : y_1 : z_1 = a\Omega_2 - x_2\Theta_2 : b\Omega_2 - y_2\Theta_2 : c\Omega_2 - z_2\Theta_2,$$

where

$$\Omega_2 = ax_2^2 + by_2^2 + cz_2^2, \quad \Theta_2 = a^2x_2 + b^2y_2 + c^2z_2,$$

and the correspondence is thus in this case also a (1, 1) quadric correspondence.

820.

ON A PROBLEM OF ANALYTICAL GEOMETRY.

[From the *Johns Hopkins University Circulars*, No. 15 (1882), p. 209.]

THE object of the present note is only to call attention to a problem of Analytical Geometry which presents itself in connexion with the reduction of an algebraical integral, and which is solved, pp. 21, 22 of Clebsch and Gordan's *Theorie der Abel'schen Functionen* (Leipzig, 1866); viz. the problem is, considering a line drawn through two given points of a curve $f=0$ of the order n, to find the equation of a curve $\Omega=0$ of the order $n-2$ passing through the remaining $n-2$ points of intersection of the line with the curve f, through the double points of f, and through as many other given points as are required for the determination of the curve. If, for instance, f is a quartic curve without double points, then Ω is the quadric curve which passes through the remaining two intersections of the line with Ω, and through three given points. Take (ξ_1, η_1, ζ_1), (ξ_2, η_2, ζ_2) the coordinates of the two given points on the curve f; (x_1, y_1, z_1), (x_2, y_2, z_2), (x_3, y_3, z_3) for the coordinates of the three given points: and write $\Omega, =(a, b, c, f, g, h\between x, y, z)^2=0$ for the equation of the required curve. In the equation $f, =(x, y, z)^4=0$, write $x, y, z=\lambda\xi_1+\mu\xi_2,\ \lambda\eta_1+\mu\eta_2,\ \lambda\zeta_1+\mu\zeta_2$: we obtain an equation originally of the fourth order in (λ, μ), but which divides by $\lambda\mu$, and which when this factor is thrown out becomes

$$\alpha\lambda^2+\beta\lambda\mu+\gamma\mu^2=0\,;$$

where

$$\begin{aligned}\alpha &= (\xi_1, \eta_1, \zeta_1)^3(\xi_2, \eta_2, \zeta_2),\quad =(\xi_2\partial_{\xi_1}+\eta_2\partial_{\eta_1}+\zeta_2\partial_{\zeta_1})f_1,\\ \beta &= (\xi_1, \eta_1, \zeta_1)^2(\xi_2, \eta_2, \zeta_2)^2,\\ \gamma &= (\xi_1, \eta_1, \zeta_1)\,(\xi_2, \eta_2, \zeta_2)^3,\quad =(\xi_1\partial_{\xi_2}+\eta_1\partial_{\eta_2}+\zeta_1\partial_{\zeta_2})f_2,\end{aligned}$$

where for shortness f_1, f_2 are written to denote $f(\xi_1, \eta_1, \zeta_1)$, $f(\xi_2, \eta_2, \zeta_2)$ respectively.

The condition as to the two points obviously is that, making the same substitution $x, y, z = \lambda\xi_1 + \mu\xi_2, \lambda\eta_1 + \mu\eta_2, \lambda\zeta_1 + \mu\zeta_2$ in the equation $\Omega = (a, \ldots \between x, y, z)^2, = 0$, we must obtain the same quadric equation in (λ, μ). We have thus two conditions, which, introducing an indeterminate multiplier θ, are expressed by the three equations

$$\begin{aligned}(a, \ldots \between \xi_1, \eta_1, \zeta_1)^2 &= \theta\alpha,\\ (a, \ldots \between \xi_1, \eta_1, \zeta_1)(\xi_2, \eta_2, \zeta_2) &= \theta\beta,\\ (a, \ldots \between \xi_2, \eta_2, \zeta_2)^2 &= \theta\gamma.\end{aligned}$$

The conditions as to the three points are obviously

$$\begin{aligned}(a, \ldots \between x_1, y_1, z_1)^2 = 0,\\ (a, \ldots \between x_2, y_2, z_2)^2 = 0,\\ (a, \ldots \between x_3, y_3, z_3)^2 = 0,\end{aligned}$$

and these equations determine the ratios of a, b, c, f, g, h. But to complete the solution the convenient course is to regard the function $\Omega, = (a, \ldots \between x, y, z)^2$ as a quantity to be determined, and consequently to join to the foregoing the equation

$$(a, \ldots \between x, y, z)^2 = \Omega;$$

we have thus seven equations from which (a, b, c, f, g, h) may be eliminated, the result being expressed by means of a determinant of the seventh order

$$\left|\begin{matrix}(x, y, z)^2, & \Omega\\ (\xi_1, \eta_1, \zeta_1)^2, & \theta\alpha\\ (\xi_1, \eta_1, \zeta_1 \between \xi_2, \eta_2, \zeta_2), & \theta\beta\\ (\xi_2, \eta_2, \zeta_2)^2, & \theta\gamma\\ (x_1, y_1, z_1)^2, & 0\\ (x_2, y_2, z_2)^2, & 0\\ (x_3, y_3, z_3)^2, & 0\end{matrix}\right| = 0,$$

viz. this is an equation of the form $A\Omega = \theta\nabla$, where A is a constant determinant of the sixth order (i.e. a determinant not involving x, y, z), ∇ a determinant of the seventh order, a quadric function of (x, y, z), obtained from the foregoing determinant by writing therein $\Omega = 0$ and $\theta = 1$: the multiplier θ is and remains arbitrary: but it is convenient to take it to be $= 1$, viz. we thus not only find the equation $\Omega = 0$, of the required conic, but we put a determinate value on the quadric function Ω itself. And this being so, it is to be remarked that, for $(x, y, z) = (\xi_1, \eta_1, \zeta_1)$, we have $\Omega = \alpha, = (\xi_2\partial_{\xi_1} + \eta_2\partial_{\eta_1} + \zeta_2\partial_{\zeta_1})f_1$: and so for $(x, y, z) = (\xi_2, \eta_2, \zeta_2)$, we have

$$\Omega = \gamma, = (\xi_1\partial_{\xi_2} + \eta_1\partial_{\eta_2} + \zeta_1\partial_{\zeta_2})f_2.$$

821.

ON THE GEOMETRICAL REPRESENTATION OF AN EQUATION BETWEEN TWO VARIABLES.

[From the *Johns Hopkins University Circulars*, No. 15 (1882), p. 210.]

An equation between two variables cannot be represented in a satisfactory manner by a curve, for this serves only to represent the corresponding *real* values of the variables: to represent the imaginary values the natural course is to represent each variable by a point in a plane, viz. the variable $z, = x + iy$, will be represented by a point the coordinates of which are the components x and y of the variable, and similarly the variable $z', = x' + iy'$, by a point the coordinates of which are the components x' and y' of the variable: the equation between the two variables then establishes a correspondence between the two variable points, or say a correspondence between the planes which contain the two points respectively: and it is this correspondence of two planes which is the proper geometrical representation of the equation between the two variables: to exhibit the correspondence we may in either of the planes draw a network of curves at pleasure, and then draw in the other plane the network of corresponding curves. This well-known theory [can be] illustrated for the case $z' = \sqrt{z^4 - 1}$; taking in the infinite half-plane y positive about the origin as centre a system of semicircles, to these correspond in the infinite plane of $x'y'$ a series of lemniscate-shaped curves: and by means of these it is easy to show in the second plane the path corresponding to a given path of the point $z, = x + iy$, in the first half-plane.

822.

ON ASSOCIATIVE IMAGINARIES.

[From the *Johns Hopkins University Circulars*, No. 15 (1882), pp. 211, 212.]

THE imaginaries (x, y) defined by the equations

$$x^2 = ax + by,$$
$$xy = cx + dy,$$
$$yx = ex + fy,$$
$$y^2 = gx + hy,$$

will not be in general associative: to make them so, we must have 8 double relations corresponding to the combinations x^3, x^2y, xyx, xy^2, yx^2, yxy, y^2x, y^3 respectively, viz. the first of these gives $x \,.\, x^2 - x^2 \,.\, x = 0$, that is, $0 = x(ax + by) - (ax + by)x$, $= b(xy - yx)$, $= b[(c - e)x + (d - f)y]$; that is, $0 = b(c - e)$ and $0 = b(d - f)$: and similarly for each of the other terms. We thus obtain apparently 16, but really only 12, relations between the 8 coefficients a, b, c, d, e, f, g, h, viz. the relations so obtained are

$$b(e - c) = 0 \text{ (twice)},\quad b(d - f) = 0,\quad g(c - e) = 0,\quad g(f - d) = 0 \text{ (twice)},$$
$$bg - ed = 0 \text{ (twice)},\quad bg - ef = 0 \text{ (twice)},$$
$$c^2 + dg - ag - ch = 0,\quad d^2 + bc - ad - bh = 0,\quad e^2 + fg - ag - eh = 0,\quad f^2 + be - af - bh = 0,$$
$$a(c - e) - cf + de = 0,\quad h(f - d) - cf + de = 0.$$

From the first four equations it appears that either $b = 0$ and $g = 0$, or else $c = e$ and $d = f$: for brevity, I attend only to the latter case, giving the commutative system

$$x^2 = ax + by,$$
$$xy = yx = cx + dy,$$
$$y^2 = gx + hy.$$

In order that this may be associative, we must still have the relations

$$bg = cd,$$
$$c^2 + dg - ag - ch = 0,$$
$$d^2 + bc - ad - bh = 0,$$

which are all three of them satisfied by $g = \dfrac{cd}{b}$, $h = \dfrac{d^2 + bc - ad}{b}$, viz. we thus have the associative and commutative system

$$x^2 = ax + by,$$
$$xy = yx = cx + dy,$$
$$y^2 = \frac{cd}{b}x + \frac{d^2 + bc - ad}{b}y.$$

I did not perceive how to identify this system with any of the double algebras of B. Peirce's Linear Associative Algebra, see pp. 120—122 of the Reprint, *American Journal of Mathematics*, t. IV. (1881); but it has been pointed out to me by Mr C. S. Peirce that my system, in the general case $ad - bc$ not $= 0$, is expressible as a mixture of two algebras of the form (a_1), see p. 120 (*l.c.*); whereas if $ad - bc = 0$, it is reducible to the form (c_2), see p. 122 (*l.c.*). The object of the present Note is to exhibit in the simple case of two imaginaries the whole system of relations which must subsist between the coefficients in order that the imaginaries may be associative; that is, the system of equations which are solved implicitly by the establishment of the several multiplication tables of the memoir just referred to.

823.

ON THE GEOMETRICAL INTERPRETATION OF CERTAIN FORMULÆ IN ELLIPTIC FUNCTIONS.

[From the *Johns Hopkins University Circulars*, No. 17 (1882), p. 238.]

I HAVE given in my *Elliptic Functions* expressions for the sn^2 of $u+\tfrac{1}{2}K$, $u+\tfrac{1}{2}iK'$, $u+\tfrac{1}{2}K+\tfrac{1}{2}iK'$; but it is better to consider the dn^2, sn^2, cn^2 of these combinations respectively, and to write the formulæ thus:

$$\text{dn}^2(u+\tfrac{1}{2}K) = k'\frac{\text{dn}\,u-(1-k')\,\text{sn}\,u\,\text{cn}\,u}{\text{dn}\,u+(1-k')\,\text{sn}\,u\,\text{cn}\,u}, \quad = k'\frac{1-k^2x-(1-k')\,y}{1-k^2x+(1-k')\,y};$$

$$\text{sn}^2(u+\tfrac{1}{2}iK') = \frac{1}{k}\frac{(1+k)\,\text{sn}\,u+i\,\text{cn}\,u\,\text{dn}\,u}{(1+k)\,\text{sn}\,u-i\,\text{cn}\,u\,\text{dn}\,u}, \quad = \frac{1}{k}\frac{(1+k)\,x+iy}{(1+k)\,x-iy};$$

$$\text{cn}^2(u+\tfrac{1}{2}K+\tfrac{1}{2}iK') = \frac{-ik'}{k}\frac{\text{cn}\,u-(k+ik')\,\text{sn}\,u\,\text{dn}\,u}{\text{cn}\,u+(k+ik')\,\text{sn}\,u\,\text{dn}\,u}, \quad = \frac{-ik'}{k}\frac{1-x-(k+ik')\,y}{1-x+(k+ik')\,y},$$

where in the last set of values x, y are used to denote $\text{sn}^2 u$ and $\text{sn}\,u\,\text{cn}\,u\,\text{dn}\,u$ respectively; and the formulæ are thus brought into connexion with the cubic curve $y^2=x(1-x)(1-k^2x)$. The curve has an inflexion at infinity on the line $x=0$; and the three tangents from the inflexion are $x=0$, $x=1$, $x=\frac{1}{k^2}$, touching the curve at the points $x, y=(0, 0)$, $(1, 0)$, $\left(\frac{1}{k^2}, 0\right)$ respectively: hence these points are sextactic points. We may from any one of them, for instance the point $(0, 0)$, draw four tangents to the curve, $(1+k)\,x+iy=0$, $(1+k)\,x-iy=0$; $(1-k)\,x+iy=0$, $(1-k)\,x-iy=0$; where the first and second of these lines form a pair, and the third and fourth of them form a pair, viz. the two tangents of a pair touch in points such that the line joining them passes through the point of inflexion: in particular, for the first-mentioned pair, the equation of the line joining the points of contact is $1+kx=0$. The linear functions belonging to a pair of tangents are precisely those which present themselves in the formulæ; thus if $T_1=(1+k)\,x+iy$, $T_2=(1+k)\,x-iy$, the second of the three formulæ is $\text{sn}^2(u+\tfrac{1}{2}K)=\frac{1}{k}\frac{T_1}{T_2}$; and the other two formulæ correspond in like manner to pairs of tangents from the sextactic points $\left(\frac{1}{k^2}, 0\right)$, and $(1, 0)$ respectively. The formulæ are connected with the fundamental equations expressing the functions sn, cn, dn as quotients of theta functions.

824.

NOTE ON THE FORMULÆ OF TRIGONOMETRY.

[From the *Johns Hopkins University Circulars*, No. 17 (1882), p. 241.]

THE equations $a = c\cos B + b\cos C$, $b = a\cos C + c\cos A$, $c = b\cos A + a\cos B$, which connect together the sides a, b, c and the angles A, B, C of a plane triangle, may be presented in an algebraical rational form, by introducing in place of the angles A, B, C the functions $\cos A + i\sin A$, $\cos B + i\sin B$, $\cos C + i\sin C$, viz. calling these $\frac{x}{w}, \frac{y}{w}, \frac{z}{w}$ respectively, or, what is the same thing, writing $2\cos A = \frac{x}{w} + \frac{w}{x}$, $2\cos B = \frac{y}{w} + \frac{w}{y}$, $2\cos C = \frac{z}{w} + \frac{w}{z}$, then the foregoing equations may be written

$$\begin{array}{llll}(-2yzw\ , & y(z^2+w^2), & z(y^2+w^2)\between a,\ b,\ c) = 0, \\ (x(z^2+w^2), & -2zxw\ , & z(x^2+w^2)\between\ \ ,, &) = 0, \\ (x(y^2+w^2), & y(x^2+w^2), & -2xyw\ \between\ \ ,, &) = 0,\end{array}$$

that is, as a system of bipartite equations linear in (a, b, c) and cubic in (x, y, z, w) respectively.

Similarly in Spherical Trigonometry, writing as above for the angles, and for the sides writing in like manner $2\cos a = \frac{\alpha}{\delta} + \frac{\delta}{\alpha}$, $2\cos b = \frac{\beta}{\delta} + \frac{\delta}{\beta}$, $2\cos c = \frac{\gamma}{\delta} + \frac{\delta}{\gamma}$, we have a system of bipartite equations separately homogeneous in regard to (x, y, z, w) and $(\alpha, \beta, \gamma, \delta)$ respectively.

825.

A MEMOIR ON THE ABELIAN AND THETA FUNCTIONS.

[Chapters I to III, *American Journal of Mathematics*, t. V. (1882), pp. 137—179; Chapters IV to VII, *ib.*, t. VII. (1885), pp. 101—167.]

THE present memoir is based upon Clebsch and Gordan's *Theorie der Abel'schen Functionen*, Leipzig, 1866 (here cited as C. and G.); the employment of differential rather than of integral equations is a novelty; but the chief addition to the theory consists in the determination which I have made for the cubic curve, and also (but not as yet in a perfect form) for the quartic curve, of the differential expression $d\Pi_{\xi\eta}$ (or as I write it $d\Pi_{12}$) in the integral of the third kind $\int_a^\beta d\Pi_{\xi\eta}$ in the final normal form (endliche Normalform) for which we have (p. 117) $\int_\xi^\eta d\Pi_{a\beta} = \int_a^\beta d\Pi_{\xi\eta}$, the limits and parametric points interchangeable. The want of this determination presented itself to me as a *lacuna* in the theory during the course of lectures on the subject which I had the pleasure of giving at the Johns Hopkins University, Baltimore, U.S.A., in the months January to June, 1882, and I succeeded in effecting it for the cubic curve; but it was not until shortly after my return to England that I was able partially to effect the like determination in the far more difficult case of the quartic curve. The memoir contains, with additional developments, a reproduction of the course of lectures just referred to. I have endeavoured to simplify as much as possible the notations and demonstrations of Clebsch and Gordan's admirable treatise; to bring some of the geometrical results into greater prominence; and to illustrate the theory by examples in regard to the cubic, the nodal quartic, and the general quartic curves respectively. The various chapters are: I, Abel's Theorem; II, Proof of Abel's Theorem; III, The Major Function; IV, The Major Function (continued); V, Miscellaneous Investigations; VI, The Nodal Quartic; VII, The Functions T, U, V, Θ. The paragraphs of the whole memoir will be numbered continuously.

CHAPTER I. ABEL'S THEOREM.

The Differential Pure and Affected Theorems. Art. Nos. 1 to 5.

1. We have a fixed curve and a variable curve, and the differential pure theorem consists in a set of linear relations between the displacements of the intersections of the two curves; in the affected theorem, a linear function of the displacements is equated to another differential expression. I state the two theorems, giving afterwards the necessary explanations.

The pure theorem is

$$\Sigma (x,\ y,\ z)^{n-3}\, d\omega = 0.$$

The affected theorem is

$$\Sigma \frac{(x,\ y,\ z)_{12}{}^{n-2}\, d\omega}{012} = -\frac{\delta\phi_1}{\phi_1} + \frac{\delta\phi_2}{\phi_2}.\ *$$

2. We have a fixed curve $f=0$, or say the curve f, or simply the fixed curve, of the order n, with δ dps, and therefore of the deficiency $\tfrac{1}{2}(n-1)(n-2)-\delta,\ =p$. The expression "the dps" means always the δ dps of f.

And we have a variable curve $\phi=0$, or say the curve ϕ, or simply the variable curve, of the order m, passing through the dps and besides meeting the fixed curve in $mn-2\delta$ variable points.

Moreover, $d\omega$ is the displacement of the current point 0, coordinates $(x,\ y,\ z)$, on the fixed curve, viz. the equation $f=0$ gives

$$\frac{df}{dx}dx + \frac{df}{dy}dy + \frac{df}{dz}dz = 0,$$

$$\frac{df}{dx}\ x + \frac{df}{dy}\ y + \frac{df}{dz}\ z = 0,$$

and we thence have

$$\frac{df}{dx} : \frac{df}{dy} : \frac{df}{dz} = y\,dz - z\,dy : z\,dx - x\,dz : x\,dy - y\,dx,$$

so that we have three equal values each of which is put $=d\omega$, viz. we write

$$\frac{y\,dz - z\,dy}{\frac{df}{dx}} = \frac{z\,dx - x\,dz}{\frac{df}{dy}} = \frac{x\,dy - y\,dx}{\frac{df}{dz}},\ = d\omega,$$

and $d\omega$ as thus defined is the displacement.

* For comparison with C. and G. observe that in the equation, p. 47, $V = \log \frac{\psi(\eta)\,\phi(\xi)}{\phi(\eta)\,\psi(\xi)},\ = \log\frac{\psi_2\phi_1}{\phi_2\psi_1}$ suppose, their ψ belongs to the upper limit and corresponds to my ϕ: the equation gives therefore $dV = -\frac{\delta\psi_1}{\psi_1} + \frac{\delta\psi_2}{\psi_2}$, agreeing with the formula in the text.

$(x, y, z)^{n-3} = 0$ is the minor curve, viz. the general curve of the order $n-3$, which passes through the dps*; and the function $(x, y, z)^{n-3}$ is the minor function.

1 and 2 are fixed points on f, called the parametric points, coordinates (x_1, y_1, z_1) and (x_2, y_2, z_2) respectively; and 012 denotes the determinant

$$\begin{vmatrix} x, & y, & z \\ x_1, & y_1, & z_1 \\ x_2, & y_2, & z_2 \end{vmatrix},$$

so that $012 = 0$ is the equation of the line joining the points 1 and 2: this line meets the fixed curve in $n-2$ other points, called the residues of 1, 2.

$(x, y, z)_{12}^{n-2} = 0$ is the major curve *quoad* the points 1 and 2; viz. this is the general curve of the order $n-2$, passing through the dps and also through the residues of 1, 2.

But further, the function $(x, y, z)_{12}^{n-2}$ is the proper major function; viz. the implicit factor of the function is so determined that, taking $0 = 1$, the current point at 1, that is, writing (x_1, y_1, z_1) for (x, y, z), the function $(x, y, z)_{12}^{n-2}$ reduces itself to the polar function $\left(x_2 \dfrac{d}{dx_1} + y_2 \dfrac{d}{dy_1} + z_2 \dfrac{d}{dz_1}\right) f_1$, afterwards written $n \,.\, 1^{n-1} 2$, of f: this implies that taking $0 = 2$, the current point at 2, the function reduces itself to the polar function $n \,.\, 12^{n-1}$.

ϕ_1 is what ϕ becomes on writing therein (x_1, y_1, z_1) for (x, y, z): and similarly ϕ_2 is what ϕ becomes on writing therein (x_2, y_2, z_2) for (x, y, z).

δ denotes differentiation in regard only to the coefficients of ϕ; viz. writing $\phi = (\text{a}, \ldots \between x, y, z)^m$ we have $\delta\phi = (d\text{a}, \ldots \between x, y, z)^m$, and similarly $\delta\phi_1$ and $\delta\phi_2 = (d\text{a}, \ldots \between x_1, y_1, z_1)^m$ and $(d\text{a}, \ldots \between x_2, y_2, z_2)^m$ respectively.

The sum Σ extends to all the variable intersections of the two curves.

3. As to the meaning of the theorems, consider first the pure theorem. The variable intersections are not all of them arbitrary points on the fixed curve: a certain number of them taken at pleasure on the fixed curve will determine the remaining variable intersections; and there are thus a certain number of integral relations between the coordinates of the variable intersections; to each such integral relation there corresponds a linear relation between the displacements $d\omega$ of these points, or say a displacement-relation. It is precisely these displacement-relations which are given by the theorem, viz. the equation

$$\Sigma\, (x, y, z)^{n-3}\, d\omega = 0$$

breaks up into as many linear relations as there are arbitrary constants in the function $(x, y, z)^{n-3}$, which equated to zero gives a curve of the order $n-3$ passing through the dps; for instance $n = 3$, $\delta = 0$, the equation gives the single relation $\Sigma d\omega = 0$; but $n = 4$, $\delta = 0$, the equation gives the three relations $\Sigma x\, d\omega = 0$, $\Sigma y\, d\omega = 0$, $\Sigma z\, d\omega = 0$.

* This definition implies that the number of dps is at most $= \frac{1}{2}(n-1)(n-2) - 1$, that is, that the fixed curve is not unicursal. But see *post*, No. 21.

4. It is of course important to show, and it will be shown, that the number of independent displacement-relations given by the theorem is equal to the number of independent integral relations between the variable intersections.

5. Observe that the pure theorem gives *all* the displacement-relations between the variable intersections; we are hereby led to see the nature of the affected theorem. Taking at pleasure on the fixed curve the sufficient number of variable intersections, the coefficients of ϕ are thereby determined in terms of the coordinates of the assumed variable intersections*, and hence the value of $-\frac{\delta\phi_1}{\phi_1}+\frac{\delta\phi_2}{\phi_2}$ is given as a linear function of the corresponding displacements $d\omega$; and, substituting this value, the affected theorem gives a linear relation between the displacements $d\omega$ of the several variable intersections. But any such linear relation must clearly be a mere linear combination of the displacement-relations $\Sigma(x, y, z)^{n-3}\,d\omega=0$ given by the pure theorem.

Examples of the Pure Theorem—The Fixed Curve a Cubic. Art. Nos. 6 to 12.

6. The pure theorem is not applicable to the case $n=2$, the fixed curve a conic: it in fact gives no displacement-relation; and this is as it should be, for the variable intersections are all of them arbitrary.

The next case is $n=3$, $\delta=0$, the fixed curve a cubic. For greater simplicity the equation is taken in Cartesian coordinates. In general for such an equation, writing in the homogeneous formulæ $z=1$, we have

$$d\omega=\frac{dx}{\frac{df}{dy}}=-\frac{dy}{\frac{df}{dx}},$$

the two values being of course equal in virtue of $\frac{df}{dx}dx+\frac{df}{dy}dy=0$; taking the former value and considering $\frac{df}{dy}$ as expressed in terms of x, let this be called P (of course, P is an irrational function of x): then we have $d\omega=\frac{dx}{P}$; and similarly $d\omega_1=\frac{dx_1}{P_1}$, &c.

The fixed curve being then a cubic, let the variable curve be a line; this meets the cubic in three points, say 1, 2, 3; and any two of these determine the line, and therefore the third point; there should therefore be one integral relation, and consequently one displacement-relation; and this is what is given by the theorem, viz. we have $\Sigma d\omega=0$, that is, $d\omega_1+d\omega_2+d\omega_3=0$, or, what is the same thing,

$$\frac{dx_1}{P_1}+\frac{dx_2}{P_2}+\frac{dx_3}{P_3}=0.$$

* The coefficients are determined, except it may be as to some constants which remain arbitrary but which disappear from the difference $-\frac{\delta\phi_1}{\phi_1}+\frac{\delta\phi_2}{\phi_2}$; this will be explained further on in the text.

The corresponding integral equation is the equation which expresses that the points 1, 2, 3 are in a line, viz. considering y_1, y_2, y_3 as given functions of x_1, x_2, x_3 respectively, this is

$$\begin{vmatrix} x_1, & y_1, & 1 \\ x_2, & y_2, & 1 \\ x_3, & y_3, & 1 \end{vmatrix} = 0,$$

or, in the notation already made use of for such a determinant, $123 = 0$.

7. This equation $d\omega_1 + d\omega_2 + d\omega_3 = 0$, where $d\omega$ denotes $\frac{dx}{P}$, has a peculiar interpretation when we consider the coefficients of the cubic as arbitrary constants, and therefore the cubic as a curve depending upon nine arbitrary constants*. In taking 1 a point on the curve, we in effect determine y_1 as a function of x_1 and the nine constants; and similarly in taking 2 a point on the curve, we determine y_2 as a function of x_2 and the nine constants; the points 1 and 2 determine the third intersection 3, and we have thus x_3 determined as a function of x_1, x_2 and the nine constants.

Considering x_3 as thus expressed, we have $dx_3 = \frac{dx_3}{dx_1} dx_1 + \frac{dx_3}{dx_2} dx_2$, an equation which must agree with $d\omega_1 + d\omega_2 + d\omega_3 = 0$, that is, with $dx_3 = -\frac{P_3}{P_1} dx_1 - \frac{P_3}{P_2} dx_2$. It follows that we have $\frac{dx_3}{dx_1} \div \frac{dx_3}{dx_2} = \frac{P_2}{P_1}$, and taking the logarithms and differentiating with regard to x_1 and x_2, we find $\frac{d}{dx_1}\frac{d}{dx_2} \log\left(\frac{dx_3}{dx_1} \div \frac{dx_3}{dx_2}\right) = 0$, a partial differential equation of the third order, independent of any particular cubic curve, and satisfied by x_3 considered as a function of x_1, x_2 and the nine constants. Observe that starting from the expression for x_3, and proceeding to the differential coefficients of the third order, we have ten equations from which the nine constants can be eliminated, that is, we ought to have a partial differential equation of the third order: and conversely that the equation for x_3, as containing nine arbitrary constants, is a complete solution of the partial differential equation: the complete solution of the partial differential equation in question is thus the equation which expresses that 3 is the remaining intersection of the line through 1 and 2 with a cubic.

8. The partial differential equation has a geometrical interpretation, or is at least very closely connected with a geometrical property. Consider three consecutive positions of the line, meeting the cubic in the points 1, 2, 3; 1′, 2′, 3′ and 1″, 2″, 3″ respectively: the three lines constitute a cubic curve: the nine points are thus the intersections of two cubic curves, or say they are an "ennead" of points: and any eight of the points thus determine uniquely the ninth point.

* This theory was communicated by me to Section A of the British Association at the York meeting. See *B. A. Report*, 1881, pp. 534, 535, [712], "A Partial Differential Equation connected with the Simplest Case of Abel's Theorem."

9. As a particular example, let the cubic be $x^3+y^3-1=0$; then $y=\sqrt[3]{1-x^3}$, and $d\omega=\dfrac{dx}{y^2}, =\dfrac{dx}{(1-x^3)^{\frac{2}{3}}}$ *; and with these values we have as before the differential relation $d\omega_1+d\omega_2+d\omega_3=0$, and the integral relation $123=0$. I give a direct verification. To find x_3, y_3 the coordinates of the third intersection, we may in the equation of the cubic write x_3, y_3, $1=\lambda x_1+\mu x_2$, $\lambda y_1+\mu y_2$, $\lambda+\mu$ respectively, and then writing for shortness $1^22=x_1{}^2x_2+y_1{}^2y_2-1$, $12^2=x_1x_2{}^2+y_1y_2{}^2-1$, we obtain for the determination of λ, μ the equation $\lambda\,.\,1^22+\mu\,.\,12^2=0$.

This being so, from the equation $123=0$ we obtain by differentiation

$$\Sigma\,\{dx_1\,(y_2-y_3)-dy_1\,(x_2-x_3)\}=0,$$

the sum consisting of three terms, the second and third of them being obtained from the one written down by the cyclical interchange of the numbers 1, 2, 3. But we have $x_1{}^2dx_1+y_1{}^2dy_1=0$, and the equation thus is

$$\Sigma\,\frac{dx_1}{y_1{}^2}\,\{y_1{}^2\,(y_2-y_3)+x_1{}^2\,(x_2-x_3)\}=0:$$

this will reduce itself to $\Sigma\,\dfrac{dx_1}{y_1{}^2}=0$, if only the three coefficients in { } are equal, that is, we ought to have

$$y_1{}^2\,(y_2-y_3)+x_1{}^2\,(x_2-x_3)=y_2{}^2\,(y_3-y_1)+x_2{}^2\,(x_3-x_1)=y_3{}^2\,(y_1-y_2)+x_3{}^2\,(x_1-x_2).$$

Comparing for instance the first and second terms, the equation is

$$-y_3\,(y_1{}^2+y_2{}^2)-x_3\,(x_1{}^2+x_2{}^2)+(x_1{}^2x_2+y_1{}^2y_2+x_1x_2{}^2+y_1y_2{}^2)=0,$$

or, as this may be written,

$$-(\lambda y_1+\mu y_2)\,(y_1{}^2+y_2{}^2)-(\lambda x_1+\mu x_2)\,(x_1{}^2+x_2{}^2)+(\lambda+\mu)\,(x_1{}^2x_2+y_1{}^2y_2+x_1x_2{}^2+y_1y_2{}^2)=0,$$

where the whole coefficient of λ is $-x_1{}^3-y_1{}^3+x_1{}^2x_2+y_1{}^2y_2$, which in fact is $x_1{}^2x_2+y_1{}^2y_2-1$, $=1^22$; and similarly the whole coefficient of μ is 12^2; the equation is thus $\lambda\,.\,1^22+\mu\,.\,12^2=0$, which is right. The first and second coefficients are thus equal, and in like manner the first and third coefficients are equal; we have thus the required result,

$$\frac{dx_1}{y_1{}^2}+\frac{dx_2}{y_2{}^2}+\frac{dx_3}{y_3{}^2}=0.$$

10. In all that follows, the cubic might be any cubic whatever, but to fix the ideas I take a particular form.

Let the cubic be $y^2-X=0$, X a cubic function $(x, 1)^3$, or say even $X=x\,.\,1-x\,.\,1-k^2x$, then $y=\sqrt{X}$, $d\omega=\dfrac{dx}{y}, =\dfrac{dx}{\sqrt{X}}$; and with these values we have the differential relation

* Writing $f=x^3+y^3-1$ we should have $\dfrac{df}{dy}=3y^2$, and therefore $d\omega=\dfrac{dx}{3y^2}$; but the $\frac{1}{3}$ enters as a common factor in all the $d\omega$'s, and it may clearly be disregarded: the value in the text, $d\omega=\dfrac{dx}{y^2}$ could of course be obtained by writing, as we may do, $f=\frac{1}{3}\,(x^3+y^3-1)$, and so in other cases.

$d\omega_1 + d\omega_2 + d\omega_3 = 0$, and the integral relation $123 = 0$. This last equation is an integral of the differential equation $d\omega_1 + d\omega_2 + d\omega_3 = 0$; as not containing any arbitrary constant, it is a particular integral.

But regard one of the three points, say 3, as a fixed point, that is, let the line pass through the fixed point 3 of the cubic, and besides meet it in the points 1 and 2. We write $d\omega_3 = 0$, and the differential equation thus is $d\omega_1 + d\omega_2 = 0$, while the integral equation is as before $123 = 0$; this equation, as containing one arbitrary constant, is the general integral of $d\omega_1 + d\omega_2 = 0$.

Let the variable curve be a conic; say the intersections with the cubic are 1, 2, 3, 4, 5, 6. Any five of these points determine the conic, and therefore the sixth point; there is thus one integral relation, the equation $123456 = 0$, which expresses that the six points are in a conic, and there should therefore be one displacement-relation, viz. this is the equation $\Sigma d\omega = 0$, that is, $d\omega_1 + d\omega_2 + d\omega_3 + d\omega_4 + d\omega_5 + d\omega_6 = 0$.

We have thus $123456 = 0$, as a particular integral of

$$d\omega_1 + d\omega_2 + d\omega_3 + d\omega_4 + d\omega_5 + d\omega_6 = 0.$$

If, however, we take 6 a fixed point on the cubic, then we have the same equation as the general integral of $d\omega_1 + d\omega_2 + d\omega_3 + d\omega_4 + d\omega_5 = 0$.

But taking also 5 a fixed point of the cubic we have as an integral of $d\omega_1 + d\omega_2 + d\omega_3 + d\omega_4 = 0$, the foregoing equation $123456 = 0$, which contains apparently two arbitrary constants; and so if we also fix the point 4, or the points 4 and 3, we have for the differential equations $d\omega_1 + d\omega_2 + d\omega_3 = 0$ and $d\omega_1 + d\omega_2 = 0$, integrals with apparently three arbitrary constants and four arbitrary constants respectively.

11. The explanation is contained in the theory of *Residuation* on a cubic curve. Take the case $d\omega_1 + d\omega_2 + d\omega_3 = 0$, with the integral $123456 = 0$, containing apparently three arbitrary constants, viz. the relation between the variable points 1, 2, 3, is here given by a construction depending on the three fixed points 4, 5, 6 on the cubic; it is to be shown that two of these points can always be regarded as no-matter-what* points. To see that this is so, take on the cubic any two no-matter-what points 4′, 5′: then according to the theory referred to, we can find on the cubic a determinate point 6′ such that the points 4′, 5′ and 6′ establish between the variable points 1, 2, 3, the same relation which is established between them by means of the points 4, 5 and 6; viz. whether in order to determine the point 3 we draw a conic through 1, 2, 4, 5 and 6, or a conic through 1, 2, 4′, 5′ and 6′, we obtain as the remaining intersection of the conic with the cubic one and the same point 3. The construction of 6′ is, through 4, 5 and 6 draw a conic meeting the cubic in any three points 1, 2, 3; through these points and 4′, 5′ draw a conic, the remaining intersection of this with the cubic will be the required point 6′, and the point 6′ thus obtained will be a

* The epithet explains, I think, itself; the point may be any point at pleasure, but it is quite immaterial what point, and for this reason it is not counted as an arbitrary point. The most simple instance is that of two constants presenting themselves in a combination such as $c + c'$: either of them may be regarded as a no-matter-what constant.

determinate point, independent of the particular conic through 4, 5 and 6 used for the construction. Thus 4 and 5 are replaceable by the no-matter-what points 4′ and 5′, or, what is the same thing, two of the points 4, 5 and 6 may be regarded as no-matter-what points, and the number of arbitrary constants is thus reduced to one. And so in other cases, all but one of the fixed points may be regarded as no-matter-what points, and the integral as containing in each case only one arbitrary constant.

But conversely, it being known that the integral of the differential equation contains but one arbitrary constant, we can thence arrive at the theory of residuation.

12. We might go on to the case where the variable curve is a cubic; there are here nine intersections; any eight of these do *not* determine the variable cubic, but they *do* determine the ninth intersection; and there is between the nine intersections one integral relation, and corresponding to it one displacement-relation $\Sigma d\omega = 0$, that is, $d\omega_1 + d\omega_2 + \ldots + d\omega_9 = 0$, given by the pure theorem. But as to this see further on, where it is shown in general that the number of independent integral relations is equal to the number of independent displacement-relations given by the theorem.

Example of the Affected Theorem—Fixed Curve a Circle. Art. Nos. 13 and 14.

13. The fixed curve is taken to be the circle $x^2 + y^2 - 1 = 0$, and the parametric points 1 and 2 to be the points (1, 0) and (0, 1) on this circle. The variable curve is taken to be a line, say the line $ax + by - 1 = 0$, meeting the circle in the points 3 and 4, coordinates (x_3, y_3) and (x_4, y_4) respectively.

Starting from the formula

$$\Sigma \frac{(x, y, 1)^0_{12}\, d\omega}{012} = -\frac{\delta\phi_1}{\phi_1} + \frac{\delta\phi_2}{\phi_2},$$

where the summation extends to the points 3 and 4, $(x, y, 1)^0_{12}$ is here a constant, $= 2 . 12$, that is, $2(x_1x_2 + y_1y_2 - 1)$, which for the points 1, 2 in question is $= -2$. We have 012 denoting the determinant

$$\begin{vmatrix} x, & y, & 1 \\ 1, & 0, & 1 \\ 0, & 1, & 1 \end{vmatrix},$$

which is $= -x - y + 1$; and $d\omega = \dfrac{dx}{2y}$. Also $\dfrac{\delta\phi_1}{\phi_1}, = \dfrac{xda + ydb}{ax + by - 1}$, is $= \dfrac{da}{a-1}$, and similarly $\dfrac{\delta\phi_2}{\phi_2}$ is $= \dfrac{db}{b-1}$. The formula thus is

$$\Sigma \frac{dx}{y(x + y - 1)} = -\frac{da}{a-1} + \frac{db}{b-1}.$$

The coefficients a and b are determined by means of the points 3 and 4, that is, they are functions of x_3, x_4; and considering them as thus expressed, then (inasmuch as there is no linear relation between the displacements $\frac{dx_3}{y_3}$ and $\frac{dx_4}{y_4}$ of the two arbitrary points 3 and 4 on the circle) the equation must become an identity in regard to the terms in dx_3 and dx_4 respectively. It only remains to verify that this is so.

14. Writing $P, Q, R = -y_3+y_4,\ x_3-x_4,\ x_3y_4-x_4y_3$; also L_3 and $L_4 = x_3+y_3-1$ and x_4+y_4-1 respectively, we have $a=P\div R$, $b=Q\div R$, and the equation is found to be

$$\frac{dx_3}{y_3L_3}+\frac{dx_4}{y_4L_4}=\frac{1}{(Q-R)(R-P)}\{(Q-R)\,dP+(R-P)\,dQ+(P-Q)\,dR\},$$

where, substituting for dy_3, dy_4 their values in terms of dx_3, dx_4, we have

$$dP,\ dQ,\ dR=\frac{1}{y_3}x_3dx_3-\frac{1}{y_4}x_4dx_4,\quad \frac{1}{y_3}y_3dx_3-\frac{1}{y_4}y_4dx_4,\quad \frac{1}{y_3}(x_3x_4+y_3y_4)\,dx_3-\frac{1}{y_4}(x_3x_4+y_3y_4)\,dx_4,$$

and with these values, and by aid of the relations

$$Q-R,\quad R-P,\quad P-Q=x_4L_3-x_3L_4,\quad y_4L_3-y_3L_4,\quad -L_3+L_4,$$

the equation is found to be

$$\frac{dx_3}{y_3L_3}+\frac{dx_4}{y_4L_4}=\frac{L_3L_4\,(x_3x_4+y_3y_4-1)}{(x_4L_3-x_3L_4)(y_4L_3-y_3L_4)}\left(\frac{dx_3}{y_3L_3}+\frac{dx_4}{y_4L_4}\right);$$

viz. this will be true if only

$$L_3L_4\,(x_3x_4+y_3y_4-1)-(x_4L_3-x_3L_4)(y_4L_3-y_3L_4)=0,$$

that is,

$$-x_4y_4L_3^2-x_3y_3L_4^2+L_3L_4\,(x_3x_4+y_3y_4+x_3y_4+x_4y_3-1)=0.$$

But from the values of L_3, L_4 we have $x_4y_4=\tfrac{1}{2}L_4^2+L_4$, $x_3y_3=\tfrac{1}{2}L_3^2+L_3$, and the coefficient of L_3L_4 is $=L_3L_4+L_3+L_4$; the equation is thus verified.

The example would perhaps have been more instructive if the points 1 and 2 had been left arbitrary points on the circle, but the working out would have been more difficult.

The Variable Intersections of the Two Curves—Number of Independent Integral Relations. Art. Nos. 15 to 19.

15. Suppose $n=3$, $\delta=0\ (p=1)$, the fixed curve a cubic; and suppose successively $m=1, 2, 3, \ldots$, the variable curve a line, conic, cubic, &c.

If $m=1$, then two points on the cubic determine the line, and consequently the remaining intersection with the cubic; hence there is one integral relation.

If $m=2$, then five points on the cubic determine the conic, and consequently the remaining intersection with the cubic; hence there is one integral relation.

If $m=3$, then eight points on the fixed cubic do not determine the variable cubic, but they do determine the ninth intersection. For draw through the eight points a no-matter-what cubic $\chi=0$, the general cubic through the eight points is $\chi+\alpha f=0$, and this meets the fixed cubic in the points $\chi=0$, $f=0$, that is, in the eight points and in one other point independent of the constant α and therefore completely determinate. Hence in this case also there is one integral relation.

So if $m=4$, then eleven points on the cubic do not determine the quartic, but they do determine the remaining intersection. For draw through the eleven points a no-matter-what quartic $\chi=0$, the general quartic through the eleven points is $\chi+(x, y, z)^1 f=0$, and this meets the cubic in the points $\chi=0$, $f=0$, that is, in the eleven points and in one other point independent of the constants of the linear function $(x, y, z)^1$, and therefore completely determinate. Hence there is one integral relation.

And so in general, the fixed curve being a cubic, then, whatever be the order of the variable curve, there is always one integral relation.

16. Suppose next $n=4$, $\delta=0$ $(p=3)$, the fixed curve a general quartic; and as before suppose successively $m=1, 2, 3, \ldots$, the variable curve a line, conic, cubic, &c.

If $m=1$, then two points on the quartic determine the line, and therefore the remaining two intersections; the number of integral relations is $=2$.

If $m=2$, then five points on the quartic determine the conic, and therefore the remaining three intersections; the number of integral relations is $=3$, and similarly if $m=3$, the number of integral relations is $=3$.

If $m=4$, then thirteen points on the fixed quartic do not determine the variable quartic, but they do determine the remaining three intersections. For draw through the thirteen points a no-matter-what quartic $\chi=0$; the general quartic through the thirteen points is $\chi+\alpha f=0$, and this meets the fixed quartic in the points $\chi=0$, $f=0$, that is, in the thirteen points and in three other points, independent of α and thus completely determinate, and the number of integral relations is $=0$; and so in general for any higher value of m, the number is still $=3$.

17. Suppose $n=5$, $\delta=0$ $(p=6)$, the fixed curve a general quintic; and as before suppose $m=1, 2, 3, \ldots$ successively.

If $m=1$, then two points on the quintic determine the line, and therefore the remaining three intersections; the number of integral relations is $=3$.

If $m=2$, then five points on the quintic determine the conic, and therefore the remaining five intersections; the number of integral relations is $=5$.

If $m=3$, then 9 points on the quintic determine the cubic, and therefore the remaining six intersections; the number of integral relations is $=6$, and so if $m=4$, or if $m=5$ or any greater number, in the first case directly, and in the other cases by consideration of the quintic $\chi+\alpha f=0$, &c., we find that the number of integral relations is always $=6$.

18. The reasoning is scarcely altered in the case where the fixed curve has dps; thus considering the general case of a fixed curve n, δ, p:

If $m=1$, then $2-\delta$ points on the fixed curve (this implies $\delta \not> 2$) determine the line, and therefore the remaining $n-2\delta-(2-\delta)$, $=n-2-\delta$ intersections; the number of integral relations is $=n-2-\delta$.

If $m=2$, then $5-\delta$ points on the fixed curve (this implies $\delta \not> 5$) determine the conic, and therefore the remaining $2n-2\delta-(5-\delta)$, $=2n-5-\delta$ intersections; the number of integral relations is $=2n-5-\delta$, and so for any value of $m \not> n-3$, and indeed for the values $n-2$ and $n-1$; here $\frac{1}{2}m(m+3)-\delta$ points on the fixed curve (this implies $\delta \not> \frac{1}{2}m(m+3)$) determine the variable curve, and therefore the remaining $mn-2\delta-\{\frac{1}{2}m(m+3)-\delta\}$, $=mn-\frac{1}{2}m(m+3)-\delta$ intersections. Hence the number of integral relations is $=mn-\frac{1}{2}m(m+3)-\delta$, that is, $=p-\frac{1}{2}(n-m-1)(n-m-2)$. And thus in the cases $m=n-2$ and $n-1$ the number is $=p$.

If $m=n$, then $\frac{1}{2}n(n+3)-1-\delta$ points on the fixed curve do not determine the variable curve, but they do determine the remaining $n^2-2\delta-\{\frac{1}{2}n(n+3)-1-\delta\}$, $=\frac{1}{2}(n-1)(n-2)-\delta$, that is, p intersections, and the number of integral relations is thus $=p$; and so, for any higher value of m, the number is still $=p$.

19. The conclusion thus is

$m \not> n-3$, the number of integral relations $=p-\frac{1}{2}(n-m-1)(n-m-2)$,

$m=$ or $>n-2$, " " $=p$.

The integral equations spoken of throughout are of course independent relations.

The Variable Intersections of the Two Curves. Number of Independent Displacement Relations given by the Pure Theorem. Art. No. 20.

20. The number of displacement-relations given by the pure theorem is = number of constants in minor function $(x, y, z)^{n-3}$, which equated to zero represents a curve through the dps, viz. this is always

$$\tfrac{1}{2}(n-1)(n-2)-\delta, \ =p.$$

But for $m>n-2$, these relations are not independent. For instance, for $n=4$, $\delta=0$, $m=1$, the displacement-relations are

$$\Sigma\,(x,\ y,\ z)^1\, d\omega=0, \text{ that is, } \Sigma x\, d\omega=0, \quad \Sigma y\, d\omega=0, \quad \Sigma z\, d\omega=0,$$

and conversely from these last equations we have $\Sigma\,(x, y, z)^1\, d\omega=0$. But in this case the variable curve is a line $ax+by+cz=0$; hence writing $(x, y, z)^1=ax+by+cz$, the equation $(x, y, z)^1=0$ is satisfied for each of the intersections of the line with the quartic, and the corresponding equation $\Sigma\,(x, y, z)^1\, d\omega=0$ is an identity. Hence the number of independent displacement-relations is $3-1$, $=2$.

So for $n=5$, $\delta=0$, $m=1$, the displacement-relations are

$$\Sigma\,(x,\ y,\ z)^2\, d\omega=0, \text{ that is, } \Sigma\,(x^2,\ y^2,\ z^2,\ yz,\ zx,\ xy)\, d\omega=0,$$

and these six equations give conversely $\Sigma(x, y, z)^2 d\omega = 0$, and in particular they give $\Sigma x(x, y, z)^1 d\omega = 0$, $\Sigma y(x, y, z)^1 d\omega = 0$, $\Sigma z(x, y, z)^1 d\omega = 0$. But if $(x, y, z)^1$ denote $ax + by + cz$, then as before we have $(x, y, z)^1 = 0$, for each of the intersections of the two curves, and the last-mentioned three equations are identities. The number of independent displacement-relations is thus $6 - 3, = 3$.

So for $n = 5$, $\delta = 0$, $m = 2$. Here if the variable curve is $\phi = (a, \ldots \between x, y, z)^2 = 0$, then taking $(x, y, z)^2 = (a, \ldots \between x, y, z)^2$, the equation $(x, y, z)^2 = 0$ is satisfied for each of the intersections of the two curves, and the corresponding equation $\Sigma(x, y, z)^2 d\omega = 0$ is an identity; the number of independent displacement-relations is $6 - 1, = 5$.

The reasoning is the same when δ is not $= 0$, and we see generally that for $m < n - 2$, or say

$$m \not> n - 3, \text{ the number of independent displacement-relations}$$

$$= p - \tfrac{1}{2}(n - m - 1)(n - m - 2);$$

while for

$$m = \text{or} > n - 2, \text{ the number is } = p;$$

since in this case the relations given by the theorem are all of them independent. It thus appears *à posteriori*, that in every case the number of independent displacement-relations given by the pure theorem is equal to the number of independent integral relations.

As to the dps of the Fixed Curve. Art. No. 21.

21. I conclude with a general remark applicable to the whole of the three chapters. There is no necessity to attend to all or indeed to any of the dps of the fixed curve. Suppose that the fixed curve has $\delta + \delta'$ dps, where δ, δ' may be either of them or each $= 0$, but attend only to the δ dps, the δ' dps being wholly disregarded, and accordingly let the expression the dps mean as before the δ dps of the fixed curve. No alteration at all is required: only, if δ' be not $= 0$, then $p = \frac{1}{2}(n-1)(n-2) - \delta$ will no longer be the deficiency. To obtain the best theorems we use all the $\delta + \delta'$ dps: but disregarding the δ' dps, we obtain theorems, as for a curve with δ dps, which are true, and may frequently be useful either in their original form or with simplifications introduced therein by afterwards taking account of the δ' dps.

CHAPTER II. PROOF OF ABEL'S THEOREM.

Preparation. Art. Nos. 22 and 23.

22. Starting from the equation $\phi = (a, \ldots \between x, y, z)^m = 0$ of the variable curve, we have

$$\frac{d\phi}{dx} dx + \frac{d\phi}{dy} dy + \frac{d\phi}{dz} dz + \delta\phi = 0,$$

$$\frac{d\phi}{dx}\, x + \frac{d\phi}{dy}\, y + \frac{d\phi}{dz}\, z \qquad = 0,$$

where $\delta\phi = (da, \ldots \between x, y, z)^m$. Let τ denote an arbitrary linear function, $= ax + by + cz$; multiply the two equations by $ax + by + cz$, $= \tau$, and $-(a\,dx + b\,dy + c\,dz)$, $= -d\tau$ respectively, and add: we obtain

$$(y\,dz - z\,dy)\left(b\frac{d\phi}{dz} - c\frac{d\phi}{dy}\right) + (z\,dx - x\,dz)\left(c\frac{d\phi}{dx} - a\frac{d\phi}{dz}\right)$$
$$+ (x\,dy - y\,dx)\left(a\frac{d\phi}{dy} - b\frac{d\phi}{dx}\right) + \tau\delta\phi = 0;$$

introducing $d\omega$, this becomes

$$d\omega\left[\frac{df}{dx}\left(b\frac{d\phi}{dz} - c\frac{d\phi}{dy}\right) + \frac{df}{dy}\left(c\frac{d\phi}{dx} - a\frac{d\phi}{dz}\right) + \frac{df}{dz}\left(a\frac{d\phi}{dy} - b\frac{d\phi}{dx}\right)\right] + \tau\delta\phi = 0,$$

or observing that a, b, c are the differential coefficients $\frac{d\tau}{dx}, \frac{d\tau}{dy}, \frac{d\tau}{dz}$, the term in [ ] is $J(f, \tau, \phi)$, or say $J(\phi, f, \tau)$, and the equation is

$$d\omega J(\phi, f, \tau) + \tau\delta\phi = 0,$$

where $J(\phi, f, \tau)$ is the Jacobian, or functional determinant

$$\begin{vmatrix} \frac{d\phi}{dx}, & \frac{d\phi}{dy}, & \frac{d\phi}{dz} \\ \frac{df}{dx}, & \frac{df}{dy}, & \frac{df}{dz} \\ \frac{d\tau}{dx}, & \frac{d\tau}{dy}, & \frac{d\tau}{dz} \end{vmatrix}, = \frac{d(\phi, f, \tau)}{d(x, y, z)};$$

and we hence have

$$d\omega = \frac{-\tau\delta\phi}{J(\phi, f, \tau)}.$$

23. The two theorems thus become

$$\Sigma\,(x, y, z)^{n-3}\,\frac{\tau\delta\phi}{J(\phi, f, \tau)} = 0,$$

$$\Sigma\,\frac{(x, y, z)_{12}^{n-2}}{012}\,\frac{-\tau\delta\phi}{J(\phi, f, \tau)} = -\frac{\delta\phi_1}{\phi_1} + \frac{\delta\phi_2}{\phi_2}.$$

But further, if in the first equation we write $\tau = z$, and in the second equation we retain τ, using it to denote the linear function 012, the equations become

$$\Sigma\,(x, y, z)^{n-3}\,\frac{z\delta\phi}{J(\phi, f)} = 0;$$

$$\Sigma\,(x, y, z)_{12}^{n-2}\,\frac{-\delta\phi}{J(\phi, f, \tau)} = -\frac{\delta\phi_1}{\phi_1} + \frac{\delta\phi_2}{\phi_2};$$

where, in the first equation, $J(\phi, f)$ denotes the Jacobian

$$\begin{vmatrix} \frac{d\phi}{dx}, & \frac{d\phi}{dy} \\ \frac{df}{dx}, & \frac{df}{dy} \end{vmatrix}, = \frac{d(\phi, f)}{d(x, y)}.$$

In these equations the only differential symbol is the δ affecting the coefficients a, b, ... of ϕ, ϕ_1, ϕ_2; the equations are, in respect to the coordinates (x, y, z) of the several variable intersections of the two curves, purely algebraical equations, which are in fact given by Jacobi's Fraction-theorem about to be explained. But for the further reduction of the affected theorem I interpose the next three articles.

The Coordinates (ρ, σ, τ). Art. Nos. 24 to 26.

24. The letter τ has just been used to denote the determinant 012: there is often occasion to consider three points 1, 2, 3 coordinates (x_1, y_1, z_1), (x_2, y_2, z_2), (x_3, y_3, z_3) respectively; and then writing ρ, σ, τ to denote the determinants 023, 031, 012 respectively, and Δ the determinant 123, we have identically

$$\Delta x = x_1\rho + x_2\sigma + x_3\tau,$$
$$\Delta y = y_1\rho + y_2\sigma + y_3\tau,$$
$$\Delta z = z_1\rho + z_2\sigma + z_3\tau,$$

which equations, regarding therein the point 0, coordinates (x, y, z), as a current point, are in fact equations for transformation from the coordinates (x, y, z) to the coordinates ρ, σ, τ belonging to the triangle of reference 123. The points 1 and 2 have been already taken to be on the fixed curve, and it will be assumed that 3 is also a point on this curve.

25. If the function f, which equated to zero gives the equation of the fixed curve, be transformed to the new coordinates (ρ, σ, τ), the coefficients of the transformed function are polar functions, each divided by ∇^n, viz. the coefficient of ρ^n is $\frac{1}{\nabla^n} 1^n$, which by reason that 1 is a point on the curve is $=0$ (and similarly the coefficients of σ^n and of τ^n are each $=0$), the coefficient of $\rho^{n-1}\sigma$ is $=\frac{1}{\nabla^n} n \,.\, 1^{n-1} 2$; that of $\rho^{n-1}\tau$ is $=\frac{1}{\nabla^n} n \,.\, 1^{n-1} 3$; that of $\rho^{n-2}\sigma^2$ is $=\frac{1}{\nabla^n} \tfrac{1}{2} n(n-1) 1^{n-2} 2^2$; and so in other cases. I write this in the form

$$f = \frac{1}{\nabla^n} (1^n = 0, \ldots \between \rho, \sigma, \tau)^n;$$

or we might also use the symbolic form

$$f = \frac{1}{\Delta^n} (\rho 1 + \sigma 2 + \tau 3)^n.$$

The terms independent of τ contain, it is clear, the factor $\rho\sigma$, and separating these terms from the others, the equation may be written

$$f = \frac{1}{\nabla_n} \rho\sigma (n \,.\, 1^{n-1} 2, \ldots \dagger)(\rho, \sigma)^{n-2} + \text{terms involving } \tau.$$

26. The equations $\tau = 0$, $(\ldots \dagger)(\rho, \sigma)^{n-2} = 0$, determine the residues of the points 1, 2, and hence the major function $(x, y, z)_{12}^{n-2}$, expressed in terms of ρ, σ, τ, and writing therein $\tau = 0$, must reduce itself to $(\ldots \dagger)(\rho, \sigma)^{n-2}$ multiplied by a constant factor which is at once found to be $= \frac{1}{\nabla_{n-2}}$; for taking the current point at 1 we have $(\rho, \sigma, \tau) = (\Delta, 0, 0)$, and the corresponding value of the major function is thus $\frac{1}{\nabla_{n-2}} n \,.\, 1^{n-1} 2 \,.\, \Delta^{n-2}$, $= n \,.\, 1^{n-2} 2$, as it ought to be. We have thus

$$(x, y, z)_{12}^{n-2} = \frac{1}{\nabla_{n-2}} (n \,.\, 1^{n-1} 2, \ldots \dagger)(\rho, \sigma)^{n-2} + \text{terms involving } \tau;$$

and we hence see that, for $\tau = 0$,

$$(x, y, z)_{12}^{n-2} = \frac{\Delta^2 f}{\rho\sigma},$$

an equation which will be useful.

The Preparation for the Affected Theorem resumed. Art. No. 27.

27. In the affected theorem instead of (x, y, z) we introduce the new coordinates (ρ, σ, τ). We have

$$J(\phi, f, \tau) = \frac{d(\phi, f, \tau)}{d(\rho, \sigma, \tau)} \frac{d(\rho, \sigma, \tau)}{d(x, y, z)},$$

where the first factor is $= \frac{d(\phi, f)}{d(\rho, \sigma)}$, say that this is $\bar{J}(\phi, f)$, the Jacobian in regard to ρ, σ: and the second factor is at once found to be $= \Delta^2$. We have consequently

$$\frac{1}{J(\phi, f, \tau)} = \frac{1}{\Delta^2 \bar{J}(\phi, f)},$$

and the equation for the affected theorem becomes

$$\Sigma (x, y; z)_{12}^{n-2} \frac{\delta\phi}{\bar{J}(\phi, f)} = -\Delta^2 \left(-\frac{\delta\phi_1}{\phi_1} + \frac{\delta\phi_2}{\phi_2} \right),$$

where $(x, y, z)_{12}^{n-2}$ is to be regarded as standing for its value in terms of (ρ, σ, τ).

Jacobi's Fraction-Theorem. Art. Nos. 28 to 31.

28. This is the extension of a well-known theorem, which, in a somewhat disguised form, may be thus written: viz. if U be any rational and integral function $(x, 1)^m$, then we have

$$\frac{1}{U} = \Sigma \frac{1}{x - x' \,.\, J(U')},$$

or, introducing an arbitrary constant A by the equation $AU=X$, say this is

$$\frac{A}{X}\left(=\frac{1}{U}\right)=\Sigma\,\frac{1}{x-x'\,.\,J(U')},$$

where U' is the same function $(x', 1)^m$ of x' that U is of x: $J(U')$, $=\dfrac{dU'}{dx'}$ is the Jacobian of U', and the summation extends to all roots x' of the equation $U'=0$: obviously this is nothing else than the formula for the decomposition of $\dfrac{1}{U}$ into simple fractions.

29. Take now $U=(x, y, 1)^m$, $V=(x, y, 1)^n$, functions of x, y of the degrees m and n respectively, and assume

$$AU+BV=X,\ \text{a function}\ (x, 1)^{mn},$$
$$CU+DV=Y,\ \quad ,,\quad (y, 1)^{mn},$$

viz. let $X=0$ and $Y=0$ be the equations obtained by elimination from $U=0$ and $V=0$ of the y and the x respectively. The forms are

$$A=(x, y^{n-1}, 1)^{mn-m},\qquad B=(x, y^{m-1}, 1)^{mn-n},$$
$$C=(x^{n-1}, y, 1)^{mn-m},\qquad D=(x^{m-1}, y, 1)^{mn-n},$$

where these equations denote the first of them that A is a rational and integral function of the degree $mn-m$ in x and y jointly, but only of the degree $n-1$ in y: and so for the other equations. It follows that

$$AD-BC=(x^{mn-1}, y^{mn-1}, 1)^{2mn-m-n}.$$

The theorem now is

$$\frac{AD-BC}{XY}=\Sigma\,\frac{1}{x-x'\,.\,y-y'\,.\,J(U', V')},$$

where U', V' are the same functions of (x', y') that U, V are of (x, y); $J(U', V')$ is the Jacobian $\dfrac{d(U', V')}{d(x', y')}$; and the summation extends to all the simultaneous roots x', y' of the equations $U=0$, $V=0$.

30. For the proof, observe that $AD-BC$ is a sum of terms of the form $x^\alpha y^\beta$ where α and β are each of them at most $=mn-1$; hence X being of the degree mn we have $\dfrac{x^\alpha}{X}=$ a sum of fractions $\dfrac{L}{x-x'}$, where x' is any root of $X=0$; and similarly $\dfrac{y^\beta}{Y}=$ a sum of fractions $\dfrac{M}{y-y'}$, where y' is any root of $Y=0$; multiplying the two expressions and taking the sum for the several terms $\lambda x^\alpha y^\beta$ of $AD-BC$ we obtain a formula

$$\frac{AD-BC}{XY}=\Sigma\,\frac{K}{x-x'\,.\,y-y'},$$

where the summation extends to all the combinations of the mn values of x' with the mn values of y. But such a formula existing, the coefficients K may be determined in the usual manner, viz. multiplying by XY and then writing $x=x'$, $y=y'$, there is on the right-hand only one term which does not vanish, and we find

$$(AD-BC)_{x'y'} = K\left(\frac{X}{x-x'}\right)_{x'}\left(\frac{Y}{y-y'}\right)_{y'}, = K\left(\frac{dX}{dx}\frac{dY}{dy}\right)_{x'y'},$$

where the factor which multiplies K does not vanish.

We distinguish the cases where (x', y') are corresponding or non-corresponding roots of $X=0$, $Y=0$; viz. corresponding roots are those for which $U=0$, $V=0$, but for non-corresponding roots these equations do not hold good; there are obviously mn pairs of corresponding roots.

In the latter case $(AD-BC)\,U=DX-BY$; $(AD-BC)\,V=-CX+AY$, and since for the values in question X, Y each vanish, but U, V do not each of them vanish, we must for these values have $AD-BC=0$, and the foregoing equation for K gives then $K=0$.

31. The formula thus is

$$\frac{AD-BC}{XY} = \Sigma\frac{K}{x-x'\,.\,y-y'},$$

where the summation now extends only to corresponding roots x', y', for which we have $U=0$, $V=0$. We have for K the foregoing expression, which, to complete the determination, we write under the form

$$AD-BC = KJ(X,\ Y)_{x'y'};$$

this is allowable, for $J(X, Y), =\dfrac{d(X, Y)}{d(x, y)}$, differs from $\dfrac{dX}{dx}\dfrac{dY}{dy}$ only by the zero term $-\dfrac{dX}{dy}\dfrac{dY}{dx}$. Moreover, differentiating the expressions for X, Y, and considering (x, y) as therein standing for a pair of corresponding roots (x', y'), the terms containing U, V will all vanish; we thus in effect differentiate as if A, B, C, D were constants, and the result is $(AD-BC)\,J(U, V)$, or say this is $(AD-BC)_{x'y'}\,J(U', V')$: hence, in the equation for K, the factor $(AD-BC)_{x'y'}$ divides out, and we have $1=KJ(U', V')$; hence the required formula is

$$\frac{AD-BC}{XY} = \Sigma\frac{1}{x-x'\,.\,y-y'}\,.\,J(U',\ V')$$

the summation extending to all the simultaneous roots (x', y') of $U=0$, $V=0$.

Homogeneous Form of the Fraction-Theorem. Art. No. 32.

32. For x, y, x', y' we write $\dfrac{x}{z}$, $\dfrac{y}{z}$, $\dfrac{x'}{z'}$, $\dfrac{y'}{z'}$; supposing that U, V now denote homogeneous functions $(x, y, z)^m$, $(x, y, z)^n$, and that we have

$$AU+BV=X,\ =(x,\ z)^{mn},\ =\alpha x^{mn}+\ldots,$$
$$CU+DV=Y,\ =(y,\ z)^{mn},\ =\beta y^{mn}+\ldots,$$

where the forms are

$$A = (x,\ y^{n-1},\ z)^{mn-m},\quad B = (x,\ y^{m-1},\ z)^{mn-n},$$
$$C = (x^{n-1},\ y,\ z)^{mn-m},\quad D = (x^{m-1},\ y,\ z)^{mn-n},$$
$$AD - BC = (x^{mn-1},\ y^{mn-1},\ z)^{2mn-m-n},$$

viz. the degree of A in (x, y, z) is $= mn - m$, but y rises only to the degree $n-1$, and so in other cases; then the theorem becomes

$$\frac{z^{m+n-2}(AD-BC)}{XY} = \Sigma \frac{z'^{m+n}}{xz'-x'z\,.\,yz'-y'z}\,.\,J(U',\ V'),$$

where $J(U', V')$ denotes the Jacobian $\dfrac{d(U',\ V')}{d(x',\ y')}$, and the summation extends to the simultaneous roots (x', y', z') of $U=0$, $V=0$.

It is proper to introduce into the formula τ', an arbitrary linear function $ax'+by'+cz'$ of (x', y', z'): observe that, in the Jacobian, (x', y', z') have always values for which $U'=0$, $V'=0$: we have therefore

$$x'\frac{dU'}{dx'} + y'\frac{dU'}{dy'} + z'\frac{dU'}{dz'} = 0,$$

$$x'\frac{dV'}{dx'} + y'\frac{dV'}{dy'} + z'\frac{dV'}{dz'} = 0,$$

and thence

$$x' : y' : z' = \frac{d(U',\ V')}{d(y',\ z')} : \frac{d(U',\ V')}{d(z',\ x')} : \frac{d(U',\ V')}{d(x',\ y')},$$

and if the expressions on the right-hand are for a moment called A', B', C', then writing $\tau' = ax'+by'+cz'$, we have $J(U',\ V',\ \tau') = aA'+bB'+cC'$, $= \dfrac{\tau'}{z'}C'$, $= \dfrac{\tau'}{z'}J(U',\ V')$, that is,

$$\frac{1}{J(U',\ V')} = \frac{\tau'}{z'J(U',\ V',\ \tau')},$$

or the equation becomes

$$\frac{z^{m+n-2}(AD-BC)}{XY} = \Sigma \frac{z'^{m+n-1}\tau'}{xz'-x'z\,.\,yz'-y'z}\,.\,J(U',\ V',\ \tau'),$$

the summation being as before.

Resulting Special Theorems. Art. Nos. 33 to 35.

33. Reverting to the Cartesian form, we have

$$\frac{xy(AD-BC)}{XY} = \Sigma \frac{1}{J(U',\ V')}\left(1+\frac{x'}{x}+\ldots\right)\left(1+\frac{y'}{y}+\ldots\right),$$

$$= \Sigma \frac{1}{J(U',\ V')}\left\{1 + H_1\left(\frac{x'}{x},\ \frac{y'}{y}\right) + H_2\left(\frac{x'}{x},\ \frac{y'}{y}\right) + \ldots\right\}$$

where H_m is the homogeneous sum of the order m,

$$H_1(u,\ v) = u+v,\ H_2(u,\ v) = u^2+uv+v^2,\ \&c.$$

The left-hand side is

$$(AD-BC)\left(\frac{1}{\alpha x^{mn-1}}+\frac{\&c.}{x^{mn}}\ldots\right)\left(\frac{1}{\beta y^{mn-1}}+\frac{\&c.}{y^{mn}}+\ldots\right)$$

and in $AD-BC$ the terms of the highest order in (x, y), say $(AD-BC)_0$ are

$$(AD-BC)_0=(xy)^{mn-m-n+1}(a, b, .., k\between x, y)^{m+n-2}.$$

There is thus on the left-hand no term which is in (x, y) of a higher degree than $-(m+n-2)$; hence on the right-hand every term of a higher degree than this in (x, y) must vanish, viz. we must have

$$0=\Sigma\frac{x'^{\alpha}y'^{\beta}}{J(U', V')} \text{ so long as } \alpha+\beta \not> m+n-3,$$

or, what is the same thing, we must have

$$0=\Sigma\frac{(x', y', 1)^{m+n-3}}{J(U', V')}, \text{ say the } (m+n-3) \text{ theorem},$$

where $(x', y', 1)^{m+n-3}$ is the arbitrary function of the degree $m+n-3$.

34. Passing to the next lower degree $-(m+n-2)$, we have

$$\frac{1}{\alpha\beta(xy)^{m+n-2}}(a, b, .., k\between x, y)^{m+n-2}=\Sigma\frac{1}{J(U', V')}H_{m+n-2}\left(\frac{x'}{x}, \frac{y'}{y}\right)$$

and if in $(a, b, .., k\between x, y)^{m+n-2}$ we consider any term $gx^{m+n-2-p}y^{m+n-2-q}$, where $p+q=m+n-2$, then we have on the left-hand the term $\frac{g}{\alpha\beta x^p y^q}$, and the corresponding term on the right-hand must be

$$\Sigma\frac{1}{J(U', V')}\frac{x'^p y'^q}{x^p y^q};$$

that is, we have

$$\frac{g}{\alpha\beta}=\Sigma\frac{x'^p y'^q}{J(U', V')}.$$

But from the foregoing expression for $(AD-BC)_0$, it appears that $(AD-BC)_0$ contains the term $gx^{mn-1-p}y^{mn-1-q}$, and it hence appears that g is the constant term of the quotient $(AD-BC)_0$ divided by $x^{mn-1-p}y^{mn-1-q}$, or as this may be written

$$g=\text{const. of }\frac{(AD-BC)_0\, x^p y^q}{(xy)^{mn-1}};$$

comparing the two values of g, we obtain

$$\text{const. of }\frac{(AD-BC)_0\, x^p y^q}{\alpha\beta(xy)^{mn-1}}=\Sigma\frac{x'^p y'^q}{J(U', V')}, \quad (p+q=m+n-2),$$

and we hence derive

$$\text{Const. of }\frac{(AD-BC)_0(x, y)^{m+n-2}}{\alpha\beta(xy)^{mn-1}}=\Sigma\frac{(x', y')^{m+n-2}}{J(U', V')},$$

where $(x, y)^{m+n-2}$ is the general function of the degree $m+n-2$, and, of course, $(x', y')^{m+n-2}$ is the same function of x', y'. The two functions may be written $(x, y, 0)^{m+n-2}$ and $(x', y', 0)^{m+n-2}$; and this being so we may on the right-hand write instead $(x', y', 1)^{m+n-2}$, for, by so doing we introduce in the numerator of the fraction new terms of an order not exceeding $m+n-3$, and by the $(m+n-3)$ theorem already obtained the sum Σ of the quotient of such terms by $J(U', V')$ is $=0$. We thus have

$$\text{Const. of } \frac{(AD-BC)_0\,(x, y, 0)^{m+n-2}}{\alpha\beta\,(xy)^{mn-1}} = \Sigma\,\frac{(x', y', 1)^{m+n-2}}{J(U', V')}, \text{ say the } (m+n-2) \text{ theorem,}$$

where $(x', y', 1)^{m+n-2}$ is the general non-homogeneous function of the degree $m+n-2$, and $(x, y, 0)^{m+n-2}$ is obtained from it by attending only to the terms of the highest degree $m+n-2$, and therein substituting x, y for x', y'.

35. We may, it is clear, in the equations for the $(m+n-3)$ and for the $(m+n-2)$ theorems respectively, omit the accents on the right-hand sides; doing this, and moreover in each equation transposing the two sides, the two special theorems are

$$\Sigma\,\frac{(x, y, 1)^{m+n-3}}{J(U, V)} = 0, \qquad (m+n-3) \text{ theorem,}$$

$$\Sigma\,\frac{(x, y, 1)^{m+n-2}}{J(U, V)} = \text{const. of } \frac{(AD-BC)_0\,(x, y, 0)^{m+n-2}}{\alpha\beta\,(xy)^{mn-1}}, \quad (m+n-2) \text{ theorem,}$$

Homogeneous Form of the Special Theorems. Art. No. 36.

36. Writing $\frac{x}{z}$, $\frac{y}{z}$ for x, y, and introducing as before the arbitrary linear function $\tau = ax+by+cz$, we at once obtain, U, V being now homogeneous functions $(x, y, z)^m$ and $(x, y, z)^n$ respectively, and the A, B, C, D being also homogeneous functions accordingly,

$$\Sigma\,\frac{z\,(x, y, z)^{m+n-3}}{J(U, V)} = 0, \qquad (m+n-3) \text{ theorem,}$$

$$\Sigma\,\frac{\tau\,(x, y, z)^{m+n-2}}{zJ(U, V, \tau)} = \text{const. of } \frac{(AD-BC)_0\,(x, y, 0)^{m+n-2}}{\alpha\beta\,(xy)^{mn-1}}, \quad (m+n-2) \text{ theorem,}$$

where the suffix 0 denotes that we are in $AD-BC$ to write $z=0$.

If in the last formula we change throughout the letters x, y, z into ρ, σ, τ (that is, consider U, V as given functions of ρ, σ, τ), but retain τ as standing for the particular function $0\rho+0\sigma+1\tau$, then the formula becomes

$$\Sigma\,\frac{(\rho, \sigma, \tau)^{m+n-2}}{\bar{J}(U, V)} = \text{const. of } \frac{(AD-BC)_0\,(\rho, \sigma, \tau)^{m+n-2}}{\alpha\beta\,(\rho\sigma)^{mn-1}}, \quad (m+n-2) \text{ theorem,}$$

where $\bar{J}(U, V)$ denotes $\frac{d(U, V)}{d(\rho, \sigma)}$, the Jacobian in regard to ρ, σ.

The effect of dps of the Curves $U=0$, $V=0$. Art. Nos. 37 and 38.

37. We must, in regard to the foregoing special theorems, consider the effect of any dps of the curves $U=0$, $V=0$.

Suppose one of the curves, say V, has a dp, but that the other curve U does not pass through it; the dp is not an intersection of U, V, and the theorems are in nowise affected.

If U passes through the dp, then the dp counts twice among the intersections of U, V; at the dp we have $J(U', V')=0$, and (to fix the ideas, attending to the $(m+n-3)$ theorem) the sum $\Sigma \dfrac{(x, y, z)^{m+n-3}}{J(U, V)}$ will contain two infinite terms; these may very well, and indeed (assuming that the theorem remains true) must have a finite sum, but except by the theorem itself, this finite sum is not calculable, and the theorem thus becomes nugatory.

If, however, the curve $(x, y, z)^{m+n-3}=0$ be a curve passing through the dp, then considering, instead, the case where the last-mentioned curve and U each approach indefinitely near to the dp of V; there are two intersections of U, V indefinitely near to each other and to the dp; at either intersection, the numerator $(x', y', z')^{m+n-3}$ and the denominator $J(U, V)$ are infinitesimals of the same order, say the first, and the fraction has a finite value; the finite values for the two intersections have not in general a zero sum, and consequently in the limit it would not be allowable to disregard the intersections belonging to the dp.

38. But if the numerator curve $(x, y, z)^{m+n-3}=0$ passes twice through the dp (that is, has there a dp), then reverting to the two consecutive intersections, at either of these the denominator $J(U, V)$ is as before an infinitesimal of the first order, but the numerator $(x, y, z)^{m+n-3}$ is an infinitesimal of the second order, and in the limit the value of the fraction is $=0$; we may in this case disregard the intersections belonging to the dp; and so in general, the curve $(x, y, z)^{m+n-3}=0$ passing twice through each dp of U which lies upon V, we have

$$\Sigma \frac{z\,(x, y, z)^{m+n-3}}{J(U, V)}=0,$$

the summation now extending to all the intersections of U, V other than the dps in question, which are to be disregarded. And the like in regard to the other theorem

$$\Sigma \frac{(x, y, z)^{m+n-2}}{J(U, V)}=\text{const. of } \frac{(AD-BC)_0\,(x, y, 0)^{m+n-2}}{\alpha\beta\,(xy)^{mn-1}}.$$

The Pure Theorem.—Completion of the Proof. Art. No. 39.

39. The theorem was reduced to

$$\Sigma \frac{z\,(x, y, z)^{n-3}\,\delta\phi}{J(\phi, f)}=0,$$

which is therefore the equation to be proved.

The $(m+n-3)$ theorem, writing therein ϕ, f in place of U, V respectively (the degrees being as before m and n), is

$$\Sigma \frac{z(x,\ y,\ z)^{m+n-3}}{J(U,\ V)} = 0.$$

Here $(x,\ y,\ z)^{m+n-3}$ is an arbitrary function of the degree $m+n-3$, and this may therefore be put $=(x, y, z)^{n-3}\,\delta\phi$, where $\delta\phi$, $=(da, \ldots \between x, y, z)^m$, is a function of the degree m; and since the curve $\phi=0$ passes always through the dps of f, and varies subject to this condition, the curve $\delta\phi=0$ will also pass through the dps; hence taking $(x,\ y,\ z)^{n-3}=0$ a curve through the dps, the curve $(x,\ y,\ z)^{n-3}\,\delta\phi=0$ is a curve passing twice through each of dps, and the $(m+n-3)$ theorem thus gives the equation which was to be proved. This completes the proof of the pure theorem

$$\Sigma\,(x,\ y,\ z)^{n-3}\,d\omega = 0.$$

The Affected Theorem.—Completion of the Proof. Art. Nos. 40 and 41.

40. The theorem was reduced to

$$\Sigma \frac{(x,\ y,\ z)_{12}{}^{n-2}\,\delta\phi}{\bar{J}(\phi,\ f)} = -\Delta^2\left(-\frac{\delta\phi_1}{\phi_1}+\frac{\delta\phi_2}{\phi_2}\right),$$

which is therefore the equation to be proved.

The $(m+n-2)$ theorem, written with $(\rho,\ \sigma,\ \tau)$ in place of $(x,\ y,\ z)$, and putting therein ϕ, f for U, V, is

$$\Sigma \frac{(\rho,\ \sigma,\ \tau)^{m+n-2}}{\bar{J}(\phi,\ f)} = \text{const. of } \frac{(AD-BC)_0\,(\rho,\ \sigma,\ 0)^{m+n-2}}{\alpha\beta\,(\rho\sigma)^{mn-1}},$$

where it will be recollected that the suffix (0) denotes that τ is to be put $=0$. Here $(\rho,\ \sigma,\ \tau)^{m+n-2}$ is an arbitrary function of the degree $m+n-2$, and this may therefore be put $=(x,\ y,\ z)_{12}{}^{n-2}\,\delta\phi$, the two factors being each of them considered as expressed in terms of $(\rho,\ \sigma,\ \tau)$; and since each of the curves $(x,\ y,\ z)_{12}{}^{n-2}=0$ and $\delta\phi=0$ passes through the dps of f, the curve $(x,\ y,\ z)_{12}{}^{n-2}\,\delta\phi=0$ is a curve passing twice through each of the dps. We have therefore

$$\Sigma \frac{(x,\ y,\ z)_{12}{}^{n-2}\,\delta\phi}{\bar{J}(\phi,\ f)} = \text{const. of } \frac{(AD-BC)_0\,(x,\ y,\ z)_{12}{}^{n-2}\,\delta\phi_0}{\alpha\beta\,(\rho\sigma)^{mn-1}},$$

where on the right-hand side $(x,\ y,\ z)_{12}{}^{n-2}$ is considered as a function of ρ, σ, τ, and we are to put therein $\tau=0$; it has been seen (No. 26) that the value is $=\dfrac{f_0\Delta^2}{\rho\sigma}$, where f_0 is what f, considered as a function of ρ, σ, τ, becomes on writing therein $\tau=0$; the right-hand side thus becomes

$$= \text{const. of } \frac{(AD-BC)_0 f_0 \Delta^2 \delta\phi_0}{\alpha\beta\,(\rho\sigma)^{mn}}.$$

41. But for $\tau = 0$, we have

$$A_0\phi_0 + B_0 f_0 = \alpha\rho^{mn},$$

$$C_0\phi_0 + D_0 f_0 = \beta\sigma^{mn},$$

and hence

$$(AD - BC)_0 f_0 = A_0\beta\sigma^{mn} - C_0\alpha\rho^{mn},$$

and the right-hand side thus becomes, say

$$-\Delta^2 \text{ const. of} \left(-\frac{A_0}{\alpha\rho^{mn}} + \frac{C_0}{\beta\sigma^{mn}}\right)\delta\phi_0.$$

But in calculating the constant of $\frac{A_0}{\alpha\rho^{mn}}\delta\phi_0$, we may suppose not only $\tau = 0$, but also $\sigma = 0$: we then have $\phi_0 = (x, y, z)^m$, $= \left(\frac{\rho}{\Delta}\right)^m (x_1, y_1, z_1)^m$, $= \left(\frac{\rho}{\Delta}\right)^m \phi_1$, and hence also $\delta\phi_0 = \left(\frac{\rho}{\Delta}\right)^m \delta\phi_1$.

Similarly, in calculating the constant of $\frac{B_0}{\beta\sigma^{mn}}\delta\phi_0$, we may suppose not only $\tau = 0$, but also $\rho = 0$: we then have $\phi_0 = (x, y, z)^m$, $= \left(\frac{\sigma}{\Delta}\right)^m (x_2, y_2, z_2)^m$, $= \left(\frac{\sigma}{\Delta}\right)^m \phi_2$, and hence $\delta\phi_0 = \left(\frac{\sigma}{\Delta}\right)^m \delta\phi_2$.

Moreover, in the equations

$$A_0\phi_0 + B_0 f_0 = \alpha\rho^{mn},$$

$$C_0\phi_0 + D_0 f_0 = \beta\sigma^{mn},$$

writing in the first equation $\sigma = 0$, we find $A_0\left(\frac{\rho}{\Delta}\right)^m \phi_1 = \alpha\rho^{mn}$, that is, $\frac{A_0}{\alpha\rho^{mn}} = \left(\frac{\Delta}{\rho}\right)^m \frac{1}{\phi_1}$; and similarly writing in the second equation $\rho = 0$, we find $C_0\left(\frac{\sigma}{\Delta}\right)^m \phi_2 = \beta\sigma^{mn}$, that is, $\frac{C_0}{\beta\sigma^{mn}} = \left(\frac{\Delta}{\sigma}\right)^m \frac{1}{\phi_2}$: and the expression thus becomes

$$= -\Delta^2\left(-\frac{\delta\phi_1}{\phi_1} + \frac{\delta\phi_2}{\phi_2}\right),$$

giving the equation which was to be proved. This completes the proof of the affected theorem

$$\Sigma\,\frac{(x, y, z)_{12}^{n-2}\,d\omega}{012} = -\frac{\delta\phi_1}{\phi_1} + \frac{\delta\phi_2}{\phi_2}.$$

CHAPTER III. THE MAJOR FUNCTION $(x, y, z)_{12}^{n-2}$.

Analytical Expression of the Function. Art. Nos. 42 to 49.

42. The function has been defined by the conditions that the curve $(x, y, z)_{12}^{n-2} = 0$ shall pass through the dps, and also through the $n-2$ residues of the parametric points 1, 2: and moreover, that on writing therein (x_1, y_1, z_1) for (x, y, z), the function shall become $= n\,.\,1^{n-1}\,2$. Obviously the function is not completely determined: calling it Ω (or when required Ω_{12}), then if Ω' be any particular form of it, the general form is $\Omega = \Omega' + (x, y, z)^{n-3}.\,012$, where $(x, y, z)^{n-3}$ is the general minor function, viz. $(x, y, z)^{n-3} = 0$ is a curve passing through the dps: the major function thus contains $\tfrac{1}{2}(n-1)(n-2) - \delta$, $= p$, arbitrary constants.

Agreeing with the definition, we have the before-mentioned equation

$$\Omega = \frac{1}{\Delta^{n-2}}(n\,.\,1^{n-1}\,2, \ldots)\!(\rho, \sigma)^{n-2} + \text{terms involving } \tau,$$

viz. from this expression for Ω it appears that the curve $\Omega = 0$ meets the line through 1, 2 in the $n-2$ residues of these points, and moreover, for $(x, y, z) = (x_1, y_1, z_1)$ and therefore $(\rho, \sigma, \tau) = (\Delta, 0, 0)$, the value of Ω is $= n\,.\,1^{n-1}\,2$.

43. We can without difficulty write down an equation determining Ω' as a function $(x, y, z)^{n-2}$, which, on putting therein $\tau = 0$, becomes equal to the foregoing expression $\dfrac{1}{\Delta^{n-2}}(n\,.\,1^{n-1}\,2, \ldots)\!(\rho, \sigma)^{n-2}$, and which is moreover such that the curve $\Omega' = 0$ passes through the dps; which being so, we have as before, $\Omega = \Omega' + (x, y, z)^{n-3}.\,012$, for the general value of Ω.

To fix the ideas, consider the particular case $n = 4$, the fixed curve a quartic: Ω', on putting therein $\tau = 0$, should become

$$= \frac{1}{\Delta^2}(4\,.\,1^3 2,\; 6\,.\,1^2 2^2,\; 4\,.\,12^3)\!(\rho, \sigma)^2;$$

and it is to be shown that this will be the case if we determine Ω', a quadric function of (x, y, z), by the equation

$$\left|\begin{array}{lc} (x\,,\, y\,,\, z\,)^2 \quad , & \Omega' \\ 1\,(x_1,\, y_1,\, z_1)^2 \quad , & 4\,.\,1^3 2 \\ 2\,(x_1,\, y_1,\, z_1)\!(x_2,\, y_2,\, z_2), & 6\,.\,1^2 2^2 \\ 1\,(x_2,\, y_2,\, z_2)^2 \quad , & 4\,.\,12^3 \\ \quad a\,,\, b\,,\, c,\; f\,,\, g\,,\, h\;, & 0 \\ \quad \vdots & \end{array}\right| = 0,$$

where the left-hand side is a determinant of seven lines and columns, the top line being x^2, y^2, z^2, $2yz$, $2zx$, $2xy$, Ω', and similarly for the second line; the third line is

$2x_1x_2$, $2y_1y_2$, $2z_1z_2$, $2(y_1z_2+y_2z_1)$, $2(z_2x_1+z_1x_2)$, $2(x_1y_2+x_2y_1)$, $6 \,.\, 1^2 2^2$, and in each of the last three lines we have six arbitrary constants followed by a 0. The equation is of the form $\square + M\Omega' = 0$, where $\square$ is a quadric function $(x, y, z)^2$, and M is a constant factor.

44. If the quartic curve has a dp, suppose at the point α, coordinates $(x_\alpha, y_\alpha, z_\alpha)$, then in order that the curve $\Omega' = 0$ may pass through the dp, we must for one of the last three lines substitute $(x_\alpha, y_\alpha, z_\alpha)^2$, 0; and so for any other dp or dps of the quartic curve. And the conditions as to the dp or dps (if any) being satisfied in this manner, we may if we please, taking $(x_\beta, y_\beta, z_\beta)$ as the coordinates of an arbitrary point β (not of necessity on the fixed curve), write any line not already so expressed, of the last three lines, in the form $(x_\beta, y_\beta, z_\beta)^2$, 0; the effect being to make the curve $\Omega' = 0$ pass through the arbitrary point β.

45. To show that the equation on putting therein $\tau = 0$ does in fact give the required value, $\Omega' = \frac{1}{\Delta^2}(4 \,.\, 1^3 2,\ 6 \,.\, 1^2 2^2,\ 4 \,.\, 12^3 \between \rho, \sigma)^2$, $= \Phi$ suppose, it is to be observed that, effecting a linear substitution upon the first six columns, the equation may be written

$$\begin{vmatrix} (\rho\,,\ \sigma\,,\ \tau\,)^2 & , & \Omega' \\ 1\,(\rho_1,\ \sigma_1,\ \tau_1)^2 & , & 4\,.\,1^3 2 \\ 2\,(\rho_1,\ \sigma_1,\ \tau_1 \between \rho_2,\ \sigma_2,\ \tau_2), & & 6\,.\,1^2 2^2 \\ 1\,(\rho_2,\ \sigma_2,\ \tau_2)^2 & , & 4\,.\,12^3 \\ a',\ b',\ c',\ f',\ g',\ h' & , & 0 \\ \vdots & & \end{vmatrix} = 0,$$

where $(\rho_1, \sigma_1, \tau_1)$, $(\rho_2, \sigma_2, \tau_2)$ are what (ρ, σ, τ) become on writing therein for (x, y, z) the values (x_1, y_1, z_1) and (x_2, y_2, z_2) respectively; viz. we have $(\rho_1, \sigma_1, \tau_1) = (\Delta, 0, 0)$; $(\rho_2, \sigma_2, \tau_2) = (0, \Delta, 0)$; the equation thus is

$$\begin{vmatrix} \rho^2, & \sigma^2, & \tau^2, & 2\sigma\tau, & 2\tau\rho, & 2\rho\sigma, & \Omega' \\ \Delta^2, & 0, & 0, & 0, & 0, & 0, & 4\,.\,1^3 2 \\ 0, & 0, & 0, & 0, & 0, & 2\Delta^2, & 6\,.\,1^2 2^2 \\ 0, & \Delta^2, & 0, & 0, & 0, & 0, & 4\,.\,12^3 \\ a', & b', & c', & f', & g', & h', & 0 \\ \vdots & & & & & & \end{vmatrix} = 0,$$

and then by another linear substitution upon the columns, the last column can be changed into $\Omega' - \Phi$, 0, 0, 0, 0, 0, 0; whence writing $\tau = 0$, the equation becomes

$$\begin{vmatrix} \rho^2, & \sigma^2, & 0, & 0, & 0, & 2\rho\sigma, & \Omega' - \Phi \\ \Delta^2, & 0, & 0, & 0, & 0, & 0, & 0 \\ 0, & 0, & 0, & 0, & 0, & 2\Delta^2, & 0 \\ 0, & \Delta^2, & 0, & 0, & 0, & 0, & 0 \\ a', & b', & c', & f', & g', & h', & 0 \\ \vdots & & & & & & \end{vmatrix} = 0,$$

or, omitting a constant factor, it is

$$\begin{vmatrix} \rho^2, & \sigma^2, & 2\rho\sigma, & \Omega'-\Phi \\ \Delta^2, & 0, & 0, & 0 \\ 0, & 0, & 2\Delta^2, & 0 \\ 0, & \Delta^2, & 0, & 0 \end{vmatrix} = 0,$$

that is, $\Omega'-\Phi=0$, or $\Omega'=\Phi, =\frac{1}{\Delta^2}(4.1^32,\ 6.1^22^2,\ 4.12^3 \rangle\!\langle \rho,\ \sigma)^2$, the required value.

46. Considering the equation for Ω' as expressed in the before-mentioned form $\square + M\Omega'=0$, the value of the constant factor M is

$$M = \begin{vmatrix} (x_1,\ y_1,\ z_1)^2 \\ (x_1,\ y_1,\ z_1 \rangle\!\langle x_2,\ y_2,\ z_2) \\ (x_2,\ y_2,\ z_2)^2 \\ a,\ b,\ c,\ f,\ g,\ h, \\ \vdots \end{vmatrix};$$

or if, instead of each line such as a, b, c, f, g, h, we have a line $(x_\alpha, y_\alpha, z_\alpha)^2$, then we have

$$M = \begin{vmatrix} (x_1,\ y_1,\ z_1)^2 \\ (x_1,\ y_1,\ z_1 \rangle\!\langle x_2,\ y_2,\ z_2) \\ (x_2,\ y_2,\ z_2)^2 \\ (x_\alpha,\ y_\alpha,\ z_\alpha)^2 \\ (x_\beta,\ y_\beta,\ z_\beta)^2 \\ (x_\gamma,\ y_\gamma,\ z_\gamma)^2 \end{vmatrix},$$

a value which is

$$= \begin{vmatrix} x_1, & y_1, & z_1 \\ x_2, & y_2, & z_2 \\ x_\alpha, & y_\alpha, & z_\alpha \end{vmatrix} \begin{vmatrix} x_1, & y_1, & z_1 \\ x_2, & y_2, & z_2 \\ x_\beta, & y_\beta, & z_\beta \end{vmatrix} \begin{vmatrix} x_1, & y_1, & z_1 \\ x_2, & y_2, & z_2 \\ x_\gamma, & y_\gamma, & z_\gamma \end{vmatrix} \begin{vmatrix} x_\alpha, & y_\alpha, & z_\alpha \\ x_\beta, & y_\beta, & z_\beta \\ x_\gamma, & y_\gamma, & z_\gamma \end{vmatrix},$$

or say this is $=12\alpha.12\beta.12\gamma.\alpha\beta\gamma$.

47. It is obvious that the foregoing process is applicable to the general case of the fixed curve of the order n with δ dps, and gives always Ω', by an equation of the foregoing form $\square + M\Omega'=0$, where $\square$ is a function $(x, y, z)^{n-2}$ of the coordinates, and M is a constant factor. Supposing that in the determinant for Ω', each of the lower lines is written in the form $(x_\alpha, y_\alpha, z_\alpha)^{n-2}$, 0, the number of the points α is $=\frac{1}{2}(n-1)(n-2)$, viz. these are the δ dps, and $\frac{1}{2}(n-1)(n-2)-\delta$, $=p$, other points α. The general expression of M is $M=12\alpha.12\beta\ldots(\alpha^{n-3}\beta^{n-3}\ldots)$, viz. equating to zero a factor such as 12α, this expresses that the point α is on the line 12; but equating to zero the last factor $(\alpha^{n-3}\beta^{n-3}\ldots)$, this expresses that the several points α, viz. the dps and the p other points α, are on a curve of the order $n-3$.

48. Preceding the case $n=4$, above considered, we have, of course, the case $n=3$, $\delta=0$, the fixed curve a cubic; the equation for Ω' is here

$$\begin{vmatrix} x, & y, & z, & \Omega' \\ x_1, & y_1, & z_1, & 3.1^22 \\ x_2, & y_2, & z_2, & 3.12^2 \\ x_a, & y_a, & z_a, & \end{vmatrix} = 0,$$

giving

$$\tfrac{1}{3}\Omega' = \frac{1^22.02\alpha + 12^2.0\alpha 1}{12\alpha},$$

or if we write herein 3 for α, this is

$$\tfrac{1}{3}\Omega' = \frac{1^22.023 + 12^2.031}{123};$$

and we have hence the general form

$$\tfrac{1}{3}\Omega = \frac{1^22.023 + 12^2.031}{123} + K.012,$$

where K is an arbitrary constant.

49. There is, however, a more simple particular solution $\tfrac{1}{3}\Omega'$ = polar function 012 ($f=x^3+y^3+z^3$, then $012 = xx_1x_2 + yy_1y_2 + zz_1z_2$), which, to avoid a confusion of notation, we may write $=\widetilde{012}$. We at once verify this; for, expressing the coordinates (x, y, z) in terms of (ρ, σ, τ), we have $\tfrac{1}{3}\Omega' = \widetilde{012}, = \frac{1}{\Delta}(1^22.\rho + 12^2.\sigma + \widetilde{123}.\tau)$, which, for $\tau=0$ becomes $=\frac{1}{\Delta}\{1^22.\rho + 12^2.\sigma\}$.

We must, of course, have an identity of the form

$$\widetilde{012} = \frac{1^22.023 + 12^2.031}{123} + K.012,$$

and to find K, writing here $0=3$, we have $K = \frac{\widetilde{123}}{123}$, or we have the identity

$$123\,\widetilde{012} - \widetilde{123}\,012 = 1^22.023 + 12^2.031.$$

Single Letter Notation for the Polar Functions of the Cubic. Art. Nos. 50 and 51.

50. The notation of single letters for the polar functions is not much required in the case of the cubic, but, in the next following case of the quartic it can hardly be dispensed with, and I therefore establish it in the case of the cubic: viz. I write

$$2^23,\ 3^21,\ 1^22 = f,\ g,\ h,\quad 23^2,\ 31^2,\ 12^2 = i,\ j,\ k;\quad \widetilde{123} = l,$$

or, what is the same thing, the expression for the cubic function f in terms of ρ, σ, τ is

$$\Delta^3.f = 3h\rho^2\sigma + 3j\rho^2\tau + 3k\rho\sigma^2 + 6l\rho\sigma\tau + 3g\rho\tau^2 + 3f\sigma^2\tau + 3i\sigma\tau^2;$$

an equation which, writing 0^3 instead of f, may also be written

$$\Delta^3 . 0^3 = (3h,\ 3j,\ 3k,\ 6l,\ 3g,\ 3f,\ 3i \between \rho^2\sigma,\ \rho^2\tau,\ \rho\sigma^2,\ \rho\sigma\tau,\ \rho\tau^2,\ \sigma^2\tau,\ \sigma\tau^2);$$

and I join to it the series of equations

$$\begin{aligned}
\Delta^2 . 0^2 1 &= (0,\ 2h,\ 2j,\ k,\ 2l,\ g \between \rho^2,\ \rho\sigma,\ \rho\tau,\ \sigma^2,\ \sigma\tau,\ \tau^2),\\
,,\quad 0^2 2 &= (h,\ 2k,\ 2l,\ 0,\ 2f,\ i \between \quad ,, \quad),\\
,,\quad 0^2 3 &= (j,\ 2l,\ 2g,\ f,\ 2i,\ 0 \between \quad ,, \quad),\\
\Delta\ .\ 01^2 &= (0,\ h,\ j \between \rho,\ \sigma,\ \tau),\\
\Delta\ \widetilde{012} &= (h,\ k,\ l \between \ ,, \ ),\\
,,\quad \widetilde{013} &= (j,\ l,\ g \between \ ,, \ ),\\
,,\quad 02^2 &= (k,\ 0,\ f \between \ ,, \ ),\\
,,\quad \widetilde{023} &= (l,\ f,\ i \between \ ,, \ ),\\
,,\quad 03^2 &= (g,\ i,\ 0 \between \ ,, \ ).
\end{aligned}$$

51. In particular, we have $\Delta . \widetilde{012} = h\rho + k\sigma + l\tau$, and the above-mentioned identity $123\,\underset{\sim}{012} - \underset{\sim}{123}\,012 = 1^2 2 . 023 + 12^2 . 031$ is simply $h\rho + k\sigma + l\tau - l\tau = h\rho + k\sigma$.

Single Letter Notation for the Polar Functions of the Quartic. Art. No. 52.

52. I write here

$$2^3 3,\ 3^3 1,\ 1^3 2 = f,\ g,\ h;\quad 23^3,\ 31^3,\ 12^3 = i,\ j,\ k;$$

$$1^2 23,\ 12^2 3,\ 123^2 = l,\ m,\ n;\quad 2^2 3^2,\ 3^2 1^2,\ 1^2 2^2 = p,\ q,\ r;$$

so that the expression for the quartic function f in terms of ρ, σ, τ is

$$\begin{aligned}
\Delta^4 . f = {} & 4h\rho^3\sigma + 4j\rho^3\tau + 6p\rho^2\sigma^2 + 12l\rho^2\sigma\tau + 6q\rho^2\tau^2\\
& + 4k\rho\sigma^3 + 12m\rho\sigma^2\tau + 12n\rho\sigma\tau^2 + 4g\rho\tau^3 + 4f\sigma^3\tau + 6r\sigma^2\tau^2 + 4i\sigma\tau^3,
\end{aligned}$$

which, putting 0^4 for f, may also be written

$$\begin{aligned}
\Delta^4 . 0^4 = {} & (4h,\ 4j;\ 6p,\ 12l,\ 6q:\ 4k,\ 12m,\ 12n,\ 4g:\ 4f,\ 6r,\ 4i \between\\
& (\rho^3\sigma,\ \rho^3\tau;\ \rho^2\sigma^2,\ \rho^2\sigma\tau,\ \rho^2\tau^2,\ \rho\sigma^3,\ \rho\sigma^2\tau,\ \rho\sigma\tau^2,\ \rho\tau^3,\ \sigma^3\tau,\ \sigma^2\tau^2,\ \sigma\tau^3);
\end{aligned}$$

and I join to it the series of equations

$$\begin{aligned}
\Delta^3 . 0^3 1 &= (0;\ 3h,\ 3j;\ 3r,\ 6l,\ 3q;\ k,\ 3m,\ 3n,\ g \between \rho^3,\ \rho^2\sigma,\ \rho^2\tau,\ \rho\sigma^2,\ \rho\sigma\tau,\ \rho\tau^2,\ \sigma^3,\ \sigma^2\tau,\ \sigma\tau^2,\ \tau^3),\\
,,\quad 0^3 2 &= (h;\ 3r,\ 3l;\ 3k,\ 6m,\ 3n;\ 0,\ 3f,\ 3p,\ i \between \quad ,, \quad),\\
,,\quad 0^3 3 &= (j;\ 3l,\ 3q;\ 3n,\ 6n,\ 3g;\ f,\ 3p,\ 3i,\ 0 \between \quad ,, \quad),
\end{aligned}$$

$$\begin{array}{lll}
\Delta^2 . 0^21^2 & = (0;\ 2h,\ 2j;\ r,\ 2l,\ q \between \rho^2;\ \rho\sigma,\ \rho\tau;\ \sigma^2,\ \sigma\tau,\ \tau^2), \\
\quad\text{,,}\quad 0^212 & = (h;\ 2r,\ 2l;\ k,\ 2m,\ n \between \quad\text{,,}\quad), \\
\quad\text{,,}\quad 0^213 & = (j;\ 2l,\ 2q;\ m,\ 2n,\ g \between \quad\text{,,}\quad), \\
\quad\text{,,}\quad 0^22^2 & = (r;\ 2k,\ 2m;\ 0,\ 2f,\ p \between \quad\text{,,}\quad), \\
\quad\text{,,}\quad 0^223 & = (l;\ 2m,\ 2n;\ f,\ 2p,\ i \between \quad\text{,,}\quad), \\
\quad\text{,,}\quad 0^23^2 & = (q;\ 2n,\ 2g;\ p,\ 2i,\ 0 \between \quad\text{,,}\quad), \\
\Delta . 01^3 & = (0,\ h,\ j \between \rho,\ \sigma,\ \tau), \\
\quad\text{,,}\quad 01^22 & = (h,\ r,\ l \between \quad\text{,,}\quad), \\
\quad\text{,,}\quad 01^23 & = (j,\ l,\ q \between \quad\text{,,}\quad), \\
\quad\text{,,}\quad 012^2 & = (r,\ k,\ m \between \quad\text{,,}\quad), \\
\quad\text{,,}\quad 0123 & = (l,\ m,\ n \between \quad\text{,,}\quad), \\
\quad\text{,,}\quad 013^2 & = (q,\ n,\ g \between \quad\text{,,}\quad), \\
\quad\text{,,}\quad 02^3 & = (k,\ 0,\ f \between \quad\text{,,}\quad), \\
\quad\text{,,}\quad 02^23 & = (m,\ f,\ p \between \quad\text{,,}\quad), \\
\quad\text{,,}\quad 023^2 & = (n,\ p,\ i \between \quad\text{,,}\quad), \\
\quad\text{,,}\quad 03^3 & = (g,\ i,\ 0 \between \quad\text{,,}\quad),
\end{array}$$

which will be convenient in the sequel.

Major Function—The Fixed Curve a Cubic. Art. No. 53.

53. It has been already seen that a simple particular form is $\frac{1}{3}\Omega' = \widetilde{012}$: and that the general form is $\Omega = \Omega' + K . 012$.

Major Function—The Fixed Curve a Quartic. Art. No. 54.

54. It is to be shown that a particular form is

$$\tfrac{1}{2}\Omega' = \frac{-\,01^3 . 02^3 + 01^22 . 012^2 + 0^212 . 1^22^2}{1^22^2}.$$

In fact, by the foregoing values of $\Delta . 01^3$, &c., the numerator of this expression, multiplied by Δ^2, is =

$$\begin{aligned}
&-(h\sigma + j\tau)(k\rho + f\tau) \\
&+(h\rho + r\sigma + l\tau)(r\rho + k\sigma + m\tau) \\
&+ r(h\rho^2 + 2r\rho\sigma + 2l\rho\tau + k\sigma^2 + 2m\sigma\tau + n\tau^2),
\end{aligned}$$

which is

$$= 2hr\rho^2 + 3r^2\rho\sigma + (hm - jk + 3lr)\rho\tau + 2kr\sigma^2 + (-fh + kl + 3mr)\sigma\tau + (-fj + lm + nr)\tau^2;$$

and this, for $\tau = 0$, becomes

$$= r(2h\rho^2 + 3r\rho\sigma + 2k\sigma^2).$$

Hence for $\tau=0$, we have

$$\Omega' = \frac{1}{\Delta^2}(4h\rho^2 + 6r\rho\sigma + 4k\sigma^2),$$

that is,

$$= \frac{1}{\Delta^2}\{4\,.\,1^3 2\,.\,\rho^2 + 6\,.\,1^2 2^2\,.\,\rho\sigma + 4\,.\,12^3\,.\,\sigma^2\};$$

and Ω' is thus a form of the major function $(x, y, z)_{12}^2$. Of course the general form is $\Omega = \Omega' + (x, y, z)^1\,.\,012$.

Syzygy of the Major Function. Art. No. 55.

55. Writing now $(x, y, z)_{12}^{n-2} = \Omega_{12}$; and taking on the fixed curve a new point 3, consider the like functions Ω_{23} and Ω_{31}: it is to be shown that we have identically

$$\Omega_{23}\,.\,031\,.\,012 + \Omega_{31}\,.\,012\,.\,023 + \Omega_{12}\,.\,023\,.\,031 - (123)^2 f = 023\,.\,031\,.\,012\,(x, y, z)^{n-3},$$

where $(x, y, z)^{n-3}$ is a *properly determined* minor function; or, considering herein 0 as a point on the fixed curve and writing therefore $f=0$, the equation is

$$\frac{\Omega_{23}}{023} + \frac{\Omega_{31}}{031} + \frac{\Omega_{12}}{012} = (x, y, z)^{n-3}.*$$

56. Write for a moment $X = \Omega_{23}\,.\,031\,.\,012 + \Omega_{31}\,.\,012\,.\,023 + \Omega_{12}\,.\,023\,.\,031$; then k being an arbitrary coefficient, we have $X - kf = 0$, a curve of the order n, passing through the points 1, 2, 3, and also through the residues of 2, 3, the residues of 3, 1, and the residues of 1, 2; in fact, at the point 1 we have $012=0$, $031=0$, and therefore $X=0$; also $f=0$; and therefore 1 is a point on the curve. Again, at any residue of 2, 3, we have $\Omega_{23}=0$, $023=0$, and therefore $X=0$; also $f=0$; and hence the residue of 2, 3 is a point on the curve.

It is next to be shown that k can be so determined that the curve $X-kf=0$ shall have a dp at each of the points 1, 2, 3. Supposing this to be so, we have the line 23 meeting the curve $X-kf=0$ in the points 2 and 3, each counting twice, and in the $n-2$ residues of 2, 3, that is, in $n+2$ points; hence the curve $X-kf=0$ must contain as part of itself the line 23, and similarly it must contain as part of itself each of the other lines 31 and 12, viz. we shall then have $X - kf = 023.031.012.(x, y, z)^{n-3}$; and from this equation observing that the curves $\Omega_{23}=0$, $\Omega_{31}=0$, $\Omega_{12}=0$ each pass through the dps, it follows that the curve $(x, y, z)^{n-3}=0$ also passes through the dps; hence, k being found to be $=(123)^2$, the theorem will be proved.

57. Taking an arbitrary point α coordinates $(x_\alpha, y_\alpha, z_\alpha)$, and writing

$$D = x_\alpha \frac{d}{dx} + y_\alpha \frac{d}{dy} + z_\alpha \frac{d}{dz},$$

* This is the differential theorem corresponding to C. and G.'s integral theorem, p. 26, viz. this is $S\xi\eta + S\eta\zeta + S\zeta\xi = I$, a sum of three integrals of the third kind = an integral of the first kind.

we have to find k, so that the curve $D(X-kf)=0$ shall pass through the point 1. Observing that $D023=\alpha 23$, &c., we have

$$\begin{aligned} D(X-kf) = & D\Omega_{23}\,.\,031\,.\,012+\Omega_{23}(031\,.\,\alpha 12+\alpha 31\,.\,012) \\ & +\alpha 23(\Omega_{31}\,.\,012+\Omega_{12}\,.\,031) \\ & +023\{\Omega_{31}\,.\,\alpha 12+\Omega_{12}\,.\,\alpha 31+D\Omega_{31}\,.\,012+D\Omega_{12}\,.\,031\} \\ & -kDf, \end{aligned}$$

and, to make the curve pass through 1, writing herein $0=1$, we have

$$0=123(\Omega_{31}{}^1\,.\,\alpha 12+\Omega_{12}{}^1\,.\,\alpha 31)-k(Df)^1,$$

where the superfix (1) denotes that we are in Ω_{31}, Ω_{12} and Df respectively to write $0=1$. We have $\Omega_{31}{}^1=n\,.\,1^{n-1}3$, $\Omega_{12}{}^1=n\,.\,1^{n-1}2$, $(Df)^1=n\,.\,1^{n-1}\alpha$, and the equation thus is

$$n\,.\,123(1^{n-1}3\,.\,\alpha 12+1^{n-1}2\,.\,\alpha 31)-kn\,.\,1^{n-1}\alpha=0.$$

But we have identically $1^{n-1}1\,.\,\alpha 23+1^{n-1}2\,.\,\alpha 31+1^{n-1}3\,.\,\alpha 12=1^{n-1}\alpha\,.\,123$, where $1^{n-1}1$, $=1^n$ is, in fact, $=0$; the factor $1^{n-1}\alpha$ thus divides out, and the equation becomes $k=(123)^2$; viz. k having this value, the curve $X-kf=0$ will have a dp at 1; and clearly by symmetry, it will also have a dp at 2, and at 3; the theorem is thus proved.

The Syzygy, Fixed Curve a Cubic. Art. No. 58.

58. The syzygy may be verified independently in the case where the fixed curve is a cubic. Observe that the syzygy, if satisfied for any particular form of Ω, will be generally satisfied; we may therefore take $\frac{1}{3}\Omega_{12}=\widetilde{012}$. Writing then

$$\frac{\frac{1}{3}\Omega_{12}}{012}=\frac{\widetilde{012}}{012},\ =\{012\}\text{ suppose,}$$

and taking 0 to be a point on the cubic curve, we ought to have $\{023\}+\{031\}+\{012\}=$ a constant; the value of this constant comes out to be $=\{123\}$, and the syzygy in its complete form thus is

$$\{023\}+\{031\}+\{012\}=\{123\}.$$

We have

$$\Delta\,\widetilde{023},\ \Delta\,\widetilde{031},\ \Delta\,\widetilde{012}=l\rho+f\sigma+i\tau,\ j\rho+l\sigma+g\tau,\ h\rho+k\sigma+l\tau,$$

and the equation thus is

$$\frac{l\rho+f\sigma+i\tau}{\rho}+\frac{j\rho+l\sigma+g\tau}{\sigma}+\frac{h\rho+k\sigma+l\tau}{\tau}-l=0;$$

this, multiplied by $\rho\sigma\tau$, becomes

$$h\rho^2\sigma+j\rho^2\tau+k\rho\sigma^2+2l\rho\sigma\tau+g\rho\tau^2+f\sigma^2\tau+i\sigma\tau^2=0,$$

which is, in fact, $\frac{1}{3}f=0$, the equation of the cubic curve.

Observe that the new symbol $\{012\}$ is, in virtue of its determinant denominator, an alternate function, $\{012\}=-\{102\}$, $\{012\}=\{120\}=\{201\}$. The syzygy is a relation between any four points 1, 2, 3, 0 of the curve, and it may be also expressed in the form

$$\{123\}-\{230\}+\{301\}-\{012\}=0.$$

The Syzygy, Fixed Curve a Quartic. Art. No. 59.

59. Taking Ω_{12} as before, we have

$$\frac{\frac{1}{2}\Omega_{12}}{012} = \frac{-01^3 . 02^3 + 01^2 2 . 012^2 + 0^2 12 . 1^2 2^2}{012 . 1^2 2^2} = \{0^2 12\} \text{ suppose:}$$

and then, taking 0 to be a point on the quartic curve, we ought to have

$$\{0^2 23\} + \{0^2 31\} + \{0^2 12\} = (x, y, z)^1, \text{ a linear function of } (x, y, z),$$

or, what is the same thing, considering the left-hand side as expressed in terms of ρ, σ, τ, the sum should be

$$= (\rho, \sigma, \tau)^1, \text{ a linear function of } (\rho, \sigma, \tau).$$

By a preceding formula we have

$$\{0^2 12\} = \frac{1}{\Delta^2 r\tau}\{2hr\rho^2 + 3r^2\rho\sigma + hm - jk + 3lr)\rho\tau$$
$$+ 2kr\sigma^2 + (-fh + kl + 3mr)\sigma\tau + (-fj + lm + nr)\tau^2\},$$

which is

$$= \frac{1}{\Delta^2}\left\{\left(3l + \frac{hm - jk}{r}\right)\rho + \left(3m + \frac{-fh + kl}{r}\right)\sigma + \left(n + \frac{-fj + lm}{r}\right)\tau\right\} + \frac{1}{\Delta^2}\,\frac{2h\rho^2 + 3r\rho\sigma + 2k\sigma^2}{\tau}.$$

And hence, forming the sum $\{0^2 23\} + \{0^2 31\} + \{0^2 12\}$, we have first a fractional part which is found to be integral, viz. this is

$$\frac{1}{\Delta^2}\left\{\frac{2f\sigma^2 + 3p\sigma\tau + 2i\tau^2}{\rho} + \frac{2g\tau^2 + 2q\tau\rho + 2j\rho^2}{\sigma} + \frac{2h\rho^2 + 3r\rho\sigma + 2k\sigma^2}{\tau}\right\},$$

$$= \frac{1}{\Delta^2\rho\sigma\tau}\{2h\rho^3\sigma + 2j\rho^3\tau + 3r\rho^2\sigma^2 + 3q\rho^2\tau^2 + 2k\rho\sigma^3 + 2g\rho\tau^3 + 2f\sigma^3\tau + 3p\sigma^2\tau^2 + 2i\sigma\tau^3\},$$

$$= \frac{1}{\Delta^2\rho\sigma\tau}\{\tfrac{1}{2}\Delta^4 f - 6l\rho^2\sigma\tau - 6m\rho\sigma^2\tau - 6n\rho\sigma\tau^2\},$$

or since $f = 0$, this is

$$= \frac{1}{\Delta^2}(-6l\rho - 6m\sigma - 6n\tau).$$

We then have integral terms which are at once deduced from the above integral terms of $0^2 12$; collecting the several terms, we find

$$\{0^2 23\} + \{0^2 31\} + \{0^2 12\} =$$

$$\frac{1}{\Delta^2}\left\{\rho\left(l + \frac{mn - gk}{p} + \frac{jn - gh}{q} + \frac{hm - jk}{r}\right) + \sigma\left(m + \frac{fn - ik}{p} + \frac{ln - hi}{q} + \frac{kl - fh}{r}\right)\right.$$
$$\left. + \tau\left(n + \frac{lm - fg}{p} + \frac{gl - ij}{q} + \frac{lm - fj}{r}\right)\right\},$$

which is the required result.

Preparation for the Conversion—The Symbol ∂. Art. Nos. 60 to 63.

60. I use ∂ as the symbol of a quasi-differentiation, viz. U being any function of (x, y, z), ∂U denotes $\frac{1}{d\omega}$ multiplied by the differential $\frac{dU}{dx}dx + \frac{dU}{dy}dy + \frac{dU}{dz}dz$; in such a differential, the increments dx, dy, dz do not in general present themselves in the combinations $ydz - zdy$, $zdx - xdz$, $xdy - ydx$; but they will do so if U is a function of the degree zero in the coordinates x, y, z (that is, if U be the quotient of two homogeneous functions of the same degree); and this being so, we can by the equations

$$\frac{ydz - zdy}{\frac{df}{dx}} = \frac{zdx - xdz}{\frac{df}{dy}} = \frac{xdy - ydx}{\frac{df}{dz}},\ = d\omega$$

get rid of the increments, and ∂U will denote a function of (x, y, z) derived in a definite manner from the function U. The symbol ∂ will be used only in the case in question of a function of the degree zero. Of course ∂_1 will denote the like operation in regard to (x_1, y_1, z_1); and so ∂_2, &c.; and we may for greater clearness write ∂_0 in place of ∂.

61. Consider then $\partial\frac{P}{Q}$, where P, Q are functions $(x, y, z)^m$ of the same degree: we have

$$\partial\frac{P}{Q} = \frac{1}{Q^2 d\omega}(QdP - PdQ),$$

and then

$$dP = \frac{dP}{dx}dx + \frac{dP}{dy}dy + \frac{dP}{dz}dz,\quad \frac{1}{m}P = \frac{dP}{dx}x + \frac{dP}{dy}y + \frac{dP}{dz}z,$$

with the like formulæ for Q. Substituting, we find

$$\partial\frac{P}{Q} = \frac{1}{mQ^2 d\omega}\left\{\frac{d(Q, P)}{d(y, z)}(ydz - zdy) + \frac{d(Q, P)}{d(z, x)}(zdx - xdz) + \frac{d(Q, P)}{d(x, y)}(xdy - ydx)\right\},$$

that is,

$$\partial\frac{P}{Q} = \frac{1}{mQ^2}\left\{\frac{df}{dx}\frac{d(Q, P)}{d(y, z)} + \&\text{c.}\right\} = \frac{1}{mQ^2}\frac{d(f, Q, P)}{d(x, y, z)},\quad = \frac{1}{mQ^2}J(f, Q, P),$$

or say

$$\partial\frac{P}{Q} = -\frac{1}{mQ^2}J(P, Q, f).$$

62. As an example, consider

$$\partial\{012\},\ = -\frac{1}{(012)^2}J(\widetilde{012}, 012, f).$$

The determinant is

$$\begin{vmatrix} \frac{d}{dx}\widetilde{012}, & y_1z_2 - y_2z_1, & \frac{df}{dx} \\ \frac{d}{dy}\widetilde{012}, & z_1x_2 - z_2x_1, & \frac{df}{dy} \\ \frac{d}{dz}\widetilde{012}, & x_1y_2 - x_2y_1, & \frac{df}{dz} \end{vmatrix},$$

and the coefficient herein of $\frac{d}{dx}\widetilde{012}$ is $(z_1x_2 - z_2x_1)\frac{df}{dz} - (x_1y_2 - x_2y_1)\frac{df}{dy}$, which is

$$= x_2\left(x_1\frac{df}{dx} + y_1\frac{df}{dy} + z_1\frac{df}{dz}\right) - x_1\left(x_2\frac{df}{dx} + y_2\frac{df}{dy} + z_2\frac{df}{dz}\right),\ = 3\,(0^21\,.\,x_2 - 0^22\,.\,x_1);$$

and so for the other terms.

The determinant is thus

$$= 3\left[0^21\left(x_2\frac{d}{dx} + y_2\frac{d}{dy} + z_2\frac{d}{dz}\right) - 0^22\left(x_1\frac{d}{dx} + y_1\frac{d}{dy} + z_1\frac{d}{dz}\right)\right]\widetilde{012}$$

say this is

$$= 3\,[0^21\,.\,\mathbb{D} - 0^22\,\mathbb{D}]\,\widetilde{012}.$$

But we have $\mathbb{D}\,\widetilde{012} = 12^2$, $\mathbb{D}\widetilde{012} = 1^22$, and the determinant is then $= 3(0^21\,.\,12^2 - 0^22\,.\,1^22)$; whence finally, writing ∂_0 instead of ∂,

$$\partial_0\,\{012\} = -3\,.\,\frac{0^21\,.\,12^2 - 0^22\,.\,1^22}{(012)^2}.$$

63. By cyclical interchange of the 0, 1, 2, we have

$$\partial_1\,\{012\} = -3\,.\,\frac{1^22\,.\,0^22 - 01^2\,.\,02^2}{(012)^2},$$

$$\partial_2\,\{012\} = -3\,.\,\frac{02^2\,.\,01^2 - 12^2\,.\,0^21}{(012)^2};$$

and thence adding, we find

$$(\partial_0 + \partial_1 + \partial_2)\,\{012\} = 0,$$

an important property which, joined to the equation before obtained,

$$\{023\} + \{031\} + \{012\} = \{123\},$$

completes the theory of the function $\{012\}$.

Conversion of the Major Function (Interchange of Limits and Parametric Points).

Art. No. 64.

64. Write in general

$$\frac{(x,\ y,\ z)_{12}^{\,n-2}}{012} = Q_{0,\,12},$$

$Q_{0,\,12}$ is an alternate function in regard to the points 1, 2 ($Q_{0,\,12} = -Q_{0,\,21}$); and in regard to the coordinates of the points 0, 1, 2, it is rational, but not integral, of the degrees $n-3$, 0, 0 respectively: it can therefore be operated upon with ∂_1 or ∂_2, but (except in the case $n = 3$) not with ∂_0.

The conversion relates not to the general major function $(x, y, z)_{12}^{n-2}$, but to this function *with the arbitrary constants properly determined*, and consists in a relation between two functions $Q_{4,\,12}$ and $Q_{1,\,45}$ (each of them a function of three out of four arbitrary points 1, 2, 4, 5 on the fixed curve), viz. the conversion is

$$\partial_1 Q_{4,\,12} = \partial_4 Q_{1,\,45},$$

an equation which may be written in four different forms, viz. we may in the form written down interchange 1, 2 and also 4, 5.*

The determination of the constants is a very peculiar one, inasmuch as it is not algebraical; viz. in the case of the cubic curve, about to be considered, it appears that $Q_{0,\,12}$ contains the term $\int_2^1 d\omega \partial_3 \{036\}$, which is a transcendental function of the coordinates of the parametric points 1 and 2.

The Conversion, Fixed Curve a Cubic. Art. No. 65.

65. We may write $Q_{0,\,12} = \{012\} + K$, where K is a constant, that is, it is independent of the point 0, but depends on the parametric points 1 and 2. I assume K to be properly determined, and give an *à posteriori* verification of the equation $\partial_1 Q_{4,\,12} = \partial_4 Q_{1,\,45}$. The value is $K = \int_2^1 d\omega \partial_3 \{036\} - \{123\}$, where 3, 6 are arbitrary points on the cubic curve, and where in the definite integral, regarded as an integral $\int U\,du$ with a current variable u, the meaning is that this variable has at the limits the values u_1, u_2 which belong to the points 1 and 2 respectively: a fuller explanation might be proper, but the investigation will presently be given in a form not depending on any integral at all.

Substituting for K its value, we have

$$Q_{0,\,12} = \{012\} + \left[\int_2^1 d\omega\, \partial_3 \{036\} - \{123\}\right],$$

or, as this may also be written,

$$= -\{023\} - \{031\} + \int_2^1 d\omega\, \partial_3 \{036\}.$$

We have thence

$$\partial_1 Q_{0,\,12} = -\partial_1 \{031\} + \partial_3 \{136\},$$

* The meaning of the property is better seen from the integral form: $Q_{0,\,12}$ is a function of the points 0, 1, 2 and $Q_{0,\,45}$ the like function of the points 0, 4, 5 such that $\int_5^4 d\omega\, Q_{0,\,12} = \int_2^1 d\omega\, Q_{0,\,45}$; which equation operated upon with $\partial_1 \partial_4$ gives the formula of the text. And there is thus the meaning (alluded to in the heading) that there exists for the integral of the third kind a canonical form (C. and G.'s endliche Normalform), such that the integral is not altered by the interchange of the limits and the parametric points. The expression for $Q_{0,\,12}$ mentioned further on in the text for the case, fixed curve a cubic, shows that in this case the canonical form of the integral of the third kind is $\int_5^4 d\omega \left[\{012\} + \left(\int_2^1 d\omega\, \partial_3 \{036\} - \{123\}\right)\right]$.

and consequently

$$\partial_1 Q_{4,\,12} = -\partial_1 \{431\} + \partial_3 \{136\},$$
$$\partial_4 Q_{1,\,45} = -\partial_4 \{134\} + \partial_3 \{436\};$$

hence, observing that $\{431\} = -\{134\}$, &c., we have

$$\partial_1 Q_{4,\,12} - \partial_4 Q_{1,\,45} = (\partial_1 + \partial_4)\{134\} + \partial_3 \{136\} - \partial_3 \{436\},$$
$$= -\partial_3 \{134\} + \partial_3 \{136\} - \partial_3 \{436\},$$

which, observing that we have $\partial_3 \{641\} = 0$, is

$$= \partial_3 (\{136\} - \{364\} + \{641\} - \{413\}),\ = 0,$$

the required theorem.

To avoid, in the proof, the use of the integral sign, we have only to consider the required function $Q_{0,\,12}$ as given by the foregoing differential formula

$$\partial_1 Q_{0,\,12} = -\partial_1 \{031\} + \partial_3 \{136\},$$

for we have then the values of $\partial_1 Q_{4,\,12}$ and $\partial_4 Q_{1,\,45}$: the rest of the proof the same as before.

The Conversion, Fixed Curve a Quartic. Art. Nos. 66 to 73.

66. We have

$$Q_{0,\,12} = \{0^2 12\} + (x, y, z)^1,$$

where $(x, y, z)^1$ is a linear function of (x, y, z), but depending also on the parametric points 1 and 2, which is to be determined so as to satisfy the conversion equation

$$\partial_1 Q_{4,\,12} = \partial_4 Q_{1,\,45}.$$

Observing that we have $\{0^2 23\} + \{0^2 31\} + \{0^2 12\} =$ a linear function of (x, y, z), the linear function $(x, y, z)^1$ of $Q_{0,\,12}$ may be taken to be $= \Theta_{0,\,12} - \{0^2 23\} - \{0^2 31\} - \{0^2 12\}$; that is, we may assume

$$Q_{0,\,12} = \{0^2 12\} + \Theta_{0,\,12} - (\{0^2 23\} + \{0^2 31\} + \{0^2 12\}),$$
$$= -\{0^2 23\} - \{0^2 31\} + \Theta_{0,\,12},$$

where $\Theta_{0,\,12}$ is a linear function of (x, y, z), but depending also on the points 1 and 2, which has to be determined. We have

$$\partial_1 Q_{0,\,12} = -\partial_1 \{0^2 31\} + \partial_1 \Theta_{0,\,12},$$

and thence

$$\partial_1 Q_{4,\,12} = -\partial_1 \{4^2 31\} + \partial_1 \Theta_{4,\,12},$$
$$\partial_4 Q_{1,\,45} = -\partial_4 \{1^2 34\} + \partial_4 \Theta_{1,\,45},$$

giving an equation for Θ,

$$\partial_1 \Theta_{4,\,12} - \partial_4 \Theta_{1,\,45} = \partial_1 \{4^2 31\} - \partial_4 \{1^2 34\};$$

here 4 is an arbitrary point of the quartic, and we may instead of it write 0, the equation thus becoming

$$\partial_1 \Theta_{0,\,12} - \partial_0 \Theta_{1,\,05} = \partial_1 \{0^2 31\} - \partial_0 \{1^2 30\}.$$

67. Of the terms on the left-hand side, the first is a linear function of (x, y, z), or say it is an integral function 0^1, and the second is a linear function of (x_1, y_1, z_1), or say it is an integral function 1^1: the given function on the right-hand side must therefore admit of expression in the form $\phi(0^1, 1, 3) - \phi(1^1, 0, 3)$, where $\phi(0^1, 1, 3)$ is a known function, integral and linear as regards the coordinates (x, y, z) of the point 0, but depending also on the points 1, 3; and $\phi(1^1, 0, 3)$ is the like known function, integral and linear as regards the coordinates (x_1, y_1, z_1) of the point 1, but depending also on the points 0, 3. Moreover, since 2 and 5 are arbitrary points entering only on the left-hand side, it is clear that $\partial_1\Theta_{0,12}$ must be independent of 2, and $\partial_0\Theta_{1,05}$ independent of 5; reverting to the cubic case, observe that here $\Theta_{0,12} = \int_2^1 d\omega\,\partial_3\{036\}$, whence $\partial_1\Theta_{0,12} = \partial_3\{136\}$, and so $\partial_0\Theta_{1,05} = \partial_3\{036\}$, and that the corresponding equation thus is $\partial_3\{136\} - \partial_3\{036\} = \partial_1\{031\} - \partial_0\{130\}$, where the left-hand side is $= \partial_3\{013\}$, and the equation itself $(\partial_0 + \partial_3 + \partial_1)\{031\} = 0$. We then have

$$\partial_1\Theta_{0,12} - \phi(0^1, 1, 3) = \partial_0\Theta_{1,02} - \phi(1^1, 0, 3),$$

where the one side is derived from the other by the interchange of the 0, 1. The solution therefore is

$$\partial_1\Theta_{0,12} - \phi(0^1, 1, 3) = X(\overline{0, 1}, 3),$$

a function which is symmetrical in regard to the points 0 and 1, and, inasmuch as the left-hand is an integral function 0^1, must itself be an integral function $(0^1, 1^1)$, that is, integral and linear as regards the coordinates (x, y, z) and (x_1, y_1, z_1) of the points 0 and 1 respectively. We thus have

$$\partial_1\Theta_{0,12} = \phi(0^1, 1, 3) + X(\overline{0, 1}, 3),$$

and thence

$$\partial_2\Theta_{0,12} = -\phi(0^1, 2, 3) - X(\overline{0, 2}, 3),$$

viz. the second of these expressions is, with its sign reversed, the same function of 2 that the first is of 1.

68. It follows that, taking a new symbol 7 for the variable of the definite integral (in the cubic case $\Theta_{0,12}$ was independent of 0, and there was nothing to prevent the use of 0 for the current point of the definite integral), we may write $\Theta_{0,12} = \int_2^1 d\omega_7 P(7, 0, 3)$, where $\partial_1 P(1, 0, 3) = \phi(0^1, 1, 3) + X(\overline{0, 1}, 3)$, an equation which implies $\partial_2 P(0, 2, 3) = \phi(0^1, 2, 3) + X(\overline{0, 2}, 3)$. But the first of these equations in P is nothing else than the first of the equations in $\Theta_{0,12}$.

69. I have succeeded in finding $\phi(0^1, 1, 3)$, but the calculation is a very tedious one, and I give only the principal steps, omitting all details. We have to bring $\partial_1 0^2 13 - \partial_0 1^2 03$ into the form $\phi(0^1, 1, 3) - \phi(1^1, 0, 3)$. From the value of

$$\{0^2 13\}, \; = \frac{-01^3 . 03^3 + 01^2 3 . 013^2 - 0^2 13 . 1^2 3^2}{013 . 1^2 3^2},$$

we find, by a process such as that of No. 62,

$$\begin{aligned}\partial_1\{0^213\} = \frac{1}{\sigma^2}\Big\{&-0^23^2.01^3 - 0^33.j\\&+\frac{1}{q}\begin{pmatrix}2.01^3[03^3.01^23-(013^2)^2]\\ \qquad +[2.0^213.013^2+1.0^23^2.01^23-3.0^21^2.0^23^2]j\end{pmatrix}\\&+\frac{1}{q^2}\begin{pmatrix}2.01^3(-01^3.03^3+013^2.01^23)g\\-2.013^2(-01^3.03^3+013^2.01^23)j\end{pmatrix}\Big\}.\end{aligned}$$

Substituting herein the values $01^3 = \frac{1}{\Delta}(h\sigma + j\tau)$, &c., we have $\frac{1}{\sigma^2}$ multiplied by a cubic function $(\rho, \sigma, \tau)^3$; writing down first the integral terms, and then the others, we have

$$\begin{aligned}\partial_1\{0^213\} = \frac{1}{\Delta^3}\Big\{&\rho\left[(-6hn+5jm)+\frac{1}{q}(4ghl-3gjr-2hij+j^2p+2jnl)+\frac{1}{q^2}(-2g^2h^2+4ghjn-2j^2n^2)\right]\\&+\sigma\left[fj-hp+\frac{1}{q}(2hil-2hn^2-3ijr+jlp+2jmn)+\frac{1}{q^2}(2ghln-2gh^2i+2hijn-2jln^2)\right]\\&+\tau\left[3jp+\frac{1}{q}(-2ghn+2gjm-2ijl-2jn^2)+\frac{1}{q^2}(2g^2hl-2ghij-2gjln+2ij^2n)\right]\Big\}\end{aligned}$$

(say this linear function of ρ, σ, τ is $=\Box$)

$$+\frac{1}{\Delta^3\sigma^2}\{\rho^3.2j^2+\rho^2\sigma(6jl-3hg)+\rho^2\tau.3jq+\rho\sigma\tau(-2gh+6jn)+\rho\tau^2.2gj+\sigma\tau^2.2ij\}.$$

70. The expression of $\partial_0\{1^203\}$ is deduced from this by the interchange of 0, 1: and I write

$$\begin{aligned}&\partial_1\{0^213\}-\partial_0\{1^203\} = \Box - *\\&+\frac{1}{\sigma^2\Delta^3\rho^3}[\rho^3\{\rho^3.2j^2+\rho^2\sigma(-3hq+6jl)+\rho^2\tau.3jq+\rho\sigma\tau(-2gh+6jn)+\rho\tau^2.2gj+\sigma\tau^2.2ij\}\\&\quad-\Delta^3\{\Delta^3.2(0^23^2)^2-\Delta^2\sigma(-3.0^32.0^23^2+6.0^33.0^223)-\Delta^2\tau.3.0^33.0^23^2\\&\quad+\Delta\sigma\tau(-2.03^3.0^32+6.0^33.023^2)+\Delta\tau^2.2.03^3.0^33-\sigma\tau^2.2i.0^33\}],\end{aligned}$$

where, and in what follows, the $*$ denotes the function immediately to the left of it, interchanging therein the 0, 1. It will be observed that the $\Box$, *quâ* linear function of (ρ, σ, τ), that is, of (x, y, z), is a term of the required function $\phi(0^1, 1, 3)$: the remaining portion has to be reduced by means of the expressions for $\Delta^2(0^23^2)$, &c., in terms of ρ, σ, τ.

71. We obtain

$$\begin{aligned}&\partial_1\{0^213\}-\partial_0\{1^203\} = \Box - *\\&+\frac{1}{\Delta^3}\{\sigma(2fj-3hp-3kq+18lm-9nr)+\tau(-3gr-3jp+9mq)\}\\&+\frac{1}{\Delta^3\rho}\{\sigma^2(12fl-12kn+18m^2-9pr)+\sigma\tau(6fg-6gk-6ir+18mn)+\tau^2.6gm\}\\&+\frac{1}{\Delta^3\rho^2}\{\sigma^3(18fm-9kp)+\sigma^2\tau(12fn-8ik+9mp)+\sigma\tau^2(4fg+6im)\}\\&+\frac{1}{\Delta^3\rho^3}\{\sigma^4.4f^2+\sigma^3\tau.6fp+\sigma^2\tau^2.4fi\}.\end{aligned}$$

The terms of the second line may be transformed as follows:

$$\frac{\sigma}{\Delta^3}(2fj - 3hp - 3kq + 18lm - 9nr)$$

$$= \tfrac{1}{2}\frac{\sigma}{\Delta^3}(2fj - 3hp - 3kq + 18lm - 9nr) - *$$

$$+ \frac{1}{\Delta^3\rho}\{\sigma^2(-12fl + 12kn - 18m^2 + 9pr) + \sigma\tau(-\tfrac{3}{2}fq + 3gk + \tfrac{3}{2}ir - \tfrac{9}{2}lp - 9mn)\}$$

$$+ \frac{1}{\Delta^3\rho^2}\{\sigma^3(-30fm + 15kp) + \sigma^2\tau(-12fn + 12ik - 18mp) + \sigma\tau^2 . -9np\}$$

$$+ \frac{1}{\Delta^3\rho^3}\{\sigma^4 . -10f^2 + \sigma^3\tau . -15fp + \sigma^2\tau^2 . -9p^2 + \sigma\tau^3 . -3ip\},$$

and

$$\frac{\tau}{\Delta^3}(-3gr - 3jp + 9mq)$$

$$= \tfrac{1}{2}\frac{\tau}{\Delta^3}(-3gr - 3jp + 9mq) - *$$

$$+ \frac{1}{\Delta^3\rho}\{\sigma\tau(-\tfrac{3}{2}fq + 3gk + \tfrac{3}{2}ir + \tfrac{3}{2}lp - 9mn) + \tau^2 . -6gm\}$$

$$+ \frac{1}{\Delta^3\rho^2}\{\sigma^2\tau(-9fn + 3ik) + \sigma\tau^2(-6fg - 6im) + \tau^3 . -3gp\}$$

$$+ \frac{1}{\Delta^3\rho^3}\{\sigma^3\tau . -3fp + \sigma^2\tau^2 . -6fi + \sigma\tau^3 . -3ip\};$$

substituting these values, the whole third line is destroyed, and we find

$$\partial_1\{0^2 13\} - \partial_0\{1^2 03\} = \square - *$$

$$+ \tfrac{1}{2} . \frac{1}{\Delta^3}\{\sigma(2fj - 3hq - 3kj + 18lm - 9nr) + \tau(-3gr - 3jp + 9mq)\} - *$$

$$+ \frac{1}{\Delta^3\rho^2}\{\sigma^3(-12fm + 6kp) + \sigma^2\tau(-9fn + 7ik - 9mp) + \sigma\tau^2(-2fg - 9np) + \tau^3 . -3gp\}$$

$$+ \frac{1}{\Delta^3\rho^3}\{\sigma^4 . -6f^2 + \sigma^3\tau . -12fp + \sigma^2\tau^2(-2fi - 9p^2) + \sigma\tau^3 . -6ip\}.$$

Ultimately the last two lines of this expression are found to be

$$= \frac{1}{\Delta^3}\{\rho(-2hn + 4jm + 2l^2 - 2qr) + \sigma(-2fj + 2hp + 2kq - 10lm + 5nr)$$
$$+ \tau(2gr + hi - jp + 7ln - 4mq)\} - *$$

so that the whole is now a sum of three linear functions of $(\rho, \sigma, \tau) . - *$.

72. Collecting the terms, we have

$$\partial_1 \{0^2 13\} - \partial_0 \{1^2 03\} =$$

$$\frac{1}{\Delta^3}\Big[\rho\Big\{(-8hn+9jm+2l^2-2qr)+\frac{1}{q}(4ghl-3gjr-2hij+j^2p+2jln)$$
$$+\frac{1}{q^2}(-2g^2h^2+4ghjn-2j^2n^2)\Big\}$$
$$+\sigma\Big\{\tfrac{1}{2}(-hp+kq-2lm+nr)+\frac{1}{q}(2hil-2hn^2-3ijr+jlp+2jmn)$$
$$+\frac{1}{q^2}(-2gh^2i+2ghln+2hijn-2jln^2)\Big\}$$
$$+\tau\Big\{\tfrac{1}{2}(gr+2hi+jp+mq+14ln)+\frac{1}{q}(-2ghn+2gjm-2ijl-2jn^2)$$
$$+\frac{1}{q^2}(2g^2hl-2ghij-2gjln+2ij^2n)\Big\}\Big]$$

$-$ *.

73. The right-hand side depends on the points 0, 1, 3 and 2: viz. we have therein $\rho = 023$, $\Delta = 123$, &c., but the left-hand side depending on only the points 0, 1 and 3, the right-hand side cannot really contain 2, and it must thus remain unaltered, if for 2 we substitute any other point on the quartic, say 6: the right-hand side may therefore be understood as a function of 0, 1, 3 and 6, viz. ρ, Δ, f, &c., will mean 063, 163, $6^3 3$, &c.: we have thus $\phi(0^1, 1, 3) =$ the above linear function with 2 thus replaced by 6; say

$$\phi(0^1, 1, 3) = \frac{1}{\Delta^3}[\rho(\ \ ) + \sigma(\ \ ) + \tau(\ \ )],$$

a given function of the points 0, 1, 3 and the arbitrary point 6, on the quartic curve; we therefore write it $\phi(0^1, 1, 3, 6)$. There is no obvious value for $X(\overline{0, 1}, 3)$ which will produce any simplification: I therefore take this function to be $=0$; and the final result is

$$Q_{0,12} = \{0^2 12\} + \Theta_{0,12} - (\{0^2 23\} + \{0^2 31\} + \{0^2 12\}),$$

where $\Theta_{0,12}$ is a function integral and linear as regards the coordinates (x, y, z) of the point 0, but transcendental as regards the parametric points 1, 2; and containing besides the arbitrary points 3, 6, of the quartic curve, its value being determined by the differential formulæ

$$\partial_1 \Theta_{0,12} = \phi(0^1, 1, 3, 6), \quad \partial_2 \Theta_{0,12} = -\phi(0^1, 2, 3, 6),$$

where $\phi(0^1, 1, 3, 6)$ is a given function as above. I do not see the meaning of the very complicated linear function of (ρ, σ, τ), nor how to reduce it to any form such as the simple one $\partial_3\{036\}$, which presents itself in the case of the cubic curve.

Cambridge, England, October 5, 1882.

CHAPTER IV. THE MAJOR FUNCTION $(x, y, z)_{12}^{n-2}$, CONTINUED.

The Conversion, Fixed Curve a Quartic, continued. Art. Nos. 74 to 82.

74. I resume the question considered *ante* Nos. 66 to 73. The general problem, where the fixed curve is any given curve whatever, has recently been solved in a very complete and elegant form by Dr Nöther, in the two notes "Zur Reduction algebraischer Differentialausdrücke auf die Normalformen" and "Ueber die algebraischen Differentialausdrücke, 2e Note," *Sitzungsb. der phys.-med. Soc. zu Erlangen*, 10 Dec. 1883 and 14 Jan. 1884. I consider here the case of the quartic curve, $n=4$, and connect his result with my former investigations.

We have the differential

$$\frac{(x, y, z)_{12}^{n-2}\, d\omega}{012}, = \frac{\Omega_{12} d\omega}{012},$$

where Ω_{12}, or as I also write it $\Omega\,(0;\ 1,\ 2;\ 3,\ 4,\ 5)$, is a rational and integral function of the degree $(n-2=)\,2$ in the current coordinates (x, y, z): it depends also on the parametric points 1, 2, which are points on the quartic, coordinates (x_1, y_1, z_1), (x_2, y_2, z_2) respectively; and on $(p=)\,3$ other points 3, 4, 5 on the quartic, coordinates (x_3, y_3, z_3), (x_4, y_4, z_4), (x_5, y_5, z_5) respectively. The curve $\Omega=0$ is a conic, which is taken to pass through the dps (none in the present case) and through the $(n-2=)\,2$ residues of the parametric points; and the function Ω is such that on writing therein (x_1, y_1, z_1) for (x, y, z) it becomes $=(n.1^{n-1}2^{*}=)\,{}_4.1^32$: viz. we have $\Omega\,(1;\,1,\,2;\ 3,\ 4,\ 5)={}_4.1^32$, which implies also $\Omega\,(2;\ 1,\ 2;\ 3,\ 4,\ 5)={}_4.12^3$: so defined, the function would contain $(p=)\,3$ arbitrary constants, but these are determined so that the curve $\Omega=0$ passes through the 3 points 3, 4, 5 on the quartic: and the function Ω, $=\Omega\,(0:\ 1,\ 2:\ 3,\ 4,\ 5)$ is thus a completely determinate function, rational and integral of the degree 2 in the coordinates (x, y, z) of the current point, and rational in the coordinates of the other five points respectively. I call to mind that 012 denotes the determinant formed with the coordinates (x, y, z), &c., of the points 0, 1, 2 respectively: the like notation is used throughout.

75. The function $\Omega\,(0;\ 1,\ 2;\ 3,\ 4,\ 5)$ is, in fact, the function Ω' of No. 43 with only the further condition in regard to the points 3, 4, 5 of the quartic; viz. Ω is the function determined by the equation

$$\begin{vmatrix} (x\,,\ y\,,\ z\,)^2 & ,\ \Omega \\ 1\,(x_1,\ y_1,\ z_1)^2 & ,\ 4\,.\,1^32 \\ 2\,(x_1,\ y_1,\ z_1\between x_2,\ y_2,\ z_2), & 6\,.\,1^22^2 \\ 1\,(x_2,\ y_2,\ z_2)^2 & ,\ 4\,.\,12^3 \\ (x_3,\ y_3,\ z_3)^2 & ,\ 0 \\ (x_4,\ y_4,\ z_4)^2 & ,\ 0 \\ (x_5,\ y_5,\ z_5)^2 & ,\ 0 \end{vmatrix} = 0:$$

* $\left(x_2\dfrac{d}{dx_1}+y_2\dfrac{d}{dy_1}+z_2\dfrac{d}{dz_1}\right)f(x_1,\ y_1,\ z_1)=n\,.\,1^{n-1}2$; see No. 2.

this is of the form

$$M\Omega + \square = 0,$$

and as appears in No. 46, M is $= 123 . 124 . 125 . 345$. Hence writing $\square\,(0;\ 1,\ 2;\ 3,\ 4,\ 5)$ for $\square$, we have

$$\Omega, = \Omega\,(0;\ 1,\ 2;\ 3,\ 4,\ 5),\ = \frac{-\square\,(0;\ 1,\ 2;\ 3,\ 4,\ 5)}{123 . 124 . 125 . 345}.$$

Hence further writing

$$Q, = Q\,(0;\ 1,\ 2;\ 3,\ 4,\ 5),\ = \frac{\Omega\,(0;\ 1,\ 2;\ 3,\ 4,\ 5)}{012},$$

so that the differential is $Qd\omega$, $= Q\,(0;\ 1,\ 2;\ 3,\ 4,\ 5)\,d\omega$, we have

$$Q = Q\,(0;\ 1,\ 2;\ 3,\ 4,\ 5),\ = \frac{-\square\,(0;\ 1,\ 2;\ 3,\ 4,\ 5)}{012 . 123 . 124 . 125 . 345},$$

which is of the form

$$Q = \frac{0^2\,\overline{12}^4\,\overline{345}^2}{0^1\,\overline{12}^4\,\overline{345}^2},\ = 0^1\,\overline{12345}^0,$$

viz. Q is a rational fraction where the numerator is of the degree 2, and the denominator of the degree 1 as regards the coordinates $(x,\ y,\ z)$ of the current point: but the numerator and denominator are each of the degree 4 as regards the coordinates of the points 1, 2 separately, and of the degree 2 as regards the coordinates of the points 3, 4, 5 separately: that is, Q is of the degree 1 as regards the coordinates $(x,\ y,\ z)$, but of the degree 0 as regards the coordinates of the points 1, 2, 3, 4, 5 separately.

76. The signification of the symbol of quasi-differentiation ∂ (applicable only to a function of the degree 0 in the coordinates to which the differentiations have reference) is explained *ante* No. 60. The function Q just mentioned is of the degree 0 in regard to the coordinates of each of the points 1, 2, 3, 4, 5; and it can thus be operated upon by the symbols $\partial_1,\ \partial_2,\ \partial_3,\ \partial_4,\ \partial_5$ respectively. Observe, in particular, that we have $\partial_1 Q\,(0;\ 1,\ 2;\ 3,\ 4,\ 5) = \overline{01}^1\,\overline{2345}^0$, viz. it is of the degree 1 in the coordinates of the points 0 and 1 respectively, but of the degree 0 in regard to the coordinates of the points 2, 3, 4, 5 respectively.

77. This being so, we may consider the function

$$\begin{aligned} H\,(0;\ 1,\ 2;\ 3,\ 4,\ 5;\ 6,\ 7,\ 8) = {} & \partial_1 Q\,(0;\ 1,\ 2;\ 3,\ 4,\ 5) \\ & + \partial_3 Q\,(1;\ 3,\ 2;\ 6,\ 4,\ 5) . \frac{045}{345} \\ & + \partial_4 Q\,(1;\ 4,\ 2;\ 3,\ 7,\ 5) . \frac{053}{453} \\ & + \partial_5 Q\,(1;\ 5,\ 2;\ 3,\ 4,\ 8) . \frac{034}{534}, \end{aligned}$$

where 6, 7, 8 are arbitrary points on the quartic; the functions

$$\partial_3 Q\,(1;\ 3,\ 2;\ 6,\ 4,\ 5),\quad \partial_4 Q\,(1;\ 4,\ 2;\ 3,\ 7,\ 5),\quad \partial_5 Q\,(1;\ 5,\ 2;\ 3,\ 4,\ 8),$$

are functions of the same form as $\partial_1 Q(0;\ 1,\ 2;\ 3,\ 4,\ 5)$, and derived from it by changing in each case the current point 0 into the parametric point 1, and by further changing in the three cases this parametric point into the points 3, 4, 5 respectively, and replacing the corresponding point 3, 4 or 5 by the new arbitrary point 6, 7 or 8. Further 045, &c., denote determinants as above; so that in H each of the last three terms is, in fact, as regards the point 0, a mere linear function of the coordinates (x, y, z) of this point.

We have $\partial_3 Q(1;\ 3,\ 2;\ 6,\ 4,\ 5) = \overline{13}^1\overline{2645}^0$, and hence this function multiplied by $\dfrac{045}{345}$ is $= \overline{01}^1\overline{23456}^0$; and so for the third and fourth terms of H: thus each of the four terms of H is $= \overline{01}^1\overline{2345678}^0$, of the degree 1 in the coordinates of the points 0 and 1 respectively, but of the degree 0 in the coordinates of the other points 2, 3, 4, 5, 6, 7, 8 respectively.

Nöther's conversion-theorem consists herein, that the function

$$H(0;\ 1,\ 2;\ 3,\ 4,\ 5;\ 6,\ 7,\ 8)$$

is unaltered by the interchange of the two points 0, 1; or putting for shortness

$$H(0;\ 1,\ 2;\ 3,\ 4,\ 5;\ 6,\ 7,\ 8) = H_1(0),$$

the theorem is

$$H_1(0) = H_0(1).$$

78. We have, No. 59,

$$\frac{\tfrac{1}{2}\Omega_{12}}{012} = \frac{-01^3 \, . \, 02^3 + 01^2 2 \, . \, 012^2 + 0^2 12 \, . \, 1^2 2^2}{012 \, . \, 1^2 2^2}, = \{0^2 12\},$$

or as for greater simplicity I write it $\qquad = 0^2 12,$

viz. $0^2 12$ is now written instead of $\{0^2 12\}$ to denote the function just given as the value of $\dfrac{\tfrac{1}{2}\Omega_{12}}{012}$, Ω_{12} is thus $= {}_2 \, . \, 012 \, . \, 0^2 12$, viz. this is a particular form of Ω_{12} satisfying the conditions that $\Omega_{12} = 0$ is a conic passing through the residues of the points 1, 2, and such that Ω_{12} on writing therein (x_1, y_1, z_1) for (x, y, z) becomes $= {}_4 \, . \, 1^3 2$: hence the general form of the function satisfying these conditions is $= {}_2 \, . \, 012 \, \{0^2 12 +$ arbitrary linear function of $(x, y, z)\}$. The before-mentioned function $\Omega(0;\ 1,\ 2;\ 3,\ 4,\ 5)$ is a function satisfying these conditions and the further conditions that the conic $\Omega = 0$ shall pass through the three points 3, 4, 5 on the quartic: these further conditions serve to determine the linear function: and we at once obtain

$$\frac{\tfrac{1}{2}\Omega(0;\ 1,\ 2;\ 3,\ 4,\ 5)}{012} = 0^2 12 - 3^2 12\,\frac{045}{345} - 4^2 12\,\frac{053}{453} - 5^2 12 \, . \, \frac{034}{534},$$

viz. the value of Ω given by this equation, on writing therein $0 = 3$, 4, or 5, becomes $= 0$ as it should do.

79. We thus have

$$Q(0;\ 1,\ 2;\ 3,\ 4,\ 5) = 0^2 12 - 3^2 12\,\frac{045}{345} - 4^2 12\,\frac{053}{453} - 5^2 12\,\frac{034}{534};$$

and Nöther's conversion-equation becomes

$$\begin{aligned}
&\partial_1 \left\{0^2 12 - 3^2 12\,\frac{045}{345} - 4^2 12\,\frac{053}{453} - 5^2 12\,\frac{034}{534}\right\} \\
&+ \partial_3 \left\{1^2 32 - 6^2 32\,\frac{145}{645} - 4^2 32\,\frac{156}{456} - 5^2 32\,\frac{164}{564}\right\} \cdot \frac{045}{345} \\
&+ \partial_4 \left\{1^2 42 - 3^2 42\,\frac{175}{375} - 7^2 42\,\frac{153}{753} - 5^2 42\,\frac{137}{537}\right\} \frac{053}{453} \\
&+ \partial_5 \left\{1^2 52 - 3^2 52\,\frac{148}{348} - 4^2 52\,\frac{183}{483} - 5^2 48\,\frac{134}{834}\right\} \cdot \frac{034}{534} \\
= \ &\partial_0 \left\{1^2 02 - 3^2 02\,\frac{145}{345} - 4^2 02\,\frac{153}{453} - 5^2 02\,\frac{134}{534}\right\} \\
&+ \partial_3 \left\{0^2 32 - 6^2 32\,\frac{045}{645} - 4^2 32\,\frac{056}{456} - 5^2 32\,\frac{064}{564}\right\} \frac{145}{345} \\
&+ \partial_4 \left\{0^2 42 - 3^2 42\,\frac{075}{375} - 7^2 42\,\frac{053}{753} - 5^2 42\,\frac{037}{537}\right\} \cdot \frac{153}{453} \\
&+ \partial_5 \left\{0^2 52 - 3^2 52\,\frac{048}{348} - 4^2 52\,\frac{083}{483} - 5^2 48\,\frac{034}{834}\right\} \cdot \frac{134}{534},
\end{aligned}$$

an equation where the functions operated on with the ∂'s are only functions such as $0^2 12$; for there is not any determinant operated upon containing the number which is the suffix of the ∂ operating upon it.

80. Taking all the terms over to the left-hand side, there are in all 32 terms: but of these $3+3$ destroy each other, and $6+6$ unite in pairs into 6 terms: there are thus in all $7+7+6$, $=20$ terms: viz. multiplying the whole equation by 345, it is found that the equation becomes

	or as this may be written	where
$345\,(\ \partial_1\, 0^2 12 - \partial_0\, 1^2 02)$	$345\ \boxed{012}$	$\boxed{012} = \partial_1\, 0^2 12 - \partial_0\, 1^2 02$, &c.
$-\,045\,(-\,\partial_3\, 1^2 32 + \partial_1\, 3^2 12)$	$-\,045\ \boxed{312}$	
$-\,053\,(-\,\partial_4\, 1^2 42 + \partial_1\, 4^2 12)$	$-\,305\ \boxed{412}$	
$-\,034\,(-\,\partial_5\, 1^2 52 + \partial_1\, 5^2 12)$	$-\,340\ \boxed{512}$	

$$-145\,(\ \partial_3\, 0^2 32-\partial_0\, 3^2 02) \qquad -145\ \boxed{032}$$

$$-153\,(\ \partial_4\, 0^2 42-\partial_0\, 4^2 02) \qquad -315\ \boxed{042}$$

$$-134\,(\ \partial_5\, 0^2 52-\partial_0\, 5^2 02) \qquad -341\ \boxed{052}$$

$$+013\,(-\partial_4\, 5^2 42+\partial_5\, 4^2 52) \qquad +301\ \boxed{452}$$

$$+014\,(-\partial_5\, 3^2 52+\partial_3\, 5^2 32) \qquad +140\ \boxed{532}$$

$$+015\,(-\partial_3\, 4^2 32+\partial_4\, 3^2 42) \qquad +015\ \boxed{342}$$

$$=0, \qquad\qquad =0,$$

viz. the equation is

$$\Sigma \pm 345\ \boxed{012}=0,$$

the nine terms which follow the first term $345\ \boxed{012}$ of the sum being obtained by the interchanges of 0, 1 (one or each) with the 3, 4, 5, each interchange giving rise to a sign $-$.

81. In obtaining the foregoing result, we have, for instance, a pair of terms

$$\partial_4\, 5^2 42\, \frac{-137\,.\,053+037\,.\,153}{537},\ =\partial_4\, 5^2 42\, \frac{-013\,.\,537}{537},\ =\partial_4\, 5^2 42\,(-013),$$

viz. this depends on the equation

$$137\,.\,053-037\,.\,153-537\,.\,013=0,$$

or say

$$-137\,.\,035+037\,.\,135+013\,.\,357=0,$$

an identity which, in a form which will be readily understood, may be written

$$\text{det.}\begin{vmatrix} 0137 \\ 013735 \end{vmatrix}=0.$$

Similarly, the two terms which contain $\partial_5\, 4^2 52$ combine into the single term $\partial_5\, 4^2 52\,(013)$: and the two new terms taken together are

$$013\,(-\partial_4\, 5^2 42+\partial_5\, 4^2 52).=301\ \boxed{452}\ .$$

82. The proof of the identity,

$$\Sigma \pm 345 \boxed{012} = 0,$$

depends on the property of the function

$$\boxed{012}\,, \;= \partial_1 0^2 12 - \partial_0 1^2 02,$$

enunciated No. 67, and proved *à posteriori* by the tedious calculation Nos. 69 to 73, viz. in No. 67, writing 2 in place of 3, this is:—$\partial_1 0^2 12 - \partial_0 1^2 02$ is equal to the difference of two functions, the first of them linear in the coordinates (x, y, z) of the point 0, but depending also on the coordinates of the points 1 and 2; the second of them linear in the coordinates (x_1, y_1, z_1) of the point 1 but depending also on the coordinates of the points 0 and 2. Or, what is the same thing, the property is

$$\boxed{012} = A_{12}x + B_{12}y + C_{12}z - (A_{02}x_1 + B_{02}y_1 + C_{02}z_1),$$

where A_{12}, B_{12}, C_{12} are functions of (x_1, y_1, z_1), (x_2, y_2, z_2), and A_{02}, B_{02}, C_{02} are the like functions of (x, y, z), (x_2, y_2, z_2).

Substituting such values in the sum $\Sigma \pm 345 \boxed{012}$, but writing down only the terms which contain x, these are

$$\begin{array}{lll} 345\,(A_{12}x - A_{02}x_1) & & \\ -045\,(A_{12}x_3 - A_{32}x_1) & -145\,(A_{32}x - A_{02}x_3) & +301\,(A_{52}x_4 - A_{42}x_5) \\ -305\,(A_{12}x_4 - A_{42}x_1) & -315\,(A_{42}x - A_{02}x_4) & +140\,(A_{32}x_5 - A_{52}x_3) \\ -340\,(A_{12}x_5 - A_{52}x_1) & -341\,(A_{52}x - A_{02}x_5) & +015\,(A_{42}x_3 - A_{32}x_4). \end{array}$$

This is

$$\begin{aligned} = \quad & A_{02}(-x_1 345 + x_3 145 + x_4 315 + x_5 341) \\ & + A_{12}(\;x\;345 - x_3 045 - x_4 305 - x_5 340) \\ & + A_{32}(\;x_1 045 - x\;145 + x_5 140 - x_4 015) \\ & + A_{42}(-x\;315 + x_3 015 + x_1 305 - x_5 301) \\ & + A_{52}(-x\;341 + x_4 301 + x_1 340 - x_3 140), \end{aligned}$$

where the coefficient of each of the A's is identically $=0$: and similarly, the terms in y and the terms in z are each $=0$. We have thus the proof of the identity

$$\Sigma \pm 345 \boxed{012} = 0,$$

that is, of the conversion-equation $H_1(0) = H_0(1)$.

The Syzygy—Fixed Curve a Quartic. Art. No. 83.

I revert to the theory of the Syzygy, *ante* No. 59.

83. We have

$$Q(0;\ 1,\ 2;\ 3,\ 4,\ 5) = 0^2 12 - 3^2 12 \frac{045}{345} - 4^2 12 \frac{053}{453} - 5^2 12 \frac{034}{534},$$

or if for convenience we take instead of 1, 2, the parametric points to be α, β coordinates $(x_\alpha,\ y_\alpha,\ z_\alpha)$ and $(x_\beta,\ y_\beta,\ z_\beta)$ respectively, then this equation is

$$Q_{\alpha\beta} = Q(0;\ \alpha,\ \beta;\ 3,\ 4,\ 5) = 0^2\alpha\beta - 3^2\alpha\beta \frac{045}{345} - 4^2\alpha\beta \frac{053}{453} - 5^2\alpha\beta \frac{034}{534}.$$

Considering a new parametric point γ, and forming the like functions $Q_{\beta\gamma}$ and $Q_{\gamma\alpha}$, it is to be shown that we have identically

$$Q_{\alpha\beta} + Q_{\beta\gamma} + Q_{\gamma\alpha} = 0.$$

To prove this, observe that, in the equation at the end of No. 59, Δ, ρ, σ, τ denote 123, 023, 031, 012 respectively. Hence writing therein α, β, γ in place of 1, 2, 3 respectively, and putting A, B, C for the coefficients $\left(\text{including therein the factor } \frac{1}{\Delta^2}\right)$ of ρ, σ, τ respectively, the equation is

$$0^2\beta\gamma + 0^2\gamma\alpha + 0^2\alpha\beta = A\,.\,0\beta\gamma + B\,.\,0\gamma\alpha + C\,.\,0\alpha\beta,$$

where A, B, C are absolute constants (functions, that is, of the coefficients of the quartic) each divided by $(\alpha\beta\gamma)^2$. We hence obtain

$$\begin{aligned}(Q_{\beta\gamma} + Q_{\gamma\alpha} + Q_{\alpha\beta})\,.\,345 = \ & 345\,(A\,.\,0\beta\gamma + B\,.\,0\gamma\alpha + C\,.\,0\alpha\beta)\\ & - 045\,(A\,.\,3\beta\gamma + B\,.\,3\gamma\alpha + C\,.\,3\alpha\beta)\\ & - 053\,(A\,.\,4\beta\gamma + B\,.\,4\gamma\alpha + C\,.\,4\alpha\beta)\\ & - 034\,(A\,.\,5\beta\gamma + B\,.\,5\gamma\alpha + C\,.\,5\alpha\beta).\end{aligned}$$

On the left-hand side the whole coefficient of A is $=0$; viz. the coefficient has the value det. $\begin{vmatrix} 0345 \\ 0345\beta\gamma \end{vmatrix}$, which is $=0$. Similarly, the whole coefficient of B is $=0$, and the whole coefficient of C is $=0$: and we have thus the required result

$$Q_{\beta\gamma} + Q_{\gamma\alpha} + Q_{\alpha\beta} = 0.$$

The syzygy is thus obtained in a more perfect form than in No. 59; viz. by considering (instead of $0^2\alpha\beta$) the new form $Q_{\alpha\beta}$, then, instead of a sum which is a linear function of the coordinates $(x,\ y,\ z)$, we obtain a sum $=0$.

The Fixed Curve a Cubic—Syzygy and Conversion. Art. Nos. 84 and 85.

84. In the case when the fixed curve is a cubic (see Nos. 58 and 64), the analogous formulæ are

$$\tfrac{1}{3}\Omega\,(0;\ 1,\ 2;\ 3) = \frac{1^22\,.\,023 + 12^2\,.\,031}{123},\quad = \widetilde{012} - \frac{\widetilde{123}}{123}\,012 \text{ (see No. 49)},$$

that is,

$$Q\,(0;\ 1,\ 2;\ 3),\ = \frac{\tfrac{1}{3}\Omega}{012},\ = \frac{\widetilde{012}}{012} - \frac{\widetilde{123}}{123},\ = \{012\} - \{123\},$$

where 1, 2 are the parametric points: 3 any other point on the cubic: the brackets { } are of course here necessary in order to distinguish {012} from the determinant 012. It will be remembered that {012} is an alternate function

$$\{012\},\ = -\{102\},\ = \{120\},\ \&c.$$

If instead of 1, 2 we take the parametric points to be α, β, coordinates $(x_\alpha,\ y_\alpha,\ z_\alpha)$ and $(x_\beta,\ y_\beta,\ z_\beta)$ respectively, then the formula is

$$Q_{\alpha\beta} = Q\,(0;\ \alpha,\ \beta;\ 3) = \{0\alpha\beta\} - \{3\alpha\beta\}.$$

Hence taking on the cubic a new point γ, coordinates $(x_\gamma,\ y_\gamma,\ z_\gamma)$ and forming the functions $Q_{\beta\gamma}$ and $Q_{\gamma\alpha}$ we have

$$\begin{aligned} Q_{\beta\gamma} + Q_{\gamma\alpha} + Q_{\alpha\beta} = {} & \{0\beta\gamma\} + \{0\gamma\alpha\} + \{0\alpha\beta\} \\ & - \{3\beta\gamma\} - \{3\gamma\alpha\} - \{3\alpha\beta\}. \end{aligned}$$

But by the formula No. 58,

$$\{0\beta\gamma\} + \{0\gamma\alpha\} + \{0\alpha\beta\} = \{\alpha\beta\gamma\};$$

hence also

$$\{3\beta\gamma\} + \{3\gamma\alpha\} + \{3\alpha\beta\} = \{\alpha\beta\gamma\}:$$

and we have thus

$$Q_{\beta\gamma} + Q_{\gamma\alpha} + Q_{\alpha\beta} = 0,$$

the syzygy for the cubic.

85. For the conversion, the definition of H is

$$\begin{aligned} H\,(0;\ 1,\ 2;\ 3,\ 6) = {} & \partial_1\,Q\,(0;\ 1,\ 2;\ 3) \\ & + \partial_3\,Q\,(1;\ 3,\ 2;\ 6), \end{aligned}$$

viz. this is

$$\begin{aligned} H_0(1) = H\,(0;\ 1,\ 2;\ 3,\ 6) = {} & \partial_1\,(\{012\} - \{123\}) \\ & + \partial_3\,(\{132\} - \{326\}), \\ = {} & \partial_1\ \{012\} - (\partial_1 + \partial_3)\,\{123\} - \partial_3\,\{326\}, \end{aligned}$$

which, in virtue of $(\partial_1 + \partial_2 + \partial_3)\,\{123\} = 0$ (see No. 63), becomes

$$H_0(1) = \partial_1\,\{012\} + \partial_2\,\{123\} - \partial_3\,\{326\}.$$

Interchanging the 0 and 1, we thence have

$$H_1(0) = \partial_0\{102\} + \partial_2\{023\} - \partial_3\{326\}.$$

Hence the difference $H_0(1) - H_1(0)$ is

$$= \partial_1\{012\} - \partial_0\{102\} + \partial_2\{123\} - \partial_2\{023\},$$

viz. this is

$$= (\partial_1 + \partial_0)\{012\} + \partial_2(\{123\} - \{230\}),$$

where the first term is

$$= - \partial_2\{012\},$$

and the whole therefore is

$$= \partial_2(\{123\} - \{230\} - \{012\})$$
$$= - \partial_2\{301\},$$

in virtue of

$$\{123\} - \{230\} + \{301\} - \{012\} = 0;$$

the whole is consequently $= 0$.

We have thus

$$H_0(1) - H_1(0) = 0,$$

the conversion-equation in the case of the cubic.

Chapter V. Miscellaneous Investigations.

The Differential Symbol dω. Art. Nos. 86 and 87.

86. The definition is

$$\frac{y\,dz - z\,dy}{\frac{df}{dx}} = \frac{z\,dx - x\,dz}{\frac{df}{dy}} = \frac{x\,dy - y\,dx}{\frac{df}{dz}} = d\omega,$$

and it hence follows that we have

$$d\omega = \frac{\begin{vmatrix} dx, & dy, & dz \\ x, & y, & z \\ \lambda, & \mu, & \nu \end{vmatrix}}{\lambda\frac{df}{dx} + \mu\frac{df}{dy} + \nu\frac{df}{dz}},$$

where (λ, μ, ν) are arbitrary constants or, if we please, arbitrary functions of (x, y, z): viz. the expression just written down is altogether independent of the values of λ, μ, ν: and is consequently equal to the value obtained by writing any two of these symbols $= 0$, that is, the expression is equal to any one of the foregoing three equal values of $d\omega$. The expression was first given by Aronhold (1863), in the memoir presently referred to.

It is to be remarked that, considering (λ, μ, ν) as the coordinates of a point, the denominator $\lambda \frac{df}{dx} + \mu \frac{df}{dy} + \nu \frac{df}{dz}$ equated to 0 is the polar $(n-1)$ thic of the point λ, μ, ν in regard to the fixed curve.

If instead of λ, μ, ν we write $bc' - b'c,\ ca' - c'a,\ ab' - a'b$, where $(a, b, c)(a', b', c')$ are constants, then the numerator is

$$= (ax + by + cz)(a'dx + b'dy + c'dz) - (a'x + b'y + c'z)(adx + bdy + cdz),$$

or introducing ρ, σ to denote the arbitrary linear functions $ax + by + cz$ and $a'x + b'y + c'z$ respectively, the numerator is $= \rho d\sigma - \sigma d\rho$: moreover, observing that a, b, c and a', b', c' are the differential coefficients of ρ, σ in regard to the coordinates (x, y, z), the denominator is $= J(f, \rho, \sigma)$; and the value of $d\omega$ is

$$d\omega = \frac{\rho d\sigma - \sigma d\rho}{J(f, \rho, \sigma)},$$

where, in accordance with a previous remark, the denominator equated to 0 is the polar $(n-1)$ thic of the intersection of the lines $\rho = 0,\ \sigma = 0$ in regard to the fixed curve.

Obviously, by taking for ρ, σ any two of the three coordinates x, y, z, we reproduce the original three forms of $d\omega$.

87. The last-mentioned form of $d\omega$ suggests the expression for this symbol in the case where the fixed curve, instead of being a plane curve, is a curve of double curvature defined by two equations $f = 0,\ g = 0$ between the four coordinates (x, y, z, w): viz. ρ, σ being now arbitrary linear functions

$$ax + by + cz + dw, \text{ and } a'x + b'y + c'z + d'w$$

of the four coordinates, the expression is

$$d\omega = \frac{\rho d\sigma - \sigma d\rho}{J(f, g, \rho, \sigma)}:$$

and by taking for ρ, σ any two of the four coordinates x, y, z, w, we have for $d\omega$ six values which must of course be equal to each other; it is easy to verify *à posteriori* that this is so.

In the case where the curve of double curvature is not the complete intersection of two surfaces, the denominator (regarded as the Jacobian of the curve and of the arbitrary planes ρ, σ) will have a definite meaning, but what this is I do not at present consider.

The last-mentioned expression for $d\omega$ will be applied further on to the case of the quadri-quadric curve $y^2 + x^2 = 1,\ z^2 + k^2x^2 = 1$.

Integral Formulæ. Art. Nos. 88 to 90.

88. In what precedes, $d\omega$ has been used as a single symbol to denote any one of the equal differential expressions

$$\frac{y\,dz - z\,dy}{\frac{df}{dx}}, = \frac{z\,dx - x\,dz}{\frac{df}{dy}}, = \frac{x\,dy - y\,dx}{\frac{df}{dz}};$$

there is no quantity ω. The expressions are of the order $-(n-3)$ in the coordinates (x, y, z), and since (x, y, z) are as to their absolute magnitudes altogether arbitrary (only their ratios being determinate), a symbol such as

$$\omega, = \int d\omega, = \int \frac{y\,dz - z\,dy}{\frac{df}{dy}},$$

would, except in the case $n = 3$, be altogether meaningless. In fact, the integral would be

$$\int \frac{z^2 d\left(\frac{y}{z}\right)}{z^{n-1}\phi\left(\frac{x}{z}, \frac{y}{z}\right)}, = \int \frac{d\left(\frac{y}{z}\right)}{z^{n-3}\phi\left(\frac{x}{z}, \frac{y}{z}\right)},$$

where $\frac{x}{z}$ is, by the equation of the fixed curve, given as a function of $\frac{y}{z}$; but the other factor z^{n-3} is an absolutely indeterminate variable value, and the expression is meaningless.

But we have integrals $\int Q d\omega$, where Q is a homogeneous function of the order $n-3$ in the coordinates (x, y, z); and, in particular, we have such integrals where (corresponding to the forms which present themselves in the differential pure and affected theorems respectively) Q is either a rational and integral function $(x, y, z)^{n-3}$, or a rational and integral function $(x, y, z)^{n-2}$ divided by a linear function $(x, y, z)^1$: for in every such case, the form of integral is

$$\int \frac{d\left(\frac{y}{z}\right)}{\phi\left(\frac{x}{z}, \frac{y}{z}\right)},$$

where $\frac{x}{z}$ is a given function of $\frac{y}{z}$, and the factor of $d\left(\frac{y}{z}\right)$ is thus a mere function of $\frac{y}{z}$. More definitely, in the integrals $\int Q d\omega$ which are considered, Q is either a minor function $(x, y, z)^{n-3}$, or it is the quotient of a major function $(x, y, z)_{12}^{n-2}$ by the linear function 012.

In the case $n=2$, there is no rational and integral function $(x, y, z)^{n-3}$, but the function may be of the form belonging to the affected theorem, viz. it is unity divided by a linear function $(x, y, z)^1$; or say the integral is $\int \frac{d\omega}{\alpha x+\beta y+\gamma z}$, where the (x, y, z) are connected by a quadric equation $(a, \ldots \between x, y, z)^2=0$: it will be shown presently that this integral is obtainable as a logarithmic function.

In the case $n=3$, we have the rational and integral function $(x, y, z)^{n-3}$, $=$ a constant, or say $=1$, so that there is here an integral $\int d\omega$: we do not call this ω, but introducing a new letter, say u, and fixing at pleasure the inferior limit of the integral, we write $u=\int d\omega$.

89. In the foregoing form $\int Q d\omega$, so long as we retain the symbol $d\omega$, there is nothing to show what is the variable in regard to which the integration is to be performed; we may, for instance, writing

$$d\omega=\frac{y^2 d\left(\frac{z}{y}\right)}{\frac{df}{dx}},$$

make it to be $\frac{y}{z}$, or in like manner to be any other of the six quotients. We thus cannot attribute a *value* to the inferior or superior limit of such an integral, but we may take the limits to be each of them a point on the fixed curve: for instance, if 1, 0 be points on the fixed curve, then the integral $\int_0^1 Q d\omega$ means the integral taken from the value at the point 0 to the value at the point 1 of the variable in regard to which the integration is performed; or when there is no expressed superior limit, then the integral is to be taken from the value for the expressed or known inferior limit to the value at the current point (x, y, z) of the variable in regard to which the integration is performed. The actual value of the integral will of course depend upon the *path* of the variable; but this is a question which is not here entered upon.

If using Cartesian Coordinates x, y, we write for instance

$$d\omega=\frac{dx}{\frac{df}{dy}}, \text{ then } \int \frac{Q dx}{\frac{df}{dy}}$$

will denote an integral $\int \phi x \, dx$ in regard to the variable x, and the inferior and superior limits will be as usual values of x; or if there is no expressed superior limit, then the integral $\int_0 \phi x \, dx$ will be the integral taken from the inferior limit x_0 to the current value x.

We may, if we please, consider the coordinates (x, y, z) as depending upon a parameter ϑ, viz. the ratios $x : y : z$ may be regarded as given functions of ϑ, and the integral $\int Q d\omega$, is then an integral $\int \Omega d\vartheta$, which taken from a constant inferior limit up to the value ϑ, belonging to a given point 1 of the curve, is a given function of ϑ_1, or say of the point 1. But except in the case of the cubic (or generally if $p = 1$), we do not have the coordinates actually given as known functions of a parameter ϑ (say they are potentially known functions of ϑ), and it is further to be noticed the functions which present themselves are functions not of a single point, but of p or more points: thus in the case of the quartic, $n = 4$, $p = 3$; we have $\int^1 x\, d\omega$, $\int^1 y\, d\omega$, $\int^1 z\, d\omega$, each standing for a given function of the parameter ϑ_1, but these integrals do not present themselves singly, but in combinations such as

$$\left(\int^1 + \int^2 + \int^3 + \int^\xi\right)(x\, d\omega,\ y\, d\omega,\ z\, d\omega),$$

say these sums of integrals are u, v, w: each of the functions u, v, w is a potentially known function of the parameters $\vartheta_1, \vartheta_2, \vartheta_3, \vartheta_\xi$ which belong to the points 1, 2, 3, ξ respectively, and is consequently regarded as a given function of these four points.

90. Consider as before, in the case of a cubic curve, the integral $u = \int d\omega$: it will presently be seen that for the general curve as given by a cubic equation $f = 0$ of any form whatever, we arrive at a form of elliptic function: but the ordinary elliptic functions sn, cn, dn connect themselves most readily with the cubic curve $y^2 = x \,.\, 1 - x \,.\, 1 - k^2 x$. We have here

$$d\omega = \frac{\tfrac{1}{2} dx}{y},\ = \frac{\tfrac{1}{2} dx}{\sqrt{x \,.\, 1 - x \,.\, 1 - k^2 x}},$$

or, in the equation $u = \int d\omega$, taking the inferior limit to be 0, say

$$u = \int_0 \frac{\tfrac{1}{2}\, dx}{\sqrt{x \,.\, 1 - x \,.\, 1 - k^2 x}},$$

an equation which determines u as a function of x, or conversely, x as a function of u. We might thence, by means of Abel's theorem as applied to the curve in question, investigate the properties of the function $x = \lambda(u)$ thus arising, and so establish the theory of elliptic functions: but it is more convenient, treating the elliptic functions as known functions, to write for λu its value; viz. to take for x as given by this equation, the value $x = \text{sn}^2 u$: we thence have $y = \text{sn}\, u \,\text{cn}\, u \,\text{dn}\, u$; viz. these values $x = \text{sn}^2 u$, $y = \text{sn}\, u \,\text{cn}\, u \,\text{dn}\, u$, satisfy the equation $y^2 = x \,.\, 1 - x \,.\, 1 - k^2 x$ of the curve, and give, moreover, $d\omega = du = \dfrac{\tfrac{1}{2}\, dx}{y}$: and we can with these values, and the formulæ for elliptic functions, verify any results given by Abel's theorem. This will be done in considerable detail: but at present I wish only to remark that the formulæ give the coordinates x, y of a point on the cubic curve expressed as one-valued functions of a parameter or argument u: but that this argument u is not a one-valued function of the coordinate x, or even of the coordinates x, y of the given point on the curve: say the argument u has not a unique value for a given point (x, y) of the curve. There are, in fact, an infinity of values $u = u_0 + 2mK + 2m'iK'$, where m, m' are any positive or negative integers: that this is so, depends on the multiplicity of values, according to the different paths of the variable, of the integral

$$u = \int_0 \frac{\frac{1}{2}\,dx}{\sqrt{x \,.\, 1-x \,.\, 1-k^2x}};$$

or, regarding the elliptic functions as known functions, it depends upon the double periodicity of these functions.

Aronhold's Quadric Integral. Art. Nos. 91 to 93.

91. I reproduce the investigation contained in Aronhold's paper "Ueber eine neue algebraische Behandlungsweise u.s.w.," *Crelle*, t. LXI. (1863), pp. 95—145. We take f the general quadric function $(a, b, c, f, g, h \between x, y, z)^2$; $\alpha x + \beta y + \gamma z$ an arbitrary linear function of x, y, z: the theorem is $\dfrac{d\varpi}{\alpha x + \beta y + \gamma z}$ = differential of logarithm of an algebraic function of (x, y, z); viz. taking (ξ, η, ζ) for the coordinates of either of the points of intersection of the line $\alpha x + \beta y + \gamma z = 0$ with the quadric $(a, \ldots \between x, y, z)^2 = 0$, and writing also

$$\Omega^2 = -(bc - f^2,\ ca - g^2,\ ab - h^2,\ gh - af,\ hf - bg,\ fg - ch \between \alpha, \beta, \gamma)^2,$$

then the theorem is

$$\frac{d\varpi}{\alpha x + \beta y + \gamma z} = \frac{1}{\Omega}\, d \,.\log \frac{(a, \ldots \between x, y, z \between \xi, \eta, \zeta)}{\alpha x + \beta y + \gamma z},$$

or, what is the same thing,

$$\int \frac{d\varpi}{\alpha x + \beta y + \gamma z} = \frac{1}{\Omega} \log \frac{(a, \ldots \between x, y, z \between \xi, \eta, \zeta)}{\alpha x + \beta y + \gamma z} + \text{const.}$$

It is to be observed, in reference to this equation, that the two sides respectively are in regard to (α, β, γ) homogeneous functions of the degree -1, and in regard to (ξ, η, ζ) homogeneous of the degree 0; viz. on the right-hand side the effect of a change in the absolute magnitudes of ξ, η, ζ, say the change into $k\xi$, $k\eta$, $k\zeta$, is merely to change by $\log k$ the constant of integration.

It is to be remarked also that the equation $(a, \ldots \between \xi, \eta, \zeta \between x, y, z) = 0$ represents the tangent to the conic at the point (ξ, η, ζ) of intersection with the line $\alpha x + \beta y + \gamma z = 0$; calling the linear function in question T, the value of the integral is $\dfrac{1}{\Omega} \log \dfrac{T}{\alpha x + \beta y + \gamma z}$; if (ξ_1, η_1, ζ_1), (ξ_2, η_2, ζ_2) are the coordinates of the two points of intersection respectively, then in passing from one of these to the other we change the sign of the radical Ω, and the two values thus are $\dfrac{1}{\Omega} \log \dfrac{T_1}{\alpha x + \beta y + \gamma z}$ and $-\dfrac{1}{\Omega} \log \dfrac{T_2}{\alpha x + \beta y + \gamma z}$. These must differ by a constant only; viz. we should have $\log \frac{T_1 T_2}{(\alpha x + \beta y + \gamma z)^2} = \text{a const.}$ And, in fact, T_1 and T_2 being the tangents to the conic f at its intersections with the line $\alpha x + \beta y + \gamma z = 0$, we have it is clear $f = \lambda T_1 T_2 + \mu (\alpha x + \beta y + \gamma z)^2$, that is, (x, y, z) referring to a point of the conic $f = 0$, we have $\frac{T_1 T_2}{(\alpha x + \beta y + \gamma z)^2} = \text{a constant}$, which is right.

92. We require the coordinates (ξ, η, ζ) of an intersection: these are determined by the equations $\alpha\xi + \beta\eta + \gamma\zeta = 0$, $(a, \ldots \between \xi, \eta, \zeta)^2 = 0$, or as these may be written

$$\begin{aligned} &\alpha\xi \qquad\qquad\qquad + \beta\eta \qquad\qquad\qquad + \gamma\zeta = 0, \\ &(a\xi + h\eta + g\zeta)\,\xi + (h\xi + b\eta + f\zeta)\,\eta + (g\xi + f\eta + c\zeta)\,\zeta = 0\,; \end{aligned}$$

we have thence ξ, η, ζ proportional to the determinants

$$\left|\begin{matrix} a\xi + h\eta + g\zeta, & h\xi + b\eta + f\zeta, & g\xi + f\eta + c\zeta \\ \alpha\quad, & \beta\quad, & \gamma \end{matrix}\right|,$$

say these determinants are $\Omega\xi$, $\Omega\eta$, $\Omega\zeta$, where Ω is a value as yet undetermined. The equations are $\gamma(h\xi + b\eta + f\zeta) - \beta(g\xi + f\eta + c\zeta) - \Omega\xi = 0$, &c., viz. these are

$$\begin{aligned} &(\gamma h - \beta g - \Omega)\,\xi + (\gamma b - \beta f \quad)\,\eta + (\gamma f - \beta c \quad)\,\zeta = 0 \\ &(\alpha g - \gamma a \quad)\,\xi + (\alpha f - \gamma h - \Omega)\,\eta + (\alpha c - \gamma g \quad)\,\zeta = 0 \\ &(\beta a - \alpha h \quad)\,\xi + (\beta h - \alpha b \quad)\,\eta + (\beta g - \alpha f - \Omega)\,\zeta = 0\,; \end{aligned}$$

eliminating (ξ, η, ζ), we have an equation which may be written

$$\left|\begin{matrix} A - \Omega, & B\quad, & C \\ A'\quad, & B' - \Omega, & C' \\ A''\quad, & B''\quad, & C'' - \Omega \end{matrix}\right| = 0,$$

that is,

$$\left|\begin{matrix} A\,, & B\,, & C \\ A', & B', & C' \\ A'', & B'', & C'' \end{matrix}\right| - \Omega\,(B'C'' - B''C' + C''A - CA'' + AB' - A'B) + \Omega^2(A + B' + C'') - \Omega^3 = 0.$$

We find very easily that the determinant and $A + B' + C''$ are each $= 0$; and the equation thus reduces itself to

$$\Omega^2 = B'C'' - B''C' + C''A - CA'' + AB' - A'B,$$

or substituting for A, B, &c., their values,

$$\Omega^2 = -(bc - f^2, \ldots \between \alpha, \beta, \gamma)^2;$$

this being so, the ratios of ξ, η, ζ are determined by means of any two of the foregoing linear equations.

93. We may now verify the theorem; in the general expression for $d\omega$ writing for λ, μ, ν the values ξ, η, ζ, the equation to be verified becomes

$$\frac{\begin{vmatrix} dx, & dy, & dz \\ x, & y, & z \\ \xi, & \eta, & \zeta \end{vmatrix}}{(\alpha x+\beta y+\gamma z)\,(a, \dots \between x, y, z \between \xi, \eta, \zeta)} = \frac{1}{\Omega}\left\{\frac{(a, \dots \between \xi, \eta, \zeta \between dx, dy, dz)}{(a, \dots \between \xi, \eta, \zeta \between x, y, z)} - \frac{\alpha dx+\beta dy+\gamma dz}{\alpha x+\beta y+\gamma z}\right\},$$

viz. this is

$$\Omega \begin{vmatrix} dx, & dy, & dz \\ x, & y, & z \\ \xi, & \eta, & \zeta \end{vmatrix} \begin{aligned} &= (a, \dots \between \xi, \eta, \zeta \between dx, dy, dz)\,.\,(\alpha x+\beta y+\gamma z) \\ &\quad - (a, \dots \between \xi, \eta, \zeta \between x, y, z)\,.\,(\alpha dx+\beta dy+\gamma dz). \end{aligned}$$

Here, on the right-hand side, the coefficient of dx is

$$\begin{aligned} &(a\xi+h\eta+g\zeta)(\alpha x+\beta y+\gamma z) \\ &-\alpha\{(a\xi+h\eta+g\zeta)x+(h\xi+b\eta+f\zeta)y+(g\xi+f\eta+c\zeta)z\}, \end{aligned}$$

which is

$$\begin{aligned} &= y\{\beta(a\xi+h\eta+g\zeta)-\alpha(h\xi+b\eta+f\zeta)\} \\ &\quad - z\{\alpha(g\xi+f\eta+c\zeta)-\gamma(a\xi+h\eta+g\zeta)\}, \\ &= y\,.\,\Omega\zeta - z\,.\,\Omega\eta, \\ &= \Omega(y\zeta-z\eta), \end{aligned}$$

which is right; and similarly, the coefficients of dy and dz have the same values on the two sides of the equation respectively.

Aronhold's Quadric Integral deduced from the Affected Theorem. Art. Nos. 94 to 98.

94. Let the fixed curve be a conic, say $f=\tfrac{1}{2}(a, b, c, f, g, h \between x, y, z)^2, =0$: and let the variable curve be a line meeting the conic in the points 3 and 4. The affected theorem is

$$\Sigma\frac{12d\omega}{012} = -\frac{\delta 134}{134}+\frac{\delta 234}{234},$$

where (x_1, y_1, z_1) and (x_2, y_2, z_2) being the coordinates of the points 1 and 2 respectively, 12 denotes the constant $(a, \dots \between x_1, y_1, z_1 \between x_2, y_2, z_2)$: and 012, &c., denote determinants as usual.

The left-hand side is here

$$12\left\{\frac{d\omega_3}{312}+\frac{d\omega_4}{412}\right\}:$$

on the right-hand side, δ refers to the variation of the constants of ϕ, that is, to the variations of the points 3 and 4; or we may write $\delta = d_3 + d_4$; the points 3, 4 are independent, and the equation, being satisfied at all, must be satisfied separately in regard to the variations of 3, and in regard to the variations of 4: we must therefore have

$$12\frac{d\omega_3}{312} = -\frac{d_3 134}{134} + \frac{d_3 234}{234},$$

and the like equation obtained herefrom by the interchange of the numbers 3 and 4.

95. The equation just written down relates to any four points 1, 2, 3, 4 of the conic; and if for 3, 4 we write 0, 3 respectively, it becomes

$$12\frac{d\omega}{012} = -\frac{d.031}{031} + \frac{d.023}{023},$$

which relates to the points 0, 1, 2, 3 of the conic: writing, as before, 023, 031, 012 $= \rho, \sigma, \tau$, this equation is

$$12\frac{d\omega}{\tau} = -\frac{d\sigma}{\sigma} + \frac{d\rho}{\rho},$$

which may be verified as follows: the equation of the conic is $f = 23.\sigma\tau + 31.\tau\rho + 12.\rho\sigma, = 0$: we have $d\omega = \dfrac{\rho\, d\sigma - \sigma\, d\rho}{\dfrac{df}{d\tau}}$, where $\dfrac{df}{d\tau} = 23.\sigma + 31\rho, = -\dfrac{12.\rho\sigma}{\tau}$, that is, $d\omega = \dfrac{12}{\tau}\left(-\dfrac{d\sigma}{\sigma} + \dfrac{d\rho}{\rho}\right)$, the equation in question.

96. We have, as a property of any four points 0, 1, 2, 3 of a conic,

$$\frac{23}{123.023} = \frac{-01}{012.031}, \text{ or say } \frac{23}{\Delta.\rho} = \frac{-01}{\sigma\tau}, \text{ that is, } \frac{23}{\Delta}\frac{\sigma}{\rho} = -\frac{01}{\tau};$$

hence considering 0 as a variable point, and differentiating the logarithms,

$$-d\log\frac{01}{\tau} = -\frac{d\sigma}{\sigma} + \frac{d\rho}{\rho},$$

and the foregoing equation $12\dfrac{d\omega}{\tau} = -\dfrac{d\sigma}{\sigma} + \dfrac{d\rho}{\rho}$ thus becomes $12\dfrac{d\omega}{\tau} = -d\log\dfrac{01}{\tau}$, or restoring for τ its value 012,

$$12\frac{d\omega}{012} = -d\log\frac{01}{012}.$$

Taking now $\alpha x + \beta y + \gamma z = 0$ for the equation of the line 012; this meets the conic in the points 1, 2, coordinates (x_1, y_1, z_1) and (x_2, y_2, z_2) respectively: and we have

$$\alpha, \beta, \gamma = y_1 z_2 - y_2 z_1,\ z_1 x_2 - z_2 x_1,\ x_2 y_1 - x_1 y_2,$$
$$12 = (a, \ldots \between x_1, y_1, z_1 \between x_2, y_2, z_2),$$

and from this last value

$$\overline{12}^2 = \{(a, \ldots \between x_1, y_1, z_1 \between x_2, y_2, z_2)\}^2 - (a, \ldots \between x_1, y_1, z_1)^2 . (a, \ldots \between x_2, y_2, z_2)^2$$

(the second term being of course $= 0$), viz. this is

$$\overline{12}^2 = -(bc - f^2, \ldots \between y_1 z_2 - y_2 z_1,\ z_1 x_2 - z_2 x_1,\ x_1 y_2 - x_2 y_1)^2$$
$$= -(bc - f^2, \ldots \between \alpha, \beta, \gamma)^2,$$

or say $12 = -\Omega$, if $\Omega^2 = -(bc - f^2, \ldots \between \alpha, \beta, \gamma)^2$ as before: and the equation thus is

$$\frac{\Omega d\omega}{\alpha x + \beta y + \gamma z} = d \log \frac{(a, \ldots \between x_1, y_1, z_1 \between x, y, z)}{\alpha x + \beta y + \gamma z},$$

or finally, writing (ξ, η, ζ) instead of (x_1, y_1, z_1) to denote the coordinates of one or other of the intersections of the line $\alpha x + \beta y + \gamma z = 0$ with the conic, the equation becomes

$$\frac{\Omega d\omega}{\alpha x + \beta y + \gamma z} = d \log \frac{(a, \ldots \between \xi, \eta, \zeta \between x, y, z)}{\alpha x + \beta y + \gamma z},$$

which is Aronhold's quadric integral.

97. (The foregoing property, which may also be written

$$\frac{23}{023 \,.\, 123} = \frac{01}{201 \,.\, 301},$$

is verified very simply in the case of four points 0, 1, 2, 3 of a circle: in fact

$$23 = x_2 x_3 + y_2 y_3 - 1, \; = \cos 23 - 1, \; = -2 \sin^2 \tfrac{1}{2} 23,$$
$$023 = 2 \sin \tfrac{1}{2} 23 \sin \tfrac{1}{2} 30 \sin \tfrac{1}{2} 02;$$

and so for the other like expressions; each side of the equation is thus reduced to $1 \div \sin \tfrac{1}{2} 02 \sin \tfrac{1}{2} 03 \sin \tfrac{1}{2} 12 \sin \tfrac{1}{2} 13$.)

98. In particular, if the conic is taken to be the circle $x^2 + y^2 - 1 = 0$, then for the coordinates $\left(\frac{\xi}{\zeta}, \frac{\eta}{\zeta}\right)$ of the intersections with the line $\alpha x + \beta y + \gamma = 0$, we have

$$\begin{aligned} \Omega\xi + \gamma\eta + \beta\zeta &= 0, \\ \gamma\xi - \Omega\eta + \alpha\zeta &= 0, \\ \beta\xi - \alpha\eta + \Omega\zeta &= 0, \end{aligned}$$

giving $\Omega^2 = \alpha^2 + \beta^2 + \gamma^2$; and then

$$\begin{aligned} \xi : \eta : \zeta &= -\beta^2 + \gamma^2 : \alpha\beta - \gamma\Omega : \alpha\gamma + \beta\Omega \\ &= -\alpha\beta - \gamma\Omega : \alpha^2 - \gamma^2 : \beta\gamma + \alpha\Omega \\ &= \alpha\gamma + \beta\Omega : \beta\gamma - \alpha\Omega : -\alpha^2 - \beta^2. \end{aligned}$$

The formula then becomes

$$\int \frac{dx}{(\alpha x + \beta\sqrt{1 - x^2} + \gamma)\sqrt{1 - x^2}} = \frac{1}{\sqrt{\alpha^2 + \beta^2 - \gamma^2}} \log \frac{\xi x + \eta\sqrt{1 - x^2} - \zeta}{\alpha x + \beta\sqrt{1 - x^2} + \gamma}:$$

or, retaining Ω, y for the values $\sqrt{\alpha^2 + \beta^2 - \gamma^2}$, and $\sqrt{1 - x^2}$, this may also be written

$$= \frac{1}{\Omega} \log \frac{\gamma(\alpha x + \beta y + \gamma) + \Omega(\beta x - \alpha y + \Omega)}{\alpha x + \beta y + \gamma}.$$

The form of the integral is still such that the value is not very readily obtainable by ordinary methods: the value just written down can of course be verified, but the verification is scarcely easier than for the original more general form.

In the very particular case $\alpha=0$, $\beta=0$, $\gamma=1$, we have $\Omega=i$; $\xi:\eta:\zeta=1:-i:0$, and the formula becomes

$$\int\frac{dx}{\sqrt{1-x^2}}=\frac{1}{i}\log(x-iy):$$

viz. this is

$$\sin^{-1}x=\tfrac{1}{2}\pi+\frac{1}{i}\log(x-i\sqrt{1-x^2}),$$

which is right; for putting $\sin^{-1}x=u$, and therefore $x=\sin u$, the equation becomes $i(u-\tfrac{1}{2}\pi)=\log(\sin u-i\cos u)$: that is, $\cos(u-\tfrac{1}{2}\pi)+i\sin(u-\tfrac{1}{2}\pi)=\sin u-i\cos u$.

Fixed Curve a Cubic: the Parametric Points 1, 2 *consecutive points on the Curve.*
Art. Nos. 99 to 106.

99. The major function $(x, y, z)^1_{12}$ is taken to be $=\dfrac{1^22\,.\,023-12^2\,.\,013}{123}$, so that, calling the differential $Qd\omega$, we have

$$Q=\frac{1^22\,.\,023-12^2\,.\,013}{123\,.\,012};$$

it is required to find what this becomes when 1, 2 are consecutive points on the curve, or what is the same thing when the line 012 is a tangent at the point 1.

I take for convenience the cubic to be $f,=\tfrac{1}{3}(x^3+y^3+z^3),=0$. The coordinates of 1 are (x_1, y_1, z_1), those of 2 are $(x_1+\delta x_1, y_1+\delta y_1, z_1+\delta z_1)$, or as for shortness I write them $(x_1+\alpha, y_1+\beta, z_1+\gamma)$, where α, β, γ are considered as infinitesimals of the first order: this being so, the denominator of Q is at once seen to be of the second order; it will appear that the numerator is of the third order; whence Q is of the first order.

100. We have

$$d\omega=\frac{y\,dz-z\,dy}{x^2},\ =\frac{z\,dx-x\,dz}{y^2}=\frac{x\,dy-y\,dx}{z^2},$$

and in analogy herewith we may write

$$\delta\omega_1=\frac{y_1\gamma-z_1\beta}{x_1^2},\ =\frac{z_1\alpha-x_1\gamma}{y_1^2},\ =\frac{x_1\beta-y_1\alpha}{z_1^2};$$

this being so, we have

$$012=\begin{vmatrix} x, & y, & z \\ x_1 & y_1 & z_1 \\ \alpha & \beta & \gamma \end{vmatrix}=(xx_1^2+yy_1^2+zz_1^2)\,\delta\omega_1=01^2\,.\,\delta\omega_1,$$

and similarly $312=31^2\,.\,\delta\omega_1$.

Moreover

$$023=\begin{vmatrix} x, & y, & z \\ x_1 & y_1 & z_1 \\ x_3 & y_3 & z_3 \end{vmatrix}+\begin{vmatrix} x\,, & y\,, & z \\ \alpha\,, & \beta\,, & \gamma \\ x_3, & y_3, & z_3 \end{vmatrix}=013+0\delta13,$$

as the second term may be written; moreover

$$1^2 2 = x_1^2(x_1+\alpha) + y_1^2(y_1+\beta) + z_1^2(z_1+\gamma), = \alpha x_1^2 + \beta y_1^2 + \gamma z_1^2,$$

$$12^2 = x_1(x_1+\alpha)^2 + y_1(y_1+\beta)^2 + z_1(z_1+\gamma)^2, = 2(\alpha x_1^2+\beta y_1^2+\gamma z_1^2) + \alpha^2 x_1 + \beta^2 y_1 + \gamma^2 z_1,$$

and hence

$$\begin{aligned} 1^2 2 \,.\, 023 - 12^2 \,.\, 013 = &\quad (\alpha x_1^2+\beta y_1^2+\gamma z_1^2)\quad (013+0\delta 13) \\ &- [2(\alpha x_1^2+\beta y_1^2+\gamma z_1^2) + (\alpha^2 x_1+\beta^2 y_1+\gamma^2 z_1)]\,013 \\ = &- \quad [(\alpha x_1^2+\beta y_1^2+\gamma z_1^2) + (\alpha^2 x_1+\beta^2 y_1+\gamma^2 z_1)]\,013 \\ &+ \quad (\alpha x_1^2+\beta y_1^2+\gamma z_1^2)\,.\,0\delta 13, \end{aligned}$$

or reducing by

$$3(\alpha x_1^2+\beta y_1^2+\gamma z_1^2) + 3(\alpha^2 x+\beta^2 y+\gamma^2 z) + (\alpha^3+\beta^3+\gamma^3) = 0,$$

this is

$$= +\tfrac{1}{3}(\alpha^3+\beta^3+\gamma^3)\,013 - (\alpha^2 x_1+\beta^2 y_1+\gamma^2 z_1)\,0\delta 13,$$

which is of the third order.

101. We may show that each of the terms contains the factor $(\delta\omega_1)^2$: we have, in fact,

$$y_1 z_1 (\delta\omega_1)^2 = \alpha\beta\frac{x_1}{y_1} - \alpha^2 - \beta\gamma\frac{x_1^2}{y_1 z_1} + \gamma\alpha\frac{x_1}{z_1},$$

$$z_1 x_1 (\delta\omega_1)^2 = \beta\gamma\frac{y_1}{z_1} - \beta^2 - \gamma\alpha\frac{y_1^2}{z_1 x_1} + \alpha\beta\frac{y_1}{x_1},$$

$$x_1 y_1 (\delta\omega_1)^2 = \gamma\alpha\frac{z_1}{x_1} - \gamma^2 - \alpha\beta\frac{z_1^2}{x_1 y_1} + \beta\gamma\frac{z_1}{y_1};$$

hence, first multiplying by α, β, γ and adding, we have

$$\begin{aligned}(\alpha y_1 z_1 + \beta z_1 x_1 + \gamma x_1 y_1)(\delta\omega_1)^2 = &\frac{\beta\gamma}{y_1 z_1}(\beta y_1^2+\gamma z_1^2) + \frac{\gamma\alpha}{z_1 x_1}(\gamma z_1^2+\alpha x_1^2) + \frac{\alpha\beta}{x_1 y_1}(\alpha x_1^2+\beta y_1^2) \\ &-\alpha\beta\gamma\left(\frac{x_1^2}{y_1 z_1} + \frac{y_1^2}{z_1 x_1} + \frac{z_1^2}{x_1 y_1}\right) \\ &-(\alpha^3+\beta^3+\gamma^3).\end{aligned}$$

But in virtue of $\alpha x_1^2+\beta y_1^2+\gamma z_1^2=0$, the first line becomes = the second line, or the two together are

$$= -2\alpha\beta\gamma\left(\frac{x_1^2}{y_1 z_1} + \frac{y_1^2}{z_1 x_1} + \frac{z_1^2}{x_1 y_1}\right),$$

which is $=0$ in virtue of $x_1^3+y_1^3+z_1^3=0$; hence the equation is

$$(\alpha y_1 z_1 + \beta z_1 x_1 + \gamma x_1 y_1)(\delta\omega_1)^2 = -(\alpha^3+\beta^3+\gamma^3),$$

the required expression for the first term.

102. Again, multiplying by x_1, y_1, z_1, and adding, we have

$$3x_1 y_1 z_1(\delta\omega_1)^2 = \frac{\beta\gamma}{y_1 z_1}(y_1^3+z_1^3-x_1^3) + \frac{\gamma\alpha}{z_1 x_1}(z_1^3+x_1^3-y_1^3) + \frac{\alpha\beta}{x_1 y_1}(x_1^3+y_1^3-z_1^3) - (\alpha^2 x_1+\beta^2 y_1+\gamma^2 z_1),$$

where, in virtue of $x_1^3+y_1^3+z_1^3=0$, the first line is

$$= -\frac{2}{x_1y_1z_1}(\beta\gamma x_1^4+\gamma\alpha y_1^4+\alpha\beta z_1^4),$$

and this again is $=-2(\alpha^2x_1+\beta^2y_1+\gamma^2z_1)$: in fact, we have identically

$$\begin{aligned} x_1y_1z_1(\alpha^2x_1+\beta^2y_1+\gamma^2z_1) &= (\alpha x_1^2+\beta y_1^2+\gamma z_1^2)(\alpha y_1z_1+\beta z_1x_1+\gamma x_1y_1) \\ &\quad -(\beta\gamma x_1+\gamma\alpha y_1+\alpha\beta z_1)(x_1^3+y_1^3+z_1^3) \\ &\quad +(\beta\gamma x_1^4+\gamma\alpha y_1^4+\alpha\beta z_1^4), \end{aligned}$$

which, in virtue of $\alpha x_1^2+\beta y_1^2+\gamma z_1^2=0$, and $x_1^3+y_1^3+z_1^3=0$, becomes

$$x_1y_1z_1(\alpha^2x_1+\beta^2y_1+\gamma^2z_1)=(\beta\gamma x_1^4+\gamma\alpha y_1^4+\alpha\beta z_1^4).$$

Hence the equation is

$$3x_1y_1z_1(\delta\omega_1)^2=-3(\alpha^2x_1+\beta^2y_1+\gamma^2z_1),$$

or finally

$$x_1y_1z_1(\delta\omega_1)^2=-\quad(\alpha^2x_1+\beta^2y_1+\gamma^2z_1),$$

the required expression for the second term.

103. Writing for shortness

$$\alpha y_1z_1+\beta z_1x_1+\gamma x_1y_1=\delta(x_1y_1z_1),$$

we have

$$1^22.023-12^2.013=\{-\tfrac{1}{3}\delta(x_1y_1z_1)013+x_1y_1z_1.0\delta13\}(\delta\omega_1)^2,$$

and hence dividing by

$$012.123,\quad =01^2.31^2.(\delta\omega_1)^2,$$

we have

$$Q=\frac{1^22.023-12^2.013}{012.123}=\frac{-\tfrac{1}{3}\delta(x_1,\ y_1,\ z_1).013+x_1y_1z_1.0\delta13}{01^2.31^2}.$$

But this can be further reduced: the numerator, multiplied by 3, is

$$=-(\alpha y_1z_1+\beta z_1x_1+\gamma x_1y_1)\begin{vmatrix} x, & y, & z \\ x_1, & y_1, & z_1 \\ x_3, & y_3, & z_3 \end{vmatrix}+3x_1y_1z_1\begin{vmatrix} x, & y, & z \\ \alpha, & \beta, & \gamma \\ x_3, & y_3, & z_3 \end{vmatrix},$$

which is

$$=\begin{vmatrix} x & , & y & , & z \\ x_1(y_1^3-z_1^3), & & y_1(z_1^3-x_1^3), & & z_1(x_1^3-y_1^3) \\ x_3 & , & y_3 & , & z_3 \end{vmatrix}\delta\omega_1,$$

where $x_1(y_1^3-z_1^3)$, $y_1(z_1^3-x_1^3)$, $z_1(x_1^3-y_1^3)$ are the coordinates of the tangential of the point 1 in regard to the cubic, viz. the point of intersection of the tangent at 1 with the cubic. The determinant may for shortness be called $0t13$; and we thus have

$$Q=\frac{1^22.023-12^2.013}{012.123}=\frac{\tfrac{1}{3}\delta\omega_1}{31^2}\frac{0t13}{01^2},$$

where observe that $01^2=0$ is the equation of the tangent at the point 0: and $0t13=0$ is the equation of the line joining the tangential of 1 with the arbitrary point 3.

104. The identity just referred to is proved very easily. Comparing on each side the coefficient of $yz_3 - y_3 z$, the factor x_1 divides out and we ought to have

$$-(\alpha y_1 z_1 + \beta z_1 x_1 + \gamma x_1 y_1) + 3 y_1 z_1 \alpha = (y_1^3 - z_1^3)\, \delta\omega_1,$$

that is,

$$(y_1^3 - z_1^3)\, \delta\omega_1 = 2\alpha y_1 z_1 - \beta z_1 x_1 - \gamma x_1 y_1 :$$

and, in fact, from $y_1^2 \delta\omega_1 = z_1 \alpha - x_1 \gamma$, $z_1^2 \delta\omega_1 = x_1 \beta - y_1 \alpha$, we have

$$(y_1^3 - z_1^3)\, \delta\omega_1 = y_1 (z_1 \alpha - x_1 \gamma) - z_1 (x_1 \beta - y_1 \alpha),$$

which is the value in question. Similarly the coefficients of $zx_3 - z_3 x$, $xy_3 - x_3 y$ are equal on the two sides; and the equation is thus verified.

105. The proof has been given in regard to the particular cubic $x^3 + y^3 + z^3 = 0$; but it might have been given for the canonical form $x^3 + y^3 + z^3 + 6lxyz = 0$: and from the invariantive form it is clear that the result in fact applies to any cubic whatever. The result is an important one: we see by it that when the points 1 and 2 are consecutive points on the curve we must, in place of the differential $Qd\omega$, which is evanescent, consider a new form $\dfrac{0t13}{01^2}\, \delta\omega_1$, where, as already remarked, the denominator represents the tangent at the point 1, and the numerator the line joining the tangential of this point with the point 3.

106. We have

$$\{023\} + \{031\} + \{012\} = \{123\},$$

or writing this in the form

$$\{012\} - \{312\} + \{023\} - \{013\} = 0,$$

suppose 2 is here the consecutive point $1 + \delta 1$; then

$$\{012\} - \{312\}, = \frac{1^2 2 \,.\, 023 - 12^2 \,.\, 013}{012 \,.\, 312}$$

becomes $= \{0t13\}\, \delta\omega_1$: we have also

$$\{023\} = \{013\} + \partial_1 \{013\}\, \delta\omega_1,$$

and the result is

$$-\{0t13\} + \partial_1\, 013 = 0,$$

that is, $\partial_1 \{013\} = \{0t13\}$. The form in the case of the cubic $x^3 + y^3 + z^3 + 6lxyz = 0$, is =

$$\frac{-\begin{vmatrix} x\ , & y\ , & z \\ x_1 (y_1^3 - z_1^3), & y_1 (z_1^3 - x_1^3), & z_1 (x_1^3 - y_1^3) \\ x_3\ , & y_3\ , & z_3 \end{vmatrix}}{3 (x_1^2 x + y_1^2 y + z_1^2 z)(x_1^2 x_3 + y_1^2 y_3 + z_1^2 z_3)},$$

i.e. the differential coefficient of $\{013\}$ in regard to the parametric point 1 is $= \{0t13\}$, the symbol for the case where the parametric line is the tangent at 1.

Fixed Curve a Cubic: the Parametric Points corresponding points. Art. Nos. 107 to 110.

107. The parametric points 1, 2 are taken to be corresponding points, that is, such that the tangents at these points meet at a point, say 3, on the cubic. We may from 3 draw two other tangents, touching the cubic, say at the points 1′ and 2′. The four points 1, 2, 1′, 2′ are then such that the lines 12, 1′2′ meet in a point, say 4, of the cubic; and moreover 3, 4 are corresponding points.

We may take $(x, y, z), =(1, 0, 0), (0, 1, 0), (0, 0, 1)$ for the coordinates of the points 1, 2, 3 respectively: $x=0$, $y=0$ are thus the equations of the lines 32, 31 respectively, and $z=0$ is the equation of the line 12, viz. we have $z=012$. Taking $x-M_1y=0$, $x-M_2y=0$, for the equations of the tangents 31′, 32′ respectively, and $\zeta=0$ for the equation of the line 1′2′ joining their points of contact, where ζ is a properly determined linear function of (x, y, z), it is to be shown that the differential $Qd\omega$ may be taken to be $\dfrac{\zeta d\omega}{z}$, and that this is $=\tfrac{1}{2}\left(\dfrac{dx}{x}-\dfrac{dy}{y}\right)$: the affected theorem thus assumes a special form, which will be noticed.

108. The cubic passes through the points $(x=0,\ z=0)$ and $(y=0,\ z=0)$, the tangents at these points being $x=0$, and $y=0$ respectively: also through the point $x=0,\ y=0$: its equation thus is

$$f, = gz^2x + 2lzxy + iz^2y + hx^2y + kxy^2, =0,$$

and writing

$$d\omega = \frac{x\,dy - y\,dx}{\dfrac{df}{dz}},$$

we have

$$\frac{df}{dz} = 2\,(gzx + lxy + izy),$$

which, from the equation of the curve written in the form

$$z\,(gzx + lxy + izy) + xy\,(hx + ky + lz) = 0,$$

or say

$$z\,(gzx + lxy + izy) + xy\zeta = 0,$$

becomes

$$= \frac{-2xy\zeta}{z};$$

and we thus have

$$d\omega = \frac{-z}{2xy\zeta}(x\,dy - y\,dx), \ = \tfrac{1}{2}\frac{z}{\zeta}\left(\frac{dx}{x} - \frac{dy}{y}\right),$$

where $\zeta = hx + ky + lz$. To find the meaning of ζ, observe that the line $x - My = 0$ meets the curve in the point $(x=0,\ y=0)$, and in two other points determined by the equation

$$z^2\,(gm + i) + 2zylM + y^2\,(hM^2 + kM) = 0;$$

this line will be a tangent if

$$(gM + i)\,(hM + k) - l^2M = 0,$$

and we then have at the point of contact $(hM+k)y+lz=0$; and writing this in the form $hx+ky+lz=0$, we see that the equation $\zeta=0$ is satisfied at the point of contact of each of the two tangents $x-M_1y=0$, $x-M_2y=0$; viz. $\zeta=0$ is the equation of the line joining the two points of contact. Moreover, from the equation of the curve written in the foregoing form

$$z(gzx+lxy+izy)+xy\zeta=0,$$

it appears that the lines $z=0$, $\zeta=0$, meet on the curve; or, what is the same thing, that the line $\zeta=0$ passes through the residue of the parametric points 1, 2.

109. The function ζ at 1 becomes $=h$, and this is the value of $3 \,.\, 1^2 2$; in fact,

$$\begin{aligned} 3 \,.\, 1^2 2 &= \left(x_2 \frac{d}{dx_1} + y_2 \frac{d}{dy_1} + z_2 \frac{d}{dz_1}\right) f_1, \quad (x_2,\ y_2,\ z_2)=(0,\ 1,\ 0), \\ &= \frac{df_1}{dy_1}, \\ &= 2lz_1x_1 + iz_1^2 + hx_1^2 + 2kx_1y_1, \quad (x_1,\ y_1,\ z_1)=(1,\ 0,\ 0), \\ &= h. \end{aligned}$$

We have thus ζ, satisfying the required conditions for the major function: and the differential $Qd\omega$ may therefore be taken to be $=\frac{\zeta}{z}d\omega$, that is, we have

$$Qd\omega = \tfrac{1}{2}\left(\frac{dx}{x}-\frac{dy}{y}\right).$$

The affected theorem thus becomes

$$\Sigma \tfrac{1}{2}\left(\frac{dx}{x}-\frac{dy}{y}\right) = -\frac{\delta\phi_1}{\phi_1}+\frac{\delta\phi_2}{\phi_2}.$$

110. The meaning of this will be better understood from the integral form. Integrating each side, and assuming that the superior limits are given by a line ϕ which cuts the cubic in the points 4, 5, 6, and the inferior limits by a line ψ which cuts the cubic in the points 7, 8, 9, we find

$$\log \frac{x_4x_5x_6}{x_7x_8x_9} - \log \frac{y_4y_5y_6}{y_7y_8y_9} = 2 \log \frac{\phi_2}{\psi_2}\frac{\psi_1}{\phi_1},$$

that is,

$$\frac{x_4x_5x_6\, y_7y_8y_9}{x_7x_8x_9\, y_4y_5y_6} = \left(\frac{\phi_2\psi_1}{\psi_2\phi_1}\right)^2,$$

where ϕ_1, ψ_1, ϕ_2, ψ_2 denote the values of the linear functions ϕ, ψ at the points 1 and 2 respectively. We have a cubic cut by the lines ϕ, ψ, x, y in the points 4, 5, 6; 7, 8, 9; 2, 2′, 3 and 1, 1′, 3 respectively: where for the moment 1′, 2′ are written to denote the points on the curve consecutive to 1 and 2 respectively. Hence, by a known theorem in transversals,

$$\left(\frac{x}{y}\right)_{456} \div \left(\frac{x}{y}\right)_{789} = \left(\frac{\phi}{\psi}\right)_{22'3} \div \left(\frac{\phi}{\psi}\right)_{11'3},$$

that is,

$$\frac{x_4x_5x_6\,y_7y_8y_9}{x_7x_8x_9\,y_4y_5y_6}=\frac{\psi_1\psi_{1'}\psi_3\,.\,\phi_2\phi_{2'}\phi_3}{\psi_2\psi_{2'}\psi_3\,.\,\phi_1\phi_{1'}\phi_3},$$

which, dividing out the $\phi_3\psi_3$, and writing 1, 2 in place of 1′, 2′, becomes

$$=\left(\frac{\phi_2\psi_1}{\phi_1\psi_2}\right)^2,$$

agreeing with the result just obtained.

Aronhold's Cubic Transformation. Art. Nos. 111 to 119.

111. This was obtained in the paper "Algebraische Reduction des Integrals $\int F(x, y)\,dx$, u. s. w.," *Berl. Monatsb.*, April, 1861, pp. 462—468. I give in the first place the analytical results, independently of the general theory, with the values for the canonical form $f, =\frac{1}{3}(x^3+y^3+z^3+6lxyz), =0$, of the cubic.

T	sextic invariant,	$=1-20l^3-8l^6$,
S	quartic „ (Aronhold's)	$=-4(l-l^4)$,
R	discriminant	$=(1+8l^3)^3$,
$P=3h\alpha^20$		$=\{-3l^2\alpha^2+(1+2l^3)\beta\gamma\}\,x+\&c.$,
$Q=f\alpha^20$		$=(\alpha^2+2l\beta\gamma)\,x+\&c.$,
$B=f\mathrm{a}^20$		$=(\mathrm{a}^2+2l\mathrm{bc})\,x+\&c.$,
$C=f\mathrm{a}0^2$		$=\mathrm{a}\,(x^2+2lyz)+\&c.$,
$D=f0^3$		$=x^3+y^3+z^3+6lxyz$,

where $\mathrm{a, b, c}=\alpha(\beta^3-\gamma^3),\ \beta(\gamma^3-\alpha^3),\ \gamma(\alpha^3-\beta^3)$.

Then we have

$$2TQ^4+6SPQ^3+8P^3Q=-R^{\frac{2}{3}}(6C^2-8BD),$$

viz. this equation, where each side is a quartic function $(x, y, z)^4$, is an identity when (α, β, γ) are connected by the equation, $f\alpha^3, =\alpha^3+\beta^3+\gamma^3+6l\alpha\beta\gamma, =0$; and further,

$$Q\,dP-P\,dQ=-R^{\frac{1}{3}}\{\mathrm{a}\,(y\,dz-z\,dy)+\mathrm{b}\,(z\,dx-x\,dz)+\mathrm{c}\,(x\,dy-y\,dx)\}.$$

Hence writing

$$\lambda=\frac{6h\alpha^20}{f\alpha^20},\ =\frac{2P}{Q},\ =\frac{2\,(\{-3l^2\alpha^2+(1+2l^3)\beta\gamma\}\,x+\&c.)}{(\alpha^2+2l\beta\gamma)\,x+\&c.},$$

we have

$$Q^4(\lambda^3-3S\lambda-2T)=R^{\frac{2}{3}}(6C^2-8BD),$$

and

$$Q^2\,d\lambda,\ =2\,(Q\,dP-P\,dQ)=-2R^{\frac{1}{3}}\{\mathrm{a}\,(y\,dz-z\,dy)+\&c.\}.$$

112. Supposing now that (x, y, z) are the coordinates of a point on the cubic, then $D=0$; and taking the square root of each side of the first equation, we may write

$$Q^2\sqrt{\lambda^3-3S\lambda-2T} = -R^{\frac{1}{3}}\sqrt{6}\,.\,C,$$
$$Q^2 d\lambda \qquad = -2R^{\frac{1}{3}}\{\mathrm{a}\,(y\,dz-z\,dy)+\&c.\}.$$

We have

$$d\omega = \frac{y\,dz-z\,dy}{x^2+2lyz} = \frac{z\,dx-x\,dz}{y^2+2lzx} = \frac{x\,dy-y\,dx}{z^2+2lxy},$$

whence

$$d\omega = \frac{\mathrm{a}\,(y\,dz-z\,dy)+\&c.}{C};$$

and we consequently have

$$\frac{d\lambda}{\sqrt{\lambda^3-3S\lambda-2T}} = \frac{2}{\sqrt{6}},$$

or, as this may also be written,

$$\frac{d\lambda}{\sqrt{4\lambda^3-12S\lambda-8T}} = \frac{1}{\sqrt{6}}\,d\omega,$$

which, if $12S$, $8T$ are put $=g_2$, g_3 respectively, takes the Weierstrassian form

$$\frac{d\lambda}{\sqrt{4\lambda^3-g_2\lambda-g_3}} = \frac{1}{\sqrt{6}}\,d\omega.$$

The conclusion is that for the cubic curve, taking λ a quotient of two linear functions of (x, y, z), the differential $d\omega$ is transformed into $d\lambda \div$ square root of a cubic function of λ: viz. we have thus a form of differential, not the same, but such as that which belongs to the ordinary theory of elliptic functions, and which has been adopted by Weierstrass as a canonical form.

113. The transformation depends on the arbitrary point (α, β, γ) of the cubic: the point (a, b, c) is the tangential of this point, viz. the point of intersection of the tangent at (α, β, γ) with the cubic: we can from (a, b, c) draw four tangents to the cubic, viz. the tangent at (α, β, γ) and three other tangents: the equations of the four tangents being $\frac{2P}{Q}$, $=\frac{6h\alpha^2 0}{f\alpha^2 0}$, $=\infty$, λ_1, λ_2, λ_3 respectively; where λ_1, λ_2, λ_3 are the roots of the equation $\lambda^3-3S\lambda-2T=0$.

Suppose for a moment that (α, β, γ) is a point not on the cubic curve, and write $A=\alpha^3+\beta^3+\gamma^3+6l\alpha\beta\gamma$. We have

$$A^2D^2+4AC^3+4B^3D-3B^2C^2-6ABCD=0,$$

for the equation of the six tangents which can be drawn from the point (α, β, γ) to the cubic: when (α, β, γ) is on the cubic, $A=0$, and the equation becomes $B^2(4BD-3C^2)=0$, where $B=0$ is the equation of the tangent at the point (α, β, γ): throwing out the factor B^2, we have $4BD-3C^2=0$ for the equation of the four tangents from (α, β, γ) to the curve; viz. the equation of the four tangents is

$$2TQ^4+6SPQ^3+8P^3Q=0,$$

or, as this may be written,

$$Q(2P-\lambda_1 Q)(2P-\lambda_2 Q)(2P-\lambda_3 Q)=0,$$

viz. the equations of the four tangents are as is mentioned above; it was, in fact, by these geometrical considerations that Aronhold obtained his results.

114. The foregoing expression for $Q\,dP-P\,dQ$, say

$$Q\,dP-P\,dQ=(1+8l^3)\{\mathrm{a}(y\,dz-z\,dy)+\mathrm{b}(z\,dx-x\,dz)+\mathrm{c}(x\,dy-y\,dx)\},$$

may be verified without difficulty. Writing for a moment

$$\begin{aligned}Q\,dP-P\,dQ=&\ (Ax+By+Cz)(Ldx+Mdy+Ndz)\\&-(Adx+Bdy+Cdz)(Lx+My+Nz)\\=&\ (BN-CM)(y\,dz-z\,dy)-\&c.;\end{aligned}$$

we have

$$\begin{aligned}BN-CM=&\ \{-3l^2\beta^2+(1+2l^3)\gamma\alpha\}(\gamma^2+2l\alpha\beta)\\&-\{-3l^2\gamma^2+(1+2l^3)\alpha\beta\}(\beta^2+2l\gamma\alpha)\\=&-6l^3\alpha\beta^3+(1+2l^3)\alpha\gamma^3+6l^3\alpha\gamma^3-(1+2l^3)\alpha\beta^3,\\=&-(1+8l^3)\alpha(\beta^3-\gamma^3)\\=&-(1+8l^3)\,\mathrm{a},\end{aligned}$$

which proves the theorem.

115. I content myself with a partial verification of the identity

$$2TQ^4+6SPQ^3-8P^3Q=-(1+8l^3)^2(6C^2-8BD).$$

Writing herein $x, y, z=1, -1, 0$, we have $D=0$, and the equation becomes

$$2TQ^4+6SPQ^3-8P^3Q+6(1+8l^3)^2C^2=0,$$

where now

$$\begin{aligned}&Q=(\alpha-\beta)(\alpha+\beta-2l\gamma),\ P=(\alpha-\beta)\{-3l^2(\alpha+\beta)-(1+2l^3)\gamma\},\\&C=\mathrm{a}+\mathrm{b}-2l\mathrm{c},\ =(\alpha-\beta)\{-\alpha\beta^2-\alpha^2\beta-\gamma^3-2l\gamma(\alpha^2+\alpha\beta+\beta^2)\},\end{aligned}$$

which, putting therein $-\gamma^3=\alpha^3+\beta^3+6l\alpha\beta\gamma$, becomes

$$=(\alpha-\beta)^2(\alpha+\beta-2l\gamma).$$

Hence writing

$$X=\alpha+\beta-2l\gamma,\quad Y=-3l^2(\alpha+\beta)-(1+2l^3)\gamma,$$

we have

$$Q,\ P,\ C=(\alpha-\beta)X,\quad(\alpha-\beta)Y,\quad(\alpha-\beta)^3X:$$

substituting these values, the factor $(\alpha-\beta)^4X$ divides out, and the equation becomes

$$2TX^3+6SX^2Y-8Y^3+6(1+8l^3)^2(\alpha-\beta)^2X=0.$$

To complete the verification, observe that we have $Y+3l^2X=-(1+8l^3)\gamma$, whence

$$-Y^3=(1+8l^3)^3\gamma^3+9(1+8l^3)^2l^2\gamma^2X+27(1+8l^3)l^4\gamma X^2+27l^6X^3,$$

and herein $-\gamma^3 = \alpha^3 + \beta^3 + 6l\alpha\beta\gamma$, whence

$$-\gamma^3 + \alpha\beta X = (\alpha+\beta)^3 = (X + 2l\gamma)^3 = X^3 + 6l\gamma X^2 + 12l^2\gamma^2 X + 8l^3\gamma^3,$$

that is,

$$-(1+8l^3)\gamma^3 = X^3 + 6l\gamma X^2 + (12l^2\gamma^2 - 3\alpha\beta) X.$$

Hence the equation to be verified becomes

$$2TX^3 + 6SX^2Y - 8\begin{Bmatrix} (1+8l^3)^2 [X^3 + 6l\gamma X^2 + (12l^2\gamma^2 - 3\alpha\beta) X] \\ -(1+8l^3)^2\, 9l^2\gamma^2 X \\ -(1+8l^3)\, 27l^4\gamma X^2 \\ -27l^6X^3 \end{Bmatrix}$$
$$+ 6(1+8l^3)^2(\alpha-\beta)^2 X = 0;$$

viz. throwing out the factor X, this is

$$\{2T - 8(1+8l^3)^2 + 216l^6\} X^2 + 6SXY - 48(1+8l^3)^2 l\gamma X + 216(1+8l^3)\, l^4\gamma X$$
$$-(1+8l^3)^2 \{96l^2\gamma^2 - 24\alpha\beta - 72l^2\gamma^2 - 6(\alpha-\beta)^2\} = 0,$$

where the last term is

$$= +6(1+8l^3)^2 \{(\alpha+\beta)^2 - 4l^2\gamma^2\},$$

viz. this is

$$= 6(1+8l^3)^2 (\alpha+\beta+2l\gamma) X,$$

and there is again the factor X which can be thrown out: the equation thus becomes

$$[2T - 8(1+8l^3)^2 + 216l^6] X + 6SY - 48(1+8l^3)^2 l\gamma + 216(1+8l^3)\, l^4\gamma$$
$$+ 6(1+8l^3)^2(\alpha+\beta+2l\gamma) = 0.$$

This may be written

$$[2T - 8(1+8l^3)^2 + 216l^6] X + 6S[-3l^2X - (1+8l^3)\gamma] - 48(1+8l^3)^2 l\gamma$$
$$+ 216(1+8l^3)\, l^4\gamma + 6(1+8l^3)^2(X + 4l\gamma) = 0,$$

or, finally, it is

$$[2T - 8(1+8l^3)^2 + 216l^6 - 18l^2S + 6(1+8l^3)^2] X$$
$$+ [-6l^{-1}S - 48(1+8l^3) + 216l^3 + 24(1+8l^3)](1+8l^3)\, l\gamma = 0;$$

substituting for T, S their values $1 - 20l^3 - 8l^6$ and $-4l + 4l^4$ respectively, the coefficients of X and $(1+8l^3)\, l\gamma$ are separately $= 0$, and the equation is thus verified.

116. The foregoing equation $\lambda = \dfrac{6ha^2 0}{fa^2 0}$, regarding therein λ as an arbitrary parameter and (x, y, z) as current coordinates, is the equation of an arbitrary line through the point (a, b, c) of the cubic: it meets the cubic in two other points depending, of course, on the value of λ; and the coordinates of either of these is thus expressible, irrationally, in terms of λ, the expressions involving the radical $\sqrt{\lambda^3 - 3S\lambda - 2T}$: from the values of x, y, z in terms of λ, we should be able to deduce the foregoing equation

$\frac{2}{\sqrt{6}}d\omega = \frac{d\lambda}{\sqrt{\lambda^3 - 3S\lambda - 2T}}$. The expressions assume a peculiarly simple form when (α, β, γ), instead of being an arbitrary point of the cubic, is a point of inflexion of the cubic; and it is easy to see *à priori* why this is so: in fact, if we assume

$$x : y : z = u + \alpha\sqrt{\Lambda} : v + \beta\sqrt{\Lambda} : w + \gamma\sqrt{\Lambda},$$

where u, v, w are linear functions and Λ a cubic function of λ; then the locus is a cubic curve, and corresponding to the value $\lambda = \infty$, we have $x : y : z = \alpha : \beta : \gamma$, viz. the curve passes through the point (α, β, γ): moreover, it can be shown that this point is an inflexion of the curve; expressions of the foregoing simple form thus only exist in the case where the point (α, β, γ) is an inflexion; and the formulæ referring to an arbitrary point (α, β, γ) of the curve are necessarily of a more complex form.

117. To work this out, we start from the foregoing equation

$$\lambda = \frac{6h\alpha^2 0}{f\alpha^2 0} = \frac{2\left(\{-3l^2\alpha^2 + (1 + 2l^3)\beta\gamma\} x + \&c.\right)}{(\alpha^2 + 2l\beta\gamma) x + \&c.}$$

which, putting therein

$$L = \lambda + 6l^2, \quad M = l\lambda - (1 + 2l^3),$$

and

$$A, B, C = L\alpha^2 + 2M\beta\gamma, \quad L\beta^2 + 2M\gamma\alpha, \quad L\gamma^2 + 2M\alpha\beta,$$

becomes $Ax + By + Cz = 0$, the equation of a line through the point

$$\text{a, b, c}, = \alpha(\beta^3 - \gamma^3), \quad \beta(\gamma^3 - \alpha^3), \quad \gamma(\alpha^3 - \beta^3),$$

as before: and we have to find the intersections of this line with the cubic

$$x^3 + y^3 + z^3 + 6lxyz = 0.$$

We have

$$C^3(x^3 + y^3) - (Ax + By)^3 - 6lC^2(Ax + By)xy = 0:$$

the cubic function contains as we know the factor $\text{b}x - \text{a}y$, and in the remaining quadric factor it is easy to calculate the coefficients of x^2 and y^2: we thus obtain the identity

$$\begin{aligned} &C^3(x^3 + y^3) - (Ax + By)^3 - 6lC^2(Ax + By)xy \\ &= (\text{b}x - \text{a}y)\{[(-\beta^2 - 6l\gamma\alpha)L^3 + 6\alpha\gamma L^2M - 8\beta^2M^3]\,x^2 \\ &\qquad + 2H \quad xy \\ &\qquad + [(-\alpha^2 - 6l\alpha\beta)L^3 + 6\beta\gamma L^2M - 8\alpha^2M^3]\,y^2\}, \end{aligned}$$

from which the as yet unknown coefficient $2H$ is to be obtained. This is most easily effected by assuming $x, y = \alpha, -\beta$; values which give

$$x^3 + y^3 = \alpha^3 - \beta^3, \quad Ax + By = L(\alpha^3 - \beta^3), \quad \text{b}x - \text{a}y = -\alpha\beta(\alpha^3 - \beta^3):$$

the whole equation becomes divisible by $\alpha^3 - \beta^3$ and, omitting this factor, we have

$$\begin{aligned} &C^3 - L^3(\alpha^3 - \beta^3)^2 + 6lC^2L\alpha\beta \\ &\qquad = \alpha\beta\,\{[2\alpha^2\beta^2 + 6l\gamma(\alpha^3 + \beta^3)]\,L^3 - 6\gamma(\alpha^3 + \beta^3)L^2M + 16\alpha^2\beta^2M^3\} + 2H\alpha^2\beta^2, \end{aligned}$$

where for C is to be substituted its value $L\gamma^2 + 2M\alpha\beta$. We may also reduce by $\alpha^3 + \beta^3 + \gamma^3 + 6l\alpha\beta\gamma = 0$. The left-hand side is

$$C^3 - L^3(\alpha^3 + \beta^3)^2 - 4\alpha^3\beta^3L^3 + 6lC^2L\alpha\beta,$$

which after reduction is found to contain the factor $\alpha\beta$; and omitting this factor and reducing also the right-hand side, the whole equation becomes

$$\begin{aligned}
&L^3(-6l\gamma^4 - 36l^2\gamma^2\alpha\beta + 4\alpha^2\beta^2) && = L^3(-6l\gamma^4 - 36l^2\gamma^2\alpha\beta + 2\alpha^2\beta^2)\\
&+ L^2M(6\gamma^4 + 24l\gamma^2\alpha\beta) && \quad + L^2M(6\gamma^4 + 36l\gamma^2\alpha\beta)\\
&+ LM^2(12\gamma^2\alpha\beta + 24l\alpha^2\beta^2) && \\
&+ M^3(8\alpha^2\beta^2) && \quad + M^3(16\alpha^2\beta^2)\\
& && \quad + 2H\alpha\beta;
\end{aligned}$$

omitting here the terms which destroy each other, the equation again divides by $\alpha\beta$ and we thus obtain the value of H; and the required identity is

$$\begin{aligned}
&C^3(x^3 + y^3) - (Ax + By)^3 - 6lC^2(Ax + By)xy\\
&\quad = (\mathrm{b}x - \mathrm{a}y)\{[(-\beta^2 - 6l\gamma\alpha)L^3 + 6\alpha\gamma L^2M - 8\beta^2M^3]x^2\\
&\qquad\qquad + [\alpha\beta L^3 - 6l\gamma^2L^2M + (6\gamma^2 + 12l\alpha\beta)LM^2 - 4\alpha\beta M^3]\,2xy\\
&\qquad\qquad + [(-\alpha^2 - 6l\beta\gamma)L^3 + 6\beta\gamma L^2M - 8\alpha^2M^3]y^2\}.
\end{aligned}$$

Hence putting for shortness

$$\begin{aligned}
\mathfrak{A} &= (\alpha^2 + 6l\beta\gamma)L^3 - 6\beta\gamma L^2M && + 8\alpha^2M^3,\\
\mathfrak{B} &= (\beta^2 + 6l\gamma\alpha)L^3 - 6\alpha\gamma L^2M && + 8\beta^2M^3,\\
\mathfrak{H} &= \alpha\beta L^3 - 6l\gamma^2L^2M + (6\gamma^2 + 12l\alpha\beta)LM^2 - 4\alpha\beta M^3, &&
\end{aligned}$$

the equation in (x, y) is

$$\mathfrak{B}x^2 - 2\mathfrak{H}xy + \mathfrak{A}y^2 = 0,$$

giving

$$\mathfrak{B}x - \mathfrak{H}y = \sqrt{\mathfrak{H}^2 - \mathfrak{A}\mathfrak{B}}\,y,$$

or say

$$x : y = \mathfrak{H} + \sqrt{\mathfrak{H}^2 - \mathfrak{A}\mathfrak{B}} : \mathfrak{B}.$$

We find without difficulty, reducing always by $\alpha^3 + \beta^3 + \gamma^3 + 6l\alpha\beta\gamma = 0$,

$$\begin{aligned}
\tfrac{1}{6}(\mathfrak{H}^2 - \mathfrak{A}\mathfrak{B}) = \;& l\gamma^4\,L^6\\
&+ (-\gamma^4 + 4l\alpha\beta\gamma^2)\,L^5M\\
&+ (6l^2\gamma^4 - 4\alpha\beta\gamma^2 + 4l\alpha^2\beta^2)\,L^4M^2\\
&+ (-4l\gamma^4 + 24l^2\alpha\beta\gamma^2 - 4\alpha^2\beta^2)\,L^3M^3\\
&+ (-2\gamma^4 - 16l\alpha\beta\gamma^2 + 24l\alpha^2\beta^2)\,L^2M^4\\
&+ (-8\alpha\beta\gamma^2 - 16l\alpha^2\beta^2)\,LM^5\\
&+ (-8\alpha^2\beta^2)\,M^6,
\end{aligned}$$

which is

$$= (lL - M)(L^3 + 6lLM^2 + 2M^3)(\gamma^2 L + 2\alpha\beta M)^2.$$

But we have

$$lL - M = 1 + 8l^3,$$

$$L^3 + 6lLM^2 + 2M^3 = (1 + 8l^3)(\lambda^3 - 3S\lambda - 2T),$$

and the equation thus is

$$\tfrac{1}{6}(\mathfrak{H}^2 - \mathfrak{A}\mathfrak{B}) = (1 + 8l^3)^2 (\lambda^3 - 3S\lambda - 2T)[(\gamma^2 + 2l\alpha\beta)\lambda + \{6l^2\gamma^2 - 2(1 + 2l^3)\alpha\beta\}]^2,$$

showing that the solution involves the radical $\sqrt{\lambda^3 - 3S\lambda - 2T}$.

118. If (α, β, γ) is the inflexion $(1, -1, 0)$, the expression for λ is here

$$\lambda = \frac{-6l^2x - 6l^2y - 2(1 + 2l^3)z}{x + y - 2lz};$$

the equation in (x, y) is

$$(L^3 + 8M^3)x^2 + (2L^3 + 24lLM^2 - 8M^3)xy + (L^3 + 8M^3)y^2 = 0,$$

or, as this may be written,

$$(L^3 + 6lLM^2 + 2M^3)(x + y)^2 + (-6lLM^2 + 6M^3)(x - y)^2 = 0,$$

say

$$(L^3 + 6lLM^2 + 2M^3)(x + y)^2 = 6M^2(lL - M)(x - y)^2;$$

viz. we thus have

$$\sqrt{\lambda^3 - 3S\lambda - 2T}\,(x + y) = M\sqrt{6}\,(x - y),$$

or, substituting for M its value,

$$x = \sqrt{6}\,\{l\lambda - (1 + 2l^3)\} + \sqrt{\lambda^3 - 3S\lambda - 2T},$$

$$y = \sqrt{6}\,\{l\lambda - (1 + 2l^3)\} - \sqrt{\lambda^3 - 3S\lambda + 2T},$$

whence also

$$z = \sqrt{6}\,(\lambda + 6l^2),$$

these values satisfying identically

$$(\lambda + 6l^2)(x + y) - 2[l\lambda - (1 + 2l^3)]z = 0,$$

and

$$x^3 + y^3 + z^3 + 6lxyz = 0.$$

119. Starting from these values we, in fact, easily obtain

$$x\,dy - y\,dx = \frac{-\sqrt{6}\,d\lambda}{\sqrt{\lambda^3 - 3S\lambda - 2T}} \times$$

$$\{l\lambda^3 + (-3 - 6l^3)\lambda^2 + (-12l^2 + 12l^5)\lambda + (-8l - 92l^4 - 8l^7)\},$$

$$z^2 + 2lxy = -2\,\{\text{Do.}\},$$

and hence

$$d\omega = \frac{x\,dy - y\,dx}{z^2 + 2lxy} = \frac{\tfrac{1}{2}\sqrt{6}\,d\lambda}{\sqrt{\lambda^3 - 3S\lambda - 2T}}.$$

The same result might of course have been obtained from the values of x, z or y, z, the factor which divides out being in each of these cases irrational.

The Cubic $y^2 = x(1-x)(1-k^2x)$. Art. Nos. 120 to 130 (*several sub-headings*).

120. The curve is a semi-cubical parabola, symmetrical in regard to the axis of x; and if, as usual, k^2 is taken to be real, positive and less than 1, then the curve consists of an oval, and an infinite portion which may be called the anguis. (See Figure.)

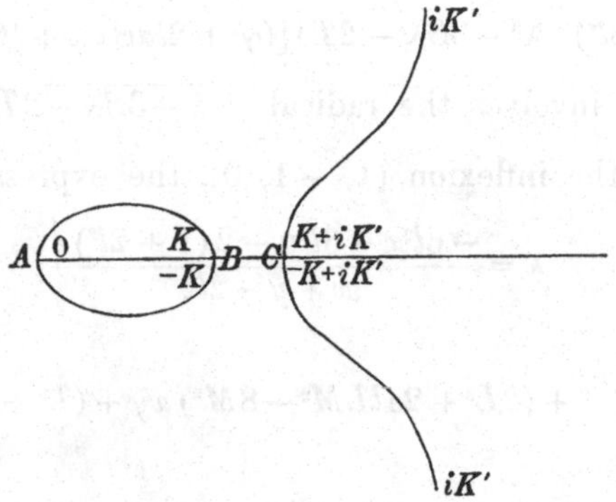

The equation is satisfied by

$$x = \text{sn}^2 u,$$
$$y = \text{sn}\, u \,\text{cn}\, u \,\text{dn}\, u.$$

Observe that the periods for these combinations of the elliptic functions are $2K$, $2iK'$; in fact,

$$\begin{aligned} \text{sn}\,(u+2K) &= -\,\text{sn}\, u, & \text{sn}\,(u+2iK') &= \ \ \text{sn}\, u,\\ \text{cn}\quad ,, \quad &= -\,\text{cn}\, u, & \text{cn}\quad ,, \quad &= -\,\text{cn}\, u,\\ \text{dn}\quad ,, \quad &= \ \ \text{dn}\, u, & \text{dn}\quad ,, \quad &= -\,\text{dn}\, u, \end{aligned}$$

whence the sn^2 and the sn . cn . dn are each unaltered by the change of u into $u+2K$ or $u+2iK'$. Hence to a given point (x, y) on the curve, the argument u is not $=$ a determinate value u_0, for it may be equally well taken to be $= u_0 + 2mK + 2m'iK'$, where m, m' are any positive or negative integers whatever: we express this by $u \equiv u_0$, or say u is congruent to u_0. But when u is thus given by a congruence $u \equiv u_0$ where u_0 has a determinate value, the point on the curve is uniquely determined. It is, however, to be noticed that a congruence $2u \equiv u_0$ does not uniquely determine the point on the curve: there are, in fact, four incongruent values of u, viz.

$$\tfrac{1}{2}u_0,\ \tfrac{1}{2}u_0 + K,\ \tfrac{1}{2}u_0 + iK',\ \tfrac{1}{2}u_0 + K + iK',$$

and the point on the curve is thus one of the four points belonging to these values of u respectively.

121. If, to fix the ideas, we select for each point of the curve one of the congruent values of the argument, we may assume for the oval, u real: at A, $u=0$; from A to B above the axis u positive and at B, $=K$; below the axis u negative and at B, $=-K$; there is thus a discontinuity, K, $-K$ at B, but the two values are congruent. For the anguis, $u = iK' +$ real value v: above the axis v positive, viz.

at infinity $v=0$, and at C, $=K$; below the axis v negative, viz. at infinity $v=0$ and at C, $=-K$: there thus is a discontinuity $iK'+K$, $iK'-K$ at C, but the two values are congruent. Observe that for points opposite to each other in regard to the axis, the arguments are, for points on the oval u, $-u$; for points on the anguis $iK'+v$, $iK'-v$: but that we have $iK'-v\equiv-(iK'-v)$.

122. The pure theorem gives for three points u_1, u_2, u_3 in a line

$$du_1+du_2+du_3=0;$$

and thence $u_1+u_2+u_3=C$. The constant C cannot have a determinate value (for if it had, then assuming the values of u_1 and u_2 at pleasure u_3 would have the determinate value $=C-u_1-u_2$), but must be given by a congruence: or, what is the same thing, assigning to C any admissible value, we must instead of $u_1+u_2+u_3=C$, write $u_1+u_2+u_3\equiv C$. Taking any particular line, for instance the tangent at A, we have $u_1, u_2, u_3=0, 0, iK'$; whence $C=iK'$; and we have $u_1+u_2+u_3\equiv iK'$, viz. this is the relation between the arguments u_1, u_2, u_3 belonging to the points of intersection of the cubic with a line: in particular, for a line at right angles to the axis, we have $u_1, u_2, u_3=\alpha, -\alpha, iK'$ or $=iK'+\beta, iK'-\beta, iK'$ (according as the line cuts the oval or the anguis): and the congruence is in each case satisfied. But I shall in general instead of $\equiv$ use the sign $=$, understanding it as in general meaning $\equiv$, and only replacing it by this sign when for clearness it seems necessary to do so.

Writing $\operatorname{sn} u_1, \operatorname{cn} u_1, \operatorname{dn} u_1=s_1, c_1, d_1$, and so in other cases, the condition in order that the three points may be in a line is

$$\begin{vmatrix} s_1^2, & s_1c_1d_1, & 1 \\ s_2^2, & s_2c_2d_2, & 1 \\ s_3^2, & s_3c_3d_3, & 1 \end{vmatrix}=0,$$

a relation which must be satisfied when the arguments are connected by the foregoing relation $u_1+u_2+u_3\equiv iK'$.

We can show from this equation alone that s_3^2 and $s_3c_3d_3$ are expressible rationally in terms of s_1^2, $s_1c_1d_1$, s_2^2, $s_2c_2d_2$; in fact, writing therein x_1, y_1 in place of s_1^2, $s_1c_1d_1$, &c., we thence have $x_3, y_3, 1=\lambda x_1+\mu x_2, \lambda y_1+\mu y_2, \lambda+\mu$, and substituting in

$$y_3^2=x_3(1-x_3)(1-k^2x_3),$$

we obtain an equation $\lambda\mu(F\lambda+G\mu)=0$, that is, $F\lambda+G\mu=0$, or say $\lambda, \mu=G, -F$, and thence

$$x_3=s_3^2=\frac{Gx_1-Fx_2}{G-F}, \quad y_3=s_3c_3d_3=\frac{Gy_1-Fy_2}{G-F}.$$

The values of F, G are easily found to be

$$F=y_1^2+2y_1y_2-x_1(1-x_1)(1-k^2x_2)-x_1(1-k^2x_1)(1-x_2)-(1-x_1)(1-k^2x_1)x_2,$$

$$G=2y_1y_2+y_2^2-x_2(1-x_2)(1-k^2x_1)-x_2(1-k^2x_2)(1-x_1)-(1-x_2)(1-k^2x_2)x_1,$$

or, as these may also be written,

$$F = -(y_1 - y_2)^2 + (x_1 - x_2)^2 \{1 + k^2 + k^2(x_1 + x_2)\} + k^2(x_1 - x_2)^2 x_1,$$
$$G = -(y_1 - y_2)^2 + (x_1 - x_2)^2 \{1 + k^2 + k^2(x_1 + x_2)\} + k^2(x_1 - x_2)^2 x_2,$$

where of course x_1, y_1, x_2, y_2 should be replaced by their values s_1^2, $s_1c_1d_1$, s_2^2, $s_2c_2d_2$. This is, in fact, the ordinary process of finding the third point of intersection of a cubic by a line which meets it in two given points.

Writing $iK' - u_3 = u$, and s, c, d for the sn, cn, dn of u, we have

$$s_3,\ c_3,\ d_3 = -\frac{1}{ks},\ \frac{id}{ks},\ \frac{ic}{s},$$

whence

$$s_3^2 = \frac{1}{k^2s^2},\quad s_3c_3d_3 = \frac{cd}{k^2s^3},$$

and the determinant equation becomes

$$\begin{vmatrix} s_1^2, & s_1c_1d_1, & 1 \\ s_2^2, & s_2c_2d_2, & 1 \\ 1\ , & \dfrac{cd}{s}\ , & k^2s^2 \end{vmatrix} = 0,$$

that is,

$$(1 - k^2s^2s_2^2)\, s_1c_1d_1 - (1 - k^2s^2s_1^2)\, s_2c_2d_2 - (s_1^2 - s_2^2)\frac{cd}{s} = 0,$$

corresponding to the relation $u = u_1 + u_2$ of the arguments. This is easily verified: we have

$$s_1 = \frac{sc_2d_2 - s_2cd}{1 - k^2s^2s_2^2},\quad s_2 = \frac{sc_1d_1 - s_1cd}{1 - k^2s^2s_1^2},\quad s = \frac{s_1^2 - s_2^2}{s_1c_2d_2 - s_2c_1d_1};$$

the equation thus becomes

$$(sc_2d_2 - s_2cd)\, c_1d_1 - (sc_1d_1 - s_1cd)\, c_2d_2 - (s_1^2 - s_2^2)\frac{cd}{s} = 0,$$

that is,

$$cd\left\{- s_2c_1d_1 + s_1c_2d_2 - \frac{s_1^2 - s_2^2}{s}\right\} = 0,$$

which is right.

The Four Tangents from a Point of the Cubic.

123. Suppose that the line is a tangent to the cubic, say the line touches the cubic at the point u, and again meets it at the point w: then instead of u_1, u_2, u_3 we have u, u, w: and the relation becomes $2u + w \equiv iK'$.

Here u being given, w is uniquely determined: viz. given the argument u of the point of contact, we have a unique value for the argument w of the tangential. But given w, we have $2u = iK' - w$; and we have thus for u the four values

$$\tfrac{1}{2}(iK' - w),\quad \text{Do.} + K,\quad \text{Do.} + iK',\quad \text{Do.} + K + iK',$$

corresponding to the four tangents which can be drawn from the point w to the cubic.

The tangents are real for a point of the anguis, and for such a point we may write $w = iK' + v$, where v is real and included between the values $\pm K$; the corresponding values of u are

$$u_1 = -\tfrac{1}{2}v, \quad u_2 = -\tfrac{1}{2}v + K, \quad u_3 = -\tfrac{1}{2}v + iK', \quad u_4 = -\tfrac{1}{2}v + K + iK':$$

the first and second of these belong to tangents to the oval, the third and fourth to tangents to the anguis. We may further distinguish a tangent according as it passes between or does not pass between the vertices B and C: say in the former case it is intermediate, and in the latter case extramediate: and we then see that, for the tangents from the point $iK' + v$ of anguis,

$u = -\tfrac{1}{2}v$	for intermediate to oval,
$u = -\tfrac{1}{2}v + K$	„ extramediate to oval,
$u = -\tfrac{1}{2}v + iK'$	„ extramediate to anguis,
$u = -\tfrac{1}{2}v + K + iK'$	„ intermediate to anguis.

124. We may make a corresponding division of the real lines which meet the curve in three real points: any such line meets the oval twice (and then of course the anguis once), or else it meets the anguis three times: and taking the arguments to be u_1, u_2, u_3, we have

$\tfrac{1}{2}(u_1 + u_2 + u_3) =$	$\tfrac{1}{2}iK'$	for intermediate line meeting oval twice,
„ $=$	$\tfrac{1}{2}iK' + K$	„ extramediate line, Do.,
„ $=$	$-\tfrac{1}{2}iK'$	„ extramediate line meeting anguis three times,
„ $=$	$-\tfrac{1}{2}iK' + K$	„ intermediate line, Do.

125. Returning to the tangents, the point $iK' + v$ may be an inflexion: we have then the point of contact of the intermediate tangent to the anguis coinciding with the point $iK' + v$; viz. $iK' + v \equiv -\tfrac{1}{2}v + K + iK'$, or say $= -\tfrac{1}{2}v \pm K + iK'$: that is, $v = \pm\tfrac{2}{3}K$; or $iK' + \tfrac{2}{3}K$ and $iK' - \tfrac{2}{3}K$ are the arguments for the real points of inflexion, above and below the axis respectively.

126. Write for a moment the equation in the form $y^2 = Bx + Cx^2 + Dx^3$; then if (α, β) be a point on the curve $(\beta^2 = B\alpha + C\alpha^2 + D\alpha^3)$, and we consider the intersections of the curve with the line $y - \beta = m(x - \alpha)$, we find for the remaining two intersections

$$B + C(x + \alpha) + D(x^2 + \alpha x + \alpha^2) = 2m\beta + m^2(x - \alpha).$$

If the line be a tangent, this will be satisfied by $x - \alpha$; the condition for this is $2m\beta = B + 2C\alpha + 3D\alpha^2$, and supposing this satisfied, then throwing out the factor $x - \alpha$ we obtain $C + D(x + 2\alpha) = m^2$, giving $Dx = m^2 - C - 2D\alpha$ for the coordinate x of the tangential of the point (α, β).

In the case of an inflexion, $x = \alpha$; and we have

$$m^2 = C + 3D\alpha, \ = \frac{(B + 2C\alpha + 3D\alpha^2)^2}{4(B\alpha + C\alpha^2 + D\alpha^3)},$$

giving for α the equation

$$3D^2\alpha^4 + 4CD\alpha^3 + 6BD\alpha - B^2 = 0,$$

or for B, C, D writing their values 1, $-(1+k^2)$, k^2, this is

$$3k^4\alpha^4 - 4k^2(1+k^2)\alpha^3 + 6k^2\alpha^2 - 1 = 0$$

for the x-coordinate of the inflexion. There is one negative root, and one or three positive roots; but only one positive root giving a real value of β, and corresponding hereto we have the two real inflexions on the anguis.

It should be possible, from the formulæ of No. 122, writing therein $(x_2, y_2) = (x_1, y_1)$, to arrive at the foregoing result, say $Dx_3 = m^2 - C - 2Dx_1$; but the functions F and G present themselves in vanishing forms, and the reduction is not immediate.

127. The general condition for an inflexion is $3u \equiv iK'$: the nine inflexions thus are $u = iK'$, inflexion at infinity, $u = iK' \pm \frac{2}{3}K$, the two real inflexions, and besides $u = \pm\frac{1}{3}iK'$, $u = \pm\frac{1}{3}iK' \pm \frac{2}{3}K$, six imaginary inflexions.

The Sextactic Points.

128. The vertices A, B, C are each of them a sextactic point: in fact, writing the equation in the form $y^2 = x - (1+k^2)x^2 + k^2x^3$, we see at once that the conic $y^2 = x - (1+k^2)x^2$ meets the curve in the point A counting six times: and there is obviously a like proof for the vertices B and C respectively. Hence, for the six intersections with any conic whatever, we have the condition

$$u_1 + u_2 + u_3 + u_4 + u_5 + u_6 \equiv 0:$$

and for the sextactic points we have the condition $6u \equiv 0$. This gives the 36 points $u = \frac{1}{6}(2mK + 2m'iK')$ or say $= \frac{1}{3}(mK + m'iK')$, $m = 0, \pm 1, \pm 2, 3$, $m' = 0, \pm 1, \pm 2, 3$: but among these are included the 9 inflexions (each of these being an improper sextactic point, the conic becoming the tangent taken twice) and there remain 27 points: among these are included the three vertices ($u = 0$, K, $iK' + K$), points of contact with the tangents from the inflexion at infinity; and of the remaining 24 points 6 are real, viz. these are the points $u = \pm\frac{1}{3}K$, $\pm\frac{2}{3}K$ on the oval, and the points $iK' \pm \frac{1}{3}K$ on the anguis: these are, in fact, the points of contact of the tangents from the real inflexions, viz. the three tangents from the inflexion $iK' + \frac{2}{3}K$ touch the oval in the points $\frac{2}{3}K$, $-\frac{1}{3}K$, and the anguis in the point $iK' - \frac{1}{3}K$; the three tangents from the inflexion $iK' - \frac{2}{3}K$ touch the oval in the points $-\frac{2}{3}K$, $\frac{1}{3}K$, and the anguis in the point $iK' + \frac{1}{3}K$.

Formulæ Relating to the Tangents from the Vertices.

129. I annex some formulæ relating to the tangents to the curve from the vertices A, B, C respectively. We have from each vertex four tangents say $\rho = 0$, $\sigma = 0$, symmetrically situate in regard to the axis, and $\rho' = 0$, $\sigma' = 0$, symmetrically situate in regard to the axis: the line joining the points of contact of ρ, σ is a line $\tau = 0$ at

right angles to the axis, and that joining the points of contact of ρ', σ' is a line $\tau'=0$ at right angles to the axis.

Vertex A, Tangents Imaginary.	Coordinates of Point of Contact.
$\rho\,,\ \sigma\,,\ \tau=y-i(1+k)\,x,\ \ y+i(1+k)\,x,\ \ kx+1\,;$	$x=-\dfrac{1}{k},\ \ y=\mp\dfrac{i(1+k)}{k},$
$\rho',\ \sigma',\ \tau'=y-i(1-k)\,x,\ \ y+i(1-k)\,x,\ \ -kx+1\,;$	$x=\ \dfrac{1}{k},\ \ y=\pm\dfrac{i(1-k)}{k}\,;$

the equation of the curve is

$$f,\ =y^2-x(1-x)(1-k^2x),\ =\rho\sigma-x\tau^2,\ =\rho'\sigma'-x\tau'^2,\ =0.$$

Vertex B, Tangents Imaginary.	Contacts.
$\rho\,,\sigma\,,\tau=y-(k-ik')(1-x),\ y+(k+ik')(1-x),\ kx-(k-ik')\,;$	$x=1-\dfrac{ik'}{k},\ y=\mp\dfrac{ik'}{k}(k+ik'),$
$\rho',\sigma',\tau'=y-(k-ik')(1-x),\ y+(k-ik')(1-x),\ kx-(k+ik')\,;$	$x=1+\dfrac{ik'}{k},\ y=\pm\dfrac{ik'}{k}(k-ik')\,;$

the equation of the curve is

$$f=\rho\sigma+(1-x)\,\tau^2,\ =\rho'\sigma'+(1-x)\,\tau'^2,\ =0.$$

Vertex C, Tangents Real.	Contacts.
$\rho\,,\ \sigma\,,\ \tau=y-\dfrac{1}{1+k'}(1-k^2x),\ \ y+\dfrac{1}{1+k'}(1-k^2x),\ \ x-\dfrac{1}{1+k'}\,;$	$x=\dfrac{1}{1+k'},\ \ y=\dfrac{\pm k'}{1+k'},$
$\rho',\ \sigma',\ \tau'=y-\dfrac{1}{1-k'}(1-k^2x),\ \ y+\dfrac{1}{1-k'}(1-k^2x),\ \ x-\dfrac{1}{1-k'}\,;$	$x=\dfrac{1}{1-k'},\ \ y=\dfrac{\pm k'}{1-k'}\,;$

the equation of the curve is

$$f,\ =\rho\sigma-(1-k^2x)\,\tau,\ =\rho'\sigma'-(1-k^2x)\,\tau',\ =0.$$

130. These linear functions ρ, σ, &c., considering therein x, y as denoting $\mathrm{sn}^2 u$, $\mathrm{sn}\,u\,\mathrm{cn}\,u\,\mathrm{dn}\,u$, respectively present themselves as the numerators and the denominators of some formulæ given No. 105 of my *Elliptic Functions* (1876), see p. 76: viz. we have

$$\mathrm{sn}^2(u+\tfrac{1}{2}K)=\frac{1}{1+k'}\,\frac{1-k^2x+(1+k')\,y}{1-k^2x+(1-k')\,y},$$

which is

$$=\frac{1}{1-k'}\,\frac{y+\dfrac{1}{1+k'}(1-k^2x)}{y+\dfrac{1}{1-k'}(1-k^2x)},\quad =\frac{1}{1-k'}\,\frac{\sigma}{\sigma'},\quad \text{(vertex C)};$$

$$\mathrm{sn}^2(u+\tfrac{1}{2}iK')=\frac{1}{k}\,\frac{(1+k)\,x+iy}{(1+k)\,x-iy},$$

which is

$$=-\frac{1}{k}\,\frac{y-i(1+k)\,x}{y+i(1+k)\,x},\qquad =-\frac{1}{k}\,\frac{\rho}{\sigma},\qquad \text{(vertex A)};$$

and

$$\mathrm{sn}^2(u+\tfrac{1}{2}K+\tfrac{1}{2}iK')=\frac{k+ik'}{k}\,\frac{1-x+(k-ik)\,y}{1-x+(k+ik)\,y},$$

which is

$$=\frac{k-ik'}{k}\frac{y+(k+ik')(1-x)}{y+(k-ik')(1-x)},\quad =\frac{k-ik'}{k}\frac{\sigma}{\sigma'},\ \text{(vertex B)}.$$

Observe here that, in the second formula, we have a pair of tangents ρ, σ which belong to a chord τ through the inflexion at ∞; but in the first and third formulæ we have tangents σ, σ' not forming such a pair. This is as it should be, for the zero and infinity of $\text{sn}^2(u+\tfrac{1}{2}iK')$ are $u=-\tfrac{1}{2}iK'$, $u=\tfrac{1}{2}iK'$, which belong to points in lineâ with the inflexion at infinity: but for $\text{sn}^2(u+\tfrac{1}{2}K)$, the zero is $u=-\tfrac{1}{2}K$, and the infinity is $u=iK'-\tfrac{1}{2}K$, which do not belong to points in lineâ with the inflexion at infinity: and the like for $\text{sn}^2(u+\tfrac{1}{2}K+\tfrac{1}{2}iK')$.

Fixed Curve a Quartic in Space, the Quadri-quadric Curve $y^2=1-x^2$, $z^2=1-k^2x^2$.
Art. Nos. 131 to 135.

131. It is assumed that k^2 is real, positive, and less than unity: the curve may be regarded as the intersection of the two cylinders

$$x^2+y^2=1,\quad k^2x^2+z^2=1;$$

but there is through it a third cylinder $y^2-k^2z^2=k'^2$. The cylinder $k^2x^2+z^2=1$, or say the horizontal cylinder, has for its section an ellipse axes $\frac{1}{k}$ and 1 respectively: and it is pierced by the cylinder $x^2+y^2=1$, or say the vertical cylinder, in two detached ovals (of double curvature) lying on opposite sides of the plane of xy: only the upper oval $ABA'B'$ is shown in the figure.

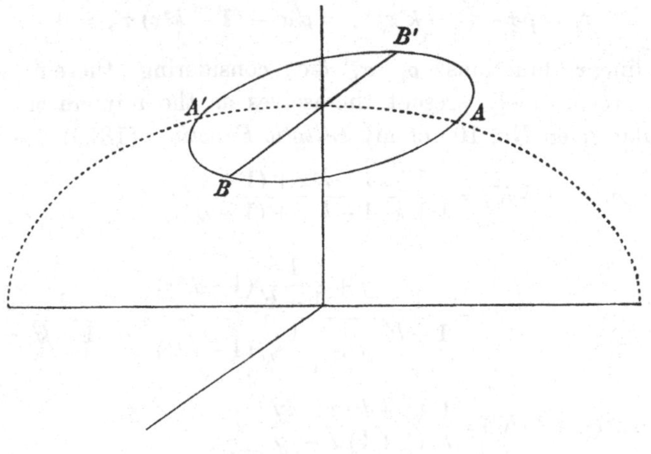

Each of the vertices A, A', B, B' is an inflexion*, viz. a point such that the osculating plane at that point meets the curve in the point counting four times. We may consider two generating lines of the horizontal cylinder, each meeting the oval in two points; the plane through the generating lines meets the curve in these

* There are in all 16 inflexions: 4 in each of the planes $x=0$, $y=0$, $z=0$, and 4 in the plane infinity.

four points, and when the generating lines come each of them to coincide with the tangent at A, we have the osculating plane meeting the curve in the point A counting four times. The like reasoning, with two generating lines of the third cylinder, shows that the vertex B is an inflexion.

132. The equations are satisfied by writing therein $x, y, z = \text{sn}\, u, \text{cn}\, u, \text{dn}\, u$: the periods are here $4K, 4iK'$: hence, at a given point on the curve, the argument is not $u =$ a determinate value u_0, but it may equally well be taken to be $= u_0 + 4mK + 4m'iK'$, where m and m' are any positive or negative integers: we express this by $u \equiv u_0$, or say u congruent to u_0. For the upper oval, u may be taken to be real, and to be $= 0$ at B, positive for the half oval BAB', and negative for the half oval $BA'B'$; having the values $+K$, $-K$ at A and A' respectively, and the discontinuity $2K$, $-2K$ at B', these two values being congruent. For the lower oval, we have $u = 2iK' +$ real value v.

For the intersections of the curve with a plane, we have

$$du_1 + du_2 + du_3 + du_4 = 0\,; \text{ whence } u_1 + u_2 + u_3 + u_4 = C\,;$$

and by taking the plane to be the osculating plane at B, we find 0 as a value of the constant, and the relation thus is $u_1 + u_2 + u_3 + u_4 \equiv 0$. Writing as before $\text{sn}\, u_1$, $\text{cn}\, u_1$, $\text{dn}\, u_1$, $= s_1, c_1, d_1$ and so in other cases, the relation between the elliptic functions is

$$\begin{vmatrix} s_1, & c_1, & d_1, & 1 \\ s_2, & c_2, & d_2, & 1 \\ s_3, & c_3, & d_3, & 1 \\ s_4, & c_4, & d_4, & 1 \end{vmatrix} = 0.$$

It is important to remark that, given three of the points, the fourth point is determined uniquely: that is, the equation really gives s_4, c_4, d_4, each as a rational function of the $s_1, c_1, d_1, s_2, c_2, d_2, s_3, c_3, d_3$.

In fact, we may write $s_4 = \lambda_1 s_1 + \lambda_2 s_2 + \lambda_3 s_3$, and similarly for c_4 and d_4, and $1 = \lambda_1 + \lambda_2 + \lambda_3$: substituting in $s_4^2 + c_4^2 - 1 = 0$, $k^2 s_4^2 + d_4^2 - 1 = 0$, we have

$$X_{23}\lambda_2\lambda_3 + X_{31}\lambda_3\lambda_1 + X_{12}\lambda_1\lambda_2 = 0,$$
$$Y_{23} \;\;,, \;\; + Y_{31} \;\;,, \;\; + Y_{12} \;\;,, \;\; = 0,$$

where

$$X_{12} = s_1 s_2 + c_1 c_2 - 1, \quad Y_{12} = k^2 s_1 s_2 + d_1 d_2 - 1, \text{ \&c.};$$

we thence have

$$\lambda_2\lambda_3 : \lambda_3\lambda_1 : \lambda_1\lambda_2 = X_{31}Y_{12} - X_{12}Y_{31} : X_{12}Y_{23} - X_{23}Y_{12} : X_{23}Y_{31} - X_{31}Y_{23}$$
$$= \quad A_1 \quad : \quad A_2 \quad : \quad A_3, \text{ suppose};$$

that is,

$$\lambda_1 : \lambda_2 : \lambda_3 = A_2 A_3 : A_3 A_1 : A_1 A_2,$$

and consequently

$$\begin{aligned} s_4, &= -\,\text{sn}\,(u_1 + u_2 + u_3), = A_2A_3s_1 + A_3A_1s_2 + A_1A_2s_3 \div (A_2A_3 + A_3A_1 + A_1A_2), \\ c_4, &= \;\;\text{cn}\,(\quad ,, \quad) \;= \;,,\; c_1 + \;,,\; c_2 + \;,,\; c_3 \div (\quad ,, \quad), \\ d_4, &= \;\;\text{dn}\,(\quad ,, \quad) \;= \;,,\; d_1 + \;,,\; d_2 + \;,,\; d_3 \div (\quad ,, \quad), \end{aligned}$$

which are the required expressions. If $u_3=0$, and consequently $s_3, c_3, d_3 = 0, 1, 1$, the resulting expressions give the sn, cn, and dn of u_1+u_2; but the expressions are in a very complicated form, not easily identifiable with the ordinary ones.

133. The determinant equation may be written

$$\begin{aligned}(s_1-s_2)(c_3d_4-c_4d_3)+(s_3-s_4)(c_1d_2-c_2d_1)\\ +(c_1-c_2)(d_3s_4-d_4s_3)+(c_3-c_4)(d_1s_2-d_2s_1)\\ +(d_1-d_2)(s_3c_4-s_4c_3)+(d_3-d_4)(s_1c_2-s_2c_1)=0,\end{aligned}$$

and, in fact, each of the three lines is separately $=0$. This appears from the following three formulæ:—

$$\frac{\operatorname{sn}(u_1+u_2)}{\operatorname{cn}(u_1+u_2)-\operatorname{dn}(u_1+u_2)}=\frac{s_1-s_2}{c_1d_2-c_2d_1},$$

$$\frac{\operatorname{sn}(u_1+u_2)}{\operatorname{cn}(u_1+u_2)+1}=\frac{c_1-c_2}{d_1s_2-d_2s_1},$$

$$\frac{\operatorname{sn}(u_1+u_2)}{\operatorname{dn}(u_1+u_2)+1}=\frac{-\frac{1}{k^2}(d_1-d_2)}{s_1c_2-s_2c_1},$$

which are themselves at once deducible from the formulæ

$$\begin{aligned}\operatorname{sn}(u_1+u_2)&=s_1^2-s_2^2,\ =-(c_1^2-c_2^2),\ =-\frac{1}{k^2}(d_1^2-d_2^2),\ \div(s_1c_2d_2-s_2c_1d_1),\\ \operatorname{cn}(u_1+u_2)&=s_1c_1d_2-s_2c_2d_1 \qquad \div \quad \text{Do.}\\ \operatorname{dn}(u_1+u_2)&=s_1d_1c_2-s_2d_2c_1 \qquad \div \quad \text{Do.}\end{aligned}$$

In fact, the numerators of $\operatorname{cn}(u_1+u_2)+\operatorname{dn}(u_1+u_2)$, $\operatorname{cn}(u_1+u_2)+1$, $\operatorname{dn}(u_1+u_2)+1$, thus become $=(s_1+s_2)(c_1d_2-c_2d_1)$, $-(c_1+c_2)(d_1s_2-d_2s_1)$, $(d_1+d_2)(s_1c_2-s_2c_1)$, respectively: so that, taking the numerator of $\operatorname{sn}(u_1+u_2)$ under its three forms successively, we have by division the formulæ in question.

134. The above three equations, putting therein $u_1+u_2=-2u_3$ and reducing the functions of $2u_3$, become

$$\begin{aligned}\frac{1}{k'^2}\frac{c_3d_3}{s_3}&=\frac{s_1-s_2}{c_1d_2-c_2d_1}, &&\text{giving } -k^2s_3^2 =\frac{(c_1-c_2)(d_1-d_2)}{(d_1s_2-d_2s_1)(s_1c_2-s_2c_1)},\\ -\frac{s_3d_3}{c_3}&=\frac{c_1-c_2}{d_1s_2-d_2s_1}, &&\quad ,, \qquad \frac{k^2}{k'^2}c_3^2=\frac{(d_1-d_2)(s_1-s_2)}{(s_1c_2-s_2c_1)(c_1d_2-c_2d_1)},\\ \frac{k^2s_3c_3}{d_3}&=\frac{d_1-d_2}{s_1c_2-s_2c_1}, &&\quad ,, \qquad -\frac{1}{k'^2}d_3^2=\frac{(s_1-s_2)(c_1-c_2)}{(c_1d_2-c_2d_1)(d_1s_2-d_2s_1)},\end{aligned}$$

equations which must of course give each of them the same value for s_3^2: the equations belong to the relation $u_1+u_2+2u_3=0$, viz. (s_3, c_3, d_3) are the coordinates of a point of contact of the tangent plane drawn through the two given points (s_1, c_1, d_1) and (s_2, c_2, d_2) of the curve.

135. Write

$$a, b, c, f, g, h = s_1 - s_2,\ c_1 - c_2,\ d_1 - d_2,\ c_1 d_2 - c_2 d_1,\ d_1 s_2 - d_2 s_1,\ s_1 c_2 - s_2 c_1,$$
$$a', b', c', f', g', h' = s_3 - s_4,\ c_3 - c_4,\ d_3 - d_4,\ c_3 d_4 - c_4 d_3,\ d_3 s_4 - d_4 s_3,\ s_3 c_4 - s_4 c_3,$$

so that (a, b, c, f, g, h) are the six coordinates of the line 12, and a', b', c', f', g', h' are the six coordinates of the line 34. The determinant equation is nothing else than the condition of the intersection of the two lines, viz. this is

$$af' + a'f + bg' + b'g + ch' + c'h = 0.$$

By what precedes, it appears that not only is this so: but that we have separately $af' + a'f = 0$, $bg' + b'g = 0$, $ch' + c'h = 0$, viz. these three equations are satisfied by the coordinates of the lines 12 and 34, which join in pairs the intersections 1, 2 and 3, 4 of the quadri-quadric curve by a plane. But this is a geometrical property depending only on the four points being in a plane: and it is thus a result of Abel's theorem that, when the arguments are such that

$$u_1 + u_2 + u_3 + u_4 = 0,$$

then not only the original equation, but each of the three equations, holds good.

The Cubic Curve $xy^2 - 2y + (1 + k^2) x - k^2 x^3 = 0$. Art. Nos. 136 and 137.

136. Writing the equation in the form

$$(xy - 1)^2 = (1 - x^2)(1 - k^2 x^2), \text{ or say } xy - 1 = \div \sqrt{1 - x^2 \,.\, 1 - k^2 x^2},$$

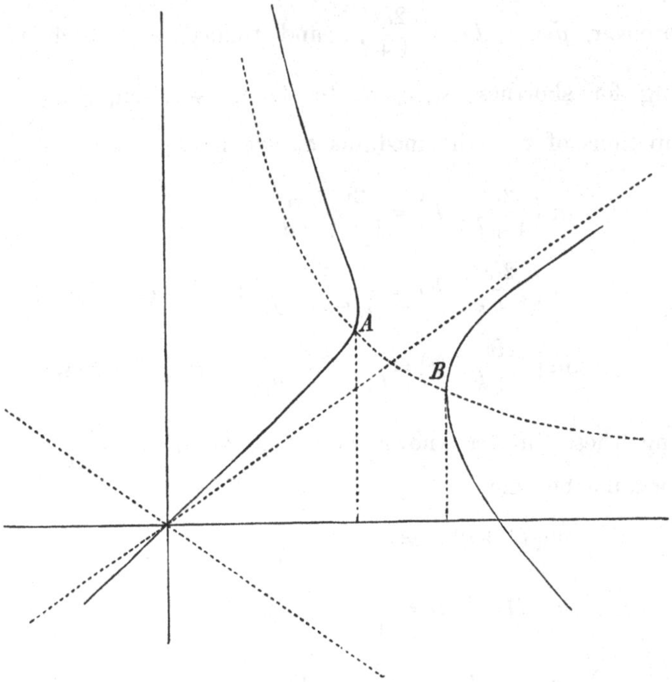

we see that the general form is as shown in the figure: the real portions of the curve lie between the values $x=-\infty,\ -\frac{1}{k};\ -1,\ +1$; and $\frac{1}{k},\ \infty$. The curve may be made to depend on elliptic functions in two different ways: we may write

$$x = \operatorname{sn} u \quad , \ = \frac{2i}{1+k}\,\frac{\operatorname{sn}_1 v}{\operatorname{cn}_1 v\,\operatorname{dn}_1 v},$$

$$y = \frac{1-\operatorname{cn} u \operatorname{dn} u}{\operatorname{sn} u}, \ = \frac{i}{1+k}\,\frac{\operatorname{sn}_1 v}{\operatorname{cn}_1 v\,\operatorname{dn}_1 v}\{1+k^2-(1-k)^2\operatorname{sn}_1^2 v\},$$

where the functions sn, cn, dn belong to the modulus k, and the functions sn_1, cn_1, dn_1 to the modulus θ, $=\frac{1-k}{1+k}$. The first mode is obvious; as to the second, observe that the formulæ give

$$y-kx = \frac{i}{1+k}\,\frac{\operatorname{sn}_1 v}{\operatorname{cn}_1 v\,\operatorname{dn}_1 v}(1-k)^2\operatorname{cn}_1^2 v, \ = \frac{i(1-k)^2}{1+k}\,\frac{\operatorname{sn}_1 v\operatorname{cn}_1 v}{\operatorname{dn}_1 v},$$

$$y+kx = \frac{i}{1+k}\,\frac{\operatorname{sn}_1 v}{\operatorname{cn}_1 v\,\operatorname{dn}_1 v}(1+k)^2\operatorname{dn}_1^2 v, \ = i(1+k)\,\frac{\operatorname{sn}_1 v\operatorname{dn}_1 v}{\operatorname{cn}_1 v},$$

whence

$$y^2-k^2x^2 = -(1-k)^2\operatorname{sn}_1^2 v;$$

and therefore

$$x(y^2-k^2x^2) = \frac{-2i(1-k)^2}{1+k}\,\frac{\operatorname{sn}_1^3 v}{\operatorname{cn}_1 v\,\operatorname{dn}_1 v},$$

which is also the value of $2y-(1+k^2)x$, as it should be.

We find, moreover, $d\omega$, $=du$, $=\frac{2idv}{1+k}$; and thence, u, v each vanishing together, $u=\frac{2iv}{1+k}$. Writing for shortness s_1, c_1, d_1 to denote $\operatorname{sn}_1 v$, $\operatorname{cn}_1 v$, $\operatorname{dn}_1 v$: that is, s_1, c_1, d_1 are the elliptic functions of v to the modulus θ: we have

$$\operatorname{sn}\left(\frac{2iv}{1+k},\ k\right) = \frac{2i}{1+k}\,\frac{s_1}{c_1 d_1},$$

$$\operatorname{cn}\left(\frac{2iv}{1+k},\ k\right) = \frac{1}{(1+k)\,c_1 d_1}\{1+k+(1-k)\,s_1^2\},$$

$$\operatorname{dn}\left(\frac{2iv}{1+k},\ k\right) = \frac{1}{(1+k)\,c_1 d_1}\{1+k-(1-k)\,s_1^2\}.$$

137. To bring these into a known form, for k write $\frac{1-k}{1+k}$, then θ is changed into k, and the formulæ become

$$\operatorname{sn}_1(1+k)\,iv = (1+k)\,i\,\frac{s}{cd},$$

$$\operatorname{cn}_1(1+k)\,iv = \frac{1}{cd}(1+ks^2),$$

$$\operatorname{dn}_1(1+k)\,iv = \frac{1}{cd}(1-ks^2),$$

where the sn_1, cn_1, dn_1 refer to the modulus θ, and s, c, d denote $\text{sn}\, v$, $\text{cn}\, v$, $\text{dn}\, v$, modulus k.

But from the formulæ, p. 63, of my *Elliptic Functions*,

$$\text{sn}(iu, k),\ \text{cn}(iu, k),\ \text{dn}(iu, k) = \frac{i\, \text{sn}(u, k')}{\text{cn}(u, k')},\ \frac{1}{\text{cn}(u, k')},\ \frac{\text{dn}(u, k')}{\text{cn}(u, k')},$$

and herein for u writing $(1+k)v$, and for k writing θ, $= \frac{1-k}{1+k}$, whence k' becomes $\frac{2\sqrt{k}}{1+k}$, $=\gamma$, and $\gamma' = \frac{1-k}{1+k}$, we have

$$\text{sn}_1(1+k)\, iv,\ \text{cn}_1(1+k)\, iv,\ \text{dn}_1(1+k)\, iv = \frac{i\, \text{sn}(\overline{1+k}v, \gamma)}{\text{cn}(\overline{1+k}v, \gamma)},\ \frac{1}{\text{cn}(\overline{1+k}v, \gamma)},\ \frac{\text{dn}(\overline{1+k}v, \gamma)}{\text{cn}(\overline{1+k}v, \gamma)};$$

and the formulæ above obtained are

$$\frac{\text{sn}(\overline{1+k}v, \gamma)}{\text{cn}(\overline{1+k}v, \gamma)} = (1+k)\frac{s}{cd},\quad \text{giving}\quad \text{sn}(\overline{1+k}v, \gamma) = \frac{(1+k)\, s}{1+ks^2},$$

$$\frac{1}{\text{cn}(\overline{1+k}v, \gamma)} = \frac{1}{cd}(1+ks^2),\quad ,,\quad \text{cn}(\overline{1+k}v, \gamma) = \frac{cd}{1+ks^2},$$

$$\frac{\text{dn}(\overline{1+k}v, \gamma)}{\text{cn}(\overline{1+k}v, \gamma)} = \frac{1}{cd}(1-ks^2),\quad ,,\quad \text{dn}(\overline{1+k}v, \gamma) = \frac{1-ks^2}{1+ks^2},$$

where, as before, s, c, d denote $\text{sn}(v, k)$, $\text{cn}(v, k)$, $\text{dn}(v, k)$: these last are, in fact, the formulæ of the second line of the table, *Elliptic Functions*, p. 183.

Fixed Curve the Cubic $y^2 = x(1-x)(1-k^2x)$: *the Function* $\{01'\theta\}$. Art. Nos. 138 to 142.

138. It was shown, No. 65, how for the affected theorem, when the fixed curve is a cubic, the form of differential is $d\omega$ multiplied by

$$\{012\} + \left[\int_2^1 d\omega\, \partial_3 \{036\} - \{123\}\right],$$

the last term being the properly determined constant K, attached to the variable term $\{012\}$, in order to obtain a standard form of integral. The object of the present articles is to show what these formulæ become for the before-mentioned form of cubic curve $y^2 = x(1-x)(1-k^2x)$, which is most directly connected with elliptic functions: and to exhibit the connexion of the formulæ with the ordinary formulæ for elliptic integrals of the second and the third kinds respectively.

139. We have, in general,

$$\{012\},\ = \frac{\widetilde{012}}{\overline{012}},\ = \left\{\begin{matrix} (1+k^2)\, x_1x_2 - (x_1+x_2) + y_1y_2 \\ + x[-1 + (1+k^2)(x_1+x_2) \\ + y(y_1+y_2) \end{matrix}\right\} \div \begin{vmatrix} x, & y, & 1 \\ x_1, & y_1, & 1 \\ x_2, & y_2, & 1 \end{vmatrix}.$$

Taking here 2, $=\theta$, for the point, coordinates x_2, $y_2=0$, 0, we have

$$\{01\theta\}=\frac{-x_1-x+(1+k^2)\,xx_1+yy_1}{xy_1-x_1y};$$

and if, retaining 1 to denote the point coordinates (x_1, y_1) belonging to the argument u_1, we write $1'$ for the point belonging to the argument $u+iK'$, then the coordinates of $1'$ are $\frac{1}{k^2x_1}$, $\frac{-y_1}{k^2x_1^2}$, and the formula becomes

$$\{01'\theta\}=\frac{x_1+k^2xx_1^2-(1+k^2)\,xx_1}{xy_1+x_1y};$$

the numerator hereof multiplied by x is $=y\,(xy_1+x_1y)-k^2x^2x_1\,(x-x_1)$, and we thence have

$$\{01'\theta\}=\frac{y}{x}-\frac{k^2xx_1\,(x-x_1)}{xy_1-x_1y},$$

which, substituting for x, y, x_1, y_1 their values in terms of u, u_1, is

$$=\frac{\operatorname{cn} u \operatorname{dn} u}{\operatorname{sn} u}-k^2 \operatorname{sn} u \operatorname{sn} u_1 \operatorname{sn}(u-u_1).$$

Operating on each side with $\frac{d}{du_1}$, $=\partial_1$, we obtain

$$\partial_1\{01'\theta\}=k^2\operatorname{sn}^2 u_1-k^2\operatorname{sn}^2(u-u_1),$$

the differentiation being, in fact, that which occurs in establishing the fundamental property of the elliptic integral of the second kind

$$Zu=u\left(1-\frac{E}{K}\right)+k^2\int_0 \operatorname{sn}^2 u\, du:$$

viz. we have

$$Zu-Zu_1-Z\,(u-u_1)=-k^2\operatorname{sn} u \operatorname{sn} u_1 \operatorname{sn}(u-u_1),$$

and thence

$$\partial_1[-k^2\operatorname{sn} u \operatorname{sn} u_1 \operatorname{sn}(u-u_1)]=-Z'u_1+Z'(u-u_1),\ =k^2\operatorname{sn}^2 u_1-k^2\operatorname{sn}^2(u-u_1).$$

Observe that, $1'$ referring to u_1+ik', the subscript 1 might be written $1'$.

The same result should of course be obtainable by the differentiation of the expression of $\{01'\theta\}$ in terms of x, y, x_1, y_1. We have

$$\partial_1 x_1=2y_1,\ \partial_1 y_1=1-2\,(1+k^2)\,x_1+3k^2x_1^2,\ =\Omega_1;$$

and we thence obtain

$$\partial_1\{01'\theta\}=\frac{k^2}{(xy_1+x_1y)^2}\left[-2\,(xy_1+x_1y)(x-2x_1)\,xy_1+xx_1\,(x-x_1)(x\Omega_1+2yy_1)\right],$$

where the term in [ ] is found to be $=x\,(xy_1+x_1y)^2-xx_1\,(x-x_1)^2$; whence the equation is

$$\partial_1\{01'\theta\}=k^2x-\frac{k^2xx_1\,(x-x_1)^2}{(xy_1+x_1y)^2},$$

giving the foregoing result.

140. To introduce into the formulæ 1 instead of 1′, we have only to write $u_1 - iK'$ instead of u_1; putting also for shortness s, c, d, s_1, c_1, d_1 for the functions of u and u_1 respectively, we thus obtain

$$\{01\theta\} = -\frac{cd}{s} + \frac{s}{s_1 \operatorname{sn}(u-u_1)},$$

where observe that, interchanging u and u_1, we have

$$\{10\theta\} = -\frac{c_1 d_1}{s_1} - \frac{s_1}{s \operatorname{sn}(u-u_1)},$$

that is, $\{10\theta\} = -\{01\theta\}$, as it should be. The formulæ may be written

$$\{01\theta\},\ = -\{10\theta\},\ = \frac{s^3 c_1 d_1 + s_1^3 cd}{sc_1 d_1 + s_1 cd}\ \frac{1}{ss_1 \operatorname{sn}(u-u_1)}:$$

and

$$\partial_1 \{01\theta\} = \frac{1}{s_1^2} - \frac{1}{\operatorname{sn}^2(u-u_1)},$$

whence

$$\partial_0 \{01\theta\} = -\frac{1}{s^2} + \frac{1}{\operatorname{sn}^2(u-u_1)};$$

we have, moreover,

$$\{012\} = \{12\theta\} + \frac{s}{s_1 \operatorname{sn}(u-u_1)} - \frac{s}{s_2 \operatorname{sn}(u-u_1)},$$

and

$$\partial_0 \{012\} = \frac{1}{\operatorname{sn}^2(u-u_1)} - \frac{1}{\operatorname{sn}^2(u-u_2)},$$

which last equation gives $(\partial_0 + \partial_1 + \partial_2)\{012\} = 0$, as it should do.

141. Supposing that the differential $d\Pi_{12}$ is defined by the equation

$$d\Pi_{12} = du\,\{012\} + du \left[\int_2^1 du\,\partial_3\,\{036\} - \{123\}\right],$$

we have

$$\int_5^4 d\Pi_{12} = \int_5^4 du\,\{012\} + \int_5^4 du \left[\int_2^1 du\,\partial_3\,\{036\} - \{123\}\right],$$

and thence

$$\begin{aligned}\partial_1 \partial_4 \int_5^4 d\Pi_{12} &= \partial_1\{124\} + \partial_3\{136\} - \partial_1\{123\},\\ &= \partial_1\{134\} - \partial_3\{316\},\\ &= \left[\frac{1}{\operatorname{sn}^2(u_1-u_3)} - \frac{1}{\operatorname{sn}^2(u_1-u_4)}\right] - \left[\frac{1}{\operatorname{sn}^2(u_3-u_1)} - \frac{1}{\operatorname{sn}^2(u_3-u_6)}\right],\\ &= \frac{1}{\operatorname{sn}^2(u_3-u_6)} - \frac{1}{\operatorname{sn}^2(u_1-u_4)},\end{aligned}$$

or establishing between the constants u_3, u_6 the relation $\text{sn}^2(u_3 - u_6) = \dfrac{K}{K-E}$, this becomes

$$= 1 - \frac{E}{K} - k^2 \,\text{sn}^2(u_4 - u_1 + iK'),$$

which is

$$= \partial_2 \partial_4 \log \frac{\Theta(u_4 - u_2 + iK')\,\Theta(u_5 - u_1 + iK')}{\Theta(u_4 - u_1 + iK')\,\Theta(u_5 - u_2 + iK')},$$

where Θ is Jacobi's theta-function, see my *Elliptic Functions*, p. 144. The expression is, in fact, $= -\partial_1 \partial_4 \log \theta(u_4 - u_1 + iK')$, $= \phi(u_4 - u_1 + iK')$, if for a moment $\phi v = \partial_v^2 \log \Theta v$. But we have

$$\log \Theta v = \log \sqrt{\frac{2k'K}{\pi}} + \tfrac{1}{2} v^2 \left(1 - \frac{E}{K}\right) - k^2 \int_0 \int_0 \text{sn}^2\, v dv^2,$$

that is, $\phi v = 1 - \dfrac{E}{K} - k^2 \,\text{sn}^2\, v$, and consequently we have $\phi(u_4 - u_1 + iK')$

$$= 1 - \frac{E}{K} - k^2 \,\text{sn}^2(u_4 - u_1 + iK').$$

142. In connexion with the same curve $y^2 = x(1-x)(1-k^2x)$, we may establish the identity

$$\frac{d}{du_1} \frac{y_1}{x_1 - x} + \frac{d}{du} \frac{y}{x_1 - x} = k^2 (x_1 - x),$$

where as before $x, y = s^2, scd$, and $x_1, y_1 = s_1^2, s_1 c_1 d_1$. We have

$$(x_1 - x)\frac{dy_1}{du_1} - y_1 \frac{dx_1}{du_1} = (s_1^2 - s^2)\{1 - 2(1+k^2)s_1^2 + 3k^2 s_1^4\} - 2s_1^2(1 - s_1^2)(1 - k^2 s_1^2)$$
$$= -s^2 - s_1^2 + 2(1+k^2)s^2 s_1^2 - 3k^2 s^2 s_1^4 + k^2 s_1^6;$$

and similarly

$$(x - x_1)\frac{dy}{du} - y\frac{dx}{du} = -s^2 - s_1^2 + 2(1+k^2)s^2 s_1^2 - 3k^2 s^4 s_1^2 + k^2 s^6.$$

The difference of the two functions on the right-hand side is $= k^2(s_1^2 - s^2)^3$; which is $= k^2(x_1 - x)^3$, and this divided by $(x_1 - x)^2$ is $= k^2(x_1 - x)$; the identity is thus verified.

Fixed Curve the Quartic $y^2 = (1 - x^2)(1 - k^2 x^2)$. Art. Nos. 143 to 145.

143. This is a curve having a tacnode at infinity on the line $x = 0$, as may be seen by writing the equation in the homogeneous form $y^2 z^2 = (z^2 - x^2)(z^2 - k^2 x^2)$; we have as it were two branches having the line infinity for a common tangent at the point in question. The equation is satisfied by $x = \text{sn}\, u$, $y = \text{cn}\, u\, \text{dn}\, u$, values which are unaltered by the change of u into $u + 4mK + 2m'iK'$, m and m' any positive or negative integers; in regard to this curve, the sign $\equiv$ is to be understood accordingly. I consider with

reference to this curve only the affected theorem, in the particular form in which it most readily connects itself with the ordinary theory of the integral of the third kind.

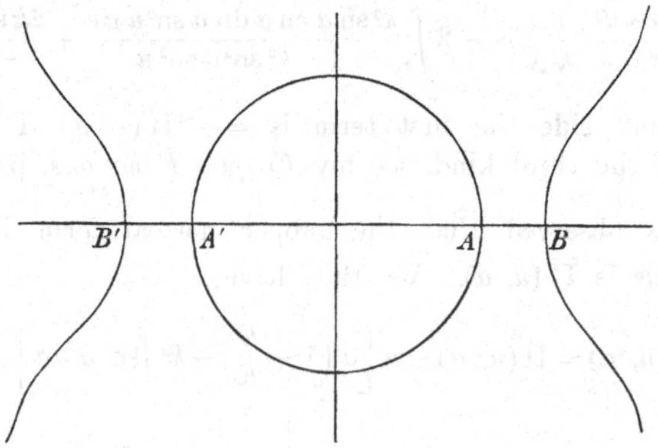

144. I consider the differential $\frac{(x, y, z)^2_{12}\, d\omega}{012}$ in the particular case where the line 012 is a line parallel to the axis of y: taking its equation to be $x - x_1 = 0$, and putting for shortness $X = \sqrt{1-x^2 . 1-k^2x^2}$, $X_1 = \sqrt{1-x_1^2 . 1-k^2x_1^2}$, the parametric points are taken to be $(x_1, \sqrt{X_1})$, $(x_1, -\sqrt{X_1})$, and the residues are the intersections with the two branches at the tacnode. The conic $(x, y, z)^2_{12} = 0$ is to satisfy the conditions of passing through the two nodes of the tacnode, and through the two residues, that is, again through the tacnode twice—in all, four conditions; and we have thus the form $z(x - \theta z) = 0$, containing the arbitrary constant θ. The major function itself is then easily determined, and putting again $z = 1$ we arrive at the form

$$\frac{\sqrt{X_1}}{x_1 - \theta}\, \frac{x - \theta}{x - x_1}\, \frac{dx}{\sqrt{X}}.$$

If the limits are taken to be two points on a line parallel to the axis of x, or what is the same thing, if the limits in regard to x are x, $-x$, we have the integral

$$\int_{-x}^{x} \frac{\sqrt{X_1}}{x_1 - \theta}\, \frac{x - \theta}{x - x_1}\, \frac{dx}{\sqrt{X}}, \quad = \int_0^x \frac{\sqrt{X_1}}{x_1 - \theta} \left(\frac{x - \theta}{x - x_1} + \frac{x + \theta}{x + x_1}\right) \frac{dx}{\sqrt{X}},$$

$$= 2\int_0^x \frac{\sqrt{X_1}}{x_1 - \theta} \cdot \frac{x^2 - x_1\theta}{x^2 - x_1^2}\, \frac{dx}{\sqrt{X}}.$$

We have

$$\frac{1}{x_1}\left(\frac{x^2}{x_1^2 - x^2} - \frac{\theta}{x_1 - \theta}\right) = \frac{x^2 - x_1\theta}{x_1^2 - x^2 . x_1 - \theta},$$

and the integral thus becomes

$$= -2\int_0^x \frac{\sqrt{X_1}\, x^2 dx}{x_1 (x_1^2 - x^2)\sqrt{X}} + 2\frac{\theta\sqrt{X_1}}{x_1(x_1 - \theta)} \int_0^x \frac{dx}{\sqrt{X}}.$$

Taking here $x = \operatorname{sn} u$, $x_1 = \operatorname{sn}(a + iK')$, $= \dfrac{1}{k \operatorname{sn} a}$, we have $dx = \operatorname{cn} u \operatorname{dn} u \, du = \sqrt{X}\, du$, and the result is

$$\int_{-x}^{x} \frac{\sqrt{X_1}}{x_1 - \theta} \frac{x - \theta}{x - x_1} \frac{dx}{\sqrt{X}} = -2 \int_0 \frac{k^2 \operatorname{sn} a \operatorname{cn} a \operatorname{dn} a \operatorname{sn}^2 u \, du}{1 - k^2 \operatorname{sn}^2 a \operatorname{sn}^2 u} + \frac{2k \operatorname{sn} a \operatorname{cn} a}{1 - k\theta \operatorname{sn} a} u,$$

where on the right-hand side the first term is $= -2\Pi(u, a)$, if $\Pi(u, a)$ be Jacobi's form of the integral of the third kind, see my *Elliptic Functions*, p. 143.

145. It is to be observed that the proper normal form is not $\Pi(u, a)$, but $\Pi(u, a) - uZa$; say this is $\overline{\Pi}(u, a)$. We then have

$$\overline{\Pi}(u, a) = \Pi(u, a) - u\left[a\left(1 - \frac{E}{K}\right) - k^2 \int \operatorname{sn}^2 a \, da\right],$$

and thence

$$\partial_u \overline{\Pi}(u, a) = \frac{k^2 \operatorname{sn} a \operatorname{cn} a \operatorname{dn} a \operatorname{sn}^2 u}{1 - k^2 \operatorname{sn}^2 a \operatorname{sn}^2 u} - a\left(1 - \frac{E}{K}\right) + k^2 \int \operatorname{sn}^2 a \, da,$$

$$\partial_a \partial_u \overline{\Pi}(u, a) = k^2 \operatorname{sn}^2 u \, \partial_a \frac{\operatorname{sn} a \operatorname{cn} a \operatorname{dn} a}{1 - k^2 \operatorname{sn}^2 a \operatorname{sn}^2 u} - \left(1 - \frac{E}{K}\right) + k^2 \operatorname{sn}^2 a;$$

or, if for shortness we write $\operatorname{sn} u, \operatorname{sn} a = s, \sigma$, this is

$$\partial_a \partial_u \overline{\Pi}(u, a) = \frac{k^2 s^2 [1 - 2(1 + k^2)\sigma^2 + 3k^2\sigma^4] + k^4 s^4 \sigma^2}{(1 - k^2 s^2 \sigma^2)^2} - \left(1 - \frac{E}{K}\right) + k^2 \sigma^2,$$

which is

$$= \frac{k^2 \{(s^2 + \sigma^2)(1 + k^2 s^2 \sigma^2) - 2(1 + k^2) s^2 \sigma^2\}}{(1 - k^2 s^2 \sigma^2)^2} - \left(1 - \frac{E}{K}\right);$$

or, this being symmetrical in regard to s, σ, we have

$$\partial_a \partial_u \overline{\Pi}(u, a) = \partial_a \partial_u \overline{\Pi}(a, u),$$

and thence by integration, and a proper determination of the constants,

$$\overline{\Pi}(u, a) = \overline{\Pi}(a, u).$$

Chapter VI. The Nodal Quartic.

Nodal Quartic; the General and Fleflecnodal Forms. Art. Nos. 146 to 148.

146. For a cubic, or other curve of deficiency 1, we are concerned with single points on the curve, and corresponding thereto with functions of a single argument (elliptic functions): but for a curve of deficiency 2, we have to consider pairs of points on the curve, and functions of two arguments: there is thus a marked change in the form of the results.

The most simple curve of deficiency 2 is the nodal quartic, $n=4$, $p=2$. Using homogeneous coordinates, the general form is $Az^2+2Bz+C=0$, where

$$A = (i,\ j,\ k\between x,\ y)^2,$$
$$B = (l,\ m,\ n,\ o\between x,\ y)^3,$$
$$C = (p,\ q,\ r,\ s,\ t\between x,\ y)^4,$$

and where we write also

$$B^2 - AC = (l^2 - ip)(x-ay)(x-by)(x-cy)(x-dy)(x-ey)(x-fy).$$

Clearly the equation of the two tangents at the node is $A=0$; and the equations of the six tangents from the node are $x-ay=0, \ldots, x-fy=0$: at the points of contact we have $Az+B=0$, viz. this is the equation of a nodal cubic, the node and the two tangents there being the same with the node and two tangents of the quartic. Hence the node counts as 6 intersections, and there are besides 6 intersections which are the points of contact of the 6 tangents respectively: say these are the points a, b, c, d, e, f: the coordinates of the point a are given by the equations

$$x : y : z = a : 1 : -\frac{B_a}{A_a}\left(= -\frac{C_a}{B_a}\right),$$

where A_a, B_a, C_a are what A, B, C become on writing therein a, 1 for x, y: and similarly for the other points.

147. An important special case occurs when $B=0$; say we have here

$$A = i(x-ey)(x-fy),$$
$$B = 0,$$
$$C = p(x-ay)(x-by)(x-cy)(x-dy),$$

or, omitting the factors i and p,

$$(x-ey)(x-fy)z^2 = (x-ay)(x-by)(x-cy)(x-dy).$$

The origin is here a fleflecnode; the tangents $x-ey=0$, $x-fy=0$ count as two of the six tangents from the node, and there remain the four tangents

$$x-cy=0,\quad x-dy=0,\quad x-ay=0,\quad x-by=0;$$

the four points of contact are the intersections of the curve with the line $z=0$.

148. The general nodal form depends on 11 constants, but by writing $\alpha x+\beta y$, $\gamma x+\delta y$, ϵz in place of x, y, z, we introduce 5 apoclastic constants, and so reduce the number to $11+1-5$, $=7$. Similarly the fleflecnodal form depends on 7 constants, but we reduce the number in like manner to $7+1-5$, $=3$: the final form might here be taken to be

$$z^2xy = (x-y)(x-by)(x-cy)(x-dy),$$

but it is more convenient to retain the original form

$$z^2(x-ey)(x-fy)=(x-ay)(x-by)(x-cy)(x-dy),$$

bearing in mind that this is reducible to the form just referred to, and thus depends virtually upon only 3 constants.

It is a general property that a curve of deficiency p greater than 1 can be by a rational transformation reduced to a curve of that deficiency depending upon $3p-3$ parameters: in particular, if $p=2$, then the form depending upon 3 parameters may be taken to be the flecnodal quartic as above: and I proceed to show how the general nodal quartic can, in fact, be reduced to this flecnodal form.

Reduction to the Flecnodal Form. Art. Nos. 149 to 152.

149. Consider the general nodal quartic $Az^2+2Bz+C=0$: take $\zeta=0$ for the equation of the line joining the points of contact of the tangents $x-ey=0$, $x-fy=0$; and then writing $x=\xi$, $y=\eta$, let the curve be transformed in the first instance from x, y, z to the new coordinates ξ, η, ζ.

Writing A_e for the value $(i, j, k \between e, 1)^2$, which A assumes on putting therein $(e, 1)$ for (x, y) respectively, and similarly A_f, B_e, B_f for the other like values, we may take

$$A_eA_f(e-f)\zeta=\begin{vmatrix} x, & y, & z \\ eA_e, & A_e, & -B_e \\ fA_f, & A_f, & -B_f \end{vmatrix},$$

$$=-x(A_eB_f-A_fB_e)+y(eA_eB_f-fB_eA_f)+z(e-f)A_eA_f,$$

say this equation is $\zeta=-\lambda x-\mu y+z$, the values of λ, μ being

$$\lambda=\frac{A_eB_f-A_fB_e}{(e-f)A_eA_f},\qquad \mu=\frac{-eA_eB_f+fB_eA_f}{(e-f)A_eA_f},$$

and therefore

$$\lambda e+\mu=-\frac{B_e}{A_e},\qquad \lambda f+\mu=-\frac{B_f}{A_f}.$$

150. From the values $\zeta, \xi, \eta=-\lambda x-\mu y+z, x, y$, we obtain $z, x, y=\zeta+\lambda\xi+\mu\eta, \xi, \eta$; and the transformed equation is

$$A'(\zeta+\lambda\xi+\mu\eta)^2+2B'(\zeta+\lambda\xi+\mu\eta)+C'=0,$$

where

$$A'=(i, j, k \between \xi, \eta)^2,$$
$$B'=(l, m, n, o \between \xi, \eta)^3,$$
$$C'=(p, q, r, s, t \between \xi, \eta)^4,$$

say this equation is $\mathfrak{A}\zeta^2+2\mathfrak{B}\zeta+\mathfrak{C}=0$, where

$$\mathfrak{A}=A',$$
$$\mathfrak{B}=A'(\lambda\xi+\mu\eta)+B',$$
$$\mathfrak{C}=A'(\lambda\xi+\mu\eta)^2+2B'(\lambda\xi+\mu\eta)+C',$$

and thence

$$\mathfrak{B}^2 - \mathfrak{A}\mathfrak{C} = B'^2 - A'C', \ = (l^2 - ip)(\xi - a\eta)(\xi - b\eta)(\xi - c\eta)(\xi - d\eta)(\xi - e\eta)(\xi - f\eta).$$

We have here $\mathfrak{B}, = A'(\lambda\xi + \mu\eta) + B'$, a cubic function $(\xi, \eta)^3$ containing the factors $\xi - e\eta$ and $\xi - f\eta$: in fact, writing $\xi, \eta = e, 1$, it becomes $A_e(\lambda e + \mu) + B_e$, which is $= 0$; and similarly writing $\xi, \eta = f, 1$, it becomes $A_f(\lambda f + \mu) + B_f$, which is $= 0$. Calling the other factor $L\xi + M\eta$, we have thus

$$\mathfrak{B} = (\xi - e\eta)(\xi - f\eta)(L\xi + M\eta),$$

and thence

$$\begin{aligned}\mathfrak{A}\mathfrak{C} &= (\xi - e\eta)^2(\xi - f\eta)^2(L\xi + M\eta)^2 - (l^2 - ip)(\xi - a\eta)(\xi - b\eta)(\xi - c\eta)(\xi - d\eta)(\xi - e\eta)(\xi - f\eta),\\ &= (\xi - e\eta)(\xi - f\eta)[(\xi - e\eta)(\xi - f\eta)(L\xi + M\eta)^2 - (l^2 - ip)(\xi - c\eta)(\xi - d\eta)(\xi - e\eta)(\xi - f\eta)].\end{aligned}$$

Hence $\mathfrak{C}$ contains the factor $(\xi - e\eta)(\xi - f\eta)$, say we have

$$\mathfrak{C} = \theta(\xi - e\eta)(\xi - f\eta)(\xi - \epsilon\eta)(\xi - \phi\eta).$$

151. In the equation $\mathfrak{A}\zeta^2 + 2\mathfrak{B}\zeta + \mathfrak{C} = 0$ of the quartic curve, writing $\zeta = 0$, we find $\mathfrak{C} = 0$, that is, $(\xi - e\eta)(\xi - f\eta)(\xi - \epsilon\eta)(\xi - \phi\eta) = 0$: but $\zeta = 0$ is the equation of the line joining the points of contact of the tangents $\xi - e\eta = 0$, $\xi - f\eta = 0$; hence $\xi - \epsilon\eta = 0$, $\xi - \phi\eta = 0$ are the lines drawn from the node to the two points ϵ, ϕ which are the residues of these two points of contact. We now have

$$\theta\mathfrak{A}(\xi - \epsilon\eta)(\xi - \phi\eta) = (\xi - e\eta)(\xi - f\eta)(L\xi + M\eta)^2 - (l^2 - ip)(\xi - a\eta)(\xi - b\eta)(\xi - c\eta)(\xi - d\eta),$$

and thence

$$\begin{aligned}0 &= (\epsilon - e)(\epsilon - f)(L\epsilon + M)^2 - (l^2 - ip)(\epsilon - a)(\epsilon - b)(\epsilon - c)(\epsilon - d),\\ 0 &= (\phi - e)(\phi - f)(L\phi + M)^2 - (l^2 - ip)(\phi - a)(\phi - b)(\phi - c)(\phi - d),\end{aligned}$$

which equations determine L and M; and then with these values of L, M, and for $\mathfrak{A}$ substituting its value $(i, j, k)(\xi, \eta)^2$, the equation must become an identity.

We have in what precedes, by the transformation $z = \zeta + \lambda\xi + \mu\eta$, $x = \xi$, $y = \eta$, passed from the form $Az^2 + 2Bz + C = 0$ to the form

$$\mathfrak{A}\zeta^2 + 2\mathfrak{B}\zeta + \mathfrak{C} = 0,$$

where

$$\begin{aligned}\mathfrak{A} &= (i, j, k)(\xi, \eta)^2,\\ \mathfrak{B} &= (\xi - e\eta)(\xi - f\eta)(L\xi + M\eta),\\ \mathfrak{C} &= \theta(\xi - e\eta)(\xi - f\eta)(\xi - \epsilon\eta)(\xi - \phi\eta),\end{aligned}$$

viz. $\mathfrak{B}$ and $\mathfrak{C}$ have here the common factor $(\xi - e\eta)(\xi - f\eta)$.

152. Assume now

$$\xi, \eta, \zeta = X, Y, \theta\frac{(X - \epsilon Y)(X - \phi Y)}{Z - LX - MY},$$

and therefore conversely

$$X, Y, Z = \xi, \eta, L\xi + M\eta + \theta\frac{(\xi - \epsilon\eta)(\xi - \phi\eta)}{\zeta};$$

then in the new coordinates (X, Y, Z) we have the equation

$$(i, j, k)(X, Y)^2 \theta^2 \left\{\frac{(X-\epsilon Y)(X-\phi Y)}{Z-LX-MY}\right\}^2$$
$$+2(X-eY)(X-fY)(LX+MY)\theta\frac{(X-\epsilon Y)(X-\phi Y)}{Z-LX-MY}$$
$$+\theta(X-eY)(X-fY)(X-\epsilon Y)(X-\phi Y)=0,$$

that is,

$$\begin{aligned}&(i, j, k)(X, Y)^2\theta(X-\epsilon Y)(X-\phi Y)\\ &+2(X-eY)(X-fY)(LX+MY)(Z-LX-MY)\\ &+\ (X-eY)(X-fY)(Z-LX-MY)^2=0,\end{aligned}$$

where the second and third lines together are

$$=(X-eY)(X-fY)\{Z^2-(LX+MY)^2\},$$

and the equation thus is

$$(X-eY)(X-fY)Z^2+\{\theta(i, j, k)(X, Y)^2(X-\epsilon Y)(X-\phi Y)$$
$$-(X-eY)(X-fY)(LX+MY)^2\}=0.$$

But the term in $\{\ \}$ is identically

$$=-(l^2-ip)(X-aY)(X-bY)(X-cY)(X-dY),$$

and the equation thus becomes

$$(X-eY)(X-fY)Z^2-(l^2-ip)(X-aY)(X-bY)(X-cY)(X-dY)=0;$$

viz. the original equation $Az^2+2Bz+C=0$ of the general nodal quartic is, by the equations

$$x, y, z=X, Y, \lambda X+\mu Y+\theta\frac{(X-\epsilon Y)(X-\phi Y)}{Z-LX-MY},$$

or conversely

$$X, Y, Z=x, y, Lx+My+\frac{\theta(x-\epsilon y)(x-\phi y)}{z-\lambda x-\mu y},$$

transformed into the fleflecnodal form as above.

It originally appeared to me that the fleflecnodal form was more easily dealt with than the general form; and I effected the transformation for this reason: there is, however, the disadvantage that the six points a, b, c, d, e, f enter into the equation unsymmetrically; and I afterwards found that the general form could be dealt with nearly as easily, and in what follows I use therefore the general form. The transformation is given as interesting for its own sake, and as an illustration of the theorem in regard to the number of constants in a curve of deficiency p.

Application of Abel's Theorem. Art. Nos. 153 to 157.

153. Taking the fixed curve to be $f, =\frac{1}{2}(Az^2+2Bz+C), =0$, we have

$$\frac{df}{dz}=Az+B=\sqrt{(x,\ y)^6},$$

if for shortness we write

$$(x,\ y)^6=B^2-AC,\ =(l^2-ip)(x-ay)(x-by)(x-cy)(x-dy)(x-ey)(x-fy),$$

and we thence have

$$d\omega=\frac{x\,dy-y\,dx}{\sqrt{(x,\ y)^6}}.$$

The minor curve $(x, y, z)^{n-3}=0$ is an arbitrary line passing through the node, that is, the point $x=0$, $y=0$; and the pure theorem thus gives the two relations $\Sigma x\,d\omega=0$, $\Sigma y\,d\omega=0$; where the summation extends to the intersections of the fixed curve $Az^2+2Bz+C=0$ with the variable curve ϕ.

The variable curve is taken to be a cubic $Az+B=(\alpha, \beta, \gamma, \delta \between x, y)^3$, or say $Az+B=\Omega$, where Ω is a given cubic function $(x, y)^3$: viz. this is a nodal cubic, the node and the two tangents there being the same with the node and the two tangents of the quartic: hence it meets the quartic in the node counting 6 times, and in 6 other points, say these are the points 1, 2, 3, 4, 5, 6: hence the differential relations are

$$x_1d\omega_1+x_2d\omega_2+x_3d\omega_3+x_4d\omega_4+x_5d\omega_5+x_6d\omega_6=0,$$

$$y_1d\omega_1+y_2d\omega_2+y_3d\omega_3+y_4d\omega_4+y_5d\omega_5+y_6d\omega_6=0.$$

154. Observe that the intersections of the cubic with the fixed curve are given by the equation $\Omega^2=B^2-AC$, or say $\Omega^2=(x, y)^6$, an equation which determines the ratio $x:y$ for the six points respectively; and the ratio $z:x$ is then determined rationally by the original equation $Az+B=\Omega$. Instead of regarding Ω as a given function, we may, if we please, take 1, 2, 3, 4 given points on the quartic: we then have four equations for the determination of the coefficients $(\alpha, \beta, \gamma, \delta)$ of the function Ω; viz. these equations may be taken to be

$$(\alpha,\ \beta,\ \gamma,\ \delta)(x_1,\ y_1)^3=\sqrt{(x_1,\ y_1)^6},$$
$$(\quad ,, \quad)(x_2,\ y_2)^3=\sqrt{(x_2,\ y_2)^6},$$
$$(\quad ,, \quad)(x_3,\ y_3)^3=\sqrt{(x_3,\ y_3)^6},$$
$$(\quad ,, \quad)(x_4,\ y_4)^3=\sqrt{(x_4,\ y_4)^6}.$$

Ω is hereby completely determined: and this being so, the remaining intersections 5 and 6 are also completely determined: there are thus between the six points 2 integral relations, which agrees with the number, $=2$, of the differential relations obtained above.

155. If we now assume

$$du = x_1 d\omega_1 + x_2 d\omega_2,\quad du' = x_3 d\omega_3 + x_4 d\omega_4,\quad du'' = x_5 d\omega_5 + x_6 d\omega_6,$$

$$dv = y_1 d\omega_1 + y_2 d\omega_2,\quad dv' = y_3 d\omega_3 + y_4 d\omega_4,\quad dv'' = y_5 d\omega_5 + y_6 d\omega_6,$$

or say

$$u,\ u',\ u'' = \int_{\alpha\beta}^{12},\ \int_{\alpha\beta}^{34},\ \int_{\alpha\beta}^{56} \frac{x\,(x\,dy - y\,dx)}{\sqrt{(x,\ y)^6}},$$

$$v,\ v',\ v'' = \int_{\alpha\beta}^{12},\ \int_{\alpha\beta}^{34},\ \int_{\alpha\beta}^{56} \frac{y\,(x\,dy - y\,dx)}{\sqrt{(x,\ y)^6}};$$

that is,

$$u = \left(\int_\alpha^1 + \int_\beta^2\right) \frac{x\,(x\,dy - y\,dx)}{\sqrt{(x,\ y)^6}},\quad v = \left(\int_\alpha^1 + \int_\beta^2\right) \frac{y\,(x\,dy - y\,dx)}{\sqrt{(x,\ y)^6}},$$

where α, β are points assumed at pleasure on the quartic: and similarly for u', v': u'', v'': then u, v are hereby determined as functions of the points 1, 2: and we may conversely regard the points 1, 2 as determined in terms of the two arguments u, v. We might, selecting any two symmetrical functions of the degree zero, for instance, $\frac{x_1}{y_1} + \frac{x_2}{y_2}$, $\frac{x_1 x_2}{y_1 y_2}$, represent them as functions $\phi(u,\ v)$, $\psi(u,\ v)$; and then $\frac{x_1}{y_1}$ and $\frac{x_2}{y_2}$ will be functions of $\phi(u,\ v)$, $\psi(u,\ v)$. But instead of this selection, it is proper to consider the ratios of six functions depending on the points a, b, c, d, e, f respectively: viz. we assume

$$\begin{array}{ccccccc} \sqrt{(x_1 - ay_1)(x_2 - ay_2)} & : & \sqrt{(x_1 - by_1)(x_2 - by_2)} & :\ldots: & \sqrt{(x_1 - fy_1)(x_2 - fy_2)} \\ = A\,(u,\ v) & : & B\,(u,\ v) & :\ldots: & F\,(u,\ v), \end{array}$$

and of course 3, 4 will be in like manner determined by means of the corresponding functions of u', v', and 5, 6 by means of the corresponding functions of u'', v''. The squared functions A^2, B^2, C^2, D^2, E^2, F^2 are proportional to given linear functions of $x_1 x_2$, $x_1 y_2 + x_2 y_1$, $y_1 y_2$, and are thus connected by three independent linear relations.

156. The differential relations then become

$$du + du' + du'' = 0,\quad dv + dv' + dv'' = 0,$$

and we have consequently

$$u + u' + u'' = I,\quad v + v' + v'' = J,$$

where I, J are constants which are determinable as definite integrals by the consideration that, when the cubic is taken to be $Az + B = 0$, the six points 1, 2, 3, 4, 5, 6 coincide with the points of contact a, b, c, d, e, f. I do not at present see my way to a proper development of this point of the theory: but in explanation of the nature of the result, I assume for the moment that by a proper determination of the inferior limits α, β, or otherwise, we may take $I = 0$, $J = 0$. We then have $u'' = -u - u'$, $v'' = -v - v'$; and the integral equations, which determine the points 5, 6 in terms of the points 1, 2 and the points 3, 4, then in effect determine the functions A, B, &c., of $-u - u'$, $-v - v'$, or say those of $u + u'$, $v + v'$ in terms of the like functions of $(u,\ v)$ and of $(u',\ v')$: viz. these equations give the addition-theory of the functions $A\,(u,\ v)$, &c.

157. We may, in the first instance, disregarding altogether the consideration of the arguments u, v, &c., attend only to the algebraic functions such as $\sqrt{(x_1 - ay_1)(x_2 - ay_2)}$, &c., of the coordinates of the pairs of points 1, 2; 3, 4, and 5, 6; and we can in regard to these develope a proper theory. This depends only on the equation $\Omega = \sqrt{(x, y)^6}$; it will be convenient to assume herein $y = 1$, and slightly modifying the form, to write it

$$(\alpha, \beta, \gamma, \delta)(x, 1)^3 = \sqrt{a - x . b - x . c - x . d - x . e - x . f - x};$$

and accordingly to consider the functions $\sqrt{a - x_1 . a - x_2}$, &c. These are called the single-letter functions A, &c., but there are certain double-letter functions AB, &c., which have also to be considered; and I will, in the first instance, show how these present themselves in connexion with the cubic curve.

Origin of the Double-Letter Functions. Art. Nos. 158 and 159.

158. The cubic curve $Az + B = \Omega$ may be taken to be a curve through two of the points of contact, say the points a, b; these will then be two out of the six points, and taking the remaining four points to be the pairs 1, 2 and 3, 4, we have single-letter functions of 3, 4 presenting themselves as double-letter functions of 1, 2. In fact, the equation of the curve is

$$Az + B = \lambda (x - ay)(x - by)(x - ky);$$

for the intersections with the quartic we have $\lambda^2 (x - ay)^2 (x - by)^2 (x - ky)^2 = \Omega^2$, or throwing out the factor $(x - ay)(x - by)$ and changing the constant λ, this is

$$(x - ay)(x - by)(x - ky)^2 - \lambda (x - cy)(x - dy)(x - ey)(x - fy) = 0;$$

and the quartic function must be a multiple of

$$(xy_1 - x_1 y)(xy_2 - x_2 y)(xy_3 - x_3 y)(xy_4 - x_4 y).$$

Putting each of the y's equal 1, we have the identity

$$(a - x)(b - x)(k - x)^2 - \lambda (c - x)(d - x)(e - x)(f - x) = \mu (x_1 - x)(x_2 - x)(x_3 - x)(x_4 - x);$$

and hence, introducing a notation which will be convenient, $a - x = \mathrm{a}$, $a - x_1 = \mathrm{a}_1$, and so in other cases, we have by giving different values to x the equations

$$\begin{aligned}
&\mathrm{a}_1 \mathrm{b}_1 \mathrm{k}_1^2 = \lambda \mathrm{c}_1 \mathrm{d}_1 \mathrm{e}_1 \mathrm{f}_1, && (a - c)(b - c)(k - c)^2 = \mu \mathrm{c}_1 \mathrm{c}_2 \mathrm{c}_3 \mathrm{c}_4, \\
&\mathrm{a}_2 \mathrm{b}_2 \mathrm{k}_2^2 = \lambda \mathrm{c}_2 \mathrm{d}_2 \mathrm{e}_2 \mathrm{f}_2, && (a - d)(b - d)(k - d)^2 = \mu \mathrm{d}_1 \mathrm{d}_2 \mathrm{d}_3 \mathrm{d}_4, \\
&\mathrm{a}_3 \mathrm{b}_3 \mathrm{k}_3^2 = \lambda \mathrm{c}_3 \mathrm{d}_3 \mathrm{e}_3 \mathrm{f}_3, && (a - e)(b - e)(k - e)^2 = \mu \mathrm{e}_1 \mathrm{e}_2 \mathrm{e}_3 \mathrm{e}_4, \\
&\mathrm{a}_4 \mathrm{b}_4 \mathrm{k}_4^2 = \lambda \mathrm{c}_4 \mathrm{d}_4 \mathrm{e}_4 \mathrm{f}_4, && (a - f)(b - f)(k - f)^2 = \mu \mathrm{f}_1 \mathrm{f}_2 \mathrm{f}_3 \mathrm{f}_4, \\
& && -\lambda (c - a)(d - a)(e - a)(f - a) = \mu \mathrm{a}_1 \mathrm{a}_2 \mathrm{a}_3 \mathrm{a}_4, \\
& && -\lambda (c - b)(d - b)(e - b)(f - b) = \mu \mathrm{b}_1 \mathrm{b}_2 \mathrm{b}_3 \mathrm{b}_4.
\end{aligned}$$

We have thus

$$\frac{(a-c)(b-c)}{(a-d)(b-d)}\left(\frac{c-k}{d-k}\right)^2 = \frac{\mathrm{c}_1\mathrm{c}_2\mathrm{c}_3\mathrm{c}_4}{\mathrm{d}_1\mathrm{d}_2\mathrm{d}_3\mathrm{d}_4},$$

and

$$\frac{\mathrm{k}_1{}^2}{\mathrm{k}_2{}^2} = \frac{\mathrm{a}_2\mathrm{b}_2\mathrm{c}_1\mathrm{d}_1\mathrm{e}_1\mathrm{f}_1}{\mathrm{a}_1\mathrm{b}_1\mathrm{c}_2\mathrm{d}_2\mathrm{e}_2\mathrm{f}_2},$$

which last equation, writing for a moment γ, $\delta = \sqrt{\mathrm{a}_2\mathrm{b}_2\mathrm{c}_1\mathrm{d}_1\mathrm{e}_1\mathrm{f}_1}$, $\sqrt{\mathrm{a}_1\mathrm{b}_1\mathrm{c}_2\mathrm{d}_2\mathrm{e}_2\mathrm{f}_2}$, gives $\dfrac{k-x_1}{k-x_2} = \dfrac{\gamma}{\delta}$, whence $k(\gamma-\delta) = x_2\gamma - x_1\delta$, and thence

$$\frac{c-k}{d-k} = \frac{\gamma\mathrm{c}_2 - \delta\mathrm{c}_1}{\gamma\mathrm{d}_2 - \delta\mathrm{d}_1} = \frac{\sqrt{\mathrm{c}_1\mathrm{c}_2}\,\{\sqrt{\mathrm{a}_2\mathrm{b}_2\mathrm{c}_2\mathrm{d}_1\mathrm{e}_1\mathrm{f}_1} - \sqrt{\mathrm{a}_1\mathrm{b}_1\mathrm{c}_1\mathrm{d}_2\mathrm{e}_2\mathrm{f}_2}\}}{\sqrt{\mathrm{d}_1\mathrm{d}_2}\,\{\sqrt{\mathrm{a}_2\mathrm{b}_2\mathrm{d}_2\mathrm{c}_1\mathrm{e}_1\mathrm{f}_1} - \sqrt{\mathrm{a}_1\mathrm{b}_1\mathrm{d}_1\mathrm{c}_2\mathrm{e}_2\mathrm{f}_2}\}},$$

or, substituting in the first equation, we have

$$\frac{\sqrt{(a-c)(b-c)}}{\sqrt{(a-d)(b-d)}} \cdot \frac{\sqrt{\mathrm{a}_2\mathrm{b}_2\mathrm{c}_2\mathrm{d}_1\mathrm{e}_1\mathrm{f}_1} - \sqrt{\mathrm{a}_1\mathrm{b}_1\mathrm{c}_1\mathrm{d}_2\mathrm{e}_2\mathrm{f}_2}}{\sqrt{\mathrm{a}_2\mathrm{b}_2\mathrm{d}_2\mathrm{c}_1\mathrm{e}_1\mathrm{f}_1} - \sqrt{\mathrm{a}_1\mathrm{b}_1\mathrm{d}_1\mathrm{c}_2\mathrm{e}_2\mathrm{f}_2}} = \frac{\sqrt{\mathrm{c}_3\mathrm{c}_4}}{\sqrt{\mathrm{d}_3\mathrm{d}_4}}.$$

159. Considering the duad DE as an abbreviation for the double triad $ABC.DEF$, the expressed duad being always accompanied by the letter F, we are thus led to the consideration of the double-letter functions

$$AB_{12} = \frac{1}{x_1 - x_2}\{\sqrt{\mathrm{a}_1\mathrm{b}_1\mathrm{f}_1\mathrm{c}_2\mathrm{d}_2\mathrm{e}_2} - \sqrt{\mathrm{a}_2\mathrm{b}_2\mathrm{f}_2\mathrm{c}_1\mathrm{d}_1\mathrm{e}_1}\}, \text{ \&c.,}$$

in connexion with the already mentioned single-letter functions $A_{12} = \sqrt{\mathrm{a}_1\mathrm{a}_2}$, &c., viz. in this notation the equation just obtained is

$$\frac{C_{34}}{D_{34}} = \sqrt{\frac{(a-c)(b-c)}{(a-d)(b-d)}}\,\frac{DE_{12}}{CE_{12}},$$

and it thus appears that, the points 3, 4 being obtained as above from the given points 1, 2, then the quotient of two of the single-letter functions of 3, 4 is a constant multiple of the quotient of two of the double-letter functions of 1, 2. Observe that the points 3, 4 are derived from 1, 2 by means of the two points a, b: we have DE standing for $ABC.DEF$, CE for $ABD.CEF$, and if the two functions were represented by ABC, ABD respectively, then the form would have been

$$\frac{C_{34}}{D_{34}} = \sqrt{\frac{(a-c)(b-c)}{(a-d)(b-d)}} \cdot \frac{ABC_{12}}{ABD_{12}},$$

which is a clearer expression of the theorem; the apparent want of symmetry of the first form arises only from the arbitrary selection of the letter F to accompany the expressed duad, and is at once removed by substituting for a duad DE the triad $ABC.DEF$ which is thereby signified. The denominator factor $x_1 - x_2$ is introduced in order to make the degree in x_1 or x_2 equal to that of the single-letter functions.

The Addition Theory. Art. Nos. 160 to 163.

160. We have the six single-letter symbols A, B, C, D, E, F; viz. $A_{12} = \sqrt{\mathrm{a}_1\mathrm{a}_2}$, &c.: and the ten double-letter symbols $AB, AC, AD, AE, BC, BD, BE, CD, CE, DE$, viz.

$$AB_{12} = \frac{1}{x_1 - x_2}\{\sqrt{\mathrm{a}_1\mathrm{b}_1\mathrm{f}_1\mathrm{c}_2\mathrm{d}_2\mathrm{e}_2} - \sqrt{\mathrm{a}_2\mathrm{b}_2\mathrm{f}_2\mathrm{c}_1\mathrm{d}_1\mathrm{e}_1}\}, \text{ \&c.,}$$

these 16 functions being connected by algebraical relations which are immediately deducible from these expressions of the functions in terms of x_1, x_2. The problem is to express the functions of 5, 6 in terms of those of 1, 2 and of those of 3, 4. The relation between the variables $x_1, x_2, x_3, x_4, x_5, x_6$ consists herein that we have $x_1, x_2, x_3, x_4, x_5, x_6$ as the roots of the equation

$$(\alpha x^3 + \beta x^2 + \gamma x + \delta)^2 - \lambda(a-x)(b-x)(c-x)(d-x)(e-x)(f-x) = 0;$$

or, what is the same thing, it consists in the identity

$$(\alpha x^3 + \beta x^2 + \gamma x + \delta)^2 - \lambda(a-x)(b-x)(c-x)(d-x)(e-x)(f-x)$$
$$- \mu(x_1 - x)(x_2 - x)(x_3 - x)(x_4 - x)(x_5 - x)(x_6 - x) = 0.$$

Again, it may be expressed by the plexus of equations

$$\left\|\begin{matrix} 1, & 1, & 1, & 1, & 1, & 1 \\ x_1, & x_2, & x_3, & x_4, & x_5, & x_6 \\ x_1^2, & x_2^2, & x_3^2, & x_4^2, & x_5^2, & x_6^2 \\ x_1^3, & x_2^3, & x_3^3, & x_4^3, & x_5^3, & x_6^3 \\ \sqrt{X_1}, & \sqrt{X_2}, & \sqrt{X_3}, & \sqrt{X_4}, & \sqrt{X_5}, & \sqrt{X_6} \end{matrix}\right\| = 0,$$

where $X_1 = (a - x_1)(b - x_1)\ldots(f - x_1)$, &c.; these are equivalent of course to two equations, and serve to determine x_5, x_6 in terms of x_1, x_2, x_3, x_4.

161. The solution is, in fact, as is given in my paper "On the addition of the double ϑ-functions," *Crelle*, t. LXXXVIII. (1880), pp. 74—81, [703]. Writing successively $x = x_1, x_2, x_3, x_4$, we have

$$\alpha x_1^3 + \beta x_1^2 + \gamma x_1 + \delta = \sqrt{\lambda}\sqrt{X_1},$$
$$\alpha x_2^3 + \beta x_2^2 + \gamma x_2 + \delta = \sqrt{\lambda}\sqrt{X_2},$$
$$\alpha x_3^3 + \beta x_3^2 + \gamma x_3 + \delta = \sqrt{\lambda}\sqrt{X_3},$$
$$\alpha x_4^3 + \beta x_4^2 + \gamma x_4 + \delta = \sqrt{\lambda}\sqrt{X_4},$$

which equations serve to determine the ratios $\alpha : \beta : \gamma : \delta$ in terms of x_1, x_2, x_3, x_4; and we have then the two like equations

$$\alpha x_5^3 + \beta x_5^2 + \gamma x_5 + \delta = \sqrt{\lambda}\sqrt{X_5},$$
$$\alpha x_6^3 + \beta x_6^2 + \gamma x_6 + \delta = \sqrt{\lambda}\sqrt{X_6},$$

which determine the symmetric functions of x_5, x_6.

If, reverting to the identity, we write therein for instance $x = a$, we find

$$\alpha a^3 + \beta a^2 + \gamma a + \delta = \sqrt{\mu}\, A_{12} A_{34} A_{56},$$

which equation when properly reduced gives the proportional value of A_{56}.

162. Calling for a moment the function on the left-hand side Ω, we have

$$\begin{vmatrix} x_1^3, & x_1^2, & x_1, & 1, & \sqrt{\lambda}\sqrt{X_1} \\ x_2^3, & x_2^2, & x_2, & 1, & \sqrt{\lambda}\sqrt{X_2} \\ x_3^3, & x_3^2, & x_3, & 1, & \sqrt{\lambda}\sqrt{X_3} \\ x_4^3, & x_4^2, & x_4, & 1, & \sqrt{\lambda}\sqrt{X_4} \\ a^3, & a^2, & a, & 1, & \Omega \end{vmatrix} = 0,$$

that is,

$$\Omega \begin{vmatrix} x_1^3, & x_1^2, & x_1, & 1 \\ x_2^3, & x_2^2, & x_2, & 1 \\ x_3^3, & x_3^2, & x_3, & 1 \\ x_4^3, & x_4^2, & x_4, & 1 \end{vmatrix} + \sqrt{\lambda} \begin{vmatrix} x_1^3, & x_1^2, & x_1, & 1, & \sqrt{X_1} \\ x_2^3, & x_2^2, & x_2, & 1, & \sqrt{X_2} \\ x_3^3, & x_3^2, & x_3, & 1, & \sqrt{X_3} \\ x_4^3, & x_4^2, & x_4, & 1, & \sqrt{X_4} \\ a^3, & a^2, & a, & 1, & 0 \end{vmatrix} = 0,$$

viz. this is

$$\begin{aligned} \Omega\, (x_1 - x_2)(x_1 - x_3)(x_1 - x_4)(x_2 - x_3)(x_2 - x_4)(x_3 - x_4) \\ = -\sqrt{\lambda}\,\{\sqrt{X_1}\,.\,x_2 - x_3\,.\,x_2 - x_4\,.\,x_2 - a\,.\,x_3 - x_4\,.\,x_3 - a\,.\,x_4 - a \\ + \sqrt{X_2}\,.\,x_3 - x_4\,.\,x_3 - a\,.\,x_3 - x_1\,.\,x_4 - a\,.\,x_4 - x_1\,.\,a - x_1 \\ + \sqrt{X_3}\,.\,x_4 - a\,.\,x_4 - x_1\,.\,x_4 - x_2\,.\,a - x_1\,.\,a - x_2\,.\,x_1 - x_2 \\ + \sqrt{X_4}\,.\,a - x_1\,.\,a - x_2\,.\,a - x_3\,.\,x_2 - x_3\,.\,x_2 - x_4\,.\,x_3 - x_4\}, \end{aligned}$$

or, as this may be written,

$$\begin{aligned} \Omega\,.\,x_1 - x_3\,.\,x_1 - x_4\,.\,x_2 - x_3\,.\,x_2 - x_4 \\ = \frac{\sqrt{\lambda}\,.\,a - x_3\,.\,a - x_4}{x_1 - x_2}\{x_2 - x_3\,.\,x_2 - x_4\,.\,a - x_2\,.\sqrt{X_1} - (x_1 - x_3\,.\,x_1 - x_4\,.\,a - x_1\,.\sqrt{X_2})\} \\ + \frac{\sqrt{\lambda}\,.\,a - x_1\,.\,a - x_2}{x_3 - x_4}\{x_4 - x_1\,.\,x_4 - x_2\,.\,a - x_4\,.\sqrt{X_3} - (x_3 - x_1\,.\,x_3 - x_2\,.\,a - x_3\,.\sqrt{X_4})\}. \end{aligned}$$

We have here $\sqrt{\lambda}\,.\,a - x_3\,.\,a - x_4 = \sqrt{\lambda}\,A^2_{34}$, and the function

$$\frac{1}{x_1 - x_2}\{x_2 - x_3\,.\,x_2 - x_4\,.\,a_2\sqrt{X_1} - (x_1 - x_3\,.\,x_1 - x_4\,.\,a_1\sqrt{X_2})\},$$

which multiplies this, is without difficulty found to be

$$= \frac{-A_{12}}{c - d\,.\,d - b\,.\,b - c}\,\Sigma\,\{c - d\,.\,B^2_{34} BE_{12}\,.\,C_{12}\,.\,D_{12}\},$$

where the summation extends to the three terms obtained by the cyclical interchange of the letters b, c, d: these being a set of three out of the five letters other than a. Similarly $\sqrt{\lambda}\,.\,a-x_1\,.\,a-x_2$ is $=\sqrt{\lambda}\,A^2_{12}$, and the function which multiplies this is

$$=\frac{-A_{34}}{c-d\,.\,d-b\,.\,b-c}\,\Sigma\,\{c-d\,.\,B^2_{12}\,.\,BE_{34}C_{14}D_{14}\}.$$

The expression for Ω thus contains the factor $A_{12}A_{34}$: but we have

$$\Omega,\ =\alpha a^3+ba^2+ca+d,\ =\sqrt{\mu}\,A_{12}A_{34}A_{56};$$

this equation contains therefore the factor $A_{12}A_{34}$, and omitting it we find

$$-\frac{\sqrt{\mu}}{\sqrt{\lambda}}(x_1-x_3\,.\,x_1-x_4\,.\,x_2-x_3\,.\,x_2-x_4)\,(c-d\,.\,d-b\,.\,b-c)\,A_{56}$$
$$=A_{34}\,\Sigma\,\{c-d\,.\,B^2_{34}BE_{12}C_{12}D_{12}\}+A_{12}\Sigma\,\{c-d\,.\,B^2_{12}BE_{34}C_{34}D_{14}\},$$

where, as before, the summations refer each to the three terms obtained by the cyclical interchange of the letters b, c, d; these being any three of the five letters other than a: and the remaining two letters e, f enter into the formulæ symmetrically. The formula thus gives for A_{56} ten values which are of course equal to each other.

By reason of the undetermined factor $\frac{\sqrt{\mu}}{\sqrt{\lambda}}$, the formula gives only the proportional value of A_{56}; viz. combining it with the like formulæ for B_{56}, &c., we have determinate values of the ratios $A_{56} : B_{56} : .. : F_{56}$. But this being understood, we regard the formula as a formula for each single-letter function of x_5, x_6 in terms of the single and double-letter functions of x_1, x_2 and of x_3, x_4 respectively.

163. We require further the expressions for the double-letter functions of x_5, x_6. Consider for example the function DE_{56}, which is

$$=\frac{1}{x_5-x_6}\{\sqrt{\mathrm{d}_5\mathrm{e}_5\mathrm{f}_5\mathrm{a}_6\mathrm{b}_6\mathrm{c}_6}-\sqrt{\mathrm{d}_6\mathrm{e}_6\mathrm{f}_6\mathrm{a}_5\mathrm{b}_5\mathrm{c}_5}\};$$

then multiplying by $A_{56}B_{56}C_{56}$, $=\sqrt{\mathrm{a}_5\mathrm{b}_5\mathrm{c}_5\mathrm{a}_6\mathrm{b}_6\mathrm{c}_6}$, we have

$$DE_{56}A_{56}B_{56}C_{56}=\frac{1}{x_5-x_6}\{\mathrm{a}_6\mathrm{b}_6\mathrm{c}_6\sqrt{X_5}-\mathrm{a}_5\mathrm{b}_5\mathrm{c}_5\sqrt{X_6}\},$$

or recollecting that $\sqrt{\lambda}\sqrt{X_5}$ and $\sqrt{\lambda}\sqrt{X_6}$ are $=\alpha x_5^3+\beta x_5^2+\gamma x_5+\delta$ and $\alpha x_6^3+\beta x_6^2+\gamma x_6+\delta$ respectively, this may be written

$$\sqrt{\lambda}\,DE_{56}A_{56}B_{56}C_{56}=\frac{1}{x_5-x_6}\{a-x_6\,.\,b-x_6\,.\,c-x_6\,.\,(\alpha x_5^3+\beta x_5^2+\gamma x_5+\delta)$$
$$-(a-x_5\,.\,b-x_5\,.\,c-x_5)\,(\alpha x_6^3+\beta x_6^2+\gamma x_6+\delta)\}.$$

Using the well-known identity

$$\alpha x_5^3+\beta x_5^2+\gamma x_5+\delta=\Sigma\,.\,\alpha a^3+\beta a^2+\gamma a+\delta\,.\,\frac{b-x_5\,.\,c-x_5\,.\,d-x_5}{b-a\,.\,c-a\,.\,d-a},$$

where the summation extends to the four terms obtained by the cyclical interchanges of the letters a, b, c, d: and the like identity for $\alpha x_6^3+\beta x_6^2+\gamma x_6+\delta$: there will be terms in $\alpha a^3+\beta a^2+\gamma a+\delta$, $\alpha b^3+\beta b^2+\gamma b+\delta$, $\alpha c^3+\beta c^2+\gamma c+\delta$, but the term in $\alpha d^3+\beta d^2+\gamma d+\delta$ will disappear of itself. After some easy reductions, the result is

$$\sqrt{\lambda}\, DE_{56} A_{56} B_{56} C_{56} = \Sigma \frac{\alpha a^3+\beta a^2+\gamma a+\delta}{b-a\,.\,c-a} B^2_{56} C^2_{56},$$

where the summation extends to the three terms obtained by the cyclical interchanges of the letters a, b, c. We have $\alpha a^3+\beta a^2+\gamma a+\delta = \sqrt{\mu}\,.\,A_{12}A_{34}A_{56}$, and similarly for the other two terms; the whole equation thus divides by $A_{56}B_{56}C_{56}$, and we find

$$-\frac{\sqrt{\mu}}{\sqrt{\lambda}} DE_{56} = \frac{1}{b-c\,.\,c-a\,.\,a-b}\left(\frac{\sqrt{\mu}}{\sqrt{\lambda}}\right)^2 . \Sigma\,(b-c\,.\,A_{12}A_{34}B_{56}C_{56}),$$

in which equation, if we imagine $\frac{\sqrt{\mu}}{\sqrt{\lambda}}A_{56}$, $\frac{\sqrt{\mu}}{\sqrt{\lambda}}B_{56}$, $\frac{\sqrt{\mu}}{\sqrt{\lambda}}C_{56}$, each replaced by its value in terms of the single and double-letter functions of x_1, x_2 and x_3, x_4, we have an equation of the form

$$-\frac{\sqrt{\mu}}{\sqrt{\lambda}}(x_1-x_3\,.\,x_1-x_4\,.\,x_2-x_3\,.\,x_2-x_4)\, DE_{56} = \frac{1}{x_1-x_3\,.\,x_1-x_4\,.\,x_2-x_3\,.\,x_2-x_4} M,$$

where M is a given rational and integral function of the single and double-letter functions of x_1, x_2 and x_3, x_4. The factor on the left-hand side has been made the same as in the formula for the single-letter functions A_{56}, &c., and to do this it was necessary to bring in on the right-hand side the factor

$$\frac{1}{x_1-x_3\,.\,x_1-x_4\,.\,x_2-x_3\,.\,x_2-x_4};$$

this disappears in the expression for the ratio of two double-letter functions; but it enters into the expression for the ratio of a single-letter to a double-letter function, and it then requires to be itself expressed in terms of the functions of x_1, x_2 and x_3, x_4: it is easy to see that we have

$$x_1-x_3\,.\,x_1-x_4\,.\,x_2-x_3\,.\,x_2-x_4 = \Sigma \frac{(A^2_{34}B^2_{12}-A^2_{12}B^2_{34})(A^2_{34}C^2_{12}-A^2_{12}C^2_{34})}{(a-b)^2(a-c)^2},$$

where the summation extends to the three terms obtained by the cyclical interchanges of the letters a, b, c: these being a set of any three out of the six letters.

We have, in what precedes, obtained the expressions for the ratios of the 16 functions $A_{56}, \ldots, F_{56}$, $AB_{56}, \ldots, DE_{56}$ in terms of the ratios of the like functions of x_1, x_2 and x_3, x_4.

CHAPTER VII. THE FUNCTIONS T, U, V, Θ.

The present chapter is substantially a reproduction of C. and G.'s seventh section, "Die Function $T_{\xi\eta}$", borrowing only from the next section the definition of the theta-function; but for greater simplicity I consider for the most part, the case, fixed curve a quartic, $n=4$, $p=3$.

Integral Form of the Affected Theorem. Art. Nos. 164 to 169.

164. Writing for shortness $\dfrac{(x, y, z)_{12}{}^{n-2}d\omega}{012}=d\Pi_{12}$, we are concerned with the integrals $\int_{a'}^{a} d\Pi_{12}$ which present themselves in connexion with the affected theorem: the notation is explained, Chap. V.; a, a' are points on the curve f; the variable may be any parameter serving for the determination of the current point, and the integral, taken from the value which belongs to the point a' to the value which belongs to the point a, is represented as above by means of the two points a, a' as limits of the integral. It is assumed that the integral is a canonical integral having the limits and the parametric points interchangeable, $\int_{a'}^{a} d\Pi_{12}=\int_{2}^{1} d\Pi_{aa'}$: see Chapter IV.

165. Writing for shortness

$$\left(\int_{a'}^{a}+\int_{b'}^{b}+\int_{c'}^{c}+\ldots\right)d\Pi_{12}=\int\begin{pmatrix}a, & b, & c, \ldots\\ a', & b', & c', \ldots\end{pmatrix}d\Pi_{12},$$

then if ϕ, ψ are curves each of the order m, the former of them intersecting the fixed curve f in the points a, b, c,..., and the latter of them intersecting the same curve in the points a', b', c',..., and if ϕ_1, ψ_1, ϕ_2, ψ_2 are what the functions ϕ, ψ become on substituting therein in place of the current coordinates the values which belong to the parametric points 1, 2 respectively; the theorem becomes

$$\int\begin{pmatrix}a, & b, & c, \ldots\\ a', & b', & c', \ldots\end{pmatrix}d\Pi_{12}=\log\frac{\phi_2\psi_1}{\phi_1\psi_2}.$$

The superior limits may be interchanged in any manner, and so also the inferior limits may be interchanged in any manner. If a superior limit coincide with an inferior limit, the two may thus be considered as belonging to an integral which will then have the value 0, and the coincident points may therefore be omitted from the expression on the left-hand side: and so in the case of any number of coincidences.

166. If the intersections of the curves ϕ, ψ and the parametric points are situate on a curve of the order m; then taking the equation of this curve to be $\phi+\lambda\psi=0$, we have simultaneously $\phi_1+\lambda\psi_1=0$, $\phi_2+\lambda\psi_2=0$; whence $\phi_2\psi_1=\psi_1\phi_2$, and the logarithmic term disappears: viz. the theorem becomes

$$\int\begin{pmatrix}a, & b, & c, \ldots\\ a', & b', & c', \ldots\end{pmatrix}d\Pi_{12}=0.$$

167. Suppose that the curves ϕ, ψ are each of them a major curve, that is, a curve of the order $n-2$ passing through the δ dps, and consequently besides meeting the curve f in $n(n-2)-2\delta$, $=2p+n-2$ points: the theorem is

$$\int \begin{pmatrix} a, & b, & c, \ldots \\ a', & b', & c', \ldots \end{pmatrix} d\Pi_{12} = \log \frac{\phi_2 \psi_1}{\phi_1 \psi_2},$$

where the numbers of the superior and of the inferior points are each $=2p+n-2$.

168. Suppose further that the curves ϕ, ψ, being major curves as above, pass each of them through the $n-2$ residues of 1, 2; they besides meet in $(n-2)(n-3)$ points, viz. these are the δ dps and $(n-2)(n-3)-\delta$ variable points: these $(n-2)(n-3)$ points lie on a minor curve, that is, a curve of the order $n-3$ passing through the dps; and the minor curve together with the parametric line 12 make together a major curve, passing through the intersections of ϕ, ψ and also through the parametric points 1, 2: viz. these points and the intersections of ϕ, ψ are situate on a curve of the order $n-2$; the logarithmic term thus vanishes. The intersections of ϕ with the fixed curve are the δ dps, the $n-2$ residues and $2p$ other points, say these are $a, b, c, \ldots, a^\times, b^\times, c^\times, \ldots$; similarly the curve ψ meets the fixed curve in the δ dps, the $n-2$ residues, and in $2p$ other points, say these are $d, e, f, \ldots, d^\times, e^\times, f^\times, \ldots$: the theorem is

$$\int \begin{pmatrix} a, & b, & c, \ldots a^\times, & b^\times, & c^\times, \ldots \\ d, & e, & f, \ldots d^\times, & e^\times, & f^\times, \ldots \end{pmatrix} d\Pi_{12} = 0,$$

where there are $2p$ superior and inferior points respectively.

169. I introduce the definitions: a minor curve meets the fixed curve in the dps and in $2p-2$ other points, called "cominors": a major curve passing through the $n-2$ residues of the points 1, 2, meets the fixed curve in the δ dps, the $n-2$ residues and in $2p$ other points, called "comajors in regard to the points 1, 2." Observe that $p-1$ of the cominors determine uniquely the remaining $p-1$ cominors; and similarly p of the comajors determine uniquely the remaining p comajors.

The foregoing theorem thus is that the sum $\int \left(\; \right) d\Pi_{12}$ is $=0$, when the superior points and the inferior points are each of them a system of comajors in regard to the parametric points 1, 2.

Fixed Curve a Quartic. Art. No. 170.

170. It would be easy to go on with the general form; but as already mentioned, I prefer to consider the case, fixed curve a quartic, $n=4$, $p=3$. A minor curve is here a line meeting the quartic in 4 points, which are "cominors"; the major curve is a conic, and if this passes through the residues of 1, 2 it besides meets the quartic in 6 points, which are "comajors in regard to the points 1, 2." Two points and their residues are cominors, but this is only by reason that $n-3=1$.

The Function T. Art. No. 171.

171. In conformity with C. and G., I introduce the functional symbol

$$T_{12} \begin{pmatrix} a, & b, & c, \ldots \\ a', & b', & c', \ldots \end{pmatrix} = \int \begin{pmatrix} a, & b, & c, \ldots \\ a', & b', & c', \ldots \end{pmatrix} d\Pi_{12},$$

so that T denotes a function of the parametric points, and of the sets of superior and inferior points respectively. The foregoing theorem for the quartic thus is

$$T_{12}\begin{pmatrix} a, & b, & c, & a^\times, & b^\times, & c^\times \\ d, & e, & f, & d^\times, & e^\times, & f^\times \end{pmatrix} = 0.$$

Observing that

$$\int_d^a + \int_{d^\times}^{a^\times} = 2\int_d^a - \int_{a^\times}^a + \int_{d^\times}^d,$$

and so in other cases, this may be written

$$2T_{12}\begin{pmatrix} a, & b, & c \\ d, & e, & f \end{pmatrix} = T_{12}\begin{pmatrix} a\ , & b\ , & c \\ a^\times, & b^\times, & c^\times \end{pmatrix} - T_{12}\begin{pmatrix} d\ , & e\ , & f \\ d^\times, & e^\times, & f^\times \end{pmatrix};$$

and if, as a definition of $T_{12}(a, b, c)$, we write

$$T_{12}(a, b, c) = T_{12}\begin{pmatrix} a\ , & b\ , & c \\ a^\times, & b^\times, & c^\times \end{pmatrix},$$

where $a^\times$, $b^\times$, $c^\times$ are the comajors of a, b, c in regard to 1, 2, then the equation is

$$2T_{12}\begin{pmatrix} a, & b, & c \\ d, & e, & f \end{pmatrix} = T_{12}(a, b, c) - T_{12}(d, e, f),$$

viz. the function of the $(2p+2=)\,8$ points 1, 2, a, b, c, d, e, f is here expressed as a difference of two functions each of $(p+2=)\,5$ points: $T_{12}(a, b, c)$ is regarded as a function of the 5 points 1, 2, a, b, c, because the remaining points $a^\times$, $b^\times$, $c^\times$ depend only on these 5 points.

The Function U. Art. Nos. 172 to 175.

172. We consider on the quartic the points ξ, μ; 1, 2, 3; and take f, f' for the cominors of 2, 3; g, g' for the cominors of 3, 1; and h, h' for the cominors of 1, 2. We write

$$T = T_{\xi\mu}(1, 2, 3),$$

$$T_1 = T_{1\mu}(\xi, f, f'),\quad T_2 = T_{2\mu}(\xi, g, g'),\quad T_3 = T_{3\mu}(\xi, h, h');$$

it is to be shown that there exists a function $U(1, 2, 3;\ \xi)$, such that

$$\delta U = \tfrac{1}{2}\{\delta_\xi T + \delta_1 T_1 + \delta_2 T_2 + \delta_3 T_3\},$$

viz. considering ξ, 1, 2, 3 as variable points on the quartic, the whole infinitesimal variation of U is the sum of these parts, where $\delta_\xi T$ is the variation of T when only ξ is varied, $\delta_1 T_1$ the variation of T_1 when only 1 is varied, and similarly for $\delta_2 T_2$ and $\delta_3 T_3$. We consider in the proof three other points 4, 5, 6 on the quartic; and taking l, l' for the cominors of 5, 6; m, m' for those of 6, 4; and n, n' for those of 4, 5, we write further

$$X_1 = T_{1\mu}(4, l, l'),\quad X_2 = T_{2\mu}(5, m, m'),\quad X_3 = T_{3\mu}(6, n, n');$$

and it then requires to be shown that

$$\tfrac{1}{2}T \qquad = \int \begin{pmatrix}123\\456\end{pmatrix} d\Pi_{\xi\mu} + \tfrac{1}{2}T_{\xi\mu}(4,\ 5,\ 6),$$

$$\tfrac{1}{2}(T_1 - X_1) = \int \begin{pmatrix}\xi 56\\423\end{pmatrix} d\Pi_{1\mu} + \log \frac{\mu 23 \,.\, 156}{\mu 56 \,.\, 123},$$

$$\tfrac{1}{2}(T_2 - X_2) = \int \begin{pmatrix}\xi 64\\531\end{pmatrix} d\Pi_{2\mu} + \log \frac{\mu 31 \,.\, 264}{\mu 64 \,.\, 123},$$

$$\tfrac{1}{2}(T_3 - X_3) = \int \begin{pmatrix}\xi 45\\612\end{pmatrix} d\Pi_{3\mu} + \log \frac{\mu 12 \,.\, 345}{\mu 45 \,.\, 123},$$

where $\mu 23$ is the determinant formed with the coordinates of the points μ, 2, 3 respectively: and so in other cases.

173. We have

$$T_{\xi\mu}\begin{pmatrix}1, & 2, & 3\\4, & 5, & 6\end{pmatrix} = \tfrac{1}{2}T_{\xi\mu}(1,\ 2,\ 3) - \tfrac{1}{2}T_{\xi\mu}(4,\ 5,\ 6),$$

that is,

$$= \tfrac{1}{2}T \qquad - \tfrac{1}{2}T_{\xi\mu}(4,\ 5,\ 6),$$

and thence the above value of $\tfrac{1}{2}T$.

The affected theorem gives

$$\int \begin{pmatrix}f, & f', & 2, & 3\\l, & l', & 5, & 6\end{pmatrix} d\Pi_{1\mu} = \log \frac{F_\mu L_1}{F_1 L_\mu},$$

where $F=0$ is the equation of the line through f, f', 2, 3; and F_1, F_μ are what the function F becomes, on substituting therein for the current coordinates the coordinates of the points 1, μ respectively. And similarly $L=0$ is the equation of the line through l, l', 5, 6; and L_1, L_μ are what the function L becomes by the same substitutions respectively. The values of F_1, F_μ are 123, $\mu 23$: those of L_1, L_μ are 156, $\mu 56$; and the logarithmic term is thus

$$= \log \frac{\mu 23 \,.\, 156}{\mu 56 \,.\, 123}.$$

We then have

$$\tfrac{1}{2}(T_1 - X_1) = \int \begin{pmatrix}\xi, & f, & f'\\4, & l, & l'\end{pmatrix} d\Pi_{1\mu}, \ = \int \begin{pmatrix}\xi, & 5, & 6\\4, & 2, & 3\end{pmatrix} d\Pi_{1\mu} + \int \begin{pmatrix}f, & f', & 2, & 3\\l, & l', & 5, & 6\end{pmatrix} d\Pi_{1\mu},$$

and in this last expression for $\tfrac{1}{2}(T_1 - X_1)$ substituting for the second term the logarithmic value just obtained, we have the required expression for $\tfrac{1}{2}(T_1 - X_1)$: and those for $\tfrac{1}{2}(T_2 - X_2)$ and $\tfrac{1}{2}(T_3 - X_3)$ are deduced by mere cyclical permutations of the letters.

174. Returning to the assumed relation $\delta U = \tfrac{1}{2}\{\delta_\xi T + \delta_1 T_1 + \delta_2 T_2 + \delta_3 T_3\}$; in order to the existence of the function U, it is only necessary to show that $T - T_1$ contains no term in 1, ξ, that is, no term depending on both these points, and that $T_1 - T_2$ contains no term in 1, 2: for then, by symmetry, the like properties hold in regard to $T - T_2$, $T - T_3$, $T_1 - T_3$, $T_2 - T_3$ respectively, and the assumed expression is a complete differential, from which the function U may be obtained by integration.

175. To show that $T - T_1$ contains no term in 1, ξ.

For T, the only term in 1, ξ is $\int_4^1 d\Pi_{\xi\mu}$,

" T_1 " " $\int_4^\xi d\Pi_{1\mu}$,

and it is to be shown that the difference of the two integrals contains no term in 1, ξ. Considering on the quartic the two new points δ, ϵ, the first integral is

$$\int \binom{1,\ \delta}{\delta,\ 4} (d\Pi_{\xi\epsilon} + d\Pi_{\epsilon\mu}),\ = \int_\delta^1 d\Pi_{\xi\epsilon} + \int_4^\delta d\Pi_{\xi\epsilon} + \int_4^1 d\Pi_{\epsilon\mu},$$

and the second is

$$\int \binom{\xi,\ \epsilon}{\epsilon,\ 4} (d\Pi_{1\delta} + d\Pi_{\delta\mu}),\ = \int_\epsilon^\xi d\Pi_{1\delta} + \int_4^\epsilon d\Pi_{\delta\mu} + \int_\mu^\xi d\Pi_{\delta\mu}.$$

Hence in the difference the only terms which can contain 1, ξ are

$$\int_\delta^1 d\Pi_{\xi\epsilon} - \int_\epsilon^\xi d\Pi_{1\delta}$$

and this is $=0$: wherefore there is not in the difference any term in 1, ξ. This proves the property for $T - T_1$. The property for $T_1 - T_2$ is proved in a similar manner.

Theorems in regard to the Function U. Art. Nos. 176 to 179.

176. Theorem (A). To prove

$$U(1,\ 2,\ 3;\ \xi) - U(1,\ 2,\ 3;\ \mu) = \tfrac{1}{2} T_{\xi\mu}(1,\ 2,\ 3), \tag{A}$$

we have

$$U(1,\ 2,\ 3;\ \xi) - U(1,\ 2,\ 3;\ \mu) = \int_\mu^\xi d_\xi U = \tfrac{1}{2}\int_\mu^\xi d_\xi T$$
$$= \tfrac{1}{2} T_{\xi\mu}(1,\ 2,\ 3) - \tfrac{1}{2} T_{\mu\mu}(1,\ 2,\ 3),$$

and $T_{\mu\mu}(1,\ 2,\ 3) = \int \binom{1,\ \ 2,\ \ 3}{1^\times\ 2^\times\ 3^\times} d\Pi_{\mu\mu}$, where $d\Pi_{\mu\mu} = 0$, viz. considering this as derived from $d\Pi_{\xi\mu}$, $= Q_{\xi\mu}\, d\omega$, by making the point ξ coincide with μ, then when ξ is indefinitely near to μ, the numerator and denominator of $Q_{\xi\mu}$ are each of them infinitesimal of the orders 3 and 2 respectively, and thus the function $Q_{\xi\mu}$ ultimately vanishes (see as to this, Chap. V. Art. Nos. 99 to 106). We have therefore $T_{\mu\mu}(1,\ 2,\ 3) = 0$, and the required theorem is proved.

177. Theorem (B). To prove

$$U(1,\ 2,\ 3;\ \xi) - U(4,\ 2,\ 3;\ \xi) = \tfrac{1}{2} T_{14}(\xi,\ f,\ f'), \tag{B}$$

where, as before, f, f' are the cominors of 2, 3, that is, 2, 3, f, f' lie on a line, we have

$$U(1,\ 2,\ 3;\ \xi) - U(4,\ 2,\ 3;\ \xi) = \int_4^1 d_1 U = \tfrac{1}{2}\int_4^1 d_1 T_1$$
$$= \tfrac{1}{2} T_{1\mu}(\xi,\ f,\ f') - \tfrac{1}{2} T_{4\mu}(\xi,\ f,\ f'),$$

the point μ is arbitrary, and it may be taken to coincide with 4; but we then have $T_{44}(\xi, f, f')=0$, and the theorem is thus proved.

178. Theorem (C). We have

$$T_{11^\times}(\xi, f, f')+T_{11^\times}(\eta, k, k')=0; \qquad \text{(C)}$$

where ξ, η, 1, 2, 3 are arbitrary points on the quartic; $1^\times$, $2^\times$, $3^\times$ are the comajors of 1, 2, 3 in regard to ξ, η, viz. the points 1, 2, 3, $1^\times$, $2^\times$, $3^\times$ lie on a conic which passes through ξ', η' the residues (or cominors) of ξ, η: f, f' are the cominors of 2, 3; and k, k' are the cominors of $2^\times$, $3^\times$.

Taking θ, θ' for the cominors of 1, $1^\times$, the four lines $\xi\eta\xi'\eta'$, $11^\times\theta\theta'$, $23ff'$ and $2^\times 3^\times kk'$ form a quartic cutting the fixed quartic in the 16 points: but of these, ξ', η', 1, 2, 3, $1^\times$, $2^\times$, $3^\times$ lie in a conic: hence the remaining 8 points θ, θ', ξ, η, f, f', k, k' lie in a conic; that is, ξ, η, f, f', k, k' lie on a conic through θ, θ', the residues of 1, $1^\times$, or they are comajors in regard to 1, $1^\times$; whence the theorem.

179. We have

	From A.	From B.
$\tfrac{1}{2}T_{11^\times}(\xi, f, f') =$	$U(\xi, f, f';\ 1) - U(\xi, f, f';\ 1^\times)$,	$= U(1, 2, 3;\ \xi) - U(1^\times, 2, 3;\ \xi)$,
$\tfrac{1}{2}T_{11^\times}(\eta, k, k') =$	$U(\eta, k, k';\ 1) - U(\eta, k, k';\ 1^\times)$,	$= U(1, 2^\times, 3^\times;\ \eta) - U(1^\times, 2, 3;\ \eta)$;

viz. we have thus two expressions for each term of the equation (C),

$$T_{11^\times}(\xi, f, f')+T_{11^\times}(\eta, k, k')=0.$$

In particular, we have Theorem (D)

$$U(1, 2, 3;\ \xi) - U(1^\times, 2, 3;\ \xi) = -U(1, 2^\times, 3^\times;\ \eta) + U(1^\times, 2^\times, 3^\times;\ \eta). \qquad \text{(D)}$$

Again, we have $T_{\xi\eta}(1, 2, 3)+T_{\xi\eta}(1^\times, 2^\times, 3^\times)=0$; where 1, 2, 3, $1^\times$, $2^\times$, $3^\times$ are comajors in regard to ξ, η: and

$$\tfrac{1}{2}T_{\xi\eta}(1, 2, 3) = U(1, 2, 3;\ \xi) - U(1, 2, 3;\ \eta),$$
$$\tfrac{1}{2}T_{\xi\eta}(1^\times, 2^\times, 3^\times) = U(1^\times, 2^\times, 3^\times, \xi) - U(1^\times, 2^\times, 3^\times, \eta);$$

whence Theorem (E),

$$U(1, 2, 3;\ \xi) - U(1, 2, 3;\ \eta) = -U(1^\times, 2^\times, 3^\times;\ \xi) + U(1^\times, 2^\times, 3^\times, \eta). \qquad \text{(E)}$$

The Function V. Art. Nos. 180 to 182.

180. It is convenient to consider U as a logarithm, say

$$-U(1, 2, 3;\ \xi) = \log V(1, 2, 3;\ \xi), \text{ or } V(1, 2, 3;\ \xi) = \exp. - U(1, 2, 3;\ \xi);$$

V, like U, is a function of the $(p+1=)\,4$ points 1, 2, 3, ξ, on the quartic.

The equation (D) thus becomes

$$\frac{V(1, 2, 3;\ \xi)}{V(1^\times, 2^\times, 3^\times;\ \eta)} = \frac{V(1^\times, 2, 3;\ \xi)}{V(1, 2^\times, 3^\times;\ \eta)},$$

where 1, 2, 3, $1^\times$, $2^\times$, $3^\times$ are comajors in regard to ξ, η: the equation shows that, in the function $\dfrac{V(1, 2, 3;\ \xi)}{V(1^\times, 2^\times, 3^\times;\ \eta)}$, we can without alteration of the value interchange a pair of points 1, $1^\times$ out of the system of comajor points; and it of course follows that we can in any manner whatever interchange these points, so as to have any three of them in the numerator-function and the remaining three in the denominator-function. In particular, we have

$$\frac{V(1, 2, 3;\ \xi)}{V(1^\times, 2^\times, 3^\times;\ \eta)} = \frac{V(1^\times, 2^\times, 3^\times;\ \xi)}{V(1, 2, 3;\ \eta)}.$$

The equation (E) becomes

$$\frac{V(1, 2, 3;\ \xi)}{V(1, 2, 3;\ \eta)} = \frac{V(1^\times, 2^\times, 3^\times;\ \eta)}{V(1^\times, 2^\times, 3^\times;\ \xi)},$$

and multiplying we find

$$V^2(1, 2, 3;\ \xi) = V^2(1^\times, 2^\times, 3^\times;\ \eta),$$

that is, $V(1, 2, 3;\ \xi) = \pm V(1^\times, 2^\times, 3^\times;\ \eta)$, the sign being determinately + or determinately −, according to the precise definition of the function V.

181. Considering η and also $1^\times$, $2^\times$, $3^\times$ as fixed points on the curve; but ξ as a variable point (that is, the parametric line $\xi\eta$ as rotating about the fixed point η), the points 1, 2, 3 are then determined as the remaining intersections with the quartic of the conic which passes through the points $1^\times$, $2^\times$, $3^\times$ and through the points ξ', η', which are the residues of ξ, η. And by the theorem just obtained it appears that, ξ, 1, 2, 3 thus varying, the function $V(1, 2, 3;\ \xi)$ remains constant. This comes to saying that V, considered as a function of the points 1, 2, 3, ξ, satisfies a certain linear partial differential equation of the first order, having a solution $V = F(u, v, w)$, an arbitrary function of u, v, w, determinate functions of the points 1, 2, 3, ξ. And if we can find u, v, w functions of these points such that they each of them remain constant when the points 1, 2, 3, ξ vary as above, then the arbitrary function of u, v, w will remain constant for the variation in question and will thus be a value of the function V.

182. It is easily seen that such functions are

$$u,\ v,\ w = \left(\int^1 + \int^2 + \int^3 - \int^\xi\right) x\,d\omega,\ y\,d\omega,\ z\,d\omega,$$

the inferior limits being given points which are regarded as absolute constants. For by the pure theorem, we have

$$\Sigma\,(x,\ y,\ z)^1\,d\omega = 0,$$

where $(x,\ y,\ z)^1$ is an arbitrary linear function, and where the summation extends to all the intersections of the quartic with any given curve. Writing

$$p = \int x\,d\omega,\ \int y\,d\omega \text{ or } \int z\,d\omega, \text{ that is, } p_1 = \int^1 x\,d\omega,\ \int^1 y\,d\omega \text{ or } \int^1 z\,d\omega,$$

the inferior limits being any absolutely fixed point on the curve, and similarly p_2, &c.; the integral form of the theorem is $\Sigma p =$ constant. And applying the theorem successively to the parametric line, and to the conic which determines the points 1, 2, 3, we have

$$p_\xi + p_\eta + p_{\xi'} + p_{\eta'} = \text{const.},$$

$$p_{\xi'} + p_{\eta'} + p_1 + p_2 + p_3 + p_{1^\times} + p_{2^\times} + p_{3^\times} = \text{const.}$$

Taking the difference of these equations, we have

$$p_1 + p_2 + p_3 - p_\xi + p_{1^\times} + p_{2^\times} + p_{3^\times} - p_\eta = \text{const.},$$

viz. the points η, $1^\times$, $2^\times$, $3^\times$ being fixed points, this is

$$p_1 + p_2 + p_3 - p_\xi = \text{const.},$$

that is, the functions u, v, w defined as above are each of them constant under the variation in question.

The Function Θ. Art. Nos. 183 and 184.

183. The function $V(1, 2, 3; \xi)$ of the $(p+1=)$ 4 points 1, 2, 3, ξ, is thus a function of the $(p=)$ 3 arguments

$$u, v, w, = \left(\int^1 + \int^2 + \int^3 - \int^\xi\right) x\, d\omega,\ y\, d\omega,\ z\, d\omega.$$

Disregarding a constant and exponential factor, we say that it is a theta-function of these arguments, and we write the result provisionally in the form

$$V(1, 2, 3; \xi) = \Theta(u, v, w),$$

the more precise definition of the theta-function being reserved for further consideration.

184. It appears by what precedes that a sum of $(p=)$ 3 integrals $\int \begin{pmatrix} abc \\ def \end{pmatrix} d\Pi_{12}$, otherwise called $T_{12}\begin{pmatrix} a, b, c \\ d, e, f \end{pmatrix}$, is in the first place expressed (see No. 171) as a difference, $= \frac{1}{2}T_{12}(a, b, c) - \frac{1}{2}T_{12}(d, e, f)$, of two functions T. Each of these is by Theorem (A) (No. 176) expressed as a difference of two functions U, that is, as the difference of the logarithms, or logarithm of the quotient, of two functions V: such function V is according to its original definition a function of $(p+1=)$ 4 points, but in such wise that the function is expressible as a function of $(p=)$ 3 arguments, and so expressed it is a Θ-function of these arguments: and the final result thus is that the sum of $(p=)$ 3 integrals $\int \begin{pmatrix} a, b, c \\ d, e, f \end{pmatrix} d\Pi_{12}$ is equal to the logarithm of a fraction, whereof the numerator and the denominator are each of them a product of two Θ-functions.

826.

NOTE ON A PARTITION-SERIES.

[From the *American Journal of Mathematics*, vol. VI. (1884), pp. 63, 64.]

PROF. SYLVESTER, in his paper, "A Constructive theory of Partitions, &c.," *American Journal of Mathematics*, vol. V. (1883), p. 282, has given the following very beautiful formula

$$(1+ax)(1+ax^2)(1+ax^3)\ldots=1+\frac{1}{1-x}(1+ax^2)\,xa+\frac{1}{1-x\,.\,1-x^2}(1+ax)(1+ax^4)\,x^5a^2$$

$$+\frac{1}{1-x\,.\,1-x^2\,.\,1-x^3}(1+ax)(1+ax^2)(1+ax^6)\,x^{12}a^3+\ldots,$$

or, as this may be written,

$$\Omega=1+P+Q(1+ax)+R(1+ax)(1+ax^2)+S(1+ax)(1+ax^2)(1+ax^3)+\ldots\qquad,$$

where

$$P=\frac{(1+ax^2)\,xa}{\mathbf{1}},\quad Q=\frac{(1+ax^4)\,x^5a^2}{\mathbf{1.2}},\quad R=\frac{(1+ax^6)\,x^{12}a^3}{\mathbf{1.2.3}},\quad S=\frac{(1+ax^8)\,x^{22}a^4}{\mathbf{1.2.3.4}},\ \&\text{c.},$$

the heavy figures **1**, **2**, **3**, **4**,... of the denominators being, for shortness, written to denote $1-x$, $1-x^2$, $1-x^3$, $1-x^4$,... respectively. The x-exponents 1, 5, 12, 22,... are the pentagonal numbers $\frac{1}{2}(3n^2-n)$.

To prove this, writing

$$P'=\frac{ax^2}{\mathbf{1}},\quad Q'=\frac{ax^3}{\mathbf{1}}+\frac{a^2x^7}{\mathbf{1.2}},\quad R'=\frac{ax^4}{\mathbf{1}}+\frac{a^2x^9}{\mathbf{1.2}}+\frac{a^3x^{15}}{\mathbf{1.2.3}},\quad S'=\frac{ax^5}{\mathbf{1}}+\frac{a^2x^{11}}{\mathbf{1.2}}+\frac{a^3x^{18}}{\mathbf{1.2.3}}+\frac{a^4x^{26}}{\mathbf{1.2.3.4}},\ \&\text{c.},$$

where the x-exponents are

$$2;\ 3,\ 3+4;\ 4,\ 4+5,\ 4+5+6;\ 5,\ 5+6,\ 5+6+7,\ 5+6+7+8;\ \&\text{c.},$$

we find without difficulty (see infrà) that

$$\begin{aligned}1+P&=(1+ax\ )(1+P'),\\ 1+P'+Q&=(1+ax^2)(1+Q'),\\ 1+Q'+R&=(1+ax^3)(1+R'),\\ 1+R'+S&=(1+ax^4)(1+S'),\ \text{\&c.};\end{aligned}$$

and hence, using Ω to denote the sum

$$\Omega=1+P+Q(1+ax)+R(1+ax)(1+ax^2)+S(1+ax)(1+ax^2)(1+ax^3)+\ldots,$$

we obtain successively

$$\begin{aligned}\Omega\div(1+ax)&=1+P'+Q+R(1+ax^2)+S(1+ax^2)(1+ax^3)+\ldots,\\ \Omega\div(1+ax)(1+ax^2)&=1+Q'+R+S(1+ax^3)+T(1+ax^3)(1+ax^4)+\ldots,\\ \Omega\div(1+ax)(1+ax^2)(1+ax^3)&=1+R'+S+T(1+ax^4)+\ldots,\end{aligned}$$

and so on. In these equations, on the right-hand sides, the lowest exponent of x is 2, 3, 4, &c., respectively, so that in the limit the right-hand side becomes $=1$, or the final equation is $\Omega=(1+ax)(1+ax^2)(1+ax^3)\ldots$; viz. we have the series represented by Ω equal to this infinite product, which is the theorem in question.

One of the foregoing identities is

$$1+R'+S=(1+ax^4)(1+S'),$$

viz. substituting for R', S, S' their values, this is

$$1+\frac{ax^4}{1}+\frac{a^2x^9}{1.2}+\frac{a^3x^{15}}{1.2.3}+\frac{(1+ax^8)\,a^4x^{22}}{1.2.3.4}=(1+ax^4)\left\{1+\frac{ax^5}{1}+\frac{a^2x^{11}}{1.2}+\frac{a^3x^{18}}{1.2.3}+\frac{a^4x^{26}}{1.2.3.4}\right\},$$

viz. this equation is

$$-ax^4+\frac{ax^4-ax^5(1+ax^4)}{1}+\frac{a^2x^9-a^2x^{11}(1+ax^4)}{1.2}$$
$$+\frac{a^3x^{15}-a^3x^{18}(1+ax^4)}{1.2.3}+\frac{(1+ax^8)\,a^4x^{22}-a^4x^{26}(1+ax^4)}{1.2.3.4},$$

that is

$$0=-ax^4+ax^4-\frac{a^2x^9}{1}+\frac{a^2x^9}{1}-\frac{a^3x^{15}}{1.2}+\frac{a^3x^{15}}{1.2}-\frac{a^4x^{22}}{1.2.3}+\frac{a^4x^{22}}{1.2.3}.$$

In the same way each of the other identities is proved.

Writing $a=-1$, we have Ω, $=\mathbf{1.2.3.4}\ldots$,

$$=1+P+Q.1+R.1.2+S.1.2.3+\ldots,$$

where

$$P=-(1+x)\,x,\quad Q=\frac{(1+x^2)\,x^5}{1},\quad R=-\frac{(1+x^3)\,x^{12}}{1.2},\ldots$$

and therefore

$$\mathbf{1.2.3.4}\ldots = 1-(1+x)\,x+(1+x^2)\,x^5-(1+x^3)\,x^{12}+\ldots,$$

which is Euler's theorem.

It might appear that the identities used in the proof would also, for this particular value $a=-1$, lead to interesting theorems; but this is found *not* to be the case: we have

$$P' = \frac{-x^2}{1},\quad Q' = \frac{-x^3}{1}+\frac{x^7}{1.2},\quad R' = \frac{-x^4}{1}+\frac{x^9}{1.2}-\frac{x^5}{1.2.3},\ \&\text{c.},$$

but the expressions in terms of these quantities for the products $\mathbf{2.3.4}\ldots$, $\mathbf{3.4}\ldots$, &c., contain denominator factors, and are thus altogether without interest; we have, for example,

$$\mathbf{2.3.4}\ldots = 1+\frac{-x^2+x^5+x^7}{1}-\frac{(1+x^3)\,x^{12}}{1}+\&\text{c.},$$

which is, with scarcely a change of form, the expression obtained from that of the original product $\mathbf{1.2.3.4}\ldots$, by division by $\mathbf{1}$, $=1-x$. And similarly as regards the products $\mathbf{3.4}\ldots$, &c.

Cambridge, June, 1883.

827.

ON THE NON-EUCLIDIAN PLANE GEOMETRY.

[From the *Proceedings of the Royal Society of London*, vol. XXXVII. (1884), pp. 82—102. Received May 27, 1884.]

1. I CONSIDER the hyperbolic or Lobatschewskian geometry: this is a geometry such as that of the imaginary spherical surface $x^2 + y^2 + z^2 = -1$; and the imaginary surface may be *bent* (without extension or contraction) into the real surface considered by Beltrami, which I will call the Pseudosphere, viz. this is the surface of revolution defined by the equations $x = \log \cot \frac{1}{2}\theta - \cos \theta$, $\sqrt{y^2 + z^2} = \sin \theta$. We have on the imaginary spherical surface imaginary points corresponding to real points of the pseudosphere, and imaginary lines (arcs of great circle) corresponding to real lines (geodesics) of the pseudosphere, and, moreover, any two such imaginary points or lines of the imaginary spherical surface have a real distance or inclination equal to the corresponding distance or inclination on the pseudosphere. Thus the geometry of the pseudosphere, using the expression straight line to denote a geodesic of the surface, is the Lobatschewskian geometry; or rather I would say this in regard to the metrical geometry, or trigonometry, of the surface; for in regard to the descriptive geometry, the statement requires (as will presently appear) some qualification.

2. I would remark that this realisation of the Lobatschewskian geometry sustains the opinion that Euclid's twelfth axiom is undemonstrable. We may imagine rational beings living in a two-dimensional space and conceiving of space accordingly, that is, having no conception of a third dimension of space; this two-dimensional space need not however be a plane, and taking it to be the pseudospherical surface, the geometry to which their experience would lead them would be the geometry of this surface, that is, the Lobatschewskian geometry. With regard to our own two-dimensional space, the plane, I have, in my Presidential Address (B.A., Southport, 1883), [784], expressed the opinion that Euclid's twelfth axiom in Playfair's form of it does not need demonstration, but is part of our notion of space, of the physical space of our experience;

the space, that is, which we become acquainted with by experience, but which is the representation lying at the foundation of all physical experience.

3. I propose in the present paper to develope further the geometry of the pseudosphere. In regard to the name, and the subject generally, I refer to two memoirs by Beltrami, "Teoria fondamentale degli spazii di curvatura costante," *Annali di Matem.*, t. II. (1868—69), pp. 232—255, and "Saggio di interpretazione della geometria non-Euclidea," *Battaglini, Giorn. di Matem.*, t. VI. (1868), pp. 284—312, both translated, *Ann. de l'École Normale*, t. VI. (1869); in the last of these, he speaks of surfaces of constant negative curvature as "pseudospherical," and in a later paper, "Sulla superficie di rotazione che serve di tipo alle superficie pseudosferiche," *Battaglini, Giorn. di Matem.*, t. X. (1872), pp. 147—160, he treats of the particular surface which I have called the pseudosphere. The surface is mentioned, Note IV. of Liouville's edition of Monge's *Application de l'Analyse à la Géométrie* (1850), and the generating curve is there spoken of as "bien connue des géomètres."

4. In ordinary plane geometry, take (fig. 1) a line Bx, and on it a point B; from B, in any direction, draw the line BA; take upon it a point A, and from

Fig. 1.

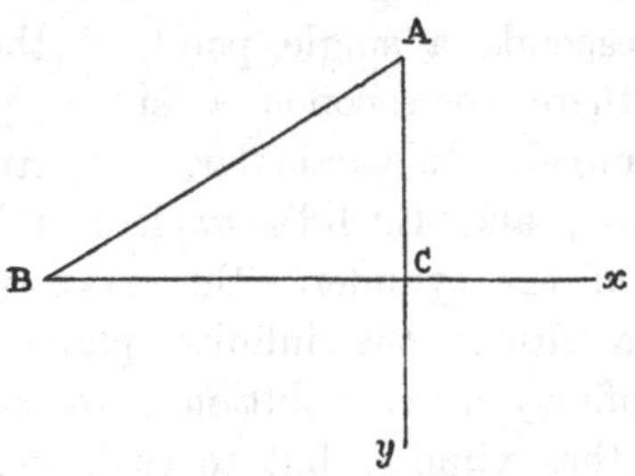

this point, at right angles to Bx, draw Ay, cutting it at C. We have thus a triangle ACB, right-angled at C; and we may denote the other angles, and the lengths of the sides, by A, B, c, a, b, respectively. In the construction of the figure, the length c and the angle B are arbitrary.

The plane is a surface which is homogeneous, isotropic, and palintropic, that is, whatever be the position of B, the direction of Bx, and the sense in which the angle B is measured, we have the same expressions for a, b as functions of c, B; these expressions, of course, are

$$a = c\cos B, \quad b = c\sin B.$$

But considering Ay as the initial line and AB, $=c$, as a line drawn from A at an inclination thereto $=A$, we have in like manner

$$b = c\cos A, \quad a = c\sin A,$$

and consequently $\cos A = \sin B$, $\sin A = \cos B$; whence $\sin(A+B) = 1$, $\cos(A+B) = 0$, and thence $A + B =$ a right angle, or $A + B + C =$ two right angles.

Hence also in any triangle ABC, drawing a perpendicular, say AD, from A to the side BC, and so dividing the triangle into two right-angled triangles, we prove that the sum $A+B+C$ of the angles is = two right angles, and we further establish the relations

$$a = b\cos C + c\cos B,\quad b = c\cos A + a\cos C,\quad c = a\cos B + b\cos A,$$

which are the fundamental formulæ of plane trigonometry; that is, we derive the metrical geometry or trigonometry of the plane from the two original equations $a = c\cos B$, $b = c\sin B$.

5. Supposing the plane bent in any manner, that is, converted into a developable surface or torse, and using the term straight line to denote a geodesic of the surface, then the straight line of the surface is in fact the form assumed, in consequence of the bending, by a straight line of the plane. The sides and angles of the rectilinear triangle ABC on the surface are equal to those of the rectilinear triangle ABC on the plane, and the metrical relations hold good without variation. But it is not *simpliciter* true that the descriptive properties of the torse are identical with those of the plane. This will be the case if the points of the plane and torse have with each other a (1, 1) correspondence, but not otherwise. For instance, consider a plane curve (such as the parabola or one branch of the hyperbola) extending from infinity to infinity, and let the torse be the cylinder having this curve for a plane section; then to each point of the plane there corresponds a single point of the cylinder; and conversely to each point of the cylinder there corresponds a single point of the plane; and the descriptive geometries are identical. In particular, two straight lines (geodesics) on the cylinder cannot inclose a space; and Euclid's twelfth axiom holds good in regard to the straight lines (geodesics) of the cylinder. But take the plane curve to be a closed curve, or (to fix the ideas) a circle; the infinite plane is bent into a cylinder considered as composed of an infinity of convolutions; to each point of the plane there corresponds a single point of the cylinder, but to each point of the cylinder an infinity of points of the plane; and the descriptive properties are in this case altered; the straight lines (geodesics) of the cylinder are helices; and we can through two given points of the cylinder draw, not only one, but an infinity of helices; any two of these will inclose a space. And even if instead of the geodesics we consider only the shortest lines, or helices of greatest inclination; yet even here for a pair of points on opposite generating lines of the cylinder, there are two helices of equal inclination, that is, two shortest lines inclosing a space. We have, in what precedes, an illustration in regard to the descriptive geometry of the pseudosphere; this is not identical with the Lobatschewskian geometry, but corresponds to it in a manner such as that in which the geometry of the surface of the circular cylinder corresponds to that of the plane.

6. The surface of the sphere is, like the plane, homogeneous, isotropic, and palintropic. We may on the spherical surface construct, as above, a right-angled triangle ABC, wherein the side c and the angle B are arbitrary; and (corresponding to the before-mentioned formulæ for the plane) we then have

$$\tan a = \tan c\cos B,\quad \sin b = \sin c\sin B,$$

whence also

$$\tan b = \tan c \cos A, \quad \sin a = \sin c \sin A.$$

We deduce from these

$$\frac{\tan^2 a}{\tan^2 c} + \frac{\sin^2 b}{\sin^2 c} = 1,$$

leading to $\cos^2 c = \cos^2 a \cos^2 b$; and then

$$\frac{\sin b}{\tan a} = \cos c \tan B, \quad \frac{\sin a}{\tan b} = \cos c \tan A,$$

giving

$$\cos a \cos b = \cos^2 c \tan A \tan B;$$

that is,

$$\tan A \tan B = \frac{1}{\cos a \cos b}, \text{ which is } > 1.$$

Hence $A + B >$ a right angle, or in the right-angled triangle ACB, the sum $A + B + C$ of the angles is > two right angles. Whence also in any triangle ABC whatever, dividing it into two right-angled triangles by means of a perpendicular let fall from an angle on the opposite side, we have the sum $A + B + C$ of the angles > two right angles. And we obtain, moreover,

$$\begin{aligned} a &= \tan^{-1}(\tan c \cos B) + \tan^{-1}(\tan b \cos C), \\ b &= \tan^{-1}(\tan a \cos C) + \tan^{-1}(\tan c \cos A), \\ c &= \tan^{-1}(\tan b \cos A) + \tan^{-1}(\tan a \cos B), \end{aligned}$$

which lead to all the formulæ of spherical trigonometry.

7. Suppose the radius of the sphere to be $1/\lambda$: then a, b, c being the lengths of the sides, the lengths in spherical measure are $\lambda a, \lambda b, \lambda c$; and we must in the formulæ instead of a, b, c write $\lambda a, \lambda b, \lambda c$ respectively. In particular, for the imaginary sphere $x^2 + y^2 + z^2 = -1$, we have $\lambda = i$, and we must instead of a, b, c write ai, bi, ci respectively. The fundamental formulæ for the right-angled triangle thus become

$$\tanh a = \tanh c \cos B, \quad \sinh b = \sinh c \sin B,$$

and these lead to all the trigonometrical formulæ, viz. any one of these is deduced from the corresponding formula of spherical trigonometry by writing therein ai, bi, ci for a, b, c respectively; or, what is the same thing, by changing the circular functions of the sides into the corresponding hyperbolic functions.

In particular, for the right-angled triangle ACB, we have

$$\tan A \tan B = \frac{1}{\cosh a \cosh b},$$

which for a and b real is < 1, that is, $A + B <$ a right angle, or $A + B + C <$ two right angles, and thence also in any triangle whatever $A + B + C <$ two right angles. But the points A, B, C of any such triangle ABC on the imaginary sphere, and the lines BC, CA, AB which connect them, are imaginary: the meaning of the proof will better appear on passing to the pseudosphere.

8. We have to consider the imaginary spherical surface as *bent* into a real surface. This is, of course, an imaginary process, as any process must be which gives a transformation of imaginary points and lines into real points and lines; but the notion is not more difficult than that of the transformation of imaginary similarity, consisting in the substitution of ix, iy, iz for x, y, z respectively. We thus pass from imaginary points of the imaginary sphere $x^2+y^2+z^2=-1$ to real points of the real sphere $x^2+y^2+z^2=1$; or, again, from imaginary points of either of the real hyperboloids $x^2+y^2-z^2=-1$, $x^2+y^2-z^2=1$, to real points of the other of the same two real hyperboloids.

9. I consider the formulæ for the flexure of the imaginary sphere $X^2+Y^2+Z^2=-1$, into the pseudosphere $x=\log\cot\tfrac{1}{2}\theta-\cos\theta$, $\sqrt{y^2+z^2}=\sin\theta$: it would be allowable to dispense with Beltrami's subsidiary variables u, v, but I prefer to collect here all the formulæ. We have

$$X=\frac{-i}{\sqrt{1-u^2-v^2}},\quad Y=\frac{u}{\sqrt{1-u^2-v^2}},\quad Z=\frac{v}{\sqrt{1-u^2-v^2}},$$

values which give $X^2+Y^2+Z^2=-1$. And observe that, taking u, v to be real magnitudes such that $u^2+v^2<1$, we have X a pure imaginary, but Y and Z each of them real. We consider on the imaginary sphere points having such coordinates X, Y, Z; any such point corresponds as will immediately appear to a real point on the pseudosphere, and (the distances and angles being the same for the pseudosphere as for the original imaginary spherical surface) it hence appears that (notwithstanding that the points on the imaginary spherical surface, and the lines joining such points, are imaginary) the distances and angles on the imaginary spherical surface are real. Also

$$\sin\theta=\frac{1-u}{\sqrt{1-u^2-v^2}},\quad \phi=\frac{v}{1-u},$$

and thence

$$iX-Y=\sin\theta,\quad iX+Y=\sin\theta\,(\phi^2+\operatorname{cosec}^2\theta),\quad Z=\sin\theta\,.\,\phi.$$

Further

$$u=\frac{\phi^2-1+\operatorname{cosec}^2\theta}{\phi^2+1+\operatorname{cosec}^2\theta},\quad v=\frac{2\phi}{\phi^2+1+\operatorname{cosec}^2\theta},$$

$$x=\log\cot\tfrac{1}{2}\theta+\cos\theta,\quad y=\sin\theta\cos\phi,\quad z=\sin\theta\sin\phi.$$

10. We have $dX^2+dY^2+dZ^2$ and $dx^2+dy^2+dz^2$ each $=\cot^2\theta\,d\theta^2+\sin^2\theta\,d\phi^2$. Writing P, $Q=iX-Y$, $iX+Y$ respectively, we in fact have

$$dX^2+dY^2+dZ^2=-dP\,dQ+dZ^2,$$

where P, Q, $Z=\sin\theta$, $\phi^2\sin\theta+\operatorname{cosec}\theta$, $\phi\sin\theta$ respectively; and thence

$$\begin{aligned}dZ&=\sin\theta\,d\phi+\phi\cos\theta\,d\theta,\\ dP&=\cos\theta\,d\theta,\\ dQ&=2\sin\theta\phi\,d\phi+(\phi^2\cos\theta-\operatorname{cosec}\theta\cot\theta)\,d\theta,\end{aligned}$$

giving the formula

$$dX^2 + dY^2 + dZ^2 = \cot^2\theta\, d\theta^2 + \sin^2\theta\, d\phi^2;$$

and then also

$$\begin{aligned} dx^2 + dy^2 + dz^2 &= dx^2 + (d \sin\theta)^2 + \sin^2\theta\, d\phi^2 \\ &= (\cos^2\theta \cot^2\theta + \cos^2\theta)\, d\theta^2 + \sin^2\theta\, d\phi^2 = \cot^2\theta\, d\theta^2 + \sin^2\theta\, d\phi^2. \end{aligned}$$

Joining to these the differential expression in u, v, we have

$$\begin{aligned} dX^2 + dY^2 + dZ^2 &= \frac{(1 - u^2 - v^2)(du^2 + dv^2) + (u\,du + v\,dv)^2}{(1 - u^2 - v^2)^2}, \\ &= \cot^2\theta\, d\theta^2 + \sin^2\theta\, d\phi^2, \\ &= dx^2 + dy^2 + dz^2, \end{aligned}$$

where the final equation $dX^2 + dY^2 + dZ^2 = dx^2 + dy^2 + dz^2$ shows that the imaginary sphere $X^2 + Y^2 + Z^2 = -1$ can be bent into the pseudosphere.

Observe that to given values of θ, ϕ there corresponds a single point on the pseudosphere, but not conversely; for if θ, ϕ be values corresponding to a given point, then corresponding to the same point we have θ, $\phi + n\pi$, where n is an arbitrary integer.

11. The geodesics of the imaginary spherical surface are, of course, its plane sections, any such section being determined by a linear equation $\alpha X + \beta Y + \gamma Z = 0$ between the coordinates X, Y, Z. Since for a point corresponding to a real point of the pseudosphere, X is a pure imaginary while Y and Z are real, we see that for a geodesic corresponding to a real geodesic of the pseudosphere we must have α a pure imaginary, β and γ real; and, in fact, writing as above, $P = iX - Y$, $Q = iX + Y$, and therefore conversely $X = \frac{1}{2}i(-P - Q)$, $Y = \frac{1}{2}(-P + Q)$, the equation $\alpha X + \beta Y + \gamma Z = 0$ becomes $(-\frac{1}{2}i\alpha - \frac{1}{2}\beta)P + (-\frac{1}{2}i\alpha + \frac{1}{2}\beta)Q + \gamma Z = 0$, which will then be of the form $AP + BQ + CZ = 0$, with real coefficients A, B, C: viz. we have

$$P,\ Q,\ Z = \sin\theta,\quad \sin\theta\,(\phi^2 + \operatorname{cosec}^2\theta),\quad \sin\theta \,.\, \phi;$$

and the equation thus is

$$A + B(\phi^2 + \operatorname{cosec}^2\theta) + C\phi = 0,$$

which is the equation for a geodesic (or straight line) on the pseudosphere. The equation $A + C\phi = 0$, that is, $\phi = \text{const.}$, is obviously that of a meridian.

12. If the geodesic pass through a given point θ_1, ϕ_1, we have, of course,

$$A + B(\phi_1^2 + \operatorname{cosec}^2\theta_1) + C\phi_1 = 0,$$

and hence also the equation of a geodesic through the two points (θ_1, ϕ_1), (θ_2, ϕ_2) is

$$\begin{vmatrix} 1, & \phi^2 + \operatorname{cosec}^2\theta, & \phi \\ 1, & \phi_1^2 + \operatorname{cosec}^2\theta_1, & \phi_1 \\ 1, & \phi_2^2 + \operatorname{cosec}^2\theta_2, & \phi_2 \end{vmatrix} = 0.$$

We may for ϕ_1, ϕ_2 write $\phi_1+2n_1\pi$, $\phi_2+2n_2\pi$ respectively, n_1, n_2 being arbitrary integers; and it would thus at first sight appear that there could be drawn through the two points a doubly infinite series of geodesics. There is, in fact, a singly infinite system of geodesics: to show how this is, write for shortness Λ, Λ_1, Λ_2, α, α_1, α_2 for $\operatorname{cosec}^2\theta$, $\operatorname{cosec}^2\theta_1$, $\operatorname{cosec}^2\theta_2$, $2n\pi$, $2n_1\pi$, $2n_2\pi$ respectively; then the equation of the geodesic through the two points may be written

$$\begin{vmatrix} 1, & (\phi+\alpha)^2+\Lambda, & \phi+\alpha \\ 1, & (\phi_1+\alpha_1)^2+\Lambda_1, & \phi_1+\alpha_1 \\ 1, & (\phi_2+\alpha_2)^2+\Lambda_2, & \phi_2+\alpha_2 \end{vmatrix}=0,$$

where the constant $\alpha=2n\pi$ may be disposed of so as to simplify the formula as much as may be: it is what I have called an apoclastic constant. Taking β an arbitrary value, this may be transformed into

$$\begin{vmatrix} 1, & (\phi+\alpha+\beta)^2+\Lambda, & \phi+\alpha+\beta \\ 1, & (\phi_1+\alpha_1+\beta)^2+\Lambda_1, & \phi_1+\alpha_1+\beta \\ 1, & (\phi_2+\alpha_2+\beta)^2+\Lambda_2, & \phi_2+\alpha_2+\beta \end{vmatrix}=0,$$

and then assuming $\alpha=\alpha_1$, $\beta=-\alpha_1$, this becomes

$$\begin{vmatrix} 1, & \phi^2+\Lambda & , \ \phi \\ 1, & \phi_1^2+\Lambda_1 & , \ \phi_1 \\ 1, & (\phi_2+\alpha_2-\alpha_1)^2+\Lambda_2, & \phi_2+\alpha_2-\alpha_1 \end{vmatrix}=0,$$

which is what the equation

$$\begin{vmatrix} 1, & \phi^2+\Lambda, & \phi \\ 1, & \phi_1^2+\Lambda_1, & \phi_1 \\ 1, & \phi_2^2+\Lambda_2, & \phi_2 \end{vmatrix}=0,$$

becomes on changing only ϕ_2 into $\phi_2+\alpha_2-\alpha_1$, that is, $\phi_2+2k_2\pi$, where k_2 is an arbitrary integer. We have thus through the two points a singly infinite series of geodesic lines; in general, only one of these is a shortest line, but for points on opposite meridians there are two equal shortest lines.

13. For the distance between two points (θ_1, ϕ_1) and (θ_2, ϕ_2) on the pseudosphere, taking (X_1, Y_1, Z_1) and (X_2, Y_2, Z_2) for the corresponding points on the imaginary sphere, and writing as above P_1, $Q_1=iX_1-Y_1$, iX_1+Y_1; P_2, $Q_2=iX_2-Y_2$, iX_2+Y_2, we have

$$\begin{aligned}\cosh\delta &= -X_1X_2-Y_1Y_2-Z_1Z_2,\\ &= \tfrac{1}{2}(P_1Q_2+P_2Q_1)-Z_1Z_2,\\ &= \sin\theta_1\sin\theta_2\left\{\tfrac{1}{2}(\phi_1^2+\operatorname{cosec}^2\theta_1)+\tfrac{1}{2}(\phi_2^2+\operatorname{cosec}^2\theta_2)-\phi_1\phi_2\right\},\\ &= \tfrac{1}{2}\sin\theta_1\sin\theta_2(\phi_2-\phi_1)^2+1+\frac{\tfrac{1}{2}(\sin\theta_2-\sin\theta_1)^2}{\sin\theta_1\sin\theta_2}.\end{aligned}$$

Observe here that, writing $\theta_2,\ \phi_2 = \theta_1 + d\theta_1,\ \phi_1 + d\phi_1$, and therefore δ small so that $\cosh\delta = 1 + \tfrac{1}{2}\delta^2$, we obtain

$$\delta^2 = \sin^2\theta_1 d\phi_1^2 + \cot^2\theta_1 d\theta_1^2,$$

agreeing with the expression for $dx^2 + dy^2 + dz^2$. If in the form first obtained we write $\Lambda_1 = \operatorname{cosec}^2\theta_1$, $\Lambda_2 = \operatorname{cosec}^2\theta_2$, we find

$$\cosh\delta = \frac{\tfrac{1}{2}}{\sqrt{\Lambda_1\Lambda_2}}\{\phi_1^2 + \Lambda_1 + \phi_2^2 + \Lambda_2 - 2\phi_1\phi_2\},$$

which is a convenient form.

In like manner, to find the mutual inclination of the two geodesics

$$A_1 + B_1(\phi^2 + \operatorname{cosec}^2\theta) + C_1\phi = 0,$$
$$A_2 + B_2(\phi^2 + \operatorname{cosec}^2\theta) + C_2\phi = 0,$$

these correspond to the plane sections

$$A_1P + B_1Q + C_1Z = 0,\quad A_2P + B_2Q + C_2Z = 0,$$

that is,

$$(A_1 + B_1)iX + (-A_1 + B_1)Y + C_1Z = 0,\quad (A_2 + B_2)iX + (-A_2 + B_2)Y + C_2Z = 0,$$

of the imaginary sphere: and we thence find

$$\cos\Omega = \frac{C_1C_2 - 2(A_1B_2 + A_2B_1)}{\sqrt{C_1^2 - 4A_1B_1}\sqrt{C_2^2 - 4A_2B_2}}.$$

14. Suppose that the two geodesics meet in the point $\theta_0,\ \phi_0$: then writing for shortness $\operatorname{cosec}^2\theta = \Lambda$, and therefore $\operatorname{cosec}^2\theta_0 = \Lambda_0$, we have

$$A_1 + B_1(\phi_0^2 + \Lambda_0) + C_1\phi_0 = 0,$$
$$A_2 + B_2(\phi_0^2 + \Lambda_0) + C_2\phi_0 = 0.$$

Suppose that the meridian through this point is

$$A_3 + B_3(\phi^2 + \Lambda) + C_3\phi = 0;$$

then $B_3 = 0$, $A_3 + C_3\phi_0 = 0$. Take $\Omega_1,\ \Omega_2$, for the inclinations to this meridian of the two geodesics respectively; then

$$\cos\Omega_1 = \frac{C_1C_3 - 2A_3B_1}{\sqrt{C_1^2 - 4A_1B_1}\,.\,C_3} = \frac{C_1 + 2B_1\phi_0}{\sqrt{C_1^2 - 4A_1B_1}},$$

whence

$$\sin\Omega_1 = \frac{2B_1\sqrt{\Lambda_0}}{\sqrt{C_1^2 - 4A_1B_1}},$$

and similarly

$$\cos\Omega_2 = \frac{C_2 + 2B_2\phi_0}{\sqrt{C_2^2 - 4A_2B_2}},$$

whence

$$\sin\Omega_2 = \frac{2B_2\sqrt{\Lambda_0}}{\sqrt{C_2^2 - 4A_2B_2}}.$$

We thence obtain

$$\cos(\Omega_1 - \Omega_2) = \frac{C_1C_2 + 2\phi_0(B_1C_2 + B_2C_1) + 4B_1B_2(\phi_0^2 + \Lambda_0)}{\sqrt{C_1^2 - 4A_1B_1}\sqrt{C_2^2 - 4A_2B_2}},$$

which is

$$= \frac{C_1C_2 - 2(A_1B_2 + A_2B_1)}{\sqrt{C_1^2 - 4A_1B_1}\sqrt{C_2^2 - 4A_2B_2}} = \cos\Omega,$$

as above, the equality of the two numerators depending on the identity

$$\{A_1 + B_1(\phi_0^2 + \Lambda_0) + C_1\phi_0\}B_2 + \{A_2 + B_2(\phi_0^2 + \Lambda_0) + C_2\phi_0\}B_1 = 0.$$

In particular, if we consider the two geodesics

$$\phi^2 + \operatorname{cosec}^2\theta - \operatorname{cosec}^2\theta_1 + C_1\phi = 0, \quad \phi = 0,$$

the second of which may be considered as representing any meridian section of the pseudosphere, and the first is an arbitrary geodesic meeting this at the point $\theta = \theta_1$, $\phi = 0$, then the formula for the inclination is

$$\cos\Omega = \frac{C_1}{\sqrt{C_1^2 + 4\operatorname{cosec}^2\theta_1}}.$$

Hence also, $\cos\Omega = 0$, or $\Omega = 90°$, if $C_1 = 0$: viz. we have $\phi^2 + \operatorname{cosec}^2\theta - \operatorname{cosec}^2\theta_1 = 0$ for the equation of the geodesic through the point $\theta = \theta_1$, $\phi = 0$, at right angles to the meridian section $\phi = 0$.

15. Consider a right-angled triangle ACB, where the points A, C are on the meridian $\phi = 0$, and write $(\theta_1, 0;\ \Lambda_1 = \operatorname{cosec}^2\theta_1)$, $(\theta_2, \phi_2;\ \Lambda_2 = \operatorname{cosec}^2\theta_2)$, $(\theta_3, 0;\ \Lambda_3 = \operatorname{cosec}^2\theta_3)$, for the points A, B, C respectively. Then if the equations are—

for the side BC, $A_1 + B_1(\phi^2 + \Lambda) + C_1\phi = 0$, we have $C_1 = 0$,

$A_1 + B_1(\phi_2^2 + \Lambda_2) = 0$, $A_1 + B_1\Lambda_3 = 0$, whence $\phi_2^2 + \Lambda_2 = \Lambda_3$;

for the side CA, $A_2 + B_2(\phi^2 + \Lambda) + C_2\phi = 0$, we have $A_2 = 0$, $B_2 = 0$;

for the side AB, $A_3 + B_3(\phi^2 + \Lambda) + C_3\phi = 0$,

we have $A_3 + B_3(\phi_2^2 + \Lambda_2) + C_3\phi_2 = 0$, $A_3 + B_3\Lambda_1 = 0$.

Observing that $\phi_1 = \phi_3 = 0$, we have

$$\cosh a = \frac{1}{2\sqrt{\Lambda_2\Lambda_3}}(\phi_2^2 + \Lambda_2 + \Lambda_3),$$

$$\cosh b = \frac{1}{2\sqrt{\Lambda_1\Lambda_3}}(\Lambda_3 + \Lambda_1),$$

$$\cosh c = \frac{1}{2\sqrt{\Lambda_1\Lambda_2}}(\phi_2^2 + \Lambda_1 + \Lambda_2);$$

or, reducing these by the relation $\phi_2^2 + \Lambda_2 = \Lambda_3$, they become

$$\cosh a = \frac{\sqrt{\Lambda_3}}{\sqrt{\Lambda_2}}, \text{ whence } \sinh a = \frac{\sqrt{\Lambda_3 - \Lambda_2}}{\sqrt{\Lambda_2}}, \ \tanh a = \frac{\sqrt{\Lambda_3 - \Lambda_2}}{\sqrt{\Lambda_3}};$$

$$\cosh b = \frac{\Lambda_1 + \Lambda_3}{2\sqrt{\Lambda_1 \Lambda_3}}, \quad ,, \quad \sinh b = \frac{\Lambda_1 - \Lambda_3}{2\sqrt{\Lambda_1 \Lambda_3}}, \ \tanh b = \frac{\Lambda_1 - \Lambda_3}{\Lambda_1 + \Lambda_3};$$

$$\cosh c = \frac{\Lambda_1 + \Lambda_3}{2\sqrt{\Lambda_1 \Lambda_2}}, \quad ,, \quad \sinh c = \frac{\sqrt{(\Lambda_1 + \Lambda_3)^2 - 4\Lambda_1\Lambda_2}}{2\sqrt{\Lambda_1 \Lambda_2}}, \ \tanh c = \frac{\sqrt{(\Lambda_1 + \Lambda_3)^2 - 4\Lambda_1\Lambda_2}}{\Lambda_1 + \Lambda_3}.$$

We have, moreover,

$$\cos B = \frac{C_1 C_3 - 2(A_1 B_3 + A_3 B_1)}{\sqrt{C_1^2 - 4A_1 B_1}\sqrt{C_3^2 - 4A_3 B_3}} = \frac{-2(A_1 B_3 + A_3 B_1)}{\sqrt{-4A_1 B_1}\sqrt{C_3^2 - 4A_3 B_3}},$$

which, writing $A_3 = -B_3\Lambda_1$ and $A_1 = -B_1\Lambda_3$, becomes

$$\cos B = \frac{B_3(\Lambda_1 + \Lambda_3)}{\sqrt{\Lambda_3}\sqrt{C_3^2 - 4A_3 B_3}};$$

or, further reducing by means of

$$\begin{aligned}\phi_2^2 (C_3^2 - 4A_3 B_3) &= B_3^2\, (\phi_2^2 + \Lambda_2 - \Lambda_1)^2 + 4\phi_2^2 B_3^2 \Lambda_1^2 \\ &= B_3^2\, \{(\phi_2^2 + \Lambda_2 - \Lambda_1)^2 + 4\phi_2^2 \Lambda_1\} \\ &= B_3^2\, \{(\Lambda_3 + \Lambda_1)^2 + 4\Lambda_1 (\Lambda_3 - \Lambda_2)\} \\ &= B_3^2\, \{(\Lambda_3 + \Lambda_1)^2 - 4\Lambda_1 \Lambda_2\},\end{aligned}$$

this becomes

$$\cos B = \frac{(\Lambda_1 + \Lambda_3)\sqrt{\Lambda_3 - \Lambda_2}}{\sqrt{\Lambda_3}\sqrt{(\Lambda_1 + \Lambda_3)^2 - 4\Lambda_1\Lambda_2}},$$

whence

$$\sin B = \frac{\sqrt{\Lambda_2}\,(\Lambda_1 - \Lambda_3)}{\sqrt{\Lambda_3}\sqrt{(\Lambda_1 + \Lambda_3)^2 - 4\Lambda_1\Lambda_2}};$$

and with these values we verify

$$\tanh a = \tanh c \,.\, \cos B,$$

$$\sinh b = \sinh c \,.\, \sin B,$$

which are the expressions for the sides BC, CA, in terms of the length BA, $= c$ and angle B, which are arbitrary. I have not thought it necessary to give the direct verification of these equations for a more general position of the right-angled triangle: we already know, and it appears *à posteriori* by the following number, that the verification really extends to any right-angled triangle whatever on the surface.

16. The pseudosphere is homogeneous, isotropic, and palintropic, viz. this is the case when bending is allowed; in other words, the surface is applicable upon itself, with three degrees of freedom. Considering any infinitesimal linear element Ax, the point A may be brought to coincide with an arbitrary point A' of the surface, and

the element Ax to lie in an arbitrary direction $A'x'$ through A'; the area about A will then coincide with the area about A'. The analytical theory is at once derived from that for the sphere, viz. we have a rectangular transformation

	X	Y	Z
X_1	α	β	γ
Y_1	α'	β'	γ'
Z_1	α''	β''	γ''

where the coefficients are such that identically

$$X_1^2 + Y_1^2 + Z_1^2 = X^2 + Y^2 + Z^2;$$

in fact, the coefficients are connected by six equations only, the system thus depending on three arbitrary parameters. If, as before, we write P_1, Q_1, P, Q, for $iX_1 - Y_1$, $iX_1 + Y_1$, $iX - Y$, $iX + Y$ respectively, then the relation is readily found to be

	P	Q	Z
P_1	$\frac{1}{2}(\alpha + i\alpha' - i\beta + \beta')$	$\frac{1}{2}(\alpha + i\alpha' + i\beta - \beta')$	$i\gamma - \gamma'$
Q_1	$\frac{1}{2}(\alpha - i\alpha' - i\beta - \beta')$	$\frac{1}{2}(\alpha - i\alpha' + i\beta + \beta')$	$i\gamma + \gamma'$
Z_1	$\frac{1}{2}(- i\alpha'' - \beta'')$	$\frac{1}{2}(- i\alpha'' + \beta'')$	γ''

this being read according to the lines only $P_1 = \frac{1}{2}(\alpha + i\alpha' - i\beta + \beta') P +$ &c., not according to the columns: in order that the coefficients may be real, we must have α, β', γ', β'', γ'', real, β, γ, α', α'' pure imaginaries.

Writing the equations in the form

	P	Q	Z
P_1	A	B	C
Q_1	A'	B'	C'
Z_1	A''	B''	C''

viz.

$$P_1 = AP + BQ + CZ, \text{ \&c.,}$$

it would be possible to deduce the equations which connect the new coefficients; but these are more easily obtained from the consideration that we must have identically $P_1Q_1 - Z_1^2 = PQ - Z^2$; the equations are thus found to be

$$A''^2 - AA' = 0, \quad B''^2 - BB' = 0, \quad C''^2 - CC' = 1,$$
$$2A''B'' - AB' - A'B = -1, \quad 2A''C'' - AC' - A'C = 0, \quad 2B''C'' - BC' - B'C = 0.$$

17. The general theory of the transformation of a quadric function into itself enables us to express the coefficients in terms of three arbitrary parameters. There is no difficulty in working out the formulæ, and we finally obtain

$$\begin{aligned}\Omega P_1 &= - (\nu+1)^2 P - \lambda^2 Q + 2\lambda(\nu+1) Z,\\ \Omega Q_1 &= - \mu^2 P - (\nu-1)^2 Q + 2\mu(\nu-1) Z,\\ \Omega Z_1 &= - \mu(\nu+1) P - \lambda(\nu-1) Q + (-1+\nu^2+\lambda\mu) Z;\end{aligned}$$

and conversely

$$\begin{aligned}\Omega P &= - (\nu-1)^2 P_1 - \lambda^2 Q_1 + 2\lambda(\nu-1) Z_1,\\ \Omega Q &= - \mu^2 P_1 - (\nu+1)^2 Q_1 + 2\mu(\nu+1) Z_1,\\ \Omega Z &= - \mu(\nu-1) P_1 - \lambda(\nu+1) Q_1 + (1+\nu^2+\lambda\mu) Z_1,\end{aligned}$$

where $\Omega = -1 + \nu^2 - \lambda\mu$: it can be at once verified that each of the two sets of formulæ does, in fact, give $P_1Q_1 - Z_1^2 = PQ - Z^2$.

18. The pseudosphere is a surface of revolution having for its meridian section the curve $x = \log\cot\frac{1}{2}\theta - \cos\theta$, $y = \sin\theta$. This is a curve symmetrical in regard to the axis of y; and we obtain the portion of it lying on the positive side of this axis, by giving to θ the series of values $\theta = 0$ to $\theta = 90°$; for $\theta = 0$, we have $y = 0$, $x = \infty$, or the axis of x is an asymptote; for $\theta = 90°$, $x = 0$, $y = 1$, the point being a cusp of the curve. The geometrical definition is that the portion of the tangent included between the curve and the axis of x has the constant length $= 1$; the inclination of

Fig. 2.

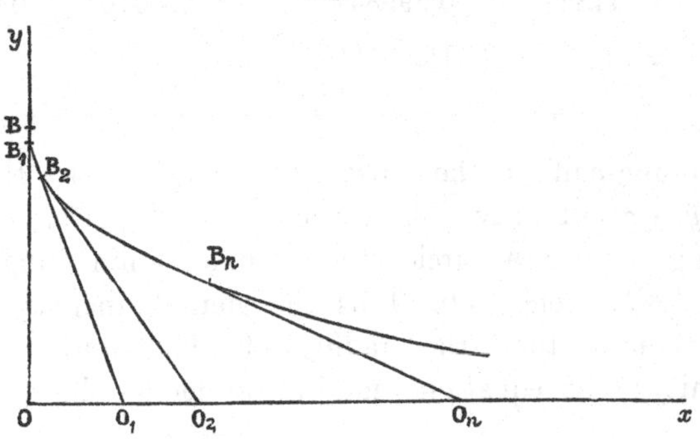

the tangent is in fact $= \theta$. We have $dx = \dfrac{\cos^2\theta\, d\theta}{\sin\theta}$, $dy = \cos\theta d\theta$; and thence $ds = \cot\theta d\theta$, and the length in question is $\dfrac{y\,ds}{dy} = 1$. The curve may be constructed graphically: take (fig. 2) the distance $BO = 1$, on OB, B_1 very near to B, and then $B_1O_1 = 1$; on

O_1B_1, B_2 very near to B_1, and then $B_2O_2 = 1$, and so on; the curve is shown on a larger scale in fig. 3, p. 235.

But the curve may also be laid down numerically; writing $\alpha = \frac{1}{2}\pi - \theta$, so that α is the inclination of the tangent to the axis of y, we have $x = \log\tan(\frac{1}{4}\pi + \frac{1}{2}\alpha) - \sin\alpha$, $y = \cos\alpha$, where $\log\tan(\frac{1}{4}\pi + \frac{1}{2}\alpha)$, the hyperbolic logarithm (which has been the signification of log throughout), is the function tabulated Tab. IV., Legendre's *Traité des Fonctions Elliptiques*, t. II. pp. 256—259.

We may hence obtain the values of the coordinates as follows:—

$\alpha = 90° - \theta$	$\log\tan(\frac{1}{4}\pi + \frac{1}{2}\alpha)$	$-\sin\alpha$	x	$y = \cos\alpha$
0	0·0000000	−0·0000000	0·0000000	1·0000000
10°	0·1754258	0·1736482	0·0017776	0·9848078
20°	0·3563785	0·3420201	0·0143584	0·9396926
30°	0·5493061	0·5000000	0·0493061	0·8660254
40°	0·7629096	0·6427876	0·1201220	0·7660444
50°	1·0106831	0·7660444	0·2446387	0·6427876
60°	1·3169578	0·8660254	0·4509324	0·5000000
70°	1·7354151	0·9396926	0·7957225	0·3563785
80°	2·4362460	0·9848078	1·4514382	0·1754258
85°	3·1313013	0·9961947	2·1351066	0·0871557
86°	3·3546735	0·9975641	2·3571019	0·0697565
87°	3·6425333	0·9986295	2·6439038	0·0523360
88°	4·0481254	0·9993908	3·0487346	0·0348995
89°	4·7413487	0·9998477	3·7415010	0·0174524
90°	∞	−1·0000000	∞	0·0000000

Attending only to one-half of the surface, we may regard the surface as standing on the circular base $y^2 + z^2 = 1$: say this circle is the equator, or the unit-circle: the horizontal section being always a circle, the radius diminishing at first rapidly and then more and more slowly from 1 to 0 as the height increases from 0 to ∞. It is hardly necessary to remark that the radius of the equator is any given length whatever, taken as unity: the equations might, of course, have been written

$$x = c\{\log\cot\tfrac{1}{2}\theta - \cos\theta\},\quad \sqrt{y^2 + z^2} = c\sin\theta,$$

but there would have been no gain of generality in this.

19. The geodesics are as already seen given by an equation

$$A + B(\phi^2 + \operatorname{cosec}^2\theta) + C\phi = 0.$$

If $B=0$, we have $A+C\phi=0$, that is, $\phi=$ const., which belongs to the meridians; if B be not $=0$, we may by a mere change of ϕ, that is, of the initial meridian, reduce the form to $A+B(\phi^2+\operatorname{cosec}^2\theta)=0$, which is the equation of a geodesic cutting at right angles the meridian $\phi=0$; writing herein $\sin\theta=\frac{1}{r}$, we have $A+B\left(\phi^2+\frac{1}{r^2}\right)=0$, which is the equation in the polar coordinates r, ϕ of the projection of the geodesic on the equatorial plane $x=0$: putting herein for greater convenience $B=-Ak^2$, we have $r^2=\frac{k^2}{1-k^2\phi^2}$: we require only such portions of the curves as lie within the unit-circle, and need therefore attend only to those for which k is not greater than 1, and in any such curve consider ϕ as extending from $\phi=0$ to $\phi=\pm\frac{\sqrt{1-k^2}}{k}$: writing this last value $=\pm\gamma$, we have $k=\frac{1}{\sqrt{1+\gamma^2}}$; if $\gamma<\pi$, that is, $k<\frac{1}{\sqrt{1+\pi^2}}$, the curve is a mere arc cutting at right angles (at the distance $r=k$ from the centre) the meridian $\phi=0$, and extending itself out on each side to meet the unit-circle in the points $\phi=\gamma$, $\phi=-\gamma$ respectively; in the case $\gamma=\pi$, that is, $k=\frac{1}{\sqrt{1+\pi^2}}$, the two points $\phi=\pm\gamma$ come together at the point $\phi=\pi$, or the curve becomes a loop; and for larger values, $k=\frac{1}{\sqrt{1+\pi^2}}$ to $\frac{1}{\sqrt{1+4\pi^2}}$, we have the two branches crossing each other on the meridian $\phi=\pi$ at the distance $r=\frac{k}{\sqrt{1-k^2\pi^2}}$ from the centre and then extending themselves in the opposite semicircles, so as to meet the unit-circle at the points $\phi=\pm\gamma$. And we have thus another critical value $k=\frac{1}{\sqrt{1+4\pi^2}}$, for which the two branches having thus crossed each other come to unite themselves at the point $\phi\,(=2\pi)=0$ of the unit-circle; and in like manner the critical values $\frac{1}{\sqrt{1+9\pi^2}}$, $\frac{1}{\sqrt{1+16\pi^2}}$, &c.: for a value of k between such limits, the branch is a spiral having a determinate number of convolutions, and the two branches cross each other always on the radii $\phi=0$ and $\phi=\pi$ respectively.

20. Let ψ denote the inclination of the radius vector to the normal, or, what is the same thing, that of the element of the circular arc to the tangent; we have $\tan\psi=\frac{dr}{rd\phi}$, and $\frac{dr}{rd\phi}=\frac{k^2\phi}{1-k^2\phi^2}$, $=r^2\phi$, that is, $\tan\psi=r^2$. At the intersection with the unit section $r=1$, and therefore $\tan\psi=\phi$; moreover putting $k=\cos\kappa$, so that the equation of the curve now is $r^2=\frac{\cos^2\kappa}{1-\phi^2\cos^2\kappa}$, then for $r=1$ we have $\phi=\tan\kappa$; and hence at the intersection with the unit-circle $\psi=\kappa$, that is, as k decreases from $k=1$, or k increases from $k=0$, the angle at which each curve cuts the unit-circle is always $=\kappa$, and thus this angle continually increases from $\kappa=0$; for $k=\frac{1}{\sqrt{1+\pi^2}}=\cos\kappa$,

and therefore $\tan\kappa = \pi$, we have $\kappa = 72° 20'$ nearly: the complement hereof $17° 40'$ is thus the angle at which each branch of the loop cuts the meridian $\phi = \pi$.

21. To obtain another datum convenient in tracing the curve, I write $\phi = \phi_0 = \tan\kappa$ for the value of ϕ at the unit-circle; and introducing for the moment the rectangular coordinates $X = r\sin\phi$, $Y = 1 - r\cos\phi$, then we easily find

$$\frac{dY}{dX} = \frac{r\sin\phi - r^3\phi\cos\phi}{r\cos\phi + r^3\phi\sin\phi};$$

and thence, for the equation of the tangent at the point on the unit-circle,

$$(y - 1 + \cos\phi_0) = \frac{\sin\phi_0 - \phi_0\cos\phi_0}{\cos\phi_0 + \phi_0\sin\phi_0}(x - \sin\phi_0).$$

For the tangent at the point of intersection with the radius $\phi = 0$, or say the apse, we have $y = 1 - \cos\kappa$; and hence, at the intersection of the two tangents,

$$x = \sin\phi_0 + \frac{\cos\phi_0 + \phi_0\sin\phi_0}{\sin\phi_0 - \phi_0\cos\phi_0}(\cos\phi_0 - \cos\kappa)$$

$$= \frac{1 - \cos\kappa\,(\cos\phi_0 + \phi_0\sin\phi_0)}{\sin\phi_0 - \phi_0\cos\phi_0},$$

which, putting therein $\phi_0 = \tan\kappa$, becomes

$$= \frac{\cos\kappa\,\{1 - \cos(\phi_0 - \kappa)\}}{\sin(\phi_0 - \kappa)} = \cos\kappa\tan\tfrac{1}{2}(\phi_0 - \kappa),$$

where ϕ_0 is given in terms of κ by the just-mentioned equation $\phi_0 = \tan\kappa$. We have $y = 1 - \cos\kappa$, $x = \cos\kappa\tan\frac{1}{2}(\phi_0 - \kappa)$, for the locus of the intersection of the two tangents; this is easily seen to be a curve having a cusp at the unit-circle.

22. Fig. 3 shows the curves for the values

$\phi_0 =$	$\tan\kappa$	$\kappa =$
$30° = \frac{1}{6}\pi$	0·5235988	27° 38′
60	1·0471976	46 19
90	1·5707963	57 31
120	2·0943941	64 29
150	2·6179939	69 5
$180 = \pi$	3·1415926	72 20

We construct and graduate the unit-circle; draw to it a tangent at 0°, and measuring off from 0 a distance equal to the semi-circumference, graduate this in like manner in equal parts 0° to 180°; then to find the curve belonging, for instance, to

$\phi_0 = 90°$, we join with the centre of the circle the point 90° of the tangent, thus determining on the unit-circle a point belonging to the angle $\kappa = 57° \, 31'$; at this point we draw parallel to the tangent a line which is the tangent at the lowest

Fig. 3.

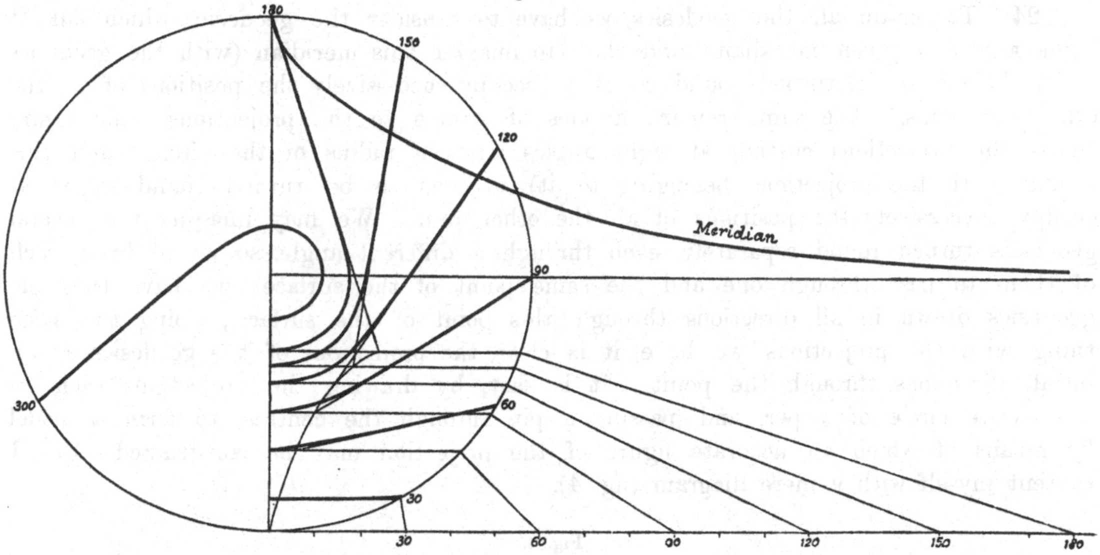

point; the curve passes through the point 90° on the unit-circle, and there cuts the circle at the angle $\kappa = 57° \, 31'$ (or, what is the same thing, the radius at the complementary angle), and we have thus the tangent at the point 90° of the unit-circle; it will be noticed that this meets the tangent at the apse at a point near to this apse, so that the arc as determined by the two tangents is for a large part of its course nearly a right line; this is still more the case for smaller values of ϕ_0 or κ, while for larger values the deviation increases, but in the neighbourhood of the unit-circle the form is always nearly rectilinear.

I show in the same figure the form of the curve for $\phi_0 = 300°$, $= 5{\cdot}2359877$, $= \tan \kappa$, that is, $\kappa = 79° \, 11'$, $r = \cos \kappa = 0{\cdot}1876670$, the value at the apse: the construction for the tangent at the unit-circle is the same as before, but in order to lay down the curve with tolerable accuracy we require also the value of r at the node on the meridian $\phi = 180°$; this is, of course, given by $r = \dfrac{\cos \kappa}{\sqrt{1 - \pi^2 \cos^2 \kappa}}$, that is, $r = \dfrac{\cos \kappa}{\cos \alpha}$, if $\pi \cos \kappa = \sin \alpha$; whence without difficulty $r = 0{\cdot}23236$, the value at the node.

23. The curves shown in the figure are projections upon the plane of the unit-circle, viz. they are the projections on this plane of the geodesics, which cut at right angles a given meridian; but bearing in mind the form of the meridian, it is easy, by means of the projection, to understand the actual forms on the surface of the pseudosphere. A point near the centre of the figure represents a point high up on the surface; and in any radius the portions near the centre are the more foreshortened in the figure, and represent greater distances on the surface. Each geodesic

cutting the meridian at right angles at the apse descends symmetrically on the two sides, reaches ultimately—it may be after many convolutions—the unit-circle; the meridian itself is a limiting or special form of geodesic. The unit-circle is not properly a geodesic, but it is an envelope of geodesics.

24. To obtain all the geodesics, we have to consider the geodesics which cut at right angles a given meridian; and then to imagine this meridian (with the geodesics which belong to it) turned round so as to occupy successively the positions of all the other meridians. The same remark applies of course to the projections; the figure shows the projections cutting at right angles a given radius of the circle; and this radius (with the projections belonging to it) is then to be turned round so as to occupy successively the positions of all the other radii. We may imagine the several geodesics turned round separately, each through a different angle, so as to bring each of them to pass through one and the same point of the surface; we have then the geodesics drawn in all directions through this point of the surface; doing the same thing with the projections, we have, it is clear, the projections of the geodesics drawn in all directions through the point. It is easy, by drawing the projections each on a separate circle of paper, and passing a pin through the centres, to form a model by means of which an accurate figure of the projection may be constructed. But I content myself with a mere diagram (fig. 4).

Fig. 4.

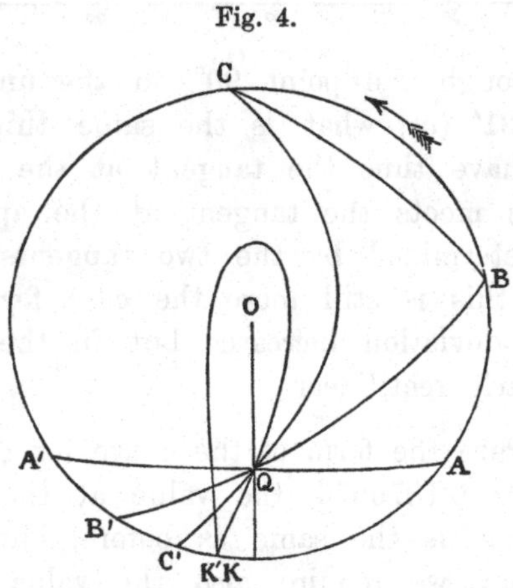

25. Taking a point Q so low down on the surface that the geodesic at right angles to the meridian through Q is a simple arc $A'A$, then imagine the two extremities A, A' each moving in the same sense round the circle, but A faster than A', so as to assume the positions B, B'; C, C'; and so on to K, K' coinciding with each other. We have the arcs $B'B$, $C'C$, and so on until we come to the loop form $K'K$: after which we have L' in advance of L, and so on to curves of any number of convolutions. Considering any two arcs—$B'B$, $C'C$—and drawing the geodesic BC which joins their extremities B and C, then any geodesic through Q intermediate to $B'B$, $C'C$, or, say, to QB, QC, will meet the arc BC; while the geodesics through Q extramediate to QB, QC *will not meet, or will only after a convolution or convolutions meet,*

the arc BC. This of course corresponds to the Lobatschewskian theory, according to which we have through a point Q to the extremities at infinity of a line BC, two distinct lines QB, QC, said to be the parallels through Q of the line BC; and which are such that any line through Q intermediate to QB, QC meets the line BC; while any line through Q extramediate to QB, QC *does not meet* the line BC.

26. It is interesting to connect the theory of the geodesics of the pseudosphere with the general theory of geodesics. Starting with the form

$$ds^2 = \cot^2\theta\, d\theta^2 + \sin^2\theta\, d\phi^2, \quad = E\, d\theta^2 + 2F\, d\theta\, d\phi + G\, d\phi^2,$$

we have $E = \cot^2\theta$, $F = 0$, $G = \sin^2\theta$; and therefore $E + 1 = \dfrac{1}{G}$, or $E = \dfrac{1}{G} - 1$, and the differential equation of the geodesic becomes

$$E\theta' \,.\, 2G_1\theta'\phi' - G\phi'(E_1\theta'^2 - G_1\phi'^2) + 2EG(\theta'\phi'' - \theta''\phi') = 0,$$

that is,

$$\phi'[(2EG_1 - GE_1)\theta'^2 + GG_1\phi'^2] + 2EG(\theta'\phi'' - \theta''\phi') = 0,$$

where

$$E_1 = \frac{dE}{d\theta}, \quad G_1 = \frac{dG}{d\theta};$$

and writing here

$$E = \frac{1}{G} - 1,$$

we have

$$E_1 = -\frac{G_1}{G^2}, \quad 2EG_1 - E_1G = G_1\left(\frac{3}{G} - 2\right).$$

Moreover, from $G = \sin^2\theta$, we find $G_1 = 2\sqrt{G\,.\,1-G}$; and the equation becomes

$$\frac{\sqrt{G}}{\sqrt{1-G}}\left[\left(\frac{3}{G} - 2\right)\theta'^2 + G\phi'^2\right]\phi' + \theta'\phi'' - \theta''\phi' = 0.$$

Introducing here G in place of θ by the equation $G = \sin^2\theta$, we have

$$\theta' = \frac{G'}{2\sqrt{G\,.\,1-G}},$$

$$\theta'' = \frac{1}{4G^{\frac{3}{2}}(1-G)^{\frac{3}{2}}}\{2G(1-G)G'' - G'^2(1-2G)\},$$

and the equation thus becomes

$$(3-2G)G'^2\phi' + 4G^3(1-G)\phi'^3 + 2G(1-G)G'\phi'' + \{-2G(1-G)G'' + (1-2G)G'^2\}\phi' = 0.$$

The whole term in ϕ' is thus $\phi'\{-2G(1-G)G'' + (4-4G)G'^2\}$, which divides by $2(1-G)$; the whole equation thus divides by $2(1-G)$, and omitting this factor, the equation becomes

$$\phi'(2G'^2 - GG'') + 2\phi'^3G^3 + \phi''GG' = 0,$$

which is simplified by introducing $H = \sin^2\theta = \frac{1}{G}$, instead of G. We, in fact, have

$$G' = \frac{-H'}{H^2},\quad G'' = \frac{-H''}{H^2} + \frac{2H'^2}{H^3},$$

and substituting these values, the equation becomes

$$\phi'\left(\frac{2H'^2}{H^4} + \frac{H''}{H^3} - \frac{2H'^2}{H^4}\right) + \frac{2}{H^3}\phi'^3 - \frac{H'}{H^3}\phi'' = 0;$$

viz. this is

$$H''\phi' + 2\phi'^3 - H'\phi'' = 0.$$

Writing herein $\phi + \alpha = \sqrt{K}$ (α an arbitrary constant), we have

$$\phi' = \frac{\frac{1}{2}K'}{\sqrt{K}},\quad \phi'' = \frac{\frac{1}{2}K''}{\sqrt{K}} - \frac{\frac{1}{4}K'^2}{K\sqrt{K}},$$

and the equation becomes

$$H''\frac{\frac{1}{2}K'}{\sqrt{K}} + \frac{\frac{1}{4}K'^3}{K\sqrt{K}} - \frac{\frac{1}{2}H'K''}{\sqrt{K}} + \frac{\frac{1}{4}H'K'^2}{K\sqrt{K}} = 0;$$

viz. this is

$$2(H''K' - H'K'')K + (K' + H')K'^2 = 0,$$

which is satisfied by $K' + H' = 0$ or $K + H = \beta$, β an arbitrary constant. Substituting for K, H their values, this is

$$(\phi + \alpha)^2 + \sin^2\theta = \beta,$$

that is,

$$\alpha^2 - \beta + (\phi^2 + \operatorname{cosec}^2\theta) + 2\alpha\phi = 0,$$

or, what is the same thing,

$$A + B(\phi^2 + \operatorname{cosec}^2\theta) + C\phi = 0,$$

where the ratios $A : B : C$ are arbitrary.

828.

A MEMOIR ON SEMINVARIANTS.

[From the *American Journal of Mathematics*, t. VII. (1885), pp. 1—25.]

Introductory. Art. Nos. 1 to 3.

1. A VERY remarkable discovery in the Theory of Seminvariants has been recently made by Capt. MacMahon, viz. considering the equation

$$0 = 1 + b\frac{x}{1} + c\frac{x^2}{1\,.\,2} + d\frac{x^3}{1\,.\,2\,.\,3} + \&c.,$$

and its roots $\alpha, \beta, \gamma, \ldots$, as defined by the identity

$$1 + b\frac{x}{1} + c\frac{x^2}{1\,.\,2} + \&c. = 1 - \alpha x\,.\,1 - \beta x\,.\,1 - \gamma x \ldots,$$

then, any symmetrical function of the roots being represented by a partition symbol in the usual manner, $1 = \Sigma\alpha$, $2 = \Sigma\alpha^2$, 11 or $1^2 = \Sigma\alpha\beta$, &c., the theorem is that any symmetric function represented by a non-unitary symbol (or symbol not containing a 1), say for shortness any non-unitary symmetric function, is a seminvariant in regard to the coefficients 1, b, c, d, e, &c.

We have for instance

$$2 = - \quad (c - b^2),$$

$$3 = -\tfrac{1}{2}(d - 3bc + 2b^3),$$

$$4 = \quad \tfrac{1}{6}(-e + 4bd + 3c^2 - 12bc + 6b^4),$$

where to verify that this is a seminvariant, observe that the value may be written

$$=\tfrac{1}{6}\{-(e-4bd+3c^2)+6(c-b^2)^2\},$$

$$22=\tfrac{1}{12}(e-4bd+3c^2),$$

&c.;

and observe further that the forms 2, 4 and 22, are connected by the identical relation

$$2.2=4+{\scriptstyle 2}.22.$$

2. We conclude that the theory of seminvariants is a part of that of symmetric functions. I take the opportunity of remarking that (the subject of the memoir being seminvariants) I use in general non-unitary symbols, even in cases where the restriction is unnecessary, and the symbols might have contained 1's: thus instead of $2.2=4+{\scriptstyle 2}.22$, the equation $1.1=2+{\scriptstyle 2}.11$ would have served equally well as an instance of an identical relation between symmetric functions; and so in general, in formulæ relating to symmetric functions, the symbols are not restricted to be non-unitary. I remark also that, for instance, instead of 4443322222, or $4^33^22^5$, I usually write 444332^5, introducing the index only for the 2; the reason is only that the 2 is often repeated a large number of times, so that the abbreviation, which I dispense with for the higher numbers, becomes convenient for the 2.

3. Reckoning the coefficients 1, b, c, d, e,... as being each of them of the degree 1, and of the weights 0, 1, 2, 3, 4,... respectively, then any symmetric function is of a degree which is equal to the highest number, and of a weight which is equal to the sum of all the numbers, in the partition-symbol. And we frequently speak of the deg. weight: thus for the function 22, the deg. weight is $=2.4$.

Multiplication of Two Symmetric Functions. Art. Nos. 4 to 17.

4. We require a theory for the multiplication of two symmetric functions. We have for instance $3.2=5+32$: for 3 denoting $\Sigma\alpha^3$, and 2 denoting $\Sigma\alpha^2$, the product contains the term α^5, and the term $\alpha^3\beta^2$, and it is thus $=\Sigma\alpha^5+\Sigma\alpha^3\beta^2$, which is $=5+32$. But multiplying for instance 2 by itself, the product contains the term α^4, and the term $\alpha^2\beta^2$ twice, and it is thus $=\Sigma\alpha^4+2\Sigma\alpha^2\beta^2$, and we thus have the before-mentioned formula $2.2=4+{\scriptstyle 2}.22$.

And so, l, m being different

$$l.m=(l+m)+lm;$$

but when $m=l$,

$$l.l\ =(2l)\quad +{\scriptstyle 2}.ll.$$

And in general, for any symmetric function $lmnpqr$..., where the numbers are all of them different, if any two of the m's become equal, we must multiply the term by 2; if any three of them become equal, we must multiply the term by 6; and so in other cases, viz. if the term becomes $l^\alpha m^\beta n^\gamma$..., we must multiply it by $[\alpha]^\alpha[\beta]^\beta[\gamma]^\gamma$...

5. We may, taking in the first instance the numbers $l, m, n, p, q, \ldots$ to be all of them different, develope an algorithm as follows:

$$\begin{array}{l|ll} l & m & \\ \hline m & & =(l+m), \\ & m & lm \end{array}\qquad \text{that is,}\quad \begin{array}{l} l\,.\,m=(l+m) \\ +lm, \end{array}$$

$$\begin{array}{ll|ll} l & m & n & \\ \hline n & & & =(l+n)\,m, \\ & n & & l\,(m+n) \\ & & n & lmn \end{array}\qquad \text{that is,}\quad \begin{array}{l} lm\,.\,n=(l+n)\,m \\ +l\,(m+n) \\ +lmn, \end{array}$$

and so in other cases;

$$\begin{array}{lll|ll} l & m & n & p & \\ \hline p & & & & =(l+p)\,mn \\ & p & & & l\,(m+p)\,n \\ & & p & & lm\,(n+p) \\ & & & p & lmnp, \end{array}\qquad\qquad \begin{array}{ll|ll} l & m & np & \\ \hline n & p & & =(l+n)\,(m+p) \\ p & n & & (l+p)\,(m+n) \\ n & & p & (l+n)\,mp \\ p & & n & (l+p)\,mn \\ & n & p & l\,(m+n)\,p \\ & p & n & l\,(m+p)\,n \\ & & np & lmnp, \end{array}$$

$$\begin{array}{lll|ll} l & m & n & pq & \\ \hline p & & q & & =(l+p)\,(m+q)\,n \\ q & p & & & (l+q)\,(m+p)\,n \\ p & & q & & (l+p)\,m\,(n+q) \\ q & & p & & (l+q)\,m\,(n+p) \\ & p & q & & l\,(m+p)\,(n+q) \\ & q & p & & l\,(m+q)\,(n+p) \\ p & & & q & (l+p)\,mnq \\ q & & & p & (l+q)\,mnp \\ & p & & q & l\,(m+p)\,q \\ & q & & p & l\,(m+q)\,p \\ & & p & q & lm\,(n+p)\,q \\ & & q & p & lm\,(n+q)\,p \\ & & & pq & lmnpq\,; \end{array}$$

and so on.

6. Observe that, if the two factors contain i numbers and j numbers respectively, $i >$ or $= j$, then in the product we have

$$\begin{array}{ll} [i]^j & \text{terms each containing } i \text{ numbers,} \\ \frac{[j]^1}{[1]^1}[i]^{j-1} & \quad\text{,,}\quad\text{,,}\quad i+1 \quad\text{,,} \\ \frac{[j]^2}{[2]^2}[i]^{j-2} & \quad\text{,,}\quad\text{,,}\quad i+2 \quad\text{,,} \\ \vdots & \\ \frac{[j]^j}{[j]^j}[i]^0, = 1, & \text{term containing } i+j \quad\text{,,} \end{array}$$

so that the whole number of terms in the product is

$$\{i,\ j\},\ = [i]^j + \frac{j}{1}[i]^{j-1} + \ldots + 1.$$

We may, if we please, take the smaller number first, and we then have

$$\{j,\ i\} = \{i,\ j\};$$

the $\{j,\ i\}$ series in fact begins with zero terms, but following these we have terms which are identical each to each with those of the $\{i,\ j\}$ series. Thus

$$\begin{aligned} \{5,\ 3\} &= [5]^3 + 3[5]^2 + 3[5]^1 + [5]^0 = 60 + 3.20 + 3.5 + 1, = 136, \\ \{3,\ 5\} &= [3]^5 + 5[3]^4 + 10[3]^3 + 10[3]^2 + 5[3]^1 + [3]^0 = 0 + 0 + 10.6 + 10.6 + 5.3 + 1, = 136 \end{aligned}$$

and it is easy to see that the general theorem can be verified in like manner.

In particular, we have Putting $i = 1$, these give

$$\begin{aligned} \{i,\ 1\} &= [i]^1 + 1 &&= i + 1 &&, \{1,\ 1\} = 2, \\ \{i,\ 2\} &= [i]^2 + 2[i]^1 + 1 &&= i^2 + i + 1 &&, \{1,\ 2\} = 3, \\ \{i,\ 3\} &= [i]^3 + 3[i]^2 + 3[i]^1 + 1 &&= i^3 + 2i + 1 &&, \{1,\ 3\} = 4, \\ \{i,\ 4\} &= [i]^4 + 4[i]^3 + 6[i]^2 + 4[i]^1 + 1 &&= i^4 - 2i^3 + 5i^2 + 1 &&, \{1,\ 4\} = 5, \\ \{i,\ 5\} &= [i]^5 + 5[i]^4 + 10[i]^3 + 10[i]^2 + 5[i]^1 + 1 &&= i^5 - 5i^4 + 15i^3 - 15i^2 + 9i + 1, &&\{1,\ 5\} = 6, \end{aligned}$$

which agree with

$$\{i,\ 1\} = i + 1.$$

Hence also the values of

{1, 1},	are 2,
{1, 2}, {2, 2},	3, 7,
{1, 3}, {2, 3}, {3, 3},	4, 13, 34,
{1, 4}, {2, 4}, {3, 4}, {4, 4},	5, 21, 73, 209,
{1, 5}, {2, 5}, {3, 5}, {4, 5}, {5, 5},	6, 31, 136, 501, 1546.

7. In forming a product $lmn \ldots pqr \ldots$, we may have equalities among the numbers $l, m, n, \ldots$ of the first symbol, and also between the numbers $p, q, r, \ldots$ of the second symbol: moreover (whether there are or are not any such equalities), we may have equalities presenting themselves between the numbers $l+p,\ m+q, \ldots$ of any symbol $(l+p)(m+q)\ldots$ on the right-hand side of the equation: and the process must be performed so as to take account of all these equalities. The actual process is best shown by an example: say we require the product 3332 . 322, which is of the deg. weight 6 . 18.

3332	322			÷12	
322		6	6552	2	1
32 2		12	6543	1	1
22 3		6	5553	6	3
32	2	12	65322	2	2
3 2	2	6	64332	2	1
2 3	2	6	55332	4	2 }
22	3	6	55332	4	2 }
2 2	3	6	54333	6	3
3	22	3	633222	12	3
3	22	1	533322	12	1 }
2	32	6	533322	12	6 }
2	32	2	433332	24	4
	322	1	3333222	144	12
		73			

8. Observe first that $\{4, 3\} = 24 + 36 + 12 + 1, = 73$. In placing all or any of the numbers of the 322 under those of the 3332, we do this in all the really distinct ways, inserting a numerical coefficient for the frequency of each way. Thus when the whole of the 322 is thus placed, the 3 may be under a 3, and the two 2's may then be under 33 or under 32: or else the 3 may be under a 2, and the two 2's must then be under 33: there are thus three ways, and these have the frequencies 6, 12, 6 respectively. For as to the first way, the 3 may be under any one of the three 3's, and for each such position of the 3, the 22 can be placed in either of two orders under the other two 3's; the frequency is thus 3 . 2, = 6. And similarly the other two frequencies are 12 and 6; the sum 6 + 12 + 6, = 24, is, the first term of $\{4, 3\}$. In like manner, when two of the numbers of the 322 are placed under the 3332, there are five ways having the frequencies 12, 6, 6, 6, 6 respectively, the sum of these frequencies is 36, which is the second term of $\{4, 3\}$. And so when one number of the 322 is placed under the 3332, there are four ways having the frequencies 3, 1, 6, 2 respectively: the sum of these frequencies is 12, which is the

third term of {4, 3}. Finally, when no number of the 322 is placed under the 3332, there is a single way, having the frequency 1, which is the last term of {4, 3}. These agreements are a useful verification for the frequencies: the sum of all the frequencies is of course = 73, the value of {4, 3}.

9. We next form the column of symmetric functions by adding to each line the 3332: to avoid accidental errors of addition, it is proper to verify for each of these that the weight is the sum, = 18, of the weights 11 and 7 of the two factors respectively. It is to be observed that the same symmetric function may present itself more than once: thus we have the functions 55332, and 533322, each of them twice. We then form a column of multiplicities: in 6552, the two 5's give the multiplicity 2; in 5553, the three 5's give 6: in 633222, the two 3's and the three 2's give 2 . 6, = 12, and so on. There is in like manner for the factor 3332 a multiplicity 6, and for the factor 322 a multiplicity 2; and these combined together give 6 . 2, = 12, viz. this is the heading ÷ 12 of the column. And then forming the products 6 . 2, 12 . 1, 6 . 6, &c., of the corresponding frequencies and multiplicities and dividing in each case by 12, we have the right-hand column of numerical coefficients.

10. The result is to be read

$$\begin{aligned} 3332 \,.\, 322 = \;& 1 \,.\, 6552 \\ & + 1 \,.\, 6543 \\ & + 3 \,.\, 5553 \\ & + \&c., \end{aligned}$$

the coefficients 2, 2 and 1, 6 of the repeated terms being of course united together, so that we have

$$\begin{aligned} & + 4 \,.\, 55332 \\ & + 7 \,.\, 533322. \end{aligned}$$

With a little practice, the operation is performed without difficulty and with very small risk of error.

11. We may apply the process to obtain analytical formulæ for certain forms of products. Consider, for instance, the product $2^\alpha \,.\, 2^\beta$, where $\alpha >$ or $= \beta$: and in the coefficients write for shortness α, instead of $[\alpha]^\alpha$ or $\Pi\alpha$, to denote $1 \,.\, 2 \,.\, 3 \ldots \alpha$. We have

$$\begin{array}{c|c|c|c} 2^\alpha & 2^\beta & \div\, \alpha \,.\, \beta & \\ \hline 2^A & 2^{\beta-A} & \dfrac{\alpha}{A \,.\, \alpha - A}\, \dfrac{\beta}{\beta - A} \,.\, 4^A 2^{\alpha+\beta-2A} \,.\, A \,.\, \alpha+\beta-2A & \dfrac{\alpha+\beta-2A}{\alpha - A \,.\, \beta - A} \end{array},$$

viz. the formula is

$$2^\alpha \,.\, 2^\beta = \Sigma \, \frac{\alpha+\beta-2A}{\alpha - A \,.\, \beta - A} \,.\, 4^A 2^{\alpha+\beta-2A},$$

where A has any integer value from β to 0; the first term is $= 1 \,.\, 4^\beta 2^{\alpha-\beta}$, and the last term is $\dfrac{\alpha+\beta}{\alpha \,.\, \beta} 2^{\alpha+\beta}$. The weight of any term is $4A + 2(\alpha+\beta-2A)$, $= 2\alpha + 2\beta$, as it should be.

In explanation as to the frequency, observe that out of the β 2's we take any A 2's and place them in any order under any A of the α 2's. The number of combinations taken is $\frac{\beta}{A\,.\,\beta-A}$, which (in A orders) gives $\frac{\beta}{\beta-A}$ sets to be placed under $\frac{\alpha}{A\,.\,\alpha-A}$ sets out of the α 2's: we thus have the foregoing coefficient of frequency $\frac{\alpha}{A\,.\,\alpha-A}\cdot\frac{\beta}{\beta-A}$, where as mentioned above α, &c. are written to denote $[\alpha]^\alpha$, &c.

12. We may in like manner find a formula for the product $3^\alpha 2^\beta\,.\,3^\gamma 2^\delta$. Taking $\gamma, =x+y+z$, any partition of γ, and $\delta, =p+q+r$, any partition of δ, and writing also $A=x$, $B=y+p$, $C=q$, we have

$$\frac{3^\alpha 2^\beta \mid 3^\gamma 2^\delta \qquad\qquad \div\,\alpha\,.\,\beta\,.\,\gamma\,.\,\delta}{3^x 2^p 3^y 2^q \mid 3^z 2^r\, M\,.\,6^A 5^B 4^C 3^{\alpha+\gamma-2A-B} 2^{\beta+\delta-B-2C}\,.\,N\,|}$$

where for the frequency, M, we have

$$\text{No. of terms } 3^x, =\frac{\gamma}{x\,.\,\gamma-x}; \qquad 3^y, =\frac{\gamma-x}{y\,.\,z},$$

$$\text{,, \quad ,, } \quad 2^p, =\frac{\delta}{p\,.\,\delta-p}; \qquad 2^q, =\frac{\delta-p}{q\,.\,r},$$

to be placed under

$$\text{sets of terms } 3^\alpha, =\frac{\alpha}{x+p\,.\,\alpha-x-p}; \quad 2^\beta, =\frac{\beta}{y+q\,.\,\beta-y-q},$$

$$\text{in orders, } =x+p; \qquad\qquad =y+q,$$

whence, multiplying, we have for the frequency

$$M=\frac{\alpha\,.\,\beta\,.\,\gamma\,.\,\delta}{x\,.\,y\,.\,z\,.\,p\,.\,q\,.\,r\,.\,\alpha-x-p\,.\,\beta-y-q};$$

and for the multiplicity, N, we have

$$\begin{aligned} N &= A\,.\,B\,.\,C\,.\,\alpha+\gamma-2A-B\,.\,\beta+\delta-B-2C, \\ &= x\,.\,B\,.\,q\,.\,\alpha+\gamma-2A-B\,.\,\beta+\delta-B-2C. \end{aligned}$$

13. The coefficient is thus

$$\begin{aligned} \frac{M\,.\,N}{\alpha\,.\,\beta\,.\,\gamma\,.\,\delta}, &= \frac{B\,.\,\alpha+\gamma-2A-B\,.\,\beta+\delta-B-2C}{y\,.\,z\,.\,p\,.\,r\,.\,\alpha-x-p\,.\,\beta-y-q}, \\ &= \frac{B\,.\,\alpha+\gamma-2A-B\,.\,\beta+\delta-B-2C}{y\,.\,p\,.\,\gamma-A-y\,.\,\alpha-A-p\,.\,\beta-C-y\,.\,\delta-C-p}, \end{aligned}$$

or, putting also

$$\begin{aligned} D &= \alpha+\gamma-2A-B, \\ E &= \beta+\delta \qquad -B-2C, \end{aligned}$$

whence

$$6A + 5B + 4C + 3D + 2E = 3(\alpha + \gamma) + 2(\beta + \delta),$$

the formula finally is

$$3^\alpha\, 2^\beta\,.\,3^\gamma\, 2^\delta = \Sigma\Lambda\,.\,6^A\, 5^B\, 4^C\, 3^D\, 2^E,$$

where

$$\Lambda = \frac{B \quad . \quad D \quad . \quad E}{y\,.\,p\,.\,\gamma - A - y\,.\,\alpha - A - p\,.\,\beta - C - y\,.\,\delta - C - p},$$

or, as this may also be written,

$$= \frac{B}{y\,.\,B - y}\,.\,\frac{D}{z\,.\,D - z}\,.\,\frac{E}{r\,.\,E - r},$$

so that the coefficient Λ is in fact the product of three binomial coefficients: it must, however, be recollected that the same term $6^A\, 5^B\, 4^C\, 3^D\, 2^E$ may occur more than once, with different coefficients Λ, so that in the final result, when these terms are united together, the numerical coefficients are not each of them of the form in question.

14. The limits of the summation are conveniently defined by means of the diagram

A	p	$\alpha - A - p$	α	$D = \alpha + \gamma - 2A - B,$
y	C	$\beta - C - y$	β	$E = \beta + \delta \qquad - B - 2C,$
z	r			*ut suprà*,
γ	δ			

viz. here the sums of the first and second lines are α and β respectively, and the sums of the first and second columns are γ and δ respectively; we have $B = y + p$, a partition of B; but the values of y, p must be such as not to render negative any one of the four terms z, r, $\alpha - A - p$, $\beta - C - y$ of the diagram: for if any one of these numbers were negative, the corresponding factorial in the denominator of Λ would be infinite and we should have $\Lambda = 0$. For any given term $6^A\, 5^B\, 4^C\, 3^D\, 2^E$ of the proper weight $3(\alpha + \gamma) + 2(\beta + \delta)$, there may be no suitable values of y, p, and the coefficient is then $= 0$: there may be a single pair of values, and the coefficient is then (as remarked above) = a product of three binomial coefficients: or there may be more than a single pair of values, and the entire coefficient of the term has not in this case a like simple form.

15. To exhibit the working of the formula, I apply it to the recalculation of the foregoing product $3332\,.\,322$ (α, β, γ, $\delta = 3$, 1, 1, 2). Properly the whole series of symbols 666, 6642, 6633, &c., of the symmetric functions of the weight 18 should be written down, and the coefficient be calculated for each of them: but I write down only those which have coefficients not $= 0$: for each of these, I take *all* the partitions

$B = y + p$, several of these giving, as will be seen, zero values, and the others giving the values already obtained for the coefficients of the several terms.

							1			2			3					
	A	B	C	D	E	B	y	p	D	$\gamma - A - y$	$\alpha - A - p$	E	$\beta - C - y$	$\delta - C - p$	1	2	3	
							2	0		−2	2		−1	2	1	0	0	
6552	1	2	0	0	1	2	1	1	0	−1	1	1	0	1	2	0	1	
							0	2		0	0		1	0	1	1	1	1.6552
6543	1	1	1	1	0	1	1	0	1	−1	2	0	−1	1	1	0	0	
							0	1		0	1		0	0	1	1	1	1.6543
6532^2	1	1	0	1	2	1	1	0	1	−1	2	2	0	2	1	0	1	
							0	1		0	1		1	1	1	1	2	2.6532^2
64332	1	0	1	2	1	0	0	0	2	0	2	1	.0	1	1	1	1	1.64332
6332^3	1	0	0	2	3	0	0	0	2	0	2	3	1	2	1	1	3	3.6332^3
							3	0		−2	3		−2	2	1	0	0	
5553	0	3	0	1	0	3	2	1	1	−1	2	0	−1	1	3	0	0	3.5553
							1	2		0	1		0	0	3	1	1	
							0	3		1	0		1	−1	1	1	0	
							2	0		−1	3		−1	2	1	0	0	2
55332	0	2	0	2	1	2	1	1	2	0	2	1	0	1	2	1	1	} 4.55332
							0	2		1	1		1	0	1	2	1	2
54333	0	1	1	3	0	1	1	0	3	0	3	0	−1	1	1	1	0	
							0	1		1	2		0	0	1	3	1	3.54333
53332^2	0	1	0	3	2	1	1	0	3	0	3	2	0	1	1	1	1	1 } 7.53332^2
							0	1		1	2		2	1	1	3	2	6
433332	0	0	1	4	1	0	0	0	4	1	3	1	0	1	1	4	1	4.433332
33332^3	0	0	0	4	3	0	0	0	4	1	3	3	1	2	1	4	3	12.33332^3

It is quite possible that abbreviations and verifications might be introduced, but the process as it stands seems to be at once less expeditious and less safe than the one first made use of.

16. Particular cases of the general formula are

$$2^\beta . 2^\delta = \Sigma\Lambda 4^C 2^E;\ E = \beta + \delta - 2C, \text{ and thence } 4C + 2E = 2(\beta + \delta),$$

$$\Lambda = \frac{E}{\beta - C . \delta - C},$$

which agrees with a result before obtained. Again,

$$3^\alpha 2^\beta . 2^\delta = \Sigma\Lambda . 5^B 4^C 3^D 2^E;\ D = \alpha - B,$$

$$E = \beta + \delta - B - 2C,$$

and thence

$$5B + 4C + 3D + 2E = 3\alpha + 2\beta + 2\delta,$$

$$\Lambda = \frac{E}{\beta - C . \delta - B - C}.$$

17. We may, in like manner with the formula for $3^\alpha 2^\beta . 3^\gamma 2^\delta$, obtain the following:

$$4^\theta 3^\alpha 2^\beta . 2^\delta = \Sigma\Lambda . 6^A 5^B 4^C 3^D 2^E,$$

where

$$D = \alpha - B,$$

$$E = 2\theta + \beta + \delta - 3A - B - 2C,$$

and thence

$$6A + 5B + 4C + 3D + 2E = 4\theta + 3\alpha + 2\beta + 2\delta,$$

and the value of the coefficient Λ is

$$\Lambda = \frac{C \quad . \quad E}{\theta - A . C - \theta + A . \theta + \beta - A - C . \theta + \delta - 2A - B - C},$$

viz. Λ is the product of two binomial coefficients: and since here a given term $6^A 5^B 4^C 3^D 2^E$ occurs once only, each numerical coefficient is actually of this form.

In the particular case $\theta = 0$, we must have $A = 0$ (this appears *à posteriori* from the denominator factor $-A$, a factorial which is infinite for any positive value of A); and we thus obtain the first given formula for $3^\alpha 2^\beta . 2^\delta$.

Capitation and Decapitation. Art. Nos. 18 to 21.

18. I explain the converse processes of Decapitation and Capitation. In any symmetric function, for instance 6552 of the degree 6, the whole coefficient of α^6 is 552, this symbol referring in the first instance to the series of remaining roots $\beta, \gamma, \delta, \ldots$; but as the series of roots is unlimited, we may ultimately replace these by $\alpha, \beta, \gamma, \ldots$, and so use 552 in its original sense. Similarly, in 6652, the whole coefficient

of α^6 is $=652$, the only difference being that, while in the former case the degree is reduced to 5, in the present case it remains $=6$. In every case we decapitate the symbol by striking out the highest number—in the case of two or more equal numbers, one only of these being struck out. Observe that by decapitation we always diminish the weight, but we do or do not diminish the degree. In a product such as 3332.322 we obtain in like manner the whole coefficient of α^6 by the decapitation of each factor, viz. the coefficient is $=332.22$: and in any equation such as that obtained above for the product 3332.322, the whole coefficient of the highest power of α must be $=0$, viz. we can by decapitation obtain a new equation of lower weight: thus from the equation in question of weight 18, we obtain the new equation of weight 12,

$$\begin{aligned} 332.22 = {} & {}_1.552 \\ & + {}_1.543 \\ & + {}_2.5322 \\ & + {}_1.4332 \\ & + {}_3.33222, \end{aligned}$$

where observe that the terms of a degree lower than 6 in the original equation give no term in the new equation. The new equation of the deg. weight 5.12 might of course be obtained independently in like manner with the original equation.

19. We capitate a symbol by prefixing to it a number which is not less than the highest number contained in it: thus 552 may be capitated into 5552, 6552, &c.: and so a product 332.22 may be capitated into 3332.222, 3332.322, &c.; moreover a single symbol may be capitated into a product, 552 into 5552.4: in fact, the capitation may be any operation such that by decapitation we reproduce the original symbol. The increase of weight may be any number not less than the degree of the original symbol: but it is usually taken to be a given number: thus for any symbol of a degree not exceeding 6, we may capitate so as to increase the weight by 6. The capitation does or does not increase the degree.

20. An identical equation may be capitated in a variety of ways, but instead of an identity we obtain only a congruence, that is, an equation which requires to be completed by the adjunction to it of proper terms of lower degrees. Thus from the above equation of the deg. weight 5.12 we may obtain

$$\begin{aligned} 3332.322 \equiv {} & {}_1.6552 \\ & + {}_1.6543 \\ & + {}_2.65322 \\ & + {}_1.64332 \\ & + {}_3.633222. \end{aligned}$$

Imagine here all the terms brought to the same side so that the form is $\Omega \equiv 0$: Ω is a function not containing α^6, for the whole coefficient of α^6 therein is precisely that function which by the equation of the deg. weight 5 . 12 is expressed to be $=0$: and hence Ω, *quâ* symmetric function cannot contain $\beta^6, \gamma^6, \ldots$; viz. Ω is a symmetric function of the degree 5 at most: the congruence $\Omega \equiv 0$ thus means, $\Omega =$ a properly determined function of the degree 5 at most. Obviously by development of the term 3332.322, that is, by substituting for this term its value as given by the equation of the deg. weight 6.18, the function of the degree 5 at most would be found to be = the sum of those terms in the expression of 3332.322, which are of a degree inferior to 6: and the congruence $\Omega \equiv 0$ thus completed would be nothing else than the equation of the deg. weight 6.18.

21. We might have capitated in a different manner: for instance

$$\begin{aligned} 4332.222 = \ & {}_1.6552 \\ & + {}_1.6543 \\ & + {}_2.65322 \\ & + {}_1.44332.2 \\ & + {}_3.33222.3, \end{aligned}$$

there being here three products requiring to be developed: on replacing them by their values, there would be found for Ω a value which would be a determinate function of a degree less than 6: and putting $\Omega =$ this value, the congruence $\Omega \equiv 0$ would be completed into an equation.

Seminvariants of a given Degree; Perpetuants, &c. Art. Nos. 22 to 38.

22. We consider now seminvariants according to their degrees; in particular, those of the degrees 2, 3, 4, 5 and 6, or say quadric, cubic, quartic, quintic and sextic seminvariants: the forms of these are $2^{\alpha+1}$, $3^{\alpha+1}2^{\beta}$, $4^{\alpha+1}3^{\beta}2^{\gamma}$, $5^{\alpha+1}4^{\beta}3^{\gamma}2^{\delta}$, $6^{\alpha+1}5^{\beta}4^{\gamma}3^{\delta}2^{\epsilon}$, where each exponent $\alpha, \beta, \gamma, \delta, \epsilon$ is a positive integer not excluding zero: the exponent of the highest number is in each case written $\alpha+1$, and has thus the value 1 at least, for otherwise the form would not be of the proper degree. The several weights are $2(\alpha+1)$, $3(\alpha+1)+2\beta$, $4(\alpha+1)+3\beta+2\gamma$, &c., or, what is the same thing, we have for the several degrees respectively

$$\begin{aligned} w-2 &= 2\alpha, \\ w-3 &= 3\alpha+2\beta, \\ w-4 &= 4\alpha+3\beta+2\gamma, \\ w-5 &= 5\alpha+4\beta+3\gamma+2\delta, \\ w-6 &= 6\alpha+5\beta+4\gamma+3\delta+2\epsilon, \end{aligned}$$

and we have for a given degree and weight as many seminvariants as there are systems of exponents satisfying the corresponding equation.

23. These numbers are at once expressible by means of a series of Generating Functions (G. F.), viz. writing for shortness **2**, **3**, **4**, **5**, **6** to denote $1-x^2$, $1-x^3$, $1-x^4$, $1-x^5$, $1-x^6$, the G. F.'s are

$$\frac{x^2}{\mathbf{2}},\ \frac{x^3}{\mathbf{2.3}},\ \frac{x^4}{\mathbf{2.3.4}},\ \frac{x^5}{\mathbf{2.3.4.5}},\ \frac{x^6}{\mathbf{2.3.4.5.6}}.$$

In fact, the number of seminvariants of a given weight is = coefficient of x^w in the corresponding G. F.; for the quadric seminvariants (or those of deg. weight $2.w$) in $\frac{x^2}{\mathbf{2}}$; for the cubic seminvariants (or those of deg. weight $3.w$) in $\frac{x^3}{\mathbf{2.3}}$; and so for the others.

24. A seminvariant of a given degree may be a sum of products (of that degree) of seminvariants of lower degrees, and of seminvariants of lower degrees: and it is in this case said to be reducible: a seminvariant which is not reducible is said to be irreducible, or otherwise to be a perpetuant. This notion of a perpetuant is due to Sylvester, see his Memoir "On Subinvariants, i.e. Semi-Invariants to Binary Quantics of an Unlimited Order," *American Journal of Mathematics*, vol. v. (1882—83), pp. 78—137 (§ 4 Perpetuants, pp. 105—118). In speaking of the number of perpetuants of a given deg. weight, we assume throughout that these are independent perpetuants, not connected by any linear relation.

25. Since the seminvariants used for the reduction of a given reducible seminvariant can themselves be expressed in terms of perpetuants, we may say more definitely that a seminvariant of a given degree, which is a sum of products (of that degree) of perpetuants of lower degrees, and of perpetuants of lower degrees, is reducible. The words "of that degree" are essential to the definition: a seminvariant may be expressible as a sum of products (of a higher degree) of perpetuants of lower degrees, and of perpetuants of lower degrees, and it is not on this account reducible: a seminvariant so expressible is said to be a "syzygant"; but as to this, see No. 49.

26. Every quadric or cubic seminvariant is obviously a perpetuant: the quadric and cubic perpetuants have thus the before-mentioned G. F.'s $\frac{x^2}{\mathbf{2}}$ and $\frac{x^3}{\mathbf{2.3}}$ respectively.

27. A reducible quartic seminvariant can only be a sum of products (2.2) of two quadric perpetuants, and of quadric perpetuants, and it is clear that no quartic seminvariant the symbol of which contains a 3 is thus expressible. If the symbol does not contain a 3, viz. when the form is $4^{\alpha+1}2^{\gamma}$, the seminvariant is reducible: we have for instance

$$4 = 2.2 - {}_2.22,$$

$$42 = 22.2 - {}_3.222, \ \&c.$$

To show that this is so in general, observe that any symmetric function $2^{\alpha+1}1^{\gamma}$, *quâ* symmetric function can be expressed as a rational and integral function of the degree $2(\alpha+1)+\gamma$, of the coefficients 1, 1^2, 1^3, &c.: instead of the roots considering their squares, we have thence an expression for the quartic seminvariant $4^{\alpha+1}2^{\gamma}$ in terms of the quadric perpetuants 2, 2^2, 2^3, &c., and such expression will be of the same degree $4(\alpha+1)+2\gamma$, as the quartic seminvariant.

It thus appears, as regards the quartic seminvariants, that whenever the symbol contains a 3, and in this case only, the seminvariant is a perpetuant: or, what is the same thing, the form of a quartic perpetuant is $4^{\alpha+1}3^{\beta+1}2^{\gamma}$: for the weight w, the number is equal to that of the sets of values α, β, γ, such that $w-7=4\alpha+3\beta+2\gamma$: or, what is the same thing, the G. F. of the quartic perpetuants is $=\dfrac{x^7}{2.3.4}$.

28. Sylvester, in the memoir referred to, obtained this result in a different manner: the quartic seminvariants of a given weight are the quartic perpetuants of that weight and also the products (of that weight) of two quadric perpetuants, the same or different: say (4) is the G. F. for the perpetuants, and (2, 2) for the products: then the G. F. for the quartic seminvariants being as already mentioned $\dfrac{x^4}{2.3.4}$, we have his equation

$$(4)+(2,\ 2)=\frac{x^4}{2.3.4}.$$

He deduces (2, 2) from the G. F. $=\dfrac{x^2}{2}$ of the quadric perpetuants and thence obtains (4), $=\dfrac{x^7}{2.3.4}$, as above.

29. Write for a moment ϕx to represent the G. F. $=\dfrac{x^2}{2}$ of the quadric perpetuants, and $A, B, C, \ldots$ to represent these quadric perpetuants: we have, in an algorithm which will be readily understood,

$$\begin{aligned}\phi x &= (A+B+C\ldots),\\ (\phi x)^2 &= (A+B+C\ldots)^2, \ = A^2+2AB+\ldots,\\ \phi x^2 &= \qquad\qquad A^2+\&c.,\end{aligned}$$

and thence

$$\tfrac{1}{2}\{(\phi x)^2+\phi x^2\} = \qquad A^2+AB+\&c.;$$

viz. the G. F. (2, 2), is

$$\tfrac{1}{2}((\phi x)^2+\phi x^2),\ =\tfrac{1}{2}\left(\frac{x^4}{2.2}+\frac{x^4}{4}\right),\ =\tfrac{1}{2}\left(\frac{x^4(1+x^2)}{2.4}+\frac{x^4(1-x^2)}{2.4}\right),\ =\frac{x^4}{2.4},$$

and we thence have

$$(4)+\frac{x^4}{2.4}=\frac{x^4}{2.3.4},$$

that is, $(4)=\dfrac{x^7}{2.3.4}$, the same result as was found above by independent considerations.

30. Sylvester established in like manner (but without the terms S which will be presently explained) the equations

$$(5)+(3,\ 2)=\frac{x^5}{2\,.\,3\,.\,4\,.\,5}+S_5,$$

$$(6)+(4,\ 2)+(3,\ 3)+(2,\ 2,\ 2)=\frac{x^6}{2\,.\,3\,.\,4\,.\,5\,.\,6}+S_6,$$

viz. here (5) is the G. F. for the quintic perpetuants, (2, 3) that for the products of the quadric and the cubic perpetuants: and similarly (6) is the G. F. for the sextic perpetuants, (4, 2) that for the products of the quadric and the quartic perpetuants, (3, 3) for the products of two cubic perpetuants, the same or different: and (2, 2, 2) for the products of three quadric perpetuants, the same or different. We have at once $(3,\ 2)=(3)\,.\,(2)$; $(4,\ 2)=(4)\,.\,(2)$; (3, 3) is found by the same process as was used for finding (2, 2), substituting only $\frac{x^3}{3}$ for $\frac{x^2}{2}$; and (2, 2, 2) is found by a like process, viz. the G. F. for A^3+A^2B+ABC is $\frac{1}{6}\{(\phi x)^3+3\phi x\,.\,\phi x^2+2\phi x^3\}$, where $\phi x=\frac{x^2}{2}$ as before, viz. this is $\frac{1}{6}\left\{\frac{x^6}{2\,.\,2\,.\,2}+\frac{3x^6}{2\,.\,4}+\frac{2x^6}{6}\right\}$: reducing to the common denominator $2\,.\,4\,.\,6$, the numerator is

$$=\tfrac{1}{6}x^6\{(1+x^2)(1+x^2+x^4)+3(1-x^6)+2(1-x^2)(1-x^4)\},$$

viz. this is $=x^6$. The several functions thus are

$$(3,\ 2)=\frac{x^5}{2\,.\,2\,.\,3},\quad (4,\ 2)=\frac{x^9}{2\,.\,2\,.\,3\,.\,4},$$

$$(3,\ 3)=\frac{x^6}{3\,.\,6},\quad (2,\ 2,\ 2)=\frac{x^6}{2\,.\,4\,.\,6}.$$

31. Mr Hammond, in regard to the equation for the quintic perpetuants, made the very important observation—see his paper "On the Solution of the Differential Equation of Sources," *American Journal of Mathematics*, t. v. (1882), pp. 218—227—that the products (3, 2) of a cubic perpetuant and a quadric perpetuant are not independent: we have between them syzygies such as $32\,.\,2-3\,.\,22\equiv 0$ (viz. the difference of the two products contains no term of the degree 5; the actual value is $=43+322$: Hammond's equation (12), p. 222), hence the necessity in the equation of a term S_5 referring to these syzygies; and he moreover obtained the expression, $S_5=\frac{x^7}{2\,.\,4}$ of the G. F. for these syzygies.

The equation gives

$$(5)=\frac{-x^7+x^{10}+x^{12}}{2\,.\,3\,.\,4\,.\,5}+S_5,$$

where of course the first term is the value of (5) which would be given by the equation without the term S_5; and substituting herein for S_5 the foregoing value, viz.

$$S_5=\frac{x^7}{2\,.\,4},\quad =\frac{x^7(1-x^3)(1-x^5)}{2\,.\,3\,.\,4\,.\,5},$$

we find

$$(5) = \frac{x^{15}}{2.3.4.5},$$

which is the correct value of the G. F. for the quintic perpetuants: the lowest quintic perpetuant is thus of the weight 15.

32. The equation for the sextic perpetuants gives

$$(6) = \frac{-x^6 - x^{13} + 2x^{16} + x^{18}}{2.3.4.5.6} + S_6,$$

which is an equation connecting (6) and S_6, the G. F.'s for the sextic perpetuants, and the sextic syzygies respectively. I have, in the investigation of the value of S_6, met with a difficulty which I have not been able to overcome: but I find that

$$S_6 = \frac{x^6 + x^{13} - 2x^{16} - x^{18} + \omega(x)}{2.3.4.5.6},$$

where $\omega(x)$ is possibly the monomial function x^{31}, but this result (which Capt. MacMahon believes to be true) is not yet completely established; it is a function containing no term lower than x^{31}. We have therefore

$$(6) = \frac{\omega(x)}{2.3.4.5.6},$$

and there is, it would appear, no sextic perpetuant of a weight lower than 31.

33. But before entering on the investigation, it is proper to further develope the theory of the quintic perpetuants. We have quintic seminvariants: 5 for weight 5; 52 for weight 7; 53 for weight 8; 54, 522 for weight 9; 55, 532 for weight 10; 542, 533, 5222 for weight 11; and so on. These are reduced by means of the products 3.2 of a cubic perpetuant and a quadric perpetuant; viz. for any given weight, we have all the products $3^\alpha 2^\beta . 2^\gamma$ of that weight. Thus for weight 5 there is only the product 3.2, and this in fact serves to reduce the seminvariant 5: we have $3.2 = 5 + 32$, and therefore $5 = 3.2 - 32$.

For the weight 7; there is only the seminvariant 52, and there are the two products 32.2 and 3.22: either of these would serve for the reduction: we have

$$\begin{aligned} 32.2 = {} & 52, & \quad 3.22 = {} & 52, \\ + {} & 43, & + {} & 322, \\ + {} & 2.322, & & \end{aligned}$$

and these two equations imply the before-mentioned syzygy, $32.2 - 3.22 \equiv 0$, in virtue of which the two reductions become equivalent; or say there remains a single equation serving for the reduction: the most simple form is $52 = 3.22 - 322$.

Weight 8: there is only the seminvariant 53, and the product 33.2: this gives the reduction.

Weight 9: seminvariants 54, 522: products $32^2.2$, 32.2^2, 3.2^3; there is between the first and last of these a syzygy; and this being satisfied, there remain two equations for the reductions.

Weight 10: seminvariants 55, 532: products 332 . 2, 33 . 22: these give the reductions.

Weight 11: seminvariants 533; 542, 5222: products 333 . 2; $32^3 . 2$, $32^2 . 2^2$, $32 . 2^3$, 32^4.

34. Observe that the seminvariants, and in like manner the products, form two classes, according as the symbols contain three odd numbers or a single odd number: these correspond separately to each other, for the development of any product will contain only seminvariants having each of them as many odd numbers as there are odd numbers in the product. Hence for the weight 11 just referred to, 333 . 2 serves for the reduction of 533; $32^3 . 2$, $32^2 . 2^2$, $32 . 2^3$, $3 . 2^4$ are connected the first and fourth of them by a syzygy, and the second and third by a syzygy; and there remain two equations serving for the reduction of the two seminvariants 542, 5222.

It is easy to show in this manner that there is no quintic perpetuant for any weight under 15; and that there is a single quintic perpetuant for the weight 15.

35. Generally the syzygies exist only for an odd weight $w = 2\beta + 3$ between the products $32^{\beta-1} . 2$, $32^{\beta-2} . 2^2, \ldots, 32 . 2^{\beta-1}$, $3 . 2^{\beta}$; viz. there is a syzygy between the first and last terms: a syzygy between the second and last but one terms; and so on. The existence of these syzygies at once appears from the principle of decapitation: decapitating $32^{\beta-1} . 2 - 3 . 2^{\beta}$, we have $2^{\beta-1} - 2^{\beta-1}$, which is identically $= 0$, hence the function contains no term of the degree 5, that is, $32^{\beta-1} . 2 - 3 . 2^{\beta} \equiv 0$; and similarly for the other pairs of terms.

There are β terms: hence in the case β even, we have $\tfrac{1}{2}\beta$ pairs of terms and therefore $\tfrac{1}{2}\beta$ syzygies: in the case β odd, there is a middle term, not connected by a syzygy with any other term, and the number of syzygies is thus $= \tfrac{1}{2}(\beta - 1)$: writing for β its value $\tfrac{1}{2}(w - 3)$, the number is $= \tfrac{1}{4}(w - 3)$, and $\tfrac{1}{4}(w - 5)$ in the two cases respectively: and it thus appears that the G. F. is $= \dfrac{x^7}{\mathbf{2 . 4}}$ as already mentioned.

36. In the case w an odd number, we have seminvariants, and in like manner products, containing respectively one odd number, three odd numbers, five odd numbers, and so on: thus $w = 15$, we have

Seminvariants.	Products.	
555	$3332^2 . 2$	3 equations,
5532	$3332 . 2^2$	
5433	$333 . 2^3$	
5332^2		
5442	$32^5 . 2$	$\tfrac{1}{2}\,6, = 3$ equations;
542^3	$32^4 . 2^2$	
52^5	$32^3 . 2^3$	
	$32^2 . 2^4$	
	$32 . 2^5$	
	$3 . 2^6$	

hence the seminvariants 555, 5532, 5433, 5332^2 with three odd numbers are not reducible, but they can be linearly expressed in terms of the 3 like products $3332^2.2$, 3332.2^2, 333.2^3, and of $4-3$, $=1$ arbitrary quantity (observe, however, that this must not be the seminvariant 5332^2, for this is in fact reducible): the seminvariants 5442, 542^2, 52^5, with one odd number, are reducible. And the like as regards any other odd value of w.

37. In the case w an even number, we have seminvariants, and in like manner products, containing respectively two odd numbers, four odd numbers, six odd numbers, and so on. Thus $w=16$, we have

Seminvariants.	Products.	
5533	33332 . 2	2 equations,
53332	3333.2^2	
5542	$332^4.2$	5 equations;
552^3	$332^3.2^2$	
5443	$332^2.2^3$	
5432^2	332.2^4	
532^4	33.2^5	

and thus the seminvariants with four odd numbers, and those with two odd numbers, are each set reducible.

38. I give in the case $w=19$ the following results: the expression for the G. F. shows that there are 2 quintic perpetuants: viz. two forms X, Y such that every quintic seminvariant of the weight 19 is expressible as a linear function of these, of products (3.2) of a cubic and a quadric perpetuant, and of forms of a degree inferior to 5, that is, quartic, cubic and quadric perpetuants. Attending only to the terms in X, Y, the actual values are:

$$
\begin{aligned}
5554 &= X,\\
5552^2 &= Y,\\
55432 &= -X,\\
55333 &= 0,\\
5532^3 &= -Y,\\
54442 &= 0,\\
54433 &= X,\\
5442^3 &= 0,\\
54332^2 &= Y,\\
542^5 &= 0,\\
533332 &= 0,\\
5332^4 &= 0,\\
52^7 &= 0,
\end{aligned}
$$

viz. of the 13 quintic seminvariants of the weight in question there are 7, which as not containing either X or Y are each of them reducible; while the remaining 6 can only be expressed as linear functions of X and Y. It would be allowable to select $5554\,(=X)$ and $5552^2\,(=Y)$ as the two representative perpetuants, but there is no particular advantage in this.

Sextic Perpetuants and Sextic Syzygies: Syzygants. Art. Nos. 39 to 51.

39. Returning now to the sextic seminvariants, these are weight 6; 6: weight 8; 62: weight 9; 63: weight 10; 64, 62^2: and so on. And they are reducible by means of the products (4, 2), (3, 3) and (2, 2, 2), that is, of a quartic perpetuant and a quadric perpetuant, of two cubic perpetuants, and of three quadric perpetuants: this last form of product existing only in the case of an even weight.

40. For weight 6, we have seminvariant 6, and the two products $3\,.\,3$ and $2\,.\,2\,.\,2$; this implies a syzygy $3\,.\,3-2\,.\,2\,.\,2\equiv 0$; and there then remains a single equation for the reduction of the seminvariant. The formulæ are

$$3\,.\,3 = 6 \qquad\qquad 2\,.\,2\,.\,2 = 6$$
$$+{}_2\,.\,33, \qquad\qquad +{}_3\,.\,22\,.\,2$$
$$\qquad\qquad\qquad -{}_3\,.\,222,$$

so that the complete syzygy, and the most simple reduction, are

$$\begin{array}{ll} \quad 3\,.\,3 & \qquad 6 = 3\,.\,3 \\ -\quad 2\,.\,2\,.\,2 & \qquad\quad -{}_2\,.\,33, \\ -{}_2\,.\,33 & \\ +{}_3\,.\,22\,.\,2 & \\ -{}_3\,.\,222 = 0, & \end{array}$$

and we might in this way verify that, for the successive weights 8, 9, 10, &c., there are no sextic perpetuants; and find for these weights respectively, the number of the sextic syzygies. But such direct investigation becomes soon impracticable.

41. I endeavour to determine the number of sextic syzygies for the weight w; and for this purpose I establish the following relation:

$$(S_6) = ((0)) + ((2))' + ((3))' + ((2,\ 2))' + ((3,\ 2))' + (S_5)' + (S_6)' - ((5))' + ((\theta))',$$

where (S_6) is the number of sextic syzygies for the weight w, or, what is the same thing, it is the coefficient of x^w in the function S_6, which is the G. F. for these syzygies: ((0)) has the value 1 for $w=6$, and the value 0 in all other cases. The accented symbols refer to the weight $w-6$; $(S_6)'$ is thus the number of sextic syzygies for this weight: and for the same weight $w-6$, ((2))′ denotes the number of quadric perpetuants, or coefficient of x^{w-6} in the function (2) which is the G. F. for these perpetuants: ((3))′ the number of cubic perpetuants, ((2, 2))′ the number of products of two quadric perpetuants, the same or different, $((3,\ 2))'$ the number of products of a cubic perpetuant and a quadric perpetuant, $(S_5)'$ the number of quintic syzygies, $((5))'$ the number of quintic perpetuants, and $((\theta))'$ a term of unascertained form which will be explained further on. Transposing the term $(S_6)'$ to the left-hand side, and passing to the generating functions, we have

$$(1-x^6)\,S_6 = (0)+(2)'+(3)'+(2,\ 2)'+(3,\ 2)'+S_5'-(5)'+(\theta)',$$

or, as this may be written,

$$= x^6\,\{1+(2)+(3)+(2,\ 2)+(3,\ 2)+(S_5)-(5)+(\theta)\},$$

where (2), (3), &c., have the values already obtained for these G. F.'s respectively: viz. writing $x^6\,(\theta) = \dfrac{\omega x}{\mathbf{2.3.4.5}}$, the equation is

$$\mathbf{6}\,.\,S_6 = x^6 + \frac{x^8}{\mathbf{2}} + \frac{x^9}{\mathbf{3}} + \frac{x^4}{\mathbf{2.4}} + \frac{x^{11}}{\mathbf{2.2.3}} + \frac{x^{13}}{\mathbf{2.4}} - \frac{x^{21}}{\mathbf{2.3.4.5}} + \frac{\omega\,(x)}{\mathbf{2.3.4.5}}.$$

42. Reducing on the right-hand side the known terms to the common denominator **2.3.4.5**, the numerator is

$$\begin{array}{lll}
x^6\,. & 1-x^2.1-x^3.1-x^4.1-x^5 = & x^6-x^8-x^9-x^{10}+x^{12}+2x^{13}+x^{14}-x^{16}-x^{17}-x^{18}+x^{20} \\
+x^8\,. & 1-x^3.1-x^4.1-x^5 & +x^8-x^{11}-x^{12}-x^{13}+x^{15}+x^{16}+x^{17}-x^{20} \\
+x^9\,. & 1-x^4.1-x^5 & +x^9-x^{13}-x^{14}+x^{18} \\
+x^{10}\,. & 1-x^3.1-x^5 & +x^{10}-x^{13}-x^{15}+x^{18} \\
+x^{11}\,. & 1+x^2.1-x^5 & +x^{11}+x^{13}-x^{16}-x^{18} \\
+x^{13}\,. & 1-x^3.1-x^5 & +x^{13}-x^{16}-x^{18}+x^{21} \\
-x^{21}\,. & & -x^{21} \\
\hline
 & & x^6+x^{13}-2x^{16}-x^{18}\ ;
\end{array}$$

whence, dividing by **6**, we have the before-mentioned formula

$$S_6 = \frac{x^6+x^{13}-2x^{16}-x^{18}+\omega\,(x)}{\mathbf{2.3.4.5.6}}$$

43. We have to prove the formula for (S_6): this symbol denotes, for the weight w, the number of syzygies between the products (4, 2), (3, 3) and (2, 2, 2): we have to consider separately the cases w odd, and w even. First, if w be odd, there are no products (2, 2, 2) and the only forms are (4, 2) and (3, 3).

I consider a particular value of w, say $w=15$. The whole series of seminvariants weight 15 is

663	555	4443	33333
654	5532	4432^2	3332^3
652^2	5442	43332	32^6;
6432	5433	432^4,	
6333	542^3		
632^3,	5332^2		
	52^5,		

and from the quartic and cubic forms we obtain the forms of the products (4, 2) and (3, 3), viz. these are

$4432 . 2$	$3333 . 3$
$443 . 2^2$	$333 . 33$
$4333 . 2$	$332^3 . 3$
$432^3 . 2$	$332^2 . 32$
$432^2 . 2^2$	$332 . 32^2$
$432 . 2^3$	$33 . 32^3$;
$43 . 2^4$,	

and (S_6) will denote the number of syzygies between these products. Now from any such syzygy, we obtain by decapitation, it may be an identity, but if not an identity, then a syzygy, of the weight $15 - 6, = 9$: and from these lower identities or syzygies we can pass back to the syzygies of the weight 15. To show how this is, I decapitate the several products, thus obtaining the forms

432	333	; viz. the distinct forms are	432	; and of these	333	occur each of them twice.
$43 . 2$	$33 . 3$		$43 . 2$		32^3	
333	32^3		333		$32^2 . 2$	
32^3	$32^2 . 2$	$33 . 3$	32^3		$32 . 2^2$	
$32^2 . 2$	$32 . 2^2$		$32^2 . 2$		$3 . 2^3$	
$32 . 2^2$	$3 . 2^3$		$32 . 2^2$			
$3 . 2^3$			$3 . 2^3$			

The forms occurring each twice are 333, 32^3, viz. these are the forms (3), or cubic perpetuants of the weight 9: and $32^2 . 2$, $32 . 2^2$, $3 . 2^3$, viz. these are the forms (3, 2) or products of a cubic perpetuant and a quadric perpetuant for the weight 9; and any form thus occurring twice gives a syzygy of the weight 15: thus 333, we capitate it with $4 . 2$ or with $3 . 3$, and so obtain the syzygy $4333 . 2 - 3333 . 3 \equiv 0$; and in like manner for each of the other forms 32^3, $32^2 . 2$, $32 . 2^2$, $3 . 2^3$. And so for any other odd weight: (S_6) contains the terms $((3))'$ and $((3, 2))'$, and for an odd value of w we may assume that S_6 contains also the terms $((2))'$ and $((2, 2))'$: for these, it is clear, vanish for any odd value of w.

44. When w is even, it appears by a similar investigation that (S_6) contains the terms $((2))'$, and $((2, 2))'$, which in this case do not vanish, and also the before-mentioned terms $((3))'$ and $((3, 2))'$: so that whether w be even or odd, (S_6) contains the terms $((2))'$, $((3))'$, $((2, 2))'$, $((3, 2))'$.

In the particular case $w = 6$, there is the sextic syzygy $3 . 3 - 2 . 2 . 2 \equiv 0$, obtained by capitation from the identity $1 - 1 = 0$; and by reason hereof, we introduce into the formula the term $((0))$, $= 1$ for $w = 6$, and $= 0$ in every other case.

In what immediately follows, I revert to the instance $w = 19$, but this now represents indifferently an odd or an even value of w, there being no distinction between the two cases.

45. Attending next to the remaining distinct terms, these are 333, 32^3 of the degree 3; 432 of the degree 4; $32^2 . 2$, $32 . 2^2$, $3 . 2^3$ of the degree 5; and 43 . 2, 33 . 3 of the degree 6. For the degrees 3 and 4, there are no syzygies: but for the degree 5 we have a syzygy: this, written as a congruence, is $32^2 . 2 - 3 . 2^3 \equiv 0$, and *quâ* quintic syzygy, it will, when completed, not contain any 5; the completed form in fact is

$$\begin{array}{rl} & 32^2 . 2 \\ - & 3 . 2^3 \\ - & 432 \\ - & {}_2 . 32^3 = 0. \end{array}$$

We can capitate this, each term with 4 . 2 or else 3 . 3, or it may be indifferently either with 4 . 2 or 3 . 3; and so obtain therefrom a syzygy of the weight 15; such a syzygy (in the congruence form) is

$$\begin{array}{rl} & 432^2 . 2^2 \\ - & 43 . 2^4 \\ - & 4432 . 2 \\ - & {}_2 . 432^3 . 2 \equiv 0 ; \end{array}$$

and it is to be observed that it is quite indifferent how the capitation is performed: if for instance the first term had been capitated into $332^2 . 32$, then in virtue of the before-obtained syzygy $432^2 . 2^2 - 332^2 . 32 \equiv 0$, the new form would be equivalent to the old one. (It is I think convenient to capitate, when this can be done, with 4 . 2; and only the other terms with 3 . 3.) Clearly the case is the same with any other odd weight; and we thus see that (S_6) contains the term $(S_5)'$.

46. But further we have, between the terms 43 . 2 and 33 . 3 of the degree 6, a syzygy $43 . 2 - 33 . 3 \equiv 0$. Completing this, there will be a term containing a 5, viz. the syzygy is

$$\begin{array}{rl} & 43 . 2 \\ - & 33 . 3 \\ - & 54 \\ - & 432 \\ - & {}_3 . 333 = 0 ; \end{array}$$

and in this form we cannot capitate it, for the quintic term 54 is not to be capitated either with 4 . 2 or with 3 . 3. But 54 is not a perpetuant: we have

$$\begin{array}{rl} 54 = & 32 . 2^2, \\ - & {}_2 . 3 . 2^3 \\ - & 432 \\ - & 32^3 \end{array} \quad \text{and thence the syzygy is} \quad \begin{array}{rl} & 43 . 2 \\ - & 33 . 3 \\ - & 32 . 2^2 \\ + & {}_2 . 3 . 2^3 \\ + & {}_3 . 333 \\ + & 32^3 = 0 ; \end{array}$$

and it can be capitated, for instance, into

$$\begin{aligned} & 443 \,.\, 2^2 \\ - \;\; & 333 \,.\, 33 \\ - \;\; & 432 \,.\, 2^3 \\ + {}_2 \,.\, & 43 \,.\, 2^4 \\ + {}_3 \,.\, & 4333 \,.\, 2 \\ + \;\; & 432^3 \,.\, 2 \equiv 0, \end{aligned}$$

the form of capitation being (for the reason mentioned above) quite immaterial. Observe that in every case where the sextic syzygy contains in the first instance any quintic seminvariants, it is assumed that each of these is expressed in terms of quintic perpetuants, as shown in No. 38; and this being done, the sextic syzygy exhibits itself as a syzygy containing, or else not containing, a quintic perpetuant or perpetuants.

47. The conclusion is that from any sextic syzygy of the weight $w-6$, *which does not contain a quintic perpetuant,* we can obtain by capitation a sextic syzygy of the weight w. The number of sextic syzygies of the weight $w-6$ is $(S_6)'$, and the number of quintic perpetuants of the same weight is $((5))'$: the former of these is (for not too large values of w) the greater; and at first sight it would appear that we can, by elimination of the quintic perpetuants, obtain from the $(S_6)'$ syzygies, $(S_6)'-((5))'$ syzygies which do not contain a quintic perpetuant: if this was always the case, we should have in (S_6) the term $(S_6)'-((5))'$, completing the series of terms, and the formula would be

$$(S_6) = ((0)) + ((2))' + ((3))' + ((2,\ 2))' + ((3,\ 2))' + (S_5)' + (S_6)' - ((5))',$$

leading to

$$S_6 = \frac{x^6 + x^{13} - 2x^{16} - x^{18}}{2 \,.\, 3 \,.\, 4 \,.\, 5 \,.\, 6}.$$

48. But this result is on the face of it wrong, for as remarked by Sylvester in the memoir referred to, from the mere fact that the sum $1+1-2-1$ of the numerator coefficients is negative, it follows that the coefficients of the development ultimately become negative; and the actual calculation showing when this happens is given by him. And it is further to be noticed that not only the formula cannot be correct beyond the point at which the coefficients become negative, but it cannot be correct beyond the point for which $(S_6)'-((5))'$ becomes negative: the sextic syzygies of the weight $w-6$ may add nothing to, but they cannot take anything away from, the number of the sextic syzygies of the weight w.

49. If for a moment we further consider these syzygies of the weight $w-6$; so long as the number of these is greater than the number of quintic perpetuants of the same weight, we can by means of them presumably express each of the quintic perpetuants in terms of sextic products, viz. in the language of Capt. MacMahon, express each quintic perpetuant as a "Sextic Syzygant." The syzygy of the weight 9, above

obtained, will serve as an example: 54 is not a quintic perpetuant, but ignoring this, it is by the syzygy in question expressible in the form

$$\begin{aligned} 54 = &\quad 43 \,.\, 2 \\ &- 33 \,.\, 3 \\ &- 432 \\ &- {}_3 \,.\, 333, \end{aligned}$$

viz. as a Sextic Syzygant, inasmuch as on the right-hand side we have terms 43.2 and 33.3, of the degree 6, which exceeds the degree 5 of the seminvariant 54 in question. Referring back to the definition of reduction, No. 25, observe that this is *not* a reduction of the seminvariant 54. It may be remarked that for the weight 19 we have 15 sextic syzygies: the number of quintic perpetuants is $=3$: so that while it is conceivable that the 15 equations might be such that they would fail to determine the 3 perpetuants, it is *primâ facie* very unlikely that this should be so. I have in fact ascertained that the equations are sufficient for the determination; that is, that (weight 19) each of the three quintic perpetuants is a sextic syzygant. So in the case $w=23$, the number of the sextic syzygies is $=28$, and that of the quintic perpetuants is $=5$; here also the 28 equations are sufficient to determine the 5 perpetuants, viz. (weight 23) each of the 5 quintic perpetuants is a sextic syzygant.

50. Supposing that for any given weight $w-6$, each of the quintic perpetuants *is* a sextic syzygant: this implies that the number of sextic syzygies $(S_6)'$ is at least equal to the number $((5))'$ of quintic perpetuants (for each expression of a quintic perpetuant as a sextic syzygant is in fact a sextic syzygy): and not only so, but it further implies that the number of the sextic syzygies, which do not contain a quintic perpetuant, is precisely equal $(S_6)'-((5))'$: for if besides the equations which serve to express the perpetuants as syzygants, we have any other sextic syzygy, then either this does not contain a quintic perpetuant, or it can (by substituting therein for every quintic perpetuant its value as a sextic syzygant) be reduced to a syzygy which does not contain any quintic perpetuant.

51. In the general case, we have $(S_6)'$ sextic syzygies of the weight $w-6$, and $((5))'$ quintic perpetuants of this weight: but it may happen that certain of the quintic perpetuants do not enter into any of the sextic syzygies; and those which enter, may do so in definite combinations: by elimination of these combinations of perpetuants we obtain (it may be) α sextic syzygies not containing any quintic perpetuant; and the remaining $(S_6)'-\alpha$ equations will then serve to express each of them a quintic perpetuant, or combination of quintic perpetuants, as a sextic syzygant. The number α is at most $=((5))'$, or taking it to be $=((5))'-((\theta))'$, the number of sextic syzygies not containing any quintic perpetuant will be $=(S_6)'-((5))'+((\theta))'$, that is, the number of sextic syzygies not containing any quintic perpetuant will be equal to the whole number $(S_6)'$ of sextic syzygies diminished by some number $((5))'-((\theta))'$, which is less than or at most equal to the whole number $((5))'$ of quintic perpetuants of the weight in question $w-6$. But as already mentioned, I have not been able to obtain the expression of the function (θ), $=\dfrac{\omega(x)}{2.3.4.5}$, which is the G. F. of the number $((\theta))'$.

Cambridge, England, 17th March, 1884.

829.

TABLES OF THE SYMMETRIC FUNCTIONS OF THE ROOTS, TO THE DEGREE 10, FOR THE FORM

$$1 + bx + \frac{cx^2}{1.2} + \ldots = (1 - \alpha x)(1 - \beta x)(1 - \gamma x) \ldots$$

[From the *American Journal of Mathematics*, t. VII. (1885), pp. 47—56.]

THE tables are derived from the tables (*b*) of my "Memoir on the Symmetric Functions of the Roots of an Equation," *Phil. Trans.*, vol. CXLVII. (1857), pp. 489—496, [147]. These refer in effect to the form $1 + bx + cx^2 + \ldots$, and we have consequently to change $b, c, d, \ldots$ into $\frac{b}{1}, \frac{c}{1.2}, \frac{d}{1.2.3}, \ldots$ respectively. Thus in the heading of the original table V (*b*), we must instead of

$$f, \quad be, \quad cd, \quad b^2d, \quad bc^2, \quad b^3c, \quad b^5,$$

write

$$\frac{f}{120}, \quad \frac{be}{24}, \quad \frac{cd}{12}, \quad \frac{b^2d}{6}, \quad \frac{bc^2}{4}, \quad \frac{b^3c}{2}, \quad \frac{b^5}{1}$$

$$= \frac{1}{120}(f, \quad 5be, \quad 10cd, \quad 20b^2d, \quad 30bc^2, \quad 60b^3c, \quad 120b^5);$$

the several columns of the original table are then multiplied by 1, 5, 10, 20, 30, 60, 120, and we thus obtain the new table with the heading

$$\frac{1}{120}(f, \quad be, \quad cd, \quad b^2d, \quad bc^2, \quad b^3c, \quad b^5).$$

In the original tables, there is a remarkable property (very easily proved) in regard to the sums of the numbers in a *column*. Thus for the table V (*b*) these sums are

$$-1, \quad +2, \quad +2, \quad -3, \quad -3, \quad +4, \quad -1,$$

where the sign is + or − according as the heading is the product of an even or an odd number of letters; and the numerical value depends only on the indices in the heading: these indices are

$$1, \quad 11, \quad 11, \quad 21, \quad 21, \quad 31, \quad 5,$$

and they give the foregoing values

$$1, \quad 2, \quad 2, \quad 3, \quad 3, \quad 4, \quad 1,$$

viz. b^3c, $= 31$ gives the value $\Pi 4 \div \Pi 3 . \Pi 1$, $= 4$; b^2d, bc^2, each $= 21$, give the value $\Pi 3 \div \Pi 2 . \Pi 1$, $= 3$; and so in other cases.

In the new tables we have a property in regard to the sums of the numbers in a *line:* viz. except for the last line of each table, where there is only a single number $+1$ or -1, this sum is always $=0$. I have given in the several tables on the right-hand of each line, the sums for the positive and the negative coefficients separately: thus V (*b*), line 1, the number ± 375 means that these sums are $+375$ and -375 respectively, the sum of all the coefficients being of course $=0$. The property is an important verification as well of the original tables (*b*) as of the new tables derived from them; and I had the pleasure of thus ascertaining that there was not a single inaccuracy in the original tables (*b*).

The symbols in the left-hand outside column of each table denote symmetric functions of the roots $\alpha, \beta, \gamma, \ldots$; $5 = \Sigma\alpha^5$, $41 = \Sigma\alpha^4\beta$, &c.: and the tables are read according to the lines: thus in table V (*b*),

$$5\ (=\Sigma\alpha^5\ ) = \tfrac{1}{120}(5f + 25be + 50cd - 100b^2d - 150bc^2 + 300b^3c - 120b^5),$$

$$41\ (=\Sigma\alpha^4\beta) = \tfrac{1}{120}(5f - \ 5be - 50cd + \ 20b^2d + \ 90bc^2 - \ 60b^3c),\ \text{\&c.}$$

I (*b*)

=	*b*	
1	−1	−1

II (*b*)

÷ 2

=	*c*	b^2	
2	−2	+2	±2
1^2	+1	+1	

III (*b*)

÷ 6

=	*d*	*bc*	b^3	
3	−3	+9	−6	±9
21	+3	−3	±3	
1^3	−1	−1		

IV (*b*)

÷ 24

=	*e*	*bd*	c^2	b^2c	b^4	
4	−4	+16	+12	−48	+24	±52
31	+4	−4	−12	+12	±16	
2^2	+2	−8	+6	±8		
21^2	−4	+4	±4			
1^4	+1	+1				

V (*b*)

÷ 120

	f	*be*	*cd*	b^2d	bc^2	b^3c	b^5	
5	−5	+25	+50	−100	−150	+300	−120	±375
41	+5	−5	−50	+20	+90	−60	±115	
32	+5	−25	+10	+40	−30	±55		
31^2	−5	+5	+20	−20	±25			
2^21	−5	+15	−10	±15				
21^3	+5	−5	±5					
1^5	−1	−1						

VI (b)

÷720

	g	bf	ce	b^2e	d^2	bcd	b^3d	c^3	b^2c^2	b^4c	b^6	
6	− 6	+36	+90	−180	+60	−720	+720	−180	+1620	−2160	+720	±3246
51	+ 6	− 6	−90	+ 30	−60	+420	−120	+180	− 720	+ 360	±996	
42	+ 6	−36	+30	+ 60	−60	+240	−240	−180	+ 180	±516		
3^2	+ 3	−18	−45	+ 90	+60	−180	0	+ 90	± 243			
41^2	− 6	+ 6	+30	− 30	+60	−180	+120	±216				
321	−12	+42	+60	− 90	−60	+ 60	±162					
2^3	− 2	+12	−30	0	+20	± 32						
31^3	+ 6	− 6	−30	+ 30	±36							
2^21^2	+ 9	−24	+15	± 15								
21^4	− 6	+ 6	± 6									
1^6	+ 1	± 1										

VII (b)

÷5040

	h	bg	cf	b^2f	de	bce	b^3e	bd^2	c^2d	b^2cd	b^4d	bc^3	b^3c^2	b^5c	b^7	
7	− 7	+49	+147	−294	+245	+1470	+1470	−980	−1470	+8820	−5880	+4410	−17640	+17640	−5040	±32781
61	+ 7	− 7	−147	+ 42	−245	+ 840	− 210	+560	+1470	−3780	+ 840	−3150	+ 6300	− 2520	±10059	
52	+ 7	−49	+ 63	+ 84	−245	+ 420	− 420	+980	− 630	−2520	+1680	+1890	− 1200	±5124		
43	+ 7	−49	−147	+294	+175	+ 210	− 630	−700	+ 210	+1260	0	− 630	±2156			
51^2	− 7	+ 7	+ 42	− 42	+245	− 315	+ 210	−560	− 420	+1680	− 840	±2184				
421	−14	+56	+ 84	−126	+ 70	− 840	+ 630	+140	+ 420	− 420	±1400					
3^21	− 7	+28	+147	−168	−175	+ 105	0	+280	− 210	± 560						
32^2	− 7	+49	− 63	− 84	+ 35	+ 210	0	−140	± 294							
41^3	+ 7	− 7	− 42	+ 42	−105	+ 315	− 210	±364								
321^2	+21	−63	−126	+168	+105	− 105	± 294									
2^31	+ 7	−35	+ 63	0	− 35	± 70										
31^4	− 7	+ 7	+ 42	− 42	± 49											
2^21^3	−14	+35	− 21	± 35												
21^5	+ 7	− 7	± 7													
1^7	− 1	− 1														

(*Concluded infrà.*)

÷ 40320

VIII (*b*)

	i	bh	cg	b^2g	df	bcf	b^2f	e^2	bde	c^2e	b^2ce	b^4e	cd^2
8	− 8	+ 64	+224	−448	+448	−2688	+2688	+280	−4480	−3360	+20160	−13440	−4480
71	+ 8	− 8	−224	+ 56	−448	+1512	− 336	−280	+2520	+3360	− 8400	+ 1680	+4480
62	+ 8	− 64	+112	+112	−448	+ 672	− 672	−280	+4480	−1680	− 5040	+ 3360	+1120
53	+ 8	− 64	−224	+448	+392	+ 168	−1008	−280	+ 280	+3360	− 7560	+ 5040	−3920
4^2	+ 4	− 32	−112	+224	−224	+1344	−1344	+420	−2240	−1680	+ 3360	0	+2240
61^2	− 8	+ 8	+ 56	− 56	+448	− 504	+ 336	+280	−2520	− 840	+ 3360	− 1680	−2800
521	−16	+ 72	+112	−168	+ 56	−1354	+1008	+560	−2800	−1680	+ 9240	− 5040	+2800
431	−16	+ 72	+448	−504	+ 56	−1680	+1344	−560	+2800	0	− 840	0	− 560
42^2	− 8	+ 64	−112	−112	+448	− 672	+ 672	−280	0	+1680	− 1680	0	−1120
3^22	− 8	+ 64	+ 56	−280	−392	+ 840	0	+280	− 280	− 840	0	0	+ 560
51^3	+ 8	− 8	− 56	+ 56	−168	+ 504	− 336	−280	+1120	+ 840	− 3360	+ 1680	±4208
421^2	+24	− 80	−168	+224	−504	+1848	−1344	+280	− 280	− 840	+ 840	± 3216	
3^21^2	+12	− 40	−252	+280	+168	− 168	0	+140	− 560	+ 420	± 1020		
32^21	+24	−136	0	+280	+336	− 504	0	−280	+ 280	± 920			
2^4	+ 2	− 16	+ 56	0	−112	0	0	+ 70	± 128				
41	− 8	+ 8	+ 56	56	+168	− 504	+ 336	±568					
321^3	32	+ 88	+224	−280	−168	+ 168	± 480						
2^31^2	−16	+ 72	−112	0	+ 56	± 128							
31^5	+ 8	− 8	− 56	+ 56	± 64								
2^21^4	+20	− 48	+ 28	± 48									
21^6	8	+ 8	± 8										
1^8	+ 1	+ 1											

÷ 40320

	b^2d^2	bc^2d	b^3cd	b^5d	c^4	b^2c^3	b^4c^2	b^6c	b^8	
8	+13440	+40320	−107520	+53760	+ 5040	−80640	+201600	−161280	+ 40320	±377344
71	− 5600	−28560	+ 36960	− 6720	− 5040	+45360	− 60480	+ 20160	±116096	
62	−10080	0	+ 26880	−13440	+ 5040	−20160	+ 10080	± 51864		
53	+ 3360	+10080	− 10080	0	− 5040	+ 5040	± 28176			
4^2	+ 2240	− 6720	0	0	+ 2520	±12352				
61^2	+ 5600	+ 8400	− 16800	+ 6720	±25208					
521	− 1120	− 5040	+ 3360	±17208						
431	− 2240	+ 680	± 6400							
42^2	+ 1120	± 3984								
	± 1800									

IX (*b*)

÷ 362880

	j	bi	ch	b^2h	dg	bcg	b^3g	ef	bdf	c^2f	b^2cf	b^4f	be^2
9	−9	+81	+324	−648	+756	−4536	+4536	+1134	−9072	−6804	+40824	−27216	−5670
81	+9	−9	−324	+72	−756	+2520	−504	−1134	+5040	+6804	−16632	+3024	+3150
72	+9	−81	+180	+144	−756	+1008	−1008	−1134	+9072	−3780	−9072	+6048	+5670
63	+9	−81	−324	+648	+756	0	−1512	−1134	0	+6804	−13608	+9072	+5670
54	+9	−81	−324	+648	−756	+4536	−4536	+1386	−1008	−756	−10584	+12096	−6930
71^2	−9	+9	+72	−72	+756	−756	+504	+1134	−5040	−1512	+6048	−3024	−3150
621	−18	+90	+144	−216	0	−2016	+1512	+2268	−5040	−3024	+16632	−9072	−8820
531	−18	+90	+648	−720	0	−2520	+2016	−252	−2520	−6048	+22680	−12096	+3780
4^21	−9	+45	+324	−360	+756	−3528	+2520	−1386	+3024	+756	−1512	0	+3150
52^2	−9	+81	−180	−144	+756	−1008	+1008	−126	−4032	0	+9072	−6048	+630
432	−18	+162	+144	−792	0	−1008	+2520	−252	+1008	+4536	−7560	0	+1260
3^3	−3	+27	+108	−216	−504	+756	0	+378	+1512	−2268	0	0	−1890
61^3	+9	−9	−72	+72	−252	+756	−504	−1124	+2016	+1512	−6048	+3024	+3150
521^2	+27	−99	−216	+288	−756	+2772	−2016	−882	+7560	+4536	−22680	+12096	−630
431^2	+27	−99	−720	+792	−756	+3276	−2520	+1638	−2520	0	+1512	0	−3150
42^21	+27	−171	+36	+360	−756	+3024	−2520	+378	−1008	−4536	+4536	0	+630
3^221	+27	−171	−468	+864	+1512	−1764	0	−882	−2016	+2268	0	0	+1890
32^3	+9	−81	+180	+144	−252	−504	0	+126	+1008	0	0	0	−630
51^4	−9	+9	+72	−72	+252	−756	+504	+504	−2016	−1512	+6048	−3024	±7389
421^3	−36	+108	+288	−360	+1008	−3528	+2520	−504	+504	+1512	−1512	±5940	
3^21^3	−18	+54	+396	−432	−252	+252	0	−252	+1008	−756	±1710		
32^21^2	−54	+270	+180	−648	−756	+1008	0	+504	−504	±1962			
2^41	−9	+63	−180	0	+252	0	0	−126	±315				
41^5	+9	−9	−72	+72	−252	+756	−504	±837					
321^4	+45	−117	−360	+432	+252	−252	±729						
2^31^3	+30	−126	+180	0	−84	±210							
31^6	−9	+9	+72	−72	±81								
2^21^5	−27	+63	−36	±63									
21^7	+9	−9	±9										
1^9	−1	−1											

(*Continued next page.*)

(Concluded infrà.)

IX (b)

	cde	e^2de	bc^2e	b^3ce	b^5e	d^3	bcd^2	b^3d^2	c^3d	b^2c^2d
9	− 22680	+ 68040	+ 102060	− 272160	+ 136080	− 5040	+ 136080	− 181440	+ 68040	− 816480
81	+ 22680	− 27720	− 71820	+ 90720	− 15120	+ 5040	− 95760	+ 60480	− 68040	+ 453600
72	+ 5040	− 50400	+ 3780	+ 60480	− 30240	+ 5040	− 65520	+ 110880	+ 37800	+ 75600
63	0	− 22680	− 34020	+ 90720	− 45360	− 10080	+ 90720	− 30240	− 22680	− 136080
54	− 2520	+ 32760	+ 11340	− 30240	0	+ 5040	− 35280	− 20160	+ 7560	+ 60480
71^2	− 13860	+ 27720	+ 18900	− 37800	+ 15120	− 5040	+ 60480	− 60480	+ 15120	− 136080
621	− 5040	+ 32760	+ 45360	− 105840	+ 45360	+ 5040	− 35280	+ 10080	− 15120	+ 60480
531	+ 2520	− 12600	− 7560	+ 7560	0	+ 5040	− 20160	+ 20160	+ 15120	− 15120
4^21	+ 2520	− 12600	+ 3780	0	0	− 5040	+ 15120	0	− 7560	± 31995
52^2	+ 7560	0	− 22680	+ 15120	0	− 5040	+ 15120	− 10080	± 29347	
432	− 10080	+ 2520	+ 7560	0	0	+ 5040	− 5040	± 24750		
3^3	+ 3780	0	0	0	0	− 1680	± 6561			
61^3	+ 6300	− 12600	− 18900	+ 37800	− 15120	± 54369				
521^2	− 6300	+ 2520	+ 11340	− 7560	± 41139					
431^2	+ 1260	+ 5040	− 3780	± 13545						
42^21	+ 2520	− 2520	± 11511							
3^221	− 1260	± 6561								
	± 1467									

÷ 40320

	b^4cd	b^6d	bc^4	b^3c^3	b^5c^2	b^7c	b^9	
9	+ 1360800	− 544320	− 204120	+ 1360800	− 2449440	+ 1632960	− 362880	± 4912515
81	− 393120	+ 60480	+ 158760	− 635040	+ 635040	− 181440	± 1507419	
72	− 302400	+ 120960	− 113400	+ 226800	− 90720	± 668511		
63	+ 90720	0	+ 68040	− 45360	± 363159			
54	0	0	− 22680	± 135855				
71^2	+ 181440	− 60480	± 327303					
621	− 30240	± 219726						
	± 79164							

X (b)

÷ 3628800

		k	bj	ci	b^2i	dh	bch	b^3h	eg	bdg	c^2g	b^2cg	b^4g
1	10	−10	+100	+450	−900	+1200	−7200	+7200	+2100	−16800	−12600	+75600	−50400
2	91	+10	−10	−450	+90	−1200	+3960	−720	−2100	+9240	+12600	−30240	+5040
3	82	+10	−100	+270	+180	−1200	+1440	−1440	−2100	+16800	−7560	−15120	+10080
4	73	+10	−100	−450	+900	+1320	−360	−2160	−2100	−840	+12600	−22680	+15120
5	64	+10	−100	−450	+900	−1200	+7200	−7200	+2940	−3360	−2520	−15120	+20160
6	5^2	+5	−50	−225	+450	−600	+3600	−3600	−1050	+8400	+6300	−37800	+25200
7	81^2	−10	+10	+90	−90	+1200	−1080	+720	+2100	−9240	−2520	+10080	−5040
8	721	−20	+110	+180	−270	−120	−2880	+2160	+4200	−8400	−5040	+27720	−15120
9	631	−20	+110	+900	−990	−120	−3600	+2880	−840	−3360	−10080	+37800	−20160
10	541	−20	+110	+900	−990	+2400	−11160	+7920	−840	−5880	−10080	+45360	−25200
11	62^2	−10	+100	−270	−180	+1200	−1440	+1440	−420	−6720	.	+15120	−10080
12	532	−20	+200	+180	−1080	−120	−1080	+3600	+4200	−15960	−5040	+37800	−25200
13	4^22	−10	+100	+90	−540	+1200	−4320	+4320	−2940	+3360	+12600	−15120	.
14	43^2	−10	+100	+450	−900	−1320	+360	+2160	−420	+10920	−5040	−7560	.
15	71^3	+10	−10	−90	+90	−360	+1080	−720	−2100	+3360	+2520	−10080	+5040
16	621^2	+30	−120	−270	+360	−1080	+3960	−2880	−1260	+12600	+7560	−37800	+20160
17	531^2	+30	−120	−990	+1080	−1080	+4680	−3600	−1260	+12600	+12600	−47880	+25200
18	4^21^2	+15	−60	−495	+540	−1800	+6120	−4320	+1890	−2520	−1260	+2520	.
19	52^21	+30	−210	+90	+450	−1080	+4320	−3600	−3780	+15120	+5040	−42840	+25200
20	4321	+60	−420	−1260	+2340	+360	+7560	−8640	+2520	−12600	−10080	+17640	.
21	3^31	+10	−70	−450	+630	+1320	−1440	.	+420	−5880	+5040	.	.
22	42^3	+10	−100	+270	+180	−1200	+1440	−1440	+2100	.	−5040	+5040	.
23	3^22^2	+15	−150	+45	+630	+720	−2520	.	−1890	+2520	+2520	.	.
24	61^4	−10	+10	+90	−90	+360	−1080	+720	+840	−3360	−2520	+10080	−5040
25	521^3	−40	+130	+360	−450	+1440	−5040	+3600	+3360	−15960	−10080	+47880	−25200
26	431^3	−40	+130	+1080	−1170	+1440	−5760	+4320	−1680	+4200	.	−2520	.
27	42^21^2	−60	+330	+180	−810	+2160	−8280	+6480	−2520	+2520	+10080	−10080	.
28	3^221^2	−60	+330	+1260	−1890	−2880	+3240	.	.	+5040	−5040	.	.
29	32^31	−40	+310	−360	−630	−240	+1800	.	+1680	−2520	.	.	.
30	2^5	−2	+20	−90	.	+240	.	.	−420	.	.	.	.

(*Continued next page.*)

X (*b*)

		f^2	bef	cdf	b^2df	bc^2f	b^3cf	b^5f	ce^2	b^2e^2	d^2e
1	10	+1260	−25200	−50400	+151200	+226800	−604800	+302400	− 31500	+ 94500	−42000
2	91	−1260	+13860	+50400	− 60480	−158760	+196560	− 30240	+ 31500	− 37800	+42000
3	82	−1260	+25200	+10080	−110880	+ 15120	+120960	− 60480	+ 6300	− 69300	+42000
4	73	−1260	+25200	− 2520	− 45360	− 68040	+181440	− 90720	+ 31500	− 94500	−46200
5	64	−1260	− 5040	+50400	− 30240	−136080	+241920	−120960	− 44100	+ 56700	− 8400
6	5^2	+2520	−18900	−37800	+ 50400	+ 75600	− 75600	.	+ 15750	+ 31500	+21000
7	81^2	+1260	−13860	−30240	+ 60480	+ 37800	− 75600	+ 30240	− 18900	+ 37800	−42000
8	721	+2520	−39060	− 7560	+ 65520	+ 90720	−211680	+ 90720	− 37800	+107100	+ 4200
9	631	+2520	− 8820	−47880	+ 75600	+136080	−287280	+120960	+ 12600	− 18900	+54600
10	541	−3780	+28980	+25200	− 60480	− 7560	+ 15120	.	+ 12600	− 50400	−33600
11	62^2	+1260	−10080	−10080	+ 50400	+ 30240	−120960	+ 60480	+ 31500	− 6300	−16800
12	532	−3780	+12600	+42840	− 20160	− 98280	+ 75600	.	− 37800	+ 6300	+ 4200
13	4^22	+1260	+ 5040	−30240	+ 10080	+ 15120	.	.	+ 6300	− 18900	+ 8400
14	43^2	+1260	−10080	+ 2520	− 15120	+ 22680	.	.	+ 6300	+ 18900	− 4200
15	71^3	−1250	+13860	+12600	− 25200	− 37800	+ 75600	− 30240	+ 18900	− 37800	+12600
16	621^2	−3780	+22680	+37800	− 95760	−128520	+287280	−120960	− 18900	+ 6300	−12600
17	531^2	+2520	−15120	−10080	+ 15120	+ 15120	− 15120	.	+ 6300	+ 12600	−12600
18	4^21^2	+1260	−10080	+ 2520	+ 10080	− 7560	.	.	− 9450	+ 18900	+12600
19	52^21	+2520	− 1260	−32760	+ 10080	+ 68040	− 45360	.	+ 6300	− 6300	+12600
20	4321	−1260	.	+12600	+ 20160	− 22680	.	.	+ 12600	− 18900	−12600
21	3^31	−1260	+ 6300	− 2520	.	.	.	.	− 6300	.	+ 4200
22	42^3	−1260	.	+10080	− 10080	.	.	.	− 6300	+ 6300	±25420
23	3^22^2	+1260	− 1260	− 5040	.	.	.	.	+ 3150	± 10860	
24	61^4	+1260	− 6300	−12600	+ 25200	+ 37800	− 75600	+ 30240	±106600		
25	521^3	−1260	+ 1260	+12600	− 5040	− 22680	+ 15120	± 85750			
26	431^3	−1260	+ 6300	− 2520	− 10080	+ 7560	± 25030				
27	42^21^2	+1260	− 1260	− 5040	+ 5040	± 28050					
28	3^221^2	+1260	− 3780	+ 2520	± 13650						
29	32^31	−1260	+ 1260	± 5050							
30	2^5	+ 252	± 512								

(*Continued next page.*)

X (*b*)

		$bcde$	b^3de	c^3e	b^2c^2e	b^4ce	b^6e	bd^3	c^2d^2	b^2cd^2	b^4d^2
1	10	+756000	−1008000	+189000	−2268000	+3780000	−1512000	+168000	+378000	−3024000	+2520000
2	91	−529200	+ 327600	−189000	+1247400	−1058400	+ 151200	−117600	−378000	+1663200	− 705600
3	82	−352800	+ 604800	+113400	+ 151200	− 756000	+ 302400	−168000	+ 25200	+1411200	+1310400
4	73	+ 37800	+ 302400	−189000	+ 680400	−1134000	+ 453600	+184800	+151200	−1209600	+ 302400
5	64	+151200	− 201600	+189000	− 453600	+ 302400	.	+ 33600	−226800	.	+ 201600
6	5^2	− 63000	− 126000	− 94500	+ 189000	.	.	− 84000	+126000	+ 252000	.
7	81^2	+327600	− 327600	+ 37800	− 340200	+ 453600	− 151200	+117600	+176400	−1058400	+ 705600
8	721	+264600	− 403200	+ 75600	− 869400	+1285200	− 453600	− 67200	−176400	+ 453600	− 100800
9	631	−189000	+ 126000	.	+ 113400	− 75600	.	−117600	+ 75600	+ 302400	− 201600
10	541	.	+ 126000	.	− 37800	.	.	+ 84000	− 25200	− 151200	.
11	62^2	−100800	.	− 75600	+ 302400	− 151200	.	+ 67200	+ 50400	− 201600	+ 100800
12	532	+ 63000	− 25200	+ 75600	− 75600	.	.	− 16800	− 50400	+ 50400	± 376520
13	4^22	+ 50400	.	− 37800	.	.	.	− 33600	+ 25200	± 143470	
14	43^2	− 37800	.	.	.	.	.	+ 16800	± 82450		
15	71^3	−151200	+ 151200	− 37800	+ 340200	− 453600	+ 151200	±788260			
16	621^2	+ 88200	− 25200	+ 37800	− 151200	+ 75600	± 600330				
17	531^2	+ 50400	− 50400	− 37800	+ 37800	± 196050					
18	4^21^2	− 37800	.	+ 18900	± 75345						
19	52^21	− 37800	+ 25200	±174990							
20	4321	+ 12600	± 88440								
		± 17920									

(*Continued next page.*)

X (b)

		bc^3d	b^3c^2d	b^5cd	b^7d	c^5	b^2c^4	b^4c^3	b^6c^2	b^8c	b^{10}	
1	10	−3024000	+15120000	−18144000	+6048000	− 226800	+5670000	−22680000	+31752000	−18144000	+ 3628800	±70872610
2	91	+2343600	− 6955200	+ 4536000	− 604800	+ 226800	−3628800	+ 9072000	− 7257600	+ 1814400	±21747460	
3	82	− 604800	− 1814400	+ 3628800	−1209600	− 226800	+2041200	− 2721600	+ 907200	± 9433840		
4	73	− 151200	+ 1814400	− 907200	.	+ 226800	− 907200	+ 453600	± 4875490			
5	64	+ 604800	− 604800	.	.	− 226800	+ 226800	± 2089630				
6	5^2	− 378000	.	.	.	+ 113400	± 921125					
7	81^2	− 529200	+ 2116800	− 2116800	+ 604800	±4721980						
8	721	+ 378000	− 756000	+ 302400	±3154550							
9	631	− 226800	+ 151200	± 1212650								
10	541	+ 75600	± 424190									
		± 712540										

		k	bj	ci	b^2i	dh	bch	b^3h	eg	bdg	c^2g	b^2cg	b^4g
31	51^5	+ 10	− 10	− 90	+ 90	− 360	+1080	− 720	− 840	+3360	+2520	−10080	+ 5040
32	421^4	+ 50	−140	−450	+ 540	−1800	+6120	−4320	+ 840	− 840	−2520	+ 2520	±10070
33	3^21^4	+ 25	− 70	−585	+ 630	+ 360	− 360	.	+ 420	−1680	+1260	± 2695	
34	32^21^3	+100	−460	−540	+1260	+1440	−1800	.	− 840	+ 840	±3640		
35	241^2	+ 25	−160	+405	.	− 480	.	.	+ 210	± 640			
36	41^6	− 10	+ 10	+ 90	− 90	+ 360	−1080	+ 720	±1180				
37	321^5	− 60	+150	+540	− 630	− 360	+ 360	±1050					
38	2^31^4	− 50	+200	−270	.	+ 120	± 320						
39	31^7	+ 10	− 10	− 90	+ 90	± 100							
40	2^21^6	+ 35	− 80	+ 45	± 80								
41	21^8	− 10	+ 10	± 10									
42	1^{10}	+ 1	+ 1										

830.

NON-UNITARY PARTITION TABLES.

[From the *American Journal of Mathematics*, t. VII. (1885), pp. 57, 58.]

In the theory of Seminvariants we are concerned with the non-unitary partitions of a number, that is, the number of ways of making up the number with the parts 2, 3, 4, ...; or what is the same thing, writing $\mathbf{2} = 1 - x^2$, $\mathbf{3} = 1 - x^3$, &c., with the Generating Functions having in their denominators the factors **2**, **3**, **4**, &c. In the present short paper, I give the developments up to x^{100} of the functions $1 \div \mathbf{2}$, $\mathbf{2}.\mathbf{3}$, $\mathbf{2}.\mathbf{3}.\mathbf{4}$, $\mathbf{2}.\mathbf{3}.\mathbf{4}.\mathbf{5}$, $\mathbf{2}.\mathbf{3}.\mathbf{4}.\mathbf{5}.\mathbf{6}$, respectively: and also of the function

$$x^6 + x^{13} - 2x^{16} - x^{18} + x^{31} \div \mathbf{2}.\mathbf{3}.\mathbf{4}.\mathbf{5}.\mathbf{6},$$

which function is (there is strong reason to believe) the G. F. for the number of sextic syzygies of a given weight: the same function without the term x^{31} occurs (p. 115) in Professor Sylvester's paper "On Subinvariants, i.e. Seminvariants to Binary Quantics of an Unlimited Order," *American Journal of Mathematics*, t. V. (1882), pp. 79—136.

In the tables, X is written to denote $x^6 + x^{13} - 2x^{16} - x^{18} + x^{31}$.

Ind. x	$1\div$ 2.3	2.3.4	2.3.4.5	2.3.4.5.6	$X\div$ 2.3.4.5.6	Ind. x	$1\div$ 2.3	2.3.4	2.3.4.5	2.3.4.5.6	$X\div$ 2.3.4.5.6
0	1	1	1	1		50	9	65	258	750	186
1	0	0	0	0		51	9	61	268	783	226
2	1	1	1	1		52	9	70	286	854	203
3	1	1	1	1		53	9	65	297	891	248
4	1	2	2	2		54	10	75	316	972	223
5	1	1	2	2		55	9	70	328	1010	270
6	2	3	3	4	1	56	10	80	348	1098	242
7	1	2	3	3	0	57	10	75	361	1144	294
8	2	4	5	6	1	58	10	85	382	1236	262
9	2	3	5	6	1	59	10	80	396	1287	319
10	2	5	7	9	2	60	11	91	419	1391	284
11	2	4	7	9	2	61	10	85	433	1443	344
12	3	7	10	14	4	62	11	96	457	1555	306
13	2	5	10	13	4	63	11	91	473	1617	371
14	3	8	13	19	6	64	11	102	598	1734	328
15	3	7	14	20	7	65	11	96	515	1802	399
16	3	10	17	26	8	66	12	108	541	1932	353
17	3	8	18	27	11	67	11	102	559	2002	427
18	4	12	22	36	13	68	12	114	587	2142	377
19	3	10	23	36	15	69	12	108	606	2223	457
20	4	14	28	47	17	70	12	120	635	2369	402
21	4	12	29	49	21	71	12	114	655	2457	490
22	4	16	34	60	22	72	13	127	686	2618	429
23	4	14	36	63	28	73	12	120	707	2709	519
24	5	19	42	78	29	74	13	133	739	2881	456
25	4	16	44	80	35	75	13	127	762	2985	552
26	5	21	50	97	36	76	13	140	795	3164	483
27	5	19	53	102	44	77	13	133	819	3276	586
28	5	24	60	120	43	78	14	147	854	3472	513
29	5	21	63	126	54	79	13	140	879	3588	620
30	6	27	71	149	53	80	14	154	916	3797	542
31	5	24	74	154	64	81	14	147	942	3927	656
32	6	30	83	180	62	82	14	161	980	4144	572
33	6	27	87	189	78	83	14	154	1008	4284	693
34	6	33	96	216	72	84	15	169	1048	4520	604
35	6	30	101	227	89	85	14	161	1077	4665	730
36	7	37	111	260	84	86	15	176	1118	4915	636
37	6	33	116	270	102	87	15	169	1149	5076	769
38	7	40	127	307	96	88	15	184	1192	5336	568
39	7	37	133	322	117	89	15	176	1224	5508	809
40	7	44	145	361	108	90	16	192	1269	5789	703
41	7	40	151	378	133	91	15	184	1302	5967	849
42	8	48	164	424	123	92	16	200	1349	6264	736
43	7	44	171	441	149	93	16	192	1384	6460	891
44	8	52	185	492	137	94	16	208	1432	6768	772
45	8	48	193	515	167	95	16	200	1469	6977	934
46	8	56	207	568	152	96	17	217	1519	7308	809
47	8	52	216	594	186	97	16	208	1557	7524	977
48	9	61	232	656	169	98	17	225	1609	7873	846
49	8	56	241	682	205	99	17	217	1649	8109	1022
						100	17	234	1883	8651	883

831.

SEMINVARIANT TABLES.

[From the *American Journal of Mathematics*, t. VII. (1885), pp. 59—73.]

THE present tables are not, I think, superseded by the tables A, pp. 149—163, contained in Capt. MacMahon's paper, "Seminvariants and Symmetric Functions," *American Journal of Mathematics*, t. VI. (1883), pp. 131—163. His order of the terms, though a very ingenious one, and giving rise to a most remarkable symmetry in the form of the tables, seems to me too artificial—and I cannot satisfy myself that it ought to be adopted in preference to the more simple one which I use: I attach also considerable importance to the employment of the simple letters b, c, d, e, &c. in place of the suffixed ones a_1, a_2, a_3, a_4, &c. There is, moreover, the question of the identification of the seminvariants with their expressions as non-unitary symmetric functions of the roots of the equation $1 + bx + \frac{cx^2}{1\,.\,2} + \&\text{c.} = 0$, which requires to be considered.

As to the form in which the tables present themselves, I remark that every seminvariant is a rational and integral function of the fundamental seminvariants

$$\mathrm{c} = (1,\ b,\ c \between -b,\ 1)^2,$$
$$\mathrm{d} = (1,\ b,\ c,\ d \between -b,\ 1)^3,$$
$$\mathrm{e} = (1,\ b,\ c,\ d,\ e \between -b,\ 1)^4,\ \&\text{c.},$$

viz. up to g, these are

c =	d =	e =	f =	g =	ce =	d² =	c³ =
c + 1	d + 1	e + 1	f + 1	g + 1			
b^2 − 1	bc − 3	bd − 4	be − 5	bf − 6			
	b^3 + 2	c^2	cd	ce	+ 1		
		b^2c + 6	b^2d + 10	d^2		+ 1	
		b^4 − 3	bc^2	b^2e + 15	− 1		
			b^3c − 10	bcd	− 4	− 6	
			b^5 + 4	c^3			+ 1
				b^3d − 20	+ 4	+ 4	
				b^2c^2	+ 6	+ 9	− 3
				b^4c + 15	− 9	− 12	+ 3
				b^6 − 5	+ 3	+ 4	− 1

and if to the value of g, which is of the weight 6, we join those of the products ce, d^2, c^3 of this same weight 6, as just written down, we have what is in effect a table of the asyzygetic seminvariants of the weight 6 and which I call the Crude Table. But we do not, in this way, obtain immediately the seminvariants of the lowest degrees: in fact, the only seminvariant containing g is given by the first column as a function $g - \ldots - 5b^6$ of the degree 6, whereas there is the seminvariant $g - 6bf + 15ce - 10d^2$ of the degree 2: to obtain this, we have to form a linear combination of the columns: the proper combination is $g + 15ce - 10d^2$, giving rise to the column g of the table (p. 278, $g = 6$). And similarly each other column of the same table is a linear combination of columns of the Crude Table: and so in every case. The process would be a very laborious one, and the tables were not, in fact, thus calculated; but we see very clearly in this manner the origin and meaning of the tables.

The mere inspection of the tables gives rise to several remarks. We see that each column begins with a non-unitary term (term without the letter b), and that it ends with a power-ending term (product wherein the last letter enters as a power)—thus, weight 8, the initial terms are

$$i,\quad cg,\quad df,\quad e^2,\quad c^2e,\quad cd^2,\quad c^4,$$

the finals are

$$e^2,\quad cd^2,\quad b^2d^2,\quad c^4,\quad b^2c^3,\quad b^4c^2,\quad b^8;$$

and it will be observed further that in this case the initial terms are all the non-unitary terms taken in order, and the corresponding final terms are all the power-ending terms taken also in order. The arrangement of the columns *inter se* is of course arbitrary, and they are, in fact, arranged so that the initial terms are the non-unitary terms taken in order—and this being so, then for each weight up to the weight 9 the final terms are the power-ending terms taken in order: but for each of the weights 10 and 11, there is a single deviation from this order; and for the weight 12, there are a great many deviations from the order.

The initial terms being in order, the broken line which bounds the tops of the columns forms a series of continually descending steps; and when the final terms are also in order, the case is the same with the broken line bounding the bottom of the columns: any deviation in the order of the final terms is shown by an ascending step or steps in the broken line bounding the bottom of the columns: thus in the table (p. 281, $k = 10$) the column cdf is longer than the next following one ce^2, and there is an ascending step accordingly.

It is to be remarked that any ascending step gives rise to a certain indeterminateness in a preceding column or columns: thus in the case just referred to, the column cdf might be replaced by any linear combination of itself with the column ce^2, it would still have the original initial and final terms cdf and b^4d^2 respectively. It would be possible to fix a standard form; we might, for instance, say that the column cdf should be that combination $cdf + 2ce^2$, which does not contain the leading term ce^2 of the ce^2-column: but I have not thought it worth while to attend to this.

It will be observed that, except in the case of an ascending step or steps, each column is completely determinate: we cannot with any column combine a preceding column, for this would give it a higher initial term: nor can we with it combine a succeeding column, for this would give it a lower final term. The numbers in the column may be taken to be without any common divisor, for any such divisor, if it existed, might be divided out: and the leading coefficient of the column may be taken to be positive.

I add certain subsidiary tables to enable the expression of any column in terms of the non-unitary symmetric functions of the roots of the equation $0 = 1 + bx + \frac{cx^2}{1.2} + \text{\&c.}$ These consist of left-hand tables and right-hand tables: the left-hand table for any weight is the original table for that weight, writing therein $b = 0$ and converting the columns into lines: thus weight $= 6$, we have

$$\begin{aligned}
&\text{col. } g = g + 15ce - 10d^2,\\
&\text{col. } ce = \qquad ce - d^2 - c^3,\\
&\text{col. } d^2 = \qquad\qquad\quad d^2 + 4c^3,\\
&\text{col. } c^3 = \qquad\qquad\qquad\quad\;\; c^3;
\end{aligned}$$

viz. these are the values of the original columns writing therein $b = 0$.

The right-hand table is the table for the same weight taken from my paper "Tables of the Symmetric Functions of the Roots, to the Degree 10, for the form $1 + bx + \frac{cx^2}{1.2} + \ldots = (1 - \alpha x)(1 - \beta x)(1 - \gamma x)\ldots$," [829], writing therein $b = 0$, and giving only those lines of the table which relate to the non-unitary symmetric functions. Thus for weight 6, we have

$$\begin{aligned}
&6\ (= \Sigma\alpha^6) &&= \tfrac{1}{720}(-6g + 90ce + 60d^2 - 180c^3),\\
&42\ (= \Sigma\alpha^4\beta^2) &&= \tfrac{1}{720}(6g + 30ce - 60d^2 - 180c^3),\\
&3^2\ (= \Sigma\alpha^3\beta^3) &&= \tfrac{1}{720}(3g - 45ce + 60d^2 + 90c^3),\\
&2^3\ (= \Sigma\alpha^2\beta^2\gamma^2) &&= \tfrac{1}{720}(-2g - 30ce + 20d^2),
\end{aligned}$$

we thus have on the one side col. g, col. ce, col. d^2 and col. c^3, and on the other side the symmetric functions 6, 42, 3^2 and 2^3, each of them expressed as a linear function of g, ce, d^2 and c^3. It follows that each of the columns can be expressed as a linear function of the symmetric functions: and conversely each of the symmetric functions as a linear function of the columns: and this being done, each of the columns is to be regarded as having its complete value as a function of b and the other letters: for the columns *quâ* seminvariants are linear functions of the symmetric functions: and assuming them to be so, they can only be the linear functions determined by the foregoing process of writing $b = 0$.

The left-hand tables are carried up to $m = 12$; the right-hand only up to $k = 10$, the limit of the tables in the memoir last referred to.

Seminvariant Tables up to ($m=12$).

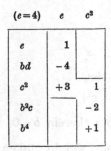

$(1=0)$	
1	+1

$(b=1)$
b

$(c=2)$	c
c	1
b^2	−1

$(d=3)$	d
d	1
bc	−3
b^3	+2

$(e=4)$	e	c^2
e	1	
bd	−4	
c^2	+3	1
b^2c		−2
b^4		+1

$(f=5)$	f	cd
f	1	
be	−5	
cd	+2	+1
b^2d	+8	−1
bc^2	−6	−3
b^3c		+5
b^5		−2

$(g=6)$	g	ce	d^2	c^3
g	+1			
bf	−6			
ce	+15	+1		
d^2	−10	−1	+1	
b^2e		−1		
bcd		+2	−6	
c^3		−1	+4	+1
b^3d			+4	
b^2c^2			−3	−3
b^4c				+3
b^6				−1

$(h=7)$	h	cf	de	c^2d
h	$+1$			
bg	-7			
cf	$+9$	$+1$		
de	-5	-1	$+1$	
b^2f	$+12$	-1		
bce	-30	-2	-3	
bd^2	$+20$	$+4$	-4	
c^2d		-1	$+3$	$+1$
b^3e		$+3$	$+2$	
b^2cd		-6	$+12$	-2
bc^3		$+3$	-9	-3
b^4d			-8	$+1$
b^3c^2			$+6$	$+8$
b^5c				-7
b^7				$+2$

$(i=8)$	i	cg	df	e^2	c^2e	cd^2	c^4
i	$+1$						
bh	-8						
cg	$+28$	$+1$					
df	-56	-3	$+3$				
e^2	$+35$	$+2$	-2	$+1$			
b^2g		-1					
bcf		$+3$	-9				
bde		-1	$+1$	-8			
c^2e		-3	$+18$	$+6$	$+1$		
cd^2		$+2$	-12		-1	$+1$	
b^3f			$+6$				
b^2ce			-15		-2		
b^2d^2			$+10$	$+16$	$+1$	-1	
bc^2d				-24	$+2$	-6	
c^4				$+9$	-1	$+4$	$+1$
b^4e					$+1$		
b^3cd					-2	$+10$	
b^2c^3					$+1$	-7	-4
b^5d						-4	
b^4c^2						$+3$	$+6$
b^6c							-4
b^8							$+1$

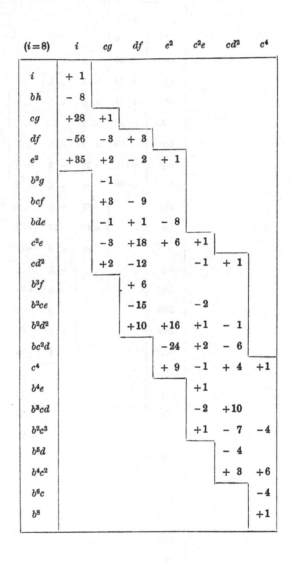

$(j=9)$	j	ch	dg	ef	c^2f	cde	d^3	c^3d
j	+ 1							
bi	− 9							
ch	+ 20	+ 2						
dg	− 28	− 7	+ 1					
ef	+ 14	+ 5	− 1	+ 1				
b^2h	+ 16	− 2						
bcg	− 56	+ 7	− 3					
bdf	+ 112	+ 22	− 2	− 4				
be^2	− 70	− 25	+ 5	− 5				
c^2f		− 27	+ 9	− 9	+ 2			
cde		+ 45	− 17	+ 32	− 5	+ 1		
d^3		− 20	+ 8	− 18	+ 3	− 1	+ 1	
b^3g			+ 2					
b^2cf			− 6	+ 24	− 4			
b^2de			+ 2	− 2	+ 5	− 1		
bc^2e			+ 6	− 51	+ 5	− 3		
bcd^2			− 4	+ 34	− 7	+ 5	− 9	
c^3d					+ 1	− 1	+ 4	+ 1
b^4f				− 12	+ 2			
b^3ce				+ 30	− 5	+ 5		
b^3d^2				− 20	− 2	− 2	+ 6	
b^2c^2d					+ 8	− 6	+ 15	− 3
bc^4					− 3	+ 3	− 12	− 3
b^5e						− 2		
b^4cd						+ 4	− 24	+ 3
b^3c^3						− 2	+ 17	+ 11
b^6d							+ 8	− 1
b^5c^2							− 6	− 15
b^7c								+ 9
b^9								− 2

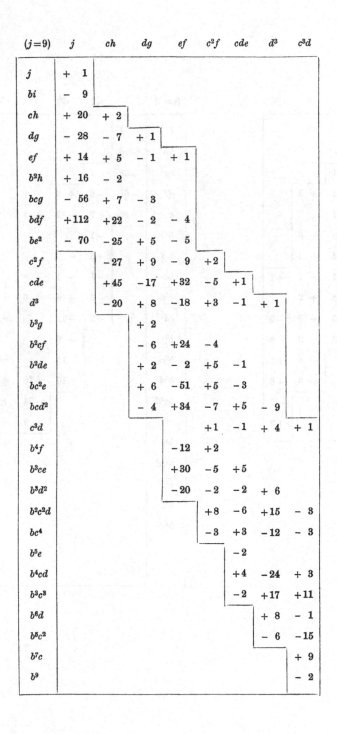

(k=10)	k	ci	dh	eg	f^2	c^2g	cdf	ce^2	d^2e	c^3e	c^2d^2	c^5
k	+ 1											
bj	− 10											
ci	+ 45	+1										
dh	−120	−4	+ 4									
eg	+210	+8	− 8	+ 16								
f^2	−126	−5	+ 5	+ 15	+ 1							
b^2i		−1										
bch		+4	−12									
bdg		−4	+ 4	− 64								
bef		+2	− 2	+ 54	−10							
c^2g		−4	+32	+ 48		+1						
cdf		+8	−64	− 60	+ 4	−3	+ 3					
ce^2		−5	+40		+16	+2	− 2	+ 1				
d^2e				+ 20	−12			− 1	+ 1			
b^2h			+ 8									
b^2cg			−28			−2						
b^2df			+56	+144	+16	+3	− 3					
b^2e^2			−35	−135	+ 9	−2	+ 2	− 1				
bc^2f				−108	−12	+3	− 9					
bcde				+180	−76	−1	+ 1	− 2	− 6			
bd^3				− 80	+48			+ 4	− 4			
c^3e					+48	−3	+18	+ 2	+ 4	+1		
c^2d^2					−32	+2	−12	− 3	+ 3	−1	+ 1	
b^4g						+1						
b^3cf						−3	+15					
b^3de						+1	− 1	+ 4	+ 4			
b^2c^2e						+3	−33	− 3	− 3	− 3		
b^2cd^2						−2	+22	− 8	+24	+2	− 2	
bc^3d								+10	−34	+2	− 6	
c^5								− 3	+ 12	−1	+ 4	+ 1
b^5f							− 6					
b^4ce							+15			+3		
b^4d^2							−10		−16	−1	+ 1	
b^3c^2d									+24	−4	+16	
b^2c^4									− 9	+2	−11	
b^6e										−1		− 5
b^5cd										+2	−14	
b^4c^3										−1	+10	+10
b^7d											+ 4	
b^6c^2											− 3	−10
b^8c												+ 5
b^{10}												− 1

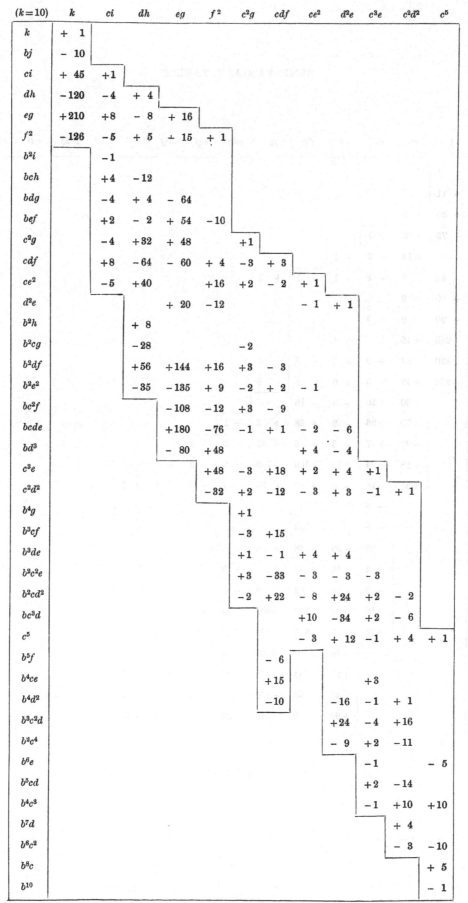

(l=11)	l	cj	di	eh	c²h	fg	cdg	cef	d²f	c³f	de²	c²de	cd³	c⁴d
l	+ 1													
bk	− 11													
cj	+ 35	+ 2												
di	− 75	− 9	+ 1											
eh	+ 90	+14	− 2	+ 1										
fg	− 42	− 7	+ 1	− 1	+ 1	+ 1								
b²j	+ 20	− 2												
bci	− 90	+ 9	− 3											
bdh	+240	+16		− 4										
beg	−420	−63	+ 9	− 2	− 5	− 5								
bf²	+252	+42	− 6	+ 6	− 6	− 6								
c²h		−30	+10	+ 3	− 16									
cdg		+70	−26	− 2	+ 58	+ 2	+ 1							
cef		−21	+ 7	− 6	+ 5	− 35	− 3	+ 1						
d²f		−56	+24	+10	−100	−100	+ 6	− 3	+ 1					
de²		+35	−15	− 5	+ 60	+ 60	− 4	+ 2	− 1	+ 1	+ 1			
b³i			+ 2											
b²ch			− 8		+ 32									
b²dg			+ 8	+20	− 48	+ 8	− 1							
b²ef			− 4	−18	+ 40		+ 3	− 1						
bc²g			+ 8	−15	− 62	− 6	− 3							
bcdf			−16	−24	+232	+408	−30	+14	− 6					
bce²			+10	+45	−205	−405	+27	−11	+ 3	− 3	− 3			
bd²e				−10	+ 20	+ 20	+ 2	− 1	+ 3	− 8	− 8			
c³f				+27	− 54	−270	+27	− 9	+ 4	−12				
c²de				−45	+ 90	+450	−45	+14	− 6	+36	+ 6	+ 1		
cd³				+20	− 40	−200	+20	− 6	+ 2	−18		− 1	+ 1	
b⁴h					− 16									

(*Continued next page.*)

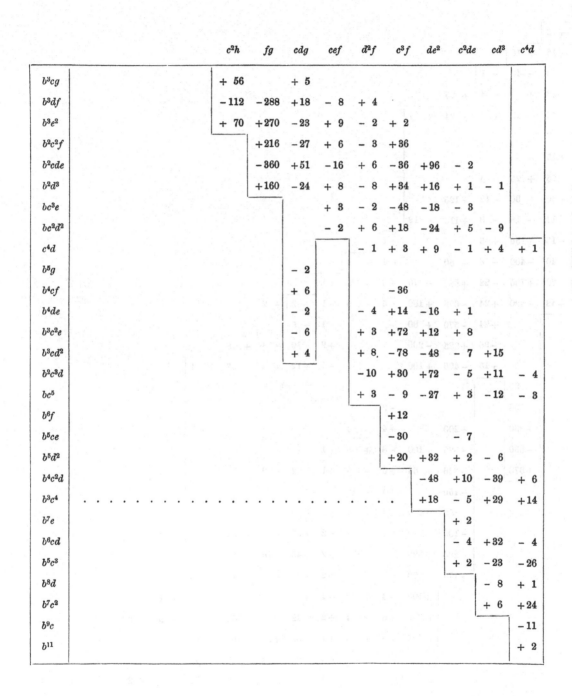

	c^2h	fg	cdg	cef	d^2f	c^3f	de^2	c^2de	cd^3	c^4d
b^3cg	+ 56		+ 5							
b^3df	− 112	− 288	+ 18	− 8	+ 4					
b^3e^2	+ 70	+ 270	− 23	+ 9	− 2	+ 2				
b^2c^2f		+ 216	− 27	+ 6	− 3	+ 36				
b^2cde		− 360	+ 51	− 16	+ 6	− 36	+ 96	− 2		
b^2d^3		+ 160	− 24	+ 8	− 8	+ 34	+ 16	+ 1	− 1	
bc^3e				+ 3	− 2	− 48	− 18	− 3		
bc^2d^2				− 2	+ 6	+ 18	− 24	+ 5	− 9	
c^4d					− 1	+ 3	+ 9	− 1	+ 4	+ 1
b^5g			− 2							
b^4cf			+ 6			− 36				
b^4de			− 2		− 4	+ 14	− 16	+ 1		
b^3c^2e			− 6		+ 3	+ 72	+ 12	+ 8		
b^3cd^2			+ 4		+ 8	− 78	− 48	− 7	+ 15	
b^2c^3d					− 10	+ 30	+ 72	− 5	+ 11	− 4
bc^5					+ 3	− 9	− 27	+ 3	− 12	− 3
b^6f						+ 12				
b^5ce						− 30		− 7		
b^5d^2						+ 20	+ 32	+ 2	− 6	
b^4c^2d							− 48	+ 10	− 39	+ 6
b^3c^4	. . . . .	. . . . .	. . . . .	. . . . .	. . . . .	. . . . .	+ 18	− 5	+ 29	+ 14
b^7e								+ 2		
b^6cd								− 4	+ 32	− 4
b^5c^3								+ 2	− 23	− 26
b^8d									− 8	+ 1
b^7c^2									+ 6	+ 24
b^9c										− 11
b^{11}										+ 2

											÷ 2										
(m=12)	m	ck	dj	ei	fh	g²	c²i	cdh	ceg	d²g	cf²	def	e³	c³g	c²df	c²e²	cd²e	d⁴	c⁴e	c³d²	c⁶
m	+1																				
bl	−12																				
ck	+66	+3																			
dj	−220	−15	+15																		
ei	+495	+40	−40	+1																	
fh	−792	−70	+70	−4	+25																
g²	+462	+42	−42	+3	−24	+1															
b²k		−3																			
bcj		+15	−45																		
bdi		−25	+25	−4																	
beh		+30	−30	+12	−125																
bfg		−14	+18	−8	+113	−12															
c²i		−15	+150	+3			+1														
cdh		+40	−400	−8	+50		−4	+4													
ceg		−70	+700	−22	+680	−70	+8	−8	+1												
cf²		+42	−420	+24	−675	+100	−5	+5	−1	+2	+2										
d²g				+24	−570	+80			−1	+5											
def				−36	+925	−200			+2	−19	−4	+18									
e³				+15	−400	+100			−1	+12	+2	−17	+1								
	. . . . .																				
b³j			+30																		
b²ci			−135				−2														
b²dh			+360		+200		+4	−4													
b²eg			−630		−525	+100	−8	+8	−1												
b²f²			+378		+336	−64	+5	−5	+1	−2	−2										
bc²h					−150		+4	−12													
bcdg					+350	−200	−4	+4	+2	−30											
bcef					−105	+20	+2	−2	−2	+37	−8	−54									
bd²f					−280	+320			−2	+46	+16	−72									
bde²					+175	−200			+2	−49	−4	+114	−12								
c³g						+100	−4	+32	−1	+20				+1							
c²df						−200	+8	−64	+2	−49	−4	+54		−3	+3						
c²e²						+125	−5	+40	+1	−32	+28	+162	−18			+1					
cd²e									−3	+91	−44	−342	+54	+4		−2	+1				
d⁴									+1	−32	+18	+135	−27	−2	−2	+1	−1	+1			
	. . . .																				
b⁴i							+1														
b³ch							−4	+20													
b³dg							+4	−4		+20											
b³ef							−2	+2		−18	+12	+36									
b²c²g							+4	−60		−15				−3							

(Continued next page.)

	c^2i	cdh	ceg	d^2g	cf^2 ÷2	def	e^3	c^3g	c^2df	c^2e^2	cd^2e	d^4	c^4e	c^3d^2	c^6
b^2cdf	−8	+120		−24	−24	+216		+6							
b^2ce^2	+5	−75		+45	−30	−360	+54	+2		−2					
b^2d^2e				−10	+20	+66	−6	−6		+2	−1				
bc^3f				+27	+12	−162		+3	−9						
bc^2de				−45	+60	−486	−180	−8	−15	+4	−6				
bcd^3				+20	−40	−252	+108	+7	+24	−4	+8	−12			
c^4e					−30	−81	+81	+1	+30	−2	+4		+1		
c^3d^2					+20	+54	−54	−2	−28	+2	−5	+8	−1	+1	
b^5h		−8													
b^4cg		+28						+3							
b^4df		−56				−144		−3	+3						
b^4e^2		+35				+135	−27			+1					
b^3c^2f						+108		−6	+24						
b^3cde						−180	+108	+9	+30	−4	+10				
b^3d^3						+80	−64	+1	−16		−4	+8			
b^2c^3e							−54	+2	−75	+2	−7		−4		
$b^2c^2d^2$							+36	−12	+6	+4	−9	+30	+3	−3	
bc^4d								+8	+48	−4	+14	−48	+2	−6	
c^6								−2	+2	+1		+16	−1	+4	+1
b^6g								−1	−16		−4				
b^5cf								+3	−21						
b^5de								−1	−15		−4				
b^4c^2e								−3	+60		+3		+6		
b^4cd^2								+2			+8	−48	−3	+3	
b^3c^3d									−40		−10	+68	−6	+22	
b^2c^5									+12		+3	−24	+3	−15	−6
b^7f									+6						
b^6ce									−15				−4		
b^6d^2									+10			+16	−23	−1	
b^5c^2d												−24	+30	−30	
b^4c^4												+9	−3	+21	+15
b^8e													+1		
b^7cd													+2	+14	−20
b^6c^3													−3	−9	
b^9d														−4	
b^8c^2														+3	+15
$b^{10}c$															−6
b^{12}															+1

Subsidiary Tables: $b=0$.

Left-hand: up to $(m=12)$.

Col.	c
c	+1

Col.	d
d	+1

Col.	e	c^2
e	+1	+3
c^2		+1

Col.	f	cd
f	+1	+2
cd		+1

Col.	g	ce	d^2	c^3
g	+1	+15	−10	
ce		+1	−1	−1
d^2			+1	+4
c^3				+1

Col.	h	cf	de	c^2d
h	+1	+9	−5	
cf		+1	−1	−1
de			+1	+3
c^2d				+1

Col.	l	cg	df	e^2	c^2e	cd^2	c^4
l	+1	+28	−56	+35			
cg		+1	−3	+2	−3	+2	
df			+3	−2	+18	−12	
e^2				+1	+6	.	+9
c^2e					+1	−1	−1
cd^2						+1	+4
c^4							+1

Col.	j	ch	dg	ef	c^2f	cde	d^3	c^3d
j	+1	+20	−28	+14				
ch		+2	−7	+5	−27	+45	−20	
dg			+1	−1	+9	−17	+8	
ef				+1	−9	+32	−18	
c^2f					+2	−5	+3	+1
cde						+1	−1	−1
d^3							+1	+4
c^3d								+1

Right-hand: up to $(k=10)$.

÷2

	c
2	−2

÷6

	d
3	−3

÷24

	e	c^2
4	−4	+12
2^2	+2	+6

÷120

	f	cd
5	−5	+50
32	+5	+10

÷720

	g	ce	d^2	c^3
6	−6	+90	+60	−180
42	+6	+30	−60	−180
3^2	+3	−45	+60	+90
2^3	−2	−30	+20	.

÷5040

	h	cf	de	c^2d
7	−7	+147	+245	−1470
52	+7	+63	−245	−630
43	+7	−147	+175	+210
32^2	−7	−63	+35	.

÷40320

	i	cg	df	e^2	c^2e	cd^2	c^4
8	−8	+224	+448	+280	−3360	−4480	+5040
62	+8	+112	−448	−280	−1680	+1120	+5040
53	+8	−224	+392	−280	+3360	−3920	−5040
4^2	+4	−112	−224	+420	−1680	+2240	+2520
42^2	−8	−112	+448	−280	+1680	−1120	.
3^22	−8	+56	−392	+280	−840	+560	
2^4	+2	+56	−112	+70			

÷362880

	j	ch	dg	ef	c^2f	cde	d^3	c^3d
9	−9	+324	+756	+1134	−6804	−22680	−5040	+68040
72	+9	+180	−756	−1134	−3780	+5040	+5040	+37800
63	+9	−324	+756	−1134	+6804	.	−10080	−22680
54	+9	−324	−756	+1386	−756	−2520	+5040	+7560
52^2	−9	−180	+756	−126	.	+7560	−5040	.
432	−18	+144	.	−252	+4536	−10080	+5040	.
3^3	−3	+108	−504	+378	−2268	+3780	−1680	.
32^3	+9	+180	−252	+126	.	.	.	.

Col.	k	ci	dh	eg	f^2	c^2g	cdf	ce^2	d^2e	c^3e	c^2d^2	c^5
Left-hand. k	+1	+45	−120	+210	−126							
ci		+1	−4	+8	−5	−4	+8	−5				
dh			+4	−8	+5	+32	−64	+40				
eg				+16	−15	+48	−60	•	+20			
f^2					+1	•	+4	+16	−12	+48	−32	
c^2g						+1	−3	+2	•	−3	+2	
cdf							+3	−2	•	+18	−12	
ce^2								+1	−1	+2	−3	−3
d^2e									+1	+4	+3	+12
c^3e										+1	−1	−1
c^2d^2											+1	+4
c^5												+1

÷3628800

	k	ci	dh	eg	f^2	c^2g	cdf	ce^2	d^2e	c^3e	c^2d^2	c^5
Right-hand. 10	−10	+450	+1320	+2100	+1260	−12600	−50400	−31500	−42000	+189000	+378000	−226800
82	+10	+270	−1320	−2100	−1260	−7560	+10080	+6300	+42000	+113400	+25200	−226800
73	+10	−450	+1320	−2100	−1260	+12600	−2520	+31500	−46200	−189000	+151200	+226800
64	+10	−450	−1200	+2940	−1260	−2520	+50400	−44100	−8400	+189000	−226800	−226800
5^2	+5	−225	−600	−1050	+2520	+6300	−37800	+15700	+21000	−94500	+126000	+113400
62^2	−10	−270	+1200	−420	+1260	•	−10080	+31500	−16800	−75600	−50400	
532	−20	+180	−120	+4200	−3780	−5040	+42840	−37800	+4200	+75600	−50400	
4^22	−10	+90	+1200	−2940	+1260	+12600	−30240	+6300	+8400	−37800	+25200	
43^2	−10	+450	−1320	−420	+1260	−5040	+2520	+6300	−4200	•		
42^3	+10	+270	−1200	+2100	−1260	−5040	+10080	−6300				
3^22^2	+15	+45	+720	−1890	+1260	+2520	−5040	+3150				
2^5	−2	−90	+240	−420	+252							

Left-hand, no right-hand.

Col.	l	cj	di	eh	fg	c^2h	cdg	cef	d^2f	de^2	c^3f	c^2de	cd^3	c^4d
l	+1	+35	−75	+90	−42									
cj		+2	−9	+14	−7	−30	+70	−21	−56	+35				
di			+1	−2	+1	+10	−26	+7	+24	−15				
eh				+1	−1	+3	−2	−6	+10	−5	+27	−45	+20	
fg					+1	−16	+58	+5	−100	+60	−54	+90	−40	
c^2h						+2	−7	+5	•	•	−27	+45	−20	
cdg							+1	−3	+6	−4	+27	−45	+20	
cef								+1	−3	+2	−9	+14	−6	
d^2f									+1	−1	+4	−6	+2	−1
de^2										+1	−12	+36	−18	+3
c^3f											+2	−5	+3	+1
c^2de												+1	−1	−1
cd^3													+1	+4
c^4d														+1

Left-hand, no right-hand.

Col.	m	ck	dj	ei	fh	g^2	c^2i	cdh	ceg	cf^2	d^2g	def	e^3	c^3g	c^2df	c^2e^2	cd^2e	d^4	c^4e	c^3d^2	c^6
m	+1	+66	−220	+495	−792	+462															
ck		+3	−15	+40	−70	+42	−15	+40	−70	+42											
dj			+15	−40	+70	−42	+150	−400	+700	−420											
ei				+1	−4	+3	+3	−8	−22	+24	+24	−36	+15								
fh					+25	−24		+50	+680	−675	−570	+925	−400								
g^2						+1			−70	+100	+80	−200	+100	−4	+8	−5					
c^2i							+1	−4	+8	−5				−4	+8	−5					
cdh								+4	−8	+5				+32	−64	+40					
ceg									+1	−1	−1	+2	−1	−1	+2	+1	−3	+1			
cf^2										+2	+5	−19	+12	+20	−49	−32	+91	−32			
d^2g											+1	−3	+2	+4	−9	−12	+27	−10	+6	−4	
def												+18	−17		+54	+162	−342	+135	−81	+54	
e^3													+1			−18	+54	−27	+81	−54	
c^3g														+1	−3				+1	−2	−2
c^2df															+3			−2	+30	−28	+2
c^2e^2																+1	−2	+1	−2	+2	+1
cd^2e																	+1	−1	+4	−5	•
d^4																		+1		+8	+16
c^4e																			+1	−1	−1
c^3d^2																				+1	+4
c^6																					+1

832.

NOTE ON AN APPARENT DIFFICULTY IN THE THEORY OF CURVES, WHEN THE COORDINATES OF A POINT ARE GIVEN AS FUNCTIONS OF A VARIABLE PARAMETER.

[From the *Messenger of Mathematics*, vol. XIV. (1885), pp. 12—14.]

SUPPOSE that the homogeneous coordinates x, y, z are given as proportional to the following functions of a parameter λ,

$$x : y : z = u + \alpha\sqrt{(\Omega)},\quad v + \beta\sqrt{(\Omega)},\quad w + \gamma\sqrt{(\Omega)},$$

where u, v, w are linear functions, Ω a cubic function, of the parameter. For the intersections of the curve with the arbitrary line $Ax + By + Cz = 0$, we have

$$Au + Bv + Cw + (A\alpha + B\beta + C\gamma)\sqrt{(\Omega)} = 0,$$

that is,

$$(Au + Bv + Cw)^2 - (A\alpha + B\beta + C\gamma)^2\,\Omega = 0,$$

a cubic equation in λ; and the curve is thus a cubic. For the value $\lambda = \infty$ we have $x : y : z = \alpha : \beta : \gamma$, or the point (α, β, γ) is a point of the curve.

Suppose now that the line $Ax + By + Cz = 0$ is an arbitrary line through the point (α, β, γ); viz. let the coefficients A, B, C satisfy the relation $A\alpha + B\beta + C\gamma = 0$; the equation for the determination of λ becomes

$$(Au + Bv + Cw)^2 = 0,$$

which equation has two equal roots, suppose $\lambda = \lambda_0$; and the meaning of this is not at once obvious.

Observe that more properly there is a root $\lambda = \infty$ which has dropped out, and that the roots are $\lambda = \infty$, $\lambda = \lambda_0$, $\lambda = \lambda_0$. The root $\lambda = \infty$ gives the point (α, β, γ), which is of course one of the intersections of the line with the curve. The two roots λ_0 give *not the same intersection* but two different intersections of the line with the curve; the line being in fact a line through the point (α, β, γ) of the curve, and which besides meets the curve in two distinct points.

To see how this is, observe that, in the general case where $A\alpha + B\beta + C\gamma$ is not $= 0$, we have λ determined by a cubic equation as above; and then taking λ equal to any root of this equation, we have further

$$Au + Bv + Cw + (A\alpha + B\beta + C\gamma)\sqrt{(\Omega)} = 0,$$

viz. the value of $\sqrt{(\Omega)}$ is hereby uniquely determined; and to each of the three values of λ, $\sqrt{(\Omega)}$, there corresponds a determinate point (x, y, z).

But suppose now $A\alpha + B\beta + C\gamma = 0$, and λ determined by the equation

$$(Au + Bv + Cw)^2 = 0,$$

giving $\lambda = \lambda_0$, as above. There is no longer an equation for the unique determination of $\sqrt{(\Omega)}$, and to the value $\lambda = \lambda_0$, there correspond the two values $\sqrt{(\Omega_0)}$, $-\sqrt{(\Omega_0)}$ of the radical: and thus to the two roots $\lambda = \lambda_0$, $\lambda = \lambda_0$ correspond the two different points

$$x : y : z = u_0 + \alpha\sqrt{(\Omega_0)} : v_0 + \beta\sqrt{(\Omega_0)} : w_0 + \gamma\sqrt{(\Omega_0)};$$

and

$$x : y : z = u_0 - \alpha\sqrt{(\Omega_0)} : v_0 - \beta\sqrt{(\Omega_0)} : w_0 - \gamma\sqrt{(\Omega_0)}.$$

It is to be added that the point (α, β, γ) is an inflexion on the curve. Write for a moment

$$u, v, w = a\lambda + f, b\lambda + g, c\lambda + h,$$

and let A, B, C be determined by the conditions

$$A\alpha + B\beta + C\gamma = 0,$$
$$Aa + Bb + Cc = 0.$$

Then the equation for the determination of λ becomes $(Af + Bg + Ch)^2 = 0$, viz. the left-hand is a mere constant, or there are the three equal roots $\lambda = \infty$; the intersections with the curve are thus the point (α, β, γ) three times; hence this point is an inflexion, the tangent being $Ax + By + Cz = 0$. The second of the two equations may be written

$$Au_\infty + Bv_\infty + Cw_\infty = 0.$$

Let λ_1 be one of the roots of the equation $\Omega = 0$; u_1, v_1, w_1 the corresponding values of u, v, w, and let A, B, C, be determined by the conditions

$$A\alpha + B\beta + C\gamma = 0,$$
$$Au_1 + Bv_1 + Cw_1 = 0.$$

The equation $(Au + Bv + Cw)^2 = 0$ for the intersections with the curve has the two equal roots $\lambda = \lambda_1$; and to each of these, since now $\sqrt{(\Omega_1)} = 0$, there corresponds the same point $x : y : z = u_1 : v_1 : w_1$; hence the line $Ax + By + Cz = 0$, or say

$$A_1x + B_1y + C_1z = 0,$$

is a tangent from the inflexion. Similarly, if λ_2, λ_3 are the other two roots of the equation $\Omega = 0$, we have $A_2x + B_2y + C_2z = 0$, $A_3x + B_3y + C_3z = 0$ for the other two tangents from the inflexion.

It would have been to some extent clearer to have represented the parameter λ as a quotient, say $\lambda = p/q$; the equations for x, y, z would then have been

$$x : y : z = (ap + fq)\sqrt{(q)} + \alpha\sqrt{(\Omega)} : (bp + gq)\sqrt{(q)} + \beta\sqrt{(\Omega)} : (cp + hq)\sqrt{(q)} + \gamma\sqrt{(\Omega)},$$

where Ω is now a homogeneous function $(p, q)^3$.

833.

ON A FORMULA IN ELLIPTIC FUNCTIONS.

[From the *Messenger of Mathematics*, vol. XIV. (1885), pp. 21, 22.]

WRITING s, c, d for the sn, cn, and dn of an argument u, and so in other cases: we have s, c, d for the coordinates of a point on the quadriquadric curve $x^2+y^2=1$, $z^2+k^2x^2=1$. Applying Abel's theorem to this curve, it appears that, if $u_1+u_2+u_3+u_4=0$, the corresponding points are in a plane; that is, the elliptic functions satisfy the relation

$$\begin{vmatrix} s_1, & c_1, & d_1, & 1 \\ s_2, & c_2, & d_2, & 1 \\ s_3, & c_3, & d_3, & 1 \\ s_4, & c_4, & d_4, & 1 \end{vmatrix} = 0.$$

This may be written

$$\begin{aligned} &(s_2-s_1)(c_3d_4-c_4d_3)+(s_4-s_3)(c_1d_2-c_2d_1) \\ &+(c_2-c_1)(d_3s_4-d_4s_3)+(c_4-c_3)(d_1s_2-d_2s_1) \\ &+(d_2-d_1)(s_3c_4-s_4c_3)+(d_4-d_3)(s_1c_2-s_2c_1)=0; \end{aligned}$$

and it may be shown that each of the three lines is, in fact, separately $=0$.

This appears from the following three formulæ:

$$\frac{\operatorname{sn}(u_1+u_2)}{\operatorname{cn}(u_1+u_2)-\operatorname{dn}(u_1+u_2)} = \frac{s_1-s_2}{c_1d_2-c_2d_1},$$

$$\frac{\operatorname{sn}(u_1+u_2)}{\operatorname{cn}(u_1+u_2)+1} = \frac{c_1-c_2}{d_1s_2-d_2s_1},$$

$$\frac{\operatorname{sn}(u_1+u_2)}{\operatorname{dn}(u_1+u_2)+1} = \frac{-\frac{1}{k^2}(d_1-d_2)}{s_1c_2-s_2c_1},$$

which are themselves at once deducible from formulæ given, p. 63, of my *Elliptic Functions*, and which may be written

$$\begin{aligned}
\operatorname{sn}(u_1+u_2) &= s_1^2 - s_2^2 = -(c_1^2 - c_2^2) = -\frac{1}{k^2}(d_1^2 - d_2^2), \quad \div (s_1c_2d_2 - s_2c_1d_1),\\
\operatorname{cn}(u_1+u_2) &= s_1c_1d_2 - s_2c_2d_1, \qquad\qquad \div \quad ,,\\
\operatorname{dn}(u_1+u_2) &= s_1d_1c_2 - s_2d_2c_1, \qquad\qquad \div \quad ,,
\end{aligned}$$

In fact, the numerators of $\operatorname{cn}(u_1+u_2) - \operatorname{dn}(u_1+u_2)$, $\operatorname{cn}(u_1+u_2)+1$, $\operatorname{dn}(u_1+u_2)+1$ thus become $=(s_1+s_2)(c_1d_2-c_2d_1)$, $-(c_1+c_2)(d_1s_2-d_2s_1)$, $(d_1+d_2)(s_1c_2-s_2c_1)$ respectively: so that, taking the numerator of $\operatorname{sn}(u_1+u_2)$ successively under its three forms, we have by division the formulæ in question. And then, if $u_1+u_2=-(u_3+u_4)$, the functions on the left-hand side become, with only a change of sign, the like functions of u_3+u_4; and we thence have the required equations

$$\frac{s_1-s_2}{c_1d_2-c_2d_1} = -\frac{s_3-s_4}{c_3d_4-c_4d_3}, \text{ \&c.}$$

834.

ON THE ADDITION OF THE ELLIPTIC FUNCTIONS.

[From the *Messenger of Mathematics*, vol. XIV. (1885), pp. 56—61.]

MR FORSYTH'S Note [*l. c.*, p. 23] on my "Formula in Elliptic Functions" has supplied a missing link, and I am now able to obtain the addition formulæ very simply from the application of Abel's theorem to the Quadriquadric Curve.

I remark that, instead of coplanar points 1, 2, 3, 4, it is advantageous to consider coresidual points 1, 2 and 3, 4; that is, pairs 1, 2 and 3, 4, which are each of them coplanar with one and the same pair of points 5, 6. The difference is as follows: for the coplanar points 1, 2, 3, 4, we have

$$du_1 + du_2 + du_3 + du_4 = 0,$$

giving

$$u_1 + u_2 + u_3 + u_4 = C,$$

and for the addition theory it is necessary to have $C = 0$; for the coresidual points, we have

$$u_1 + u_2 + u_5 + u_6 = C, \quad u_3 + u_4 + u_5 + u_6 = C;$$

and thence $u_1 + u_2 = u_3 + u_4$, irrespectively of the value of C.

As to the general theory of a curve in space, observe that, when this is a complete intersection of two surfaces

$$f(x, y, z, w) = 0, \quad g(x, y, z, w) = 0,$$

then at the point (x, y, z, w), if

$$(x + dx, \ y + dy, \ z + dz, \ w + dw)$$

are the coordinates of the consecutive point, the six coordinates of the tangent line are

$$y\,dz - z\,dy, \quad z\,dx - x\,dz, \quad x\,dy - y\,dx, \quad x\,dw - w\,dx, \quad y\,dw - w\,dy, \quad z\,dw - w\,dz.$$

But considering the line as the intersection of the two tangent planes

$$\frac{df}{dx}X+\frac{df}{dy}Y+\frac{df}{dz}Z+\frac{df}{dw}W=0,$$

and

$$\frac{dg}{dx}X+\frac{dg}{dy}Y+\frac{dg}{dz}Z+\frac{dg}{dw}W=0,$$

the six coordinates are

$$\frac{d(f,g)}{d(x,w)},\ \frac{d(f,g)}{d(y,w)},\ \frac{d(f,g)}{d(z,w)},\ \frac{d(f,g)}{d(y,z)},\ \frac{d(f,g)}{d(z,x)},\ \frac{d(f,g)}{d(x,y)},$$

so that the six quotients

$$(y\,dz-z\,dy)\Big/\frac{d(f,g)}{d(x,w)},\ \text{\&c.},$$

are equal to each other, and may be put $=d\omega$.

Considering any two quadric surfaces, there is in general a system of four conjugate points, or points such that in regard to each of the quadrics the polar plane of any one of the points is the plane through the other three points. And then taking $x=0$, $y=0$, $z=0$, $w=0$ for the equations of the faces of the tetrahedron formed by the four points, the equations of the quadric surfaces will be of the form

$$\text{a}x^2+\text{b}y^2+\text{c}z^2+\text{d}w^2=0,$$
$$\text{a}'x^2+\text{b}'y^2+\text{c}'z^2+\text{d}'w^2=0;$$

we then have the six quotients

$$(y\,dz-z\,dy)/(\text{ad}'-\text{a}'\text{d})\,xw,\ \text{\&c.},$$

equal to each other, and each $=d\omega$. Here $d\omega$ is homogeneous of the degree zero in the coordinates (x, y, z, w), or, what is the same thing, it is a differential $F\left(\frac{x}{w}\right)d\frac{x}{w}$, say it is $=du$; and taking the integrals always from one and the same fixed point on the curve, we have each point of the curve corresponding to a determinate value of a parameter u.

Supposing that u_1, u_2, u_5, u_6 are the values of u, belonging to any four coplanar points 1, 2, 5, 6; then, by Abel's theorem, $du_1+du_2+du_5+du_6=0$; that is, we have

$$u_1+u_2+u_5+u_6=C,$$

as the condition in order that the four points 1, 2, 5, 6 may be coplanar; similarly, we have

$$u_3+u_4+u_5+u_6=C,$$

as the condition in order that the four points 3, 4, 5, 6 may be coplanar; and we have therefore

$$u_1+u_2=u_3+u_4,$$

as the condition that the two pairs of points 1, 2 and 3, 4 may be coresidual.

The points 1, 2, 5, 6 are coplanar, hence the line 56 meets the line 12, say in the point A; and the points 3, 4, 5, 6 are coplanar, hence the line 56 meets the line 34, say in the point B. We can, through the curve and any arbitrary point in space, draw a quadric surface

$$(\mathrm{a}+\lambda\mathrm{a}')x^2+(\mathrm{b}+\lambda\mathrm{b}')y^2+(\mathrm{c}+\lambda\mathrm{c}')z^2+(\mathrm{d}+\lambda\mathrm{d}')w^2=0.$$

Hence we have such a quadric surface through the point A; and this surface, passing through 5 and 6, will contain the line 56, and therefore also the point B; hence, passing through 3 and 4, it will contain the line 34; viz. we have the lines 12, 34 as generating lines, obviously of the same kind, on the last-mentioned quadric surface. I say that if, on such a surface, that is, on any surface

$$Ax^2+By^2+Cz^2+Dw^2=0,$$

we have

$$(a,\ b,\ c,\ f,\ g,\ h),\quad (a',\ b',\ c',\ f',\ g',\ h'),$$

the coordinates of two generating lines of the same kind, then

$$\frac{a}{f}=\frac{a'}{f'},\quad \frac{b}{g}=\frac{b'}{g'},\quad \frac{c}{h}=\frac{c'}{h'}.$$

This is at once seen to be the case; for, taking θ an arbitrary parameter, we have for the equations of a generating line

$$\{x\sqrt{(A)}+iy\sqrt{(B)}\}+\theta\{z\sqrt{(C)}+iw\sqrt{(D)}\}=0,$$
$$\theta\{x\sqrt{(A)}-iy\sqrt{(B)}\}-\ \{z\sqrt{(C)}-iw\sqrt{(D)}\}=0,$$

and the coordinates $(a,\ b,\ c,\ f,\ g,\ h)$ of this line are

$$i\sqrt{(AD)}(1-\theta^2),\quad \sqrt{(BD)}(-1-\theta^2),\quad i\sqrt{(CD)}\,2\theta,$$
$$i\sqrt{(BC)}(-1+\theta^2),\quad \sqrt{(CA)}(1+\theta^2),\quad i\sqrt{(AB)}(-2\theta);$$

that is, the quotients $\frac{a}{f},\ \frac{b}{g},\ \frac{c}{h}$ are each of them independent of θ; and they have consequently their values unaltered when for the original line we substitute any other generating line of the same kind. Or, to prove the statement in a different manner, the equation of the quadric surface through the line $(a,\ b,\ c,\ f,\ g,\ h)$ is

$$aghx^2+bhfy^2+cfgz^2+abcw^2=0;$$

hence, if this contains the line $(a',\ b',\ c',\ f',\ g',\ h')$, we must have

$$agh\ :\ bhf\ :\ cfg\ :\ abc=a'g'h'\ :\ b'h'f'\ :\ c'f'g'\ :\ a'b'c',$$

equations which give either

$$af'+a'f=0,\quad bg'+b'g=0,\quad ch'+c'h=0,$$

or else

$$af'-a'f=0,\quad bg'-b'g=0,\quad ch'-c'h=0.$$

In the former case, the two lines are generating lines of different kinds; in the latter, they are generating lines of the same kind.

Now, considering (a, b, c, f, g, h) as the coordinates of the line 12 and (a', b', c', f', g', h') as those of the line 34, the equations just obtained are

$$\frac{y_1z_2 - y_2z_1}{x_1w_2 - x_2w_1} = \frac{y_3z_4 - y_4z_3}{x_3w_4 - x_4w_3}, \quad \frac{z_1x_2 - z_2x_1}{y_1w_2 - y_2w_1} = \frac{z_3x_4 - z_4x_3}{y_3w_4 - y_4w_3}, \quad \frac{x_1y_2 - x_2y_1}{z_1w_2 - z_2w_1} = \frac{x_3y_4 - x_4y_3}{z_3w_4 - z_4w_3}.$$

Of course the equations hold good if, instead of the two lines, we have one and the same line; the equations

$$\mathrm{a}x^2 + \mathrm{b}y^2 + \mathrm{c}z^2 + \mathrm{d}w^2 = 0, \quad \mathrm{a}'x^2 + \mathrm{b}'y^2 + \mathrm{c}'z^2 + \mathrm{d}'w^2 = 0,$$

considering therein x^2, y^2, z^2, w^2 as coordinates, may be regarded as the equations of a line; and thus the points $(x_1^2, y_1^2, z_1^2, w_1^2)$, &c., will be four points on a line. And we have thus

$$\frac{y_1^2z_2^2 - y_2^2z_1^2}{x_1^2w_2^2 - x_2^2w_1^2} = \frac{y_3^2z_4^2 - y_4^2z_3^2}{x_3^2w_4^2 - x_4^2w_3^2}, \text{ \&c.,}$$

equations which are, by means of the foregoing set, converted into

$$\frac{y_1z_2 + y_2z_1}{x_1w_2 + x_2w_1} = \frac{y_3z_4 + y_4z_3}{x_3w_4 + x_4w_3}, \quad \frac{z_1x_2 + z_2x_1}{y_1w_2 + y_2w_1} = \frac{z_3x_4 + z_4x_3}{y_3w_4 + y_4w_3}, \quad \frac{x_1y_2 + x_2y_1}{z_1w_2 + z_2w_1} = \frac{x_3y_4 + x_4y_3}{z_3w_4 + z_4w_3}.$$

If for x, y, z, w we write s, c, d, 1, then the equations are

$$\frac{c_1d_2 - c_2d_1}{s_1 - s_2} = \frac{c_3d_4 - c_4d_3}{s_3 - s_4}, \quad \frac{d_1s_2 - d_2s_1}{c_1 - c_2} = \frac{d_3s_4 - d_4s_3}{c_3 - c_4}, \quad \frac{s_1c_2 - s_2c_1}{d_1 - d_2} = \frac{s_3c_4 - s_4c_3}{d_3 - d_4},$$

$$\frac{c_1d_2 + c_2d_1}{s_1 + s_2} = \frac{c_3d_4 + c_4d_3}{s_3 + s_4}, \quad \frac{d_1s_2 + d_2s_1}{c_1 + c_2} = \frac{d_3s_4 + d_4s_3}{c_3 + c_4}, \quad \frac{s_1c_2 + s_2c_1}{d_1 + d_2} = \frac{s_3c_4 + s_4c_3}{d_3 + d_4},$$

where s_1, c_1, d_1 are the sn, cn and dn of u_1, &c.; and where the relation between the arguments is $u_1 + u_2 = u_3 + u_4$.

In particular, if $u_4 = 0$, we have $s_4, c_4, d_4 = 0, 1, 1$; and then writing u for u_3, and consequently s, c, d for s_3, c_3, d_3, the relation between the arguments is $u = u_1 + u_2$; and we have

$$\frac{c_1d_2 - c_2d_1}{s_1 - s_2} = \frac{c - d}{s}, \quad \frac{d_1s_2 - d_2s_1}{c_1 - c_2} = \frac{s}{1 - c}, \quad \frac{s_1c_2 - s_2c_1}{d_1 - d_2} = \frac{s}{d - 1},$$

$$\frac{c_1d_2 + c_2d_1}{s_1 + s_2} = \frac{c + d}{s}, \quad \frac{d_1s_2 + d_2s_1}{c_1 + c_2} = \frac{s}{1 + c}, \quad \frac{s_1c_2 + s_2c_1}{d_1 + d_2} = \frac{s}{d + 1}.$$

The last two pairs give

$$\frac{1 + c}{1 - c} = \frac{(d_1s_2 - d_2s_1)(c_1 + c_2)}{(d_1s_2 + d_2s_1)(c_1 - c_2)}, \quad \frac{d + 1}{d - 1} = \frac{(s_1c_2 - s_2c_1)(d_1 + d_2)}{(s_1c_2 + s_2c_1)(d_1 - d_2)},$$

that is,

$$c = \frac{s_1c_1d_2 - s_2c_2d_1}{s_1c_2d_2 - s_2c_1d_1}, \qquad d = \frac{s_1d_1c_2 - s_2d_2c_1}{s_1c_2d_2 - s_2c_1d_1}.$$

These last values give

$$c-d=\frac{(s_1+s_2)(c_1d_2-c_2d_1)}{s_1c_2d_2-s_2c_1d_1},\quad c+d=\frac{(s_1-s_2)(c_1d_2+c_2d_1)}{s_1c_2d_2-s_2c_1d_1};$$

and then, from the given value of either $\frac{c-d}{s}$ or $\frac{c+d}{s}$, we obtain

$$s=\frac{s_1^2-s_2^2}{s_1c_2d_2-s_2c_1d_1};$$

viz. the resulting equations thus are

$$s=\frac{s_1^2-s_2^2}{s_1c_2d_2-s_2c_1d_1},\quad c=\frac{s_1c_1d_2-s_2c_2d_1}{s_1c_2d_2-s_2c_1d_1},\quad d=\frac{s_1d_1c_2-s_2d_2c_1}{s_1c_2d_2-s_2c_1d_1},$$

which are one of the four sets given (p. 63) of my *Elliptic Functions*; it may be noticed that they have the advantage of not containing k explicitly, and the disadvantage of becoming vanishing fractions for $u_1=u_2$. To obtain the ordinary forms, we have only to multiply the numerators and denominators each by $s_1c_2d_2+s_2c_1d_1$; the denominator thus becomes $=(s_1^2-s_2^2)(1-k^2s_1^2s_2^2)$, and each of the numerators has, or acquires, the factor $s_1^2-s_2^2$, so that this factor divides out.

835.

ON CARDAN'S SOLUTION OF A CUBIC EQUATION.

[From the *Messenger of Mathematics*, vol. XIV. (1885), pp. 96, 97.]

IT is interesting to see how the solution comes out when one root of the equation is known. Say the equation is $x^3 + qx - r = 0$, where $a^3 - qa - r = 0$, that is, $r = a^3 + qa$.

Solving in the usual manner, we have

$$x = y + z, \quad y^3 + z^3 - r + (y + z)(3yz + q) = 0,$$

whence

$$y^3 + z^3 = r,$$
$$yz = -\tfrac{1}{3}q;$$

and thence

$$(y^3 - z^3)^2 = r^2 + \tfrac{4}{27}q^3, \quad = a^6 + 2qa^4 + q^2a^2 - \tfrac{4}{27}q^3, \quad = (a^2 + \tfrac{4}{3}q)(a^2 + \tfrac{1}{3}q)^2;$$

or say

$$y^3 - z^3 = (a^2 + \tfrac{1}{3}q)\sqrt{(a^2 + \tfrac{4}{3}q)};$$

and therefore

$$8y^3 = 4a^3 + 4qa + (4a^2 + \tfrac{4}{3}q)\sqrt{(a^2 + \tfrac{4}{3}q)}, \quad = \{a + \sqrt{(a^2 + \tfrac{4}{3}q)}\}^3,$$
$$8z^3 = 4a^3 + 4qa - (4a^2 + \tfrac{4}{3}q)\sqrt{(a^2 + \tfrac{4}{3}q)}, \quad = \{a - \sqrt{(a^2 + \tfrac{4}{3}q)}\}^3;$$

where observe that the essential step is the expression of the irrational functions as perfect cubes: that the functions are the cubes of $a \pm \sqrt{(a^2 + \tfrac{4}{3}q)}$ respectively is seen to be true; but if we were to attempt to find a cube root $\alpha + \beta\sqrt{(a^2 + \tfrac{4}{3}q)}$ by an algebraical process, we should be thrown back upon the original cubic equation.

Writing then ω for an imaginary cube root of unity, we have

$$2y = (1, \ \omega \ \text{ or } \ \omega^2)\{a + \sqrt{(a^2 + \tfrac{4}{3}q)}\},$$
$$2z = (1, \ \omega^2 \ \text{ or } \ \omega\ )\{a - \sqrt{(a^2 + \tfrac{4}{3}q)}\};$$

and then

$$x = y + z = a, \text{ or } = -\tfrac{1}{2}a \pm \tfrac{1}{2}(\omega - \omega^2)\sqrt{(a^2 + \tfrac{4}{3}q)},$$

where $\omega - \omega^2 = i\sqrt{3}$; the last two roots are of course the roots of the quadric equation $x^2 + ax + a^2 + q = 0$, which is obtained by throwing out the factor $x - a$ from the given equation $x^3 + qx - r = 0$.

Cambridge, Sep. 17, 1884.

836.

ON THE QUATERNION EQUATION $qQ - Qq' = 0$.

[From the *Messenger of Mathematics*, vol. XIV. (1885), pp. 108—112.]

I CONSIDER the equation $qQ - Qq' = 0$, where q, q' are given quaternions, and Q is a quaternion to be determined. Obviously a condition must be satisfied by the given quaternions; for, substituting in the given equation for q, q', Q their values, say $w + ix + jy + kz$, $w' + ix' + jy' + kz'$, and $W + iX + jY + kZ$ respectively, and equating to zero the scalar part and the coefficients of i, j, k, we have four equations linear in W, X, Y, Z, and then eliminating these quantities, we have the condition in question. Supposing the condition satisfied, the ratios of W, X, Y, Z are then completely determined, and the required quaternion Q is thus determinate except as to a scalar factor, or say Q is = product of an arbitrary scalar into a determinate quaternion expression.

It might, at first sight, appear that the condition is that the given quaternions shall have their tensors equal, $Tq = Tq'$; for the equation gives $Tq \,.\, TQ - TQ \,.\, Tq' = 0$, that is, $TQ(Tq - Tq') = 0$. But we cannot thence infer, and it is not true, that the condition is $Tq - Tq' = 0$; the formula does not give the required condition at all, but the conclusion to which it leads is that, when the condition is satisfied, then in general (that is, unless $Tq - Tq' = 0$) the required quaternion is an imaginary quaternion (or, as Hamilton calls it, a biquaternion) having its tensor $TQ = 0$. In the particular case where the given quaternions are such that $Tq - Tq' = 0$, then the required quaternion Q is determined less definitely, viz. it becomes = product of an arbitrary scalar into a not completely determined quaternion expression; and it is thus in general such that TQ is not $= 0$. In explanation, observe that, for the particular case in question, the four linear equations for W, X, Y, Z reduce themselves to two independent relations, and they give therefore for the ratios of W, X, Y, Z expressions involving an arbitrary parameter A; these expressions cannot, it is clear, be deduced from the determinate expressions which belong to the general case.

Instead of directly working out the condition in the manner indicated above, I present the investigation in a synthetic form as follows:

Taking v, v' for the vector parts of the two given quaternions, so that $q = w + v$, $q' = w' + v'$, I write for shortness

$$\begin{aligned}
\theta &= w - w', \\
\alpha &= v^2 + v'^2, \quad = -x^2 - y^2 - z^2 - x'^2 - y'^2 - z'^2, \\
\beta &= v^2 - v'^2, \quad = -x^2 - y^2 - z^2 + x'^2 + y'^2 + z'^2, \\
D &= -\theta(\alpha - \theta^2), \\
A &= \beta - \theta^2, \\
B &= \beta + \theta^2;
\end{aligned}$$

so that θ, α, β, D, A, B are all of them scalars. With these I form a quaternion $Q = (D + Av)(D + Bv')$; I say that we have identically

$$qQ - Qq' = \{D - v \,.\, v'^2 + v' \,.\, v^2 + vv' \,.\, \theta\}\,(\theta^4 - 2\alpha\theta^2 + \beta^2).$$

It of course follows that, if $\theta^4 - 2\alpha\theta^2 + \beta^2 = 0$, then $qQ - Qq' = 0$, viz. $\theta^4 - 2\alpha\theta^2 + \beta^2 = 0$ is the condition $\Omega = 0$, for the existence of the required quaternion; and this condition being satisfied, then (omitting the arbitrary scalar factor) the value of the quaternion is $Q = (D + Av)(D + Bv')$, a value giving $T(Q) = 0$, that is, $W^2 + X^2 + Y^2 + Z^2 = 0$. If $\theta = 0$ (that is, $w = w'$), then the condition becomes $\beta = 0$, that is, $x^2 + y^2 + z^2 - x'^2 - y'^2 - z'^2 = 0$; and these two conditions being satisfied, Q ceases to have the determinate value given by the foregoing formula: it has a value involving an arbitrary parameter, and is no longer such that $W^2 + X^2 + Y^2 + Z^2 = 0$.

The identical equation is at once verified; we have

$$\begin{aligned}
qQ - Qq' &= (w + v)\,Q - Q\,(w' - v') \\
&= \theta\ (D^2 + DAv + DBv' + ABvv') \\
&\quad + v\ (D^2 + DAv + DBv' + ABvv') \\
&\quad - (D^2 + DAv + DBv' + ABvv')\,v' \\
&= \theta D^2 + DAv^2 - DBv'^2 \\
&\quad + v\ (DA\theta + D^2 - ABv'^2) \\
&\quad + v'\ (DB\theta + ABv^2 - D^2) \\
&\quad + vv'\,(\theta AB + DB - DA).
\end{aligned}$$

The first line is here $= D\{D\theta + Av^2 + Bv'^2\}$, viz. the term in $\{\ \}$ is

$$\begin{aligned}
&-(\alpha - \theta^2)\,\theta^2 + (\beta - \theta^2)\,v^2 - (\beta + \theta^2)\,v'^2, \\
&= -\alpha\theta^2 + \theta^4 + \beta \,.\, \beta - \theta^2 \,.\, \alpha, \\
&= \theta^4 - 2\alpha\theta^2 + \beta^2;
\end{aligned}$$

and similarly each of the other lines contains the factor $\theta^4 - 2\alpha\theta^2 + \beta^2$, and the equation is thus seen to hold good.

The tensor of Q is $=(D^2 - A^2v^2)(D^2 - B^2v'^2)$; and we have

$$D^2 - A^2v^2 = (\alpha - \theta^2)^2\,\theta^2 - v^2(\beta - \theta^2),$$
$$= \theta^6 - (2\alpha + v^2)\,\theta^4 + (\alpha^2 + 2\beta v^2)\,\theta^2 - v^2\beta^2,$$

which, observing that

$$\alpha^2 + 2\beta v^2 = \beta^2 + 2\alpha v^2 \quad \text{is} \quad = (\theta^2 - v^2)(\theta^4 - 2\alpha\theta^2 + \beta^2);$$

and similarly

$$D^2 - B^2v'^2 \qquad \text{is} \quad = (\theta^2 - v'^2)(\theta^4 - 2\alpha\theta^2 + \beta^2);$$

hence the tensor is

$$TQ = (\theta^2 - v^2)(\theta^2 - v'^2)(\theta^4 - 2\alpha\theta^2 + \beta^2)^2,$$

which, in virtue of $\theta^4 - 2\alpha\theta^2 + \beta^2 = 0$, is $= 0$.

The particular case is when $Tq - Tq' = 0$, that is, $w^2 - v^2 - w'^2 + v'^2 = 0$, or say $w^2 - w'^2 = v^2 - v'^2$, that is, $\theta(w + w') = \beta$. Combining with this the general condition $\theta^4 - 2\alpha\theta^2 + \beta^2 = 0$, we find $\theta^2\{\theta^2 - 2\alpha + (w + w')^2\} = 0$, or in the second factor, for θ and α substituting their values, we have $2\theta^2(w^2 - v^2 + w'^2 - v'^2) = 0$, that is, $2\theta^2(Tq + Tq') = 0$. Attending to the assumed relation $Tq - Tq' = 0$, the second factor can only vanish if $Tq = 0$, $Tq' = 0$; hence, disregarding this more special case, the factor which vanishes must be the first factor, that is, $\theta = 0$; or the equations $Tq - Tq' = 0$ and $\theta^4 - 2\alpha\theta^2 + \beta^2 = 0$ give $\beta = 0$ and $\theta = 0$, that is, we have as already mentioned $w - w' = 0$, and $x^2 + y^2 + z^2 = x'^2 + y'^2 + z'^2$, viz. the given quaternions have their scalars equal, and the squares of their vectors also equal. The equation here is $vQ - Qv' = 0$; and writing $v^2 = v'^2 = -p^2$, we see at once that a solution is

$$Q = (-p^2 + vv') + A(v + v'),$$

where A is an arbitrary scalar; in fact, with this value of Q, we have at once vQ and Qv' each

$$= -p^2(v + v') + A(-p^2 + vv');$$

and the equation $vQ - Qv' = 0$ is thus satisfied. The value of the tensor is easily found to be

$$TQ = 2(A^2 + p^2)(p^2 + xx' + yy' + zz'),$$

which is not $= 0$.

In accordance with a remark in the introductory paragraphs, the solution

$$Q = -p^2 + vv' + A(v + v')$$

is not comprised in the general solution. As to this, observe that, in the case in question $\theta = 0$, $\beta = 0$, we have from the general theorem the form

$$Q = \left(-\frac{\alpha\theta}{\beta} + v\right)\left(-\frac{\alpha\theta}{\beta} + v'\right);$$

that is,

$$Q = \frac{\alpha^2\theta^2}{\beta^2} + vv' - \frac{\alpha\theta}{\beta}(v + v');$$

in the condition $\theta^4 - 2\alpha\theta^2 + \beta^2 = 0$, writing $\theta = 0$, we have $\frac{\beta^2}{\theta^2} = 2\alpha$, or for α writing its value $= -2p^2$, we have $\alpha^2 = 4p^4$, $\frac{\theta^2}{\beta^2} = -\frac{1}{4p^2}$, and thence $\frac{\alpha^2\theta^2}{\beta^2} = -p^2$, and $\frac{\alpha\theta}{\beta} = \lambda p$, if λ denote the $\sqrt{(-1)}$ of ordinary algebra. The resulting formula is thus

$$Q = -p^2 + vv' - \lambda p\,(v + v'),$$

which corresponds to the determinate value $-\lambda p$ of the constant A.

The foregoing solution $Q = -p^2 + vv' + A\,(v + v')$ may be easily identified with that given pp. 124, 125 of Tait's *Elementary Treatise on Quaternions*; the case there considered is that of a real quaternion, and it was therefore assumed that the two conditions $w = w'$, and $x^2 + y^2 + z^2 = x'^2 + y'^2 + z'^2$, were each of them satisfied.

The theory of quaternions is, as is well known, identical with that of matrices of the second order; the identity is, in effect, established by the remark and footnote "Linear Associative Algebra," *American Journal of Mathematics*, t. IV. (1881), p. 132. Writing x, y, z, w for Peirce's imaginaries i, j, k, l, these have the multiplication table

	x	y	z	w
x	x	y	0	0
y	0	0	x	y
z	z	w	0	0
w	0	0	z	w

Then if λ, $= \sqrt{(-1)}$, be the imaginary of ordinary algebra, and i, j, k the quaternion imaginaries, the relations between i, j, k and x, y, z, w are

$$\begin{aligned} x &= \tfrac{1}{2}(\ \ 1 - \lambda i), &&\text{or conversely} & 1 &= \ \ x + w\,, \\ y &= \tfrac{1}{2}(\ \ j - \lambda k), &&\quad\text{,,} & i &= \lambda\,(x - w), \\ z &= \tfrac{1}{2}(-j - \lambda k), &&\quad\text{,,} & j &= \ \ (y - z), \\ w &= \tfrac{1}{2}(\ \ 1 + \lambda i), &&\quad\text{,,} & k &= \lambda\,(y + z)\,; \end{aligned}$$

and we can thus at once express a quaternion as a linear function of the x, y, z, w, or a linear function of the x, y, z, w as a quaternion. And we then consider

$ax+by+cz+dw$ as denoting the matrix $\begin{vmatrix} a, & b \\ c, & d \end{vmatrix}$, we obtain for the product of two matrices the ordinary formula

$$\begin{vmatrix} a, & b \\ c, & d \end{vmatrix} \cdot \begin{vmatrix} a', & b' \\ c', & d' \end{vmatrix} = \begin{vmatrix} (a,\ b) \\ (c,\ d) \end{vmatrix} \begin{matrix} (a',\ c'), & (b',\ d') \\ ,, & ,, \\ ,, & ,, \end{matrix};$$

viz. we have

$$(ax+by+cz+dw)(a'x+b'y+c'z+d'w) = aa'x^2 + bc'yz + \&c.,$$
$$= (aa'+bc')\,x + (ab'+bd')\,y + (ca'+dc')\,z + (cb'+dd')\,w,$$

in accordance with the formula for the product of the two matrices. Observe that, writing

$$A + Bi + Cj + Dk = A\,(x+w) + B\lambda\,(x-w) + C\,(y-z) + D\lambda\,(y+z)$$
$$= ax + by + cz + dw,$$

we have

$$a,\ d,\ b,\ c = A + B\lambda, \quad A - B\lambda, \quad C + D\lambda, \quad -C + D\lambda,$$

and thence

$$ad - bc = A^2 + B^2 + C^2 + D^2,$$

so that the determinant of the matrix corresponds to the tensor of the quaternion.

837.

ON THE SO-CALLED D'ALEMBERT CARNOT GEOMETRICAL PARADOX.

[From the *Messenger of Mathematics*, vol. XIV. (1885), pp. 113, 114.]

THE present note has reference to Prof. Sylvester's paper on this subject [*l. c.*, pp. 92—96]. I cannot admit that D'Alembert and Carnot raised a well-founded objection "to the then and even now too prevalent interpretation of the meaning of the geometrical positive and negative": it appears to me that the objection was not a well-founded one.

Consider through the origin K an indefinite line $t'Kt$, and measure off from K in the sense Kt a distance equal to the positive quantity α, and let m be the extremity of the distance thus measured off. There is not in the ordinary theory any reason why the distance Km should be $=+\alpha$ rather than $=-\alpha$; it is $=+\alpha$, if Kt be the positive sense of the line through K, and it is $=-\alpha$ if Kt' be the positive sense of the line through K; if it be undetermined which of the two is the positive sense, then the distance Km is $=\pm\alpha$, the sign being essentially indeterminate.

The problem is from a point K outside a given circle to draw a line Kmm' such that the intercepted portion mm' within the circle has a given value c.

Supposing that the line from K to the centre meets the circle in the points A, B at the distances $KA = a$, $KB = b$; then if $Km = r$, we have $ab = r(c+r)$, or $r = -\frac{1}{2}c \pm \sqrt{(\frac{1}{4}c^2 + ab)}$; viz. we have for r, not simultaneously but alternatively, the positive value $-\frac{1}{2}c + \sqrt{(\frac{1}{4}c^2 + ab)}$, and the negative value $-\frac{1}{2}c - \sqrt{(\frac{1}{4}c^2 + ab)}$, the latter of these being the greatest in absolute magnitude; say the values are $+\rho_1$ and $-\rho_2$. We may with either of these values construct the point m; viz. we obtain m as one of the intersections of the given circle with the circle centre K and radius ρ_1, or else with the circle centre K and radius $-\rho_2$ (that is, radius ρ_2); and attending to the intersections on the same side of the line from K to the centre, it happens that the two points m thus determined are on one and the same line $t'Kt$; but there is no *à priori* reason why the positive senses should be the same, and they are in fact opposite to each other, in the two cases respectively; in the one case we measure off the distance ρ_1 in the sense Kt, in the other case the distance $-\rho_2$ in the sense Kt'; that is, we in fact measure off the positive distances $+\rho_1$, and $+\rho_2$, in one and the same sense Kt; thus obtaining for the point m one or the other extremity of a determinate secant through K.

The best illustration is I think in the elementary problem of finding the perpendicular distance of a given line from the origin. Let $Ax + By + C = 0$ be the equation of the given line: and first let a line be drawn *in a determinate sense*, say at the inclination θ to the positive part of the axis of x, to meet the given line. Taking r for the distance from the origin of the point of intersection, we have, for the coordinates of the point of intersection, $x, y = r\cos\theta, r\sin\theta$; and thence

$$r(A\cos\theta + B\sin\theta) + C = 0,$$

that is,

$$r = \frac{-C}{A\cos\theta + B\sin\theta},$$

a perfectly determinate value. But the perpendicular on the given line may be considered as drawn in one or the other of two opposite senses; that is, we have at pleasure

$$\cos\theta,\ \sin\theta = \frac{A}{\sqrt{(A^2+B^2)}},\ \frac{B}{\sqrt{(A^2+B^2)}},$$

or else

$$= \frac{-A}{\sqrt{(A^2+B^2)}},\ \frac{-B}{\sqrt{(A^2+B^2)}};$$

and thence $r = \dfrac{-C}{\sqrt{(A^2+B^2)}}$, or else $r = \dfrac{+C}{\sqrt{(A^2+B^2)}}$; that is, the perpendicular distance is $= \dfrac{\pm C}{\sqrt{(A^2+B^2)}}$, with the essentially indeterminate sign $\pm$, because the distance may be considered as drawn from the origin in one or the other of the two opposite senses.

838.

ON THE TWISTED CUBICS UPON A QUADRIC SURFACE.

[From the *Messenger of Mathematics*, vol. XIV. (1885), pp. 129—132.]

A CUBIC (twisted cubic) on a quadric surface meets each generator of the one kind twice, and each generator of the other kind once (Salmon, *Solid Geometry*, Ed. 4, p. 301). There are thus on the surface two kinds of cubics, viz. distinguishing for convenience the generating lines as generators and directors, a cubic may meet each generator twice and each director once; or it may meet each generator once and each director twice. And two cubic curves are accordingly of the same kind or of different kinds.

Consider for example the quadric surface $xw - yz = 0$, and the cubic

$$x : y : z : w = 1 : \phi : \phi^2 : \phi^3.$$

If we call the line $(x = ky,\ kw = z)$ a generator, and the line $(x = kz,\ kw = y)$ a director, for the intersections with a generator we have $1 = k\phi,\ k\phi^3 = \phi^2$; equations with the common root $k\phi = 1$, or there is a single intersection; for the intersections with a director, $1 = k\phi^2,\ k\phi^3 = \phi$, equations with the common roots $k\phi^2 = 1$, or there are two intersections.

Consider on the same quadric surface the cubic

$$x : y : z : w = \theta - \alpha . \theta - \beta . \theta - \gamma : \theta - \alpha . \theta - \epsilon . \theta - \zeta : \theta - \delta . \theta - \beta . \theta - \gamma : \theta - \delta . \theta - \epsilon . \theta - \zeta ;$$

or as, for shortness, I write these,

$$x : y : z : w = \alpha\beta\gamma : \alpha\epsilon\zeta : \delta\beta\gamma : \delta\epsilon\zeta,$$

viz. α is written to denote $\theta - \alpha$, and so in other cases; for the intersections with a generator, we have $\alpha\beta\gamma = k\alpha\epsilon\zeta,\ k\delta\epsilon\zeta = \delta\beta\gamma$, equations having the two common roots

$\beta\gamma = k\epsilon\zeta$; or there are two intersections with the generator. And in like manner there is one intersection with the director. The cubic

$$x : y : z : w = \alpha\beta\gamma : \alpha\epsilon\zeta : \delta\beta\gamma : \delta\epsilon\zeta$$

is thus a cubic of different kind from

$$x : y : z : w = 1 : \phi : \phi^2 : \phi^3.$$

And in the same manner it appears that the cubic

$$x : y : z : w = \alpha\beta\gamma : \delta\beta\gamma : \alpha\epsilon\zeta : \delta\epsilon\zeta$$

is of the same kind with

$$x : y : z : w = 1 : \phi : \phi^2 : \phi^3.$$

The two cubics of different kinds intersect in 5 points; the two cubics of the same kind in only 4 points. In fact, for the intersections with the cubic

$$x : y : z : w = 1 : \phi : \phi^2 : \phi^3,$$

we have $xz - y^2 = 0$, $yw - z^2 = 0$, and substituting herein the θ-values,

$$x : y : z : w = \alpha\beta\gamma : \alpha\epsilon\zeta : \delta\beta\gamma : \delta\epsilon\zeta,$$

we find

$$\alpha\delta\beta^2\gamma^2 - \alpha^2\epsilon^2\zeta^2 = 0, \quad \alpha\delta\epsilon^2\zeta^2 - \delta^2\beta^2\gamma^2 = 0,$$

equations with the common five roots $\delta\beta^2\gamma^2 - \alpha\epsilon^2\zeta^2 = 0$; whereas for the θ-values $x : y : z : w = \alpha\beta\gamma : \delta\beta\gamma : \alpha\epsilon\zeta : \delta\epsilon\zeta$, we have $\alpha^2\beta\gamma\epsilon\zeta - \delta^2\beta^2\gamma^2 = 0$, $\delta^2\beta\gamma\epsilon\zeta - \alpha^2\epsilon^2\zeta^2 = 0$, equations with the common four roots $\alpha^2\epsilon\zeta - \delta^2\beta\gamma = 0$.

I remark in passing that the equations just obtained have each of them a root $\theta = \infty$, but this would have been avoided if the θ-equations had been taken in the slightly altered form

$$x : y : z : w = \alpha\beta\gamma : b\alpha\epsilon\zeta : c\delta\beta\gamma : bc\delta\epsilon\zeta,$$

and

$$x : y : z : w = \alpha\beta\gamma : c\delta\beta\gamma : b\alpha\epsilon\zeta : bc\delta\epsilon\zeta,$$

b, c being arbitrary constants; and the special value is thus of no importance.

The two cubics intersecting in 5 points constitute the complete intersection of a quadric surface and a cubic surface; and conversely when a quadric surface and a cubic surface intersect in two cubics, then the cubics must have 5 intersections, and are thus by what precedes cubics of different kinds in regard to the quadric surface. In fact, considering two cubics meeting in 5 points, we can, through these 5 points, through 2 points assumed at pleasure on the first cubic, and through 2 points assumed at pleasure on the second cubic, draw a determinate quadric surface: this meets each of the two cubics in $5 + 2$ points, and consequently it entirely contains the cubic; viz. we have through the two cubics a determinate quadric surface. Again, through the 5 points, through 5 points at pleasure on the first cubic and through 5 points at pleasure on the second cubic, we may draw a cubic surface; this meets each cubic in $5+5$ points, and therefore it entirely contains the cubic; we have thus a cubic surface through the two cubics. The cubic surface has been subjected only to $5+5+5=15$ conditions, and the equation thus contains homogeneously $20-15$, $=5$ constants; viz. if $\Theta=0$ be the quadric surface through the two cubics the general form is $(\alpha x+\beta y+\gamma z+\delta w)\Theta+\lambda U=0$; disregarding the term in Θ, we have thus, in fact, a determinate cubic $U=0$.

Taking as above the first cubic as given by the equations

$$x : y : z : w = 1 : \phi : \phi^2 : \phi^3,$$

and the second cubic as given by the equations

$$x : y : z : w = \alpha\beta\gamma : \alpha\epsilon\zeta : \delta\beta\gamma : \delta\epsilon\zeta,$$

the equation of the cubic surface which contains the two cubics will be of the form

$$(Ax+By+Cz+Dw)(yw-z^2)+(A''x+B''y+C''z+D''w)(xz-y^2)=0.$$

Substituting herein the θ-values and omitting the common factor $\alpha\epsilon^2\zeta^2-\delta\beta^2\gamma^2$, we have

$$\delta(A\alpha\beta\gamma+B\alpha\epsilon\zeta+C\delta\beta\gamma+D\delta\epsilon\zeta)-\alpha(A''\alpha\beta\gamma+B''\alpha\epsilon\zeta+C''\delta\beta\gamma+D''\delta\epsilon\zeta)=0,$$

which is of the fourth order in θ. We may assume $C''=0$, and $D''=0$; in fact, the terms in C'' and D'' might be got rid of by the substitutions

$$z(xz-y^2)=y(xw-yz)-x(yw-z^2),$$

$$w(xz-y^2)=z(xw-yz)-y(yw-z^2);$$

we then have 6 coefficients A, B, C, D, A'', B'', and equating to zero the terms in θ^0, θ^1, θ^2, θ^3, θ^4, the ratios of these are determined by the five equations: and we have thus the required cubic surface.

Starting from either cubic considered as a curve on the given quadric surface; if through the centre we describe a cubic surface, this meets the quadric surface in a second cubic and the two cubics intersect each other in 5 points. In particular, if the one cubic is $x : y : z : w=1 : \phi : \phi^2 : \phi^3$, it appears by what precedes that the other cubic may be taken to be

$$x : y : z : w=\alpha\beta\gamma : \alpha\epsilon\zeta : \delta\beta\gamma : \delta\epsilon\zeta,$$

intersecting the former in the 5 points given by the equation $\delta\beta^2\gamma^2-\alpha\epsilon^2\zeta^2=0$. The five points are of course points of contact of the quadric surface and the cubic surface.

A very simple instance is the following. The quadric surface $xw-yz=0$ and the cubic surface

$$y(xz-y^2)+z(yw-z^2)=0$$

intersect in the two cubics

$$x : y : z : w=1 : \phi : \phi^2 : \phi^3 \text{ and } x : y : z : w=1 : \theta^2 : \theta : \theta^3.$$

The two cubics meet in the 5 points

$$\theta = (0,\ \infty,\ 1,\ \omega,\ \omega^2);\quad \phi = (0,\ \infty,\ 1,\ \omega^2,\ \omega),$$

or, what is the same thing, in the 5 points

$$(x,\ y,\ z,\ w) = (1,\ 0,\ 0,\ 0),\quad (0,\ 0,\ 0,\ 1),\quad (1,\ 1,\ 1,\ 1),\quad (1,\ \omega,\ \omega^2,\ 1),\quad (1,\ \omega^2,\ \omega,\ 1),$$

where ω is an imaginary cube root of unity.

I was anxious to work out this result not for its own sake, but as an illustration of a point which first arises as to octics: a unicursal octic curve is not in general situate on any surface lower than a quartic surface; if on the octic we take at pleasure 33 points, we may through these draw two quartic surfaces $U=0$, $V=0$, each of which will entirely contain the curve: the two surfaces meet in a second octic also unicursal, and the two curves intersect in 34 points, which are points of contact of the two quartic surfaces. We cannot, by adjoining to the given octic any curve of an inferior to its own, obtain a complete intersection of two surfaces: the most simple case is when we adjoin to the given octic as above another unicursal octic, and thus obtain a complete intersection of two quartic surfaces. It is to be observed that the second curve is determined completely and uniquely by the given curve: considering the former curve as given by the expressions of the coordinates in terms of a parameter ϕ, the determination of the like expressions for the latter curve would appear to be a problem of very great difficulty.

839.

ON THE MATRICAL EQUATION $qQ - Qq' = 0$.

[From the *Messenger of Mathematics*, vol. XIV. (1885), pp. 176—178.]

I CONSIDER the matrical equation $qQ - Qq' = 0$, where q, q' are given matrices $\begin{vmatrix} a, & b \\ c, & d \end{vmatrix}$, $\begin{vmatrix} a', & b' \\ c', & d' \end{vmatrix}$ and Q, $= \begin{vmatrix} A, & B \\ C, & D \end{vmatrix}$ is a matrix which has to be determined. As remarked in the paper "On the Quaternion Equation $qQ - Qq' = 0$," *Messenger*, t. XIV. (1885), pp. 108—112, [836] the question for matrices is equivalent to that for quaternions: for a matrix $\begin{vmatrix} a, & b \\ c, & d \end{vmatrix}$ may be regarded as a quaternion

$$\tfrac{1}{2}a\,(1 - \lambda i) + \tfrac{1}{2}b\,(j + \lambda k) + \tfrac{1}{2}c\,(-j - \lambda k) + \tfrac{1}{2}d\,(1 + \lambda i),$$

or (omitting the factor $\tfrac{1}{2}$) as the quaternion

$$(a + d) - \lambda\,(a - d)\,i + (b - c)\,j - \lambda\,(b + c)\,k,$$

where λ, $= \sqrt{(-1)}$, is the imaginary of ordinary algebra. Hence considering q, q' as denoting the quaternions

$$(a + d) - \lambda\,(a - d)\,i + (b - c)\,j - \lambda\,(b + c)\,k,$$
$$(a' + d') - \lambda\,(a' - d')\,i + (b' - c')\,j - \lambda\,(b' + c')\,k,$$

we can, if a certain condition is satisfied, find a quaternion Q such that $qQ - Qq' = 0$; say this is $Q = W + iX + jY + kZ$; putting this

$$= \tfrac{1}{2}\,\{(A + D) - \lambda\,(A - D)\,i + (B - C)\,j - \lambda\,(B + C)\,k\},$$

we find

$$Q = \begin{vmatrix} A, & B \\ C, & D \end{vmatrix}, \quad = \begin{vmatrix} W + \lambda X, & Y + \lambda Z \\ -Y + \lambda Z, & W - \lambda X \end{vmatrix},$$

for the required matrix Q; this being an indeterminate matrix, such that $AD - BC = 0$.

But it is better to solve directly the matrical equation

$$\begin{vmatrix} a, & b \\ c, & d \end{vmatrix} \begin{vmatrix} A, & B \\ C, & D \end{vmatrix} - \begin{vmatrix} A, & B \\ C, & D \end{vmatrix} \begin{vmatrix} a', & b' \\ c', & d' \end{vmatrix} = 0,$$

viz.

$$\begin{array}{r|cc} & (A, C), & (B, D) \\ \hline (a, b) & ,, & ,, \\ (c, d) & ,, & ,, \end{array} - \begin{array}{r|cc} & (a', c'), & (b', d') \\ \hline (A, B) & ,, & ,, \\ (C, D) & ,, & ,, \end{array} = 0,$$

that is,

$$\begin{aligned} Aa + Cb - (Aa' + Bc') &= 0, \\ Ba + Db - (Ab' + Bd') &= 0, \\ Ac + Cd - (Ca' + Dc') &= 0, \\ Bc + Dd - (Cb' + Dd') &= 0, \end{aligned}$$

or, what is the same thing,

$$\left(\begin{array}{cccc} a-a', & -c', & b, & 0, \\ -b', & a-d', & 0, & b, \\ c, & 0, & d-a', & -c', \\ 0, & c, & -b', & d-d', \end{array}\right)\!\!\!\;\Big\backslash\!\!\!\;(A, B, C, D)=0,$$

viz. we have (A, B, C, D) connected by these four linear equations: viz. the necessary condition is that the determinant formed out of the matrix which here presents itself shall be $=0$.

After some reductions, and putting for shortness

$$\nabla = ad - bc, \quad \nabla' = a'd' - b'c',$$

this is found to be

$$(\nabla - \nabla')^2 + \{\nabla(a'+d') - \nabla'(a+d)\}(a'+d'-a-d) = 0;$$

which is the condition for the existence of a solution. This condition being satisfied, the four equations will be equivalent to three independent equations, which serve to determine A, B, C, D; and, assuming the absolute value of A, we find

$$\begin{aligned} A &= \qquad\qquad\quad -(a'+d'-a-d), \\ c'B &= \nabla - \nabla' + a'(a'+d'-a-d), \\ bC &= \nabla - \nabla' + a\ (a'+d'-a-d), \\ bc'D &= (d'-a)\nabla - (d-a')\nabla' - aa'(a'+d'-a-d), \end{aligned}$$

values which give

$$-bc'(AD-BC) = (\nabla - \nabla')^2 + \{\nabla(a'+d') - \nabla'(a+d)\}(a'+d'-a-d), \ =0,$$

viz. in the case where the given matrices satisfy the above-mentioned condition, the components A, B, C, D of the required matrix have determinate values which are such that $AD-BC=0$.

If we have $\nabla-\nabla'=0$, that is, $ad-bc=a'd'-b'c'$, then the condition becomes $a'+d'=a+d$; and these two conditions being satisfied, the four equations reduce themselves to two independent equations, and the ratios $A:B:C:D$ have no longer determinate values; the linear equations may in fact be written

$$\left(\begin{array}{cccc} 0, & b, & -c', & a-a' \\ -b, & 0, & -a+d', & b' \\ c', & a'-d, & 0, & -c \\ d-d', & -b', & c, & 0 \end{array}\right)(D,\ C,\ B,\ A)=0,$$

which are of the form

$$\left(\begin{array}{cccc} 0, & \mathrm{h}, & -\mathrm{g}, & \mathrm{a} \\ -\mathrm{h}, & 0, & \mathrm{f}, & \mathrm{b} \\ \mathrm{g}, & -\mathrm{f}, & 0, & \mathrm{c} \\ -\mathrm{a}, & -\mathrm{b}, & -\mathrm{c}, & 0 \end{array}\right)(D,\ C,\ B,\ A)=0,$$

where the coefficients a, b, c, f, g, h are such that $\mathrm{af}+\mathrm{bg}+\mathrm{ch}=0$; and the four equations are thus equivalent to two independent equations. To obtain a symmetrical solution, assume a relation

$$D\xi+C\eta+B\zeta+A\omega=0,$$

with arbitrary coefficients $\xi,\ \eta,\ \zeta,\ \omega$; we then find

$$\begin{aligned} D &= \quad . \quad \mathrm{c}\eta-\mathrm{b}\zeta+\mathrm{f}\omega, \\ C &= -\mathrm{c}\xi \quad . \quad +\mathrm{a}\zeta+\mathrm{g}\omega, \\ B &= \mathrm{b}\xi-\mathrm{a}\eta \quad . \quad +\mathrm{h}\omega, \\ A &= -\mathrm{f}\xi-\mathrm{g}\eta-\mathrm{h}\zeta \quad . \ , \end{aligned}$$

viz. the complete theorem is that, if the matrices

$$q=\left|\begin{array}{cc} a, & b \\ c, & d \end{array}\right|, \quad q'=\left|\begin{array}{cc} a', & b' \\ c', & d' \end{array}\right|$$

are such that $ad-bc=a'd'-b'c'$, $a+d=a'+d'$, then writing

$$\begin{array}{ll} \mathrm{a}=\ a-a', =-d+d', & \mathrm{f}=-a+d', =-a'+d, \\ \mathrm{b}=\ b', & \mathrm{g}=-c', \\ \mathrm{c}=-c, & \mathrm{h}=\ b, \end{array}$$

values which satisfy

$$\mathrm{af}+\mathrm{bg}+\mathrm{ch}=0,$$

and taking $\xi,\ \eta,\ \zeta,\ \omega$ arbitrary, we have for the matrix Q, which is such that $Qq-Qq'=0$, the expression

$$Q=\left|\begin{array}{cc} -\mathrm{f}\xi-\mathrm{g}\eta-\mathrm{h}\zeta, & \mathrm{b}\xi-\mathrm{a}\eta \quad . \quad +\mathrm{h}\omega \\ -\mathrm{c}\xi \quad . \quad +\mathrm{a}\zeta+\mathrm{g}\omega, & \mathrm{c}\eta-\mathrm{b}\zeta+\mathrm{f}\omega \end{array}\right|,$$

depending really on one arbitrary parameter, viz. we may without loss of generality put any two of the coefficients $\xi,\ \eta,\ \zeta,\ \omega$ equal to 0.

840.

ON MASCHERONI'S GEOMETRY OF THE COMPASS.

[From the *Messenger of Mathematics*, vol. XIV. (1885), pp. 179—181.]

I HAVE not seen the original of Mascheroni's work, *La Geometria del Compasso*, 8°, Pavia (1797), but only the French translation, *Géométrie du Compas*, par L. Mascheroni, traduite de l'Italien par A. M. Carette, 2 ed. 8° Paris, 1828 (Author's Preface, pp. 5—24, and pp. 25—328, with 14 plates containing 108 figures). The title expresses the notion of the work: a straight line is given by means of two points thereof, but is not allowed to be actually drawn; and the problem is, with the compass alone to perform all the constructions of Geometry. Observe that, for the purpose of demonstration, any lines may be imagined to be drawn, and such lines are in fact shown as dotted lines in the figures; but this does not in any wise interfere with the fundamental postulate that the constructions are to be performed with the compass only.

Assuming, then, that a line is in every case given by means of two points thereof, the leading questions are those considered Book VII. "on the intersections of straight lines with circular arcs and with each other," viz. they are (1) to find the intersections of a given line and circle; (2) to find the intersection of two given lines. But in the first problem it is necessary to consider two cases; (*A*) the general case, where the line does not pass through the centre of the circle, and (*B*) the particular, but actually more difficult, case, where the line passes through the centre of the circle. It is assumed that the centre of the circle is given: if it is not given, it can be found as afterwards mentioned.

It will be convenient to establish the definition of counter-points in regard to a given line: two points, which are such that the line joining them is bisected at right angles by the given line, are said to be counter-points in regard to that line.

(1), *A*. Consider a line given by means of two points *P*, *Q*; and a given circle, centre *C*. With centre *P* and radius *PC* describe a circle, and with centre *Q* and radius QC a circle; these meet in C and in a second point D which will be the counter-point of C in regard to the line PQ; hence with centre D and radius = that of given circle, describing a circle, this meets the given circle in two points which are the intersections of the given circle by the line PQ.

The construction fails if the line PQ passes through the centre of the given circle, for then the two auxiliary circles touch at C, or we have D coincident with C. We have therefore considered this case; Q may be taken to be the centre C of the circle; so that we have:—

(1), *B*. The problem is, given a point P, and a circle, centre C: to find the intersections of the circle by the line CP. With centre P, and an arbitrary radius, describe a circle meeting the given circle in points A, B (which are of course counter-points in regard to line CP); the circle is divided into two arcs AB and $360° - AB$, and the problem is to bisect each of these two arcs; for the points of bisection are obviously the required intersections of CP with the circle.

Hence we have:—

(1), *C*. To bisect a given arc AB of a circle, centre C (fig. 1). With centre A and radius AC describe a circle; and with centre B, and equal radius BC describe a circle; and on these circles respectively, find the points D, E such that $CD = CE = AB$.

Fig. 1.

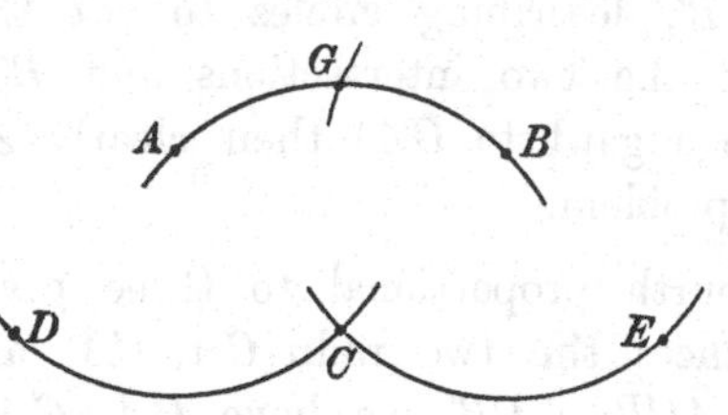

With centre E and radius EA, describe a circle; and with centre D, and equal radius DB, describe a circle; and let these two circles meet in F; then with centre E (or D) and radius CF describe a circle: this will meet the given circle in the required middle point G of the arc AB.

The proof depends on the theorem that in a rhombus the sum of the squares of the diagonals is = sum of the squares of the four side. We have a rhombus $ACEB$, and therefore

$$(AE)^2 + (BC)^2 = 2(AB)^2 + 2(AC)^2,$$

that is,

$$(AE)^2 = 2(AB)^2 + \text{rad.}^2.$$

But by construction, $EF^2 = (AE)^2$; therefore

$$EF^2 = 2(AB)^2 + \text{rad.}^2, \quad = 2(CE)^2 + \text{rad.}^2.$$

Hence in the right-angled triangle FCE, we have

$$(CF)^2 = (EF)^2 - CE^2, \quad = (CE)^2 + \text{rad.}^2;$$

and by construction, $(EG)^2 = (CF)^2$; that is,

$$(EG)^2 = (CE)^2 + \text{rad.}^2, \ = (CE)^2 + (CG)^2;$$

viz. CEG is a right-angled triangle; that is, CG is at right angles to BCE, or the line CGF bisects the arc AB in G.

For the second problem, (2), to find the intersection of two given lines, we require the solution of the problem to find a fourth proportional to three given distances, and this immediately depends on the following problem.

Problem. In two concentric circles to place chords subtending equal angles. If AB (fig. 2) be a chord of the one circle, then with centres A and B respectively, and

Fig. 2.

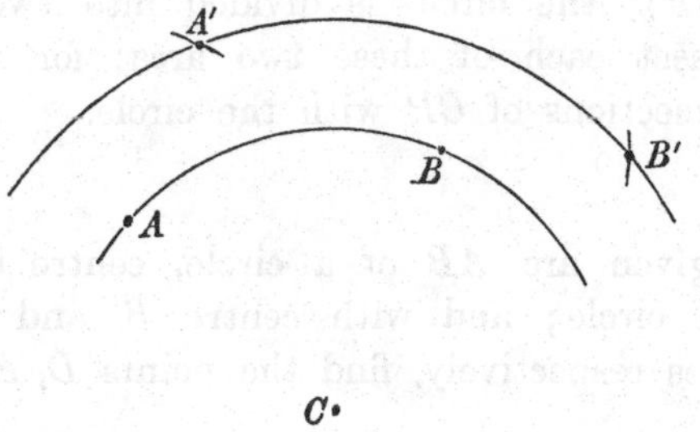

an arbitrary radius $AA' = BB'$, describing circles to cut the larger circle in the points A' and B' (A' is either of the two intersections and B' is the intersection lying in regard to CB as A' lies in regard to CA) then clearly $\angle ACB = \angle A'CB'$. And hence also we have the following problem.

Problem. To find a fourth proportional to three given distances. We have only to take as the given distances the two radii CA, CA' and the chord AB; and then from the similar triangles ACB, $A'CB'$, we have $CA : CA' :: AB : A'B'$, viz. $A'B'$ is a fourth proportional to CA, CA', AB.

We have now the solution of

(2) To construct the intersection of two given lines AB and CD (fig. 3). Find

Fig. 3.

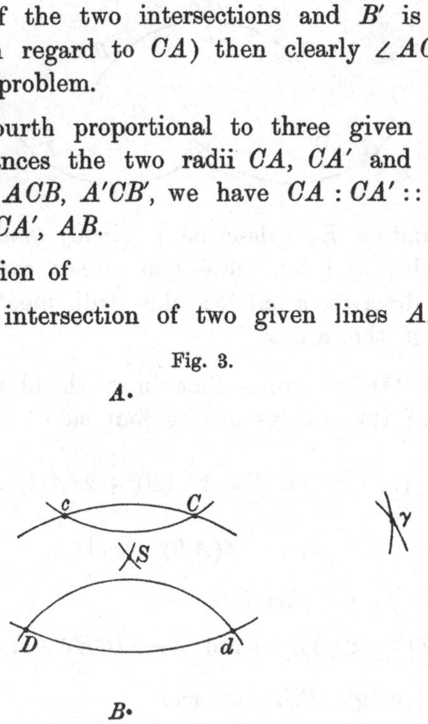

c the counter-point of C, and d the counter-point of D, in regard to the line AB:

and find γ such that the distances $C\gamma$, $d\gamma$ are $= Dd$, DC respectively; γ is in a line with cC, and we have $C\gamma dD$ a parallelogram. Find cS a fourth proportional to $c\gamma$, cd, cC, and with centres c, C respectively and radii each $= cS$ describe circles cutting in the point S; this will be the required intersection of the two lines. In fact, the required point S will be the intersection of the two lines CD, cd: supposing these lines each of them drawn, and also the lines $cC\gamma$ and $d\gamma$, we have DC parallel to $d\gamma$, that is, the triangles $cd\gamma$, cSC are similar to each other or $c\gamma : cd :: cC : cS$: viz. the distance cS having been found by this proportion, and the point S found as the intersection of the two circles, centres c and C respectively, the point S so determined is the required point of intersection of the given lines AB and CD.

If a circle be given without its centre being known, then taking any three points A, B, C on the circle, and a pair of counter-points D, E of the line AB, and also a pair of counter-points F, G of the line AC, we have obviously the centre of the given circle as the intersection of the lines DE and FG; and the centre can thus be found with the compass only.

It is proper to remark that the problems considered in the present paper are those connecting the theory with ordinary geometry, not the problems which are most readily and elegantly solved with the compass only: a large collection of these are contained in the work, and in particular the twelfth book contains some interesting *approximate* solutions of the problems of the quadrature of the circle, the duplication of the cube, and other problems not solvable by ordinary geometry.

Cambridge, March 19, 1885.

841.

ON A DIFFERENTIAL OPERATOR.

[From the *Messenger of Mathematics*, vol. XIV. (1885), pp. 190, 191.]

WRITE $X = 1 + bx + cx^2 + \ldots, = (1-\alpha x)(1-\beta x)(1-\gamma x)\ldots$; then by Capt. MacMahon's theorem, any non-unitary function of the roots $\alpha, \beta, \gamma, \ldots$ is reduced to zero by the operation

$$\Delta, = \partial_b + b\partial_c + c\partial_d + \ldots;$$

for instance, if

$$(2), = \Sigma\alpha^2 = b^2 - 2c,$$

we have

$$\Delta(b^2 - 2c) = 2b + b(-2), = 0.$$

We have

$$\Delta X = x + bx^2 + cx^3 + \ldots = xX;$$

and writing, moreover, $X', = b + 2cx + 3dx^2 + \&c.$, for the derived function of X, then

$$\Delta X' = 1 + 2bx + 3cx^2 + \ldots = (xX)'.$$

We can hence shew that $\Delta\left(\dfrac{X'}{X} - b\right) = 0$; the value is, in fact,

$$\frac{\Delta X'}{X} - \frac{X'\Delta X}{X^2} - \Delta b, \text{ that is, } \frac{(xX)'}{X} - \frac{X'xX}{X^2} - 1,$$

which is

$$= \frac{X + xX'}{X} - \frac{xX'}{X} - 1, = 0.$$

This is right, for $\dfrac{X'}{X}$ is a sum of non-unitary symmetric functions of the roots; in fact,

$$\frac{X'}{X} = \Sigma\frac{-\alpha}{1-\alpha x} = -(1) - (2)x - (3)x^2 - \&c.,$$

or since $b = -(1)$, this is

$$\frac{X'}{X} - b = -(2)x - (3)x^2 - \&c.,$$

a sum of non-unitary functions of the roots.

842.

ON THE VALUE OF $\tan(\sin\theta)-\sin(\tan\theta)$.

[From the *Messenger of Mathematics*, vol. XIV. (1885), pp. 191, 192.]

THE following equation is given p. 59 of the *Lady's and Gentleman's Diary* for 1853:

$$\tan(\sin\theta)-\sin\tan\theta=\tfrac{1}{30}\theta^7+\tfrac{29}{756}\theta^9+\&c.$$

Write in general

$$X=\theta+A\theta^3+B\theta^5+C\theta^7+D\theta^9+\ldots,$$

$$Y=\theta+A'\theta^3+B'\theta^5+C'\theta^7+D'\theta^9+\ldots.$$

Then, as far as θ^9, we have

$$\begin{aligned}
&X+A'X^3+B'X^5+C'X^7+D'X^9\\
=\ &\theta+A\theta^3+B\theta^5+C\theta^7+D\theta^9\\
&+A'\{\theta^3+3A\theta^5+3(A^2+B)\theta^7+(A^3+6AB+3C)\theta^9\}\\
&+B'\{\theta^5+5A\theta^7+(10A^2+5B)\theta^9\}\\
&+C'\{\theta^7+7A\theta^9\}\\
&+D'\{\theta^9\}\\
=\ &+\theta\\
&+\theta^3(A+A')\\
&+\theta^5(B+3AA'+B')\\
&+\theta^7(C+3A^2A'+3A'B+5AB'+C')\\
&+\theta^9(D+A^3A'+6AA'B+3A'C+10A^2B'+5BB'+7AC+D'),
\end{aligned}$$

and hence

$$X + A'X^3 + B'X^5 + C'X^7 + D'X^9$$
$$- Y - AY^3 - BY^5 - CY^7 - DY^9$$
$$= \theta^7 \{3AA'(A - A') + 2(AB' - A'B)\}$$
$$+ \theta^9 \{AA'(A^2 - A'^2) + 6AA'(B - B') + 4(AC' - A'C) + 10(A^2B' - A'^2B)\}.$$

Now let

$$X = \sin\theta = \tfrac{1}{6}\theta^3 + \tfrac{1}{120}\theta^5 - \tfrac{1}{5040}\theta^7 + \ldots; \quad A = -\tfrac{1}{6},\ B = \tfrac{1}{120},\ C = -\tfrac{1}{5040},$$
$$Y = \tan\theta = \tfrac{1}{3}\theta^3 + \tfrac{2}{15}\theta^5 + \tfrac{17}{315}\theta^7 - \ldots; \quad A' = \tfrac{1}{3},\ B' = \tfrac{2}{15},\ C' = \tfrac{17}{315};$$

we have therefore

$$AA' = -\tfrac{1}{18},\quad A - A' = -\tfrac{1}{2},\quad A + A' = \tfrac{1}{6},\quad B - B' = -\tfrac{1}{8},$$
$$AB' - A'B = -\tfrac{1}{40},\quad AC' - A'C = -\tfrac{1}{112},\quad A^2B' - A'^2B = \tfrac{1}{360}.$$

Hence

$$\text{coeff. } \theta^7 = \tfrac{1}{12} - \tfrac{1}{20},\ = \tfrac{1}{30},$$
$$\text{coeff. } \theta^9 = \tfrac{1}{216} + \tfrac{1}{24} - \tfrac{1}{28} + \tfrac{1}{36},\ = \tfrac{29}{756};$$

and the required equation is thus verified.

843.

ON THE QUADRIQUADRIC CURVE IN CONNEXION WITH THE THEORY OF ELLIPTIC FUNCTIONS.

[From the *Mathematische Annalen*, t. XXV. (1885), pp. 152—156.]

I CALL to mind that, if we have on a line two points $(\alpha, \beta, \gamma, \delta)$, $(\alpha', \beta', \gamma', \delta')$, and through the line two planes

$$Ax + By + Cz + Dw = 0,\quad A'x + B'y + C'z + D'w = 0,$$

then

$$\begin{aligned}&\beta\gamma' - \beta'\gamma : \gamma\alpha' - \gamma'\alpha : \alpha\beta' - \alpha'\beta : \alpha\delta' - \alpha'\delta : \beta\delta' - \beta'\delta : \gamma\delta' - \gamma'\delta\\ &= AD' - A'D : BD' - B'D : CD' - C'D : BC' - B'C : CA' - C'A : AB' - A'B;\end{aligned}$$

and that, putting each of these two equal sets of ratios

$$= a : b : c : f : g : h,$$

then the quantities (a, b, c, f, g, h), which it is easy to see satisfy the relation $af + bg + ch = 0$, are said to be the "six coordinates" of the line: as only the ratios of the six quantities are material, and as the last-mentioned equation establishes a single relation between these ratios, the system of the six coordinates contains four arbitrary ratios or parameters, for the determination of the particular line. See my paper "On the six coordinates of a line," *Camb. Phil. Trans.* t. XI. (1869), pp. 290—323, [435].

I consider for a moment the quadric surface

$$x^2 + y^2 + z^2 + w^2 = 0,$$

and I proceed to show that, if (a, b, c, f, g, h) are the coordinates of a generating line on the surface, then either $a = f$, $b = g$, $c = h$, or else $a = -f$, $b = -g$, $c = -h$; the one or other system of equations according as the line belongs to the one or other system of generating lines.

We satisfy the equation by

$$x + iy + \theta z + \theta iw = 0,$$

$$x - iy - \frac{1}{\theta} z + \frac{1}{\theta} iw = 0,$$

where θ is an arbitrary parameter: hence these equations determine a generating line of the surface: and the coordinates of this line are

$$\begin{array}{cccccc} a, & b, & c, & f, & g, & h \\ = -i\left(\theta - \frac{1}{\theta}\right), & -\left(\theta + \frac{1}{\theta}\right), & zi, & i\left(\theta - \frac{1}{\theta}\right), & \theta + \frac{1}{\theta}, & -zi; \end{array}$$

viz. these values give $a = -f$, $b = -g$, $c = -h$.

Similarly we satisfy the equation by

$$x + iy + \phi z - \phi iw = 0,$$

$$x - iy - \frac{1}{\phi} z - \frac{1}{\phi} iw = 0,$$

where ϕ is an arbitrary parameter: hence these equations also determine a generating line of the surface: and the coordinates of this line are

$$\begin{array}{cccccc} a, & b, & c, & f, & g, & h \\ = i\left(\phi - \frac{1}{\phi}\right), & \phi + \frac{1}{\phi}, & -zi, & i\left(\phi + \frac{1}{\phi}\right), & \phi + \frac{1}{\phi}, & -zi; \end{array}$$

viz. these values give $a = f$, $b = g$, $c = h$.

If for x, y, z, w we write $x\sqrt{p}$, $y\sqrt{q}$, $z\sqrt{r}$, $w\sqrt{s}$ respectively, then we have the theorem that, for the quadric surface

$$px^2 + qy^2 + rz^2 + sw^2 = 0,$$

the coordinates (a, b, c, f, g, h) of a generating line are such that

$$\frac{a}{f} = \pm\sqrt{\frac{ps}{qr}}, \quad \frac{b}{g} = \pm\sqrt{\frac{qs}{rp}}, \quad \frac{c}{h} = \pm\sqrt{\frac{rs}{pq}},$$

the signs being all + or all −, according as the line belongs to one or other of the systems of generating lines.

Take (a', b', c', f', g', h') for the coordinates of an arbitrary line, and write $p, q, r, s = a'g'h', b'h'f', c'f'g', a'b'c'$; the quadric surface is

$$a'g'h'x^2 + b'h'f'y^2 + c'f'g'z^2 + a'b'c'w^2 = 0,$$

which is a surface having the line (a', b', c', f', g', h') for a generating line. To verify this, observe that for the line in question we have

$$\begin{aligned} & h'y - g'z + a'w = 0, \\ -h'x \quad . \quad & + f'z + b'w = 0, \\ g'x + f'y \quad . \quad & + c'w = 0, \\ -a'x - b'y - c'z \quad . \quad & = 0, \end{aligned}$$

equivalent of course to two independent equations: these give $h'x = f'z + b'w$, $h'y = g'z - a'w$, values which substituted in the quadric equation satisfy it identically. And for the last-mentioned quadric surface we have the theorem that, if (a, b, c, f, g, h) are the coordinates of a generating line, then

$$\frac{a}{f} = \pm\frac{a'}{f'},\quad \frac{b}{g} = \pm\frac{b'}{g'},\quad \frac{c}{h} = \pm\frac{c'}{h'},$$

where obviously the sign + belongs to a generating line of the same system with the line (a', b', c', f', g', h'), and the sign − to a line of the other system.

Taking the sign −, we thus see that if (a, b, c, f, g, h), (a', b', c', f', g', h') are the coordinates of lines of the two systems respectively, then

$$af' + a'f = 0,\quad bg' + b'g = 0,\quad ch' + c'h = 0;$$

where observe that the resulting equation

$$af' + a'f + bg' + b'g + ch' + c'h = 0,$$

is the condition which expresses that the two lines meet each other.

Consider now the quadriquadric curve

$$\begin{aligned} U_1 &= Ax^2 + By^2 + Cz^2 + Dw^2 = 0, \\ U_1' &= A'x^2 + B'y^2 + C'z^2 + D'w^2 = 0; \end{aligned}$$

and let (a, b, c, f, g, h) and (a', b', c', f', g', h') be the coordinates of two lines meeting each other, and each meeting the quadric curve twice: or, again, let these be lines joining in pairs the four intersections of the curve by an arbitrary plane: or, again, let them be the nodal lines of the binodal quartic cone having an arbitrary vertex and passing through the curve. The two lines are generating lines, belonging to the two systems respectively, of a properly determined quadric surface $U + \lambda U' = 0$ passing through the curve: and by what precedes, we have

$$af' + a'f = 0,\quad bg' + b'g = 0,\quad ch' + c'h = 0,$$

the fundamental theorem which I wished to establish.

Writing in the equations $w = 1$, we have in particular the quadriquadric curve $y^2 = 1 - x^2$, $z^2 = 1 - k^2x^2$, equations which are satisfied by $x = \operatorname{sn} u$, $y = \operatorname{cn} u$, $z = \operatorname{dn} u$. Consider on the curve four points belonging to the arguments u_1, u_2, u_3, u_4 respectively;

and write for shortness s_1, c_1, d_1 for the sn, cn, and dn of u_1; and similarly for u_2, u_3 and u_4. It appears by Abel's theorem that the condition in order that the four points may be in a plane is $u_1+u_2+u_3+u_4=0$; viz. when this equation is satisfied we have

$$\begin{vmatrix} s_1, & c_1, & d_1, & 1 \\ s_2, & c_2, & d_2, & 1 \\ s_3, & c_3, & d_3, & 1 \\ s_4, & c_4, & d_4, & 1 \end{vmatrix},$$

or writing

$$\begin{array}{cccccc} a, & b, & c, & f, & g, & h \\ =c_1d_2-c_2d_1, & d_1s_2-d_2s_1, & s_1c_2-s_2c_1, & s_1-s_2, & c_1-c_2, & d_1-d_2, \\ a', & b', & c', & f', & g', & h' \\ =c_3d_4-c_4d_3, & d_3s_4-d_4s_3, & s_3c_4-s_4c_3, & s_3-s_4, & c_3-c_4, & d_3-d_4, \end{array}$$

this equation is

$$af'+a'f+bg'+b'g+ch'+c'h=0;$$

or by what precedes, it appears that not only is this so, but that we have separately

$$af'+a'f=0, \quad bg'+b'g=0, \quad ch'+c'h=0,$$

viz. it follows from Abel's theorem that, when the arguments u_1, u_2, u_3, u_4 are connected by the equation $u_1+u_2+u_3+u_4=0$, then each of these three equations holds good.

I assume in particular $u_4=0$, $u_3=-u$, so that $u=u_1+u_2$; we have

$$\begin{array}{cccccc} a, & b, & c, & f, & g, & h \\ =c_1d_2-c_2d_1, & d_1s_2-d_2s_1, & s_1c_2-s_2c_1, & s_1-s_2, & c_1-c_2, & d_1-d_2, \\ a', & b', & c', & f', & g', & h' \\ =\quad c-d, & s, & -s, & -s, & c-1, & d-1, \end{array}$$

and the three equations become

$$\frac{c-d}{s}=\frac{c_1d_2-c_2d_1}{s_1-s_2}, \quad \frac{-s}{c-1}=\frac{d_1s_2-d_2s_1}{c_1-c_2}, \quad \frac{s}{d-1}=\frac{s_1c_2-s_2c_1}{d_1-d_2},$$

equations which may also be written

$$\frac{c-d}{s}=\frac{c_1d_2-c_2d_1}{s_1-s_2}, \quad \frac{-c+1}{s}=\frac{c_1-c_2}{d_1s_2-d_2s_1}, \quad \frac{d-1}{s}=\frac{d_1-d_2}{s_1c_2-s_2c_1},$$

so that, adding these three equations, we must have an identity.

Representing the second and third of the last-mentioned equations by

$$\frac{-c+1}{s}=C, \quad \frac{d-1}{s}=D,$$

we have

$$c = 1 - Cs, \quad d = 1 + Ds,$$

from either of which we can obtain s rationally; viz. the first equation gives

$$1 - s^2 = 1 - 2Cs + C^2s^2,$$

that is, $s = \frac{2C}{1 + C^2}$, whence

$$c = \frac{1 - C^2}{1 + C^2}, \quad d = \frac{1 + C^2 + 2CD}{1 + C^2};$$

and similarly from the second equation $s = -\frac{2D}{k^2 + D^2}$, whence also

$$c = \frac{k^2 + D^2 - 2CD}{k^2 + D^2}, \quad d = \frac{k^2 - D^2}{k^2 + D^2}.$$

Substituting for C its value, the first-mentioned value of s becomes

$$s = \frac{(c_1 - c_2)(d_1 s_2 - d_2 s_1)}{1 - k^2 s_1^2 s_2^2 - c_1 c_2 - d_1 s_1 d_2 s_2},$$

which is a form that can be easily verified: we in fact have

$$s_1 = \text{sn}(u_1 + u_2) = \frac{s_1 c_2 d_2 + s_2 c_1 d_1}{1 - k^2 s_1^2 s_2^2}, \quad = \frac{s_1^2 - s_2^2}{s_1 c_2 d_2 - s_2 c_1 d_1}, \quad = \frac{s_1 c_1 d_2 + s_2 c_2 d_1}{c_1 c_2 + s_1 d_1 s_2 d_2}, \quad = \frac{s_1 d_1 c_2 + s_2 d_2 c_1}{d_1 d_2 + k^2 s_1 c_1 s_2 c_2},$$

see my *Elliptic Functions*, p. 63: or calling these values

$$= \frac{N_1}{D_1} = \frac{N_2}{D_2} = \frac{N_3}{D_3} = \frac{N_4}{D_4},$$

the above value is

$$s = \frac{N_1 - N_3}{D_1 - D_3},$$

which is right.

Cambridge, 28 *June*, 1884.

844.

ON A THEOREM RELATING TO SEMINVARIANTS.

[From the *Quarterly Journal of Pure and Applied Mathematics*, vol. XX. (1885), pp. 212, 213.]

I FIND among my papers the following example of a general theorem relating to seminvariants.

Write

$$\Theta = b\partial_a + c\partial_b + \ d\partial_c + \ e\partial_d + \ f\partial_e + \ g\partial_f + \ h\partial_g + \ i\partial_h,$$
$$\Omega = \qquad c\partial_b + 2d\partial_c + 3e\partial_d + 4f\partial_e + 5g\partial_f + 6h\partial_g + 7i\partial_h,$$

then from a seminvariant S of the degree δ and weight w, operating upon it with $\Theta - \delta b$ and $\Omega - 2wb$, we derive two seminvariants, each of them of the degree $\delta + 1$ and weight $w + 1$; we obtain a combination of these by operating on S with $2w(\Theta - \delta b) - \delta(\Omega - 2wb)$, that is, with $2w\Theta - \delta\Omega$, and this combination is of the weight $w + 1$, but only of the degree δ; viz. operating with $2w\Theta - \delta\Omega$, we obtain a new seminvariant of the same degree but of a weight increased by unity.

The example relates to the under-mentioned seminvariant S of the degree 3 and weight 7:

	$S=$
h	+ 1
bg	− 7
cf	+ 9
de	− 5
b^2f	+ 12
bce	− 30
bd^2	+ 20
	± 42

	ΘS	$-3bS$	Sum	ΩS	$-14bS$	Sum	$14\Theta S$	$-3\Omega S$			$\div 110$
i	1		1	+ 7		+ 7	+ 14	− 21	+ 7	0	
bh	− 5	− 3	− 8	− 42	− 14	− 56	− 70	+ 126	− 56	0	
cg	+ 2		+ 2	+ 38		+ 38	+ 28	− 114	+ 196	+ 110	+ 1
df	+ 4		+ 4	− 2		− 2	+ 56	+ 6	− 392	− 330	− 3
e^2	− 5		− 5	− 15		− 15	− 70	+ 45	+ 245	+ 220	+ 2
b^2g	+ 5	+ 21	+ 26	+ 60	+ 98	+ 158	+ 70	− 180		− 110	− 1
bcf	+ 3	− 27	− 24	− 96	− 126	− 222	+ 42	+ 288		+ 330	+ 3
bde	+ 5	+ 15	+ 20	+ 60	+ 70	+ 130	+ 70	− 180		− 110	− 1
c^2e	− 30		− 30	− 30		− 30	− 420	+ 90		− 330	− 3
cd^2	+ 20		+ 20	+ 20		+ 20	+ 280	− 60		+ 220	+ 2
b^3f		− 36	− 36		− 168	− 168					
bce		+ 90	+ 90		+ 420	+ 420					
b^2d^2		− 60	− 60		− 280	− 280					
	± 40	± 126	± 163	± 185	± 588	± 773	± 560	± 555	± 448	± 880	± 8

The columns show ΘS (where observe that, to operate with the $b\partial_a$ of Θ, we restore the proper power of a, reading S as being $= +1a^2h - 7abg + \&c.$, and putting ultimately $a = 1$) and $-3bS$, and the sum of these, which is a seminvariant, degree 4, weight 8; also ΩS and $-14bS$, and the sum of these, which is a seminvariant, degree 4, weight 8. They also show $14\Theta S$ and $-3\Omega S$, the sum of which would be a seminvariant, degree 3, and weight 8; instead of giving this sum, I have added a column equal to $+7(i - 8bh + 28cg - 56df + 35e^2)$, and given the sum of the three columns which will of course be a seminvariant of the same degree and weight; the coefficients contain all of them the factor 110, and, throwing this out, we have in the last column the seminvariant $cg - 3df + \ldots + 2cd^2$ of the degree 3 and weight 8, derived by the foregoing direct process from the given seminvariant S of the degree 3 and weight 7.

845.

ON THE ORTHOMORPHOSIS OF THE CIRCLE INTO THE PARABOLA.

[From the *Quarterly Journal of Pure and Applied Mathematics*, vol. XX. (1885), pp. 213—220.]

It is remarked by Schwarz (see his Memoir "Ueber einige Abbildungsaufgaben," *Crelle*, t. LXX. (1869), pp. 105—120; p. 115), that the circle $x'^2+y'^2-1=0$ can be orthomorphosed into the parabola $y^2=4(1-x)$ by the equation

$$\sqrt{(x'+iy')}=\tan\tfrac{1}{4}\pi\sqrt{(x+iy)}:$$

viz. this equation establishes a (1, 1) relation between the points interior to the circle and those interior to the parabola, which, quà relation between $x'+iy'$ and $x+iy$, will be orthomorphic, that is, infinitesimal elements of the one area will correspond to similar infinitesimal elements of the other area. The diameter $y'=0$ of the circle is trans-

Fig. 1.

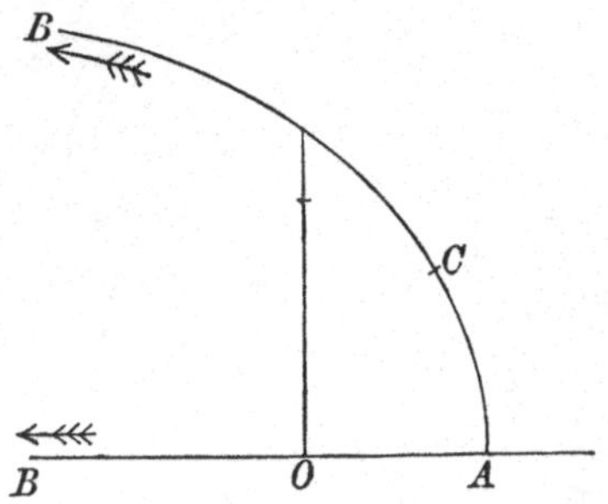

Fig. 2.

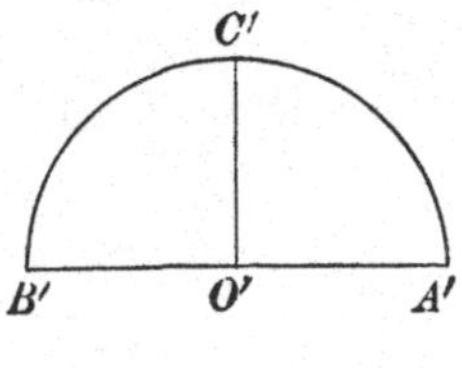

formed into the axis $y=0$ of the parabola, and figs. 1, 2 are symmetrical on the two sides of these lines respectively; we may therefore consider only the transformation of the upper semicircle into the upper half of the parabola; we have (see figs. 1 and 2) A' corresponding to the vertex A, O' to the focus O, and B' to the point at infinity ($x = -\infty$) on the axis of the parabola; the semicircular arc $A'C'B'$ corresponds to the infinite half-arc ACB of the parabola, the highest point C' corresponding to a point C between the vertex and the semi-latus rectum.

Fig. 3.

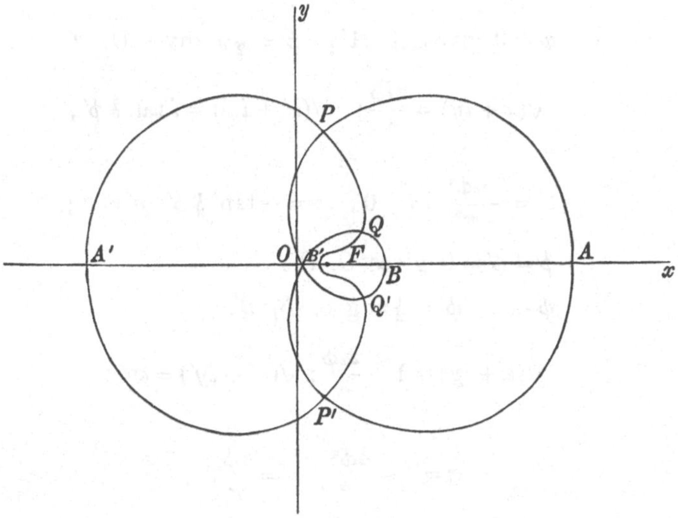

We may divide the circle by concentric circles and by radii; the corresponding curves for the parabola will be ovals and radials from the focus, the curves of each system being transcendental curves. Or we may divide the parabola by means of two systems of confocal parabolas; the corresponding curves for the circle will be (see fig. 3) two systems of orthotomic limaçons, those of the one system having B' for a crunode, and those of the other system having B' for an acnode.

To show that the circle thus corresponds to the parabola, it is only necessary to write $\sqrt{(x+iy)} = 1 + qi$, that is, $x = 1 - q^2$, $y = 2q$, implying $y^2 = 4(1-x)$, or (x, y) is a point on the parabola; and we then obtain for $x' + iy'$ a value of the form $\cos\theta' + i\sin\theta'$, that is, (x', y') is a point on the circle $x'^2 + y'^2 = 1$; but in reference to what follows, I give the proof in a somewhat more artificial form.

Writing log tan to denote the hyperbolic logarithm of the tangent, then if ϕ, ϕ' are such that

$$\phi = \log\tan(\tfrac{1}{4}\pi + \tfrac{1}{2}\phi'),$$

this equation, as is known, may also be presented in the forms

$$i\phi' = \log\tan(\tfrac{1}{4}\pi + \tfrac{1}{2}i\phi),$$
$$i\tan\tfrac{1}{2}\phi' = \tan\tfrac{1}{2}i\phi,$$

or, what is the same thing,

$$\tan\tfrac{1}{2}\phi' = \tanh\tfrac{1}{2}\phi,$$

where observe that, as ϕ' increases from 0 to $\frac{1}{2}\pi$, ϕ increases from 0 to ∞, and that always $\phi > \phi'$.

We can now establish as follows the three correspondences AO with $A'O'$, OB with $O'B'$, and ACB with $A'C'B'$.

(1) $$\sqrt{(x+iy)} = \frac{2\phi}{\pi}; \quad \sqrt{(x'+iy')} = \tan\tfrac{1}{2}\phi';$$

that is,

$$x = \frac{4\phi^2}{\pi^2}, \; y = 0; \quad x' = \tan^2\tfrac{1}{2}\phi', \; y' = 0;$$

$$\phi = 0 \text{ gives } A, A'; \quad \phi' = \tfrac{1}{2}\pi \text{ gives } O, O'.$$

(2) $$\sqrt{(x+iy)} = \frac{2i\phi}{\pi}; \quad \sqrt{(x'+iy')} = i\tan\tfrac{1}{2}\phi';$$

that is,

$$x = -\frac{4\phi^2}{\pi^2}, \; y = 0; \quad x' = -\tan^2\tfrac{1}{2}\phi', \; y' = 0;$$

$$\phi = \phi' = 0 \text{ gives } O, O';$$

$$\phi = \infty, \; \phi' = \tfrac{1}{2}\pi \text{ give } B, B'.$$

(3) $$\sqrt{(x+iy)} = 1 + \frac{2i\phi}{\pi}; \; \sqrt{(x'+iy')} = e^{i\phi'};$$

that is,

$$x = 1 - \frac{4\phi^2}{\pi^2}, \quad y = \frac{4\phi}{\pi},$$

and therefore

$$y^2 = 4(1-x),$$

the parabola; and

$$x' = \cos 2\phi', \; y' = \sin 2\phi',$$

and therefore

$$x'^2 + y'^2 = 1,$$

the circle;

$$\phi = \phi' = 0 \text{ gives } A, A'; \quad \phi = \infty, \; \phi' = \tfrac{1}{2}\pi \text{ give } B, B'.$$

Observe that for points in $O'A'$, $O'B'$, equidistant from O', we have $x' = \tan^2\frac{1}{2}\phi'$, $x' = -\tan^2\frac{1}{2}\phi'$; and corresponding hereto we have points in OA, OB on the axis of the parabola, the values of x being $x = \frac{4\phi'^2}{\pi^2}$, $x = -\frac{4\phi^2}{\pi^2}$, viz. the negative value is always the greater.

Observe further that, to the points $(x', y') = (\cos 2\phi', \sin 2\phi')$ and $(x', y') = (-\tan^2\frac{1}{2}\phi', 0)$ on the circle, and on the radius OB', correspond the points

$$(x, y) = \left(1 - \frac{4\phi^2}{\pi^2}, \; \frac{2\phi}{\pi}\right), \text{ and } (x, y) = \left(-\frac{4\phi^2}{\pi^2}, \; 0\right),$$

on the parabola and on the axis OB respectively; the axial distance of these two points is $\left(1 - \frac{4\phi^2}{\pi^2}\right) + \frac{4\phi^2}{\pi^2} = 1$, the radius of the circle, or the distance OA of the vertex and focus of the parabola; this is a rather curious theorem.

The function log tan (45° + ½ arg.) is tabulated, Legendre, *Théorie des Fonctions Elliptiques*, t. II. table iv. pp. 256—259, viz. writing as above ϕ' for the argument, we have hereby the value of ϕ, $= \log\tan(\frac{1}{4}\pi + \frac{1}{2}\phi')$, for every value of ϕ' from 0° to 90° at intervals of 30′ and to 12 decimals, and with fifth differences. Observe that ϕ' is thus given in degrees and minutes as a circular arc, ϕ as a number; it is convenient to have ϕ and ϕ' each as arcs, or each as numbers, the conversion being of course at once made by means of a table of the lengths of circular arcs, and I have calculated the Table which I give at the end of the present paper.

I had previously calculated for the correspondence of $A'O'$, $O'B'$ and $A'C'B'$ with AO, OB and ACB, the following values, at irregular intervals suited to the construction of a figure.

Circle ∠ $A'O'P'$	Parabola Ordinate PM
0°	0
30	·3372
60	·6994
90	1·1220
120	1·6768
150	2·5817
160	3·1018
170	3·9869
172	4·2713
174	4·6378
176	5·1542
178	6·0258
179	6·9195
180°	∞

Circle $O'A'$ $x'=$	Circle $O'B'$ $x'=$	Parabola OA $x=$	Parabola OB $x=$
0	0	0	0
·1	− ·1	·152	− ·173
·2	− ·2	·287	− ·375
·3	− ·3	·407	− ·611
·4	− ·4	·515	− ·943
·5	− ·5	·615	− 1·257
·6	− ·6	·704	− 1·724
·7	− ·7	·787	− 2·368
·8	− ·8	·853	− 3·386
·9	− ·9	·935	− 5·368
1·0	− 1·0	1·000	− ∞

Write in general

$$\sqrt{(x+iy)} = p + qi, \ = \frac{2}{\pi}(\psi + i\phi);$$

viz. considering x, y as given, find thence p, q by the equations $x = p^2 - q^2$, $y = 2pq$: and then writing $\frac{1}{2}\pi p = \psi$, $\frac{1}{2}\pi q = \phi$, we have

$$\sqrt{(x'+iy')} = \tan\tfrac{1}{4}\pi(p+qi) = \tan(\tfrac{1}{2}\psi + \tfrac{1}{2}i\phi) = \frac{P+iQ}{1-iPQ},$$

where

$$P = \tan \tfrac{1}{4}\pi p, \quad = \tan \tfrac{1}{2}\psi,$$

$$Q = \frac{1}{i} \tan \tfrac{1}{4}\pi q i, \quad = \tanh \tfrac{1}{4}\pi q, \quad = \frac{1}{i} \tan \tfrac{1}{2}\phi i, \quad = \tanh \tfrac{1}{2}\phi\,;$$

whence, if we introduce ϕ' connected with ϕ by the equation $\phi = \log \tan (\frac{1}{4}\pi + \frac{1}{2}\phi')$ as before, we have $Q = \tan \frac{1}{2}\phi'$, and the formula is

$$\sqrt{(x' + iy')} = \frac{P + iQ}{1 - iPQ}, \quad P = \tan \tfrac{1}{2}\psi, \quad Q = \tan \tfrac{1}{2}\phi',$$

giving the values of x', y'.

It is clear that we have

$$\sqrt{(x' + iy')} = \frac{P - iQ}{1 + iPQ},$$

and thence

$$\sqrt{(x'^2 + y'^2)} = \frac{P^2 + Q^2}{1 + P^2Q^2}.$$

Hence, to the circle $x'^2 + y'^2 = c'^2$, corresponds in the parabola the curve

$$P^2 + Q^2 = c'(1 + P^2Q^2),$$

where $p + iq = \sqrt{(x + iy)}$, $P = \tan \frac{1}{4}\pi p$, $Q = \tanh \frac{1}{4}\pi q$. This is a complicated transcendental equation, and I do not see any convenient way of tracing the curve. The set of curves satisfy the differential equation

$$\frac{P\,dP + Q\,dQ}{P^2 + Q^2} = \frac{PQ\,(Q\,dP + P\,dQ)}{1 + P^2Q^2},$$

that is,

$$P\,dP\,(1 - Q^4) + Q\,dQ\,(1 - P^4) = 0,$$

where dP, dQ are given in terms of dp, dq by the equations

$$\frac{dP}{1 + P^2} = \tfrac{1}{4}\pi\,dp, \quad \frac{dQ}{1 - Q^2} = \tfrac{1}{4}\pi\,dq.$$

We have $y = 2pq$, and thence at the highest points, or summits of the several curves, $q\,dp + p\,dq = 0$. Combining these equations, we have

$$dP : dQ = p\,(1 + P^2) : -q\,(1 - Q^2);$$

and thence

$$pP\,(1 + Q^2) - qQ\,(1 - P^2) = 0,$$

as an equation to the locus of the summits in question. If p and q are small, then putting for a moment $\frac{1}{4}\pi = m$ for shortness, we have

$$P = mp + \tfrac{1}{3}m^3p^3, \quad Q = mq - \tfrac{1}{3}m^3q^3,$$

and the equation becomes

$$p^2\,(1 + \tfrac{1}{3}m^2p^2)\,(1 + m^2q^2) - q^2\,(1 - \tfrac{1}{3}m^2q^2)\,(1 - m^2p^2) = 0,$$

that is,

$$p^2 - q^2 + \tfrac{1}{3}m^2(p^4 + 6p^2q^2 + q^4) = 0;$$

writing this in the form

$$p^2 - q^2 + \frac{\pi^2}{48}[(p^2 - q^2)^2 + 8p^2q^2] = 0,$$

we find $x + \dfrac{\pi^2}{48}(x^2 + 2y^2) = 0$, or say $y^2 = -\dfrac{24}{\pi^2}x$, as the locus of the summits in the neighbourhood of the focus O, viz. the summits lie all of them, as might have been expected, on the left-hand (or negative side) of the focus.

I have constructed for the parabola, to the scale $1 = 1\frac{1}{2}$ inch, as accurately as the data enable, the figure corresponding to the concentric circles and the radii of the circle.

Resuming the equation $\sqrt{(x + iy)} = p + iq$, that is, $x = p^2 - q^2$ and $y = 2pq$, we have ($p =$ const.), the curves $y^2 = 4p^2(p^2 - x)$, and ($q =$ const.), $y^2 = 4q^2(q^2 + x)$, which are two systems of confocal parabolas, cutting each other at right angles; the curves of the former set all of them interior to the given parabola, those of the latter set of course cutting it at right angles.

Corresponding hereto in the circle, writing for a moment

$$\sqrt{(x' + iy')} = p' + iq',$$

we have

$$p' + iq' = \frac{P + iQ}{1 - iPQ},$$

whence

$$p' = P(1 - q'Q), \quad q' = Q(1 + p'P);$$

whence eliminating P and Q successively, we find

$$p'^2 + q'^2 - p'\left(P - \frac{1}{P}\right) - 1 = 0,$$

$$p'^2 + q'^2 - q'\left(Q + \frac{1}{Q}\right) + 1 = 0;$$

say, for shortness, these are $p'^2 + q'^2 - 2mp' + 1 = 0$, and $p'^2 + q'^2 - 2nq' - 1 = 0$. Introducing the polar coordinates r', θ', we have

$$x' + iy' = r'(\cos\theta' + i\sin\theta'),$$

and thence

$$p',\ q' = \sqrt{(r')}\cos\tfrac{1}{2}\theta', \quad \sqrt{(r')}\sin\tfrac{1}{2}\theta';$$

the equations thus become

$$r' - 1 - 2m\sqrt{(r')}\cos\tfrac{1}{2}\theta' = 0,$$

$$r' + 1 - 2n\sqrt{(r')}\sin\tfrac{1}{2}\theta' = 0,$$

which belong to two limaçons having each of them B' for a node and O for a focus, and which of course cut each other at right angles; see fig. 4, which is a mere

Fig. 4.

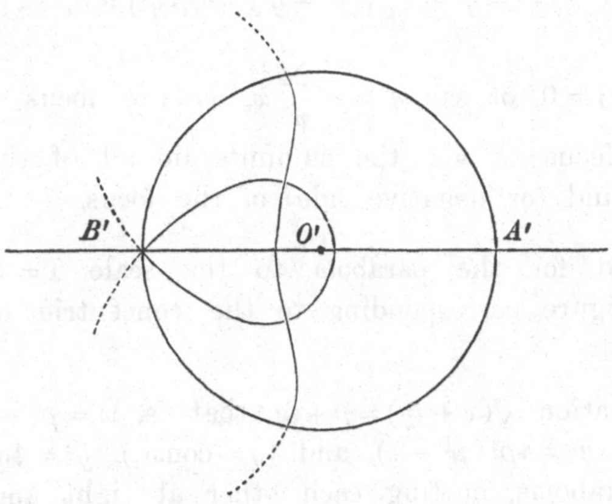

diagram. In fact, omitting for convenience the accents, but recollecting always that the curves belong to the figure of the circle, the first equation gives

$$(r-1)^2 = 4m^2 r \cos^2 \tfrac{1}{2}\theta', \quad = 2m^2 r (1+\cos\theta) = 2m^2 (r+x),$$

that is,

$$(r^2+1-2m^2 x)^2 = 4(m^2+1)^2 r^2.$$

Transforming to the point B' as origin, we must write $x+1$ for x, and then

$$r^2 = (x-1)^2 + y^2;$$

the equation thus becomes

$$\{x^2+y^2-2(m^2+1)x+2(m^2+1)\}^2 = 4(m^2+1)^2(x^2+y^2+2x+1),$$

that is,

$$\{x^2+y^2-2(m^2+1)x\}^2 \qquad = 4m^2(m^2+1)(x^2+y^2),$$

or say

$$(x^2+y^2)^2 - 4(m^2+1)(x^2+y^2)x + 4(m^2+1)(x^2-m^2y^2) = 0,$$

showing that the point B' is a crunode, the tangents there being $x = \pm my$.

Writing in the equation $y=0$, we have $x^2=0$, the crunode, and

$$x = 2\sqrt{(m^2+1)}\{\sqrt{(m^2+1)} \pm m\};$$

the product of these two values is $=4(m^2+1)$, that is, $=\left(P+\dfrac{1}{P}\right)^2$, viz. one value is greater, the other less, than 2. We see also that

$$2\sqrt{(m^2+1)}\{\sqrt{(m^2+1)}-m\},$$

the smaller root, is greater than 1. The curve corresponding to a parabola $y^2 = 4p^2(p^2 - x)$ is thus a crunodal limaçon, the crunode at B', and the loop lying wholly within the circle. Moreover, the loop includes within itself the centre O' of the circle.

The other curve is in like manner

$$\{r^2 + 1 + 2n^2x\}^2 = 4(n^2 - 1) r^2,$$

viz. transforming to the origin B', and therefore putting $x - 1$ for x, and

$$r^2 = (x-1)^2 + y^2,$$

the equation is

$$\{x^2 + y^2 + 2(n^2 - 1)x - 2(n^2 - 1)\}^2 = 4(n^2 - 1)^2(x^2 + y^2 - 2x + 1),$$

that is,

$$\{x^2 + y^2 + 2(n^2 - 1)x\}^2 \qquad = 4n^2(n^2 - 1)(x^2 + y^2),$$

or say

$$(x^2 + y^2)^2 + 4(n^2 - 1)(x^2 + y^2)x - 4(n^2 - 1)(x^2 + n^2y^2) = 0,$$

viz. the point B' is an acnode with the imaginary tangents $x = \pm iny$.

Writing in the equation $y = 0$, we have $x^2 = 0$ the acnode, and

$$x = -\surd(n^2 - 1)\{\surd(n^2 - 1) \pm n\},$$

where

$$\surd(n^2 - 1) = \tfrac{1}{2}\sqrt{\left(Q - \frac{1}{Q}\right)^2}$$

is real and may be taken to be positive; there is thus one root

$$-\surd(n^2 - 1)\{\surd(n^2 - 1) + n\},$$

which is negative; the other root, say

$$\surd(n^2 - 1)\{n - \surd(n^2 - 1)\},$$

is positive and less than 1; the curve is thus an acnodal limaçon, having B' for the acnode and cutting the axis outside the circle on the negative side of B', and inside the circle between B' and O'.

Taking B' as origin, the equation of the circle is $x^2 + y^2 - 2x = 0$, and we hence find for the intersection of the limaçon with the circle $n^2x = 2(n^2 - 1)$, that is, $x = 2\left(1 - \frac{1}{n^2}\right)$; whence also $y^2 = \frac{4}{n^2}\left(1 - \frac{1}{n^2}\right)$, values which belong to a real intersection; for $q = 0$, we have $Q = 0$, $n = \infty$, and therefore $(x, y) = (2, 0)$, viz. the intersection is at B'; for $q = \infty$, we have $Q = 1$, $n = 1$, and therefore $(x, y) = (0, 0)$, viz. the intersection is at A', results which are obviously right. Observe that for the other limaçon we have $x = 2\left(1 + \frac{1}{m^2}\right)$, $y = -\frac{4}{m^2}\left(1 + \frac{1}{m^2}\right)$, viz. there is no real intersection with the circle.

The Table above referred to.

$\phi = \log\tan(45^\circ + \frac{1}{2}\phi')$.				$\phi = \log\tan(45^\circ + \frac{1}{2}\phi')$.			
ϕ'		ϕ		ϕ'		ϕ	
0°	·0	·0	0°	46°	·8028515	·9062755	51°55′·546
1	0174533	0174542	1 0′·003	47	·8203047	·9316316	53 22 ·714
2	0349066	0349137	2 0 ·024	48	·8377580	·9574669	54 51 ·529
3	0523599	0523838	3 0 ·082	49	·8552113	·9838079	56 22 ·082
4	0698132	0698699	4 0 ·195	50	·8726646	1·0106832	57 54 ·473
5	0872665	0873774	5 0 ·381	51	·8901179	1·0381235	59 28 ·806
6	1047198	1049117	6 0 ·657	52	·9075712	1·0661617	61 5 ·194
7	1221730	1224781	7 1 ·152	53	·9250245	1·0948335	62 43 ·761
8	1396263	1400822	8 1 ·568	54	·9424778	1·1241772	64 24 ·637
9	1570796	1577296	9 2 ·269	55	·9509311	1·1542346	66 7 ·967
10	1745329	1754258	10 3 ·069	56	·9773844	1·1850507	67 53 ·904
11	1919862	1931766	11 4 ·092	57	·9948377	1·2166748	69 42 ·620
12	2094395	2109867	12 5 ·323	58	1·0122910	1·2491606	71 34 ·298
13	2268928	2288650	13 6 ·670	59	1·0297443	1·2825668	73 29 ·140
14	2443461	2468145	14 8 ·486	60	1·0471976	1·3169579	75 27 ·368
15	2617994	2648422	15 10 ·460	61	1·0646508	1·3524048	77 29 ·225
16	2792527	2829545	16 12 ·726	62	1·0821041	1·3889860	79 34 ·639
17	2967060	3011577	17 15 ·304	63	1·0995574	1·4267882	81 44 ·937
18	3141593	3194583	18 18 ·217	64	1·1170107	1·4659083	83 59 ·421
19	3316126	3378629	19 21 ·487	65	1·1344640	1·5064542	86 18 ·808
20	3490659	3563785	20 25 ·139	66	1·1519173	1·5485472	88 43 ·513
21	3665191	3750121	21 29 ·197	67	1·1693706	1·5923237	91 14 ·006
22	3839724	3937710	22 33 ·685	68	1·1868239	1·6379387	93 50 ·819
23	4014257	4126626	23 39 ·630	69	1·2042772	1·6855685	96 34 ·558
24	4188790	4316947	24 44 ·057	70	1·2217305	1·7354152	99 25 ·918
25	4363323	4508753	25 49 ·995	71	1·2391838	1·7877120	102 25 ·701
26	4537856	4702127	26 56 ·472	72	1·2566371	1·8427300	105 34 ·839
27	4712389	4897154	28 3 ·518	73	1·2740904	1·9007867	108 54 ·423
28	4886922	5093923	29 11 ·162	74	1·2915436	1·9622572	112 25 ·744
29	5061455	5292527	30 19 ·437	75	1·3089969	2·0275894	116 10 ·339
30	5235988	5493061	31 28 ·375	76	1·3264502	2·0973240	120 10 ·069
31	5410521	5695627	32 38 ·012	77	1·3439035	2·1721218	124 27 ·205
32	5585054	5900329	33 48 ·383	78	1·3613568	2·2528027	129 4 ·566
33	5759587	6107275	34 59 ·527	79	1·3788101	2·3404007	134 5 ·705
34	5934119	6316581	36 11 ·480	80	1·3962634	2·4362460	139 35 ·197
35	6108652	6528366	37 24 ·287	81	1·4137167	2·5420904	145 39 ·063
36	6283185	6742755	38 37 ·988	82	1·4311700	2·6603061	152 25 ·459
37	6457718	6959880	39 52 ·631	83	1·4486233	2·7942190	160 5 ·818
38	6632251	7179880	41 8 ·261	84	1·4660766	2·9487002	168 56 ·885
39	6806784	7402901	42 24 ·930	85	1·4835299	3·1313013	179 24 ·621
40	6981317	7629095	43 42 ·690	86	1·5009832	3·3546735	192 12 ·518
41	7155850	7858630	45 1 ·597	87	1·5184364	3·6425334	208 42 ·107
42	7330383	8091672	46 21 ·712	88	1·5358897	4·0481254	231 56 ·416
43	7504916	8328406	47 43 ·095	89	1·5533430	4·7413488	271 39 ·557
44	7679449	8569026	49 5 ·814	90	1·5707963	∞	∞
45	7853982	·8813736	50 29 ·939				

846.

A VERIFICATION IN REGARD TO THE LINEAR TRANSFORMATION OF THE THETA-FUNCTIONS.

[From the *Quarterly Journal of Pure and Applied Mathematics*, vol. XXI. (1886), pp. 77—84.]

THE notation is that of Smith's* "Memoir on the Theta and Omega Functions," differing from Hermite's only in a factor, $\vartheta_{m,n}$ (Smith) $= i^{mn}\,\vartheta_{m,n}$ (Hermite); and it is in consequence of this that the factor $i^{\mu\nu-mn}$ occurs in the expression presently given for δ. Hermite's Memoir "Sur quelques formules relatives à la transformation des fonctions elliptiques" appeared in *Liouville*, t. III. (1858), pp. 27—36.

Writing

$$\omega = \frac{c + d\Omega}{a + b\Omega},$$

where $ad - bc = 1$, the formula of transformation is

$$\vartheta_{\mu,\nu}\left\{(a + b\Omega)\frac{\pi x}{h},\ \Omega\right\} = C \exp\left\{-i\pi b(a + b\Omega)\frac{x^2}{h^2}\right\}\vartheta_{m,n}\left(\frac{\pi x}{h},\ \omega\right),$$

where

$$m = a\mu + b\nu + ab,$$

$$n = c\mu + d\nu + cd.$$

We have, according as b is positive or negative,

$$C = \frac{\delta H}{\sqrt{\{-i(a + b\Omega)\}}},\ \text{ or } C = \frac{\delta H}{\sqrt{\{i(a + b\Omega)\}}},$$

where for each case the square root is to be taken in such wise that the real part is positive; δ, H are eighth roots of unity; viz. in each case

$$\delta = \exp\{-\tfrac{1}{4}\pi i(ac\mu^2 + 2bc\mu\nu + bd\nu^2 + 2abc\mu + 2abd\nu + ab^2c)\}\, i^{\mu\nu-mn},$$

[* H. J. S. Smith, *Collected Mathematical Papers*, vol. II., pp. 415—621.]

and, according as b is positive or negative,

$$H=\left(\frac{b}{a}\right)i^{-\frac{1}{2}a},\qquad a \text{ odd}, \qquad\Big|\qquad H=\left(\frac{-b}{-a}\right)i^{\frac{1}{2}a},\qquad a \text{ odd},$$

$$H=\left(\frac{a}{b}\right)i^{-\frac{1}{2}a-\frac{1}{2}(a-1)(b-1)},\quad b \text{ odd}, \qquad\Big|\qquad H=\left(\frac{-a}{-b}\right)i^{\frac{1}{2}a-\frac{1}{2}(a+1)(b+1)},\quad b \text{ odd},$$

where $\left(\frac{a}{b}\right)$, &c., are the Legendrian symbols as generalised by Jacobi; if a and b are each of them odd, the formulæ for the cases a odd and b odd respectively are equivalent to each other, and either may be used indifferently.

To fix the ideas, I assume throughout that b is positive and odd; the proper formulæ thus are

$$C=\frac{\delta H}{\sqrt{\{-i(a+b\Omega)\}}},$$

δ as above, and

$$H=\left(\frac{a}{b}\right)i^{-\frac{1}{2}a-\frac{1}{2}(a-1)(b-1)}.$$

I will also, for greater convenience, assume that c is odd; ad is consequently even, viz. the numbers a and d are each of them even, or else one is odd and the other even.

The equation $\omega=\dfrac{c+d\Omega}{a+b\Omega}$ gives $\Omega=\dfrac{c-a\omega}{-d+b\omega}$, and thence

$$a+b\Omega=\frac{-1}{-d+b\omega}.$$

Comparing the expressions for ω, Ω, it appears that we may interchange ω and Ω, writing also $-d,\ b,\ c,\ -a$ for $a,\ b,\ c,\ d$; and changing the other letters, we may for

$$a,\ b,\ c,\quad d,\ \omega,\ \Omega,\ \mu,\ \nu,\ m,\ n,$$

write

$$-d,\ b,\ c,\ -a,\ \Omega,\ \omega,\ m,\ n,\ m',\ n',$$

where

$$m'=-dm+bn-bd,$$

$$n'=\quad cm-an-ac.$$

The formula thus becomes

$$\vartheta_{m,n}\left\{(-d+b\omega)\frac{\pi x}{h},\ \omega\right\}=C'\exp\left\{-i\pi b(-d+b\omega)\frac{x^2}{h^2}\right\}\vartheta_{m',n'}\left(\frac{\pi x}{h},\ \Omega\right),$$

or, for x writing $(a+b\Omega)x$, this is

$$\vartheta_{m,n}\left\{-\frac{\pi x}{h},\ \omega\right\}=C'\exp\left\{i\pi(a+b\Omega)\frac{x^2}{h^2}\right\}\vartheta_{m',n'}\left\{(a+b\Omega)\frac{\pi x}{h},\ \Omega\right\},$$

or, observing that the left-hand side is $=(-)^{mn}\vartheta\left(\frac{\pi x}{h},\ \omega\right)$, we have

$$\vartheta_{m',\,n'}\left\{(a+b\Omega)\frac{\pi x}{h},\ \Omega\right\}=\frac{(-)^{mn}}{C'}\exp\left\{-i\pi(a+b\Omega)\frac{x^2}{h^2}\right\}\vartheta_{m,\,n}\left(\frac{\pi x}{h},\ \omega\right),$$

an equation which is of the same form as the original equation of transformation, and is to be identified with it.

We have

$$\begin{aligned} m' &= -dm+bn-bd = -d(a\mu+b\nu+ab) && = -\mu+bd(-a+c-1),\\ &\qquad\qquad\qquad\quad +b(c\mu+d\nu+cd)-bd, && \\ n' &= \ \ cm-an-ac = \ \ c(a\mu+b\nu+ab) && = -\nu+ac(\ \ b-d-1);\\ &\qquad\qquad\qquad\quad -a(c\mu+d\nu+cd)-ac, && \end{aligned}$$

values which may be written

$$\begin{aligned} m' &= 2P-\mu, \text{ where } P=\tfrac{1}{2}bd(-a+c-1),\\ n' &= 2Q-\nu, \quad\ \text{,,}\quad\ \ Q=\tfrac{1}{2}ac(\ \ b-d-1), \end{aligned}$$

P and Q being integers. In fact, P is an integer if b or d is even, and if they are each of them odd, then from the equation $ad-bc=1$, a and c will be one of them odd, the other even, and $-a+c-1$ will be even; and similarly for Q.

The left-hand side is

$$\vartheta_{2P-\mu,\,2Q-\nu}\left\{(a+b\Omega)\frac{\pi x}{h},\ \Omega\right\},$$

which is

$$=(-)^{\mu(Q-\nu)}\vartheta_{\mu,\,\nu}\left\{(a+b\Omega)\frac{\pi x}{h},\ \Omega\right\},$$

and the equation finally is

$$(-)^{\mu(Q-\nu)}\vartheta_{\mu,\,\nu}\left\{(a+b\Omega)\frac{\pi x}{h},\ \Omega\right\}=\frac{(-)^{mn}}{C'}\exp\left\{-i\pi(a+b\Omega)\frac{x^2}{h^2}\right\}\vartheta_{m,\,n}\left(\frac{\pi x}{h},\ \omega\right).$$

Comparing this with the original equation

$$\vartheta_{\mu,\,\nu}\left\{(a+b\Omega)\frac{\pi x}{h},\ \Omega\right\}=C\exp\left\{-i\pi(a+b\Omega)\frac{x^2}{h^2}\right\}\vartheta_{m,\,n}\left(\frac{\pi x}{h},\ \omega\right),$$

the two equations will be identical if only

$$CC'=(-)^{-\mu Q+\mu\nu-mn}.$$

We have

$$C=\frac{\delta H}{\sqrt{\{-i(a+b\Omega)\}}},\quad C'=\frac{\delta' H'}{\sqrt{\{-i(-d+b\omega)\}}},$$

where the square roots are taken in such wise that the real part is positive; hence

$$CC'=\frac{\delta\delta' HH'}{\sqrt{(+1)}},$$

where $\sqrt{(+1)}$ means

$$\sqrt{\{-i(a+b\Omega)\}}\,.\,\sqrt{\{-i(-d+b\omega)\}},$$

the last-mentioned two square roots being as just explained; and we have moreover

$$\delta = \exp\{-\tfrac{1}{4}i\pi(\ \ ac\mu^2 + 2bc\mu\nu + bd\nu^2 + 2abc\mu + 2abd\nu + ab^2c)\}\ i^{\mu\nu-mn},$$
$$\delta' = \exp\{-\tfrac{1}{4}i\pi(-dcm^2 + 2bcmn - abn^2 - 2dbcm + 2abdn - db^2c)\}\ i^{mn-m'n'},$$

viz. the value of δ' is obtained from that of δ by the change $a, b, c, d, \mu, \nu, m, n$ into $-d, b, c, -a, m, n, m', n'$.

Representing these by

$$\delta = \exp\{-\tfrac{1}{4}(i\pi)\Delta\}\ i^{\mu\nu-mn},\quad \delta' = \exp\{-\tfrac{1}{4}(i\pi)\Delta'\}\ i^{mn-m'n'},$$

we have

$$\delta\delta' = \exp\{-\tfrac{1}{4}i\pi(\Delta+\Delta')\}\ i^{\mu\nu-m'n'}.$$

But

$$m'n' = (2P-\mu)(2Q-\nu),\ = 4PQ - 2P\nu - 2Q\mu + \mu\nu,$$

that is,

$$\mu\nu - m'n' = -4PQ + 2P\nu + 2Q\mu;$$

or, omitting the term divisible by 4,

$$i^{\mu\nu-m'n'} = i^{2P\nu+2Q\mu},\ = (-)^{P\nu+Q\mu}.$$

To calculate $\Delta+\Delta'$, we have

$$dm - bn = \mu + bd(\ \ a - c),$$
$$-cm + an = \nu + ac(-b+d),$$

and thence

$$-cdm^2 + (ad+bc)mn - abn^2 = \ \ \mu\nu + \mu ac(-b+d) + \nu bd(a-c) + abcd(d-b)(a-c)$$
$$(-ad+bc)mn \ \ = -ac\mu^2 - bd\nu^2 - (ad+bc)\mu\nu - \mu ac(b+d) - \nu bd(a+c) - abcd,$$

consequently

$$-cdm^2 + 2bcmn - abn^2 = -ac\mu^2 - bd\nu^2 - 2bc\mu\nu - 2abc\mu - 2bcd\nu + abcd(2bc - ab - cd);$$

also

$$-2bcdm \ = \ -2abcd\mu - 2b^2cd\nu - 2ab^2cd,$$
$$2abdn \ = \ 2abcd\mu + 2abd^2\nu + 2abcd^2,$$
$$-db^2c \ = \ -db^2c,$$

whence, adding, we obtain

$$\Delta' = -ac\mu^2 - 2bc\mu\nu - bd\nu^2 - 2abc\mu + (-2bcd - 2b^2cd + 2abd^2)\nu$$
$$+ abcd(2bc - ab - cd - 2b + 2d) - db^2c,$$

and, adding to this the expression of Δ, we find

$$\Delta + \Delta' = (2abd - 2bcd - 2b^2cd + 2abd^2)\nu + ab^2c - db^2c + abcd(2bc - ab - cd - 2b + 2d).$$

The coefficient of ν is $= 2bd(a-c-bc+ad) = 2bd(a-c+1)$, which is $= -4P\nu$. Hence writing

$$\Theta = \tfrac{1}{4}abcd(2bc-ab-cd-2b+2d) + \tfrac{1}{4}b^2c(a-d),$$

where observe that 4Θ, but not in every case Θ itself, is an integer, we have $\Delta + \Delta' = -4P\nu + 4\Theta$, and consequently

$$\delta\delta' = (-)^{\Theta+P\nu}(-)^{P\nu+Q\mu}, = (-)^{\Theta+2P\nu+Q\mu},$$

or, omitting the even number $2P\nu$,

$$\delta\delta' = (-)^{\Theta+Q\mu}.$$

Observe that $(-)^{\Theta}$ denotes, and it might properly have been written, $\exp i\pi\Theta$. The foregoing equation

$$\frac{\delta\delta' HH'}{\sqrt{(+1)}} = (-)^{-Q\mu+\mu\nu-mn}$$

becomes thus

$$(-)^{\Theta+Q\mu}\frac{HH'}{\sqrt{(+1)}} = (-)^{-Q\mu+\mu\nu-mn},$$

that is,

$$\frac{HH'}{\sqrt{(+1)}} = (-)^{-\Theta-2Q\mu+\mu\nu-mn},$$

where the even term $-2Q\mu$ may be omitted. We have moreover

$$mn = ac\mu^2 + (ad+bc)\mu\nu + bd\nu^2 + ac(b+d)\mu + bd(a+c)\nu + abcd,$$

$$-\mu\nu = \quad (-ad+bc)\mu\nu,$$

and thence

$$mn - \mu\nu = ac(\mu^2-\mu) + 2bc\mu\nu + bd(\nu^2-\nu) + ac(b+d+1)\mu + bd(a+c+1)\nu + abcd,$$

where each term is even; hence $mn - \mu\nu$ is even, and we have simply

$$\frac{HH'}{\sqrt{(+1)}} = (-)^{-\Theta},$$

where

$$\Theta = \tfrac{1}{4}abcd(ab+cd-2bc+2b-2d) - \tfrac{1}{4}b^2c(a-d).$$

I write $M = \tfrac{1}{4}b(a-d)$, then

$$\Theta - M = \tfrac{1}{4}abcd(ab+cd-2bc+2b-2d) - \tfrac{1}{4}(bc+1)b(a-d),$$

where the second term is $= -\tfrac{1}{4}abd(a-d)$, and we have therefore

$$\Theta - M = \tfrac{1}{4}abd(abc + c^2d - 2bc^2 + 2bc - 2cd - a + d).$$

I assume, as above, that b and c are each of them odd; therefore ad is even. I suppose, first, that ad divides by 4, then $\tfrac{1}{4}abd$ is an integer, and in the expression of $\Theta - M$, omitting even numbers, we have

$$\Theta - M = \tfrac{1}{4}abd(abc + c^2d - a + d),$$

which, putting therein $bc = ad - 1$, becomes

$$= \tfrac{1}{4}abd\,(a^2d + c^2d - 2a + d),$$
$$= \tfrac{1}{4}abd\,\{a^2d + (c^2 - c)\,d + (c+1)\,d - 2a\},$$

where inside the $\{\ \}$ each term is even; hence $\Theta - M$ is even.

Next, if ad is even but not divisible by 4, then $bc = ad - 1$, which is $\equiv 1$ (mod. 4), thus b and c are

$$= 4\sigma + 1 \text{ and } 4\tau + 1,$$

or else

$$= 4\sigma - 1 \text{ and } 4\tau - 1,$$

and, moreover, $bc = 4\theta + 1$, and $c^2 = 4\phi + 1$; hence

$$\Theta - M = \tfrac{1}{2}b\,.\,\tfrac{1}{2}ad\,\{a\,(4\theta + 1) + d\,(4\phi + 1) - 2b\,(4\phi + 1) + 2\,(4\theta + 1) - 2cd - a + d\},$$

or, omitting even numbers, that is, inside the $\{\ \}$ numbers which contain the factor 4, this is

$$= \tfrac{1}{4}abd\,(a + d - 2b + 2 - a + d),$$
$$= \tfrac{1}{2}abd\,(d - b + 1 - cd),$$
$$= \tfrac{1}{2}abd\,\{d\,(1 - c) + 1 - b\},$$

or, since each term within the $\{\ \}$ is even, we have in this case also $\Theta - M$ even. And this being so, the foregoing equation for HH' becomes

$$\frac{HH'}{\sqrt{(+1)}} = (-)^{-M}, \ = (-)^{\frac{1}{2}b\,(d-a)}, \text{ or say } = i^{\frac{1}{2}b\,(d-a)}.$$

The values of H, H', refer each of them to a positive odd value of b, and they thus are

$$H = \left(\frac{a}{b}\right) \quad i^{-\frac{1}{2}a - \frac{1}{2}(a-1)\,(b-1)},$$

$$H' = \left(-\frac{d}{b}\right) i^{\frac{1}{2}d - \frac{1}{2}(-d-1)\,(b-1)};$$

hence

$$HH' = \left(\frac{a}{b}\right)\left(\frac{-d}{b}\right) i^{-\frac{1}{2}(a-d) - \frac{1}{2}(a-d-2)\,(b-1)} = \left(\frac{a}{b}\right)\left(\frac{-d}{b}\right) i^{b-1+\frac{1}{2}b\,(d-a)},$$

or, since

$$\left(\frac{-d}{b}\right) = (-)^{\frac{1}{2}(b-1)}\left(\frac{d}{b}\right) = i^{b-1}\left(\frac{d}{b}\right),$$

and $2\,(b-1)$ divides by 4, this is

$$HH' = \left(\frac{a}{b}\right)\left(\frac{d}{b}\right) i^{\frac{1}{2}b\,(d-a)}.$$

Also $\left(\frac{a}{b}\right)\left(\frac{d}{b}\right) = \left(\frac{ad}{b}\right)$, but from the equation $ad - bc = 1$, or $\frac{ad}{b} = c + \frac{1}{b}$, we have $\left(\frac{ad}{b}\right) = \left(\frac{1}{b}\right) = 1$; whence

$$HH' = i^{\frac{1}{2}b\,(d-a)}.$$

We have $\omega = x + iy$, where y is positive; hence

$$-d + b\omega = -d + bx + iby = \alpha + 2by, \text{ if } \alpha = -d + bx;$$

hence

$$a + b\Omega = \frac{-1}{\alpha + iby} = \frac{-\alpha + iby}{\alpha^2 + \beta^2 y^2},$$

$$-i(-d - b\Omega) = by - i\alpha = R(\cos\theta + i\sin\theta),$$

$$-i(\ \ a + b\Omega) = \frac{by + i\alpha}{\alpha^2 + \beta^2 y^2} = \frac{R(\cos\theta - i\sin\theta)}{\alpha^2 + \beta^2 y^2},$$

where R is positive; and $\cos\theta$ is positive since y is positive, and thus θ lies between $\frac{1}{2}\pi$ and $-\frac{1}{2}\pi$. Hence

$$\sqrt{\{-i(-d + b\Omega)\}} = \sqrt{R}(\cos\tfrac{1}{2}\theta + i\sin\tfrac{1}{2}\theta),$$

$\cos\frac{1}{2}\theta$ being positive,

$$\sqrt{\{-i(\ \ a + b\Omega)\}} = \frac{\sqrt{R}(\cos\frac{1}{2}\theta - i\sin\frac{1}{2}\theta)}{\sqrt{(\alpha^2 + \beta^2 y^2)}},$$

and thus

$$\sqrt{\{-i(a + b\Omega)\}}\sqrt{\{-i(-d + b\Omega)\}} = \frac{R}{\sqrt{(\alpha^2 + \beta^2 y^2)}} = +1,$$

that is, $\sqrt{(+1)} = +1$, and we thus have, as we should do,

$$\frac{HH'}{\sqrt{(+1)}} = i^{\frac{1}{2}b(d-a)},$$

the equation which was to be verified.

847.

ON THE THEORY OF SEMINVARIANTS.

[From the *Quarterly Journal of Pure and Applied Mathematics*, vol. XXI. (1886), pp. 92—107.]

IN my "Mémoire sur les Hyperdéterminants," *Crelle*, t. XXX. (1846), pp. 1—37, [13, 14, 16*], I gave an investigation for the number of and relations between the quartic invariants of a given binary quantic: the very same formulæ apply to the cubic covariants of a given binary quantic, or what is the same thing to the cubic seminvariants of a given weight, and I propose at present (considering the formulæ in this point of view) to further develope the theory in regard to the solution of the systems of linear equations obtained in the memoir in question. But I first reproduce the investigation as it stands: I recall that the notation is the ordinary hyperdeterminant notation: for instance,

$$U=\tfrac{1}{2}(a,\ b,\ c\between x_1,\ y_1)^2,\quad V=\tfrac{1}{2}(a,\ b,\ c\between x_2,\ y_2)^2.$$

$$\begin{aligned}T\overline{12}^2UV &= (\partial_{x_1}\partial_{y_2}-\partial_{y_1}\partial_{x_2})^2\,UV\\ &= (\xi_1\eta_2-\eta_1\xi_2)^2\,UV\\ &= \xi_1^2U\,.\,\eta_2^2V-2\xi_1\eta_1U\,.\,\xi_2\eta_2V+\eta_1^2U\,.\,\xi_2^2V\\ &= \quad a\,.\,c \quad - \quad 2b\,.\,b \quad + \quad c\,.\,a\\ &= 2\,(ac-b^2);\end{aligned}$$

and that, when the variables do not disappear, each set $(x_1,\ y_1)$, $(x_2,\ y_2)$, ... is ultimately replaced by $(x,\ y)$, so that these are the variables of any covariant.

The investigation* is as follows:

"...We pass to the derivatives of the fourth degree, considering the forms in which all the differential coefficients are of the same order. It is easy to see that we may write

$$\begin{aligned}\square\,UVWX &= (\overline{12}\,.\,\overline{34})^\alpha\,(\overline{13}\,.\,\overline{42})^\beta\,(\overline{14}\,.\,\overline{23})^\gamma\,UVWX\\ &= D_{\alpha,\,\beta,\,\gamma}UVWX \text{ or } D_{\alpha,\,\beta,\,\gamma}.\end{aligned}$$

[* This Collection, vol. I., p. 104: see also, vol. I., p. 117.]

Writing for shortness

$$\overline{12}.\overline{34}=\mathfrak{A},\quad \overline{13}.\overline{42}=\mathfrak{B},\quad \overline{14}.\overline{23}=\mathfrak{C},$$

we have

$$D_{\alpha,\beta,\gamma}=\mathfrak{A}^\alpha\mathfrak{B}^\beta\mathfrak{C}^\gamma UVWX.$$

Suppose $U=V=W=X$, and consider the derivatives which correspond to the same value f of α, β, γ: we have to find how many of these derivatives are independent, and to express the others in terms of these. Since the functions are equal after the differentiations, we may before the differentiations interchange in any manner the symbolic numbers 1, 2, 3, 4: this gives

$$D_{\alpha,\beta,\gamma}=D_{\beta,\gamma,\alpha}=D_{\gamma,\alpha,\beta}=(-)^f D_{\alpha,\gamma,\beta}=(-)^f D_{\gamma,\beta,\alpha}=(-)^f D_{\beta,\alpha,\gamma}.$$

But we have identically

$$\mathfrak{A}+\mathfrak{B}+\mathfrak{C}=0.$$

Multiplying by $\mathfrak{A}^a\mathfrak{B}^b\mathfrak{C}^c$, and applying it to the product $UVWX$, we have

$$D_{a+1,b,c}+D_{a,b+1,c}+D_{a,b,c+1}=0,$$

which, putting $a+b+c=f-1$, gives a system of equations between the derivatives $D_{\alpha,\beta,\gamma}$ for which $\alpha+\beta+\gamma=f$. Reducing these by the conditions just obtained, suppose that Θf denotes the number of partitions of f into three parts, zero included, and the permutations of the parts being disregarded, we have a number Θf of derivatives $D_{\alpha,\beta,\gamma}$, and a number $\Theta(f-1)$ of linear relations between these derivatives. There remains therefore a number $\Theta f-\Theta(f-1)$ of independent derivatives: but when f is an even number, we have included among these the functions $D_{f,0,0}$, that is, $\overline{12}^f.\overline{34}^f UVWX$, which evidently reduces itself to

$$\overline{12}^f UV.\overline{34}^f WX,\ \text{ or }\ B_f(U,V).B_f(W,X),$$

that is to say, to the square of $B_f(U,V)$. We must therefore diminish by unity this number $\Theta f-\Theta(f-1)$, when f is even. Let $E\left(\frac{a}{b}\right)$ denote the integer part of the fraction $\frac{a}{b}$: it can be shown that the required number is equal to

$$E\left(\frac{f}{6}\right)\ \text{ or }\ E\left(\frac{f+3}{6}\right)$$

according as f is even or odd. Giving to f the six forms

$$6g,\quad 6g+1,\quad 6g+2,\quad 6g+3,\quad 6g+4,\quad 6g+5,$$

we obtain for the independent derivatives of the fourth degree the corresponding numbers

$$g,\quad g,\quad g,\quad g+1,\quad g,\quad g+1;$$

there is for example a single derivative for each of the orders 3, 5, 6, 7, 8, 10, two for each of the orders 9, 11, 12, 13, 14, 16, &c.

We may take for independent derivatives, when f is even, the terms of the series $D_{f-3,3,0}$, $D_{f-6,6,0}$, ..., and when f is odd, those of the series $D_{f-1,1,0}$, $D_{f-4,4,0}$, $D_{f-7,7,0}\ldots$: continuing each series until the last term in which the first suffix exceeds or is equal to the second suffix. We have in each case the right number of terms; for example, in the case $f=9$, we take for independent derivatives the two functions D_{810} and D_{540}, and we form the equations

$$\begin{aligned}
D_{900}+D_{810}+D_{801}&=0, & D_{621}+D_{531}+D_{522}&=0,\\
D_{810}+D_{720}+D_{711}&=0, & D_{540}+D_{450}+D_{441}&=0,\\
D_{720}+D_{630}+D_{621}&=0, & D_{531}+D_{441}+D_{432}&=0,\\
D_{711}+D_{621}+D_{612}&=0, & D_{522}+D_{432}+D_{423}&=0,\\
D_{630}+D_{540}+D_{531}&=0, & D_{432}+D_{342}+D_{333}&=0,
\end{aligned}$$

which are to be reduced by means of the formulæ

$$D_{900}=-D_{900}=0, \qquad D_{810}=-D_{801}, \ \&c.$$

We will presently give the solution of these equations; but beginning with the second order and going successively to the ninth order, we form easily the following table:

$$\begin{aligned}
D_{200}&=B_2^2, \\
D_{110}&=-\tfrac{1}{2}B_2^2. \\
\\
D_{300}&=0, \\
D_{210}&, \\
D_{111}&=0. \\
\\
D_{400}&=B_4^2, \\
D_{310}&=-\tfrac{1}{2}B_4^2, \\
D_{220}&=\tfrac{1}{2}B_4^2, \\
D_{211}&=0. \\
\\
D_{500}&=0, \\
D_{410}&, \\
D_{320}&=-D_{410}, \\
D_{311}&=0, \\
D_{221}&=0, \\
\\
D_{600}&=B_6^2, \\
D_{510}&=-\tfrac{1}{2}B_6^2, \\
D_{420}&=\tfrac{1}{6}B_6^2-\tfrac{2}{3}D_{330}, \\
D_{411}&=\tfrac{1}{3}B_6^2+\tfrac{2}{3}D_{330}, \\
D_{330}&, \\
D_{321}&=-\tfrac{1}{6}B_6^2-\tfrac{1}{3}D_{330}, \\
D_{222}&=\tfrac{1}{3}B_6^2+\tfrac{2}{3}D_{330}.
\end{aligned}$$

$$\begin{aligned}
D_{700}&=0, \\
D_{610}&, \\
D_{520}&=-D_{610}, \\
D_{511}&=0, \\
D_{430}&=D_{610}, \\
D_{421}&=0, \\
D_{331}&=0, \\
D_{322}&=0. \\
\\
D_{800}&=B_8^2, \\
D_{710}&=-\tfrac{1}{2}B_8^2, \\
D_{620}&=\tfrac{1}{6}B_8^2-\tfrac{2}{3}D_{530}, \\
D_{611}&=-\tfrac{1}{3}B_8^2-\tfrac{2}{3}D_{530}, \\
D_{530}&, \\
D_{521}&=-\tfrac{1}{12}B_8^2-\tfrac{1}{3}D_{530}, \\
D_{440}&=-\tfrac{1}{30}B_8^2-\tfrac{16}{15}D_{530}, \\
D_{431}&=\tfrac{1}{30}B_8^2+\tfrac{1}{15}D_{530}, \\
D_{422}&=\tfrac{2}{15}B_8^2+\tfrac{4}{15}D_{530}, \\
D_{332}&=-\tfrac{1}{15}B_8^2-\tfrac{2}{15}D_{530}.
\end{aligned}$$

$$\begin{aligned}
D_{900}&=0, \\
D_{810}&, \\
D_{720}&=-D_{810}, \\
D_{711}&=0, \\
D_{630}&=\tfrac{1}{2}D_{810}-\tfrac{1}{2}D_{540}, \\
D_{621}&=\tfrac{1}{2}D_{810}-\tfrac{1}{2}D_{540}, \\
D_{540}&, \\
D_{531}&=-\tfrac{1}{2}D_{810}-\tfrac{1}{2}D_{540}, \\
D_{522}&=0, \\
D_{441}&=0, \\
D_{432}&=\tfrac{1}{2}D_{810}+\tfrac{1}{2}D_{540}, \\
D_{333}&=0.
\end{aligned}$$

For any value of f except $f=2$, 3, or 4, the table commences, for f even and f odd respectively, in the following manner:

$$\begin{array}{llll} D_{f,0,6} = & B_f^2, & D_{f,0,0} = & 0, \\ D_{f-1,1,0} = & -\tfrac{1}{2} B_f^2, & D_{f-1,1,0}, & \\ D_{f-2,2,0} = & \tfrac{1}{6} B_f^2 - \tfrac{2}{3} D_{f-3,3,0}, & D_{f-2,2,0} = & - D_{f-1,1,0}, \\ D_{f-2,1,1} = & \tfrac{2}{3} B_f^2 + \tfrac{3}{2} D_{f-3,3,0}, & D_{f-2,1,1} = & 0, \end{array}$$

but beyond this I do not know the law of the series."

Before going further, I remark that if $\mathfrak{A}$, $\mathfrak{B}$, $\mathfrak{C}$, instead of the foregoing values, denote respectively

$$\begin{aligned} \mathfrak{A} &= (x\partial_{x_1} + y\partial_{y_1})\,\overline{23}, && = (x\xi_1 + y\eta_1)(\xi_2\eta_3 - \xi_3\eta_2), \\ \mathfrak{B} &= (x\partial_{x_2} + y\partial_{y_2})\,\overline{31}, && = (x\xi_2 + y\eta_2)(\xi_3\eta_1 - \xi_1\eta_3), \\ \mathfrak{C} &= (x\partial_{x_3} + y\partial_{y_3})\,\overline{12}, && = (x\xi_3 + y\eta_3)(\xi_1\eta_2 - \xi_2\eta_1), \end{aligned}$$

then identically

$$\mathfrak{A} + \mathfrak{B} + \mathfrak{C} = 0,$$

and the same theory applies to the cubic derivatives $\mathfrak{A}^\alpha\mathfrak{B}^\beta\mathfrak{C}^\gamma . UVW$, that is, to the cubic covariants, or, attending only to the coefficients of the highest powers of x, to the cubic seminvariants.

Or, we may use the Clebschian notation, for instance

$$(a, b, c \between x, y)^2 = (a_0x + a_1y)^2 = (b_0x + b_1y)^2 = \&\text{c.},$$

that is,

$$\begin{aligned} a_0^2 &= b_0^2 = \ldots = a, \\ a_0a_1 &= b_0b_1 = \ldots = b, \\ a_1^2 &= b_1^2 = \ldots = c, \end{aligned}$$

viz. in the language of Prof. Sylvester, $a_0, a_1, b_0, b_1, \ldots$ are here *umbræ*. The invariant is

$$\begin{aligned} (a_0b_1 - a_1b_0)^2 &= a_0^2b_1^2 - 2a_0b_0a_1b_1 + b_0^2a_1^2, \\ &= ac \quad - 2b\,.\,b \quad + ac, \\ &= 2(ac - b^2), \end{aligned}$$

and so in other cases. To apply this directly to the theory of seminvariants, it is more convenient to write

$$(1, b, c \between x, y)^2 = (x + \alpha y)^2 = (x + \beta y)^2 = \&\text{c.},$$

that is,

$$\begin{aligned} \alpha &= \beta = \ldots = b, \\ \alpha^2 &= \beta^2 = \ldots = c, \end{aligned}$$

where $\alpha, \beta, \ldots$ are *umbræ*.

The invariant is

$$\begin{aligned} (\alpha - \beta)^2 &= \alpha^2 - 2\alpha\beta + \beta^2 \\ &= c - 2b\,.\,b + c \\ &= 2(c - b^2); \end{aligned}$$

or, to take the case of a cubic function

$$(1,\ b,\ c,\ d) = (x+\alpha y)^3 = (x+\beta y)^3 = (x+\gamma y)^3,$$

that is,

$$\alpha = \beta = \gamma = b,$$
$$\alpha^2 = \beta^2 = \gamma^2 = c,$$
$$\alpha^3 = \beta^3 = \gamma^3 = d,$$

a seminvariant is

$$\begin{aligned}(\alpha-\beta)^2(\alpha-\gamma) &= \alpha^3 - 2\alpha^2\beta + \alpha\beta^2 - \alpha^2\gamma + 2\alpha\beta\gamma - \beta^2\gamma,\\ &= d - 2bc \ + bc \ - bc \ + 2b^3 \ - bc,\\ &= d - 3bc \ + 2b^3,\end{aligned}$$

and so in other cases. Writing here

$$\mathfrak{A} = \beta - \gamma,\quad \mathfrak{B} = \gamma - \alpha,\quad \mathfrak{C} = \alpha - \beta,$$

the general form of a cubic seminvariant of the weight ω is now $\mathfrak{A}^p\mathfrak{B}^q\mathfrak{C}^r$, or say $D_{p,\,q,\,r}$, where $p+q+r=\omega$ (the new letters p, q, r being used for the exponents, since α, β, γ are now employed in a different sense; and since f will be required as a coefficient, I have also written ω instead of f for the weight). We have, as before,

$$\mathfrak{A} + \mathfrak{B} + \mathfrak{C} = 0,$$

and we again see that the theory as to the number of and relations between the cubic seminvariants is identical with that of the quartic invariants. Observe also that for ω odd, $D_{\omega,\,0,\,0}$ is $=0$, while for ω even, $D_{\omega,\,0,\,0}$, $=\mathfrak{A}^\omega$, is a quadric seminvariant, which is of course not to be reckoned among the proper cubic seminvariants; this exactly corresponds to the B_f^2 which is not a proper quartic invariant.

The choice of the functions $D_{f,\,0,\,0}$, $D_{f-3,\,3,\,0}$, ..., or $D_{f-1,\,1,\,0}$, $D_{f-4,\,4,\,0}$, ..., for expressing in terms of them the other functions has its advantages, but also its disadvantages; in what follows, I in effect replace them by $D_{f,\,0,\,0}$, $D_{f-2,\,2,\,0}$, ... and $D_{f-1,\,1,\,0}$, $D_{f-3,\,3,\,0}$, ... respectively. And instead of considering the whole series of the functions $D_{p,\,q,\,r}$ for a given weight ω, I consider only those of the form $D_{p,\,q,\,0}$; these form a single series $D_{\omega,\,0,\,0}$, $D_{\omega-1,\,1,\,0}$, $D_{\omega-2,\,2,\,0}$, ...; and for shortness, I call them A, B, C, D, &c. respectively, viz. I write

$$A,\ B,\ C,\ D\,... = (\mathfrak{A}^\omega,\ \mathfrak{A}^{\omega-1}\mathfrak{B},\ \mathfrak{A}^{\omega-2}\mathfrak{B}^2,\ \mathfrak{A}^{\omega-3}\mathfrak{B}^3, ...).$$

Suppose first that ω is odd; we have

$$D_{\omega-q-r,\,q,\,r} = (-)^\omega D_{\omega-q-r,\,r,\,q};$$

and, in particular,

$$D_{\omega-2q,\,q,\,q} = - D_{\omega-2q,\,q,\,q};$$

that is,

$$D_{\omega-2q,\,q,\,q} = 0;$$

that is,

$$\mathfrak{A}^\omega = 0,\quad \mathfrak{A}^{\omega-2}\mathfrak{B}\mathfrak{C} = 0,\quad \mathfrak{A}^{\omega-4}\mathfrak{B}^2\mathfrak{C}^2 = 0, ...$$

Hence

$$A = 0;$$

moreover

$$B + C = \mathfrak{A}^{\omega-2}\mathfrak{B}(\mathfrak{A}+\mathfrak{B}) = -\mathfrak{A}^{\omega-2}\mathfrak{B}\mathfrak{C} = 0\,;$$

and similarly $C + 2D + E = 0$, &c.; that is, we have

$$\begin{array}{llll} A & & & = 0, \\ B + C & & & = 0, \\ & C + 2D + E & & = 0, \\ & & D + 3E + 3F + G & = 0, \\ \text{\&c.} & & & \end{array}$$

The readiest way of solving these is to express the other functions in terms of B, D, F, H, &c.; viz. we thus have

	A	B	C	D	E	F	G	H	I	J	K	L	M
B	0	1	− 1	0	1	0	− 3	0	17	0	− 155	0	2073
D				1	− 2	0	5	0	− 28	0	255	0	− 3410
F						1	− 3	0	14	0	− 126	0	1683
H								1	− 4	0	30	0	− 396
J										1	− 5	0	55
L												1	− 6

read according to the columns

$$\begin{aligned} A &= 0, \\ B &= B, \\ C &= -B, \\ D &= D, \\ E &= B - 2D, \\ F &= F, \\ G &= -3B + 5D - 3F, \\ &\;\vdots \end{aligned}$$

and, by way of verification of the numbers, observe that the sum of the numbers in a column is 1 and − 1 alternately.

The formulæ are true for any odd value of ω whatever; but they require an explanation, viz. for any finite value of ω, the terms B, D, F,... are not all of them arbitrary. For any given value of ω the number of arbitrary terms is in fact given by what precedes, and it is also known by Captain MacMahon's theorem, viz. the number of cubic seminvariants is = that of the non-unitary symmetric functions $3^\alpha 2^\beta$ of the proper weight $3\alpha + 2\beta = \omega$; thus for $\omega = 9$, the forms are 3222 and 333; so that there should be only two arbitrary terms.

We have here, writing for shortness 900, 810, &c., instead of D_{900}, D_{810}, &c. respectively,

$$\begin{aligned}
900 &= 0,\\
810 &= B,\\
720 &= -B,\\
630 &= D,\\
540 &= B - 2D,\\
450 &= F,\\
360 &= -3B + 5D - 3F,\\
270 &= H,\\
180 &= 17B - 28D + 14F - 4H,\\
090 &= L.
\end{aligned}$$

But we have $900 = 0$, $810 + 180 = 0$, $720 + 270 = 0$, $630 + 360 = 0$, $540 + 450 = 0$; that is,

$$\begin{aligned}
L &= 0,\\
18B - 28D + 14F - 4H &= 0,\\
-B + H &= 0,\\
-3B + 6D - 3F &= 0,\\
B - 2D + F &= 0,
\end{aligned}$$

all satisfied by $F = -B + 2D$, $H = B$, $L = 0$. The proper course is to stop at the equation for 540, viz. the system is

$$\begin{aligned}
900 &= 0,\\
810 &= B,\\
720 &= -B,\\
630 &= D,\\
540 &= B - 2D,
\end{aligned}$$

equations which may be considered as expressing the several functions 900, 810, 720, 630, 540 in terms of $B = 810$ and $D = 630$. They agree, as they should do, with a foregoing result,

$$\begin{aligned}
&900, = 0,\\
&810,\\
&720, = -810,\\
&630, = \tfrac{1}{2}810 - \tfrac{1}{2}540,\\
&540,
\end{aligned}$$

and it would of course be possible to express the remaining forms 711, 621, 522, 441, 432, and 333 in terms of $B = 810$ and $D = 630$. But observe that the speciality of the present process is that, instead of the whole series of forms 900, 810, 720, 711, 630, 621, 540, 531, 522, 441, 432, 330, we work only with the forms 900, 810, 720, 630, 540, for which the last element is $=0$.

A more simple solution of the system of linear equations in $A, B, C, \ldots$ is the following:

	A	B	C	D	E	F	G	H	I	J	K	L	M
x	0	1	−1	1	−1	1	−1	1	−1	1	−1	1	−1
y				1	−2	3	−4	5	−6	7	−8	9	−10
z						1	−3	6	−10	15	−21	28	−36
w								1	−4	10	−20	35	−56
t										1	−5	15	−35
u												1	−6

read

$$\begin{aligned} A &= 0, \\ B &= x, \\ C &= -x, \\ D &= x + y, \\ E &= -x - 2y, \\ F &= x + 3y + z, \\ G &= -x - 4y - 3z, \end{aligned}$$

where $x, y, z, \ldots$ are arbitrary; the numbers in the table are the binomial coefficients with the signs + and − alternately.

We may, it is clear, express $x, y, z, \ldots$ in terms of $B, D, F, \ldots$, viz. we thus have

$$\begin{aligned} x &= B, \\ y &= -B + D, \\ z &= 2B - 3D + F, \\ &\vdots \end{aligned}$$

and substituting these values, we reproduce the foregoing expressions of $A, B, C, \ldots$ in terms of $B, D, F, \ldots$.

For a given finite value of ω, we have of course the same number of arbitrary coefficients $x, y, z, \ldots$ as there were of arbitrary coefficients $B, D, F, \ldots$; thus for $\omega = 9$, only x and y are arbitrary, and the remaining coefficients z, &c., are determined in terms of these. Or, what is the same thing, we stop with the equation $E = -x - 2y$, next preceding an equation with the non-arbitrary coefficient z.

For the case ω even, I write

$$A+2B=A',$$
$$B+2C=B',$$
$$C+2D=C',$$
&c.;

and I say that the new functions $A', B', C', \ldots$ are related together precisely in the same manner as are the functions $A, B, C, \ldots$ belonging to an odd weight ω; viz. we have

$$A' \qquad\qquad = 0,$$
$$B' + C' \qquad = 0,$$
$$C' + 2D' + E' = 0,$$
$$\vdots$$

which being so, the theory is included in what precedes, and there is no occasion to consider it separately.

To prove this, observe that ω being even, we have

$$D_{\omega-q-r,q,r} = D_{\omega-q-r,r,q},$$

that is,

$$\mathfrak{A}^{\omega-q-r}\mathfrak{B}^q\mathfrak{C}^r = \mathfrak{A}^{\omega-q-r}\mathfrak{B}^r\mathfrak{C}^q,$$

whence, in particular,

$$\mathfrak{A}^{\omega-1}\mathfrak{B} = \mathfrak{A}^{\omega-1}\mathfrak{C},\ \mathfrak{A}^{\omega-3}\mathfrak{B}^2\mathfrak{C} = \mathfrak{A}^{\omega-3}\mathfrak{B}\mathfrak{C}^2,\ \&c.;$$

we thus have

$$A' = \mathfrak{A}^{\omega} + 2\mathfrak{A}^{\omega-1}\mathfrak{B},\ = \mathfrak{A}^{\omega} + \mathfrak{A}^{\omega-1}\mathfrak{B} + \mathfrak{A}^{\omega-1}\mathfrak{C},\ = \mathfrak{A}^{\omega-1}(\mathfrak{A}+\mathfrak{B}+\mathfrak{C}),$$

which is $=0$. Similarly

$$B'+C' = B+3C+2D = \mathfrak{A}^{\omega-1}\mathfrak{B} + 3\mathfrak{A}^{\omega-2}\mathfrak{B}^2 + 2\mathfrak{A}^{\omega-3}\mathfrak{B}^3,$$
$$= \mathfrak{A}^{\omega-3}\mathfrak{B}(\mathfrak{A}^2 + 3\mathfrak{A}\mathfrak{B} + 2\mathfrak{B}^2),\ = \mathfrak{A}^{\omega-3}\mathfrak{B}(\mathfrak{A}+\mathfrak{B})(\mathfrak{A}+2\mathfrak{B}),$$

or, since

$$\mathfrak{A}+\mathfrak{B}+\mathfrak{C}=0,$$

this is

$$= -\mathfrak{A}^{\omega-3}\mathfrak{B}\mathfrak{C}(\mathfrak{A}+2\mathfrak{B}),$$

which, in virtue of

$$\mathfrak{A}^{\omega-3}\mathfrak{B}^2\mathfrak{C} = \mathfrak{A}^{\omega-3}\mathfrak{B}\mathfrak{C}^2,$$

is

$$= -\mathfrak{A}^{\omega-3}\mathfrak{B}\mathfrak{C}(\mathfrak{A}+\mathfrak{B}+\mathfrak{C}),$$

which is $=0$. And similarly

$$C'+2D'+E'=0,\ \&c.$$

I come to a new part of the theory: to fix the ideas, consider the weight $\omega=27$; and write also

$$a_0, a_1, a_2, a_3, \ldots = 1, b, c, d, e, f, g, h, i, j, k, l, m, \ldots,$$

using, if we please, the suffixed a for the higher terms. The cubic seminvariants are 32^{12}, $3^3 2^9$, $3^5 2^6$, $3^7 2^3$, 3^9, viz. the number of them is $=5$; we have thus the 5 terms B, D, F, H, J. Here $B=(\beta-\gamma)^{26}(\gamma-\alpha)=(\alpha-\beta)^{26}(\beta-\gamma)$, or, interchanging the α and β, $=(\alpha-\beta)^{26}(\alpha-\gamma)$, there is an initial term $\alpha^{27}=a_{27}$, and a final term $\alpha^{13}\beta^{13}\gamma=bn^2$, or we may write

$$B=(\alpha-\beta)^{26}(\alpha-\gamma)=a_{27}+bn^2;$$

and similarly for the other forms. We have thus

$$\begin{aligned}
B &= (\alpha-\beta)^{26}(\alpha-\gamma)^1 = a_{27}+bn^2,\\
D &= (\alpha-\beta)^{24}(\alpha-\gamma)^3 = a_{27}+dm^2,\\
F &= (\alpha-\beta)^{22}(\alpha-\gamma)^5 = a_{27}+fl^2,\\
H &= (\alpha-\beta)^{20}(\alpha-\gamma)^7 = a_{27}+hk^2,\\
J &= (\alpha-\beta)^{18}(\alpha-\gamma)^9 = a_{27}+j^3,
\end{aligned}$$

where the initial term $\alpha^{27}=a_{27}$ has in each case the coefficient unity, but the final terms have each of them the proper numerical coefficient.

It must be possible to form linear combinations

$$\begin{aligned}
&B &&= a_{27} + bn^2,\\
&(B,\ D) &&= a_{25}(c+b^2)+dm^2,\\
&(B,\ D,\ F) &&= a_{23}(e+c^2)+fl^2,\\
&(B,\ D,\ F,\ H) &&= a_{21}(g+d^2)+hk^2,\\
&(B,\ D,\ F,\ H,\ J) &&= a_{19}(i+e^2)+j^3,
\end{aligned}$$

where $c+b^2$, $e+c^2$, $g+d^2$, $i+e^2$ denote the seminvariants

$$c-b^2,\ e-4bd+3c^2,\ g-6bf+15ce-10d^2,$$
$$i-8bh+28cg-56df+35e^2,$$

respectively. The expressions indicated by their initial and final terms $a_{27}+bn^2$, $a_{25}c+dm^2$, &c., are what I call "columns;" see my paper "Seminvariant Tables," *American Journal of Mathematics*, t. VII. (1885), pp. 59—73, [831], where, however, the theory is not by any means completely developed.

As to this, observe that B, D are each of them of the form $a_{27}+a_{26}b+a_{25}(c,\ b^2)+\dots$, the proper combination in order to eliminate a_{27} is obviously $B-D$, the term in $a_{26}b$ will then disappear of itself, and the terms in $a_{25}(c,\ b^2)$ will combine into a term $a_{25}(c-b^2)$, viz. in the function *quâ* seminvariant the coefficient of the highest letter a_{25} will be the seminvariant $c-b^2$. Similarly, in the combination B, D, F, the two coefficients will be determinable so that there are no terms in a_{27}, a_{26}, a_{25} or a_{24}, and this being so, the term in a_{23} will be $a_{23}(e-4bd+3c^2)$, viz. in the function, *quâ* seminvariant, the coefficient of the highest letter a_{23} will be the seminvariant $e-4bd+3c^2$. And similarly for the remaining two linear combinations.

As a partial verification, I write

$$B = (\alpha-\beta)^{26}(\alpha-\gamma)$$
$$= \{\alpha^{26}+\beta^{26}-26(\alpha^{25}\beta+\alpha\beta^{25})+325(\alpha^{24}\beta^2+\alpha^2\beta^{24})\ldots\}(\alpha-\gamma),$$

or, retaining only the terms which have an index at least $=25$, this is

$$\begin{aligned} B = \;& \alpha^{27}+\alpha\beta^{26} \quad -26(\alpha^{26}\beta+\alpha^2\beta^{25})+325\alpha^{25}\beta^2 \\ & -(\alpha^{26}+\beta^{26})\gamma-26(\alpha^{25}\beta+\alpha\beta^{25})\gamma, \\ = \;& a_{27}+\quad a_{26}b \\ & -26a_{26}b \;-\; 26a_{25}c \\ & \qquad\qquad +325a_{25}c \\ & -\;\; 2a_{26}b \;+\; 52a_{25}b^2 \\ = \;& a_{27}-27a_{26}b+\quad a_{25}(299c+52b^2); \end{aligned}$$

and similarly

$$D = (\alpha-\beta)^{24}(\alpha-\gamma)^3$$
$$= \{\alpha^{24}+\beta^{24}-24(\alpha^{23}\beta+\alpha\beta^{23})+276(\alpha^{22}\beta^2+\alpha^2\beta^{22})\ldots\}(\alpha^3-3\alpha^2\gamma+3\alpha\gamma^2-\gamma^3),$$

or, retaining only the terms which have an index at least $=25$, this is

$$\begin{aligned} D = \;& \alpha^{27} \quad -24\alpha^{26}\beta \quad +276\alpha^{25}\beta^2 \\ & -3\alpha^{26}\gamma+72\alpha^{25}\beta\gamma \\ & +3\alpha^{25}\gamma^2 \\ = \;& a_{27} \quad -24a_{26}b \quad +276a_{25}c \\ & \qquad -\;\; 3a_{26}b \;+\; 72a_{25}b^2 \\ & \qquad\qquad\qquad +\;\; 3a_{25}c \\ = \;& a_{27} \quad -27a_{26}b \;+\; a_{25}(279c+72b^2), \end{aligned}$$

and thence

$$B-D = a_{25}(20c-20b^2),$$

viz. the coefficient of a_{25} is $=20(c-b^2)$.

But instead of the foregoing forms B, D, F, H, J, we may start with five other forms which are in fact linear combinations of these, viz. these are the forms

$$\begin{aligned} X &= (\alpha-\beta)^{26}(\alpha+\beta-2\gamma) &&= a_{27} \quad + bn^2, \\ Y &= (\alpha-\beta)^{24}(\alpha+\beta-2\gamma)(\alpha-\gamma)(\beta-\gamma) &&= a_{26}(c+b^2)+dm^2, \\ Z &= (\alpha-\beta)^{22}(\alpha+\beta-2\gamma)(\alpha-\gamma)^2(\beta-\gamma)^2 &&= a_{25}(c+b^2)+fl^2, \\ W &= (\alpha-\beta)^{20}(\alpha+\beta-2\gamma)(\alpha-\gamma)^3(\beta-\gamma)^3 &&= a_{24}(e+c^2)+hk^2, \\ T &= (\alpha-\beta)^{18}(\alpha+\beta-2\gamma)(\alpha-\gamma)^4(\beta-\gamma)^4 &&= a_{23}(e+c^2)+j^3, \end{aligned}$$

where as well the initial as the final terms should have their proper numerical coefficients.

As to this, observe that in Y we have a term $\alpha^{26}(\beta-\gamma)$ which is $=0$, and similarly a term $\beta^{25}(\alpha-\gamma)$ which is $=0$; the highest powers are thus α^{25} and β^{25}, giving a term in a_{25} which can only be $a_{25}(c-b^2)$. In Z we have the terms $\alpha^{25}(\beta-\gamma)^2$, and $\beta^{25}(\alpha-\gamma)^2$, giving the term $a_{25}(c-b^2)$; in W we have the terms $\alpha^{24}(\beta-\gamma)^3$ and $\beta^{24}(\alpha-\gamma)^3$ which are each $=0$, the highest powers thus are α^{23}, β^{23} giving a term in a_{23} which can only be $a_{23}(e-4bd+3c^2)$; and in T we have the terms $\alpha^{23}(\beta-\gamma)^4$ and $\beta^{23}(\alpha-\gamma)^4$, which give the term $a_{23}(e-4bd+3c^2)$.

The combinations which give the columns thus are

$$\begin{aligned}
X \quad &= a_{27} \qquad + bn^2, \\
(Y) \quad &= a_{25}(c+b^2)+dm^2, \\
(Y,\ Z) \quad &= a_{23}(e+c^2)+fl^2, \\
(Y,\ Z,\ W) \quad &= a_{21}(g+d^2)+hk^2, \\
(Y,\ Z,\ W,\ T) &= a_{19}(i+e^2)+j^3.
\end{aligned}$$

I start with the form

$$u=(Y,\ Z,\ W,\ T)=Y+\lambda Z+\mu W+\nu T.$$

We know that it is possible to determine the coefficients λ, μ, ν, in suchwise that the linear function shall be of the proper form $a_{19}(i+e^2)+j^3$; or, what is the same thing, that there shall be no term with an index exceeding 19; the conditions to be satisfied are apparently more than three, but of course the number of independent conditions must be $=3$. We have, in particular, to get rid of all the terms of the second degree which precede $a_{19}i$, viz. the coefficients of $a_{25}c$, $a_{24}d$, $a_{23}e$, $a_{22}f$, $a_{21}g$, $a_{20}h$ must all of them vanish. Now the terms of u which produce terms of the second degree are the terms containing any two but not all three of the symbolic letters α, β, γ, say the biliteral terms; we have for instance $\alpha^{25}\beta^2$, $\alpha^{25}\gamma^2$, $\beta^{25}\alpha^2$, $\beta^{25}\gamma^2$ (there are no terms $\gamma^{25}\alpha^2$ or $\gamma^{25}\beta^2$, since the highest power of γ is γ^9), each of which is $=a_{25}c$. But in the development of u, we may in any biliteral term by an interchange of the letters α, β, γ make the letter of highest index to be α, and the other letter to be β; thus the several terms $\alpha^{25}\beta^2$, $\alpha^{25}\gamma^2$, $\beta^{25}\alpha^2$, $\beta^{25}\gamma^2$ may be each of them converted into $\alpha^{25}\beta^2$, and so in other cases. Imagining this to be done, the term in $a_{25}c$ is simply the term in $\alpha^{25}\beta^2$, and in like manner the term in $a_{24}d$ is the term in $\alpha^{24}\beta^3$, and so in other cases; the condition as to the disappearance of the terms $a_{25}c, \ldots, a_{20}h$ is the condition that the terms in $\alpha^{25}\beta^2$, $\alpha^{24}\beta^3$, $\alpha^{23}\beta^4$, $\alpha^{22}\beta^5$, $\alpha^{21}\beta^6$, $\alpha^{20}\beta^7$ shall all of them disappear. And if, in the function u transformed in the foregoing manner, we write for convenience $\alpha=1$, so that u has become a function of β only (and observe that u will contain the factor β^2), then the condition is that the terms in β^2, β^3, β^4, β^5, β^6, β^7 shall all of them disappear. The conditions may in fact be written $u=0$; viz. it is assumed that u is in the first instance transformed into a function of β as just explained, and the equation is then to be understood as denoting that the coefficients of u are to be so determined that as many as possible of the terms (beginning with that which contains the lowest power of β) shall each of them vanish.

Consider any term of u, for instance the term

$$Y=(\alpha-\beta)^{24}(\alpha+\beta-2\gamma)(\alpha-\gamma)(\beta-\gamma).$$

To obtain the biliteral terms in the proper form 1° we write therein $1=0$, 2° we write $\beta=0$, changing γ into β, 3° we write $\alpha=0$, changing β, γ into α, β; we thus obtain

$$(\alpha-\beta)^{24}(\alpha+\beta)\,\alpha\beta,$$
$$\alpha^{24}(\alpha-2\beta)(\alpha-\beta)(\ -\beta),$$
$$\alpha^{24}(\alpha-2\beta)(\ -\beta)(\alpha-\beta);$$

viz. the second and third terms are identical; the first term requires, however, the factor 2 (for any term $\alpha^p\beta^q$ is accompanied by a corresponding term $\alpha^q\beta^p$), so that, omitting this factor throughout, the terms are

$$(\alpha-\beta)^{24}(\alpha+\beta)\,\alpha\beta-\alpha^{24}(\alpha-2\beta)(\alpha-\beta)\,\beta,$$

or, putting therein $\alpha=1$, the terms are

$$(1-\beta)^{24}(1+\beta)\,\beta-(1-2\beta)(1-\beta)\,\beta.$$

Similarly from

$$Z=(\alpha-\beta)^{22}(\alpha+\beta-2\gamma)(\alpha-\gamma)^2(\beta-\gamma)^2,$$

we have

$$(1-\beta)^{22}(1+\beta)\,\beta^2+(1-2\beta)(1-\beta)^2\beta^2,$$

and the like as regards W and T. The result thus is

$$(1+\beta)\,\beta\,(1-\beta)^{24}+\lambda\,(1-\beta)^{22}\beta^2+\mu\,(1-\beta)^{20}\beta^3+\nu\,(1-\beta)^{18}\beta^4$$
$$-(1-2\beta)(1-\beta)\,\beta\,\{1-\lambda\beta\,(1-\beta)+\mu\beta^2(1-\beta)^2-\nu\beta^3(1-\beta)^3\}=0,$$

or, as this may be written,

$$\frac{(1+\beta)(1-\beta)^{23}}{1-2\beta}\left\{1+\lambda\frac{\beta}{(1-\beta)^2}+\mu\frac{\beta^2}{(1-\beta)^4}+\nu\frac{\beta^3}{(1-\beta)^6}\right\}$$
$$-1+\lambda(\beta-\beta^2)-\mu\,(\beta-\beta^2)^2+\nu\,(\beta-\beta^2)^3=0,$$

viz. each side is to be expanded in ascending powers of β, and the coefficients λ, μ, ν are then to be determined so that as many terms as possible shall vanish.

Expanding, we have

$$(1-20\beta+190\beta^2-1138\beta^3+4808\beta^4-15178\beta^5+\ldots)$$
$$\times\{1+\beta\lambda+\beta^2(2\lambda+\mu)+\beta^3(3\lambda+4\mu+\nu)+\beta^4(4\lambda+10\mu+6\nu)+\beta^5(5\lambda+20\mu+21\nu)+\ldots\}$$
$$-1+\beta\lambda+\beta^2(-\lambda-\mu)+\beta^3(2\mu+\nu)+\beta^4(-3\nu)+\beta^5.\,3\nu+\ldots=0,$$

or finally, as far as β^5, the equation is

$$0=0+\beta\,(2\lambda-20)+\beta^2(-19\lambda+190)+\beta^3(153\lambda-14\mu+2\nu-1138)$$
$$+\beta^4(-814\lambda+119\mu-17\nu+4808)+\beta^5(3027\lambda-558\mu+94\nu-15178).$$

We have in the first place $\lambda=10$, which makes the coefficient of β^2 to be $=0$; and we then have

$$-\ 14\mu+\ 2\nu+\ \ 392=0,\ \text{or say}\quad 7\mu-\ \ \nu-\ 196=0,$$
$$119\mu-17\nu-\ 3332=0,\qquad 119\mu-17\nu-3332=0,$$
$$-\ 558\mu+94\nu+15092=0,\qquad 272\mu-47\nu-7546=0,$$

where the second equation is equivalent to the first: the three equations give $\mu=\frac{833}{25}$, $\nu=\frac{931}{25}$. I have since found that proceeding one step further, that is, to β^6, the coefficient is $-8310\lambda+1791\mu-363\nu+36942$, viz. putting this $=0$, and for λ substituting its value $=10$, we have $597\mu-121\nu=15386$, an equation which is in fact satisfied by the values just obtained for λ and μ. For the function (Y, Z, W) we have the same equations with $\nu=0$; and therefore $\lambda=10$, $\mu=28$; and for the function (Y, Z) the same equations with $\mu=0$, $\nu=0$; and therefore $\lambda=10$. The linear combinations thus are

$$\begin{aligned}
X \qquad\qquad\qquad &= a_{27} \qquad\qquad + bn^2,\\
Y \qquad\qquad\qquad &= a_{25}(c+b^2)+dm^2,\\
Y+\ 10Z \qquad\qquad &= a_{23}(e+c^2)+fl^2,\\
Y+\ 10Z+\ 28W \qquad &= a_{21}(g+d^2)+hk^2,\\
25Y+250Z+823W+931T &= a_{19}(i+e^2)+j^3,
\end{aligned}$$

where remark that in $Y+10Z$, by means of the coefficient 10, we make to disappear the two terms $a_{25}c$, $a_{24}d$; in the next function, by means of the coefficients 10 and 28, the four terms $a_{25}c$, $a_{24}d$, $a_{23}e$, $a_{22}f$; and in the last function, by means of the three coefficients, the six terms $a_{25}c$, $a_{24}d$, $a_{23}e$, $a_{22}f$, $a_{21}g$ and $a_{20}h$.

It is possible that a larger number of terms will disappear; but if this is so, the general form shown by the combinations on p. 353 will require modification.

The investigation applies without alteration to any odd weight ω whatever; the condition for the determination of the coefficients $\lambda, \mu, \nu, \ldots$ is obviously

$$\frac{(1+\beta)(1-\beta)^{\omega-4}}{1-2\beta}\left\{1+\lambda\frac{\beta}{(1-\beta)^2}+\mu\frac{\beta^2}{(1-\beta)^4}+\nu\frac{\beta^3}{(1-\beta)^6}+\ldots\right\}$$
$$-1+\lambda(\beta-\beta^2)-\mu(\beta-\beta^2)^2+\nu(\beta-\beta^2)^4-\ldots=0.$$

For the case of an even weight ω, the series of functions is

$$\begin{aligned}
X &= (\alpha-\beta)^{\omega},\\
Y &= (\alpha-\beta)^{\omega-2}(\alpha-\gamma)\ (\beta-\gamma),\\
Z &= (\alpha-\beta)^{\omega-4}(\alpha-\gamma)^2(\beta-\gamma)^2,
\end{aligned}$$

and the condition for the determination of the coefficients $\lambda, \mu, \nu, \ldots$ is in like manner found to be

$$(1-\beta)^{\omega-3}\left\{1+\lambda\frac{\beta}{(1-\beta)^2}+\mu\frac{\beta^2}{(1-\beta)^4}+\nu\frac{\beta^3}{(1-\beta)^6}+\ldots\right\}$$
$$-1+\lambda(\beta-\beta^2)-\mu(\beta-\beta^2)^2+\nu(\beta-\beta^2)^3+\ldots=0,$$

which is to be understood as explained above.

848.

ON THE TRANSFORMATION OF THE DOUBLE-THETA FUNCTIONS.

[From the *Quarterly Journal of Pure and Applied Mathematics*, vol. XXI. (1886), pp. 142—178.]

I PROPOSE to reproduce Hermite's Memoir "Sur la théorie de la transformation des fonctions Abéliennes," *Comptes Rendus*, t. XL. (1855), pp. 249,..., 784, with some changes of notation and developments. Hermite's functions are even or odd according as we have $\mu q + \nu p$ even or odd; viz. his characteristic is $\begin{pmatrix}\mu, & \nu\\ q, & p\end{pmatrix}$, or the letters p, q, are misplaced; I write, therefore, r instead of p, so as to have the characteristic $\begin{pmatrix}\mu, & \nu\\ q, & r\end{pmatrix}$; and then for symmetry it is necessary to interchange the suffixes 2, 3 and the letters c, d; the invariant function of the periods, instead of being as with him

$$\omega_0 v_3 - \omega_3 v_0 + \omega_1 v_2 - \omega_2 v_1,$$

must be taken to be

$$\omega_0 v_2 - \omega_2 v_0 + \omega_1 v_3 - \omega_3 v_1.$$

Moreover, I write A, B for his G, G', so as, instead of

$$(G,\ H,\ G' \between x,\ y)^2,$$

to have in the expressions of the theta-functions the quadric function $(A,\ H,\ B \between x,\ y)^2$; and I alter the arrangement of the memoir so as to separate more completely the preliminary theory from the theory of the transformation.

GENERAL THEORY. Art. Nos. 1 to 21 (*several sub-headings*).

The functions $\Pi\,\{K$, indef. or def.$\}$.

1. Consider a function

$$\Pi \begin{pmatrix}\mu, & \nu\\ q, & r\end{pmatrix} (x,\ y)\,(A,\ H,\ B)\,\{K, \text{ indef. or def.}\},$$

having: the characteristic $\begin{pmatrix}\mu, \nu\\ q, r\end{pmatrix}$, where the characters μ, ν, q, r are positive or negative integers, which may be taken to have each of them only the values (0, 1) at pleasure, so that the number of functions is 2^4, $=16$;

the arguments (x, y);

the parameters, or conjoint quarter-periods, (A, H, B);

and the potency K, a positive integer;

and which is either indefinite or definite, as will be explained; the function, moreover, contains linearly certain arbitrary constants, the number of them depending on the value of K, as will be explained.

The function may be written $\Pi(x, y)$ or in any other less abbreviated form which may be convenient.

2. The function $\Pi(x, y)$ $\{K$ indef.$\}$ is defined by the following four equations:

$$\begin{aligned}\Pi(x+1, y) &= (-)^{\mu}\, \Pi(x, y),\\ \Pi(x, y+1) &= (-)^{\nu}\, \Pi(x, y),\\ \Pi(x+A, y+H) &= (-)^{q}\, \Pi(x, y) \exp. -2\pi K(2x+A),\\ \Pi(x+H, y+B) &= (-)^{r}\, \Pi(x, y) \exp. -2\pi K(2y+B),\end{aligned}$$

and the function $\Pi(x, y)$ $\{K$ def.$\}$ by the same equations, together with the following fifth equation,

$$\Pi(-x, -y) = (-)^{\mu q+\nu r}\, \Pi(x, y);$$

viz. the definite function is an even function or else an odd function of the arguments according as $\mu q+\nu r$ is even or odd. We may call $\mu q+\nu r$ the index; and the function is then even or odd according as the index is even or odd.

It is perhaps worth noticing that it would be allowable to define a function $\Pi(x, y)$ $\{K$, skew def.$\}$, by the corresponding relation

$$\Pi(-x, -y) = -(-)^{\mu q+\nu r}\, \Pi(x, y),$$

but I do not propose here to develope this notion.

3. The four equations give rise to the following one,

$$\Pi(x+a_0+Aa_2+Ha_3,\ y+a_1+Ha_2+Ba_3)$$
$$= (-)^{\mu a_0+\nu a_1+q a_2+r a_3}\, \Pi(x, y) \times \exp. -2\pi K\{2a_2x+2a_3y+(A, H, B\between a_2, a_3)^2\},$$

where a_0, a_1, a_2, a_3 are any positive or negative integers (zero not excluded), and which single equation, in fact, includes the preceding four equations.

4. In regard to the parameters it is to be observed that, if $A, H, B = A_0+i\alpha$, $H_0+i\eta$, $B_0+i\beta$, we must have (α, η, β) a determinate positive quadratic form; viz. this is the necessary and sufficient condition for the convergence of the series for the development of the function.

5. The number of arbitrary constants is for the indefinite function $=K^2$; but for the definite function it is, when K is odd, $=\frac{1}{2}(K^2+1)$; and when K is even, it is $=\frac{1}{2}K^2$, except in the case of a characteristic $\begin{pmatrix}0, & 0\\ q, & r\end{pmatrix}$, when it is $=\frac{1}{2}(K^2+4)$.

In particular, for $K=1$, there is only a single arbitrary constant, which is a mere factor of the function; taking it to be $=1$, as presently explained, we have the 16 theta-functions.

6. The function $\Pi(x, y)$ is developed in a series of exponentials, in the form

$$\Pi(x, y) = \Sigma\, (-)^{mq+nr} A_{m,\,n} \exp.\, i\pi \left\{(2m+\mu)\, x + (2n+\nu)\, y + \frac{1}{4K}(A, H, B\between 2m+\mu,\ 2n+\nu)^2\right\},$$

where m and n have each of them all positive and negative integer values (zero not excluded) from $-\infty$ to ∞. In fact, substituting this series in the four equations, they are all of them satisfied if only

$$A_{m+K,\,n} = A_{m,\,n}\,;\ \ A_{m,\,n+K} = A_{m,\,n}\,.$$

Consequently the following K^2 coefficients remain arbitrary, viz. those with the suffixes

	$0,\ 1, \ldots K-1$;		
0	,,	,,	,,
1			
⋮			
$K-1$	,,	,,	,,

and we have for $\Pi(x, y)$ a sum of K^2 terms, each a determinate series multiplied into one of the arbitrary coefficients $A_{0,\,0}$, $A_{0,\,1}$, &c. The indefinite function thus contains, as already mentioned, K^2 arbitrary constants.

7. Substituting in the fifth equation, we have for the definite function the further condition

$$A_{-m-\mu,\,-n-\nu} = A_{m,\,n},$$

which it is clear will be satisfied generally if only it is satisfied by the coefficients in the foregoing set of K^2 coefficients.

8. In the case K odd, we thus reduce the number of arbitrary coefficients to $\frac{1}{2}(K^2+1)$; the mode in which this takes place is best seen by an example. Suppose $K=3$, so that $A_{m+3,\,n} = A_{m,\,n}$; $A_{m,\,n+3} = A_{m,\,n}$. For the coefficients of the indefinite function, the suffixes are

00, 01, 02,

10, 11, 12,

20, 21, 22.

And if we suppose $\mu=0$, $\nu=1$, then the new condition is $A_{-m,-n-1}=A_{m,n}$, viz. writing down only the suffixes, we thus obtain

$0,\ 0=0,\ -1$	$1,\ 0=-1,\ -1$	$2,\ 0=-2,\ -1,$
$0,\ 1=0,\ -2$	$1,\ 1=-1,\ -2$	$2,\ 1=-2,\ -2,$
$0,\ 2=0,\ -3$	$1,\ 2=-1,\ -3$	$2,\ 2=-2,\ -3,$

that is,

$0,\ 0=0,\ 2$	$1,\ 0=2,\ 2$	$2,\ 0=1,\ 2,$
$0,\ 1=0,\ 1$	$1,\ 1=2,\ 1$	$2,\ 1=1,\ 1,$
$0,\ 2=0,\ 0$	$1,\ 2=2,\ 0$	$2,\ 2=1,\ 0,$

viz. one of these equations 0, 1 = 0, 1 is an identity, but the other equations occur each twice; or we have four equations, each of them an equality between two out of the remaining 8 coefficients; the number of arbitrary coefficients is thus $1+\tfrac{1}{2}(9-1), =5$; and so in general the number is

$$1+\tfrac{1}{2}(K^2-1),\quad =\tfrac{1}{2}(K^2+1).$$

9. When K is even, it is necessary to distinguish between the case $(\mu, \nu)=(0, 0)$ and the remaining three cases $(\mu, \nu)=(1, 0)$, $(0, 1)$ or $(1, 1)$. In the former case, the relation between the coefficients is $A_{-m,-n}=A_{m,n}$; there are four identities, $0, 0=0, 0$; $0, \tfrac{1}{2}K=0, \tfrac{1}{2}K$; $\tfrac{1}{2}K, 0=\tfrac{1}{2}K, 0$; $\tfrac{1}{2}K, \tfrac{1}{2}K=\tfrac{1}{2}K, \tfrac{1}{2}K$; and the remaining K^2-4 equations occur each twice, that is, we have $\tfrac{1}{2}(K^2-4)$ equations, each of them an equality between two of the remaining K^2-4 coefficients; the number of arbitrary coefficients is thus $4+\tfrac{1}{2}(K^2-4), =\tfrac{1}{2}(K^2+4)$.

In the latter case, there are no identities and the K^2 equations occur each twice, that is, we have $\tfrac{1}{2}K^2$ equations, each of them an equality between two of the K^2 coefficients; and we thus have $\tfrac{1}{2}K^2$ arbitrary coefficients.

10. Recapitulating, it thus appears that

for an indefinite function, the number of coefficients $=K^2$;
for a definite function, the number $=\tfrac{1}{2}(K^2+1)$, K odd;
$=\tfrac{1}{2}K^2$, K even, and
$(\mu, \nu)=(1, 0)$, $(0, 1)$ or $(1, 1)$;
$=\tfrac{1}{2}(K^2+4)$, K even, and
$(\mu, \nu)=(0, 0)$.

The Theta-functions.

11. In the particular case $K=1$, the distinction between the indefinite function and the definite function disappears, and we have instead of $\Pi(x, y)$, the theta-functions $\Theta(x, y)$, satisfying the four equations

$$\begin{aligned}
\Theta(x+1, y) &= (-)^{\mu}\,\Theta(x, y),\\
\Theta(x, y+1) &= (-)^{\nu}\,\Theta(x, y),\\
\Theta(x+A, y+H) &= (-)^{q}\,\Theta(x, y)\ \text{exp.} -i\pi(2x+A),\\
\Theta(x+H, y+B) &= (-)^{r}\,\Theta(x, y)\ \text{exp.} -i\pi(2y+B),
\end{aligned}$$

and the fifth equation

$$\Theta(-x,\ -y) \qquad = (-)^{\mu q+\nu r}\,\Theta(x,\ y)\,;$$

as before, $\mu q+\nu r$ is the index of the function.

The four equations are all of them included in the following one:

$$\Theta(x+a_0+Aa_2+Ha_3,\ y+a_1+Ha_2+Ba_3) = (-)^{\mu a_0+\nu a_1+q a_2+r a_3}\,\Theta(x,\ y)$$
$$\times \exp. -i\pi\,\{2a_2x+2a_3y+(A,\ H,\ B\between a_2,\ a_3)^2\},$$

where $a_0,\ a_1,\ a_2,\ a_3$ are each of them any positive or negative integer, zero not excluded.

Moreover, we take $A_{0,0}=1$, and the value of the function thus is

$$\Theta(x,\ y) = \Sigma\,(-)^{mq+nr}\exp.\,i\pi\,\{(2m+\mu)\,x+(2n+\nu)\,y+\tfrac{1}{4}(A,\ H,\ B\between 2m+\mu,\ 2n+\nu)^2\}.$$

12. The sum of two characteristics is the characteristic obtained by taking the sums of the component terms or characters,

$$\begin{pmatrix}\mu, & \nu\\ q, & r\end{pmatrix}+\begin{pmatrix}\mu', & \nu'\\ q', & r'\end{pmatrix}=\begin{pmatrix}\mu+\mu', & \nu+\nu'\\ q+q', & r+r'\end{pmatrix};$$

and similarly for any number of characteristics. I use the sign $=$, but this properly denotes a congruence, mod. 2; and the like as regards the indices.

The sum of two identical characteristics, or generally of any number of characteristics taken each of them any even number of times, is $=\begin{pmatrix}0, & 0\\ 0, & 0\end{pmatrix}$. And this characteristic $\begin{pmatrix}0, & 0\\ 0, & 0\end{pmatrix}$ may be called the characteristic 0.

It should be observed, that the index of the sum is not in general equal to the sum of the indices. To make it so, we must have, for two characteristics,

$$(\mu+\mu')(q+q')+(\nu+\nu')(r+r') = \mu q+\nu r+\mu' q'+\nu' r',$$

that is,

$$\mu q'+\mu' q+\nu r'+\nu' r=0\,;$$

and there is obviously a like formula for the case of more than two characteristics.

Two or more characteristics, such that they have the sum of the indices equal to the index of the sum, are said to be "in direct relation" or "directly related" to each other. The sum of the indices and the index of the sum may differ by unity and we then have the inverse relation; but I do not propose to consider this.

13. Consider any number K of theta-functions, of the same arguments and parameters, but with the same or different characteristics. The product of these functions is in general a function $\Pi(x,\ y)$ {K indef.}, having a characteristic which is $=$ the sum of the characteristics of the theta-functions. In fact, from the four

equations of the theta-functions, we at once obtain for their products $\Pi(x, y)$ the four equations

$$\begin{aligned}
\Pi(x+1, y) &= (-)^{\Sigma\mu}\,\Pi(x, y),\\
\Pi(x, y+1) &= (-)^{\Sigma\nu}\,\Pi(x, y),\\
\Pi(x+A, y+H) &= (-)^{\Sigma q}\,\Pi(x, y)\,\exp. - i\pi K(2x+A),\\
\Pi(x+H, y+B) &= (-)^{\Sigma r}\,\Pi(x, y)\,\exp. - i\pi K(2y+B),
\end{aligned}$$

which proves the theorem.

14. But if the indices are in direct relation to each other, then we have further

$$\Pi(-x, -y) = (-)^{\Sigma\mu\Sigma q+\Sigma\nu\Sigma r}\,\Pi(x, y),$$

and the product is thus a function $\Pi(x, y)\,\{K, \text{def.}\}$.

15. Take the square of a theta-function, the characteristic is $=\begin{pmatrix}0, & 0\\ 0, & 0\end{pmatrix}$ or 0, and we have also twice the index $=0$; viz. the theta-function is in direct relation with itself. Hence the squared function is a function $\Pi\begin{pmatrix}0, & 0\\ 0, & 0\end{pmatrix}(x, y)\,\{2, \text{def.}\}$, and as such it contains linearly $\tfrac{1}{2}(2^2+4)$, $=4$ arbitrary constants. Hence, taking any five squares, since each of them is a function of the form in question, it follows that the squares of the 5 theta-functions are connected by a linear relation.

Göpel's relation between 4 theta-functions.

16. We may in a variety of ways (in fact, in 60 ways, as will presently be shown) select four theta-functions, all of them even, or else two of them even and two odd (that is, having the sum of their indices $=0$), such that the sum of their characteristics is $=0$; for instance, the functions may be

$$P' = \begin{pmatrix}0, & 0\\ 1, & 1\end{pmatrix},\quad P'' = \begin{pmatrix}0, & 0\\ 0, & 0\end{pmatrix},\quad S' = \begin{pmatrix}1, & 1\\ 0, & 1\end{pmatrix},\quad S'' = \begin{pmatrix}1, & 1\\ 1, & 0\end{pmatrix},$$

indices

$$0\quad,\qquad 0\quad,\qquad 1\quad,\qquad 1\ .$$

The functions are thus in direct relation, and the product of the four functions is a function $\Pi\begin{pmatrix}0, & 0\\ 0, & 0\end{pmatrix}\{4, \text{def.}\}$. But obviously any one of the functions taken four times, or any two of them taken each twice, are in like manner four functions in direct relation, or the fourth powers P'^4, P''^4, S'^4, S''^4, and the squared products $P'^2P''^2$, $S'^2S''^2$, $P'^2S'^2$, $P'^2S''^2$, $P''^2S'^2$, $P''^2S''^2$, are in like manner each of them a function $\Pi\begin{pmatrix}0, & 0\\ 0, & 0\end{pmatrix}\{4, \text{def.}\}$, viz. we have thus in all $1+4+6$, $=11$ such functions. But the Π function contains only $\tfrac{1}{2}(4^2+4)$, $=10$ arbitrary constants; hence there must be a linear relation between the 11 powers and products, and this is Göpel's relation.

17. Starting with any two characteristics a, b at pleasure, the remaining characteristics form seven pairs, such that

$$a+b=c+d=e+f=g+h=i+j=k+l=m+n=o+p;$$

but among the seven pairs we have only three, suppose $c+d$, $e+f$, $g+h$, which are such that (a, b, c, d), (a, b, e, f), (a, b, g, h) are each of them either all even or else two even and two odd; that is, starting with any pair (a, b), we have these three tetrads having each of them the required property. The number of pairs (a, b) is $\frac{1}{2}16.15$, $=120$; and we thence derive 120×3, $=360$ tetrads; but each such tetrad is of course derivable from any one of the six pairs contained in it; or the number of distinct tetrads is $\frac{1}{6}360$, $=60$, viz. we have, as mentioned above, 60 Göpel-tetrads.

The four functions Π_0, Π_1, Π_2, Π_3.

18. We consider four theta-functions θ_0, θ_1, θ_2, θ_3, which are such that to the modulus 2, the sum of the characters is $\equiv 0$, and also the sum of the indices is $\equiv 0$; taking the characters to be

$$\begin{pmatrix}\mu, & \nu\\ q, & r\end{pmatrix},\quad \begin{pmatrix}\mu', & \nu'\\ q', & r'\end{pmatrix},\quad \begin{pmatrix}\mu'', & \nu''\\ q'', & r''\end{pmatrix},\quad \begin{pmatrix}\mu''', & \nu'''\\ q''', & r'''\end{pmatrix},$$

and writing throughout $=$ for $\equiv$ (mod. 2), we have

$$\mu+\mu'+\mu''+\mu'''=0,$$
$$\nu+\nu'+\nu''+\nu'''=0,$$
$$q+q'+q''+q'''=0,$$
$$r+r'+r''+r'''=0,$$
$$\mu q+\nu r+\mu' q'+\nu' r'+\mu'' q''+\nu'' r''+\mu''' q'''+\nu''' r'''=0.$$

Writing for shortness $(01)=\mu q'+\mu' q+\nu r'+\nu' r$, and so in other cases; and further $(01)+(02)+(12)=(012)$, &c., then substituting for μ''', ν''', q''', r''' their values from the first four equations, we deduce $(012)=0$; and similarly $(013)=0$, $(023)=0$, $(123)=0$.

19. Consider now a product $\theta_0^{\mathrm{a}}\theta_1^{\mathrm{b}}\theta_2^{\mathrm{c}}\theta_3^{\mathrm{d}}$, where $\mathrm{a}+\mathrm{b}+\mathrm{c}+\mathrm{d}$ is $=$ a given odd number k; the characteristic is

$$\begin{pmatrix}\mu\mathrm{a}+\mu'\mathrm{b}+\mu''\mathrm{c}+\mu'''\mathrm{d}, & \nu\mathrm{a}+\nu'\mathrm{b}+\nu''\mathrm{c}+\nu'''\mathrm{d}\\ q\mathrm{a}+q'\mathrm{b}+q''\mathrm{c}+q'''\mathrm{d}, & r\mathrm{a}+r'\mathrm{b}+r''\mathrm{c}+r'''\mathrm{d}\end{pmatrix},$$

and it hence follows that the index is

$$=\mathrm{a}(\mu q+\nu r)+\mathrm{b}(\mu' q'+\nu' r')+\mathrm{c}(\mu'' q''+\nu'' r'')+\mathrm{d}(\mu''' q'''+\nu''' r''').$$

In fact, forming the index in question, we have first terms in a^2, b^2, c^2, d^2, which upon writing therein a, b, c, d for these values respectively ($\mathrm{a}^2=\mathrm{a}$, &c.) give the required value; we have therefore only to show that the sum of the remaining terms in ab, &c., is $=0$. These terms are

$$\text{ab}\,(01)+\text{ac}\,(02)+\text{ad}\,(03)+\text{bc}\,(12)+\text{bd}\,(13)+\text{cd}\,(23),$$

and writing herein $\text{a}+\text{b}+\text{c}+\text{d}=1$, and thence $\text{ad}=\text{a}(1-\text{a}-\text{b}-\text{c})$, $=\text{ab}+\text{ac}$, and similarly $\text{bd}=\text{ab}+\text{bc}$, $\text{cd}=\text{ac}+\text{bc}$, the terms in question become

$$=\text{ab}\,(013)+\text{ac}\,(023)+\text{bc}\,(123),$$

that is, they become $=0$.

20. We thus see that the function $\theta_0^{\text{a}}\theta_1^{\text{b}}\theta_2^{\text{c}}\theta_3^{\text{d}}$ has an index which is

$$=\text{a ind}\,\theta_0+\text{b ind}\,\theta_1+\text{c ind}\,\theta_2+\text{d ind}\,\theta_3.$$

Consider separately four products $\theta_0^{\text{a}}\theta_1^{\text{b}}\theta_2^{\text{c}}\theta_3^{\text{d}}$, in which the exponents a, b, c, d satisfy successively the relations (always to modulus 2)

$$\begin{array}{ll} \text{b}+\text{d}=0, & \text{c}+\text{d}=0,\\ \text{b}+\text{d}=1, & \text{c}+\text{d}=0,\\ \text{b}+\text{d}=0, & \text{c}+\text{d}=1,\\ \text{b}+\text{d}=1, & \text{c}+\text{d}=1. \end{array}$$

Combining herewith the relation $\text{a}+\text{b}+\text{c}+\text{d}=1$, it follows that the exponents a, b, c, d are

$$\begin{array}{llll} =\text{d}+1, & \text{d}\;, & \text{d}\;, & \text{d}\;,\\ \text{d}\;, & \text{d}+1, & \text{d}\;, & \text{d}\;,\\ \text{d}\;, & \text{d}\;, & \text{d}+1, & \text{d}\;,\\ \text{d}\;, & \text{d}\;, & \text{d}\;, & \text{d}+1, \end{array}$$

in the four cases respectively. Then substituting these values, the characteristics become

$$\begin{pmatrix}\mu, & \nu\\ q, & r\end{pmatrix},\quad \begin{vmatrix}\mu', & \nu'\\ q', & r'\end{vmatrix},\quad \begin{vmatrix}\mu'', & \nu''\\ q'', & r''\end{vmatrix},\quad \begin{vmatrix}\mu''', & \nu'''\\ q''', & r'''\end{vmatrix},$$

viz. the four products have the same characters as θ_0, θ_1, θ_2, θ_3 respectively; and in like manner recollecting that

$$\text{ind}\,\theta_0+\text{ind}\,\theta_1+\text{ind}\,\theta_2+\text{ind}\,\theta_3=0,$$

we see that the four products have the same indices as θ_0, θ_1, θ_2, θ_3 respectively.

More generally write Π_0, Π_1, Π_2, $\Pi_3=\Sigma\theta_0^{\text{a}}\theta_1^{\text{b}}\theta_2^{\text{c}}\theta_3^{\text{d}}$, where for the four cases respectively the exponents a, b, c, d satisfy the conditions already referred to; then Π_0, Π_1, Π_2, Π_3 have the same characteristics, and the same indices, as θ_0, θ_1, θ_2, θ_3 respectively.

21. It can be shown that each of the functions Π contains $\tfrac{1}{2}(k^2+1)$ constants. It will be recollected that we have between θ_0, θ_1, θ_2, θ_3 an equation of the form

$$0=(\theta_0^4,\ \theta_1^4,\ \theta_2^4,\ \theta_3^4,\ \theta_0^2\theta_1^2,\ \theta_0^2\theta_2^2,\ \theta_0^2\theta_3^2,\ \theta_1^2\theta_2^2,\ \theta_1^2\theta_3^2,\ \theta_2^2\theta_3^2,\ \theta_0\theta_1\theta_2\theta_3);$$

this serves to express, say θ_3^4, in lower powers of θ_3; and by successive applications of this equation, we can reduce $\Sigma\theta_0^{a}\theta_1^{b}\theta_2^{c}\theta_3^{d}$ to a form in which d has only one of the values 0, 1, 2 or 3; we do not by this transformation alter the suffix of Π, viz. a term originally of the form Π_0, Π_1, Π_2 or Π_3, will by the transformation give rise only to terms which are of the same form Π_0, Π_1; Π_2, Π_3 (as the case may be). The number of constants in Π_0 is thus

$$= \text{number of partitions of } k \text{ into four parts a, b, c, d,}$$

under the conditions

$$\text{d} = 0 \text{ or } 2; \text{ a odd, b, c each even,}$$

$$\text{d} = 1 \text{ or } 3; \text{ a even, b, c each odd,}$$

where, in reckoning the partitions, the order of the parts is taken into account: the partitions are thus as follows

$$\begin{aligned} \text{d} = 0, \quad & (\text{a} - 1) + \text{b} + \text{c} = k - 1, \\ \text{d} = 1, \quad & \text{a} + (\text{b} - 1) + (\text{c} - 1) = k - 3, \\ \text{d} = 2, \quad & (\text{a} - 1) + \text{b} + \text{c} = k - 3, \\ \text{d} = 3, \quad & \text{a} + (\text{b} - 1) + (\text{c} - 1) = k - 5, \end{aligned}$$

where the parts a or (a − 1), b or (b − 1), c or (c − 1), as the case may be, are all of them even; hence, writing $k' = \tfrac{1}{2}(k-1)$, the cases are

$$\text{a}' + \text{b}' + \text{c}' = k', \; k' - 1, \; k' - 1, \text{ or } k' - 2,$$

where the a′, b′, c′ are odd or even (zero not excluded) at pleasure; as already mentioned, the order of the parts is taken into account: thus the particulars of 3 would be

300,	210,	120,	030,	No. is 10, $= \tfrac{1}{2}4.5$.
	201,	111,	021,	
		102,	012,	
			033.	

Hence, in the four cases respectively, the numbers are

$$\tfrac{1}{2}(k'^2 + 3k' + 2),$$
$$\tfrac{1}{2}(k'^2 + k'),$$
$$\tfrac{1}{2}(k'^2 + k'),$$
$$\tfrac{1}{2}(k'^2 - k'),$$

giving a total

$$= \tfrac{1}{2}(4k'^2 + 4k' + 2), = \tfrac{1}{2}\{(2k' + 1)^2 + 1\},$$

that is, $= \tfrac{1}{2}(k^2 + 1)$. And similarly the number is $= \tfrac{1}{2}(k^2 + 1)$ in the other three cases respectively.

PREPARATION FOR THE TRANSFORMATION. Art. Nos. 22 to 44 (*several sub-headings*).

The Hermitian quartic matrix.

22. Observing that, for the adopted form of the Π- or Θ-functions, the periods

$$\begin{array}{ll} \omega_0,\ \upsilon_0 \text{ are} & 1,\ 0, \\ \omega_1,\ \upsilon_1 & 0,\ 1, \\ \omega_2,\ \upsilon_2 & A,\ H, \\ \omega_3,\ \upsilon_3 & H,\ B, \end{array}$$

so that we have

$$\omega_0\upsilon_2 - \omega_2\upsilon_0 + \omega_1\upsilon_3 - \omega_3\upsilon_1, \quad = H - 0 + 0 - H, \ = 0,$$

we have to consider the automorphic transformation of the bilinear form

$$\omega_0\upsilon_2 - \omega_2\upsilon_0 + \omega_1\upsilon_3 - \omega_3\upsilon_1.$$

23. We write

$$(\omega_0,\ \omega_1,\ \omega_2,\ \omega_3) = \left(\begin{array}{llll} a_0, & a_1, & a_2, & a_3 \\ b_0, & b_1, & b_2, & b_3 \\ c_0, & c_1, & c_2, & c_3 \\ d_0, & d_1, & d_2, & d_3 \end{array}\right)(\Omega_0,\ \Omega_1,\ \Omega_2,\ \Omega_3),$$

$$(\upsilon_0,\ \upsilon_1,\ \upsilon_2,\ \upsilon_3) = \left(\begin{array}{llll} a_0, & a_1, & a_2, & a_3 \\ b_0, & b_1, & b_2, & b_3 \\ c_0, & c_1, & c_2, & c_3 \\ d_0, & d_1, & d_2, & d_3 \end{array}\right)(\Upsilon_0,\ \Upsilon_1,\ \Upsilon_2,\ \Upsilon_3),$$

and the coefficients are assumed to be such that we have identically

$$\omega_0\upsilon_2 - \omega_2\upsilon_0 + \omega_1\upsilon_3 - \omega_3\upsilon_1 = k\,(\Omega_0\Upsilon_2 - \Omega_2\Upsilon_0 + \Omega_1\Upsilon_3 - \Omega_3\Upsilon_1),$$

where k is in the sequel taken to be a positive integer. We obtain by direct substitution the value of $\omega_0\upsilon_2 - \omega_2\upsilon_0 + \omega_1\upsilon_3 - \omega_3\upsilon_1$ in the following form:

	Ω_0	Ω_1	Ω_2	Ω_3
Υ_0	$a_0c_0 - c_0a_0 + b_0d_0 - d_0b_0$	$a_1c_0 - c_1a_0 + b_1d_0 - d_1b_0$	$a_2c_0 - c_2a_0 + b_2d_0 - d_2b_0$	$a_3c_0 - c_3a_0 + b_3d_0 - d_3b_0$
Υ_1	$a_0c_1 - c_0a_1 + b_0d_1 - d_0b_1$	$a_1c_1 - c_1a_1 + b_1d_1 - d_1b_1$	$a_2c_1 - c_2a_1 + b_2d_1 - d_2b_1$	$a_3c_1 - c_3a_1 + b_3d_1 - d_3b_1$
Υ_2	$a_0c_2 - c_0a_2 + b_0d_2 - d_0b_2$	$a_1c_2 - c_1a_2 + b_1d_2 - d_1b_2$	$a_2c_2 - c_2a_2 + b_2d_2 - d_2b_2$	$a_3c_2 - c_3a_2 + b_3d_2 - d_3b_2$
Υ_3	$a_0c_3 - c_0a_3 + b_0d_3 - d_0b_3$	$a_1c_3 - c_1a_3 + b_1d_3 - d_1b_3$	$a_2c_3 - c_2a_3 + b_2d_3 - d_2b_3$	$a_3c_3 - c_3a_3 + b_3d_3 - d_3b_3$

viz. equating this to its value $k(\Omega_0\Upsilon_2 - \Omega_2\Upsilon_0 + \Omega_1\Upsilon_3 - \Omega_3\Upsilon_1)$, we have 4 identities and 12 equations which are, in fact, 6 equations occurring each twice. We have thus six equations, which are the conditions in order that the matrix may be the matrix of automorphic transformation of the bilinear form $\omega_0 v_2 - \omega_2 v_0 + \omega_1 v_3 - \omega_3 v_1$. The six equations may be written

$$\begin{aligned}(ac+bd)_{01} &= 0,\\ (ac+bd)_{02} &= k,\\ (ac+bd)_{03} &= 0,\\ (ac+bd)_{12} &= 0,\\ (ac+bd)_{13} &= k,\\ (ac+bd)_{23} &= 0,\end{aligned}$$

viz. the first of these equations is $a_0c_1 - a_1c_0 + b_0d_1 - b_1d_0 = 0$, and so in other cases. It is convenient to remark that each interchange of two letters a and c, b and d, also interchange of the suffixes 0 and 2, produces a change of sign; thus the second equation may be written $(ca+db)_{02} = -k$, or $(ca+db)_{20} = k$.

24. The inverse matrix is found to be

$$\begin{pmatrix} a_0, & a_1, & a_2, & a_3 \\ b_0, & b_1, & b_2, & b_3 \\ c_0, & c_1, & c_2, & c_3 \\ d_0, & d_1, & d_2, & d_3 \end{pmatrix}^{-1} = \frac{1}{k}\begin{pmatrix} c_2, & d_2, & -a_2, & -b_2 \\ c_3, & d_3, & -a_3, & -b_3 \\ -c_0, & -d_0, & a_0, & b_0 \\ -c_1, & -d_1, & a_1, & b_1 \end{pmatrix};$$

and the determinant of each of the matrices in this formula is $= k^2$.

We have thus

$$k(\Omega_0, \Omega_1, \Omega_2, \Omega_3) = \begin{pmatrix} c_2, & d_2, & -a_2, & -b_2 \\ c_3, & d_3, & -a_3, & -b_3 \\ -c_0, & -d_0, & a_0, & b_0 \\ -c_1, & -d_1, & a_1, & b_1 \end{pmatrix}(\omega_0, \omega_1, \omega_2, \omega_3);$$

and the like formula for the Υ, v. Substituting these values in the equation

$$k(\Omega_0\Upsilon_2 - \Omega_2\Upsilon_0 + \Omega_1\Upsilon_3 - \Omega_3\Upsilon_1) = \omega_0 v_2 - \omega_2 v_0 + \omega_1 v_3 - \omega_3 v_1,$$

we obtain 6 new equations, which are in a different form the conditions for the automorphic transformation.

The 6 new equations may be written

$$\begin{aligned}(02+13)_{cd} &= 0,\\ (02+13)_{ac} &= k,\\ (02+13)_{bc} &= 0,\\ (02+13)_{ad} &= 0,\\ (02+13)_{bd} &= k,\\ (02+13)_{ab} &= 0,\end{aligned}$$

viz. these equations are

$$c_0d_2 - c_2d_0 + c_1d_3 - c_3d_1 = 0, \ \&c.$$

25. It is worth while to show how the foregoing formula for the inverse matrix comes out. Take for instance the diagonal minor

$$\begin{vmatrix} b_1, & b_2, & b_3 \\ c_1, & c_2, & c_3 \\ d_1, & d_2, & d_3 \end{vmatrix},$$

this is

$$= c_1 (db)_{23} + c_2 (db)_{31} + c_3 (db)_{12},$$

which is

$$= c_1 (ac)_{23} + c_2 \{k + (ac)_{31}\} + c_3 (ac)_{12},$$
$$= c_2 k,$$

since the remaining terms destroy each other. And dividing by the determinant, which is $= k^2$, we have the term $c_2 \div k$ of the inverse matrix.

The Symmetrical Hermitian Matrix.

26. We may consider a symmetrical Hermitian matrix, say the matrix

$$\begin{pmatrix} \mathfrak{A}, & \mathfrak{H}, & \mathfrak{G}, & \mathfrak{L} \\ \mathfrak{H}, & \mathfrak{B}, & \mathfrak{F}, & \mathfrak{M} \\ \mathfrak{G}, & \mathfrak{F}, & \mathfrak{C}, & \mathfrak{N} \\ \mathfrak{L}, & \mathfrak{M}, & \mathfrak{N}, & \mathfrak{D} \end{pmatrix},$$

viz. we have

$$\begin{aligned} &\mathfrak{A}\mathfrak{C} - \mathfrak{G}^2 + \mathfrak{H}\mathfrak{N} - \mathfrak{L}\mathfrak{F} = \phi, \\ &\mathfrak{H}\mathfrak{N} - \mathfrak{L}\mathfrak{F} + \mathfrak{B}\mathfrak{D} - \mathfrak{M}^2 = \phi, \\ &\mathfrak{A}\mathfrak{F} - \mathfrak{G}\mathfrak{H} + \mathfrak{H}\mathfrak{M} - \mathfrak{B}\mathfrak{L} = 0, \\ &\mathfrak{A}\mathfrak{N} - \mathfrak{L}\mathfrak{G} + \mathfrak{H}\mathfrak{D} - \mathfrak{L}\mathfrak{M} = 0, \\ &\mathfrak{C}\mathfrak{H} - \mathfrak{F}\mathfrak{G} + \mathfrak{B}\mathfrak{N} - \mathfrak{M}\mathfrak{F} = 0, \\ &\mathfrak{G}\mathfrak{N} - \mathfrak{C}\mathfrak{L} + \mathfrak{F}\mathfrak{D} - \mathfrak{M}\mathfrak{N} = 0. \end{aligned}$$

The characteristic property is that, effecting a Hermitian transformation, we have a new symmetrical Hermitian matrix

$$\begin{pmatrix} \mathfrak{A}, & \mathfrak{H}, & \mathfrak{G}, & \mathfrak{L} \\ \mathfrak{H}, & \mathfrak{B}, & \mathfrak{F}, & \mathfrak{M} \\ \mathfrak{G}, & \mathfrak{F}, & \mathfrak{C}, & \mathfrak{N} \\ \mathfrak{L}, & \mathfrak{M}, & \mathfrak{N}, & \mathfrak{D} \end{pmatrix} (a_0 x + a_1 y + a_2 z + a_3 w,\ b_0 x + \ldots,\ c_0 x + \ldots,\ d_0 x + \ldots)^2$$

$$= \begin{pmatrix} \mathfrak{A}', & \mathfrak{H}', & \mathfrak{G}', & \mathfrak{L}' \\ \mathfrak{H}', & \mathfrak{B}', & \mathfrak{F}', & \mathfrak{M}' \\ \mathfrak{G}', & \mathfrak{F}', & \mathfrak{C}', & \mathfrak{N}' \\ \mathfrak{L}', & \mathfrak{M}', & \mathfrak{N}', & \mathfrak{D}' \end{pmatrix} (x, y, z, w)^2.$$

In fact, the new matrix is

$$(\mathfrak{A}'\ldots) = \left(\begin{array}{cccc} a_0, & b_0, & c_0, & d_0 \\ a_1, & b_1, & c_1, & d_1 \\ a_2, & b_2, & c_2, & d_2 \\ a_3, & b_3, & c_3, & d_3 \end{array}\right) (\mathfrak{A}\ldots) \left(\begin{array}{cccc} a_0, & a_1, & a_2, & a_3 \\ b_0, & b_1, & b_2, & b_3 \\ c_0, & c_1, & c_2, & c_3 \\ d_0, & d_1, & d_2, & d_3 \end{array}\right),$$

where the first factor, *quâ* transposed matrix, is Hermitian. Hence the product is a symmetrical $k^2\phi$ matrix.

27. Write

$$\Delta_0 = H_0^2 - A_0 B_0 - (\eta^2 - \alpha\beta),\quad \theta = \alpha\beta - \eta^2,$$
$$\delta = 2H_0\eta - \alpha B_0 - \beta A_0,$$

and consider the following matrix

$$\left(\begin{array}{cccc} \beta, & -\eta, & -\eta H_0 + \beta A_0, & -\eta B_0 + \beta H_0 \\ -\eta, & \alpha, & -\eta A_0 + \alpha H_0, & -\eta H_0 + \alpha B_0 \\ -\eta H_0 + \beta A_0, & -\eta A_0 + \alpha H_0, & +\alpha\Delta_0 - \delta A_0, & -\delta H_0 + \eta\Delta_0 \\ -\eta B_0 + \beta H_0, & -\eta H_0 + \alpha B_0, & -\delta H_0 + \eta\Delta_0, & +\beta\Delta_0 - \delta B_0 \end{array}\right)(x, y, z, w)^2$$

$$= (\beta, -\eta, \alpha(x + A_0 z + H_0 \omega_1 y + H_0 z + B_0 w)^2 - (\eta^2 - \alpha\beta)(\alpha z^2 + 2\eta z w + \beta w^2);$$

it is easily shown that this is a symmetrical θ^2-matrix. In fact, representing it for a moment by

$$\left(\begin{array}{cccc} \mathfrak{A}, & \mathfrak{H}, & \mathfrak{G}, & \mathfrak{L} \\ \mathfrak{H}, & \mathfrak{B}, & \mathfrak{F}, & \mathfrak{M} \\ \mathfrak{G}, & \mathfrak{F}, & \mathfrak{C}, & \mathfrak{N} \\ \mathfrak{L}, & \mathfrak{M}, & \mathfrak{N}, & \mathfrak{D} \end{array}\right),$$

so that $\mathfrak{A} = \beta$, $\mathfrak{B} = \alpha$, $\mathfrak{H} = -\eta$, &c., we have

$$\mathfrak{A}\mathfrak{C} - \mathfrak{G}^2 + \mathfrak{H}\mathfrak{N} - \mathfrak{L}\mathfrak{F} = \beta(\alpha\Delta_0 - \delta A_0) - (-\eta H_0 + \beta A_0)^2$$
$$- \eta(-\delta H_0 + \eta\Delta_0) - (-\eta B_0 + \beta H_0)(-\eta A_0 + \alpha H_0),$$

which is

$$= \alpha\beta\Delta_0 - \beta\delta A_0 - \eta^2 H_0^2 + 2\beta\eta A_0 H_0 - \beta^2 A_0^2 + \eta\delta H_0 - \eta^2\Delta_0 - \eta^2 A_0 B_0 + \eta\alpha B_0 H_0 + \eta\beta A_0 H_0 - \alpha\beta H_0^2;$$

or, for the two terms $-\beta\delta A_0 + \eta\delta H_0$, $= (\eta H_0 - \beta A_0)\,\delta$, substituting the value

$$(\eta H_0 - \beta A_0)(2H_0\eta - \alpha B_0 - \beta A_0),$$

the whole is found to be

$$= (\alpha\beta - \eta^2)(\Delta_0 - H_0 + A_0 B_0),\ \text{that is,}\ = (\alpha\beta - \eta^2)^2 = \theta^2.$$

And, similarly, $\mathfrak{H}\mathfrak{N} - \mathfrak{L}\mathfrak{F} + \mathfrak{B}\mathfrak{D} - \mathfrak{M}^2$ is found to be $= \theta^2$; and the remaining four combinations of terms to be each of them $= 0$; the matrix is thus a symmetrical θ^2-matrix.

28. It is to be added that the diagonal minors and the determinant

$$\mathfrak{A},\quad \mathfrak{A}\mathfrak{B} - \mathfrak{H}^2,\quad \mathfrak{A}\mathfrak{B}\mathfrak{C} + \text{\&c.},\quad \mathfrak{A}\mathfrak{B}\mathfrak{C}\mathfrak{D} + \text{\&c.},$$

have respectively the values β, θ, $\alpha\theta^2$, θ^4; viz. if α, β, θ are positive, then these are all positive; or the last-mentioned quadric function is a definite positive form.

The general matrix resumed: Arithmetical theory.

29. The matrix containing, as before, the parameter k, may be called a k-matrix; if k be $=1$, it is a unit-matrix. From the fundamental equation

$$k(\Omega_0\Upsilon_2 - \Omega_2\Upsilon_0 + \Omega_1\Upsilon_3 - \Omega_3\Upsilon_1) = \omega_0 v_2 - \omega_2 v_0 + \omega_1 v_3 - \omega_3 v_1,$$

it at once follows that, compounding a k-matrix with a k'-matrix, we have a kk'-matrix; and in particular, compounding a k-matrix with a unit-matrix, we have a k-matrix.

30. The symbols a_0, a_1, b_0, &c., and k, have thus far been arbitrary magnitudes, but we now take them to be integers; and we consider in particular the case where k is a positive odd prime. The number of k-matrices is of course infinite, but if we regard as equivalent any two such matrices which are derivable one from the other by post-multiplication by a unit-matrix (viz. U being a unit-matrix, the matrices M and $M.U$ are regarded as equivalent), then the number of distinct k-matrices is finite, and $=1+k+k^2+k^3$.

The first step is to show that we can by post-multiplication, by a properly determined unit-matrix, reduce the k-matrix to the form

$$\left(\begin{array}{cccc} a_0, & a_1, & a_2, & a_3 \\ 0, & b_1, & b_2, & b_3 \\ 0, & 0, & c_2, & 0 \\ 0, & 0, & d_2, & d_3 \end{array}\right),$$

these values being such as to satisfy identically two out of the six conditions; the remaining conditions present themselves under the two equivalent forms

$$a_0c_2 = k,\ \ b_1d_3 = k,\ \ a_0b_2 + a_1b_3 - a_3b_1 = 0,\ \ a_0d_2 + a_1d_3 = 0,$$

and

$$a_0c_2 = k,\ \ b_1d_3 = k,\ \ a_1c_2 + b_1d_2 \qquad = 0,\ \ -a_3c_2 + b_2d_3 - b_3d_2 = 0.$$

Hence a_0, $c_2 = 1$, k, or k, 1; and b_1, $d_3 = 1$, k, or k, 1; so that, combining these pairs of values, we have four different types of matrix, each type depending on the coefficients a_1, d_2, a_2, b_2, a_3, b_3, connected together by two equations. But the forms of the same type are not distinct from each other, and we have to determine for each of the four types a system of non-equivalent forms comprised therein, and such that from these, by post-multiplication by a unit-matrix as before, the other forms of the type can be obtained. This final system is:—

$$\text{I.}\ \left|\begin{array}{cccc} 1, & 0, & 0, & 0 \\ 0, & 1, & 0, & 0 \\ 0, & 0, & k, & 0 \\ 0, & 0, & 0, & k \end{array}\right|, \qquad \text{II.}\ \left|\begin{array}{cccc} 1, & 0, & 0, & 0 \\ 0, & k, & 0, & i \\ 0, & 0, & k, & 0 \\ 0, & 0, & 0, & 1 \end{array}\right|,$$

$$\text{III.}\ \left|\begin{array}{cccc} k, & i, & i', & 0 \\ 0, & 1, & 1, & 0 \\ 0, & 0, & 1, & 0 \\ 0, & 0, & -i, & k \end{array}\right|, \qquad \text{IV.}\ \left|\begin{array}{cccc} k, & 0, & i', & i \\ 0, & k, & i, & i'' \\ 0, & 0, & 1, & 0 \\ 0, & 0, & 0, & 1 \end{array}\right|,$$

where i, i', i'' are integers, having each of them any one of the values 0, 1, 2, ..., $k-1$, viz. there are in the four types 1, k, k^2, k^3 forms respectively, and the number of forms is thus $=1+k+k^2+k^3$, as already mentioned. I abstain from the further details of the proof.

There is obviously a like theory for which, in place of post-multiplication, we have pre-multiplication; viz. here the matrices M and $U.M$ are regarded as equivalent.

31. Any two k-matrices are reducible one to the other by a combined pre- and post-multiplication; viz. we have always $M' = U.M.U'$, where M, M' are any two k-matrices, and U, U' properly determined unit-matrices; and in particular, M' being any given k-matrix, this is expressible in the foregoing form, where M denotes the principal matrix

$$\left(\begin{array}{cccc} 1, & 0, & 0, & 0 \\ 0, & 1, & 0, & 0 \\ 0, & 0, & k, & 0 \\ 0, & 0, & 0, & k \end{array}\right).$$

Congruence theorems, k an odd number.

32. Taking k an odd number, and using throughout $=$ instead of $\equiv$ (mod. 2), we have the following congruences:

$$(a_0a_2 + a_1a_3,\ b_0b_2 + b_1b_3,\ c_0c_2 + c_1c_3,\ d_0d_2 + d_1d_3)$$

$$= \left(\begin{array}{cccc} a_0, & a_1, & a_2, & a_3 \\ b_0, & b_1, & b_2, & b_3 \\ c_0, & c_1, & c_2, & c_3 \\ d_0, & d_1, & d_2, & d_3 \end{array}\right)\!\!\left(a_2c_2 + b_2d_2,\ a_3c_3 + b_3d_3,\ a_0c_0 + b_0d_0,\ a_1c_1 + b_1d_1\right);$$

or, conversely,

$$(a_2c_2 + b_2d_2,\ a_3c_3 + b_3d_3,\ a_0c_0 + b_0d_0,\ a_1c_1 + b_1d_1)$$

$$= \left(\begin{array}{cccc} c_2, & d_2, & -a_2, & -b_2 \\ c_3, & d_3, & -a_3, & -b_3 \\ -c_0, & -d_0, & a_0, & b_0 \\ -c_1, & -d_1, & a_1, & b_1 \end{array}\right)\!\!\left(a_0a_2 + a_1a_3,\ b_0b_2 + b_1b_3,\ c_0c_2 + c_1c_3,\ d_0d_2 + d_1d_3\right),$$

where observe that on the right-hand side the signs $-$ may be changed into $+$; in fact, to the modulus 2, we have for any integer value whatever $-p \equiv +p$.

33. The first congruence is

$$a_0a_2 + a_1a_3 = a_0(a_2c_2 + b_2d_2) + a_1(a_3c_3 + b_3d_3) + a_2(a_0c_0 + b_0d_0) + a_3(a_1c_1 + b_1d_1),\ \text{say } X = Y,$$

and to verify this, I see no other method than that of considering separately all the combinations of even and odd values of a_0, a_1, a_2, a_3; viz. we have the 16 cases

Case	a_0,	a_2,	a_1,	a_3	X,	Y
1	0	0	0	0	does not exist	
2	0	0	0	1	0	0
3	0	0	1	0	0	0
4	0	0	1	1	1	1
5	0	1	0	0	0	0
6	0	1	0	1	0	0
7	0	1	1	0	0	0
8	0	1	1	1	1	1
9	1	0	0	0	0	0
10	1	0	0	1	0	0
11	1	0	1	0	0	0
12	1	0	1	1	1	1
13	1	1	0	0	1	1
14	1	1	0	1	1	1
15	1	1	1	0	1	1
16	1	1	1	1	0	0

viz. the X column gives in each case the value of $X, = a_0a_2 + a_1a_3$, and then it has to be shown that Y has the same value. The coefficients satisfy the conditions

$$\begin{aligned}
a_0c_2 - a_2c_0 + a_1c_3 - a_3c_1 &= 1,\\
b_0d_2 - b_2d_0 + b_1d_3 - b_3d_1 &= 1,\\
a_0b_2 - a_2b_0 + a_1b_3 - a_3b_1 &= 0,\\
a_0d_2 - a_2d_0 + a_1d_3 - a_3d_1 &= 0,\\
b_0c_2 - b_2c_0 + b_1c_3 - b_3c_1 &= 0,\\
c_0d_2 - c_2d_0 + c_1d_3 - c_3d_1 &= 0,
\end{aligned}$$

so that from the first equation we cannot have a_0, a_1, a_2, a_3 each $= 0$, or the first case does not exist. As to the remaining cases, it is easy to see that they group themselves as follows: 2, 3, 5, 9: 4, 13: 6, 7, 10, 11: 8, 12, 14, 15: 16: the proof being substantially the same for the several cases in the same group.

34. Case 2. We have $Y = b_1 d_1$ which should be $=0$, and in fact, the six equations give $b_1 = 0$, $d_1 = 0$, whence $Y = 0$.

Case 4. We have $Y = c_1 + c_3 + b_1 d_1 + b_3 d_3$ which should $=1$, and in fact, the six equations give $c_3 - c_1 = 1$, $b_3 - b_1 = 0$, $d_3 - d_1 = 0$; whence $c_1 + c_3 = 1$, $b_1 d_1 + b_3 d_3 = 0$, and therefore $Y = 1$.

Case 6. Here $Y = b_0 d_0 + b_1 d_1$ which should $=0$; and in fact, the six equations give $b_0 + b_1 = 0$, $d_0 + d_1 = 0$, whence $Y = 0$.

Case 16. Here $Y = c_0 + c_1 + c_2 + c_3 + b_0 d_0 + b_1 d_1 + b_2 d_2 + b_3 d_3$, which should be $=0$; the six equations give

$$-c_0 - c_1 + c_2 + c_3 = 1,\ \ -b_0 - b_1 + b_2 + b_3 = 0,$$
$$-d_0 - d_1 + d_2 + d_3 = 0,\ \text{and}\ \ b_0 d_2 - b_2 d_0 + b_1 d_3 - b_3 d_1 = 1.$$

Writing

$$b_0 + b_2 = b_1 + b_3 \ \text{and}\ \ d_0 + d_2 = d_1 + d_3,$$

we find

$$b_0 d_0 + b_2 d_2 + b_0 d_2 + b_2 d_0 = b_1 d_1 + b_3 d_3 + b_1 d_3 + b_3 d_1,$$

that is,

$$b_0 d_0 + b_1 d_1 + b_2 d_2 + b_3 d_3 = b_0 d_2 + b_2 d_0 + b_1 d_3 + b_3 d_1,\ =1;$$

and $c_0 + c_1 + c_2 + c_3 = 1$, whence $Y = 0$.

35. Case 8. This is the only case of any difficulty: we have

$$Y = c_1 + c_3 + b_0 d_0 + b_1 d_1 + b_3 d_3,$$

which should be $=1$. The six equations give

$$-c_0 + c_3 - c_1 = 1,\ \ -b_0 + b_3 - b_1 = 0,\ \ -d_0 + d_3 - d_1 = 0,$$

or, say

$$c_0 = -1 - c_1 + c_3,\ \ b_0 = -b_1 + b_3,\ \ d_0 = -d_1 + d_3;$$

or, substituting these values and omitting even terms

$$Y = c_1 + c_3 - b_1 d_3 - b_3 d_1.$$

The remaining three of the six equations are

$$b_0 d_2 - b_2 d_0 + b_1 d_3 - b_3 d_1 = 1,\ \ b_0 c_2 - b_2 c_0 + b_1 c_3 - b_3 c_1 = 0,\ \ c_0 d_2 - c_2 d_0 + c_1 d_3 - c_3 d_1 = 0;$$

or, substituting for b_0, c_0, d_0 their values, these become

$$\begin{aligned}
(1 + c_1 - c_3)\, b_2 + (b_3 - b_1)\, c_2 \qquad\qquad\qquad &= -b_1 c_3 + b_3 c_1,\\
(d_1 - d_3)\, b_2 + \qquad\qquad (b_3 - b_1)\, d_2 &= 1 - b_1 d_3 + b_3 d_1,\\
(d_1 - d_3)\, c_2 + (-1 - c_1 + c_3)\, d_2 &= -c_1 d_3 + c_3 d_1;
\end{aligned}$$

we can from these equations eliminate b_2, c_2, d_2; viz. from the first and third equations eliminating c_2, we have

$$(1 + c_1 - c_3)\{(d_3 - d_1)\, b_2 + (b_1 - b_3)\, d_2\} = (d_3 - d_1)(-b_1 c_3 + b_3 c_1) + (b_3 - b_1)(-c_1 d_3 + c_3 d_1),$$

and this, by means of the second equation, becomes

$$(1+c_1-c_3)(-1+b_1d_3-b_3d_1)=(d_3-d_1)(-b_1c_3+b_3c_1)+(b_3-b_1)(-c_1d_3+c_3d_1),$$

viz. reducing, this is

$$-c_1+c_3-1+b_1d_3-b_3d_1=0;$$

and in virtue of it we have $Y=c_1+c_3-b_1d_3-b_3d_1=1$.

It can be further shown that, to the modulus 2 (k being, as before, odd), we have

$$(a_0a_2+a_1a_3)(c_0c_2+c_1c_3)+(b_0b_2+b_1b_3)(d_0d_2+d_1d_3)=0,$$
$$(a_0c_0+b_0d_0)(a_2c_2+b_2d_2)+(a_1c_1+b_1d_1)(a_3c_3+b_3d_3)=0.$$

To prove the first equation, write for a moment

$$\Omega=(a_0a_2+a_1a_3)(c_0c_2+c_1c_3),\quad \Omega'=(b_0b_2+b_1b_3)(d_0d_2+d_1d_3),$$
$$X=(a_0c_1-a_1c_0)(a_2c_3-a_3c_2),\quad X'=(b_0d_1-b_1d_0)(b_2d_3-b_3d_2);$$

then in virtue of the equations

$$a_0c_1-a_1c_0+b_0d_1-b_1d_0,$$
$$a_2c_3-a_3c_2+b_2d_3-b_3d_2,$$

we have $X=X'$. But we have identically

$$\Omega-X=(a_0c_2+a_1c_3)(a_2c_0+a_3c_1),$$

and from the equation

$$a_0c_2-a_2c_0+a_1c_3-a_3c_1=1,$$

that is, $a_2c_0+a_3c_1=-1+a_0c_2+a_1c_3$, we have $\Omega-X=0$; and similarly $\Omega'-X'=0$, that is, $\Omega-\Omega'=X-X'=0$; we have thus the required equation $\Omega+\Omega'=0$. In a similar manner the second equation may be verified.

36. Write

$$\mu'=\mu a_0+\nu a_1+qa_2+ra_3+a_0a_2+a_1a_3,$$
$$\nu'=\mu b_0+\nu b_1+qb_2+rb_3+b_0b_2+b_1b_3,$$
$$q'=\mu c_0+\nu c_1+qc_2+rc_3+c_0c_2+c_1c_3,$$
$$r'=\mu d_0+\nu d_1+qd_2+rd_3+d_0d_2+d_1d_3.$$

It is to be shown that to the modulus 2 we have

$$\mu'q'+\nu'r'=\mu q+\nu r.$$

In fact, forming the value of $\mu'q'+\nu'r'$, we have first a constant term (term without μ, ν, q, r) which vanishes; next writing $\mu^2=\mu$, the whole term in μ is

$$a_0c_0+b_0d_0+a_0(c_0c_2+c_1c_3)+b_0(d_0d_2+d_1d_3)+c_0(a_0a_2+a_1a_3)+d_0(b_0b_2+b_1b_3),$$

which also vanishes; and similarly the terms in ν, q, r each of them vanish; there remain only the terms in $\mu\nu$, μq, &c. The coefficient of μq is $a_0c_2+a_2c_0+b_0d_2+b_2d_0$, which (always to the modulus 2) is $= a_0c_2 - a_2c_0 + b_0d_2 - b_2d_0$, that is, it is $=1$; and similarly the coefficient of νr is $=1$; and in like manner the coefficients of the other terms are each $=0$; we have thus the required congruence $\mu' q' + \nu' r' = \mu q + \nu r$.

The quintic matrix.

37. I consider a quintic matrix composed of the foregoing coefficients $(a, b, c, d)_{0, 1, 2, 3}$, viz. the matrix contained in a linear transformation which I write as follows:

$$(T', P', Q', R', S') = \begin{array}{c|ccccc|} & 01 & 21 & 2.02 & 03 & 32 \\ \hline ab & . & . & . & . & . \\ cb & . & . & . & . & . \\ ac & . & . & -k & . & . \\ ad & . & . & . & . & . \\ dc & . & . & . & . & . \end{array} (T, P, Q, R, S),$$

read

$$T' = (ab)_{01}\, T + (ab)_{21}\, P + 2\,(ab)_{02}\, Q + (ab)_{03}\, R + (ab)_{32}\, S,$$

where as before $(ab)_{01} = a_0b_1 - a_1b_0$, &c., and so in other cases; in particular, observe that, in the expression of Q', the term involving Q is $\{-k + 2\,(ac)_{02}\}\, Q$ which, in virtue of the relation $(ac + bd)_{02} = k$, may also be written $\{(ac)_{02} - (bd)_{02}\}\, Q$. I notice that in Hermite's paper, p. 366, the term is in effect written without the $-k$, $= 2\,(ac)_{02}\, Q$; the correction of this erratum and of a corresponding one, p. 366, is made p. 787 at the conclusion of the memoir.

38. The matrix is automorphic for the form $T^2 - PR - TS$, viz. we have identically

$$Q'^2 - P'R' - T'S' = k^2 (Q^2 - PR - TS).$$

As a partial verification, consider in $Q'^2 - P'R' - T'S'$ the term containing Q^2. The coefficient of Q^2 is

$$\{-k + 2\,(ac)_{02}\}^2 - 4\,(cb)_{02}\,(ad)_{02} - 4\,(ab)_{02}\,(dc)_{02},$$

where the first term is

$$\{(ac)_{02} - (bd)_{02}\}^2, \text{ which is } = \{(ac)_{02} + (bd)_{02}\}^2 - 4\,(ac)_{02}\,(bd)_{02} = k^2 - 4\,(ac)_{02}\,(bd)_{02}.$$

Hence, observing that we have

$$(ac)_{02}\,(bd)_{02} + (cb)_{02}\,(ad)_{02} + (ab)_{02}\,(dc)_{02} = 0,$$

the whole coefficient is $= k^2$, as it should be; and in like manner the verification may be effected for any other term.

39. We require the following formulæ:

$$\left(\begin{array}{ccc} -c_0, & a_0, & b_0 \\ -c_1, & a_1, & b_1 \\ -c_2, & a_2, & b_2 \\ -c_3, & a_3, & b_3 \end{array}\right)(T', P', Q') = k \left(\begin{array}{ccccc} ., & b_1, & -b_0, & ., & -b_3 \\ ., & ., & b_1, & -b_0, & -b_2 \\ -b_1, & ., & -b_2, & -b_3, & . \\ b_0, & b_2, & b_3, & ., & . \end{array}\right)(T, P, Q, R, S),$$

that is,

$$-c_0T'+a_0P'+b_0Q'=k(b_1P-b_0Q-b_3S),\ \&c.,$$

and

$$\left(\begin{array}{ccc} -d_0, & a_0, & b_0 \\ -d_1, & a_1, & b_1 \\ -d_2, & a_2, & b_2 \\ -d_3, & a_3, & b_3 \end{array}\right)(T',\ Q',\ R')=k\left(\begin{array}{ccccc} ., & -a_1, & a_0, & ., & a_3 \\ ., & ., & -a_1, & a_0, & -a_2 \\ a_1, & ., & a_2, & a_3, & . \\ -a_0, & -a_2, & -a_3, & ., & . \end{array}\right)(T,\ P,\ Q,\ R,\ S).$$

From the first set, multiplying the first, third and fourth equations by T, P, Q respectively and adding, and again multiplying the second, third and fourth equations by T, Q, R, and adding, we obtain

$$-(c_0T+c_2P+c_3Q)\,T'+(a_0T+a_2P+a_3Q)\,P'+(b_0T+b_2P+b_3Q)\,Q'=\quad kb_3(Q^2-PR-TS),$$
$$-(c_1T+c_2Q+c_3R)\,T'+(a_1T+a_2Q+a_3R)\,P'+(b_1T+b_2Q+b_3R)\,Q'=-kb_2(Q^2-PR-TS);$$

and similarly, from the second set of equations, we obtain

$$-(d_0T+d_2P+d_3Q)\,T'+(a_0T+a_2P+a_3Q)\,Q'+(b_0T+b_2P+b_3Q)\,R'=-ka_3(Q^2-PR-TS),$$
$$-(d_1T+d_2Q+d_3R)\,T'+(a_1T+a_2Q+a_3R)\,Q'+(b_1T+b_2Q+b_3R)\,R'=\quad ka_2(Q^2-PR-TS).$$

40. We have the inverse system

$$\begin{array}{rc|ccccc|l} & & dc & ad & 2bd & cb & ab & \\ (T,\ P,\ Q,\ R,\ S)= & 32 & . & . & . & . & . & (T',\ P',\ Q',\ R',\ S'), \\ & 03 & . & . & . & . & . & \\ & 13 & . & . & -k & . & . & \\ & 21 & . & . & . & . & . & \\ & 01 & . & . & . & . & . & \end{array}$$

read

$$T=(dc)_{32}\,T'+(ad)_{32}\,P'+2\,(bd)_{32}\,Q'+(cb)_{32}\,R'+(ab)_{32}\,S';$$

and in particular observe that, in the expression for Q, the term containing Q' is $\{-k+2\,(bd)_{13}\}\,Q'$, where, in virtue of $(ac+bd)_{13}=k$, the coefficient of Q' is also $=(bd)_{13}-(ac)_{13}$.

41. These equations give

$$\left(\begin{array}{ccc} a_0, & a_2, & a_3 \\ b_0, & b_2, & b_3 \\ c_0, & c_2, & c_3 \\ d_0, & d_2, & d_3 \end{array}\right)(T,\ P,\ Q)=k\left(\begin{array}{ccccc} d_3, & ., & -a_3, & -b_3, & . \\ -c_3, & a_3, & b_3, & ., & . \\ ., & d_3, & -c_3, & ., & b_3 \\ ., & ., & d_3, & -c_3, & -a_3 \end{array}\right)(T',\ P',\ Q',\ R',\ S'),$$

and

$$\left(\begin{array}{ccc} a_1, & a_2, & a_3 \\ b_1, & b_2, & b_3 \\ c_1, & c_2, & c_3 \\ d_1, & d_2, & d_3 \end{array}\right)\!\!\between\!(T,\ Q,\ R) = k\left(\begin{array}{ccccc} -d_2, & ., & a_2, & b_2, & . \\ c_2, & -a_2, & -b_2, & ., & . \\ ., & -d_2, & c_2, & ., & -b_2 \\ ., & ., & -d_2, & c_2, & a_2 \end{array}\right)\!\!\between\!(T',\ P',\ Q',\ R',\ S').$$

From the first set of equations, multiplying the first, third and fourth equations by $-S'$, $-R'$, $+Q'$ respectively, and adding, and again multiplying the second, third and fourth equations by $-S'$, Q', $-P'$ respectively, and adding, we deduce

$$T(-a_0S' - c_0R' + d_0Q') + P(-a_2S' - c_2R' + d_2Q') \\ + Q(-a_3S' - c_3R' + d_3Q') = -kc_3(Q'^2 - P'R' - T'S'),$$

$$T(-b_0S' + c_0Q' - d_0P') + P(-b_2S' + c_2Q' - d_2P') \\ + Q(-b_3S' + c_3Q' - d_3P') = \quad kd_3(Q'^2 - P'R' - T'S');$$

and in like manner, from the second set of equations,

$$T(-a_1S' - c_1R' + d_1Q') + Q(-a_2S' - c_2R' + d_2Q') \\ + R(-a_3S' - c_3R' + d_3Q') = -kd_2(Q'^2 - P'R' - T'S'),$$

$$T(-b_1S' + c_1Q' - d_1P') + Q(-b_2S' + c_2Q' - d_2P') \\ + R(-b_3S' + c_3Q' - d_3P') = \quad kc_2(Q'^2 - P'R' - T'S').$$

42. Assume that T, P, Q, R, S and T', P', Q', R', S' are linearly connected as above; and write

$$T : P : Q : R : S = 1 : A : H : B : H^2 - AB,$$
$$T' : P' : Q' : R' : S' = 1 : A' : H' : B' : H'^2 - A'B',$$

equations which establish a like relation between $1, A, H, B, H^2 - AB$, and $1, A', H', B', H'^2 - A'B'$. Observe that these forms are admissible since, if

$$T^2 - PR - QS = 0,$$

then also

$$Q'^2 - P'R' - T'S' = 0.$$

The quantities A, H, B were taken as the parameters of a theta-function; viz. taking

$$A,\ H,\ B = A_0 + i\alpha,\ H_0 + i\eta,\ B_0 + i\beta,$$

then $(\alpha, \eta, \beta\between x, y)^2$ must be a positive form (or what is the same thing, α and $\alpha\beta - \eta^2$ must be positive). If A', H', B' are also the parameters of a theta-function, then writing

$$A',\ H',\ B' = A_0' + i\alpha',\ H_0' + i\eta',\ B_0' + i\beta',$$

$(\alpha', \eta', \beta'\between x, y)^2$ must also be a positive form. It can be shown that, A', H', B' being determined as above, the former condition implies the latter one; viz. if the form $(\alpha, \eta, \beta\between x, y)^2$ be positive, then also the form $(\alpha', \eta', \beta'\between x, y)^2$ will be positive.

Write

$$\begin{aligned}
A &= A_0 + i\alpha, \quad A' = A_0' + i\alpha', \\
B &= B_0 + i\beta, \quad B' = B_0' + i\beta', \\
H &= H_0 + i\eta, \quad H' = H_0' + i\eta'. \\
\Delta &= \Delta_0 + i\delta, \\
\Delta_0 &= H_0^2 - A_0B_0 - (\eta^2 - \alpha\beta), \\
\delta &= 2H_0\eta - B_0\alpha - A_0\beta,
\end{aligned}$$

and let the quintic matrix in the foregoing transformation

$$(T', P', Q', R', S') = (M)(T, P, Q, R, S)$$

be represented by

	t	p	q	r	s
0					
1					
2					
3					
4					

43. It is to be shown that $\alpha' x^2 + 2\eta' xy + \beta' y^2$ is definite and positive.

We require α', β', η'; we have

$$A_0' + i\alpha' = \frac{(t_1,\ p_1,\ q_1,\ r_1,\ s_1)(1,\ A_0 + i\alpha,\ H_0 + i\eta,\ B_0 + i\beta,\ \Delta_0 + i\delta)}{(t_0,\ p_0,\ q_0,\ r_0,\ s_0)(\qquad\qquad\qquad ,, \qquad\qquad\qquad)} = \frac{M_1 + N_1 i}{M_0 + N_0 i},$$

and thence

$$\alpha' = \frac{M_0N_1 - M_1N_0}{M_0^2 + N_0^2} = \frac{1}{\square}(M_0N_1 - M_1N_0).$$

Now

$$\begin{aligned}
M_0N_1 - M_1N_0 = \quad & (t_0p_1 - t_1p_0)\,\alpha \\
& + (t_0q_1 - t_1q_0)\,\eta \\
& + (t_0r_1 - t_1r_0)\,\beta \\
& + (t_0s_1 - t_1s_0)\,\delta \\
& + (p_0q_1 - p_1q_0)\,(A_0\eta - H_0\alpha) \\
& + (p_0r_1 - p_1r_0)\,(A_0\beta - B_0\alpha) \\
& + (p_0s_1 - p_1s_0)\,(A_0\delta - \Delta_0\alpha) \\
& + (q_0r_1 - q_1r_0)\,(H_0\beta - B_0\eta) \\
& + (q_0r_1 - q_1r_0)\,(H_0\beta - B_0\eta) \\
& + (q_0s_1 - q_1s_0)\,(H_0\delta - \Delta_0\eta) \\
& + (r_0s_1 - r_1s_0)\,(B_0\delta - \Delta_0\beta);
\end{aligned}$$

and so for η', β' with only the change of t_1, p_1, q_1, r_1, s_1 into t_2, p_2, q_2, r_2, s_2 and t_3, p_3, q_3, r_3, s_3 respectively; we find, omitting a factor $\frac{k}{\square}$ throughout,

	$\alpha' =$	$\eta' =$	$\beta' =$
α	$b_1^2,$	$-a_1b_1,$	$a_1^2,$
η	$-2b_0b_1,$	$a_0b_1+a_1b_0,$	$-2a_0a_1,$
β	$b_0^2,$	$-a_0b_0,$	$a_0^2,$
δ	$-(b_0b_2+b_1b_3),$	$\frac{1}{2}(a_0b_2+a_2b_0+a_1b_3+a_3b_1),$	$-(a_0a_2+a_1a_3),$
$A_0\eta - H_0\alpha$	$-2b_1b_2,$	$a_1b_2+a_2b_1,$	$-2a_1a_2,$
$H_0\beta - B_0\eta$	$+2b_0b_3,$	$-(a_0b_3+a_3b_0),$	$+2a_0a_3,$
$A_0\beta - B_0\alpha$	$b_0b_2-b_1b_3,$	$-\frac{1}{2}(a_0b_2+a_2b_0-a_1b_3-a_3b_1),$	$a_0a_2-a_1a_3,$
$A_0\delta - \Delta_0\alpha$	$-b_2^2,$	$a_2b_2,$	$-a_2^2,$
$H_0\delta - \Delta_0\eta$	$-2b_2b_3,$	$+a_2b_3+a_3b_2,$	$-2a_2a_3,$
$B_0\delta - \Delta_0\beta$	$-b_3^2,$	$a_3b_3,$	$-a_3^2,$

and hence

$$\alpha'x^2+2\eta'xy+\beta'y^2$$

$$=\frac{k}{\square}\left(\begin{array}{llll} \beta, & -\eta, & -\eta H_0+\beta A_0, & -\eta B_0+\beta H_0 \\ -\eta, & \alpha, & -\eta A_0+\alpha H_0, & -\eta H_0+\alpha B_0 \\ -\eta H_0+\beta A_0, & -\eta A_0+\alpha H_0, & +\alpha\Delta_0-\delta A_0, & -\delta H_0+\eta\Delta_0 \\ -\eta B_0+\beta H_0, & -\eta H_0+\alpha B_0, & \delta H_0-\eta\Delta_0, & +\beta\Delta_0-\delta B_0 \end{array}\right)(b_0x-a_0y,\ b_1x-a_1y,\ b_2x-a_2y,\ b_3x-a_3y)^2.$$

The right-hand is here definite and positive (*suprà* No. 28), hence also the left-hand is definite and positive.

44. As a specimen of the work, observe that we have

$$\begin{aligned} t_0p_1-t_1p_0 &= (ab)_{01}(cb)_{21}-(ab)_{21}(cb)_{01} \\ &= b_1\{b_0(ac)_{21}+b_2(ac)_{10}+b_1(ac)_{02}\} \\ &= b_1[-b_0(bd)_{21}-b_2(bd)_{10}+b_1\{-(bd)_{02}+k\}] \\ &= b_1^2k, \end{aligned}$$

since the remaining terms destroy each other. Dividing by $\square$, and then omitting the factor $\frac{k}{\square}$, we have thus the term αb_1^2 in the foregoing expression for α'.

THE TRANSFORMATION. Art. Nos. 45 to 53.

45. Consider $(a, b, c, d)_{0,1,2,3}$, the components of a quartic matrix as above, and also T, P, Q, R, S; T', P', Q', R', S' connected as already mentioned; and write for shortness

$$z_0=a_0x+b_0y,\quad z_1=a_1x+b_1y,$$
$$z_2=a_2x+b_2y,\quad z_3=a_3x+b_3y,$$

and consider the function

$$\Pi(x,\ y) = \Theta(z_0 + Az_2 + Hz_3,\ z_1 + Hz_2 + Bz_3)\ \text{exp.}\ i\pi\{(z_0z_2 + z_1z_3) + (A,\ H,\ B\between z_2,\ z_3)^2\},$$

where Θ denotes the function

$$\Theta\begin{pmatrix}\mu, & \nu\\ q, & r\end{pmatrix}(A,\ H,\ B).$$

It is to be shown that Π is a function

$$\Pi\begin{pmatrix}\mu', & \nu'\\ q', & r'\end{pmatrix}(x,\ y)(A',\ H',\ B')(K,\ \text{def.}),$$

where the new parameters A', H', B' are given in terms of the original parameters A, H, B by the equations

$$1 : A : H : B : H^2 - AB = T : P : Q : R : S,$$
$$1 : A' : H' : B' : H'^2 - A'B' = T' : P' : Q' : R' : S',$$

and where μ', ν', q', r' have the values

$$\mu' = \mu a_0 + \nu a_1 + qa_2 + ra_3 + a_0a_2 + a_1a_3,$$
$$\nu' = \mu b_0 + \nu b_1 + qb_2 + rb_3 + b_0b_2 + b_1b_3,$$
$$q' = \mu c_0 + \nu c_1 + qc_2 + rc_3 + c_0c_2 + c_1c_3,$$
$$r' = \mu d_0 + \nu d_1 + qd_2 + rd_3 + d_0d_2 + d_1d_3;$$

viz. it is to be shown that the function Π, as above defined, satisfies the fundamental equations

$$\Pi(x+1,\ y) = (-)^{\mu'}\,\Pi(x,\ y),$$
$$\Pi(x,\ y+1) = (-)^{\nu'}\,\Pi(x,\ y),$$
$$\Pi(x + A',\ y + H') = (-)^{q'}\,\Pi(x,\ y)\ \text{exp.} - 2\pi k\,(2x + A'),$$
$$\Pi(x + H',\ y + B') = (-)^{r'}\,\Pi(x,\ y)\ \text{exp.} - 2\pi k\,(2y + B'),$$

and

$$\Pi(-x,\ -y) = (-)^{\mu'\nu' + q'r'}\,\Pi(x,\ y).$$

46. It is proper in the first place to show how it is that A', H', B' are capable of being the parameters of the new function.

Write for shortness

$$X = z_0 + Az_2 + Hz_3,$$
$$Y = z_1 + Hz_2 + Bz_3,$$

and suppose x changed into $x+1$: we have z_0, z_1, z_2, z_3 increased by a_0, a_1, a_2, a_3 respectively; and thence X, Y increased by $a_0 + Aa_2 + Ha_3$, $a_1 + Ha_2 + Ba_3$; the theta-function is thus changed into

$$\Theta(X + a_0 + Aa_2 + Ha_3,\quad Y + a_1 + Ha_2 + Ba_3),$$

viz. the arguments are increased by multiples of the quarter-periods $(1,\ 0,\ A,\ H)$ and $(0,\ 1,\ H,\ B)$.

And, similarly, by the change of y into $y+1$, the theta-function is changed into

$$\Theta(X+b_0+Ab_2+Hb_3,\ Y+b_1+Hb_2+Bb_3),$$

or the arguments are increased by integer multiples of the quarter-periods.

47. But suppose next that x, y are changed into $x+A'$, $y+B'$: we have z_0, z_1, z_2, z_3 increased by

$$A'a_0+H'b_0,\quad A'a_1+H'b_1,\quad A'a_2+H'b_2,\quad A'a_3+H'b_3,$$

and thence

$$\begin{aligned} &X \text{ increased by } A'(a_0+Aa_2+Ha_3)+H'(b_0+Ab_2+Hb_3),\ =c_0+Ac_2+Hc_3,\\ &Y \quad\text{,,}\quad\text{,,}\quad A'(a_1+Ha_2+Ba_3)+H'(b_1+Hb_2+Bb_3),\ =c_1+Hc_2+Bc_3,\end{aligned}$$

since these equalities are the before-mentioned equations

$$\begin{aligned} P'(Ta_0+Pa_2+Qa_3)+Q'(Tb_0+Pb_2+Qb_3)-T'(Tc_0+Pc_2+Qc_3)=0,\\ P'(Ta_1+Qa_2+Ra_3)+Q'(Tb_1+Qb_2+Rb_3)-T'(Tc_1+Qc_2+Rc_3)=0.\end{aligned}$$

It thus appears that, by the change of x, y into $x+A'$, $y+H'$, the theta-function is changed into

$$\Theta(X+c_0+Ac_2+Hc_3,\ Y+c_1+Hc_2+Bc_3),$$

viz. we have again the arguments increased by integer multiples of the quarter-periods; and in like manner, by the change of x, y into $x+H'$, $y+B'$, the theta-function is changed into

$$\Theta(X+d_0+Ad_2+Hd_3,\ Y+d_1+Hd_2+Bd_3),$$

or the arguments are again increased by integer multiples of the quarter-periods.

48. We have to complete the verifications, first for the equation

$$\Pi(x+1,\ y)=(-)^{\mu'}\,\Pi(x,\ y).$$

Reverting to the definition of Π, we have

$$\begin{aligned} &(-)^{\mu'}\,\Pi(x+1,\ y)\div\Pi(x,\ y)\\ &=(-)^{-\mu a_0-\nu a_1-qa_2-ra_3}\ \text{exp.}-i\pi\{a_0a_2+a_1a_3\};\end{aligned}$$

$$\begin{aligned} \Theta(X+a_0+Aa_2+Ha_3,\ Y+a_1+Ha_2+Ba_3)\div\Theta(X,\ Y)\\ =\text{exp.}\ i\pi\{a_0z_2+a_2z_0+a_1z_3+a_3z_1+a_0a_2+a_1a_3\}\times\\ \text{exp.}\ i\pi\{2(Aa_2+Ha_3)z_2+2(Ha_2+Ba_3)z_3+(A,\ H,\ B)(a_2,\ a_3)^2\};\end{aligned}$$

the second line of this is

$$=(-)^{\mu a_0+\nu a_1+qa_2+ra_3}\ \text{exp.}\ i\pi\{-2a_2X-2a_3Y-(A,\ H,\ B)(a_2,\ a_3)^2\},$$

and the right-hand side of the equation will thus be $=1$ if only the whole argument of the exponential be $=0$; that is, omitting the terms which destroy each other and the common factor $i\pi$, if only

$$\begin{aligned} &-2a_2X-2a_3Y\\ &+a_0z_2+a_2z_0+a_1z_3+a_3z_1\\ &+2(Aa_2+Ha_3)z_2+2(Ha_2+Ba_3)z_3=0.\end{aligned}$$

Substituting herein for X, Y their values

$$a_0 + Aa_2 + Ha_3, \quad a_1 + Ha_2 + Ba_3,$$

the equation becomes

$$a_0 z_2 - a_2 z_0 + a_1 z_3 - a_3 z_1 = 0,$$

and finally substituting herein for z_0, z_1, z_2, z_3 their values, the coefficient of x is identically $= 0$, and the coefficient of y is

$$a_0 b_2 - a_2 b_0 + a_1 b_3 - a_3 b_1,$$

which is $(02 + 13)_{ab}$, and is thus $= 0$. This completes the proof of the equation

$$\Pi(x+1, y) = (-)^{\mu'} \Pi(x, y);$$

and we have of course a precisely similar proof for the equation

$$\Pi(x, y+1) = (-)^{\nu'} \Pi(x, y).$$

49. For the next equation, writing for shortness α_0, α_1, α_2, α_3 for $A'a_0 + H'b_0$, $A'a_1 + H'b_1$, $A'a_2 + H'b_2$, $A'a_3 + H'b_3$, so that z_0, z_1, z_2, z_3 are increased by α_0, α_1, α_2, α_3 respectively, we have

$$\begin{aligned} &(-)^{-q'} \Pi(x + A', y + H') \div \Pi(x, y) \\ &= (-)^{-\mu c_0 - \nu c_1 - q c_2 - r c_3} \exp. -i\pi (c_0 c_2 + c_1 c_3); \\ &\Theta(X + c_0 + Ac_2 + Hc_3, Y + c_1 + Hc_2 + Bc_3) \div \Theta(x, y) \\ &= \exp. i\pi \{\alpha_0 z_2 + \alpha_2 z_0 + \alpha_1 z_3 + \alpha_3 z_1 + \alpha_0 \alpha_2 + \alpha_1 \alpha_3\} \times \\ &\quad \exp. i\pi \{2(A\alpha_2 + H\alpha_3) z_2 + 2(H\alpha_2 + B\alpha_3) z_3 + (A, H, B)(\alpha_2, \alpha_3)^2\}. \end{aligned}$$

The second line is here

$$= (-)^{\mu c_0 + \nu c_1 + q c_2 + r c_3} \exp. i\pi \{-2c_2 X - 2c_3 Y - (A, H, B)(c_2, c_3)^2\},$$

and the whole expression should be

$$= \exp. i\pi \{-2kx - kA'\}.$$

Hence bringing these terms over to the left-hand side, the equation will be satisfied if only the whole argument of the exponential be $= 0$; viz. omitting the factor $i\pi$, the equation will be satisfied if only

$$\begin{aligned} &\alpha_0 z_2 + \alpha_2 z_0 + \alpha_1 z_3 + \alpha_3 z_1 + \alpha_0 \alpha_2 + \alpha_1 \alpha_3 - c_0 c_2 - c_1 c_3 \\ &\quad + 2(A, H, B)(\alpha_2, \alpha_3)(z_2, z_3) + (A, H, B)(\alpha_2, \alpha_3)^2 \\ &\quad - 2c_2 X - 2c_3 Y \qquad - (A, H, B)(c_2, c_3)^2 + k(2x + A') = 0. \end{aligned}$$

Substituting here for X, Y their values

$$z_0 + Az_2 + Hz_3, \quad z_1 + Hz_2 + Bz_3,$$

and attending to the values

$$a_0 x + b_0 y, \quad a_1 x + b_1 y, \quad a_2 x + b_2 y, \quad a_3 x + b_3 y$$

of z_0, z_1, z_2, z_3, the equation contains a term in x, a term in y, and a constant term; and we may consider these terms separately.

50. The term in x will vanish if

$$\alpha_0\alpha_2 + \alpha_2\alpha_0 + \alpha_3\alpha_1 + \alpha_1\alpha_3 + 2(A\alpha_2 + H\alpha_3)\alpha_2 + 2(H\alpha_2 + B\alpha_3)\alpha_3$$
$$- 2c_2(a_0 + Aa_2 + Ha_3) - 2c_3(a_1 + Ha_2 + Ba_3) + 2k = 0,$$

and substituting herein for α_0, α_1, α_2, α_3 their values

$$A'a_0 + H'b_0,\quad A'a_1 + H'b_1,\quad A'a_2 + H'b_2,\quad A'a_3 + H'b_3,$$

the equation becomes

$$A'[2a_0a_2 + 2a_1a_3 + 2(Aa_2 + Ha_3)a_2 + 2(Ha_2 + Ba_3)a_3]$$
$$+ H'[a_0b_2 + a_2b_0 + a_1b_3 + a_3b_1 + 2(Aa_2 + Ha_3)b_2 + 2(Ha_2 + Ba_3)b_3]$$
$$- 2c_2(a_0 + Aa_2 + Ha_3) - 2c_3(a_1 + Ha_2 + Ba_3) + 2k = 0,$$

or observing that the first and second lines may be expressed in the form

$$A'[2a_2(a_0 + Aa_2 + Ha_3) + 2a_3(a_1 + Ha_2 + Ba_3)]$$
$$+ H'[2b_2(a_0 + Aa_2 + Ha_3) + 2b_3(a_1 + Ha_2 + Ba_3)] - (a_0b_2 - a_2b_0 + a_1b_3 - a_3b_1),$$

where the term $a_0b_2 - a_2b_0 + a_1b_3 - a_3b_1$, that is, $(02 + 13)_{ab}$, is $= 0$, and may therefore be omitted, the whole equation, omitting the factor 2, which divides out, is

$$(A'a_2 + H'b_2 - c_2)(a_0 + Aa_2 + Ha_3) + (A'a_3 + H'b_3 - c_3)(a_1 + Ha_2 + Ba_3) + k = 0,$$

an equation which, in the form

$$(-c_2T' + a_2P' + b_2Q')(a_0T + a_2P + a_3Q) + (-c_3T' + a_3P' + b_3Q')(a_1T + a_2Q + a_3R) + kTT' = 0,$$

has been above shown to be true. The term in x thus vanishes.

51. The term in y will vanish if

$$\alpha_0b_2 + \alpha_2b_0 + \alpha_1b_3 + \alpha_3b_1 + 2(A\alpha_2 + H\alpha_3)b_2 + 2(H\alpha_2 + B\alpha_3)b_3$$
$$- 2c_2(b_0 + Ab_2 + Hb_3) - 2c_3(b_1 + Hb_2 + Bb_3) = 0;$$

and this is in a similar manner reduced to

$$(A'a_2 + H'b_2 - c_2)(b_0 + Ab_2 + Hb_3) + (A'a_3 + H'b_3 - c_3)(b_1 + Hb_2 + Bb_3) = 0,$$

an equation which, in the form

$$(-c_2T' + a_2P' + b_2Q')(b_0T + b_2P + b_3Q) + (-c_3T' + a_3P' + b_3Q')(b_1T + b_2Q + b_3R) = 0,$$

has been above shown to be true. The term in y thus vanishes.

52. It only remains to show that the constant term also vanishes, viz. that we have

$$\alpha_0\alpha_2 + \alpha_1\alpha_3 + (A,\ H,\ B)(\alpha_2,\ \alpha_3)^2$$
$$- c_0c_2 - c_1c_3 - (A,\ H,\ B)(c_2,\ c_3)^2 + kA' = 0.$$

Substituting here for $\alpha_0, \alpha_1, \alpha_2, \alpha_3$, their values

$$= A'a_0 + H'b_0,\quad A'a_1 + H'b_1,\quad A'a_2 + H'b_2,\quad A'a_3 + H'b_3,$$

the equation becomes

$$\begin{aligned}
A'^2(a_0a_2 + a_1a_3) + A'H'&(a_0b_2 + a_2b_0 + a_1b_3 + a_3b_1) + H'^2(b_0b_2 + b_1b_3)\\
&+ \;\; A\,[A'^2a_2^2 + 2A'H'a_2b_2 \qquad\qquad + H'^2b_2^2\,]\\
&+ 2H\,[A'^2a_2a_3 + A'H'(a_2b_3 + a_3b_2) \quad + H'^2b_2b_3]\\
&+ \;\; B\,[A'^2a_3^2 + 2A'H'a_3b_3 \qquad\qquad + H'^2b_3^2\,]\\
&- c_0c_2 - c_1c_3 - (A,\ H,\ B\between c_2,\ c_3)^2 + kA' = 0.
\end{aligned}$$

This may be written

$$\begin{aligned}
A'^2\,&\{a_2(a_0 + Aa_2 + Ha_3) + a_3(a_1 + Ha_2 + Ba_3)\}\\
+ A'H'\,&\{b_2(a_0 + Aa_2 + Ha_3) + b_3(a_1 + Ha_2 + Ba_3)\\
&\quad + a_2(b_0 + Ab_2 + Hb_3) + a_3(b_1 + Hb_2 + Bb_3)\}\\
+ \;\; H'^2\,&\{b_2(b_0 + Ab_2 + Hb_3) + b_3(b_1 + Hb_2 + Bb_3)\}\\
&- c_2(c_0 + Ac_2 + Hc_3) - c_3(c_1 + Hc_2 + Bc_3) + kA' = 0.
\end{aligned}$$

Writing herein

$$\begin{aligned}
c_0 + Ac_2 + Hc_3 &= (a_0 + Aa_2 + Ha_3)\,A' + (b_0 + Ab_2 + Hb_3)\,H',\\
c_1 + Hc_2 + Bc_3 &= (a_1 + Ha_2 + Ba_3)\,A' + (b_1 + Hb_2 + Bb_3)\,H',
\end{aligned}$$

equations which are true in virtue of

$$\begin{aligned}
-(c_0T + c_2P + c_3Q)\,T' + (a_0T + a_2P + a_3Q)\,P' + (b_0T + b_2P + b_3Q)\,Q' = 0,\\
-(c_1T + c_2Q + c_3R)\,T' + (a_1T + a_2Q + a_3R)\,P' + (b_1T + b_2Q + b_3R)\,Q' = 0,
\end{aligned}$$

the equation becomes

$$\begin{aligned}
&(a_0 + Aa_2 + Ha_3)\,A'\,(A'a_2 + H'b_2 - c_2)\\
&+ (a_1 + Ha_2 + Ba_3)\,A'\,(A'a_3 + H'b_3 - c_3)\\
&+ (b_0 + Ab_2 + Hb_3)\,H'\,(A'a_2 + H'b_2 - c_2)\\
&+ (b_1 + Hb_2 + Bb_3)\,H'\,(A'a_3 + H'b_3 - c_3) + kA' = 0,
\end{aligned}$$

that is,

$$\begin{aligned}
A'\,&[(a_0 + Aa_2 + Ha_3)(-c_2 + A'a_2 + H'b_2) + (a_1 + Ha_2 + Ba_3)(-c_3 + A'a_3 + H'b_3) + k]\\
+ H'\,&[(b_0 + Ab_2 + Hb_3)(-c_2 + A'a_2 + H'b_2) + (b_1 + Hb_2 + Bb_3)(-c_3 + A'a_3 + H'b_3)] = 0\,;
\end{aligned}$$

and we have the coefficients of A' and H' each $=0$, in virtue of

$$\begin{aligned}
(a_0T + a_2P + a_3Q)(-c_2T' + a_2P' + b_2Q') + (a_1T + a_2Q + a_3R)(-c_3T' + a_3P' + b_3Q') + kTT' &= 0,\\
(b_0T + b_2P + b_3Q)(-c_2T' + a_2P' + b_2Q') + (b_1T + b_2Q + b_3R)(-c_3T' + a_3P' + b_3Q') \qquad &= 0.
\end{aligned}$$

53. This completes the proof of the equation

$$\Pi(x+A', \; y+H') = (-)^{q'} \Pi(x, y) \exp. -i\pi k(2x+A');$$

the proof of the remaining equation

$$\Pi(x+H', \; y+B') = (-)^{r'} \Pi(x, y) \exp. -i\pi k(2y+B'),$$

is of course precisely similar.

It has already been seen that $\mu' q' + \nu' r' = \mu q + \nu r$; we have

$$\Theta(-X, -Y) = (-)^{\mu q+\nu r} \Theta(X, Y),$$

and thence

$$\Pi(-x, -y) = (-)^{\mu q+\nu r} \Pi(x, y),$$

that is,

$$\Pi(-x, -y) = (-)^{\mu' q'+\nu' r'} \Pi(x, y),$$

which is the last of the equations which should be satisfied by the function $\Pi(x, y)$.

RECAPITULATION, AND FINAL FORM. Art. Nos. 54 to 58.

54. Recapitulating, we have a Hermitian k-matrix (k an odd prime)

$$\begin{pmatrix} a_0, & a_1, & a_2, & a_3 \\ b_0, & b_1, & b_2, & b_3 \\ c_0, & c_1, & c_2, & c_3 \\ d_0, & d_1, & d_2, & d_3 \end{pmatrix},$$

susceptible of $(1+k+k^2+k^3)$ forms; further, writing for shortness

$$z_0 = a_0 x + b_0 y,$$
$$z_1 = a_1 x + b_1 y,$$
$$z_2 = a_2 x + b_2 y,$$
$$z_3 = a_3 x + b_3 y,$$

and then

$$X = z_0 + A z_2 + H z_3,$$
$$Y = z_1 + H z_2 + B z_3,$$

and assuming

$$\Pi(x, y) = \Theta\begin{pmatrix} \mu, & \nu \\ q, & r \end{pmatrix}(X, Y)(A, H, B) \exp. i\pi \{(z_0 z_2 + z_1 z_3) + (A, H, B)(z_2, z_3)^2\},$$

then $\Pi(x, y)$ satisfies the equations

$$\Pi(x+1, \; y) = (-)^{\mu'} \Pi(x, y),$$
$$\Pi(x, \; y+1) = (-)^{\nu'} \Pi(x, y),$$
$$\Pi(x+A', \; y+H') = (-)^{q'} \Pi(x, y) \exp. -i\pi k(2x+A'),$$
$$\Pi(x+H', \; y+B') = (-)^{r'} \Pi(x, y) \exp. -i\pi k(2y+B'),$$
$$\Pi(-x, -y) = (-)^{\mu' q'+\nu' r'} \Pi(x, y),$$

and it is consequently a function

$$\Pi\begin{pmatrix}\mu', & \nu' \\ q', & r'\end{pmatrix}(x,\ y)(A',\ H',\ B')\ \{K,\ \text{def.}\},$$

and as such, contains linearly $\frac{1}{2}(k^2+1)$ constants.

55. The values of the new parameters are given as above, viz. T', P', Q', R', S' being linear functions of T, P, Q, R, S (the coefficients in these relations being given functions of the coefficients $(a,\ b,\ c,\ d)_{0,\,1,\,2,\,3}$ of the Hermitian matrix), then

$$1 : A : H : B : H^2 - AB = T : P : Q : R : S,$$

$$1 : A' : H' : B' : H'^2 - A'B' = T' : P' : Q' : R' : S',$$

which represent the required relations.

56. We consider four such functions $\Pi(x,\ y)$, derived from Θ-functions having respectively the characteristics

$$\begin{pmatrix}\mu_0, & \nu_0 \\ q_0, & r_0\end{pmatrix},\quad \begin{pmatrix}\mu_1, & \nu_1 \\ q_1, & r_1\end{pmatrix},\quad \begin{pmatrix}\mu_2, & \nu_2 \\ q_2, & r_2\end{pmatrix},\quad \begin{pmatrix}\mu_3, & \nu_3 \\ q_3, & r_3\end{pmatrix},$$

being a (0123) system; and having consequently the characteristics

$$\begin{pmatrix}\mu_0', & \nu_0' \\ q_0', & r_0'\end{pmatrix},\quad \begin{pmatrix}\mu_1', & \nu_1' \\ q_1', & r_1'\end{pmatrix},\quad \begin{pmatrix}\mu_2', & \nu_2' \\ q_2', & r_2'\end{pmatrix},\quad \begin{pmatrix}\mu_3', & \nu_3' \\ q_3', & r_3'\end{pmatrix},$$

which also form a (0123) system; say the four are Π_0, Π_1, Π_2, Π_3.

This being so, consider four theta-functions

$$\theta(x,\ y)(A',\ H',\ B'),$$

having respectively the characteristics

$$\begin{pmatrix}\mu_0', & \nu_0' \\ q_0', & r_0'\end{pmatrix},\quad \begin{pmatrix}\mu_1', & \nu_1' \\ q_1', & r_1'\end{pmatrix},\quad \begin{pmatrix}\mu_2', & \nu_2' \\ q_2', & r_2'\end{pmatrix},\quad \begin{pmatrix}\mu_3', & \nu_3' \\ q_3', & r_3'\end{pmatrix},$$

which as already mentioned form a (0123) system; form with these the four sums

$$\Sigma\theta_0^{\mathrm{a}}\theta_1^{\mathrm{b}}\theta_2^{\mathrm{c}}\theta_3^{\mathrm{d}},$$

where in each case $\mathrm{a}+\mathrm{b}+\mathrm{c}+\mathrm{d}=k$, but in the first sum a, in the second sum b, in the third sum c, and in the fourth sum d is of contrary parity to the other three letters (even, if they are odd; odd, if they are even), say the four sums are Σ_0, Σ_1, Σ_2, Σ_3 respectively; each sum depends linearly on $\frac{1}{2}(k^2+1)$ constants.

The general transformation theorem then is

$$\Sigma_0 = \Pi_0,\quad \Sigma_1 = \Pi_1,\quad \Sigma_2 = \Pi_2,\quad \Sigma_3 = \Pi_3;$$

each of which equations in fact represents $\frac{1}{2}(k^2+1)$ equations; viz. on the left-hand side we assign to the constants $\frac{1}{2}(k^2+1)$ systems of values at pleasure, then to each such system there corresponds on the right-hand side a determinate system of values of the $\frac{1}{2}(k^2+1)$ constants.

57. The results may be presented in a more symmetrical form. Writing ξ, η, ζ, ω for the foregoing x, y, z, w, we may consider

$$\Theta\begin{pmatrix}\mu, & \nu\\ q, & r\end{pmatrix}(\xi+A'\zeta+H'\omega,\ \eta+H'\zeta+B'\omega)(A',\ H',\ B')$$

as a function of the four arguments ξ, η, ζ, ω, and express it by

$$\Phi\begin{pmatrix}\mu, & \nu\\ q, & r\end{pmatrix}(\xi,\ \eta,\ \zeta,\ \omega)(A',\ H',\ B').$$

Writing then

$$x=\xi+A'\zeta+H'\omega,$$
$$y=\eta+H'\zeta+B'\omega,$$
$$X,\ Y,\ Z,\ W=\left(\begin{array}{c|cccc|} & a_0, & b_0, & c_0, & d_0\\ & a_1, & b_1, & c_1, & d_1\\ & a_2, & b_2, & c_2, & d_2\\ & a_3, & b_3, & c_3, & d_3\end{array}\right)(\xi,\ \eta,\ \zeta,\ \omega),$$

we have

$$\begin{aligned}z_1+Az_2+Hz_3&=a_0x+b_0y+A(a_2x+b_2y)+H(a_3x+b_3y)\\&=(a_0+Aa_2+Ha_3)(\xi+A'\zeta+H'\omega)\\&\quad+(b_0+Bb_2+Hb_3)(\eta+H'\zeta+B'\omega),\end{aligned}$$

and also $X+AZ+HW=$

$$\begin{aligned}a_0\xi+b_0\eta+c_0\zeta+d_0\omega\quad&=\quad(a_0+Aa_2+Ha_3)\,\xi\\+A(a_2\xi+b_2\eta+c_2\zeta+d_2\omega)\quad&\quad\ +(b_0+Ab_2+Hb_3)\,\eta\\+H(a_3\xi+b_3\eta+c_3\zeta+d_3\omega)\quad&\quad\ +(c_0+Ac_2+Hc_3)\,\zeta\\&\quad\ +(d_0+Ad_2+Hd_3)\,\omega,\end{aligned}$$

or since, as above,

$$A'(a_0+Aa_2+Ha_3)+H'(b_0+Ab_2+Hb_3)=c_0+Ac_2+Hc_3,$$
$$H'(a_0+Aa_2+Ha_3)+B'(b_0+Ab_2+Hb_3)=d_0+Ad_2+Hd_3,$$

we find

$$z_0+Az_2+Hz_3=X+AZ+HW;$$

and in like manner another equation, viz. we have

$$z_0+Az_2+Hz_3=X+AZ+HW,$$
$$z_1+Hz_2+Bz_3=Y+HZ+BW.$$

Moreover

$$
\begin{aligned}
& z_0 z_2 + z_1 z_3 + (A,\ H,\ B\between z_2,\ z_3)^2 \\
= \ & z_2 (z_0 + A z_2 + H z_3) = \quad (a_2 x + b_2 y)(X + AZ + HW) \\
& + z_3 (z_1 + H z_2 + B z_3) \quad + (a_3 x + b_3 y)(Y + HZ + BW) \\
& \qquad = \ a_2 (\xi + A'\zeta + H'\omega)(X + AZ + HW) \\
& \qquad + b_2 (\eta + H'\zeta + B'\omega)(\quad " \quad) \\
& \qquad + a_3 (\xi + A'\zeta + H'\omega)(Y + HZ + BW) \\
& \qquad + b_3 (\eta + H'\zeta + B'\omega)(\quad " \quad),
\end{aligned}
$$

which *quà* function of X, Y, Z, W may be called χ.

58. And we then have the final result in the following form, viz.

$$\text{exp. } i\pi\chi \,.\, \Phi\begin{pmatrix}\mu, & \nu \\ q, & r\end{pmatrix}(X,\ Y,\ Z,\ W)(A,\ H,\ B),$$

in each of the four forms, is a homogeneous function of the order k of the corresponding four forms

$$\Phi\begin{pmatrix}\mu, & \nu \\ q, & r\end{pmatrix}(\xi,\ \eta,\ \zeta,\ \omega)(A',\ H',\ B').$$

849.

ON THE INVARIANTS OF A LINEAR DIFFERENTIAL EQUATION.

[From the *Quarterly Journal of Pure and Applied Mathematics*, vol. XXI. (1886), pp. 257—261.]

CONSIDER the equation of the second order

$$\frac{d^2y}{dx^2} + 2p\frac{dy}{dx} + qy = 0;$$

if we effect a transformation of the dependent variable y, say $y = f.Y$, where f is an arbitrary function of x, we obtain a new equation

$$\frac{d^2Y}{dx^2} + 2P\frac{dY}{dx} + QY = 0,$$

where

$$P = p + \frac{f'}{f},$$

$$Q = q + 2p\frac{f'}{f} + \frac{f''}{f},$$

(the accents denoting differentiation in regard to x); and we thence establish the identity

$$Q - P^2 - P' = q - p^2 - p',$$

viz. we have $q - p^2 - p'$ a function possessing this invariantive property, but, as remarked by Mr Harley, it is the analogue rather of a seminvariant than of an invariant; I will call it an α-seminvariant of the differential equation. The class of function was considered long ago by Sir James Cockle, and more recently by Mr Malet; see Mr Harley's paper "Professor Malet's Classes of Invariants identified with Sir James Cockle's Criticoids," *Proc. R. Soc.*, vol. XXXVIII. (1884), pp. 45—57.

Effecting for the same equation a transformation of the independent variable, say $x = \phi$, an arbitrary function of X, we obtain a new equation

$$\frac{d^2y}{dX^2} + 2P\frac{dy}{dX} + Qy = 0,$$

where

$$P = p\phi_1 - \tfrac{1}{2}\frac{\phi_2}{\phi_1},$$
$$Q = q\phi_1^2,$$

(the subscript numbers denoting differentiation in regard to X). There is not in the present case any function possessing a like invariantive property, say there is not any β-seminvariant; but such functions exist for differential equations of the third and higher orders, and have been considered by Sir James Cockle and Mr Malet. It is to be noticed that, attempting to obtain a β-seminvariant, we are led to the equation

$$Q - P^2 - P_1 = \phi_1^2(q - p^2 - p') + \tfrac{1}{2}\left\{\frac{\phi_3}{\phi_1} - \tfrac{3}{2}\left(\frac{\phi_2}{\phi_1}\right)^2\right\},$$

which but for the last term would be invariantive. It is curious to find this last term presenting itself in the form of a Schwarzian derivative.

Passing to the differential equation of the third order

$$\frac{d^3y}{dx^3} + 3p\frac{d^2y}{dx^2} + 3q\frac{dy}{dx} + r = 0,$$

here the transformation $y = f.Y$ of the dependent variable gives the new equation

$$\frac{d^3Y}{dx^3} + 3P\frac{d^2Y}{dx^2} + 3Q\frac{dY}{dx} + R = 0,$$

where

$$P = p + \frac{f'}{f},$$
$$Q = q + 2p\frac{f'}{f} + \frac{f''}{f},$$
$$R = r + 3q\frac{f'}{f} + 3p\frac{f''}{f} + \frac{f'''}{f};$$

and we thence derive as well the before-mentioned α-seminvariant $q - p^2 - p'$, as also a new α-seminvariant $r - 3pq + 2p^3 - p''$.

Again, effecting the transformation $x = \phi$ of the independent variable, we have the new equation

$$\frac{d^3y}{dX^3} + 3P\frac{d^2y}{dX^2} + 3Q\frac{dy}{dX} + R = 0,$$

where

$$P = p\phi_1 - \frac{\phi_2}{\phi_1},$$
$$Q = q\phi_1^2 - p\phi_2 + \left(\frac{\phi_2}{\phi_1}\right)^2 - \tfrac{1}{3}\frac{\phi_3}{\phi_1},$$
$$R = r\phi_1^3.$$

We thence obtain the identities

$$\frac{P_1+2P^2-3Q}{R^{\frac{2}{3}}}=\frac{p'+2p^2-3q}{r^{\frac{2}{3}}},$$

$$\frac{R_1+3PR}{R^{\frac{4}{3}}}=\frac{r'+3pr}{r^{\frac{4}{3}}};$$

so that we have $\dfrac{p'+2p^2-3q}{r^{\frac{2}{3}}}$ and $\dfrac{r'+3pr}{r^{\frac{4}{3}}}$ as β-seminvariants of the differential equation of the third order.

But these are by no means the best conclusions; it is shown by M. Halphen in his great Memoir, "Mémoire sur la réduction des équations différentielles linéaires aux formes intégrables," *Sav. Etrang.* t. XXVIII. (1884), pp. 1—297, see p. 127, that there is a function invariantive in regard to each of the two transformations, and which is thus an invariant, viz. this is the function

$$p''-3(q'-2pp')+2(r-3pq+2p^3);$$

this is for the first transformation unaltered, viz. it is

$$=P''-3(Q'-2PP')+2(R-3PQ+2P^3),$$

and for the second transformation it is only altered by the factor ϕ_1^{-3}, viz. it is

$$=\phi_1^{-3}\{P_2-3(Q_1-2PP_1)+2(R-3PQ+2P^3)\}.$$

It is interesting to directly verify this last result. Performing the differentiations in regard to X, we find without difficulty

$$P_2=p_2\phi_1+2p_1\phi_2+p\phi_3-\frac{\phi_4}{\phi_1}+\frac{3\phi_2\phi_3}{\phi_1^2}-\frac{2\phi_2^3}{\phi_1^3},$$

$$Q_1-2PP_1=(q_1-2pp_1)\phi_1^2+p_1\phi_2-2p^2\phi_1\phi_2+2q\phi_1\phi_2+p\phi_3-\frac{\phi_4}{\phi_1}+\frac{\phi_2\phi_3}{\phi_1^2},$$

$$R-3PQ+2Q^3=(r-3pq+2p^3)\phi_1^3+p\phi_3-3p^2\phi_1\phi_2+3q\phi_1\phi_2-\frac{\phi_2\phi_3}{\phi_1^2}+\frac{\phi_2^2}{\phi_1^3};$$

and thence

$$P_2-3(Q_1-2PP_1)+2(R-3PQ+2Q^3)=p_2\phi_1-3(q_1-2pp_1)\phi_1^2+2(r-3pq+2p^3)\phi_1^3-p_1\phi_2.$$

But, introducing herein the derived functions in regard to x, we have

$$p_1=p'\phi_1,\quad q_1=q'\phi_1,\quad p_2=p'\phi_2+p''\phi_1^2,$$

whence

$$p_2\phi_1-p_1\phi_2=p'''\phi_1^3;$$

and the right-hand side becomes

$$=\phi_1^3\{p''-3(q'-2pp')+2(r-3pq+2p^3)\},$$

which is the required result.

We have thus

$$p'' - 3(q' - 2pp') + 2(r - 3pq + 2p^3)$$

as an "invariant" of the differential equation

$$\frac{d^3y}{dx^3} + 3p\frac{d^2y}{dx^2} + 3q\frac{dy}{dx} + r = 0.$$

It is to be remarked that this is *not* what M. Halphen calls a "differential invariant;" he uses this expression not in regard to a differential equation, but in regard to a curve defined by an equation between the coordinates (x, y), and the differential invariant is a function of the derivatives y'', y''', ... which is either $=0$ in virtue of the equation of the curve, or else, being put $=0$, it determines certain singularities of the curve. Thus y'' is a differential invariant; the equation $y''=0$ determines the points of inflexion. Again

$$\tfrac{1}{3}\left(\frac{y'''}{y''}\right)'' - \tfrac{2}{9}\frac{y'''}{y''}\left(\frac{y'''}{y''}\right)' + \tfrac{4}{27}\left(\frac{y'''}{y''}\right)^3$$

is a differential invariant, vanishing identically if the variables (x, y) are connected by any quadric equation whatever; it is thus the differential invariant of a conic.

This last differential invariant is intimately connected with the above-mentioned invariant

$$p'' - 3(q' - 2pp') + 2(r - 3pq + 2p^3),$$

viz. writing with M. Halphen

$$p = -\tfrac{1}{3}\frac{y'''}{y''},\ q = 0,\ r = 0,$$

we have

$$p'' - 3(q' - 2pp') + 2(r - 3pq + 2p^3) = p'' + 6pp' + 4p^3$$

$$= -\left\{\tfrac{1}{3}\left(\frac{y'''}{y''}\right)'' - \tfrac{2}{9}\frac{y'''}{y''}\left(\frac{y'''}{y''}\right)' + \tfrac{4}{27}\left(\frac{y'''}{y''}\right)^3\right\}.$$

It is moreover noticed by him that, writing

$$y'',\ y''',\ y'''',\ y''''' = 2a,\ 6b,\ 24c,\ 20d,$$

respectively, the function in { } becomes

$$= \frac{20}{a^2}(a^2d - 3abc + 2b^3);$$

the form under which he had previously obtained the differential invariant of the conic. As remarked by Sylvester, it is mentioned pp. 19 and 20 in Boole's *Differential Equations* (Cambridge, 1859), that the general differential equation of a conic was obtained by Monge in the form

$$9y''^2y''''' - 45y''y'''y'''' + 40(y''')^3 = 0,$$

which (representing the differential coefficients as just mentioned) becomes $a^2d - 3abc + 2b^3 = 0$, but putting them $=a, b, c, d$ respectively, it becomes $9a^2d - 45abc + 40b^3 = 0$; the last-mentioned form presented itself to Sylvester in his theory of Reciprocants.

850.

ON LINEAR DIFFERENTIAL EQUATIONS.

[From the *Quarterly Journal of Pure and Applied Mathematics*, vol. XXI. (1886), pp. 321—331.]

1. THE researches of Fuchs, Thomé, Frobenius, Tannery, Floquet, and others, relate to linear differential equations of the form

$$p_0 \frac{d^m y}{dx^m} + p_1 \frac{d^{m-1} y}{dx^{m-1}} + \ldots + p_m y = 0,$$

or say

$$(p_0, p_1, \ldots, p_m \between \frac{d}{dx}, 1)^m y = 0,$$

where $p_0, p_1, \ldots, p_m$ are rational and integral functions of the independent variable x; any common factor of all the functions could of course be thrown out, and it is therefore assumed that the functions have no common factor. It is to be throughout understood that x and y denote complex magnitudes, which may be regarded as points in the infinite plane; viz. $x, = \xi + i\eta$, is the point the coordinates whereof are ξ, η: and similarly for y.

2. Suppose $x - a$ is *not* a factor of p_0, the point $x = a$ is in this case said to be an ordinary point in regard to the differential equation; and let $y_0, y_1, y_2, \ldots, y_{m-1}$ be arbitrary constants denoting the values of $y, \frac{dy}{dx}, \frac{d^2y}{dx^2}, \ldots, \frac{d^{m-1}y}{dx^{m-1}}$ for the point $x = a$. We can from the differential equation, and the equations derived therefrom by successive differentiations in regard to x, obtain the values for $x = a$ of the subsequent differential coefficients $\frac{d^m y}{dx^m}, \frac{d^{m+1}y}{dx^{m+1}}, \ldots$, say these are $y_m, y_{m+1}, \ldots$; viz. the value of each of these quantities will be determined as a linear function of $y_0, y_1, \ldots, y_{m-1}$; and we thus have a development of y in positive integer powers of $x - a$, viz. this is

$$y = y_0 + y_1 (x - a) + \frac{1}{1 \cdot 2} y_2 (x - a)^2 + \ldots,$$

which will in fact be a sum of m determinate series multiplied by $y_0, y_1, y_2, \ldots, y_{m-1}$ respectively; say the form is

$$y = y_0X_0 + y_1X_1 + \ldots + y_{m-1}X_{m-1},$$

where $X_0, X_1, \ldots, X_{m-1}$ are each of them a series of positive integer powers of $x-a$. Each of these series will be convergent for values of x sufficiently near to a, or say for points x within the domain of the point a; and since $y_0, y_1, \ldots, y_{m-1}$ are arbitrary, each series separately will satisfy the differential equation; and we have thus m particular integrals of the differential equation.

3. Suppose next that $x-a$ is a factor of p_0, the point $x=a$ is in this case said to be a singular point in regard to the differential equation. The foregoing process of development fails, as leading to infinite values of $\frac{d^my}{dx^m}, \frac{d^{m+1}y}{dx^{m+1}}, \ldots$; and we have to consider the developments of y which belong to the neighbourhood of the point $x=a$, or say to the domain of the singular point $x=a$. This has to be done separately in regard to each of the singular points, and among these we have, it may be, the point ∞. To decide whether ∞ is or is not a singular point, we may in the differential equation write $x=1/t$; and then transforming to this new variable, and throwing out any common factor, the coefficients of the transformed equation will be rational and integral functions of t, without any common factor; say these are $P_0, P_1, \ldots, P_m$; if t is not a factor of P_0, then ∞ will be an ordinary point of the original equation; but if t is a factor of P_0, then ∞ will be a singular point of the original equation.

4. In considering the singular point $x=a$, we may, it is clear, transform the equation to this point as origin, and it is convenient to do this; supposing it done, we have an equation wherein $x=0$ is a singular point, viz. p_0 contains x as a factor, and we have to consider the developments of y which belong to the domain of this singular point $x=0$.

It is convenient to change the form of the differential equation by dividing the whole equation by the first coefficient p_0, and then expanding each of the quotients $\frac{p_1}{p_0}, \frac{p_2}{p_0}, \ldots, \frac{p_m}{p_0}$ in a series of ascending powers of x. The new form is

$$\frac{d^my}{dx^m} + p_1\frac{d^{m-1}y}{dx^{m-1}} + \ldots + p_my = 0,$$

or say

$$P(y) = (1, p_1, \ldots, p_m \between \frac{d}{dx}, 1)^m y = 0,$$

where $p_1, p_2, \ldots, p_m$ now denote each of them a series (finite or infinite) of integer powers of x, but containing only a finite number of negative powers of x.

5. Such an equation frequently admits of a "regular integral" of the form $y = x^\rho E(x)$, where $E(x)$ is a series of positive integer powers $E_0 + E_1x + E_2x^2 + \ldots$, ($E_0$ not $=0$, for this would imply a different value of ρ)*. To determine whether

* The expression regular integral is afterwards used in a more general sense, see *post*, No. 10.

this is so, we substitute in the differential equation for y the value in question $x^\rho E(x)$, thus obtaining a series

$$\Omega_0 x^{\rho-\theta} + \Omega_1 x^{\rho-\theta+1} + \dots,$$

(where θ is a determinate positive integer depending on the negative powers of x in the equation); the coefficients $\Omega_0, \Omega_1, \dots$ are functions of ρ of an order not exceeding m, and contain also the coefficients $E_0, E_1, E_2, \dots$ linearly; in particular, Ω_0 contains E_0 as a factor, say its value is $= E_0 \Pi_0$. The series should vanish identically. Supposing that Π_0 contains ρ, then we have $\Pi_0 = 0$, an equation of an order not exceeding m for the determination of ρ. For any root $\rho = \rho_0$ of this equation, E_0 remains arbitrary and may be taken $= 1$; the equations $\Omega_1 = 0, \Omega_2 = 0, \dots$ then serve to determine the ratios to E_0 of the remaining coefficients $E_1, E_2, \dots$; and we thus have the solution $y = x^{\rho_0}(1 + E_1 x + E_2 x^2 + \dots)$, where ρ_0 and the coefficients have determinate values.

6. I stop to notice a curious form of illusory solution; the assumed form of solution is

$$y = x^\rho (\dots + E_{-2} x^{-2} + E_{-1} x^{-1} + E_0 + E_1 x + \dots),$$

the series being a double series extending both ways to infinity, or say a back-and-forward series; we have here a series of equations

$$\dots \Omega_{-2} = 0, \quad \Omega_{-1} = 0, \quad \Omega_0 = 0, \quad \Omega_1 = 0, \dots,$$

which leave ρ undetermined, but determine the ratios of the several coefficients to one of these coefficients, say E_0; or taking this $= 1$, we have a solution

$$y = x^\rho (\dots + E_{-2} x^{-2} + E_{-1} x^{-1} + 1 + E_1 x + E_2 x^2 + \dots)$$

where the coefficients are determinate functions of the arbitrary symbol ρ. Such a series is in general divergent for all values of the variable, and thus is altogether without meaning. As a simple instance, take the differential equation $\frac{dy}{dx} - y = 0$, which is satisfied by

$$y = \left\{ \dots (\rho - 1)\rho x^{\rho-2} + \rho x^{\rho-1} + x^\rho + \frac{x^{\rho+1}}{\rho+1} + \dots \right\};$$

see my paper, Cayley, Note on Riemann's paper, "Versuch einer allgemeinen Auffassung der Integration und Differentiation," *Werke*, pp. 331—344; *Math. Ann.* t. XVI. (1880), pp. 81, 82), [751].

7. A more general form of integral is Thomé's "normal elementary integral," $y = e^w x^\rho E(x)$, where w is $=$ a finite series $\frac{c_{\alpha-1}}{x^{\alpha-1}} + \dots + \frac{c_1}{x}$ of negative powers of x (α a positive integer, $= 2$ at least). To discover whether such a form exists, observe that, writing for a moment $\frac{dy}{dx} = y'$, and so for the other symbols, we have $\frac{y'}{y} = w' + \frac{\rho}{x} + \frac{E'(x)}{E(x)}$,

where the last term $\frac{E'(x)}{E(x)}$ may be expanded as a series of positive integer powers of x, so that, writing $\frac{y'}{y}=z$, we have

$$z=-\frac{(\alpha-1)\,c_{\alpha-1}}{x^\alpha}-\ldots-\frac{c_1}{x^2}+\frac{\rho}{x}+G_0+G_1x+\ldots;$$

we can, by introducing into the differential equation the new variable $z\left(=\frac{y'}{y}\right)$ in place of y, obtain for z a differential equation (not a linear one) of the order $m-1$, and this should be satisfied by the series in question, viz. a series containing a finite number of negative powers of x; we endeavour to determine the first term $\frac{-(\alpha-1)c_{\alpha-1}}{x^\alpha}$, or say $\frac{A}{x^\alpha}$, where the exponent α is a positive integer which is to be determined, and A is not $=0$; this is done by a well-known process; we write in the equation $z=\frac{A}{x^\alpha}$, and (if possible to do so) determine α (a positive integer $=2$ at least) by the condition that two or more of the terms shall have the same negative index $-p$ preceding in order all the other indices, viz. the absolute value of p must be greater than the absolute value of any other negative index; we then equate to zero the whole coefficient of the term in question x^{-p}, and thus obtain a value not $=0$ of A. And having thus obtained the first term $\frac{A}{x^\alpha}$, we can, by assuming the form

$$z=\frac{A}{x^\alpha}+\frac{B}{x^{\alpha-1}}+\ldots+\frac{K}{x^2}+\frac{L}{x}+\ldots$$

with indeterminate coefficients, and substituting in the equation, find the remaining coefficients $B,\ldots,K,L$; we then have

$$c_{\alpha-1}=\frac{-A}{\alpha-1},\ \ldots,\ c_1=-K,$$

giving the value $\frac{c_{\alpha-1}}{x^{\alpha-1}}+\ldots+\frac{c_1}{x}$ of w; instead of determining the subsequent coefficients from the z-equation, we may, from the original equation, writing therein $y=e^wY$, obtain an equation for $Y(=e^{-w}y)$ which will be linear of the order m, and will admit of a regular integral $Y=x^\rho E(x)$. Or we may, going on a step further with the z-equation, find therefrom the coefficient L which is $=\rho$.

8. As an example, take the equation

$$\frac{d^2y}{dx^2}+\left(\frac{3}{x}-\frac{1}{x^2}\right)\frac{dy}{dx}+\frac{1}{x^2}y=0,$$

which has the elementary normal integral

$$y=e^{-\frac{1}{x}}\frac{1}{x}.$$

To investigate this, writing $\frac{y'}{y}=z$, we have $\frac{y''}{y}=z'+z^2$, and thence the z-equation

$$\frac{dz}{dx}+z^2+\left(\frac{3}{x}-\frac{1}{x^2}\right)z+\frac{1}{x^2}=0.$$

Substituting herein the value $z=\frac{A}{x^a}$, we obtain the function

$$-\frac{A\alpha}{x^{\alpha+1}}+\frac{A^2}{x^{2\alpha}}-\frac{A}{x^{\alpha+2}}+\frac{1}{x^2},$$

and here the value $\alpha=2$ gives the two terms $\frac{A^2}{x^4}$ and $\frac{-A}{x^4}$, each with a negative index -4, the absolute value whereof is greater than the absolute values of the other indices; we then have $A^2-A=0$, giving a value $A=1$, which is not $=0$; and we have thus a first term $z=\frac{1}{x^2}$; the complete value is, as it happens, the finite series $z=\frac{1}{x^2}-\frac{1}{x}$; and z being $=\frac{y'}{y}$, we thence obtain the integral in question, $y=e^{-\frac{1}{x}}\frac{1}{x}$.

9. Returning to the question of the regular integrals, we may consider the function

$$P(x^\rho),\ =(1,\ p_1,\ldots,\ p_m\between\frac{d}{dx},\ 1)^m\,x^\rho,$$

say this is the so-called "determinirende Function," but I will call it the Indicial function; this is a function of (x,ρ); we have therein terms arising from

$$\left(\frac{d}{dx}\right)^m x^\rho,\quad \left(\frac{d}{dx}\right)^{m-1}x^\rho,\ldots,$$

viz. these are equal to $[\rho]^m x^{\rho-m}$, $[\rho]^{m-1}x^{\rho-m-1},\ldots$ respectively, but these will be multiplied by powers of x contained in the coefficients $p_1, p_2, \ldots, p_m$ respectively; the function is thus of the degree m as regards ρ, but the coefficient of any power of x is of the degree m, or of some inferior degree, according to the terms $[\rho]^m$, $[\rho]^{m-1},\ldots$ which enter into the coefficient. The coefficient of the lowest power of x in the indicial function may be termed the indicial coefficient, or simply the indicial; and the equation obtained by equating this coefficient to zero the indicial equation. By what precedes, the indicial is a function of ρ, which is of the degree m at most, but which may be of any inferior degree, or even be an absolute constant. Hence, considering a regular integral $x^\rho E(x)$, the index ρ is determined by the indicial equation, and to each root of this equation there corresponds in general a regular integral $x^\rho E(x)$; the number of regular integrals is thus at most $=m$.

10. Suppose, first, that the indicial equation is of the degree m, and that the roots of this equation are all unequal; there are, in this case, m values, say $\rho_1, \rho_2, \ldots, \rho_m$ of ρ, and each of these gives rise to a regular integral $x^\rho E(x)$, viz. there will be corresponding to each value of ρ a series of positive integer powers $E(x)$, which will be

a convergent series for sufficiently small values of x. Some of these series may be finite series, viz. this will be the case if the indicial equation has roots differing from each other by integer values (but a finite series may of course be regarded as convergent); and in the case of two or more equal roots, it is necessary to extend the notion of a regular integral by including under it series multiplied into positive integer powers of $\log x$; but, with this modification, the conclusion holds good; if the indicial equation be of the degree m, the number of regular integrals will always be $= m$, viz. there will be m integrals involving series $E(x)$, or, it may be, $E(x)(\log x)^k$, where each series $E(x)$ is a convergent series for sufficiently small values of x.

11. But suppose the indicial equation is of an order $m-\gamma$ inferior to m, and to fix the ideas suppose that the roots are all unequal. There are in this case $m-\gamma$ values, say $\rho_1, \rho_2, \ldots, \rho_{m-\gamma}$ of ρ, and each of these will apparently give rise to a regular integral $x^\rho E(x)$; but it may be that in any such case the series $E(x)$ is a series divergent for all values, however small, of the variable x, and that we thus have in appearance only, but not in reality, a regular integral; or say the integral is illusory. The conclusion is that the indicial equation being of a degree $m-\gamma$ inferior to m, the number of regular integrals is at most $= m-\gamma$; but it may have any less value than $m-\gamma$, or even there may be no regular integral. It is hardly necessary to remark that, if the indicial be an absolute constant, or say if the degree of the indicial equation be $= 0$, then there is no value of ρ, and consequently no regular integral.

12. By way of illustration, consider first the differential equation

$$\frac{d^2y}{dx^2} + \frac{1}{x}\frac{dy}{dx} + \frac{1}{x}y = 0,$$

for which the degree of the indicial equation is equal to the order of the equation $(=2)$; viz. the indicial function is

$$= \rho(\rho-1)x^{\rho-2} + \rho x^{\rho-2} + x^{\rho-1},$$

and the indicial equation is thus $\rho(\rho-1)+\rho = 0$, that is, $\rho^2 = 0$, viz. there are here two roots, each $= 0$.

It is convenient not in the first instance to write $\rho = 0$, but to substitute in the equation the value

$$y = Ax^\rho + Bx^{\rho+1} + Cx^{\rho+2} + \ldots.$$

We thus find

	$x^{\rho-2}$,	$x^{\rho-1}$,	$x^\rho,\ldots$
$0=$	$\rho\,.\,\rho-1\,.\,A$,	$\rho+1\,.\,\rho\,.\,B$,	$\rho+2\,.\,\rho+1\,.\,C$
	$+\quad \rho A$,	$+\rho+1\,.\,B$,	$+\rho+2\,.\,C$
		$+A$,	$+B$,

viz. we have the equations

$$\begin{aligned} \rho^2 A \quad &= 0, \\ (\rho+1)^2 B + A &= 0, \\ (\rho+2)^2 C + B &= 0, \end{aligned}$$

where observe that in each equation the function of ρ, which multiplies the posterior coefficient, is of a degree superior to that which multiplies the other coefficient. The first equation gives $\rho=0$; A arbitrary, but say $A=1$; the values of $B, C, \ldots$ are then

$$-\frac{1}{(\rho+1)^2},\quad +\frac{1}{(\rho+1)^2(\rho+2)^2},\quad -\frac{1}{(\rho+1)^2(\rho+2)^2(\rho+3)^2},\ldots$$

giving rise to a series which in this form, and when ρ is put equal to its value, $\rho=0$, is at once seen to be convergent (even for indefinitely large values of x, but this is an accident; it would have been enough if the series had been convergent for sufficiently small values), viz. the series is

$$y=1-\frac{x}{1^2}+\frac{x^2}{1^2.\,2^2}-\frac{x^3}{1^2.\,2^2.\,3^2}+\ldots.$$

There is another integral

$$y=\left(1-\frac{x}{1^2}+\frac{x^2}{1^2.\,2^2}-\ldots\right)\log x+2x-\tfrac{3}{4}x^2+\tfrac{11}{108}x^3-\tfrac{25}{3456}x^4+\ldots,$$

involving $\log x$ and a second series which is also convergent.

13. But consider next the before-mentioned equation

$$\frac{d^2y}{dx^2}+\left(\frac{3}{x}-\frac{1}{x^2}\right)\frac{dy}{dx}+\frac{1}{x^2}y=0,$$

for which the degree of the indicial equation is equal to order $-1\,(=1)$; in fact, here the indicial function is

$$=\rho(\rho-1)\,x^{\rho-2}+3\rho x^{\rho-2}-\rho x^{\rho-3}+x^{\rho-2},$$

and the indicial is thus $=-\rho$, giving the index $\rho=0$. There is thus at most one regular integral $E(x)$, but this is in fact an illusory one. Substituting in the equation the value

$$y=Ax^\rho+Bx^{\rho+1}+Cx^{\rho+2}+\ldots,$$

we find

	$x^{\rho-3}$,	$x^{\rho-2}$,	$x^{\rho-1}$,	x^{ρ}.
$0=$		$\rho(\rho-1)A$,	$(\rho+1)\rho B$,	$(\rho+2)(\rho+1)C$,
	$-\rho A$,	$-(\rho+1)B$,	$-\ (\rho+2)C$,	$-\ (\rho+3)D$,
		$+3\rho A$,	$+3(\rho+1)B$,	$+3(\rho+2)C$,
		$+\ A$,	$+\ B$,	$+\ C$,

and we thus have the equations

$$\begin{aligned}\rho A &= 0,\\ (\rho+1)^2A-(\rho+1)B&=0,\\ (\rho+2)^2B-(\rho+2)C&=0,\\ (\rho+3)^2C-(\rho+3)D&=0,\end{aligned}$$

where observe that in each equation the function of ρ, which multiplies the posterior coefficient, is of a degree inferior to that which multiplies the other coefficient. The first equation then gives $\rho=0$; A arbitrary, or say $=1$: the other coefficients B, C,... are then found to be

$$\rho+1, \quad (\rho+1)(\rho+2), \quad (\rho+1)(\rho+2)(\rho+3), \ldots$$

or substituting for ρ its value $=0$, the series is

$$1+1x+1\,.\,2\,.\,x^2+1\,.\,2\,.\,3x^3+\ldots,$$

which is divergent for any value whatever, however small, of x; or the integral is as mentioned an illusory one.

14. The last-mentioned differential equation is *reducible*, viz. we have

$$P(y)=\left\{\frac{d^2}{dx^2}+\left(\frac{3}{x}-\frac{1}{x^2}\right)\frac{d}{dx}+\frac{1}{x^2}\right\}y=\left(\frac{d}{dx}+\frac{2}{x}\right)\left(\frac{d}{dx}+\frac{1}{x}-\frac{1}{x^2}\right)y, \text{ say } =QDy,$$

where of course the order of the factors Q, D is material.

It is a general theorem that the indicial, belonging to the product P, is of a degree equal to the sum of the degrees of the indicials, belonging to the factors Q, D respectively; and, in fact, the indicial of P is (as was seen) of the degree 1; and the indicials of Q, D are of the degrees 1, 0 respectively. But, moreover, if the indicial of P is of a degree $m-\gamma$, less than m the order of the equation (in the present case $m=2$, $m-\gamma=1$), then, in order that the equation may have $m-\gamma$ regular integrals, it is a necessary and sufficient condition that P shall be decomposable into a product QD, where the factors Q, D (being functions of the same form as P) are of the orders γ and $m-\gamma$ respectively, and where, moreover, the *second* factor D shall have an indicial of the degree $m-\gamma$, and consequently the *first* factor Q an indicial of the degree zero. In the present case, it is the second factor which has an indicial of degree zero; and thus the condition is not satisfied. And accordingly the equation ought not to have $(m-\gamma=)$ 1 regular integral; and we have seen that there is in fact no regular integral. The investigation serves to show in what sense it is that there is no regular integral, or generally in what sense it is that the number of regular integrals may be less than $m-\gamma$.

15. The theorem may, it appears to me, be stated in the more general form: the necessary and sufficient condition that the differential equation $P(y)=0$ of the order m, but having an indicial of the degree $m-\gamma$, shall have $m-\gamma-\delta$ regular integrals is that the function P shall be a product of the form $P=QMD$, where the orders of Q, M, D are $\gamma+\delta-\theta$, θ, $m-\gamma-\delta$ respectively, and the degrees of their indicials δ, 0, $m-\gamma-\delta$ respectively; or, to denote this in a compendious manner, say

$$\overset{m}{\underset{m-\gamma}{P}} = \overset{\gamma+\delta-\theta}{\underset{\delta}{Q}}\ \overset{\theta}{\underset{0}{M}}\ \overset{m-\gamma-\delta}{\underset{m-\gamma-\delta}{D}};$$

θ is an integer which is not $=0$, and which is at most $=\gamma$, for otherwise the function

Q of the order $\gamma+\delta-\theta$ could not have an indicial of the degree δ. If there is no regular integral, then $m=\gamma+\delta$, D is a mere constant or say unity, and the formula is

$$\underset{m-\gamma}{\overset{m}{P}} = \underset{m-\gamma}{\overset{m-\theta}{Q}}\ \underset{0}{\overset{\theta}{M}},$$

where θ is not $=0$, and is at most $=m-\gamma$.

16. For differential equations of the form $P(y)=0$ above considered, Floquet gave a theorem, which as afterwards remarked by him is not generally true, viz. this was $P=ABC\ldots$, a product of linear factors of the form $\frac{d}{dx}+\Sigma_{-\infty}^{\infty} C_i x^i$ (i an integer). The question of decomposition is considered in my next following paper "On Linear Differential Equations (the Theory of Decomposition)," [851], and by means of the formulæ there given (and by the process indicated *ante* No. 7), it could be decided whether any particular differential equation admits of a decomposition $P=ABC\ldots$, where the linear factors are functions *not* of the form just referred to, but of the form $\frac{d}{dx}+\Sigma C_i x^i$, in which the series contains only a finite number of negative powers, say this is a decomposition into linear regular factors.

851.

ON LINEAR DIFFERENTIAL EQUATIONS (THE THEORY OF DECOMPOSITION).

[From the *Quarterly Journal of Pure and Applied Mathematics*, vol. XXI. (1886), pp. 331—335.]

1. In the theory of linear differential equations the question arises, to decompose a quantic $\left(*\right)\left(\frac{d}{dx}, 1\right)^n$ into linear factors: we have, for instance, a differential equation

$$\frac{d^2y}{dx^2} + p\frac{dy}{dx} + qy = 0,$$

which is to be expressed in the form

$$\left(\frac{d}{dx} + \alpha\right)\left(\frac{d}{dx} + \beta\right)y = 0,$$

or say we have to express $\frac{d^2}{dx^2} + p\frac{d}{dx} + q$ as a product of linear factors

$$\left(\frac{d}{dx} + \alpha\right)\left(\frac{d}{dx} + \beta\right).$$

The problem is analogous to, but wholly distinct from, that of the resolution of an algebraic equation: using accents to denote differentiation in regard to x, the relation (in the simple case just referred to) between the coefficients (p, q) and the roots (α, β) is (not $p = \alpha + \beta$, $q = \alpha\beta$, but in place thereof) $p = \alpha + \beta$, $q = \alpha\beta + \beta'$; and it thus appears that there is the important distinction that, in the present problem, the order of the factors is not indifferent.

2. The problem may be solved when the general solution of the differential equation is known, or, what is the same thing, by means of n particular solutions

of the equation; it is not here considered in this point of view, but the intention is to treat it directly by means of the relations between the coefficients and the roots: thus in the above case, $p = \alpha + \beta$, $q = \alpha\beta + \beta'$, we may (1) find first β and then α, viz. eliminating α we have $\beta' + \beta(p - \beta) - q = 0$, β determined by a differential equation (not linear) of the first order; and then $\alpha = p - \beta$; or we may (2) find first α and then β, viz. eliminating β, we have $\alpha' - p' + \alpha(\alpha - p) + q = 0$, α determined by a differential equation (not linear) of the first order, and then $\beta = p - \alpha$.

In the case of a product

$$\left(\frac{d}{dx} + \alpha\right)\left(\frac{d}{dx} + \beta\right)\left(\frac{d}{dx} + \gamma\right)\ldots$$

of more than two factors, the roots might be determined in any order; but the two orders which naturally present themselves and which will be alone considered are, say, the reverse order $(\ldots, \gamma, \beta, \alpha)$ and the direct order $(\alpha, \beta, \gamma, \ldots)$.

3. *The Reverse Order.* Writing D for $\frac{d}{dx}$, we assume

$$\begin{aligned} D + \alpha &= (1, p_1 \between D, 1), \\ (D+\alpha)(D+\beta) &= (1, p_2, q_2 \between D, 1)^2, \\ (D+\alpha)(D+\beta)(D+\gamma) &= (1, p_3, q_3, r_3 \between D, 1)^3, \end{aligned}$$

and so on. Then using accents to denote differentiation, we find

$$\begin{array}{ll}
(1) & p_1 = \alpha. \\
(2) & p_2 = \beta + p_1, \\
 & q_2 = \beta' + p_1\beta. \\
(3) & p_3 = \gamma + p_2, \\
 & q_3 = 2\gamma' + p_2\gamma + q_2, \\
 & r_3 = \gamma'' + p_2\gamma' + q_2\gamma. \\
(4) & p_4 = \delta + p_3, \\
 & q_4 = 3\delta' + p_3\delta + q_3, \\
 & r_4 = 3\delta'' + 2p_3\delta' + q_3\delta + r_3, \\
 & s_4 = \delta''' + p_3\delta'' + q_3\delta' + r_3\delta,
\end{array}$$

where the law is obvious: thus in the last set of equations, the several columns (after the first) contain the factors p_3, q_3, r_3, s_3 respectively; and there is in each column a head term with the coefficient unity: omitting these head terms, we have in the several columns the sets $(\delta, 3\delta', 3\delta'', \delta''')$, $(\delta, 2\delta', \delta'')$, (δ, δ'), δ, of derivatives of the root δ.

4. We may from each set of equations determine in order the coefficients of lower rank contained in the equations, and the last equation of each set then gives an equation independent of these coefficients of lower rank. Thus set (2) gives

$$\begin{array}{ll} p_1 = -\beta & 0 = \beta' - \beta^2 \\ \quad + p_2; & \quad + p_2\beta \\ & \quad - p_2. \end{array}$$

Set (3) gives

$$\begin{array}{lll} p_2 = -\gamma & q_2 = -2\gamma' + \gamma^2 & 0 = \gamma'' - 3\gamma\gamma' + \gamma^3 \\ \quad + p_3; & \quad - p_3\gamma & \quad + p_3(\gamma' - \gamma^2) \\ & \quad + q_3; & \quad + q_3\gamma \\ & & \quad - r_3. \end{array}$$

Set (4) gives

$$\begin{array}{lll} p_3 = -\delta & q_3 = -3\delta' + \delta^2 & r_3 = -3\delta'' + 5\delta\delta' - \delta^3 \\ \quad + p_4; & \quad - p_4\delta & \quad + p_4(-2\delta' + \delta^2) \\ & \quad + q_4; & \quad - q_4\delta \\ & & \quad + r_4; \end{array}$$

$$\begin{aligned} 0 = \ & \delta''' - 4\delta\delta'' + 6\delta^2\delta' - 3\delta'^2 - \delta^4 \\ & + p_4(\delta'' - 3\delta\delta' + \delta^3) \\ & + q_4(\delta' - \delta^2) \\ & + r_4\delta \\ & - s_4; \end{aligned}$$

and so on.

5. Thus suppose (p_4, q_4, r_4, s_4) are given, we have δ determined by a differential equation (not linear) of the third order; and δ being known, p_3, q_3, r_3 are also known; then γ is determined by a differential equation (not linear) of the second order, and γ being known, p_2, q_2 are also known; then β is determined by a differential equation (not linear) of the first order, and β being known, p_1, that is, α is also known.

Comparing the last equations of each set, that is, the equations for the determination of $\beta, \gamma, \delta, \ldots$ respectively, it will be observed that they depend on the single series of derivatives

$$\begin{aligned} & -1, \\ & \phi, \\ & \phi' - \phi^2, \\ & \phi'' - 3\phi\phi' + \phi^3, \\ & \phi''' - 4\phi\phi'' + 6\phi^2\phi' - 3\phi'^2 - \phi^4, \end{aligned}$$

where, calling the successive terms $A, B, C, D, E, \ldots$, we have

$$B = A' - \phi A,\ C = B' - \phi B,\ D = C' - \phi C,\ E = D' - \phi D.$$

6. *Direct Order.* Considering the product

$$\left(\frac{d}{dx} + \alpha\right)\left(\frac{d}{dx} + \beta\right)\left(\frac{d}{dx} + \gamma\right)\left(\frac{d}{dx} + \delta\right)$$

of four factors, and assuming

$$\begin{aligned} D+\delta &= (1,\ p_1 \between D,\ 1), \\ (D+\gamma)(D+\delta) &= (1,\ p_2,\ q_2 \between D,\ 1)^2, \\ (D+\beta)(D+\gamma)(D+\delta) &= (1,\ p_3,\ q_3,\ r_3 \between D,\ 1)^3, \\ (D+\alpha)(D+\beta)(D+\gamma)(D+\delta) &= (1,\ p_4,\ q_4,\ r_4,\ s_4 \between D,\ 1)^4, \end{aligned}$$

we have the sets of equations

$$\begin{aligned} &(1)\quad p_1 = \delta. \\ &(2)\quad p_2 = \gamma + p_1, \\ &\qquad\ q_2 = \gamma p_1 + p_1'. \\ &(3)\quad p_3 = \beta + p_2, \\ &\qquad\ q_3 = \beta p_2 + p_2' + q_2, \\ &\qquad\ r_3 = \beta q_2 + q_2'. \\ &(4)\quad p_4 = \alpha + p_3, \\ &\qquad\ q_4 = \alpha p_3 + p_3' + q_3, \\ &\qquad\ r_4 = \alpha q_3 + q_3' + r_3, \\ &\qquad\ s_4 = \alpha r_3 + r_3'. \end{aligned}$$

7. I stop to remark that there would have been a convenience in considering the product

$$\ldots\left(\frac{d}{dx}+\delta\right)\left(\frac{d}{dx}+\gamma\right)\left(\frac{d}{dx}+\beta\right)\left(\frac{d}{dx}+\alpha\right);$$

for the first, second, third, &c., sets of equations would then have contained α, β, γ, &c., respectively; but for better comparison with the equations of the reverse order, I have preferred not to make this change of notation.

8. The set (1) gives

$$\delta - p_1 = 0.$$

Set (2) gives

$$p_1 = -\gamma + p_2, \quad 0 = \begin{array}{l} \gamma' + \gamma^2 \\ - p_2\gamma - p_2' \\ + q_2. \end{array}$$

Set (3) gives

$$p_2 = -\beta + p_3, \quad q_2 = \begin{array}{l} \beta' + \beta^2 \\ - p_3\beta - p_3' \\ + q_3, \end{array} \quad 0 = \begin{array}{l} \beta'' + 3\beta\beta' + \beta^3 \\ - p_3(\beta' + \beta^2) - 2p\beta_3 - p_3'' \\ + q_3\beta + q_3' \\ - r_3. \end{array}$$

Set (4) gives

$$\begin{aligned} p_3 &= -\alpha + p_4, \quad q_3 = \alpha' + \alpha^2 - p_4\alpha - p' + p_4, \\ r_3 &= -(\alpha'' + 3\alpha\alpha' + \alpha^3) + p_4(\alpha^2 + \alpha') + 2p_4'\alpha + p_4'' - q_4\alpha - q_4' + r_4, \end{aligned}$$

$$\begin{aligned} 0 = \ & \alpha''' + 4\alpha\alpha'' + 6\alpha^2\alpha' + 3\alpha'^2 + \alpha^4 \\ & - p_4(\alpha'' + 3\alpha\alpha' + \alpha^3) - p_4'(3\alpha' + 3\alpha^2) - 3p_4''\alpha - p_4''' \\ & + q_4(\alpha' + \alpha^2) + 2q_4'\alpha + q_4'' \\ & - r_4\alpha - r_4' \\ & + s_4, \end{aligned}$$

which differ in form from the equations belonging to the reverse order in containing the derived functions p_4', p_4'', p_4''', q_4', ... of the coefficients.

9. Taking p_4, q_4, r_4, s_4 as known, we have α determined by a differential equation (not linear) of the third order; and α being known, we know p_3, q_3, r_3. We then have β determined by a differential equation (not linear) of the second order; and β being known, we know p_2, q_2. We then have γ determined by a differential equation (not linear) of the first order; and γ being known, we know p_1, that is, δ.

852.

NOTE SUR LE MÉMOIRE DE M. PICARD "SUR LES INTÉGRALES DE DIFFÉRENTIELLES TOTALES ALGÉBRIQUES DE PREMIÈRE ESPÈCE."

[From the *Bulletin des Sciences Mathématiques*, 2me Sér., t. x. (1886), pp. 75—78.]

On peut présenter l'analyse sur laquelle est fondé le Mémoire sous une forme plus symétrique en introduisant dès le commencement les fonctions homogènes.

Soit $f=(*)(x, y, z, t)^m$ une fonction du degré m des variables x, y, z, t, lesquelles seront toujours liées par l'équation $f=0$: écrivons aussi $\frac{df}{dx}, \frac{df}{dy}, \frac{df}{dz}, \frac{df}{dt} = X, Y, Z, T$, de manière que X, Y, Z, T sont des fonctions du degré $m-1$: et soient A, B, C, D des fonctions chacune du degré $m-3$ et Q une fonction du degré $m-4$, telles que $AX+BY+CZ+DT=Qf$ identiquement; donc, en supposant $f=0$, on aura

$$AX+BY+CZ+DT=0.$$

On vérifie sans peine que l'expression

$$d\Omega = \begin{vmatrix} \lambda, & \mu, & \nu, & \rho \\ A, & B, & C, & D \\ x, & y, & z, & t \\ dx, & dy, & dz, & dt \end{vmatrix} \div (\lambda X + \mu Y + \nu Z + \rho T)$$

est indépendante des valeurs de λ, μ, ν, ρ, et ainsi égale à chacune des quatre expressions $d\Omega_x$, $d\Omega_y$, $d\Omega_z$, $d\Omega_t$,

$$= \frac{1}{X} \begin{vmatrix} B, & C, & D \\ y, & z, & t \\ dy, & dz, & dt \end{vmatrix}, \quad -\frac{1}{Y} \begin{vmatrix} C, & D, & A \\ z, & t, & x \\ dz, & dt, & dx \end{vmatrix},$$

$$\frac{1}{Z}\begin{vmatrix} D, & A, & B \\ t, & x, & y \\ dt, & dx, & dy \end{vmatrix}, \quad -\frac{1}{T}\begin{vmatrix} A, & B, & C \\ x, & y, & z \\ dx, & dy, & dz \end{vmatrix},$$

respectivement.

Cela étant, soit

$$d\Omega_t = -\frac{1}{T}\begin{vmatrix} A, & B, & C \\ x, & y, & z \\ dx, & dy, & dz \end{vmatrix} = \text{une différentielle totale};$$

en écrivant pour un moment $x, y, z = x't, y't, z't$, et en dénotant par f' la fonction $(*)(x', y', z', 1)^m$, et de même par $X', Y', Z', T', A', B', C'$ les valeurs correspondantes de X, Y, Z, T, A, B, C, les variables x', y', z' seront liées par l'équation $f' = 0$, ce qui donne

$$X'dx' + Y'dy' + Z'dz' = 0;$$

et l'on voit sans peine que l'expression de $d\Omega_t$ se réduit à

$$-\frac{1}{T'}\begin{vmatrix} A', & B', & C' \\ x', & y', & z' \\ dx', & dy', & dz' \end{vmatrix},$$

fonction de la forme

$$F'dx' + G'dy' + H'dz',$$

qui ne contient que les variables x', y', z'. Donc, en omettant les accents, il est permis de prendre $t = \text{const.}$, ce qui donne

$$X\,dx + Y\,dy + Z\,dz = 0;$$

et avec cette relation entre les différentielles dx, dy, dz, de faire que $d\Omega_t = F\,dx + G\,dy + H\,dz$ soit une différentielle totale: cela donne la condition

$$X\left(\frac{dG}{dz} - \frac{dH}{dy}\right) + Y\left(\frac{dH}{dx} - \frac{dF}{dz}\right) + Z\left(\frac{dF}{dy} - \frac{dG}{dx}\right) = 0,$$

ou enfin

$$X\left(\frac{d}{dz}\,\frac{Cx - Az}{T} - \frac{d}{dy}\,\frac{Ay - Bx}{T}\right)$$

$$+ Y\left(\frac{d}{dx}\,\frac{Ay - Bx}{T} - \frac{d}{dz}\,\frac{Bz - Cy}{T}\right)$$

$$+ Z\left(\frac{d}{dy}\,\frac{Bz - Cy}{T} - \frac{d}{dx}\,\frac{Cx - Az}{T}\right) = 0.$$

On a d'abord un terme $\mathfrak{A} \div T$, où

$$\begin{aligned}\mathfrak{A} = \quad & X\left[-2A + x\left(\frac{dA}{dx}+\frac{dB}{dy}+\frac{dC}{dz}\right)-\left(x\frac{dA}{dx}+y\frac{dA}{dy}+z\frac{dA}{dz}\right)\right]\\ +\,& Y\left[-2B + y\left(\frac{dA}{dx}+\frac{dB}{dy}+\frac{dC}{dz}\right)-\left(x\frac{dB}{dx}+y\frac{dB}{dy}+z\frac{dB}{dz}\right)\right]\\ +\,& Z\;\;-2C + z\left(\frac{dA}{dx}+\frac{dB}{dy}+\frac{dC}{dz}\right)-\left(x\frac{dC}{dx}+y\frac{dC}{dy}+z\frac{dC}{dz}\right)\Big],\end{aligned}$$

ou, en réduisant,

$$\begin{aligned}\mathfrak{A} =& -2(AX+BY+CZ)\\ &+(Xx+Yy+Zz)\left(\frac{dA}{dx}+\frac{dB}{dy}+\frac{dC}{dz}\right)\\ &-(m-3)(AX+BY+CZ)\\ &+t\left(X\frac{dA}{dt}+Y\frac{dB}{dt}+Z\frac{dC}{dt}\right)\\ =&\quad (m-1)DT\\ &-Tt\left(\frac{dA}{dx}+\frac{dB}{dy}+\frac{dC}{dz}+\frac{dD}{dt}\right)\\ &+t\left(X\frac{dA}{dt}+Y\frac{dB}{dt}+Z\frac{dC}{dt}+T\frac{dD}{dt}\right);\end{aligned}$$

puis un terme $\mathfrak{B} \div T^2$, où

$$\begin{aligned}\mathfrak{B} = & -X\left[(Cx-Az)\frac{dT}{dz}-(Ay-Bx)\frac{dT}{dy}\right]\\ & -Y\left[(Ay-Bx)\frac{dT}{dx}-(Bz-Cy)\frac{dT}{dz}\right]\\ & -Z\left[(Bz-Cy)\frac{dT}{dy}-(Cx-Az)\frac{dT}{dx}\right],\end{aligned}$$

ou, en réduisant,

$$\begin{aligned}\mathfrak{B} = \quad & \frac{dT}{dx}\left[(AX+BY+CZ)\,x - A\,(xX+yY+zZ)\right]\\ +\,& \frac{dT}{dy}\left[(AX+BY+CZ)\,y - B\,(xX+yY+zZ)\right]\\ +\,& \frac{dT}{dz}\left[(AX+BY+CZ)\,z - C\,(xX+yY+zZ)\right]\\ =\,& -DT\left(x\frac{dT}{dx}+y\frac{dT}{dy}+z\frac{dT}{dz}\right)\\ &+tT\left(A\frac{dT}{dx}+B\frac{dT}{dy}+C\frac{dT}{dz}\right)\\ =\,& -(m-1)DT^2\\ &+Tt\left(A\frac{dX}{dt}+B\frac{dY}{dt}+C\frac{dZ}{dt}+D\frac{dT}{dt}\right).\end{aligned}$$

Donc, en réunissant les deux parties, $0 = \mathfrak{A} + \frac{\mathfrak{B}}{T}$, c'est-à-dire,

$$\begin{aligned} 0 &= - Tt\left(\frac{dA}{dx} + \frac{dB}{dy} + \frac{dC}{dz} + \frac{dD}{dt}\right) \\ &\quad + t\frac{d}{dt}(AX + BY + CZ + DT) \\ &= - Tt\left(\frac{dA}{dx} + \frac{dB}{dy} + \frac{dC}{dz} + \frac{dD}{dt}\right) \\ &\quad + tQT, \end{aligned}$$

ou, en omettant le facteur tT, on obtient enfin

$$0 = Q - \left(\frac{dA}{dx} + \frac{dB}{dy} + \frac{dC}{dz} + \frac{dD}{dt}\right),$$

c'est-à-dire que les fonctions A, B, C, D sont telles que

$$AX + BY + CZ + DT = \left(\frac{dA}{dx} + \frac{dB}{dy} + \frac{dC}{dz} + \frac{dD}{dt}\right) f;$$

et, cela étant, l'expression générale $d\Omega$, et de même chacune des expressions $d\Omega_x$, $d\Omega_y$, $d\Omega_z$, $d\Omega_t$, sera égale à une différentielle totale.

Cambridge, le 8 *janvier* 1886.

853.

NOTE ON A FORMULA FOR $\Delta^n 0^i/n^i$ WHEN n, i ARE VERY LARGE NUMBERS.

[From the *Proceedings of the Royal Society of Edinburgh*, t. XIV. (1887), pp. 149—153.]

THE following formula

$$\frac{\Delta^n 0^i}{n^i} = e^{-nq}\left[1 + \left(\frac{i+1-2n}{2n}\right)q - \left(\frac{n+i+2}{2}\right)q^2\right], \quad q = e^{-(i+1)/n},$$

is given by Laplace (*Théorie Analytique des Probabilités*, 2nd ed., Paris, 1814, p. 195) as an approximate value of $\Delta^n 0^i/n^i$, when n and i are very large numbers, and is applied immediately afterwards to the case where i is of the order $n\log n$. As remarked by Professor Tait, it is certainly not applicable to the case where i is of the order n; for taking $i = An$, where A is a given number however large, then q is indefinitely near to the very small value e^{-A}, but nevertheless the last term $-\frac{1}{2}(n+i+2)q^2$, by taking n sufficiently large, may be made as large as we please, and the value would thus come out negative. It is thus necessary that i should be at least of the order $n\log n$; but it may be of any higher order.

Writing for greater convenience $r = ne^{-i/n}$ (where r is not very large), then $nq = re^{-1/n} = r(1-X)$, if $X = 1 - e^{-1/n}$; and the formula becomes

$$\frac{\Delta^n 0^i}{n^i} = e^{-r(1-X)}\left[1 + \frac{i+1-2n}{2n}e^{-1/n}\frac{r}{n} - \frac{n+i+2}{2}e^{-2/n}\frac{r^2}{n^2}\right].$$

Here $X = \frac{1}{n} - \frac{1}{1.2}\frac{1}{n^2} + \frac{1}{1.2.3}\frac{1}{n^3} + \&c.$, and the exponential $e^{rX} = 1 + rX + \frac{r^2X^2}{1.2} + \ldots$ is thus also expansible in negative powers of n; the formula becomes

$$\frac{\Delta^n 0^i}{n^i} = e^{-r}\left(1 + rX + \frac{r^2X^2}{1.2} + \ldots\right)\left[1 + \frac{i+1-2n}{2n}\,.\,e^{-1/n}\frac{r}{n} - \frac{n+i+2}{2}e^{-2/n}\frac{r^2}{n^2}\right];$$

viz. putting for X its value,

$$= e^{-r}\Big\{1$$

$$+ r\left(\frac{i+1-2n}{2n^2}e^{-1/n}+1-e^{-1/n}\right)$$

$$+ r^2\left(\frac{-n-i-2}{2n^2}e^{-2/n}+\frac{i+1-2n}{2n^2}(1-e^{-1/n})e^{-1/n}+\tfrac{1}{2}(1-e^{-1/n})^2\right)$$

$$+ \text{\&c.}\Big\};$$

or finally, expanding $e^{-1/n}$ and taking the whole result as far as $\frac{1}{n^2}$, the coefficient of r is

$$\left(-\frac{1}{n}+\frac{i+1}{2n^2}\right)\left(1-\frac{1}{n}\right)+\frac{1}{n}-\frac{1}{2n^2},\quad =\frac{1+\frac{1}{2}i}{n^2};$$

the coefficient of r^2 is

$$\left(-\frac{1}{n}+\frac{-2-i}{n^2}\right)\left(1-\frac{2}{n}\right)+\left(-\frac{1}{n}+\frac{4i}{2n^2}\right)\frac{1}{n}+\frac{1}{2n^2},\quad =-\frac{\frac{1}{2}}{n}+\frac{-\frac{1}{2}-\frac{1}{2}i}{n^2};$$

whence the formula becomes

$$\frac{\Delta^n 0^i}{n^2}=e^{-r}\left\{1+r\frac{1+\frac{1}{2}i}{n^2}+\frac{r^2}{1.2}\left(-\frac{1}{n}+\frac{-1-i}{n^2}\right)+\ldots\right\}.$$

It seems to me that the correct result up to this order of approximation is

$$\frac{\Delta^n 0^i}{n^2}=e^{-r}\left\{1+r\frac{\frac{1}{2}i}{n^2}+\frac{r^2}{1.2}\left(-\frac{1}{n}+\frac{-i}{n^2}\right)\right\}.$$

My investigation is as follows: we have

$$\frac{\Delta^n 0^i}{n^i}=1-\frac{n}{1}\left(1-\frac{1}{n}\right)^i+\frac{n.n-1}{1.2}\left(1-\frac{2}{n}\right)^i+\ldots,$$

the series being a finite one; but the number of terms is very large. But observe that, however large n is, we can take i so large that the second term $n\left(1-\frac{1}{n}\right)^i$ may be as small as we please; taking this term to be of moderate amount, say $=r_1$, the subsequent terms will be not very different from $\frac{r_1^2}{1.2}$, $\frac{r_1^3}{1.2.3}$, ..., and the approximate value is $1-r_1+\frac{r_1^2}{1.2}-$&c., which is a convergent series having its sum $=e^{-r_1}$. To work this properly out, I represent the successive terms by $r_1, \frac{r_2}{1.2}, \frac{r_3}{1.2.3}, \ldots$, so that the series is

$$=1-r_1+\frac{r_2}{1.2}-\frac{r_3}{1.2.3}+\ldots$$

Taking r a value at pleasure not very different from r_1, and multiplying by

$$(1=)\, e^{-r}.\, e^{r}=e^{-r}.\left(1+r+\frac{r^2}{1.2}+\ldots\right),$$

the sum is

$$=e^{-r}.\left\{1+(r-r_1)+\frac{1}{1.2}(r^2-2rr_1+r_2)\right.$$

$$\left.+\frac{1}{1.2.3}(r^3-3r^2r_1+3rr_2-r_3)+\ldots\right\}.$$

Assuming now $r=ne^{-i/n}$, we have

$$r_1=n\left(1-\frac{1}{n}\right)^i=ne^{i\log\left(1-\frac{1}{n}\right)}=r(1+X_1),$$

where $X_1=e^{-\frac{1}{2}\frac{i}{n^2}-\frac{1}{3}\frac{i}{n^3}-\ldots}$; and similarly

$$r_2=n.n-1.\left(1-\frac{2}{n}\right)^i=n^2\left(1-\frac{1}{n}\right)e^{i\log\left(1-\frac{2}{n}\right)}$$

$$=\left(1-\frac{1}{n}\right)r^2(1+X_2),$$

where $X_2=e^{-\frac{1}{2}\frac{4i}{n^2}-\frac{1}{3}\frac{8i}{n^3}-\ldots}$; also

$$r_3=n.n-1.\left(1-\frac{2}{n}\right)^i=n^2\left(1-\frac{1}{n}\right).e^{i\log\left(1-\frac{2}{n}\right)}$$

$$=\left(1-\frac{1}{n}\right)\left(1-\frac{2}{n}\right)r^3(1+X_3),$$

where $X_3=e^{-\frac{1}{2}\frac{9i}{n^2}-\frac{1}{3}\frac{27i}{n^3}-\ldots}$, and so on. It is now easy to calculate the successive terms $r-r_1$, $r^2-2rr_1+r_2$, &c.; and it is to be observed that, in the parts independent of the X's, we have only terms divided by n, n^2, or higher powers of n: thus in $r^4-4r^3r_1+6r^2r_2-4r^3r_3+r_4$, we have r^4 multiplied by

$$1-4+6\left(1-\frac{1}{n}\right)-4\left(1-\frac{1}{n}\right)\left(1-\frac{2}{n}\right)+\left(1-\frac{1}{n}\right)\left(1-\frac{2}{n}\right)\left(1-\frac{3}{n}\right)$$

$$=\frac{3}{n^2}-\frac{6}{n^3}.$$

We thus obtain the formula

$$\begin{aligned}
\frac{\Delta^n 0^i}{n^i} = e^{-r}\Big\{1 \\
&+ r \left(\qquad -1X_1\right) \\
&+ \frac{r^2}{1.2}\left\{-\frac{1}{n} \qquad -2X_1 + \left(1-\frac{1}{n}\right)X_2\right\} \\
&+ \frac{r^3}{1.2.3}\left\{-\frac{2}{n^2} \qquad -3X_1 + 3\left(1-\frac{1}{n}\right)X_2 - \left(1-\frac{1}{n}\right)\left(1-\frac{2}{n}\right)X_3\right\} \\
&+ \frac{r^4}{1.2.3.4}\left\{-\frac{3}{n^2}-\frac{6}{n^3} - 4X_1 + 6\left(1-\frac{1}{n}\right)X_2 - 4\left(1-\frac{1}{n}\right)\left(1-\frac{2}{n}\right)X_3\right. \\
&\qquad \left. + \left(1-\frac{1}{n}\right)\left(1-\frac{2}{n}\right)\left(1-\frac{3}{n}\right)X_4\right\} + \ldots, \\
&\qquad \vdots \qquad\qquad \vdots \qquad\qquad \vdots\Big\},
\end{aligned}$$

where $r = ne^{-i/n}$ as above, and X_1, X_2, ... have the above-mentioned values, the exponentials being expanded in negative powers of n.

Writing

$$X_1 = \frac{-\frac{1}{2}i}{n^2}, \quad X_2 = \frac{-2i}{n^2},$$

we have

$$\frac{\Delta^n 0^i}{n^2} = e^{-r}\left\{1 + r\frac{\frac{1}{2}i}{n^2} + \frac{r^2}{2}\left(-\frac{1}{n} + \frac{-i}{n^2}\right)\right\},$$

which is the foregoing approximate value.

854.

AN ALGEBRAICAL TRANSFORMATION.

[From the *Messenger of Mathematics*, vol. XV. (1886), pp. 58, 59.]

THE following algebraical transformation occurs in a paper by Hermite "On the theory of the Modular Equations," *Comptes Rendus*, t. XLVIII. (1859), p. 1100.

Writing $q = 1 - 2u^8$, $l = 1 - 2v^8$, then in the transformation of the fifth order, the modular equation was expressed by Jacobi in the form

$$\Omega, = (q-l)^6 - 256(1-q^2)(1-l^2)\{16ql(9-ql)^2 + 9(45-ql)(q-l)^2\}, = 0;$$

and if we write herein $q = 1 - 2x$, $l = \dfrac{x+1}{x-1}$, or, what is the same thing, establish between q, l the relation $q - l = 3 + ql$, that is, between u, v the relation $v^8 = 1 \div (1 - u^8)$, then the function Ω becomes

$$\Omega = \frac{64}{(1-x)^6}\{(x^2-x+1)^3 + 2^7(x^2-x)^2\}\{(x^2-x+1)^3 + 2^7 . 3^3(x^2-x)^2\};$$

or, what is the same thing, the equation $\Omega = 0$ gives for $\dfrac{(x^2-x+1)^3}{(x^2-x)^2}$ the values -2^7 and $-2^7 . 3^3$.

We, in fact, have

$$q - l = 3 + ql = \frac{2(x^2-x+1)}{1-x},$$

$$1 - q^2 = 4x(1-x),\quad 1 - l^2 = \frac{-4x}{(1-x)^2},$$

and therefore

$$(1-q^2)(1-l^2) = \frac{-16x^2}{1-x}.$$

Hence

$$\Omega = \frac{64}{(1-x)^6}\left[(x^2-x+1)^6 + 64(1-x)^5 \times x^2 \{16ql(9-ql)^2 + 9(45-ql)(3+ql)^2\}\right];$$

and, putting for a moment $ql = \theta - 3$, the term in $\{\ \}$ is found to be

$$= 7\theta^3 + 34560\theta - 6912;$$

viz. this is

$$= \frac{56(x^2-x+1)^2}{(1-x)^3} + \frac{6912(x^2-x+1)}{1-x} - 6912,$$

$$= \frac{8}{(1-x)^3}\{7(x^2-x+1)^3 + 864(x-1)^2(x^2-x+1) + 864(x-1)^3\},$$

$$= \frac{8}{(1-x)^3}\{7(x^2-x+1)^3 + 864(x^2-x)^2\}.$$

Hence

$$\Omega = \frac{64}{(1-x)^6}\left[(x^2-x+1)^6 + 512(x^2-x)^2\{7(x^2-x+1)^3 + 864(x^2-x)^2\}\right],$$

which is

$$= \frac{64}{(1-x)^6}\{(x^2-x+1)^3 + 2^7(x^2-x)^2\}\{(x^2-x+1)^3 + 2^7 . 3^3(x^2-x)^2\}.$$

855.

SOLUTION OF $(a, b, c, d) = (a^2, b^2, c^2, d^2)$.

[From the *Messenger of Mathematics*, vol. XV. (1886), pp. 59—61.]

IT is required to find four quantities (no one of them zero) which are in some order or other equal to their squares, say

$$(a, b, c, d) = (a^2, b^2, c^2, d^2).$$

Supposing that the required quantities (a, b, c, d) are the roots of the biquadratic equation

$$x^4 + px^3 + qx^2 + rx + s = 0,$$

(s not $= 0$), then the function

$$(x^4 + qx^2 + s)^2 - (px^3 + rx)^2 \text{ must be } = x^8 + px^6 + qx^4 + rx^2 + s;$$

and we have thus the conditions

$$2q - p^2 = p, \quad 2s + q^2 - 2pr = q, \quad 2qs - r^2 = r, \quad s^2 = s,$$

the last of which (since s is not $= 0$) gives $s = 1$, and the others then become

$$2q = p^2 + p, \quad 2(pr - 1) = q^2 - q, \quad 2q = r^2 + r;$$

viz. regarding p, q, r as the coordinates of a point in space, this is determined as the intersection of three quadric surfaces, and the number of solutions is thus $= 8$.

We, in fact, have $2q = p^2 + p = r^2 + r$; that is, $p^2 + p = r^2 + r$, or $(p - r)(p + r + 1) = 0$; hence $r = p$ or $r = -1 - p$.

First, if $r = p$; here $2q = p^2 + p$, $2(p^2 - 1) = q^2 - q$: the last equation multiplied by 4 gives

$$8(p^2 - 1) = (p^2 + p)(p^2 + p - 2), \; = p(p^2 - 1)(p + 2),$$

that is, $p^2 - 1 = 0$ or $p^2 + 2p - 8 = 0$.

If $p^2-1=0$, then either $p=1$, giving $q=1$, $r=1$, and hence the equation is $x^4+x^3+x^2+x+1=0$; or else $p=-1$, giving $q=0$, $r=-1$, and hence the equation is $x^4-x^3-x+1=0$, that is, $(x-1)^2(x^2+x+1)=0$.

If $p^2+2p-8=0$, then either $p=2$, giving $q=3$, $r=2$, and hence the equation is $x^4+2x^3+3x^2+2x+1$, that is,

$$(x^2+x+1)^2=0;$$

or else $p=-4$, giving $q=6$, $r=-4$, and hence the equation is $x^4-4x^3+6x^2-4x+1=0$, that is, $(x-1)^4=0$.

Secondly, if $r=-1-p$; here

$$2q=p^2+p,\quad 2(-p^2-p-1)=q^2-q;$$

the last equation multiplied by 4 gives

$$8(-p^2-p-1)=(p^2+p)(p^2+p-2),$$

that is,

$$p^4+2p^3+7p^2+6p+8=0, \text{ or } (p^2+p+4)(p^2+p+2)=0.$$

If $p^2+p+4=0$, then $p=\frac{1}{2}\{-1\pm i\sqrt{(15)}\}$, whence

$$r=\tfrac{1}{2}\{-1\pm i\sqrt{(15)}\},\ 2q=p^2+p,\ =-4, \text{ or } q=-2;$$

and the equation is

$$x^4+\tfrac{1}{2}\{-1\pm i\sqrt{(15)}\}x^3-2x^2+\tfrac{1}{2}\{-1\pm i\sqrt{(15)}\}x+1=0.$$

If $p^2+p+2=0$, then $p=\frac{1}{2}\{-1\pm i\sqrt{(7)}\}$; whence

$$r=\tfrac{1}{2}\{-1\pm i\sqrt{(7)}\},\ 2q=p^2+p,\ =-2, \text{ or } q=-1;$$

and the equation is

$$x^4+\tfrac{1}{2}\{-1\pm i\sqrt{(7)}\}x^3-x^2+\tfrac{1}{2}\{-1\pm i\sqrt{(7)}\}x+1=0,$$

that is,

$$(x-1)[x^3+\tfrac{1}{2}\{1\pm i\sqrt{(7)}\}x^2+\tfrac{1}{2}\{-1\pm i\sqrt{(7)}\}x-1]=0.$$

We thus see that the eight equations are

$$\begin{array}{ll}
1 & (x-1)^4=0,\\
1 & (x^2+x+1)^2=0,\\
1 & (x-1)^2(x^2+x+1)=0,\\
1 & x^4+x^3+x^2+x+1=0,\\
2 & (x-1)\{x^3+\frac{1}{2}(1\pm i\sqrt{7})x^2+\frac{1}{2}(-1\pm i\sqrt{7})x-1\}=0,\\
2 & x^4+\frac{1}{2}(-1\pm i\sqrt{15})x^3-2x^2+\frac{1}{2}(-1\mp i\sqrt{15})x+1=0,\\
\hline
8 &
\end{array}$$

and it hence appears that, writing γ, ϵ, θ to denote respectively an imaginary cube root, fifth root, and seventh root of unity, then the values of (a, b, c, d) are

$$\begin{array}{cccc} 1, & 1, & 1, & 1; \\ \gamma, & \gamma, & \gamma^2, & \gamma^2; \\ 1, & 1, & \gamma, & \gamma^2; \\ \epsilon, & \epsilon^2, & \epsilon^3, & \epsilon^4; \\ \epsilon\gamma, & \epsilon^3\gamma^2, & \epsilon^4\gamma, & \epsilon^2\gamma^2; \\ \epsilon^2\gamma, & \epsilon^4\gamma^2, & \epsilon^3\gamma, & \epsilon\gamma^2; \\ 1, & \theta, & \theta^2, & \theta^4; \\ 1, & \theta^3, & \theta^6, & \theta^5; \end{array}$$

viz. for each of these systems we have the required relation

$$(a, b, c, d) = (a^2, b^2, c^2, d^2).$$

It may be noticed that out of the eight equations we have the following three which are irreducible:—

$$x^4 + x^3 + x^2 + x + 1 = 0,$$
$$x^4 + \tfrac{1}{2}(-1 + i\sqrt{15})\,x^3 - 2x^2 + \tfrac{1}{2}(-1 - i\sqrt{15})\,x + 1 = 0,$$
$$x^4 + \tfrac{1}{2}(-1 - i\sqrt{15})\,x^3 - 2x^2 + \tfrac{1}{2}(-1 + i\sqrt{15})\,x + 1 = 0.$$

Each of these is an Abelian equation, viz. the roots are of the form

$$a,\ \theta(a),\ \theta^2(a),\ \theta^3(a),\ (= a,\ a^2,\ a^4,\ a^8),$$

where $\theta^4(a) = a$, not identically but in virtue of the value of a, viz. we have $\theta^4(a) = a^{16} = a$, in virtue of $a^{15} = 1$: (in the first equation $a^5 = 1$, and therefore $a^{15} = 1$; in each of the other two, a^{15} is the lowest power which is $= 1$).

In the first equation, we have evidently

$$x^4 + x^3 + x^2 + x + 1$$

as the irreducible factor of $x^5 - 1$.

The second and third equations combined together give

$$(x^4 - \tfrac{1}{2}x^3 - 2x^2 - \tfrac{1}{2}x + 1)^2 + \tfrac{15}{4}(x^3 - x)^2 = 0;$$

that is,

$$x^8 - x^7 + x^5 - x^4 + x^3 - x + 1 = 0,$$

where the left-hand side is the irreducible factor of $x^{15} - 1$.

856.

NOTE ON A CUBIC EQUATION.

[From the *Messenger of Mathematics*, vol. XV. (1886), pp. 62—64.]

CONSIDER the cubic equation

$$x^3 + 3cx + d = 0;$$

then effecting upon this the Tschirnhausen-Hermite transformation

$$y = xT_1 + (x^2 + 2c)\,T_2,$$

the resulting equation in y is

$$\begin{aligned} &y^3 + 3y\,(cT_1^2 + dT_1T_2 - c^2T_2^2) \\ &\quad + dT_1^3 - 6c^2T_1^2T_2 - 3cdT_1T_2^2 - (d^2 + 2c^3)\,T_2^3 = 0, \end{aligned}$$

and this will be

$$y^3 + 3cy + d = 0,$$

if only

$$\begin{aligned} c &= cT_1^2 + dT_1T_2 - c^2T_2^2, \\ d &= dT_1^3 - 6c^2T_1^2T_2 - 3cdT_1T_2^2 - (d^2 + 2c^3)\,T_2^3, \end{aligned}$$

equations which give

$$(d^2 + 4c^3) = (d^2 + 4c^3)\,(T_1^3 + 3cT_1T_2^2 + dT_2^3)^2,$$

viz. assuming that $d^2 + 4c^3$ not $= 0$, this is

$$1 = T_1^3 + 3cT_1T_2^2 + dT_2^3.$$

Hence the coefficients T_1, T_2 being such as to satisfy these relations, the equation in z is identical with the equation in x; or, what is the same thing, if α, β, γ are the roots of the equation in x, then we have between these roots the relations

$$\begin{aligned} \beta &= \alpha T_1 + (\alpha^2 + 2c)\,T_2, \\ \gamma &= \beta T_1 + (\beta^2 + 2c)\,T_2, \\ \alpha &= \gamma T_1 + (\gamma^2 + 2c)\,T_2, \end{aligned}$$

viz. the general cubic equation $x^3+3cx+d=0$, adjoining thereto the radicals T_1, T_2 may be regarded as an Abelian equation.

In particular, if $c, d=-1, 1$, then we may write $T_1=0$, $T_2=1$; the cubic equation is here

$$x^3+3x-1=0,$$

and the roots α, β, γ are such that $\beta=\alpha^2-2$, $\gamma=\beta^2-2$, $\alpha=\gamma^2-2$; in fact, taking θ a primitive ninth root of unity, $\theta^6+\theta^3+1=0$; we have $\alpha, \beta, \gamma=\theta+\theta^8, \theta^2+\theta^7, \theta^4+\theta^5$; values which satisfy $x^3+3x-1=0$, and the relations in question.

The same question may be considered from a different point of view. Take the transforming equation to be

$$y=A+Bx+Cx^2,$$

then assuming that the values of y corresponding to the values $x=\alpha, \beta, \gamma$ are β, γ, α respectively, we have

$$\beta=A+B\alpha+C\alpha^2,$$
$$\gamma=A+B\beta+C\beta^2,$$
$$\alpha=A+B\gamma+C\gamma^2,$$

and the transforming equation thus is

$$\begin{vmatrix} y, & 1, & x, & x^2 \\ \beta, & 1, & \alpha, & \alpha^2 \\ \gamma, & 1, & \beta, & \beta^2 \\ \alpha, & 1, & \gamma, & \gamma^2 \end{vmatrix}=0.$$

This may also be written

$$\begin{aligned}(\beta-\gamma)(\gamma-\alpha)(\alpha-\beta)\{y-\tfrac{1}{2}(\alpha+\beta+\gamma+x)\}\\ = \beta^2\gamma^2+\gamma^2\alpha^2+\alpha^2\beta^2-\tfrac{1}{2}(\beta^3\gamma+\beta\gamma^3+\gamma^3\alpha+\gamma\alpha^3+\alpha^3\beta+\alpha\beta^3)\\ +x\{\alpha^3+\beta^3+\gamma^3-\tfrac{1}{2}(\beta^2\gamma+\beta\gamma^2+\gamma^2\alpha+\gamma\alpha^2+\alpha^2\beta+\alpha\beta^2)\}\\ +x^2\{\beta\gamma+\gamma\alpha+\alpha\beta-(\alpha^2+\beta^2+\gamma^2)\}.\end{aligned}$$

We have

$$(\beta-\gamma)^2(\gamma-\alpha)^2(\alpha-\beta)^2=\frac{-27}{a^4}(a^2d^2+4ac^3+4b^3d-3b^2c^2-6abcd),$$
$$=\frac{-27}{a^4}\Delta,$$

or, say

$$(\beta-\gamma)(\gamma-\alpha)(\alpha-\beta)=\frac{3(\omega-\omega^2)}{a^2}\sqrt{(\Delta)},$$

if Δ be the discriminant, and ω an imaginary cube root of unity, $\{(\omega-\omega^2)^2=-3\}$.

The remaining functions of α, β, γ are of course expressible rationally in terms of the coefficients: we have

$$\begin{aligned}
\Sigma\beta^2\gamma^2 &= \frac{1}{a^2}(-6bd+9c^2),\\
\Sigma\beta^3\gamma &= \frac{1}{a^3}(-3abd-18ac^2+27b^2c),\\
\Sigma\alpha^3 &= \frac{1}{a^3}(-3a^2d+27abc-27b^3),\\
\Sigma\beta^2\gamma &= \frac{1}{a^2}(3ad-9bc),\\
\Sigma\beta\gamma &= \frac{3c}{a},\\
\Sigma\alpha^2 &= \frac{1}{a^2}(9b^2-6ac),
\end{aligned}$$

and the final result is

$$\begin{aligned}
\tfrac{1}{3}(\omega-\omega^2)\sqrt{(\Delta)}\,\{2a(y+x)+3b\} = &\quad -abd+4ac^2-3b^2c\\
&+x\;(-a^2d+7abc-6b^3)\\
&+x^2(\;\;2a^2c-2ab^2);
\end{aligned}$$

viz. we have thus an automorphic transformation of the equation

$$ax^3+3bx^2+3cx+d=0.$$

857.

ANALYTICAL GEOMETRICAL NOTE ON THE CONIC.

[From the *Messenger of Mathematics*, vol. xv. (1886), p. 192.]

TAKE (X, Y, Z) the coordinates of a point on the conic $yz + zx + xy = 0$, so that $YZ + ZX + XY = 0$; clearly (Y, Z, X) and (Z, X, Y) are the coordinates of two other points on the same conic; I say that the three points are the vertices of a triangle circumscribed about the conic

$$x^2 + y^2 + z^2 - 2yz - 2zx - 2xy = 0.$$

In fact, the equation of one of the sides is

$$\begin{vmatrix} x, & y, & z \\ X, & Y, & Z \\ Y, & Z, & X \end{vmatrix} = 0,$$

say this is $AX + BY + CZ = 0$, where $A, B, C = XY - Z^2, YZ - X^2, XZ - Y^2$; and the condition in order that this side may touch the conic

$$x^2 + y^2 + z^2 - 2yz - 2zx - 2xy = 0$$

is

$$BC + CA + AB = 0.$$

But we have

$$\begin{aligned} BC + CA + AB &= Y^2Z^2 + Z^2X^2 + X^2Y^2 - X(Y^3 + Z^3) - Y(Z^3 + X^3) - Z(X^3 + Y^3) \\ &\qquad + X^2YZ + XY^2Z + XYZ^2 \\ &= (YZ + ZX + XY)(-X^2 - Y^2 - Z^2 + YZ + ZX + XY) = 0; \end{aligned}$$

and similarly for the other two sides. The point (X, Y, Z) is an arbitrary point on the conic $yz + zx + xy = 0$; and we thus see that we have a singly infinite series of triangles each inscribed in this conic and circumscribed about the conic

$$x^2 + y^2 + z^2 - 2yz - 2zx - 2xy = 0.$$

858.

COMPARISON OF THE WEIERSTRASSIAN AND JACOBIAN ELLIPTIC FUNCTIONS.

[From the *Messenger of Mathematics*, vol. XVI. (1887), pp. 129—132.]

THE Weierstrassian function $\sigma(u)$ corresponds of course with Jacobi's $H(u)$, but it is worth while to establish the actual formulæ of transformation.

Writing, for a moment,

$$\omega = \omega_1 + iv_1,$$
$$\omega' = \omega_2 + iv_2,$$

it is convenient to assume $\omega_1 v_2 - \omega_2 v_1$ positive; we then have

$$2(\eta\omega' - \eta'\omega) = +\pi i;$$

in particular, this will be the case if $\omega = \omega_1$, $\omega' = iv_2$, where ω_1, v_2 are each positive.

To reduce the periods into the Jacobian form, we may assume

$$\omega = \lambda K,$$
$$\omega' = \lambda i K',$$

(where observe that, if as usual k, K, K' are each real and positive, and if as above $\omega = \omega_1$, $\omega' = iv_2$, ω_1 and v_2 positive, then also λ will be real and positive). We have

$$\frac{iK'}{K} = \frac{\omega'}{\omega}, \text{ or say } q = e^{-\frac{\pi K'}{K}} = e^{\frac{i\pi\omega'}{\omega}},$$

which determines first q, and thence k, K, K', as functions of $\frac{\omega'}{\omega}$, and we then have $\lambda = \frac{\omega}{K} = \frac{-i\omega'}{K'}$, either of which equations gives λ as a function of ω, ω'. Conversely, starting with k, λ, the original equations give the values of ω, ω'; those of η, η' will be determined presently.

The form of relation is at once seen to be

$$H(u) = Ae^{Bu^2}\sigma(\lambda u),$$

and observing that, for u small, we have $H(u) = \sqrt{\left(\frac{2kk'K}{\pi}\right)}\,u$ and $\sigma(\lambda u) = \lambda u$, we have $A = \frac{1}{\lambda}\sqrt{\left(\frac{2kk'K}{\pi}\right)}$; I first write down and afterwards verify the value of B, viz. this is $= -\frac{1}{2}\frac{\lambda^2\eta}{\omega}$; and the formula thus is

$$H(u) = \frac{1}{\lambda}\sqrt{\left(\frac{2kk'K}{\pi}\right)}\, e^{-\frac{1}{2}\frac{\lambda^2\eta}{\omega}u^2}\,\sigma(\lambda u).$$

In fact, for u writing successively $u + 2K$, and $u + 2iK'$, we obtain

$$\frac{H(u+2K)}{H(u)} = e^{-\frac{\lambda^2\eta}{\omega}2K(u+K)}\,\frac{\sigma(\lambda u + 2\omega)}{\sigma(\lambda u)},$$

$$\frac{H(u+2iK')}{H(u)} = e^{-\frac{\lambda^2\eta}{\omega}2iK'(u+iK')}\,\frac{\sigma(\lambda u + 2\omega')}{\sigma(\lambda u)},$$

which should be satisfied in virtue of

$$\frac{H(u+2K)}{H(u)} = -1, \qquad \frac{\sigma(\lambda u + 2\omega)}{\sigma(\lambda u)} = -e^{2\eta(\lambda u+\omega)},$$

$$\frac{H(u+2iK')}{H(u)} = -e^{-\frac{i\pi}{K}(u+iK')}, \qquad \frac{\sigma(\lambda u + 2\omega')}{\sigma(\lambda u)} = -e^{2\eta'(\lambda u+\omega')};$$

viz. we ought to have

$$0 = -\frac{\lambda^2\eta}{\omega}2K\ \ (u + \ K\ ) + 2\eta\ (\lambda u + \omega\ )$$

$$-\frac{i\pi}{K}(u + iK') = -\frac{\lambda^2\eta}{\omega}2iK'(u + iK') + 2\eta'(\lambda u + \omega').$$

The first of these is

$$0 = -\frac{\lambda K}{\omega}(u + K) + \left(4 + \frac{\omega}{\lambda}\right),$$

that is,

$$0 = (-1 + 1)(u + K);$$

and the second is

$$0 = \left(\frac{\frac{1}{2}iK}{\lambda K} - \frac{\lambda iK'\eta}{\omega}\right)(u + iK') + \eta'\left(u + \frac{\omega'}{\lambda}\right).$$

viz. for $i\pi$ writing $2(\eta\omega' - \eta'\omega)$, this is

$$0 = \left(\frac{\eta\omega' - \eta'\omega}{\omega} - \frac{\eta\omega'}{\omega} + \eta'\right)(u + iK'),$$

and the two equations are thus each of them an identity.

The Weierstrassian function $\wp(u)$ is defined as

$$= -\frac{d^2}{du^2}\log\sigma(u);$$

or, what is the same thing, we have

$$\wp(u) = -\frac{d}{du}\frac{\sigma'(u)}{\sigma(u)}.$$

Hence

$$\wp(\lambda u) = -\frac{1}{\lambda}\frac{d}{du}\frac{\sigma'(\lambda u)}{\sigma(\lambda u)}.$$

But

$$\frac{H'(u)}{H(u)} = -\frac{\lambda^2\eta u}{\omega} + \frac{\lambda\sigma'(\lambda u)}{\sigma(\lambda u)},$$

or

$$\wp(\lambda u) = -\frac{\eta}{\omega} - \frac{1}{\lambda^2}\frac{d}{du}\frac{H'(u)}{H(u)}.$$

But from the equation $\sqrt{k}\,\text{sn}\,u = \frac{H(u)}{\Theta(u)}$, we obtain

$$\frac{d}{du}\frac{H'(u)}{H(u)} = Z'(u) - \frac{1}{\text{sn}^2 u} + k^2\,\text{sn}^2 u, \quad = 1 - \frac{E}{K} - \frac{1}{\text{sn}^2 u},$$

and consequently

$$\wp(\lambda u) = -\frac{\eta}{\omega} - \frac{1}{\lambda^2}\left(1 - \frac{E}{K}\right) + \frac{1}{\lambda^2\,\text{sn}^2 u},$$

where, on the right-hand side, expanding in ascending powers of u, the constant term is

$$= -\frac{\eta}{\omega} - \frac{1}{\lambda^2}\left(1 - \frac{E}{K}\right) + \frac{1}{\lambda^2}\tfrac{1}{3}(1 + k^2).$$

But in the function $\wp(\lambda u)$ this constant term is $= 0$, and we thus have

$$\frac{\eta}{\omega} = \frac{1}{\lambda^2}\left\{\tfrac{1}{3}(1 + k^2) - 1 + \frac{E}{K}\right\};$$

and then, since

$$\frac{\eta}{\omega} - \frac{\eta'}{\omega'} = \frac{\frac{1}{2}\pi i}{\omega\omega'} = \frac{\frac{1}{2}\pi}{\lambda^2 KK'},$$

we have

$$\frac{\eta'}{\omega'} = -\frac{\frac{1}{2}\pi}{\lambda^2 KK'} + \frac{1}{\lambda^2}\left\{\tfrac{1}{3}(1 + k^2) - 1 + \frac{E}{K}\right\};$$

or, as these equations may also be written,

$$\eta = \frac{K}{\lambda}\left\{\tfrac{1}{3}(1 + k^2) - 1 + \frac{E}{K}\right\},$$

$$\eta' = \frac{-\frac{1}{2}\pi i}{\lambda K} + \frac{2K'}{\lambda}\left\{\tfrac{1}{3}(1 + k^2) - 1 + \frac{E}{K}\right\},$$

which are the values of η, η'. And we then have

$$\wp(\lambda u) = -\frac{1}{\lambda^2}\tfrac{1}{3}(1 + k^2) + \frac{1}{\lambda^2\,\text{sn}^2 u},$$

the equation connecting $\wp(\lambda u)$ and $\text{sn}\,u$.

859.

ON THE COMPLEX OF LINES WHICH MEET A UNICURSAL QUARTIC CURVE.

[From the *Proceedings of the London Mathematical Society*, vol. XVII. (1886), pp. 232—238.]

THE curve is taken to be that determined by the equations

$$x : y : z : w = 1 : \theta : \theta^3 : \theta^4,$$

viz. it is the common intersection of the quadric surface $\Theta = 0$, and the cubic surfaces $P = 0$, $Q = 0$, $R = 0$, where

$$\Theta = xw - yz, \quad P = x^2z - y^3, \quad Q = xz^2 - y^2w, \quad R = z^3 - yw^2.$$

Writing (a, b, c, f, g, h) as the six coordinates of a line, viz.

$$(a, b, c, f, g, h) = (\beta z - \gamma y, \gamma x - \alpha z, \alpha y - \beta x, \alpha w - \delta x, \beta w - \delta y, \gamma w - \delta z),$$

if $(\alpha, \beta, \gamma, \delta)$, (x, y, z, w) are the coordinates of any two points on the line; then, if the line meet the curve, we have

$$\begin{array}{rl} . \quad h\theta - g\theta^3 + a\theta^4 &= 0, \\ -h \quad . \quad + f\theta^3 + b\theta^4 &= 0, \\ g - f\theta \quad . \quad + c\theta^4 &= 0, \\ -a - b\theta - c\theta^3 \quad . &= 0, \end{array}$$

from which four equations (equivalent, in virtue of the identity $af + bg + ch = 0$, to two independent equations), eliminating θ, we have the equation of the complex. The form may, of course, be modified at pleasure by means of the identity just referred to, but one form is

$$\Omega, = a^4 - b^3h + bf^2g + cg^3 - acfh + 2c^2h^2 - 4a^2ch + af^3 - a^3f = 0,$$

as may be verified by substituting therein the values $a = -b\theta - c\theta^3$, $g = f\theta - c\theta^4$, $h = f\theta^3 + b\theta^4$. The last-mentioned equation is thus the equation of the complex in question, in terms of the six coordinates (a, b, c, f, g, h).

If for the six coordinates we substitute their values, $\beta z - \gamma y$, &c., we obtain $\Omega, = (x, y, z, w)^4 (\alpha, \beta, \gamma, \delta)^4 = 0$, which, regarded as an equation in (x, y, z, w), is the equation of the cone, vertex $(\alpha, \beta, \gamma, \delta)$, passing through the quartic curve; this equation should evidently be satisfied if only Θ, P, Q, R are each $=0$, viz. Ω must be a linear function of (Θ, P, Q, R); and by symmetry, it must be also a linear function of $(\Theta_0, P_0, Q_0, R_0)$, where

$$\Theta_0 = \alpha\delta - \beta\gamma, \quad P_0 = \alpha^2\gamma - \beta^3, \quad Q_0 = \alpha\gamma^2 - \beta^2\delta, \quad R_0 = \gamma^3 - \beta\delta^2,$$

viz. the form is $\Omega, = (\Theta, P, Q, R)(\Theta_0, P_0, Q_0, R_0)$, an expression with coefficients which are of the first or second degree in (x, y, z, w) and also of the first or second degree in $(\alpha, \beta, \gamma, \delta)$.

To work this out, I first arrange in powers and products of (α, δ), (β, γ), expressing the quartic functions of (x, y, z, w) in terms of (Θ, P, Q, R), as follows:

$\Omega =$

	a^4	$-b^3h$	$+bf^2g$	$+cg^3$	$-acfh$	$+2c^2h^2$	$-4a^2ch$	$+af^3$	$-a^3f$		
α^4										0	
$\alpha^3\delta$		$-z^4$	$+yzw^2$							$-z^4+yzw^2$	$-zR$
$\alpha^2\delta^2$			$-2xyzw$			$+2y^2z^2$				$-2xyzw+2y^2z^2$	$-2yz\Theta$
$\alpha\delta^3$			$+x^2yz$	$-y^4$						$+x^2yz-y^4$	$+yP$
δ^4										0	
$\alpha^3\beta$			$-zw^3$					$+zw^3$		0	
$\alpha^2\beta\delta$			$+2xzw^2$		$+yz^2w$			$-3xzw^2$		$-xzw^2+yz^2w$	$-zw\Theta$
$\alpha\beta\delta^2$			$-x^2zw$	$+3y^3w$	$-xyz^2$	$-4xyz^2$		$+3x^2zw$		$+2x^2zw+3y^3w-5xyz^2$	$+2xz\Theta-3yQ$
$\beta\delta^3$				$+xy^3$				$-x^3z$		$+xy^3-x^3z$	$-xP$
$\alpha^3\gamma$		$+z^3w$						$-yw^3$		$+z^3w-yw^3$	$+wR$
$\alpha^2\gamma\delta$		$+3xz^3$	$-xyw^2$		$-y^2zw$	$-4y^2zw$		$+3xyw^2$		$+3xz^3+2xyw^2-5y^2zw$	$+2yw\Theta+3zQ$
$\alpha\gamma\delta^2$			$+2x^2yw$		$+xyz^2$			$-3x^2yw$		$-x^2yw+xyz^2$	$-xy\Theta$
$\gamma\delta^3$				$-x^3y$				$+x^3y$		0	
$\alpha^2\beta^2$										0	
$\alpha^2\beta\gamma$			$+xw^3$		$-yzw^2$					$+xw^3-yzw^2$	$+w^2\Theta$
$\alpha^2\gamma^2$		$-3xz^2w$			$+y^2w^2$	$+2y^2w^2$				$-3xz^2w+3y^2w^2$	$-3wQ$
$\alpha\beta^2\delta$				$-3y^2w^2$	$-xz^2w$		$+4yz^3$			$-3y^2w^2-xz^2w+4yz^3$	$-4z^2\Theta+3wQ$
$\alpha\beta\gamma\delta$			$-2x^2w^2$		$+2xyzw$	$+8xyzw$	$-8y^2z^2$			$-2x^2w^2+10xyzw-8y^2z^2$	$+(-2xw+8yz)\Theta$
$\alpha\gamma^2\delta$		$-3x^2z^2$			$-xy^2w$		$+4y^3z$			$-3x^2z^2-xy^2w+4y^3z$	$-4y^2\Theta-3xQ$
$\beta^2\delta^2$				$-3xy^2w$	$+x^2z^2$	$+2x^2z^2$				$-3xy^2w+3x^2z^2$	$+3xQ$
$\beta\gamma\delta^2$			$+x^3w$		$-x^2yz$					$+x^3w-x^2yz$	$+x^2\Theta$
$\gamma^2\delta^2$										0	
$\alpha\beta^3$				$+yw^3$					$-z^3w$	$+yw^3-z^3w$	$-wR$
$\alpha\beta^2\gamma$					$+xzw^2$		$-4yz^2w$		$+3yz^2w$	$+xzw^2-yz^2w$	$+zw\Theta$
$\alpha\beta\gamma^2$					$-xyw^2$	$-4xyw^2$	$+8y^2zw$		$-3y^2zw$	$-5xyw^2+5y^2zw$	$-5yw\Theta$
$\alpha\gamma^3$		$+3x^2zw$					$-4y^3w$		$+y^3w$	$+3x^2zw-3y^3w$	$+3wP$
$\beta^3\delta$				$+3xyw^2$			$-4xz^3$		$+xz^3$	$+3xyw^2-3xz^3$	$-3xR$
$\beta^2\gamma\delta$					$-x^2zw$	$-4x^2zw$	$+8xyz^2$		$-3xyz^2$	$-5x^2zw+5xyz^2$	$-5xz\Theta$
$\beta\gamma^2\delta$					$+x^2yw$		$-4xy^2z$		$+3xy^2z$	$+x^2yw-xy^2z$	$+xy\Theta$
$\gamma^3\delta$		$+x^3z$							$-xy^3$	$+x^3z-xy^3$	$+xP$
β^4	$+z^4$			$-xw^3$						$+z^4-xw^3$	$+zR-w^2\Theta$
$\beta^3\gamma$	$-4yz^3$						$+4xz^2w$			$-4yz^3+4xz^2w$	$+4z^2\Theta$
$\beta^2\gamma^2$	$+6y^2z^2$					$+2x^2w^2$	$-8xyzw$			$+2x^2w^2-8xyzw+6y^2z^2$	$+(2xw-6yz)\Theta$
$\beta\gamma^3$	$-4y^3z$						$+4xy^2w$			$-4y^3z+4xy^2w$	$+4y^2\Theta$
γ^4	$+y^4$	$-x^3w$								$+y^4-x^3w$	$-yP-x^2\Theta$

Collecting the terms multiplied by P, Q, R, Θ, respectively, we have

$$\begin{aligned}
\Omega = \ & P\{y\alpha\delta^3 - x\beta\delta^3 + 3w\alpha\gamma^3 + x\gamma^3\delta - y\gamma^4\} \\
& + Q\{-3y\alpha\beta\delta^2 + 3z\alpha^2\gamma\delta - 3w\alpha^2\gamma^2 + 3w\alpha\beta^2\delta - 3x\alpha\gamma^2\delta + 3x\beta^2\delta^2\} \\
& + R\{-z\alpha^3\delta + w\alpha^3\gamma - w\alpha\beta^3 - 3x\beta^3\delta + z\beta^4\} \\
& + \Theta\{-2yz\alpha^2\delta^2 - zw\alpha^2\beta\delta + 2xz\alpha\beta\delta^2 + 2yw\alpha^2\gamma\delta - xy\alpha\gamma\delta^2 \\
& \quad + w^2\alpha^2\beta\gamma - 4z^2\alpha\beta^2\delta + (-2xw + 8yz)\alpha\beta\gamma\delta - 4y^2\alpha\gamma^2 + x^2\beta\gamma\delta^2 \\
& \quad + zw\alpha\beta^2\gamma - 5yw\alpha\beta\gamma\delta - 5xz\beta^2\gamma\delta + xy\beta\gamma^2\delta \\
& \quad - w^2\beta^4 + 4z^2\beta^3\gamma + (2xw - 6yz)\beta^2\gamma^2 + 4y^2\beta\gamma^3 - x^2\gamma^4\},
\end{aligned}$$

which may be written as follows:—

$$\begin{aligned}
\Omega = \ & P\{y(\alpha\delta^3 - \gamma^4) + x(\gamma^3\delta - \beta\delta^3)\} && + P(3w\alpha\gamma^3) \\
& + Q\{3x(\beta^2\delta^2 - \alpha\gamma^2\delta) + 3w(\alpha\beta^2\delta - \alpha^2\gamma^2)\} && + Q(3z\alpha^2\gamma\delta - 3y\alpha\beta\delta^2) \\
& + R\{-z(\alpha^3\delta - \beta^4) + w(\alpha^3\gamma - \alpha\beta^3)\} && + R(-3x\beta^3\delta) \\
& + \Theta\{zw(-\alpha^2\beta\delta + \alpha\beta^2\gamma) && \\
& \quad + xz\,2(\alpha\beta\delta^2 - \beta^2\gamma\delta) && + \Theta(-3xz\beta^2\gamma\delta) \\
& \quad + yw\,2(\alpha^2\gamma\delta - \alpha\beta\gamma^2) && + \Theta(-3yw\alpha\beta\gamma^2) \\
& \quad + xy(-\alpha\gamma\delta^2 + \beta\gamma^2\delta) && \\
& \quad + xw\,2(-\alpha\beta\gamma\delta + \beta^2\gamma^2) && \\
& \quad + yz(-2\alpha^2\delta^2 + 8\alpha\beta\gamma\delta - 6\beta^2\gamma^2) && \\
& \quad + x^2(\beta\gamma\delta^2 - \gamma^4) && \\
& \quad + y^2\,4(-\alpha\gamma^2\delta + \beta\gamma^3) && \\
& \quad + z^2\,4(-\alpha\beta^2\delta + \beta^3\gamma) && \\
& \quad + w^2(\alpha^2\beta\gamma - \beta^4)\}, &&
\end{aligned}$$

in which all the terms contained in the { } admit of expression in terms of P_0, Q_0, R_0, Θ_0; the remaining six terms not included within { } may be written

$$3wP\alpha(\gamma^3 - \beta\delta^2) + 3(wP - yQ)\alpha\beta\delta^2 - 3\Theta xz\beta^2\gamma\delta,$$
$$-3xR\delta(\beta^3 - \alpha^2\gamma) + 3(-xR + zQ)\alpha^2\gamma\delta - 3\Theta yw\alpha\beta\gamma^2;$$

which, observing that $wP - yQ = xz\Theta$, and $-xR + zQ = yw\Theta$, are

$$-3wP\alpha(\gamma^3 - \beta\delta^2) + 3xz\Theta(\alpha\beta\delta^2 - \beta^2\gamma\delta),$$
$$-3xR\delta(\beta^3 - \alpha^2\gamma) + 3yw\Theta(\alpha^2\gamma\delta - \alpha\beta\gamma^2).$$

The expression thus becomes

$$\begin{aligned}
\Omega = \quad & P\,.\quad x(\gamma^3\delta-\beta\delta^3) && = x\delta R_0\\
& \quad + y(\alpha\delta^3-\gamma^4) && = y(-\gamma R_0+\delta^2\Theta)\\
& \quad + 3w\alpha(\gamma^3-\beta\delta^2) && = 3w\alpha R_0\\
+ & Q\,.-3x(\beta^2\delta^2-\alpha\gamma^2\delta) && = -3x\delta Q_0\\
& \quad + 3w(\alpha\beta^2\delta-\alpha^2\gamma^2) && = -3w\alpha Q_0\\
+ & R\,.-3x\delta(\beta^3-\alpha^2\gamma) && = 3x\delta P_0\\
& \quad - z(\alpha^3\delta-\beta^4) && = z(-\beta P_0-\alpha^2\Theta_0)\\
& \quad + w(\alpha^3\gamma-\alpha\beta^3) && = w\alpha P_0\\
+ & \Theta\,.\quad zw(-\alpha^2\beta\delta+\alpha\beta^2\gamma) && = -zw\alpha\beta\Theta_0\\
& \quad + 5xz(\alpha\beta\delta^2-\beta^2\gamma\delta) && = 5xz\beta\delta\Theta_0\\
& \quad + 5yw(\alpha^2\gamma\delta-\alpha\beta\gamma^2) && = 5yw\alpha\gamma\Theta_0\\
& \quad + xy(-\alpha\gamma\delta^2+\beta\gamma^2\delta) && = -xy\gamma\delta\Theta_0\\
& \quad + 2xw(-\alpha\beta\gamma\delta+\beta^2\gamma^2) && = -2xw\beta\gamma\Theta_0\\
& \quad + yz(-2\alpha^2\delta^2+8\alpha\beta\gamma\delta-6\beta^2\gamma^2) && = -2yz(\alpha\delta-3\beta\gamma)\Theta_0\\
& \quad + x^2(\beta\gamma\delta^2-\gamma^4) && = -x^2\gamma R_0\\
& \quad + 4y^2(-\alpha\gamma^2\delta+\beta\gamma^3) && = -4y^2\gamma^2\Theta_0\\
& \quad + 4z^2(-\alpha\beta^2\delta+\beta^3\gamma) && = -4z^2\beta^2\Theta_0\\
& \quad + w^2(\alpha^2\beta\gamma-\beta^4) && = w^2\beta P;
\end{aligned}$$

and we thus finally obtain

$$\begin{aligned}
\Omega = \quad & PR_0(3\alpha w-\gamma y+\delta x)\\
& + RP_0(3\delta x-\beta z+\alpha w)\\
& + P\Theta_0\,.\,\delta^2 y\\
& + R\Theta_0\,.-\alpha^2 z\\
& + P_0\Theta\,.\,\beta w^2\\
& + R_0\Theta\,.-\gamma x^2\\
& - QQ_0\;.-3(\alpha w+\delta x)\\
& + \Theta\Theta_0\{-\alpha\beta zw-\gamma\delta xy+5\beta\delta xz+5\alpha\gamma yw-2\beta\gamma xw-2\alpha\delta yz\\
& \qquad -4\gamma^2y^2+6\beta\gamma yz-4\beta^2z^2\},
\end{aligned}$$

viz. $\Omega=0$ is the equation of the cone, vertex $(\alpha, \beta, \gamma, \delta)$, which passes through the quartic curve $x : y : z : w = 1 : \theta : \theta^3 : \theta^4$. As regards the symmetry of this expression, it is to be remarked that, changing (x, y, z, w) and $(\alpha, \beta, \gamma, \delta)$ into (w, z, y, x) and $(\delta, \gamma, \beta, \alpha)$ respectively, we change (Θ, P, Q, R) and $(\Theta_0, P_0, Q_0, R_0)$ into $(\Theta, -R, -Q, -P)$ and $(\Theta_0, -R_0, -Q_0, -P_0)$, respectively, and so leave Ω unaltered. Again, interchanging (x, y, z, w) and $(\alpha, \beta, \gamma, \delta)$, we interchange (Θ, P, Q, R) and $(\Theta_0, P_0, Q_0, R_0)$, and so leave Ω unaltered.

860.

ON BRIOT AND BOUQUET'S THEORY OF THE DIFFERENTIAL EQUATION $F\left(u, \dfrac{du}{dz}\right)=0$.

[From the *Proceedings of the London Mathematical Society*, vol. XVIII. (1887), pp. 314—324.]

I CONSIDER the theory of a differential equation of the first order as developed in Briot and Bouquet's *Théorie des fonctions doublement périodiques et en particulier des fonctions elliptiques* (8°, 1st ed., Paris, 1859), but I make some substantial variations in the mode of treatment.

I remark that, writing $u=x$, $\dfrac{du}{dz}=y$, I make the theory to depend altogether upon that of the curve $F(x, y)=0$; viz. the form of my result is, in order that a differential equation of the first order $F\left(u, \dfrac{du}{dz}\right)=0$ may have a monotropic integral, the curve $F(x, y)=0$ must satisfy certain conditions. Briot and Bouquet give their result (Theorem IV., p. 296) under a somewhat different form, as follows; viz. changing only the word "monodrome" into "monotrope," their theorem is:—

"Pour qu'une équation différentielle du premier ordre de la forme

$$\left(\frac{du}{dz}\right)^m + f_1(u)\left(\frac{du}{dz}\right)^{m-1} + \ldots\ldots + f_m(u) = 0$$

admette une intégrale monotrope: 1° les coefficients $f_1(u)$, $f_2(u)$, ..., $f_m(u)$ doivent être des polynômes entiers en u et, au plus, le premier du second degré, le second du quatrième degré, ..., le dernier du degré $2m$; 2° quand pour une certaine valeur de u l'équation a une racine multiple différente de zéro, $\dfrac{du}{dz}$ doit rester monotrope par rapport à u; 3° quand pour une certaine valeur u_1 de u, l'équation a une racine multiple égale à zéro, le premier terme du développement de $\frac{du}{dz}$ suivant les puissances croissantes de $(u-u_1)^{1/m}$ doit avoir l'exposant $1-\frac{1}{n}$, si cet exposant est plus petit que l'unité; 4° enfin l'équation différentielle, que l'on déduit de la première en posant $u=\frac{1}{v}$, doit offrir pour $v=0$ les mêmes caractères.—Ces conditions sont suffisantes."

I notice that this may be regarded as a statement referring to the curve

$$F(x, y), = y^m + f_1(x)\, y^{m-1} + \ldots + f_m(x) = 0,$$

but that in 2° a notion of monotropy is assumed, for y must be a monotropic function of x; and that 4° in effect introduces the new curve $F\left(\frac{1}{x}, -\frac{y}{x^2}\right)=0$.

1. We have between the variables u and z a differential equation of the first order $F\left(u, \frac{du}{dz}\right)=0$, F a rational and integral function of u and $\frac{du}{dz}$. The equation determines u as a function of z, and we wish to know whether u is a monotropic function of z. It will not be so if we have a tropical point, viz. a point $z=c$, such that, in the neighbourhood thereof, the value of u is given by a series $u-a=B(z-c)^\beta+\ldots$, in ascending powers of $z-c$, involving fractional powers of $z-c$. Conversely, if there is no tropical point, then u will be a monotropic function of z; but we must consider and exclude not only tropical points at a finite distance, but also the point at infinity, viz. there must not be any development $u=Bz^\beta+\ldots$, in descending powers, involving fractional powers of z.

2. A one-valued function (*eindeutige Function*, or *fonction uniforme* or *bien déterminée*) is monotropic; since there is only one value, the function must, after a circuit described by the point z, recover its original value. But, conversely, a monotropic function is one-valued, that is, no two or more valued function can be monotropic. For, suppose a two-valued function of z, having the values Z_1 and Z_2, is monotropic; there is by hypothesis no tropical point, that is, no point $z=c$, such that the function Z_1, after a circuit described by the point z, instead of recovering its original value, assumes the value Z_2; and the like as to Z_2. But, this being so, it is meaningless to regard Z_1 and Z_2 as two values of the same function; they must be considered as two distinct and separate functions, each of them a one-valued monotropic function. And it thus appears that the two notions, monotropic and one-valued, are in fact equivalent.

3. In what precedes, u and z are regarded as complex variables geometrically represented by means of two real points in their own planes respectively. But we now write $\frac{du}{dz}=y$, $u=x$, whence $F(x, y)=0$, regarding x, y as the coordinates, real or imaginary, of a point of the curve represented by this equation. And this curve has, moreover, to be considered from a non-projective point of view; we have to distinguish between, and separately consider, finite points (x and y each of them

finite), and the points at infinity (x and y either or each of them infinite). And it is moreover necessary to distinguish between, and separately consider, points on the axis of x (or say axial points), and points off the axis of x (say non-axial points). Taking, for convenience, a to denote a finite value which may be $=0$, b a finite value which is not $=0$ (or say, a not $=\infty$, b not $=\infty$ or 0), we have thus the six cases, 1° (a, b); 2° $(a, 0)$; 3° (a, ∞); 4° (∞, b); 5° $(\infty, 0)$; 6° (∞, ∞). I regard also the axes of x and y as horizontal and vertical respectively, and speak of the tangent or element of the curve as horizontal, vertical, or inclined, accordingly.

4. In the neighbourhood of any given point of the curve, we have $y=$ a series of powers of x, the form of the series being different for different classes of points. I write $P(x)$ to denote a series (finite or infinite) $A+Bx+Cx^2+\ldots$, in ascending powers of x; it is to be throughout understood that the leading coefficient A is not $=0$. Of course, $P(x-a)$, $P(x-a)^{1/n}$, $P(x^{-1})$, &c., will denote the like series $A+B(x-a)+\ldots$, $A+B(x-a)^{1/n}+\ldots$, $A+Bx^{-1}+\ldots$, &c. The coefficients after the leading coefficient A may any or all of them vanish; thus $P(x-a)^{1/n}$ extends to denote the series $P(x-a)$ of integer powers, but naturally the former notation is not used unless the series contains fractional powers. We can, by means of the symbol P, express for any given point of the curve the form of the expansion in the neighbourhood thereof; and, attaching to the point the expansion which belongs to it, we may, for instance, speak of a point $y=P(x-a)$; viz. if b is the constant term of the series, then this is a non-axial point (a, b), which is an ordinary (non-singular) point of the curve, and for which the element is not vertical; a like point with the element vertical would be a point $y=b+(x-a)^{\frac{1}{2}}P(x-a)^{\frac{1}{2}}$. Observe that, in the case of a multiple point, there is a separate expansion for each branch through the point, and in thus attaching an expansion to the point we regard the point as belonging to one of these branches; in dealing with a multiple point, it is necessary to consider separately all the different expansions, that is, all the branches through the point. It is clear that for a point (a, b), $(a, 0)$, or (a, ∞) the expression for y depends on $P(x-a)$, or $P(x-a)^{1/n}$; while, for a point (∞, b), $(\infty, 0)$, or (∞, ∞), it depends on $P(x^{-1})$ or $P(x^{-1/n})$; viz. in the two cases respectively, we have a series in ascending powers of $x-a$, or in descending powers of x.

5. In regard to any given point of the curve, substituting for y, x their values $\dfrac{du}{dz}$, u, and thence forming the reciprocal of $\dfrac{du}{dz}$, we obtain $\dfrac{dz}{du}=$ a series in u; viz. this is a series in ascending powers of $(u-a)$, or in descending powers of u. Integrating, we have $z-c=$ a series in u, which series may or may not contain a logarithmic term $\log u$ or $\log(u-a)$; and, reverting the series, we obtain $u=$ a series in $(z-c)$. This result, applicable to the neighbourhood of the point $u=a$, may contain only integer powers of $z-c$, and be accordingly of the form $u=$ a one-valued series in z; and we then say that the point on the curve is a "permissive" point. Or it may contain fractional powers of $z-c$, and be accordingly of the form $u=$ a more-than-one-valued series in z; and we then say that the point on the curve is a "prohibitive" point. The necessary and sufficient condition in order that u may be

a monotropic function of z is:—the points of the curve must be all of them permissive points; or, what is the same thing, there must not be any prohibitive point.

6. I form a complete table of permissive points, as follows:—

n a positive integer $=2$ at least. $\epsilon=0$, 1, or -1,

(a, b), $y = P(x-a)$,

$(a, 0)$, $y=(x-a)P(x-a)$ or $(x-a)^2 P(x-a)$, $\quad y=(x-a)^{1+\epsilon/n} P(x-a)^{1/n}$,

(a, ∞),

(∞, b), $y=P(x^{-1})$,

$(\infty, 0)$,

(∞, ∞), $y = xP(x^{-1})$ or $x^2P(x^{-1})$, $\quad y = x^{1+\epsilon/n} P\left(\frac{1}{x^{1/n}}\right)$,

viz. it is only a point (a, b), $(a, 0)$, (∞, b) or (∞, ∞) which may be permissive; a point (a, ∞) or $(\infty, 0)$ is prohibitive.

7. A point $y=P(x-a)$ is permissive. We have

$$\frac{du}{dz} = P(u-a),$$

$$\frac{dz}{du} = P(u-a),$$

$$z-c=(u-a)P(u-a),$$
$$u-a=(z-c)P(z-c),$$

the required result. The several steps will be readily understood; the reciprocal of a series $P(u-a)$ is a series of the like form $P(u-a)$; integrating this in regard to u as a series in $u-a$ (no constant being added on the right-hand side), we obtain a series $(u-a)P(u-a)$; and finally, $z-c$ being equal to this expression, we obtain, by reversion, a like expression $u-a=(z-c)P(z-c)$.

We might, for convenience, have written x, u, z in place of $x-a$, $u-a$, $z-c$, respectively. The proof would have run

$$y=P(x),\ \frac{du}{dz}=P(u),\ \frac{dz}{du}=P(u),\ z=uP(u),\ u=zP(z),$$

meaning $u-a=(z-c)P(z-c)$, as above; and, in what follows, this is done accordingly.

It is worth while to make an instance in which the integration can be performed—say we consider the point (0, 1) of the curve $y=\frac{1+x+x^2}{1+2x}$, which point is $y=\frac{1+x+x^2}{1+2x}=P(x)$. We have

$$\frac{du}{dz}=\frac{1+u+u^2}{1+2u},\ \frac{dz}{du}=\frac{1+2u}{1+u+u^2},$$

$$z=\log(1+u+u^2),\text{ or } u+u^2=(e^z-1),$$

giving a series $u=zP(z)$; or the point is a permissive point, as in the general case. We have here $u+u^2=(e^z-1)$, viz. u is not a monotropic function of z; but this is by reason of a prohibitive point $(-\tfrac{1}{2}, \infty)$.

8. A point $y=(x-a)P(x-a)$ is permissive.

Writing as above u, z in place of $u-a$, $z-c$, we have

$$\frac{du}{dz}=uP(u),\quad \frac{dz}{du}=\frac{1}{u}P(u),\ \text{or say}\ m\frac{dz}{du}=\frac{1}{u}+Pu,$$

whence

$$mz=\log u+uPu,$$

that is,

$$ue^{uPu}=e^{mz},\ \text{or say}\ uP(u)=e^{mz},$$

whence

$$u=e^{mz}P(e^{mz}),\ \text{a one-valued series.}$$

I take a particular instance, the point (0, 0) for the curve $y=\dfrac{x-\frac{1}{3}x^3}{1-x^2}$; here

$$\frac{du}{dz}=\frac{u-\frac{1}{3}u^3}{1-u^2},\ \text{or}\ \frac{dz}{du}=\frac{1-u^2}{u-\frac{1}{3}u^3},$$

$$z=\log(u-\tfrac{1}{3}u^3),\ \text{or}\ u-\tfrac{1}{3}u^3=e^z;$$

the point is permissive. The equation just obtained shows that u is not a monotropic function of z; but this is by reason of the prohibitive points $(\pm 1, \infty)$.

9. A point $y=(x-a)^2P(x-a)$ is permissive. Here

$$\frac{du}{dz}=u^2P(u),\quad \frac{dz}{du}=\frac{1}{u^2}P(u),\ =\frac{A}{u^2}+\frac{B}{u}+P(u),$$

whence

$$z=-\frac{A}{u}+B\log u+uP(u),$$

where the logarithmic term occurs or does not occur, according as B is not or is $=0$. In the former case, we have

$$e^{z/B}=ue^{-\frac{A}{B}u+uP(u)};$$

in the latter case,

$$z=-\frac{A}{u}+uP(u);$$

and in each case we can, by reversion, express u as a one-valued series in z. We may take as examples the two curves $y=\dfrac{x^2}{1-x}$ and $y=\dfrac{x^2}{1-x^2}$; in each case the point (0, 0) is a permissive point, but, by reason of the prohibitive points $(1, \infty)$ and $(\pm 1, \infty)$, u is not a monotropic function of z.

10. As to the other forms of permissive points, the proofs depend upon those for the forms already considered. Thus for

$$y=(x-a)^{1+\epsilon/n}P(x-a)^{1/n}, \text{ or } \frac{du}{dz}=u^{1+\epsilon/n}P(u^{1/n}),$$

writing $v=u^{1/n}$, we find

$$\frac{dv}{dz}=u^{(1+\epsilon)/n}P(u^{1/n}), \;=v^{1+\epsilon}P(v),$$

that is,

$$\frac{dv}{dz}=P(v),\; vP(v), \text{ or } v^2P(v);$$

in each case v = one-valued series in z, and thence $u=v^n$ = one-valued series in z.

Similarly for

$$y=P(x^{-1}),\;\; y=xP(x^{-1}),\;\; y=x^2P(x^{-1}),$$

or say

$$y=x^{1+\epsilon}P(x^{-1});$$

here

$$\frac{du}{dz}=u^{1+\epsilon}P(u^{-1}),$$

or, writing $v=u^{-1}$, we have

$$\frac{dv}{dz}=u^{\epsilon-1}P(u^{-1}), \;=v^{1-\epsilon}P(v);$$

viz. this is

$$\frac{dv}{dz}=P(v),\; vP(v), \text{ and } v^2P(v);$$

in each case v = one-valued series in z, and thence $u=v^{-1}$ = one-valued series in z.

And finally, for

$$y=x^{1+\epsilon/n}P\left(\frac{1}{x^{1/n}}\right);$$

here

$$\frac{du}{dz}=u^{1+\epsilon/n}P\left(\frac{1}{u^{1/n}}\right),$$

or, writing $v=u^{-\epsilon/n}$, we have

$$\frac{dv}{dz}=u^{(\epsilon-1)/n}P\left(\frac{1}{u^{1/n}}\right), \;=v^{1-\epsilon}P(v);$$

that is,

$$\frac{dv}{dz}=P(v),\; vP(v), \text{ or } v^2P(v),$$

as before; v = one-valued series in z, and thence $u=v^{-n}$ = one-valued series in z.

11. There is no difficulty in showing that every point of the curve, not being a permissive point as above, is a prohibitive point. Thus for a point (a, b), if the series for $y-b$ contain any fractional power, say, if we have $y=b+(x-a)^{\alpha}$, then

$$\frac{du}{dz}=b+u^{\alpha},\;\; \frac{dz}{du}=\frac{1}{b}-\frac{1}{b^2}u^{\alpha}+\dots,\;\; z=\frac{u}{b}-\frac{1}{(\alpha+1)\,b^2}u^{\alpha+1}+\dots,$$

whence, reverting, $u=bz$ + series containing the fractional power $z^{\alpha+1}$.

Again, a point $y=(x-a)^m P(x-a)$, $m=3$ or any higher integer value, is prohibitive. Attending only to the leading term, we have $\frac{du}{dz}=u^m$, whence $\frac{dz}{du}=u^{-m}$ or $z=u^{1-m}$, whence, reverting, $u=z^{1/(1-m)}$, which is a fractional power; and this is also the case if m be a negative integer. And similarly, in other cases, the series for u will always contain fractional powers of z.

12. In order that u may be a monotropic function of z, the points of the curve must be all of them permissive, or, what is the same thing, there must be no prohibitive point; we have to consider the geometrical signification of this condition. If we attend first to the ordinary (or non-singular) points of the curve, a non-axial point (a, b) must be a point $y=P(x-a)$, viz. there must be no such point with a vertical element.

An axial point $(a, 0)$ may be $y=(x-a)P(x-a)$, viz. the element may be inclined; or the point may be $y=(x-a)^2 P(x-a)$, viz. the element may be horizontal (but in this case there must be an ordinary or two-pointic contact only); or the point may be $y=(x-a)^{\frac{1}{2}} P(x-a)^{\frac{1}{2}}$, viz. the element may be vertical.

The point must not be (a, ∞), viz. there must be no asymptote parallel to (or coincident with) the axis of y.

The point may be (∞, b), $y=P(x^{-1})$, that is, there may be an ordinary, or osculating, asymptote parallel to the axis of x. But there is no point $(\infty, 0)$, that is, there must be no asymptote coincident with the axis of x.

There may be points (∞, ∞), $y=xP(x^{-1})$, that is, ordinary or osculating asymptotes inclined to the axes; and there may also be points (∞, ∞), $y=x^2P(x^{-1})$, that is, asymptotic parabolas of vertical axis.

13. We have, moreover, conditions as to the singular points; thus, every non-axial multiple point (a, b) must be a point with each branch of the form $y=P(x-a)$, that is, each branch must be an ordinary (non-cuspidal) branch, and must be non-vertical. The axial multiple or singular points $(a, 0)$ may have ordinary (non-cuspidal) branches $y=(x-a)P(x-a)$ and $y=(x-a)^2 P(x-a)$, inclined or horizontal, or else $y=(x-a)^{\frac{1}{2}}P(x-a)^{\frac{1}{2}}$, vertical; and there may also be cuspidal branches $y=(x-a)P(x-a)^{1/n}$, $y=(x-a)^{1+1/n}P(x-a)^{1/n}$, and $y=(x-a)^{1-1/n}P(x-a)^{1/n}$, inclined, horizontal, or vertical.

There must be no singular points (a, ∞); any such point is, in fact, excluded by the condition, no asymptote parallel to or coincident with the axis of y.

There may be multiple points (∞, b), but not $(\infty, 0)$; viz. as above, there may be asymptotes parallel to, but not coincident with, the axis of x; but in this case, the several branches must be each of them an ordinary branch $y=P(x^{-1})$.

Finally, there may be multiple or singular points (∞, ∞), but each branch must be either an ordinary branch $y=xP(x^{-1})$ or $y=x^2P(x^{-1})$, or a cuspidal branch

$$y=xP\left(\frac{1}{x^{1/n}}\right),\quad y=x^{1+1/n}P\left(\frac{1}{x^{1/n}}\right),\quad \text{or}\quad y=x^{1-1/n}P\left(\frac{1}{x^{1/n}}\right).$$

14. The enumeration seems long and difficult. But for a given curve we have, in fact, only to see that there are no ordinary non-axial points (a, b) of vertical element, and to further examine the finite singular points (a, b) and $(a, 0)$, and also the several infinite branches as giving ordinary or else singular points (a, ∞), (∞, b), $(\infty, 0)$, or (∞, ∞), viz. it has to be seen for each infinite branch that the expansion is of the proper form.

15. Writing the equation of the curve in the form

$$f_0(x)\, y^m + f_1(x)\, y^{m-1} + \ldots + f_m(x) = 0,$$

where $f_0, f_1, \ldots, f_m$ denote rational and integral functions of x, then, if a be a root of the equation $f_0(x) = 0$, there will be on the curve a point (a, ∞), which is prohibitive; $f_0(x)$ is therefore a constant, or, taking it to be $= 1$, the form must be

$$y^m + f_1(x)\, y^{m-1} + \ldots + f_m(x) = 0,$$

where $f_1, f_2, \ldots, f_m$ denote rational and integral functions of x.

It is to be shown that the degrees of these functions are equal respectively to 2, 4, ..., m at most. Supposing that the degrees are 2, 4, ..., m; then for the points at infinity we find an expansion $y = Ax^\omega + \ldots$, in descending powers of x, by substitution in an equation

$$y^m + \alpha_1 x^2 y^{m-1} + \alpha_2 x^4 y^{m-2} + \ldots + \alpha_n x^{2m} = 0;$$

whence $\omega = 2$, and A is determined by an equation of the order m; there are thus m branches $y = x^2 P(x^{-1})$, corresponding to permissive points. If, however, any one of the functions f be of a higher degree, then it may be shown (but the formal proof is not easily given) that there will be at any rate a branch $y = Ax^\omega + \&c.$, where ω has a value > 2, and we have thus a prohibitive point; hence the degrees must be as stated.

16. As an example of a differential equation with a monotropic solution, take

$$\left(\frac{du}{dz}\right)^3 + 3\left(\frac{du}{dz}\right)^2 + 27u^2 - 4 = 0;$$

we have here the curve

$$y^3 + 3y^2 + 27x^2 - 4 = 0,$$

Fig. 1.

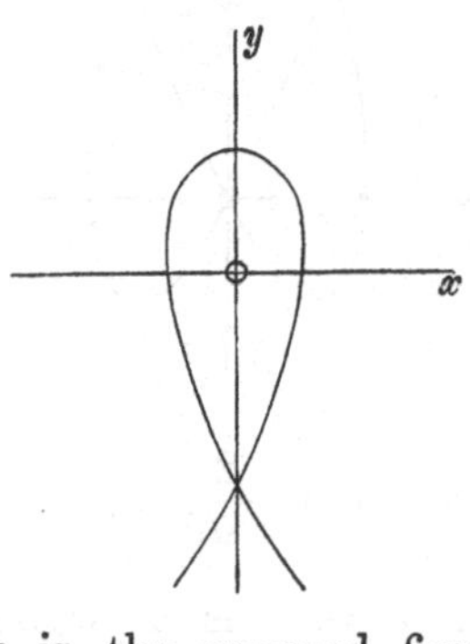

which is a nodal cubic, as shown in the annexed figure. To put the node in evidence, the equation may be written in the form $(y+2)^2(y-1) + 27x^2 = 0$; viz. the node is the point $(0, -2)$, and the equation of the tangents is $(y+2)^2 - 27x^2 = 0$. The curve besides meets the line $x=0$ in the point $y=1$, where the tangent is horizontal, and it meets the axis $y=0$ in the points $x = \pm\dfrac{2}{3\sqrt{3}}$, at each of which the tangent is vertical; these two points, as lying on the axis $y=0$, are thus each of them permissive; and they are the *only* points where the tangent is vertical (in fact, the tangents from the point at infinity on the line $x=0$ are the two lines $x = \pm\dfrac{2}{3\sqrt{3}}$, and the line infinity counting twice, as a tangent at an inflexion).

For the points at infinity, we have $y = -3x^{\frac{2}{3}} - 1 - \frac{1}{3}x^{-\frac{2}{3}} + \&c.$, which is of the form $y = x^{\frac{2}{3}}P\left(\dfrac{1}{x^{\frac{1}{3}}}\right)$, included in $y = x^{1-1/n}P\left(\dfrac{1}{x^{1/n}}\right)$, and the point is thus permissive; there is no prohibitive point, and the differential equation has a monotropic solution. This is, in fact, the rational solution $u = (z-c) - (z-c)^3$, or say $u = z - z^3$; the curve is thus given by the two equations $x = z - z^3$, $y = 1 - 3z^2$.

17. As a second example, take the differential equation

$$\left(\frac{du}{dz}\right)^3 - \left(\frac{du}{dz}\right)^2 + 4u^2 - 27u^4 = 0.$$

We have here the curve

$$y^3 - y^2 + 4x^2 - 27x^4 = 0,$$

which is a trinodal quartic curve, as shown in the figure. There is a node at the origin with the tangents $y^2 - 4x^2 = 0$, and, writing the equation in the form

$$(3y+1)(3y-2)^2 - (27x^2 - 2)^2 = 0,$$

Fig. 2.

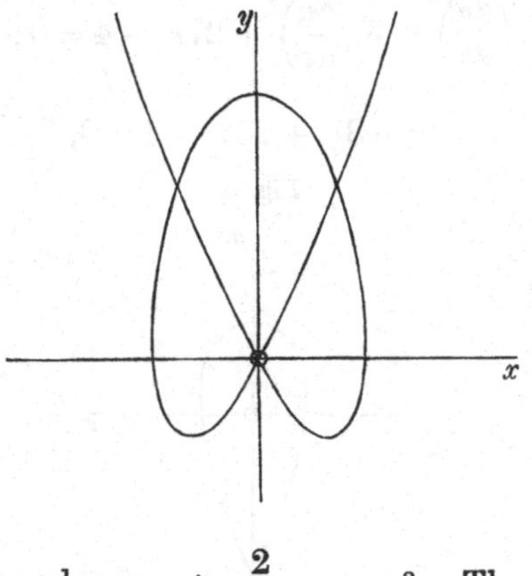

we have for the other two nodes $x = \pm\dfrac{2}{3\sqrt{3}}$, $y = \frac{2}{3}$. The lowest points are given by $x = \pm\dfrac{2}{3\sqrt{3}}$, $y = -\frac{1}{3}$, viz. the line $y = -\frac{1}{3}$ is a horizontal tangent. Writing $x=0$, we have $y^3 - y^2 = 0$, that is, $y^2 = 0$, the node at the origin, and $y = 1$, the height of the loop; and, writing $y = 0$, we have $4x^2 - 27x^4 = 0$, that is, $x^2 = 0$, the node at the origin, and $x = \pm\frac{2}{3\sqrt{3}}$, the other two intersections with the axis of x. The tangents at these two points are vertical; as being on the axis, they are thus permissive points. And they are the *only* points with a vertical tangent; in fact, the point ($x = 0$, $y = \infty$) is a point of the curve, with the line ∞ as an osculating tangent (4-pointic intersection); hence the tangents from the point $x = 0$, $y = \infty$ are the line ∞ counting four times, and the lines $x = \pm\frac{2}{3\sqrt{3}}$ touching at the points $\left(\pm\frac{2}{3\sqrt{3}}, 0\right)$ as above. For the infinite branches, we have $y = 3x + \ldots$, which is of the right form $y = x^{1+1/n} P\left(\frac{1}{x^{1/n}}\right)$. We thus see that there is no prohibitive point; and the differential equation has a monotropic solution accordingly.

Writing z in place of $z - c$, the solution in fact is

$$u = \frac{e^z - e^{-z}}{e^z + e^{-z}} - \left(\frac{e^z - e^{-z}}{e^z + e^{-z}}\right)^3;$$

hence, putting $\theta = \frac{e^z - e^{-z}}{e^z + e^{-z}}$, the curve is given by the two equations

$$x = \theta - \theta^3, \quad y = (1 - 3\theta^2)(1 - \theta^2), \quad = 1 - 4\theta^2 + 3\theta^4:$$

the monotropic function u satisfies the differential equation.

18. $F\left(u, \frac{du}{dz}\right) = 0$ must be either a rational function, a singly periodic function, or a doubly periodic function of z; say the forms are

$$u = P(z), \quad u = P(e^{gz}),$$

and

$$u = P\{\mathrm{sn}(gz, k)\} + \mathrm{cn}(gz, k)\,\mathrm{dn}(gz, k)\,Q\{\mathrm{sn}(gz, k)\},$$

where P, Q denote rational functions. I do not at present consider the criteria (such as are given in Briot and Bouquet's Theorem v., p. 301) for determining by means of the curve which is the form of the integral. I remark, however, that in the first and second cases the curve is unicursal, while in the third case it is bicursal; or say that, according as the deficiency is $= 0$ or 1, the integral is rational or simply periodic, or else it is doubly periodic. Moreover, the curve being unicursal, we can express the coordinates as equal to rational functions $P(\theta)$, $Q(\theta)$ of a parameter θ; and, being bicursal, we can express them as functions of the form

$$P(\theta) + P_1(\theta)\sqrt{1 - \theta^2 \,.\, 1 - k^2\theta^2}, \quad Q(\theta) + Q_1(\theta)\sqrt{1 - \theta^2 \,.\, 1 - k^2\theta^2},$$

of the parameter θ; and supposing the coordinates (x, y), that is, u and $\frac{du}{dz}$, thus expressed, it should be easy to establish the relation of θ with z, e^{gz} or $\mathrm{sn}(gz, k)$ in the three cases respectively.

861.

NOTE ON A FORMULA RELATING TO THE ZERO-VALUE OF A THETA-FUNCTION.

[From *Crelle's Journal der Mathem.*, t. C. (1887), pp. 87, 88.]

I HAD some difficulty in verifying for the case of a single theta-function, a formula given in Herr Thomae's paper "Beitrag zur Theorie der ϑ-Functionen," *Crelle's Journal*, vol. LXXI. (1870), pp. 201—222. The formula in question (see p. 216) is given as follows:

$$(11)\qquad \vartheta(0,\ 0, \ldots 0) = \sqrt{\frac{|A_\lambda^{(\lambda')}|}{(2\pi i)^p}}\sqrt[4]{\text{Discr.}\,(0,\ 0, \ldots 0)\ \text{Discr.}\,(0,\ 0, \ldots 0)},$$

but the denominator factor should I think be $(\pi i)^p$ instead of $(2\pi i)^p$. Making this alteration, then in the case of a single theta-function, $p = 1$, and the function belongs to the radical

$$\sqrt{x - k_1 \,.\, x - k_2 \,.\, x - k_3 \,.\, x - k_4},$$

where

$$(k_1,\ k_2,\ k_3,\ k_4) = \left(-\frac{1}{k},\ -1,\ +1,\ +\frac{1}{k}\right).$$

The determinant $|A_\lambda^{(\lambda')}|$ is a single term $= A$, and the formula becomes

$$\vartheta(0) = \sqrt{\frac{A}{\pi i}}\sqrt[4]{(k_3 - k_1)(k_4 - k_2)},$$

where $k_3 - k_1$, $k_4 - k_2$ are each $= 1 + \frac{1}{k}$, and we have therefore

$$\vartheta(0) = \sqrt{\frac{A}{\pi i}\left(1 + \frac{1}{k}\right)};$$

also A denotes the integral

$$\int_{k_1}^{k_2} \frac{dx}{\sqrt{x - k_1 \,.\, x - k_2 \,.\, x - k_3 \,.\, x - k_4}},\ = \int_{-\frac{1}{k}}^{-1} \frac{k\,dx}{\sqrt{1 - x^2 \,.\, 1 - k^2x^2}}$$

$$= \int_1^{\frac{1}{k}} \frac{k\,dx}{\sqrt{1 - x^2 \,.\, 1 - k^2x^2}} = ikK',$$

and the formula thus is

$$\vartheta(0) = \sqrt{\frac{K'(1+k)}{\pi}}.$$

But observe that, in the theta-function as defined by the equation

$$\vartheta(x) = \Sigma e^{am^2+2mx},$$

a is used to denote the value

$$a = a_1'B, \quad = \frac{2\pi}{A} B,$$

where A is the above-mentioned integral, and B is the integral

$$B = \int_{k_2}^{k_3} \frac{dx}{\sqrt{x-k_1\,.\,x-k_2\,.\,x-k_3\,.\,x-k_4}}, \quad = \int_{-1}^{1} \frac{k\,dx}{\sqrt{1-x^2\,.\,1-k^2x^2}} = 2kK,$$

which value must however be taken negatively, viz. we must write $B = -2kK$, and we then have

$$a = -\frac{2\pi K}{K'},$$

viz. writing as usual

$$q = e^{-\frac{\pi K'}{K}}, \quad r = e^{-\frac{\pi K}{K'}},$$

the e^a of the theta-function is not $=q$, but it is $=r^2$; and the zero-value $\vartheta(0)$ is $= 1 + 2r^2 + 2r^8 + 2r^{18} + \ldots$. The equation thus is

$$1 + 2r^2 + 2r^8 + 2r^{18} + \ldots = \sqrt{\frac{K'(1+k)}{\pi}},$$

which is right; in fact, writing k' in place of k, and consequently K, q in place of K', r respectively, the equation becomes

$$1 + 2q^2 + 2q^8 + 2q^{18} + \ldots = \sqrt{\frac{K(1+k')}{\pi}};$$

we have

$$1 + 2q \;+ 2q^4 + 2q^9 \;+ \ldots = \sqrt{\frac{2K}{\pi}},$$

and changing q into q^2, then (*Fund. Nova*, p. 92) K is changed into $\dfrac{K(1+k')}{2}$, and we have the formula in question. As a verification for small values of q, observe that we have

$$\frac{2K}{\pi} = 1 + 4q + 4q^2, \quad \frac{1+k'}{2} = 1 - 4q + 16q^2,$$

and thence

$$\frac{K(1+k')}{\pi} = 1 + 4q^2 \text{ or } \sqrt{\frac{K(1+k')}{\pi}} = 1 + 2q^2.$$

Cambridge, 12 February, 1886.

862.

NOTE ON THE THEORY OF LINEAR DIFFERENTIAL EQUATIONS.

[From *Crelle's Journal der Mathem.*, t. C. (1887), pp. 286—295.]

1. I CONSIDER a linear differential equation

$$p_0 \frac{d^m y}{dx^m} + p_1 \frac{d^{m-1} y}{dx^{m-1}} + \ldots + p_m y = 0,$$

where $p_0, p_1, \ldots, p_m$ are rational and integral functions of x, having no common factor: as usual, x is a complex magnitude represented by a point, and we consider the integrals belonging to a singular point $x = a$ of the differential equation. An integral may be a regular integral, or it may be what Thomé calls a normal elementary integral: the theory of these integrals (which I would rather call subregular integrals) requires, I think, further examination.

2. I retain x as the independent variable, but for shortness use t to denote $\frac{1}{x-a}$: and I take as the dependent variable z, $= \frac{1}{y}\frac{dy}{dx}$: we have then z determined by a non-linear differential equation of the order $m-1$, and from any value of z we derive a corresponding value $e^{\int z dx}$ of y.

3. To obtain the z-equation, using for shortness accents to denote differentiation in regard to x, we have $z = \frac{y'}{y}$, and thence $z' = \frac{y''}{y} - \left(\frac{y'}{y}\right)^2$, that is, $z^2 + z' = \frac{y''}{y}$: hence also $2zz' + z'' = \frac{y'''}{y} - \frac{y'y''}{y^2}$, that is,

$$z(z^2 + z') + 2zz' + z'' = \frac{y'''}{y}, \text{ or finally } z^3 + 3zz' + z'' = \frac{y'''}{y},$$

and so on; viz. the values of $\frac{y'}{y}, \frac{y''}{y}, \frac{y'''}{y}, \ldots$ are z, $z^2 + z'$, $z^3 + 3zz' + z''$, ...; generally for

$\frac{y^{(n)}}{y}$ the first term is z^n and the last term is $z^{(n-1)}$. If, to fix the ideas, the y-equation is

$$p_0 y''' + p_1 y'' + p_2 y' + p_3 = 0,$$

then, dividing by y, the z-equation is

$$p_0 (z^3 + 3zz' + z'') + p_1 (z^2 + z') + p_2 z + p_3 = 0;$$

and similarly, from the y-equation of the order m, we derive a z-equation of the order $m-1$,

$$p_0 \{z^m + \ldots + z^{(m-1)}\} + \ldots + p_{m-2} (z^2 + z') + p_{m-1} z + p_m = 0.$$

4. In this equation, each coefficient is in the first instance a rational and integral function of x, or, what is the same thing, of $x - a$; writing t to denote $\frac{1}{x-a}$, each coefficient is thus the sum of a finite number of negative integer powers of t; hence multiplying the whole equation by a proper positive power of t, the several coefficients become each of them the sum of a finite number of positive powers of t, or arranging in descending powers of t, say the general form is $p_r = \alpha_r t^\lambda + \beta_r t^{\lambda-1} + \ldots + \kappa_r$, where λ is a positive integer which may be $=0$, and which has for each coefficient its proper value. We wish to satisfy the z-equation by a value z = descending series in t, viz. the form is $z = At^\alpha + A't^{\alpha-\alpha'} + \ldots$, with powers which are integral or fractional, but where the number of positive powers is finite. The theory is almost identical with that of the solution in like form of the algebraical equation

$$p_0 z^m + p_1 z^{m-1} + \ldots + p_{m-1} z + p_m = 0,$$

or say the equation $U = 0$, where U is a rational and integral function of z and t, of the degree m as regards z.

5. Considering z and t as Cartesian coordinates, the equation in question, $U = 0$, represents a geometrical curve such for any given value of the abscissa t, the ordinate z has m values: the curve is of the order m at least, but it may be of a higher order $m + \kappa$; when this is so, there is at infinity on the axis $t = 0$, a κ-tuple point K, and thus any line parallel to the axis $t = 0$, intersects the curve in the point K counting as κ intersections, and in m other points, which are the points belonging to the m values of the ordinate z. Taking $s = 1$, and for z, t writing $\frac{z}{s}$, $\frac{t}{s}$ respectively, we have between the trilinear coordinates z, t, s an equation $U' = 0$ of the order $m + \kappa$, where $s = 0$ is the equation of the line infinity: and the curve has a κ-tuple point at the point K for which $t = 0$, $s = 0$.

6. Starting from the equation $U = 0$, for an arbitrary value of t we have m values of z all of them different. These may be developed, each of them as a descending series in t, and the development is effected in the usual manner; viz. considering for the moment t as very large, then z is in general very large and it has a leading term At^α, which is determined by writing in the equation $z = At^\alpha$, and then finding α in such wise that there are two or more exponents equal to each other and greater

(that is, nearer $+\infty$) than any other exponent: we have thus a highest power of t, the whole coefficient for which must vanish, viz. we obtain an equation of two or more terms giving for A a value or values not $=0$; in the term At^{α} in question, α is in general positive, but it may be $=0$, or be negative. The leading term At^{α} being found in this manner, the law for the exponents of the subsequent terms is frequently at once apparent; but, if necessary, we write $z = At^{\alpha} + A't^{\alpha-\alpha'}$, and determine in like manner the exponent $\alpha - \alpha'$ and the coefficient A'. Proceeding in this manner until the law of the exponent becomes apparent, and then by the method of indeterminate coefficients, we finally arrive at a series

$$z = At^{\alpha} + A't^{\alpha-\alpha'} + A''t^{\alpha-\alpha'-\alpha''} + \ldots,$$

in descending powers of t: the exponents are integral or fractional, but the number of positive exponents is always finite. There is either a single series or it may be two or more series; but the number of series is at most $=m$, and when it is less than m, then the coefficients in the several series or some of them will contain radicals, and by giving to each radical its different values, the system of series will determine m different values of z.

7. The curve meets the line infinity ($s=0$) in the point K counting as κ intersections, and in m other points, some or all of which may coincide with K; the forms of the series depend on the configuration of the m points, or say on the relation of the curve to the line infinity. Thus suppose $\kappa=0$, and further that the m points are all of them distinct, that is, let the curve be a curve of the order m meeting the line infinity in m distinct points, or, what is the same thing, having m asymptotes, no two of them parallel; we have in this case m series, each of the form

$$z = At + B + \frac{C}{t} + \frac{D}{t^2} + \ldots,$$

where the several coefficients A have distinct values.

8. In the foregoing case A is determined by an equation of the order m, having unequal roots; and taking for A any root at pleasure, the remaining coefficients $B, C, \ldots$ are each of them linearly determined. If the equation has two equal roots, this may correspond to the case of two parallel asymptotes (the curve has here a node at infinity); the coefficients $B, C, \ldots$ will in this case depend on a quadric radical, and by giving to this radical its two values, we obtain the two series

$$z = At + B + \frac{C}{t} + \ldots,$$

$$z = At + B' + \frac{C'}{t} + \ldots,$$

corresponding to the two equal roots of the equation. But if instead of a node at infinity we have the line infinity a tangent to the curve, then instead of the two parallel asymptotes, we have an asymptotic parabola; the series assumes a new form, viz. it contains terms in $t^{\frac{1}{2}}, t^{-\frac{1}{2}}, \ldots$; the coefficients of the integer powers have determinate values, but those of the powers $t^{\frac{1}{2}}, t^{-\frac{1}{2}},\ldots$ contain as factor a quadric radical, and giving to this radical its two values, we have two series

$$z = At + C + \ldots + t^{\frac{1}{2}}\left(B + \frac{D}{t} + \ldots\right),$$

$$z = At + C + \ldots - t^{\frac{1}{2}}\left(B + \frac{D}{t} + \ldots\right).$$

9. The like considerations apply in the case where the line infinity passes through higher multiple points of the curve, or has with it a contact or contacts other than a single ordinary contact: in every case, reckoning as distinct series those obtained by attaching to each radical its different values, the number of distinct series will be $= m$ precisely.

10. Reverting now to the differential equation

$$p_0\{z^m + \ldots + z^{(m-1)}\} + \ldots + p_{m-2}(z^2 + z') + p_{m-1}z + p_m = 0,$$

it is to be observed that the very same process is applicable to the determination of z in the form of a descending series $z = At^\alpha + A't^{\alpha-\alpha'} + \ldots$; moreover, it frequently happens that the determination of the leading term At^α is made by means of the algebraical equation

$$p_0z^m + \ldots + p_{m-2}z^2 + p_{m-1}z + p_m = 0,$$

viz. this is so whenever the value of α is greater than 1: in fact, writing $z = At^\alpha\ (\alpha > 1)$, then in the sets of terms $\{z^m + \ldots + z^{(m-1)}\}, \ldots, (z^3 + 3zz' + z''),\ (z^2 + z')$, the first terms $z^m, \ldots, z^3, z^2$ are each of them of a higher degree than any other term in the same brackets, so that attending only to the terms of highest degree the sets may be reduced to $z^m, \ldots, z^3, z^2$. But in the determination of the subsequent exponents and coefficients, the omitted terms or some of them would of course come into play: and it is at least not obvious that we can, for the forms of the series which satisfy the differential equation, employ geometrical considerations such as those which were used in regard to the series satisfying the algebraical equation.

11. Considering as above the coefficients $p_0, p_1, \ldots, p_m$ as rational and integral functions of t, it is easy to determine the degrees of these coefficients in such wise that the exponent α of the leading term At^α shall have a given value. Suppose first, this given value is $\alpha = \text{a}$, a positive integer greater than 1: the degrees may be $\theta,\ \theta + \text{a},\ \theta + 2\text{a}, \ldots,\ \theta + m\text{a}$, ($\theta = 0$ or any positive integer at pleasure); and not only so, but if we write

$$p_0 = L_0t^\theta + \ldots,\quad p_1 = L_1t^{\theta+\text{a}} + \ldots,\quad p_m = L_mt^{\theta+m\text{a}} + \ldots,$$

then writing $z = At^{\text{a}}$, clearly the highest power in the equation is $t^{\theta+m\text{a}}$, and equating the coefficient hereof to zero, we have

$$L_0A^m + L_1A^{m-1} + \ldots + L_{m-1}A + L_m = 0,$$

an equation of the degree m for the determination of the coefficient of the leading term At^{a} (a > 1). Assuming that the roots are all unequal, we have m series each of the form

$$z = At^{a} + Bt^{a-1} + \ldots + K + \frac{L}{t} + \ldots.$$

12. Suppose a = 1, that is, let the leading term be At; here the degrees may be θ, $\theta+1$, $\theta+2, \ldots, \theta+m$, ($\theta = 0$ or any positive integer as before), but we have a different form for the A-equation. In fact, here ($z = At$) in the sets of terms $\{z^m + \ldots + z^{(m-1)}\}, \ldots, (z^3 + 3zz' + z''), (z^2 + z')$, the terms in each brackets are of the same degree, viz. the degrees are $m, \ldots, 3, 2$; and not only so, but we have $z^2 + z' = (A^2 + A)\,t^2$, that is, $= [A]^2\, t^2$, $z^3 + 3zz' + z'' = [A]^3\, t^3, \ldots$. Hence if

$$p_0 = L_0 t^{\theta} + \ldots, \quad p_1 = L_1 t^{\theta+1} + \ldots, \quad p_m = L_m t^{\theta+m} + \ldots,$$

the A-equation is

$$L_0 [A]^m + L_1 [A]^{m-1} + \ldots + L_{m-1} [A]^1 + L_m = 0,$$

an equation of the degree m as before; assuming that the roots are unequal, we have here m series each of the form

$$z = At + B + \frac{C}{t} + \frac{D}{t^2} + \ldots,$$

being, as will presently appear, the case of m regular integrals for the y-equation.

13. In either of the last-mentioned cases, the formula for the A-equation holds good if any two or more of the coefficients $p_0, p_1, \ldots, p_m$ are of the degrees aforesaid, the others of them being of inferior degrees; we have only to put $=0$ the L's which belong to the coefficients of inferior degrees. If neither L_0 nor L_m be $=0$, the equation is still an equation of the order m, with m effective roots; but if any of the coefficients L at the beginning of the equation vanish we thus introduce roots $=\infty$: and if any of the coefficients L at the end of the equation vanish, we thus introduce roots $=0$: the number of effective roots is $= m -$ the number of roots ∞ or 0; the case requires further consideration. If the equation has equal roots, then from what precedes, it is easy to infer that we may have distinct series containing as in the general case only integer powers of t, or we may have series beginning with At^{a} or At, but containing also fractional powers of t.

14. If the leading term is $z = At^{\alpha}$, $\alpha =$ a, a value < 1, then in the sets of terms $\{z^m + \ldots + z^{(m-1)}\}, \ldots, (z^3 + 3zz' + z''), (z^2 + z')$ the last terms $z^{(m-1)}, \ldots, z'', z'$ are each of them of a higher degree than any other term in the same brackets, so that attending only to the terms of the highest degree the sets may be reduced to $z^{(m-1)}, \ldots, z'', z'$, or we may consider the equation

$$p_0 z^{(m-1)} + p_1 z^{(m-2)} + \ldots + p_{m-2} z' + p_{m-1} z + p_m = 0.$$

In particular, this is the case for $z = At^{-a}$, a being a positive integer; in this case, the degrees of $p_0, p_1, \ldots, p^{m-1}, p_m$ may be $\theta + \mathrm{a} - m + 1$, $\theta + \mathrm{a} - m + 2, \ldots, \theta + \mathrm{a}$, θ; and this being so, the leading coefficient A will be determined by a linear equation. But the case is, in fact, that of a non-singular point $x = a$: and I do not here further consider it.

15. Reverting to the case where the series is

$$z = At + B + \frac{C}{t} + \frac{D}{t^2} + \ldots,$$

and substituting for t its value $\frac{1}{x-a}$, or for greater convenience $\frac{1}{x}$, this is

$$\frac{1}{y}\frac{dy}{dx} = \frac{A}{x} + B + Cx + Dx^2 + \ldots,$$

whence

$$\log y = A \log x + Bx + \tfrac{1}{2}Cx^2 + \ldots,$$

or passing to the value of y, but expanding the exponential of $Bx + Cx^2 + \ldots$, this is

$$y = x^A(1 + B'x + C'x^2 + \ldots),$$

viz. we have here a regular integral. By what precedes, it appears that, for the values $p_0 = L_0 x^{-\theta} + \ldots$, $p_1 = L_1 x^{-\theta-1} + \ldots$, $p_m = L_m x^{-\theta-m} + \ldots$, the exponent A is determined by the equation

$$L_0 [A]^m + L_1 [A]^{m-1} + \ldots + L_{m-1} [A]^1 + L_m = 0,$$

which is the equation for that purpose considered by Fuchs, and is what I would call the indicial equation.

The logarithmic forms which belong to the case of equal roots may be deduced from the case of unequal roots, by supposing two or more of the roots to approach each other continuously and become ultimately equal.

16. Similarly, if the series be

$$z = Ft^{\mathrm{a}} + \ldots + Kt^2 + At + B + \frac{C}{t} + \ldots,$$

(a being a positive integer $= 2$ at least), then writing as above $t = \frac{1}{x}$, this is

$$\frac{1}{y}\frac{dy}{dx} = \frac{F}{x^{\mathrm{a}}} + \ldots + \frac{K}{x^2} + \frac{A}{x} + B + Cx + \ldots,$$

whence, if for shortness $w = -\frac{F}{\mathrm{a}-1}\frac{1}{x^{\mathrm{a}-1}} - \ldots - \frac{K}{x}$, we have

$$\log y = w + A \log x + Bx + \tfrac{1}{2}Cx^2 + \ldots,$$

or passing to the value of y, but expanding as before the exponential of $Bx + \frac{1}{2}Cx^2 + \ldots$, this is

$$y = e^w x^A (1 + B'x + C'x^2 + \ldots),$$

a subregular integral. By what precedes, it appears that, if

$$p_0 = L_0 x^{-\theta} + \ldots,\ p_1 = L_1 x^{-\theta-\mathrm{a}} + \ldots,\ p_m = L_m x^{-\theta-m\mathrm{a}} + \ldots,$$

then the equation for the leading coefficient F (previously represented by A) is

$$L_0 F^m + L_1 F^{m-1} + \ldots + L_{m-1} F + L_m = 0;$$

but I do not see any direct or simple way of obtaining the corresponding equation for the exponent A of this subregular integral.

17. To illustrate the case of a series with fractional powers, and also in order to work out a simple example with some completeness, I consider the differential equation of the second order $\frac{d^2y}{dx^2} + p_2 y = 0$, giving the z-equation $z^2 + z' + p_2 = 0$; p_2 is a rational and integral function of $t\left(=\frac{1}{x}\right)$, and we have two cases according as the order of the function is even or odd.

18. Suppose first that the order is even, and for convenience taking it to be $= 4$, and assuming the value of p_2 accordingly, I consider the equation

$$z^2 + z' = \alpha t^4 + \beta t^3 + \gamma t^2 + \delta t + \epsilon.$$

Here the form is at once seen to be

$$z = At^2 + Bt + C + \frac{D}{t} + \frac{E}{t^2} + \ldots,$$

and observing that we have

$$z' = -t^2 \frac{dz}{dt} = -2At^3 - Bt^2 + D + \frac{2E}{t} + \ldots,$$

we find

$$\begin{array}{ll} A^2 & = \alpha, \\ 2AB & -2A = \beta, \\ 2AC + B^2 & -\ B = \gamma, \\ 2AD + 2BC & +0C = \delta, \\ 2AE + 2BD + C^2 & +\ D = \epsilon, \\ 2AF + 2BE + 2CD + 2E & = 0, \\ \vdots & \end{array}$$

equations which give

$$A = \sqrt{\alpha},$$

$$B = 1 + \frac{\beta}{2\sqrt{\alpha}},$$

$$C = \frac{\gamma}{2\sqrt{\alpha}} - \frac{\beta}{4\alpha} - \frac{\beta^2}{8\alpha\sqrt{\alpha}},$$

$$D = \frac{\delta}{2\sqrt{\alpha}} - \frac{\gamma}{2\alpha} + \frac{1}{4\alpha\sqrt{\alpha}}(\beta - \beta\gamma) + \frac{\beta^2}{4\alpha^2} + \frac{\beta^3}{16\alpha^2\sqrt{\alpha}},$$

$$E = \frac{\epsilon}{2\sqrt{\alpha}} - \frac{3\delta}{4\alpha} + \frac{1}{4\alpha\sqrt{\alpha}}(3\gamma - \beta\delta - \tfrac{1}{2}\gamma^2) + \frac{1}{4\alpha^2}(-\tfrac{3}{2}\beta + 3\beta\gamma)$$

$$+ \frac{1}{16\alpha^2\sqrt{\alpha}}(-\tfrac{17}{2}\beta^2 + 3\beta^2\gamma) - \frac{\beta^3}{4\alpha^3} - \tfrac{5}{128}\frac{\beta^4}{\alpha^3\sqrt{\alpha}};\ \&c.$$

We then have

$$\frac{1}{y}\frac{dy}{dx} = \frac{A}{x^2} + \frac{B}{x} + C + Dx + Ex^2 + \ldots,$$

and consequently

$$y = e^{-\frac{A}{x}}\, x^B\, e^{Cx + \frac{1}{2}Dx^2 + \frac{1}{3}Ex^3 + \ldots},$$

where the second exponential is to be expanded in the form $1 + C'x + D'x^2 + \ldots$; there are of course two values, say y_1 and y_2, corresponding to the values $+\sqrt{\alpha}$ and $-\sqrt{\alpha}$ of the radical respectively.

19. Secondly, if the order be odd, and for convenience taking it to be $= 3$, and assuming the value of p_2 accordingly, I consider the equation

$$z^2 + z' = \beta t^3 + \gamma t^2 + \delta t + \epsilon.$$

Here the series is at once seen to be

$$z = At^{\frac{3}{2}} + Bt + Ct^{\frac{1}{2}} + D + Et^{-\frac{1}{2}} + \ldots:$$

we have

$$z' = -\tfrac{3}{2}At^{\frac{5}{2}} - Bt^2 - \tfrac{1}{2}Ct^{\frac{3}{2}} + \tfrac{1}{2}Et^{\frac{1}{2}} + \ldots,$$

and we thence derive the equations

$$\begin{aligned}
&A^2 &&= \beta,\\
&2AB &- \tfrac{3}{2}A &= 0,\\
&2AC + B^2 &- B &= \gamma,\\
&2AD + 2BC &- \tfrac{1}{2}C &= 0,\\
&2AE + 2BD + C^2 &&= \delta,\\
&2AF + 2BE + 2CD + \tfrac{1}{2}E &&= 0,\\
&\vdots
\end{aligned}$$

(where in the next equation we have on the right-hand side ϵ, and in each of the subsequent equations we have on the right-hand side 0). We find

$$\begin{aligned}
A &= \sqrt{\beta},\\
B &= \tfrac{3}{4},\\
C &= \frac{1}{2\sqrt{\beta}}\left(\gamma + \tfrac{3}{16}\right),\\
D &= -\frac{1}{4\beta}\left(\gamma + \tfrac{3}{16}\right),\\
E &= \frac{1}{2\beta\sqrt{\beta}}\left(\delta + \tfrac{9}{32}\gamma - \tfrac{1}{4}\gamma^2 + \tfrac{63}{1024}\right), \text{ \&c.,}
\end{aligned}$$

where observe that the coefficients A, C, E, ... contain as a factor $\dfrac{1}{\sqrt{\beta}}$, but the coefficients B, D, ... are rational.

We have

$$z = \frac{1}{y}\frac{dy}{dx} = \frac{A}{x^{\frac{3}{2}}} + \frac{B}{x} \quad + \frac{C}{x^{\frac{1}{2}}} + D + Ex^{\frac{1}{2}} + Fx + \ldots,$$

giving

$$\log y = -\frac{2A}{x^{\frac{1}{2}}} + B\log x + 2Cx^{\frac{1}{2}} + Dx + \tfrac{2}{3}Ex^{\frac{3}{2}} + \tfrac{1}{2}Fx^2 + \ldots;$$

and thence passing to the value of y, but expanding the exponential of $2Cx^{\frac{1}{2}} + Dx + \ldots$, we have say the integral

$$y_1 = e^{-\frac{2A}{x^{\frac{1}{2}}}} x^B (1 + C'x^{\frac{1}{2}} + D'x + E'x^{\frac{3}{2}} + \ldots),$$

and then also, changing the sign of $\sqrt{\beta}$, the integral

$$y_2 = e^{\frac{2A}{x^{\frac{1}{2}}}} \cdot x^B (1 - C'x^{\frac{1}{2}} + D'x - E'x^{\frac{3}{2}} + \ldots).$$

Writing for a moment

$$Q = e^{-\frac{2A}{x^{\frac{1}{2}}}},$$

$$M = 1 + D'x + F'x^2 + \ldots,$$

$$N = C' + E'x + G'x^2 + \ldots,$$

the two integrals are $x^B Q(M + Nx^{\frac{1}{2}})$, and $x^B Q^{-1}(M - Nx^{\frac{1}{2}})$: these may be replaced by their sum and difference, viz. these are

$$y_1 + y_2 = Mx^B(Q + Q^{-1}) + Nx^{B+\frac{1}{2}}(Q - Q^{-1}),$$

$$y_1 - y_2 = Mx^B(Q - Q^{-1}) + Nx^{B+\frac{1}{2}}(Q + Q^{-1}),$$

where $Q + Q^{-1}$ is a rational function of x, and $Q - Q^{-1}$ is $= x^{\frac{1}{2}}$ multiplied by a rational function of x; hence $y_1 + y_2$ is $= x^B$ multiplied by a rational function of x, but $y_1 - y_2$ is $= x^{B+\frac{1}{2}}$ multiplied by a rational function of x.

Cambridge, 28 *April* 1886.

863.

NOTE ON THE THEORY OF LINEAR DIFFERENTIAL EQUATIONS.

[From *Crelle's Journal der Mathem.*, t. CI. (1887), pp. 209—213.]

THE theorem V. given by Fuchs in the memoir "Zur Theorie der linearen Differentialgleichungen mit veränderlichen Coefficienten," *Crelle's Journal*, t. LXVIII. (1868), pp. 354—385 (see p. 374) for the purpose of deciding whether the integrals belonging to a group of roots of the "determinirenden Fundamentalgleichung" (or as I call it, the Indicial equation) do or do not involve logarithms, may I think be exhibited in a clearer form.

Starting from the differential equation

$$P(y), \quad = p_0 \frac{d^m y}{dx^m} + p_1 \frac{d^{m-1} y}{dx^{m-1}} + \ldots + p_m y, \quad = 0,$$

of the order m, then if X be any function of x not satisfying the differential equation, we can at once form a differential equation of the order $m+1$, satisfied by all the solutions of the differential equation, and having also the solution $y = X$; the required equation is in fact

$$\partial_x P(y) \,.\, P(X) - P(y) \,.\, \partial_x P(X) = 0.$$

This I call the augmented equation.

I recall that the equation $P(y) = 0$, considered by Fuchs, is an equation having for each singular point $x = a$, m regular integrals, viz. the coefficients $p_0, p_1, \ldots, p_m$ have the forms $q_0 (x-a)^m$, $q_1 (x-a)^{m-1}, \ldots, q_m$, where $q_0, q_1, \ldots, q_m$ are rational and integral functions of $x - a$, q_0 not vanishing for $x = a$, and the other functions $q_1, q_2, \ldots, q_m$ not in general vanishing for $x = a$. Writing $y = (x-a)^\theta$, we obtain

$$P(x-a)^\theta = I(\theta)(x-a)^\theta + \text{higher powers of } (x-a),$$

where $I(\theta)$, the coefficient of the lowest power of $(x-a)$, is a function of θ of the order m, which I call the indicial coefficient; and equating it to zero, we have $I(\theta) = 0$, the determinirende Fundamentalgleichung, or Indicial equation, being an equation of the order m. If the roots of this equation are such that no two of them are equal or differ only by an integer number, then we have m particular integrals each of them of the form

$$y=(x-a)^r + \text{higher powers of } (x-a),$$

where r is any root of the indicial equation: but if we have in the indicial equation a group of λ roots $r_1, r_2, \ldots, r_\lambda$, such that the difference of each two of them is either zero or an integer, then the integrals which correspond to these roots involve or may involve logarithms; in particular, if any two of the roots are equal, the integrals for the group will involve logarithms.

Consider now the differential equation $P(y)=0$ in reference to the singular point $x=a$ as above, and writing $X=(x-a)^\epsilon f$ where ϵ is in the first instance arbitrary, and f is a rational and integral function of $x-a$ not vanishing for $x=a$, we form the augmented equation which, observing that we have in general $P(X)=(x-a)^\epsilon Q$, Q a rational and integral function of $x-a$ not vanishing for $x=a$, and dividing the whole equation by $(x-a)^{\epsilon-1}$, may be written

$$\partial_x P(y) \,.\, (x-a)\, Q - P(y)\, \{\epsilon Q + (x-a)\, \partial_x Q\} = 0,$$

an equation of the same form as the original equation (but of the order $m+1$ instead of m), and having an indicial equation

$$(\theta-\epsilon)\, I(\theta)=0.$$

In fact, writing as before $y=(x-a)^\theta$, we have in $\partial_x P(y) . (x-a)\, Q$ the term of lowest order $\theta I(\theta)\, Q_0 (x-a)^\theta$ and in $P(y) . \epsilon Q$ the term of lowest order $\epsilon I(\theta)\, Q_0 (x-a)^\theta$, whereas in $P(y)(x-a)\, \partial_x Q$ the term of lowest order is $(x-a)^{\theta+1}$; the indicial equation is thus as just found.

If however ϵ be equal to a root of the indicial equation $I(\theta)=0$, then instead of $P(X)=(x-a)^\epsilon Q$, we have $P(X)=(x-a)^\mu Q$, where the index μ is $=\epsilon +$ a positive integer, and where the value of the difference $\mu-\epsilon$ may depend upon the determination of the function f in the expression $(x-a)^\epsilon f$. The indicial equation for the augmented equation is in this case $(\theta-\mu)\, I(\theta)=0$.

If the indicial equation $I(\theta)=0$ of the given differential equation has a group of roots $r_1, r_2, \ldots, r_\lambda$, the difference of any two of these roots being zero or an integer, then taking $\epsilon=$ any one of these roots, the augmented equation will have a group of roots $(\mu, r_1, r_2, \ldots, r_\lambda)$.

If any two of the roots $r_1, r_2, \ldots, r_\lambda$ are equal, the group of integrals $u_1, u_2, \ldots, u_\lambda$ will involve logarithms: the question only arises when these roots are unequal, and taking them to be so, the theorem V. is in effect as follows: "If by taking $\epsilon=$ some one of the roots $r_1, r_2, \ldots, r_\lambda$, and by a proper determination of the function f we can make μ to be $=$ one of the same roots $r_1, r_2, \ldots, r_\lambda$, then the group of integrals $u_1, u_2, \ldots, u_\lambda$ will involve logarithms; but if μ cannot be made $=$ one of the roots $r_1, r_2, \ldots, r_\lambda$, then the group of integrals will be free from logarithms."

As an example, I consider the equation

$$P(y)=(x^2-x^4)\frac{d^2y}{dx^2}-2x^3\frac{dy}{dx}-(n^2+n)\, y=0.$$

This is Legendre's equation $(1-x^2)\frac{d^2y}{dx^2}-2x\frac{dy}{dx}+(n^2+n)y=0$, with $\frac{1}{x}$ substituted for x, so that, instead of a singular point $x=\infty$, there may be a singular point $x=0$. Attending to the singular point $x=0$, we have $P(x^\theta)=(\theta^2-\theta-n^2-n)x^\theta+$ higher powers, so that the indicial equation $I(\theta)=0$ is $\theta^2-\theta-n^2-n=0$, that is, $(\theta+n)(\theta-n-1)=0$, or we have the roots $-n$, $n+1$, which differ by an integer, and thus form a group, if n be $=$ an integer, or be $=$ an integer $-\frac{1}{2}$; to fix the ideas, say that the roots are $-p$, $p+1$ or else $-p+\frac{1}{2}$, $p+\frac{1}{2}$ where p is a positive integer.

Writing for greater convenience $x^\epsilon f=x^\epsilon+F$, where F is a sum of powers of x higher than ϵ, we find without difficulty

$$P(x^\epsilon f)=x^\epsilon\{(\epsilon+n)(\epsilon-n-1)-(\epsilon^2+\epsilon)x^2+(x^2-x^4)x^{-\epsilon}F''-(n^2+n)x^{-\epsilon}F\}$$

which, so long as ϵ remains arbitrary, is of the form $x^\epsilon Q$, $Q=(\epsilon+n)(\epsilon-n-1)+$ powers of x; if however ϵ be a root of the indicial equation, for instance, if $\epsilon=-n$, then the expression in brackets $\{\ \}$ contains at any rate the factor x, so that the form is $P(x^{-n}f)=x^\mu Q$, where μ is $=-n+1$ at least; we can however, by a proper determination of the function f, make μ acquire a larger value.

For instance, suppose $-n$, $n+1=-2$, 3; $\epsilon=-n=-2$, and assume

$$x^\epsilon f=x^{-2}+Bx^{-1}+Cx^0+Dx^1+Ex^2+Fx^3+Gx^4+\ldots$$

To calculate $P(x^\epsilon f)$, we have

	x^{-2}	x^{-1}	x^0	x^1	x^2	x^3	x^4	$x^5\ldots$
	6	$2B$	$0C$	$0D$	$2E$	$6F$	$12G$	$20H\ldots$
			-6	$-2B$	$-0C$	$-0D$	$-2E$	$-6F$
			$+4$	$+2B$	$-0C$	$-2D$	$-4E$	$-6F$
	-6	$-6B$	$-6C$	$-6D$	$-6E$	$-6F$	$-6G$	$-6H.$
$P(x^\epsilon f)=$	0	$-4B$	$-6C$	$-6D$	$-4E$	$0F$	$6G$	$14H\ldots$
			-2			$-2D$	$-6E$	$-12F\ldots$

Hence if B not $=0$, we have $\mu=-1$; if $B=0$, $-6C-2$ not $=0$, we have $\mu=0$; if $B=0$, $-6C-2=0$, D not $=0$, we have $\mu=1$; if $B=0$, $-6C-2=0$, $D=0$, but E not $=0$, we have $\mu=2$; if $B=0$, $-6C-2=0$, $D=0$, $E=0$, then the coefficient of x^3, $=0F-2D$, is $=0$, and we have not $\mu=3$, but $\mu=4$ at least, viz. μ will be $=4$, if $6G-6E=0$, that is, if $G=0$; but leaving F arbitrary, we can by giving proper values to the subsequent coefficients H, I, &c., make μ to be $=5$ or any larger integer value. The values of μ are thus $=-1, 0, 1, 2, 4, 5, \ldots$, and we see that the group $(\mu, -n, n+1)$, that is, $(\mu, -2, 3)$, does not in any case contain two equal indices. Starting from the value $\epsilon=3$, the value of μ is >3, and thus here also the group $(\mu, -2, 3)$ does not contain two equal indices.

The conclusion from the theorem thus is that the integrals u_1, u_2, belonging to the roots -2, 3, do not involve logarithms: and in precisely the same manner, it appears that the integrals, belonging to the two roots $-p$, $p+1$ (p any positive integer), do not involve logarithms: this is right, for the integrals are, in fact, the Legendrian functions of the first and second kinds P_p and Q_p, with only $\frac{1}{x}$ written therein instead of x.

Similarly, if for instance $-n,\ n+1=-\tfrac{1}{2},\ \tfrac{3}{2}$, then, if $\epsilon=-n=-\tfrac{1}{2}$, assuming

$$x^{\epsilon}f = x^{-\frac{1}{2}} + Bx^{\frac{1}{2}} + Cx^{\frac{3}{2}} + Dx^{\frac{5}{2}} + Ex^{\frac{7}{2}} + \dots,$$

we have

$$\begin{array}{rccccc}
 & x^{-\frac{1}{2}} & x^{\frac{1}{2}} & x^{\frac{3}{2}} & x^{\frac{5}{2}} & x^{\frac{7}{2}} \dots \\
\hline
 & \frac{3}{4} & -\frac{1}{4}B & \frac{3}{4}C & \frac{15}{4}D & \frac{35}{4}E \\
 & & & -\frac{3}{4} & +\frac{1}{4}B & -\frac{3}{4}C \\
 & & & +1 & -B & -3C \\
 & -\frac{3}{4} & -\frac{3}{4}B & -\frac{3}{4}C & -\frac{3}{4}D & -\frac{3}{4}E \\
\hline
P(x^{\epsilon}f) = & 0 & -B & 0C & 3D & 8E \dots \\
 & & & +\frac{1}{4} & -\frac{3}{4}B & -\frac{15}{4}C \dots
\end{array}$$

We have here if B not $=0$, $\mu=\tfrac{1}{2}$; but if $B=0$, then we cannot in any way make the coefficient of $x^{\frac{3}{2}}$ to vanish, and consequently $\mu=\tfrac{3}{2}$. With this last value of μ, the group $(\mu,\ -n,\ n+1)$, that is, $(\mu,\ -\tfrac{1}{2},\ \tfrac{3}{2})$, becomes $(\tfrac{3}{2},\ -\tfrac{1}{2},\ \tfrac{3}{2})$ which contains two equal roots, and the conclusion from the theorem thus is that the integrals u_1, u_2, corresponding to the roots $-\tfrac{1}{2}$, $\tfrac{3}{2}$, involve logarithmic values. And similarly in general the integrals u_1, u_2, corresponding to the roots $-p+\tfrac{1}{2}$, $p+\tfrac{1}{2}$ (p any positive integer), involve logarithmic values: this also is right.

The examples exhibit the true character of the theorem, and show I think that it is a less remarkable one than would at first sight appear: in fact, in working them out, we really ascertain by an actual substitution whether the differential equation can be satisfied by series of powers only, without logarithms. Thus for $n=2$ as above, it appears that the equation is satisfied by the series

$$y = x^{-2} + Bx^{-1} + Cx^{0} + Dx^{1} + Ex^{2} + Fx^{3} + Gx^{4} + Hx^{5} + \dots,$$

where

$$B=0,\ C=-\tfrac{1}{3},\ D=0,\ E=0,\ F=F,\ G=0,\ H=-\tfrac{6}{7}F, \dots,$$

that is, by

$$y = x^{-2} + \tfrac{1}{3} + F(x^{3} - \tfrac{6}{7}x^{5} + \dots),$$

in other words, that we have the two particular integrals $y=x^{-2}+\tfrac{1}{3}$, and $y=x^{3}-\tfrac{6}{7}x+\dots$, belonging to the two roots -2, 3 respectively.

Similarly, when $n=\tfrac{1}{2}$, we cannot satisfy the equation by a series

$$y = x^{-\frac{1}{2}} + Bx^{\frac{1}{2}} + Cx^{\frac{3}{2}} + Dx^{\frac{5}{2}} + \dots;$$

for in order to satisfy the equation, we must have $B=0$, $C=\infty$; there is thus no series of powers $y=x^{-\frac{1}{2}}+Cx^{\frac{3}{2}}+\dots$, corresponding to the root $-\tfrac{1}{2}$: but there is a series $y=x^{\frac{3}{2}}+kx^{\frac{7}{2}}+\dots$ corresponding to the root $\tfrac{3}{2}$; and thus the integrals u_1, u_2, corresponding to these roots $-\tfrac{1}{2}$, $\tfrac{3}{2}$, involve logarithms.

Cambridge, 23 March 1887.

864.

ON RUDIO'S INVERSE CENTRO-SURFACE.

[From the *Quarterly Journal of Pure and Applied Mathematics*, vol. XXII. (1887), pp. 156—158.]

DR F. RUDIO, in an inaugural dissertation "Ueber diejenigen Flächen deren Krümmungsmittelpunktsflächen confokale Flächen zweiten Grades sind," Berlin, 1880, and *Crelle's Journal*, t. XCV., p. 240, has determined the surfaces having for their centro-surface (i.e., the locus of centres of curvature) the aggregate of the confocal quadric surfaces

$$\frac{x^2}{a-\lambda}+\frac{y^2}{b-\lambda}+\frac{z^2}{c-\lambda}=1,$$

$$\frac{x^2}{a-\mu}+\frac{y^2}{b-\mu}+\frac{z^2}{c-\mu}=1,$$

or, what is the same thing, the surfaces orthotomic to the common tangents of these two surfaces. He obtains, as the final result of an elegant analytical investigation, the following formulæ:

$$x=\sqrt{(a-\lambda)}\sqrt{\left(\frac{a-u\,.\,a-v}{a-b\,.\,a-c}\right)},$$

$$y=\sqrt{(b-\lambda)}\sqrt{\left(\frac{b-u\,.\,b-v}{b-c\,.\,b-a}\right)},$$

$$z=\sqrt{(c-\lambda)}\sqrt{\left(\frac{c-u\,.\,c-v}{c-a\,.\,c-b}\right)},$$

$$U=\sqrt{\left(\frac{a-u\,.\,b-u\,.\,c-u}{\lambda-u\,.\,\mu-u}\right)},$$

$$V=\sqrt{\left(\frac{a-v\,.\,b-v\,.\,c-v}{\lambda-v\,.\,\mu-v}\right)},$$

$$\xi\frac{U-V}{x}=1-\frac{b-\lambda\,.\,c-\lambda}{\lambda-u\,.\,\lambda-v}-\frac{a-\mu}{a-u\,.\,a-v}\,UV,$$

$$\eta\frac{U-V}{y}=1-\frac{c-\lambda\,.\,a-\lambda}{\lambda-u\,.\,\lambda-v}-\frac{b-\mu}{b-u\,.\,b-v}\,UV,$$

$$\zeta\frac{U-V}{z}=1-\frac{a-\lambda\,.\,b-\lambda}{\lambda-u\,.\,\lambda-v}-\frac{c-\mu}{c-u\,.\,c-v}\,UV,$$

(values which are such that $\xi^2+\eta^2+\zeta^2=1$),

$$\rho=\tfrac{1}{2}\left(\int\frac{du}{U}+\int\frac{dv}{V}\right)+C.$$

And then

$$x'=x+\rho\xi,\quad y'=y+\rho\eta,\quad z'=z+\rho\zeta,$$

viz. the equations give x, y, z, U, V, ξ, η, ζ, each of them as a function of two independent parameters u, v; ρ is a function of u, v and of the arbitrary constant C; hence, giving to C any assumed value, we have x', y', z' each of them a function of the two arbitrary parameters u, v; that is, x', y', z' are the coordinates of a point on a surface, one of a system of parallel surfaces (corresponding to the different values of C) which are the surfaces in question.

Observe that u, v are the elliptic coordinates of the point (x, y, z) on the first of the two confocal surfaces, and that ξ, η, ζ are the cosine-inclinations of one of the tangents from this point to the other confocal surface; so that, if ρ were left arbitrary, the equations $x'=x+\rho\xi$, $y'=y+\rho\eta$, $z'=z+\rho\zeta$ would be the equations of a common tangent of the two confocal surfaces; but ρ, as determined, is the distance of the point x, y, z from the point x', y', z' on the required surface. The expression for ρ involves hyper-elliptic integrals of the first species, which are the same as those which present themselves in the determination of the geodesic lines upon either of the two confocal surfaces.

I have, in quoting these remarkable results, written for greater simplicity a, b, c instead of the author's a^2, b^2, c^2.

Cambridge, Dec. 20, 1886.

865.

ON MULTIPLE ALGEBRA.

[From the *Quarterly Journal of Pure and Applied Mathematics*, vol. XXII. (1887), pp. 270—308.]

1. I REPRODUCE a passage from my Presidential Address, British Association, Southport, 1883.

"Outside of ordinary mathematics we have some theories which must be referred to: algebraical, geometrical, logical. It is, as in many other cases, difficult to draw the line: we do in ordinary mathematics use symbols not denoting quantities, which we nevertheless combine in the way of addition and multiplication, $a+b$ and ab, and which may be such as not to obey the commutative law $ab=ba$; in particular, this is or may be so in regard to symbols of operation; and it could hardly be said that any development whatever of the theory of such symbols of operation did not belong to ordinary algebra. But I do separate from ordinary algebra the system of multiple algebra or linear associative algebra developed in the valuable memoir by the late Benjamin Peirce, "Linear Associative Algebra" (1870, reprinted 1881 in the *American Journal of Mathematics*, vol. IV., with notes and addenda by his son, C. S. Peirce): we here consider symbols A, B, &c., which are linear functions of a determinate number of letters or units i, j, k, l, &c., with coefficients which are ordinary analytical magnitudes real or imaginary, viz. the coefficients are in general of the form $x+iy$, where i is the before-mentioned imaginary, or $\sqrt{(-1)}$ of ordinary analysis. The letters i, j, k, &c., are such that every binary combination i^2, ij, ji, &c., (ij not in general $=ji$), is equal to a linear function of the letters, but under the restriction of satisfying the associative law; viz. for each combination of three letters $ij.k$ is $=i.jk$, so that there is a determinate and unique product of three or more letters; or, what is the same thing, the laws of combination of the units i, j, k,... are defined by a multiplication table giving the values of i^2, ij, ji, &c.; the original units may be replaced by linear functions of these units, so as to give rise for the units finally adopted to a multiplication table of the most simple form; and it is very remarkable how frequently in these simplified forms we have nilpotent or idempotent symbols ($i^2=0$ or $i^2=i$ as the case may be), and symbols i, j, such that $ij=ji=0$; and consequently how simple are the forms of the multiplication tables which define the several systems respectively.

"I have spoken of this multiple algebra before referring to various geometrical theories of earlier date, because I consider it as the general analytical basis, and the true basis, of these theories. I do not realise to myself directly the notions of the addition or multiplication of two lines, areas, rotations, or other geometrical, kinematical, or mechanical entities; and I would formulate a general theory as follows: consider any such entity as determined by the proper number of parameters $a, b, c, \ldots$ (for instance, in the case of a finite line given in magnitude and position, these might be the length, the coordinates of one end, and the direction-cosines of the line considered as drawn from this end); and represent it by or connect it with the linear function $ai+bj+ck+\&c.$, formed with these parameters as coefficients and with a given set of units i, j, k, &c. Conversely, any such linear function represents an entity of the kind in question. Two given entities are represented by two linear functions; the sum of these is a like function representing an entity of the same kind, which may be regarded as the sum of the two entities; and the product of them (taken in a determined order, when the order is material) is an entity of the same kind, which may be regarded as the product (in the same order) of the two entities. We thus establish by definition the notion of the sum of the two entities, and that of the product (in a determinate order, when the order is material) of the two entities. The value of the theory in regard to any kind of entity would of course depend on the choice of a system of units $i, j, k, \ldots$, with such laws of combination as would give a geometrical or kinematical or mechanical significance to the notions of the sum and product as thus defined.

"Among the geometrical theories referred to, we have a theory (that of Argand, Warren, and Peacock) of imaginaries in plane geometry; Sir W. R. Hamilton's very valuable and important theory of quaternions; the theories developed in Grassmann's *Ausdehnungslehre*, 1844 and 1862; Clifford's theory of biquaternions, and recent extensions of Grassmann's theory to non-Euclidian space by Mr Homersham Cox. These different theories have of course been developed, not in anywise from the point of view from which I have been considering them, but from the points of view of their several authors respectively."

2. The present paper is in a great measure the development of the views contained in the foregoing extract; but, instead of establishing *ab initio* a linear function $ai+bj+ck+\ldots$ as above, I deduce this, as will be seen from the notion of addition.

3. If $x, y, \ldots$ denote ordinary (real or imaginary) analytical magnitudes, which (as such) are susceptible of addition and multiplication, and for each of these operations are commutative and associative, then we may consider a multiple symbol $(x, y, \ldots)$ involving any given number of letters, to fix the ideas say (x, y), susceptible of addition and multiplication according to determinate laws

$$(x, y)+(x', y')=(P, Q), \quad (x, y)(x', y')=(X, Y),$$

where P, Q, X, Y are given functions of x, y, x', y'. For greater simplicity the law of addition is taken to be

$$(x, y) + (x', y') = (x + x', y + y'),$$

so that as regards addition the multiple symbols are commutative and associative. But this is or is not the case for multiplication, according to the form of the given functions X, Y; for instance, if

$$(x, y)(x', y') = (xx' - yy', xy' + yx'),$$

then in regard to multiplication the symbols will be commutative and associative. But if

$$(x, y, z)(x', y', z') = (yz' - y'z, zx' - z'x, xy' - x'y),$$

then the symbols will be associative, but not commutative.

4. I remark here that we are in general concerned with symbols of a given multiplicity, double symbols (x, y), triple symbols (x, y, z), n-tuple symbols $(x_1, x_2, \ldots, x_n)$, as the case may be, and that as well the product as the sum is a symbol *ejusdem generis*, and consequently of the same multiplicity, with the component symbols; this is to be assumed throughout in the absence of an express statement to the contrary. It is, moreover, proper to narrow the notion of multiplication by restricting it to the case where the terms $(X, Y, \ldots)$ of the product are linear functions of the terms $(x, y, \ldots)$ and $(x', y', \ldots)$ of the component symbols respectively; any other form $(X, Y, \ldots)$ is better designated *not as a product*, but as a combination (or by some other name) of the component symbols $(x, y, \ldots)$ and $(x', y', \ldots)$.

5. I assume, moreover, that, if m be any ordinary analytical magnitude, this may be multiplied into a multiple symbol $(x, y, \ldots)$, according to the law

$$m(x, y, \ldots) = (mx, my, \ldots).$$

6. As a consequence of this last assumption and of the assumed law of addition, we have for instance

$$\begin{aligned}(x, y, z) &= (x, 0, 0) + (0, y, 0) + (0, 0, z)\\ &= x(1, 0, 0) + y(0, 1, 0) + z(0, 0, 1);\end{aligned}$$

that is, using single letters i, j, k for the multiple symbols $(1, 0, 0)$, $(0, 1, 0)$, $(0, 0, 1)$ respectively, we have

$$(x, y, z) = xi + yj + zk,$$

where the letters i, j, k, thus standing for determinate multiple symbols, may be termed "extraordinaries." Each extraordinary may be multiplied into any ordinary symbol x, and is commutative therewith, $xi = ix$; moreover, each extraordinary may be multiplied into itself, or into another extraordinary, according to laws which are, in fact, determined by means of the assumed law of multiplication of the original multiple symbols; and, conversely, the law of multiplication of the extraordinaries determines that of the original multiple symbols; thus, if

$$(x, y)(x', y') = (xx' - yy', xy' + yx'),$$

then

$$(ix + jy)(ix' + jy') = i(xx' - yy') + j(xy' + yx'),$$

and also

$$= i^2xx' + ijxy' + jiyx' + j^2yy',$$

which expressions will agree together if, and only if, $i^2 = i$, $ij = j$, $ji = j$, $j^2 = -i$, or, as these equations may be written,

	i	j
i	i	j
j	j	$-i$

And so in general we have a multiplication table giving each square or product as a homogeneous linear function of all or any of the extraordinaries.

7. And, conversely, from this multiplication table of the extraordinaries i, j, we have

$$(ix + jy)(ix' + jy') = i(xx' - yy') + j(xy' + yx'),$$

that is,

$$(x, y)(x', y') = (xx' - yy', xy' + yx')$$

the originally assumed formula of multiplication.

In the example just given, we have $i^2 = i$, $ij = ji = j$, viz. the symbol i comports itself like unity, and may be put $= 1$; we have $(x, y) = x + jy$, with the multiplication table

	1	j
1	1	j
j	j	-1

or simply the equation $j^2 = -1$; and then

$$(x + jy)(x' + jy') = xx' - yy' + j(xy' + yx');$$

and it is convenient to regard 1 as an extraordinary, and speak of the system of extraordinaries 1, j.

8. The separate terms $x, y, \ldots$, whatever be their number, may always without loss of generality be arranged in a line; but it may be convenient to arrange them in a different form, for instance, in that of a square; we may have for instance symbols $\begin{vmatrix} x, & y \\ z, & w \end{vmatrix}$ with the laws of combination

$$\begin{vmatrix} x, & y \\ z, & w \end{vmatrix} + \begin{vmatrix} x', & y' \\ z', & w' \end{vmatrix} = \begin{vmatrix} x + x', & y + y' \\ z + z', & w + w' \end{vmatrix},$$

$$\begin{vmatrix} x, & y \\ z, & w \end{vmatrix} \cdot \begin{vmatrix} x', & y' \\ z', & w' \end{vmatrix} = \begin{vmatrix} xx' + yz', & xy' + yw' \\ zx' + wz', & zy' + ww' \end{vmatrix},$$

where observe that the multiplication is not commutative.

9. A multiple symbol may be connected with a geometrical or physical entity of any kind: viz. any such entity, depending on a number of parameters susceptible of analytical magnitude, to fix the ideas say on two parameters (x, y), may be connected with the multiple symbol (x, y). We cannot in general directly conceive the addition or multiplication of such entities, but the assumed laws of combination of the multiple symbols in effect serve as definitions of the operations. Thus

$$(x, y)+(x', y')=(x+x', y+y'); \quad (x, y).(x', y')=(xx'-yy', xy'+yx'):$$

or the sum of the entities whose parameters are x, y and x', y' is by definition the entity whose parameters are $x+x', y+y'$; and the product of the same entities is by definition the entity whose parameters are $xx'-yy', xy'+yx'$. The entities are thus, as regards addition, commutative and associative; but, as regards multiplication they are or are not commutative, or associative, according to the assumed law of multiplication of the multiple symbols.

10. If, as above, the multiple symbol be represented by means of extraordinaries, $(x, y)=ix+jy$, then the extraordinaries i, j, *quà* multiple symbols (1, 0), (0, 1), are themselves special entities of the kind in question, their laws of combination (as such special entities) being included in the general laws for the combination of such entities; or, if we please, these general laws being derived from the assumed laws for the special entities. Thus, if $ix+jy$ represent the point whose coordinates are x, y, then i represents the point whose coordinates are (1, 0), and j the point whose coordinates are (0, 1).

11. It has been assumed that the multiple symbols combined together, and their sum and product, are symbols of one and the same multiplicity, say the types of combination are

$$(x, y, z)+(x', y', z')=(x+x', y+y', z+z'); \quad (x, y, z)(x', y', z')=(X, Y, Z);$$

if this were not so, if for instance we had

$$(x, y, z)(x', y', z')=\text{a double symbol } (X, Y),$$

then this given equation for the multiplication of two triple symbols would not serve as a definition for the multiplication of double symbols, or of a double and a triple symbol, and new definitions would be required. We might, however, have symbols of indefinite multiplicity $(x, y, z, w, \ldots)$, including within them all finite multiplicities, viz. (x, y) meaning $(x, y, 0, 0, \ldots)$, and so in other cases, these being combined into symbols of like indefinite multiplicity

$$(x, y, z, \ldots)+(x', y', z', \ldots)=(x+x', y+y', z+z', \ldots);$$
$$(x, y, z, \ldots)(x', y', z', \ldots)=(X, Y, Z, \ldots);$$

for instance, the law of multiplication might be

$$(x, y, z, \ldots)(x', y', z', \ldots)=(xx', xy'+yx', xz'+yy'+zx', \ldots),$$

and we could hereby combine symbols of any finite multiplicities (the same or different) whatever

12. Another peculiarity may be noticed: suppose that, in general,

$$(x, y, z, w)+(x', y', z', w')=(x+x', y+y', z+z', w+w');$$

$$(x, y, z, w)(x', y', z', w')=(X, Y, Z, W);$$

then, as a particular case hereof, we have the addition-equation

$$(x, y, 0, 0)+(x', y', 0, 0)=(x+x', y+y', 0, 0)$$

and it may very well be that for the same two symbols the multiplication-equation may take the form

$$(x, y, 0, 0)(x', y', 0, 0)=(0, 0, Z, W),$$

(Z, W each of them a function of x, y, x', y'); viz. here, in a different point of view, the component symbols $(x, y, 0, 0)$ and $(x', y', 0, 0)$, are double systems of a certain kind $\{x, y\}$, $\{x', y'\}$, and the product is a double system of a different kind $[Z, W]$, or it might have been $(0, Y, Z, W)$, = a triple system $[Y, Z, W]$. Of course, all peculiarity disappears when we revert from the particular symbols $(x, y, 0, 0)$ to the general symbols (x, y, z, w), from which we may regard them as derived.

13. The peculiarity just referred to presents itself naturally when we regard the symbols as representing geometrical or physical entities; it may very well be that the product of two entities of a certain kind is taken to be an entity of a different kind (for instance, the product of two points to be a line), and that the analytical theory of the multiple symbol is constructed in order to such a relation; and it may further be that only the symbol $(x, y, 0, 0)$ or $\{x, y\}$ is interpreted with reference to an entity of the one kind, and only the symbol $(0, 0, Z, W)$, or $[Z, W]$ is interpreted by reference to an entity of the other kind, without any interpretation at all being given to the general symbol (x, y, z, w); thus the two forms $(x, y, 0, 0)$, and $(0, 0, Z, W)$, naturally present themselves as double symbols $\{x, y\}$ and $[Z, W]$ of different kinds.

14. In further illustration, suppose the symbols represented as linear functions of extraordinaries,

$$(x, y, z, w)=x\epsilon_1+y\epsilon_2+z\eta_1+w\eta_2,$$

with a multiplication table for the four extraordinaries ϵ_1, ϵ_2, η_1, η_2. We then have $(x, y, 0, 0)$, $=\{x, y\}=x\epsilon_1+y\epsilon_2$; the ϵ_1, ϵ_2 have no proper multiplication table of their own, but the squares and products ϵ_1^2, $\epsilon_1\epsilon_2$, $\epsilon_2\epsilon_1$, ϵ_2^2 are each given as a linear function of η_1, η_2; hence we have

$$(x\epsilon_1+y\epsilon_2)(x'\epsilon_1+y'\epsilon_2)=Z\eta_1+W\eta_2,$$

which is a symbol $(0, 0, Z, W)$, or $[Z, W]$, of a different kind; it may, in completion of the theory, be assumed that in like manner η_1, η_2 have no proper multiplication table of their own, but that the squares and products η_1^2, $\eta_1\eta_2$, $\eta_2\eta_1$, η_2^2 are each given as a linear function of ϵ_1, ϵ_2, leading to a relation

$$(z\eta_1+w\eta_2)(z'\eta_1+w'\eta_2)=X\epsilon_1+Y\epsilon_2,$$

which is a symbol $(X, Y, 0, 0)$ or $\{X, Y\}$ of the first kind; it may further happen that each of the several products $\epsilon\eta$, $\eta\epsilon$ is $=0$, in which case

$$(x\epsilon_1 + y\epsilon_2)(z\eta_1 + w\eta_2) = 0, \quad (z\eta_1 + w\eta_2)(x\epsilon_1 + y\epsilon_2) = 0.$$

For the complete theory, we require the multiplication table of the four extraordinaries ϵ_1, ϵ_2, η_1, η_2. The geometrical interpretation may be that $x\epsilon_1 + y\epsilon_2$ represents a point, $z\eta_1 + w\eta_2$ a line (with or without any geometrical interpretation of a symbol $x\epsilon_1 + y\epsilon_2 + z\eta_1 + w\eta_2$), the product of two points is a line, the product of two lines is a point; that of a line and point is $=0$, (see *post* Grassmann, where however the system is a somewhat different one).

15. I do not propose in the present paper to consider the subject of multiple algebra from an analytical point of view; and I will make only a few remarks and give some references. The general theory of associative linear forms is treated in a very satisfactory manner in Peirce's Memoir (1870) above referred to, only it is assumed throughout *ab initio* that the forms are associative. In a Note "On Associative Imaginaries," *Johns Hopkins Circulars*, No. 15 (1882), [822], and more fully in the paper "On Double Algebra," *Proc. Lond. Math. Soc.*, t. xv. (1884), pp. 185—197, [814], starting from the assumed equations $x^2 = ax + by$, $xy = cx + dy$, $yx = ex + fy$, $y^2 = gx + hy$, between the extraordinaries x, y, I considered under what conditions the algebra of these symbols was in fact associative. And in a paper "On the 8-square Imaginaries," *American Journal of Mathematics*, t. iv. (1881), pp. 293—296, [773], I showed that the extraordinaries 0, 1, 2, 3, 4, 5, 6, 7, connected with Euler's theorem of the 8 squares, are of necessity *non-associative*. I do not know that anything else has been done in regard to non-associative algebras: and it thus appears that the question has hardly been discussed at all. Matrices are associative: I shall have again to refer to them.

16. The object of the present paper is to discuss in detail, from the point of view explained in the extract from my British Association Address, the different theories in regard to geometrical or other entities. I do this under various headings.

The $i = \sqrt{(-1)}$ *of Analysis and Analytical Geometry.* Art. Nos. 17—26.

17. We may, of course, consider the i as an extraordinary. We have a double symbol (x, y), combining according to the laws

$$(x, y) + (x', y') = (x + x', y + y'),$$
$$(x, y)(x', y') = (xx' - yy', xy' + yx');$$

and then, introducing the extraordinaries k, i, and writing

$$(x, y) = kx + iy,$$

we have

$$(kx + iy)(kx' + iy') = k^2xx' + kixy' + ikyx' + i^2yy';$$

whence

$$k^2 = k, \quad ki = ik = i, \quad i^2 = -k,$$

or the multiplication table is

	k	i
k	k	i
i	i	$-k$

But the conditions are satisfied by, and we accordingly assume, $k=1$; and there then remains only the condition $i^2=-1$. Compare herewith Sir W. R. Hamilton, "Theory of Conjugate Functions, or Algebraical Couples; with a Preliminary and Elementary Essay on Algebra as a Science of pure time" (*Trans. R. I. Acad.*, t. XVII., 1833—35). I refer to this paper in my Address, and remark upon it that I cannot appreciate the manner in which the author connects, with the notion of time, his algebraical couple or imaginary magnitude $a+bi, =a+b\sqrt{(-1)}$ as written in the memoir.

18. But we have, in fact, passed out of this view, and have come to regard $a+bi$ as an ordinary analytical magnitude; viz. in every case an ordinary symbol represents or may represent such a magnitude, and the magnitude (and as a particular case thereof, the symbol i) is commutable with the extraordinaries of any system of multiple algebra. And similarly in analytical geometry, without seeking for any real representation, we deal with imaginary points, lines, &c., that is, with points, lines, &c., depending on parameters of the foregoing form $a+bi$.

$\sqrt{(-1)}$ *denotes Perpendicularity.* Art. Nos. 19—26.

19. I give a list of the earlier works and memoirs, marking with an asterisk those which I have not seen.

Wallis. *De Algebrâ Tractatus,* 1685 Anglice editus; 1693. Chapters 66, 67, 68, and 69 have respectively the titles: De quadratis negativis eorumque radicibus dictis Imaginariis;—Eorundem exemplificatio in Geometriâ;—Effectiones geometricæ his accommodatæ; Aliæ quæ huc spectant constructiones geometricæ. I quote a single sentence from Chap. 67: "Prout igitur cum æquationis quadraticæ radix prodit negativa, dicendum verbi gratiâ punctum B pro eo statu haberi non posse ut supponitur in expositâ AC prorsum; posse tamen retrorsum ab A in eâdem rectâ—hic vero (de radice quadrati negativi) dicendum, non haberi quidem posse punctum B ut erat suppositum in AC rectâ, vel ante vel retro: posse tamen (in eodem plano) suprà rectam illam": viz. the distance represented by the square root of a negative quantity cannot be measured in the line, forwards or backwards; but can be measured (in the same plane) above the line—or, as appears elsewhere, at right angles to the line, either in the plane, or in a plane at right angles thereto.

Forcenex. "Réflexions sur les quantités imaginaires," *Mel. de Turin,* t. I. (1759), pp. 113—146. The author, after stating that there was no geometrical construction for imaginary quantities, proceeds "Cependant, pour conserver une certaine analogie avec les quantités négatives, un auteur dont nous avons un cours d'algèbre d'ailleurs fort estimable a prétendu les devoir prendre sur une ligne perpendiculaire à celle où l'on les avait supposées, si par exemple &c.," the example which he considers being as follows: On a given line AB to find a point P, such that the product of the distances AP, BP is $=\frac{1}{2}(AB)^2$. Taking $AB=2a$ and $AP=x$, then $x(2a-x)=2a^2$, that is, $(x-a)^2=-a^2$, or, $x=a+a\surd(-1)$, an imaginary value; the condition is, however, satisfied by a real point P *off* the line AB

$$(AM=MP=\tfrac{1}{2}AB=a),\ \text{(fig. 1)},$$

and hence the author in question infers that the imaginary value $a\surd(-1)$ is represented by the line MP at right angles to AB. Forcenex objects to this: if the

Fig. 1.

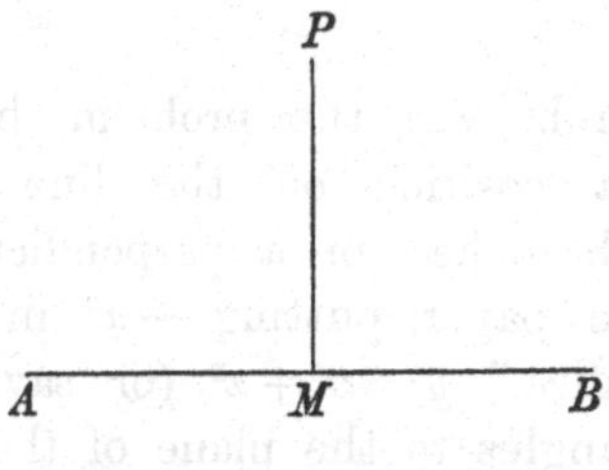

point P may be taken off the axis of x, then the condition $AP\,.\,BP=2a^2$ gives for the locus of P a certain curve; and there is no reason why the foregoing solution $x=a+a\surd(-1)$ should represent one point rather than another of this curve. As to this see *post* (Peacock).

***Truel,** Henri Dominique, 1786.

***Suremain-de-Missery.** *Théorie purement algébrique des quantités imaginaires*, Paris, 1801; it is referred to in the letter by Servois, *infra.*

20. **Buée.** "Mémoire sur les quantités imaginaires," *Phil. Trans.*, 1806, pp. 23—88.

I give an extract: "Du signe $\surd(-1)$. Je mets en titre, du signe $\surd(-1)$ et non de la quantité imaginaire ou de l'unité imaginaire, parce que $\surd(-1)$ est un signe particulier joint à l'unité réelle, non une quantité particulière: c'est un nouvel adjectif joint au substantif ordinaire et non un nouveau substantif.

"Mais que veut dire ce signe? il n'indique ni une addition ni une soustraction, ni une suppression ni une opposition par rapport aux signes + et −: une quantité accompagnée par $\surd(-1)$ n'est ni additive ni subtractive, ni égale à zéro. La qualité marquée par $\surd(-1)$ n'est opposée ni à celle qu'indique +, ni à celle marquée par −; qu'est-elle donc?

"Pour la découvrir supposons trois lignes égales AB, AC, AD (fig. 2) qui partent toutes du point A. Si je désigne la ligne AB par $+1$, la ligne AC sera -1, et la ligne AD qui est une moyenne proportionnelle entre $+1$ et -1, sera nécessairement $\surd(-1^2)$ ou plus simplement $\surd(-1)$. Ainsi $\surd(-1)$ est le signe de la PERPENDICULARITÉ: donc la propriété caractéristique est *que tous les points de la perpendiculaire sont également éloignés de points placés à égale distance de part et d'autre de son pied.* Le signe $\sqrt{(-1)}$ signifie tout cela et il est le seul qui l'exprime.

Fig. 2.

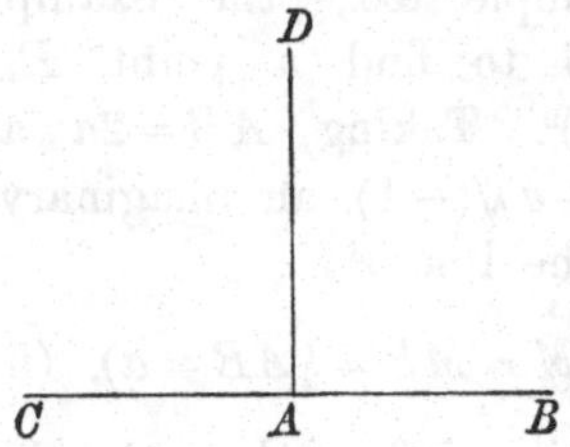

"*Ce signe mis devant a (a signifiant une ligne ou une surface) veut donc dire qu'il faut donner à a une situation perpendiculaire à celle qu'on lui donnerait si l'on avait simplement a ou* $-a$."

And he, in fact, gives (Prob. VI.) the problem before referred to, and finds no difficulty in assigning to P a position off the line AB. I notice also that Buée considers a curve as having branches in a perpendicular plane; thus for the circle $x^2+y^2=a^2$ in the plane of the paper, putting $-x^2$ in place of x^2, there are branches, or, as he calls it, "une appendice," $y^2=a^2+x^2$ (or say $y^2=a^2+z^2$), being a rectangular hyperbola, in a plane at right angles to the plane of the paper.

21. Other writings are:

Argand. *Essai sur une manière de représenter les quantités imaginaires.* Paris, 1806. Partly reproduced in Argand's memoir under the same title presently referred to.

Francais, J. F. "Nouveaux principes de géométrie de position et interprétation géométrique des symboles imaginaires." *Gergonne Annales,* t. IV. 1813—14, pp. 61—72.

Argand. "Essai sur une manière, &c." *Ibid.* pp. 133—148.

Francais, J. F. "Lettre au Rédacteur." *Ibid.* pp. 222—227.

Servois. "Lettre au Rédacteur." *Ibid.* pp. 228—235.

This last is *against* the theory, with a running commentary of notes *in favour* by Gergonne.

Francais, J. F. "Autre lettre." *Ibid.* pp. 364—366.

Lacroix. "Note transmise à M. Vecten"; it calls attention to Buée's memoir. *Ibid.* p. 367.

***Mourey.** *Vraie théorie des quantités négatives et des quantités prétendues imaginaires,* 1828, reprinted 1861.

22. **Warren, J.** *A treatise on the geometrical representation of the square roots of negative quantities.* 8vo. Cambridge, 1828, pp. 1—154.

Peacock, Preface to *Algebra*, 1830, speaks of Warren's work as "distinguished for great originality and for extreme boldness in the use of definitions." There is no Preface or Introduction; the first Chapter is entitled Definitions, Addition, Subtraction, Proportion, Multiplication, Division, Fractions, and Raising of Powers. I quote certain articles as follows:

(1) All straight lines drawn in a given direction from a given point are represented in *length* and *direction* by algebraic quantities; and in the following treatise whenever the word *quantity* is used it is to be understood as signifying *a line.*

(2) Def. The given point from which the straight lines are measured is called the *origin.*

(3) Def. The *sum* of two quantities is the diagonal of the parallelogram whose sides are the two quantities.

(12) Def. The first of four quantities is said to have to the second the same ratio which the third has to the fourth: when the first has in *length* to the second the same ratio which the third has in *length* to the fourth, according to Euclid's definition; and also the angle at which the fourth is inclined to the third is equal to the angle at which the second is inclined to the first, and is measured in the same direction.

(17) Def. Unity is a positive quantity arbitrarily assumed, from a comparison of which the values of other quantities are obtained.

(18) Def. If there be three quantities such that unity is to the first as the second is to the third, then the third is called the *product* which arises from the *multiplication* of the second by the first.

23. The signification of these definitions may be thus expressed. Let the line, drawn from an origin O to the point whose rectangular coordinates are x, y, be called the line $x+iy$; of course the line, length unity in the direction of the axis of x, is the line 1. Then, observing that, putting $x,\ y=r\cos\theta,\ r\sin\theta$, and $x',\ y'=r'\cos\theta',\ r'\sin\theta'$, we have $(x+iy)(x'+iy')=rr'\{\cos(\theta+\theta')+i\sin(\theta+\theta')\}$, the two definitions are:

The *sum* of the lines $x+iy$ and $x'+iy'$ is the line

$$x+x'+i(y+y');$$

and the *product* of the same lines is the line

$$(x+iy)(x'+iy');$$

viz. Warren in effect establishes by definition the notions of the sum and product of two lines, and that by connecting the line with a linear symbol $x+iy$.

It may be right to remark that the general notion of a proportion, used as above to define multiplication, is geometrically the more simple of the two, and that the condition for the proportion $OA : OB = OC : OD$ may be thus expressed: the lengths of the lines are proportional, and the inclination OA to OB is equal to the inclination OC to OD, these inclinations being in the same sense.

24. **Peacock, G.** *A Treatise on Algebra.* 8vo. Cambridge, 1830 (one Vol., pp. 5—38 and 1—685).

—— "Report on the Recent Progress and Present State of Certain Branches of Analysis." *B. A. Report*, (1833), pp. 185—352.

—— *A Treatise on Algebra.* Vol. I. Arithmetical Algebra; Vol. II. On Symbolical Algebra and its Application to the Geometry of Position. 8vo. Cambridge, 1842 and 1845.

The statement of the theory is substantially identical with that of the earlier writers, thus *Algebra*, (1830), p. 362, No. 439, is:—"We have shown on a former occasion that, if a designated a line in one direction, $-a$ must designate a line in the direction opposite to it, or making an angle with the former equal to two right angles or 180 degrees; but inasmuch as $a\{\sqrt{(-1)}\}^2=-a$, it follows that the *double affectation* of the line a with the sign $\sqrt{(-1)}$ produces a result represented by $-a$, and is consequently equivalent to its transfer through 180°: it follows therefore that its single affectation with the sign $\sqrt{(-1)}$ is equivalent to its transfer through half that angle, or through 90°; or, in other words, if a represents a line, $a\sqrt{(-1)}$ will represent a line at right angles to it." But further, pp. 666, 667, the Author considers the before-mentioned problem (for convenience, I write $2a$ instead of a) in the form "to divide a line $2a$ into two such parts that the rectangle contained by them may be $=b^2$"; he takes x, y for the two parts, and in the case where $b^2>a^2$, finds the two parts $x=a\pm\sqrt{(-1)}\sqrt{(b^2-a^2)}$, $y=a\mp\sqrt{(-1)}\sqrt{(b^2-a^2)}$, which he interprets as meaning the two parts are AC, CB, where C is a point off the line AB.

25. It seems to me that, except in Warren, the defect in the exposition of the theory is that it is not made clear, that the theory is in fact a theory of lines in a plane through a given point, with assumed definitions for the addition and for the multiplication of lines; thus in Peacock's problem "to divide a line a into two such parts that the product of them is $=b^2$," the real meaning is "given a line OA, $=a$, and a line OB, $=b^2$ (a and b^2 real and positive, so that each of these lines is in the given direction Ox), to find lines OP, OQ such that their sum may be equal to the given line OA, and their product equal to the given line OB." To solve it, take the two lines to be $x+iy=z$, and $x'+iy'$, $=z'$; then, working with z and z', we have $z+z'=2a$, $zz'=b^2$, giving a solution which may be written in the two forms

$$\{z=a+\ \sqrt{(a^2-b^2)},\quad z'=a-\ \sqrt{(a^2-b^2)}\},$$

and

$$\{z=a+i\sqrt{(b^2-a^2)},\quad z'=a-i\sqrt{(b^2-a^2)}\},$$

viz. for $a>b$, the two lines lie each of them in the line Ox, while for $b>a$, they are inclined at equal angles on opposite sides of the line Ox. Or if we work with the real values x, y, x', y', then the same two equations give $x+x'=2a$, $y+y'=0$, $xx'-yy'=b^2$, $xy'+yx'=0$; viz. writing $y'=-y$, then we have $x+x'=2a$, $xx'+y^2=b^2$, $y(x'-x)=0$, whence either $x=x'$ or else $y=0$, and we have the same two solutions as before. Observe that the second condition is $zz'=b^2$, not (as Forcenex understood it) the product of the lengths OP, $OQ=b^2$; and his objection is thus answered. But I cannot but consider that the form of enunciation, "to divide the given line into two parts, such that their product shall be $= b^2$," does not express the meaning with sufficient clearness.

26. I have referred to Buée's notion of a curve as having branches in a perpendicular plane—this notion is generalised and developed in the papers:—

Gregory, D. F. "On the existence of branches of curves in several planes." *Camb. Math. Journ.*, t. I. (1839), pp. 259—266 (2nd Ed., pp. 284—292).

Walton. "On the general interpretation of equations between two variables in analytical geometry." *Do.* t. II. (1840), pp. 103—113.

—— "On the general theory of multiple points." *Do.* pp. 155—167.

And see also:—

Gregory's *Examples of the Processes of the Differential and Integral Calculus.* 8vo. Cambridge, 1841, pp. 172 et seq.;

and the criticism on these papers:—

Salmon. *A Treatise on the Higher Plane Curves.* 8vo. Dublin, 1852, The Note on Imaginary Points and Curves, pp. 301—306.

Geometrical representation of $x + iy$: *Gauss.* Art. Nos. 27—29.

27. The theory is that established by **Gauss** in the paper "Theoria Residuorum biquadraticorum Commentatio Secunda," *Comm. Gott. Recent.*, t. VII. (1832); *Werke*, t. II. pp. 95—148; viz. Gauss remarks that, in the same way that a real quantity x may be represented by means of a point on a line, so an imaginary (or complex) quantity $x + iy$ may be represented by means of a point the abscissa of which is $= x$, and its ordinate is $= y$. Taking m, m' as given complex quantities, he speaks of the points answering to the complex quantities mm', m, m', 1 as forming a *proportion* (thus in effect defining the product of two points M, M'). But he defers to another occasion all further discussion of the theory.

28. The direction in which the theory has been developed is as follows: Considering two complex values z, $= x + iy$, and Z, $= X + iY$, connected by an equation $\phi(Z, z) = 0$, then we have z represented as above by means of a point in a plane, and Z represented in like manner by means of another point in a different plane; to any given value of z or Z there corresponds a determinate number of values of Z or z; that is, to any given real point of either plane, there corresponds a determinate number of real points in the other plane: the equation thus establishes a correspondence between the points of the two planes; and it is known that this is an orthomorphic correspondence, viz. to any indefinitely small area of the one plane, there correspond in the other plane indefinitely small areas each of them of the same shape as the area in the first plane. We have thus a solution of the geometrical problem of orthomorphic projection; and in the analytical point of view, the theory is of great importance as exhibiting the relation to each other of complex variables connected by an equation.

29. It is to be remarked that, although the correspondence is primarily a correspondence between real points of the two planes respectively, yet in the ulterior development of the theory of this correspondence it may become necessary to consider imaginary points in the two planes. In view hereto, there would be some propriety in replacing the i of the z, Z by a like symbol I; say we have $z=x+Iy$, $Z=X+IY$ (where the x, y, X, Y may now be ordinary complex magnitudes $u+vi$; $i^2=-1$ as before: $I^2=-1$, and $Ii=iI$, viz. I is a square root of -1, commutable with the ordinary imaginary i). But this in passing.

The Barycentric Calculus; Möbius. Art. Nos. 30, 31.

30. **Möbius,** *Der barycentrische Calcul*, 8vo. Leipzig, 1827, reprinted in Vol. I. of the *Gesammelte Werke*, Leipzig, 1885.

The theory applies to points in a line, in a plane, or in space; but it will be sufficient to consider the case of points in a plane. The idea is that, taking in the plane any three points A, B, C as fundamental points, then regarding these as loaded with the proper positive or negative weights p, q, r, any other point P of the plane may be regarded as the centre of gravity of these weights; say we have $P=pA+qB+rC$, as a representation of the point P; p, q, r are, in the first instance, given positive or negative real values (but they may be taken to be imaginary values), P, A, B, C are symbols denoting points. Observe that p, q, r are, in fact, trilinear coordinates: if for a moment ξ, η, ζ are the perpendicular distances of P from the sides of the triangle, and α, β, γ the perpendicular distances of the opposite vertices from the same sides respectively, then p, q, r are proportional to ξ/α, η/β, ζ/γ, or, what is the same thing, they are proportional to PBC/Δ, PCA/Δ, PAB/Δ, where PBC, &c., denote the three triangles and Δ the triangle ABC. It would be allowable to take p, q, r *equal* to these values, which would give $p+q+r=1$. Möbius does this very nearly, for he writes $p+q+r+s=0$, and then writing $-(p+q+r)D$ instead of P he has

$$pA+qB+rC+sD=0;$$

that is,

$$D=-\frac{p}{s}A-\frac{q}{s}B-\frac{r}{s}C,$$

where

$$-\frac{p}{s}-\frac{q}{s}-\frac{r}{s}=1:$$

writing x, y, z, instead of these quantities, I take therefore as the representation of the point, $P=xA+yB+zC$, where $x+y+z=1$.

31. The symbols A, B, C may be regarded as extraordinaries; but it can hardly be said that Möbius so regards them, for he does not establish the notion of the multiplication of points, nor consequently any multiplication table for these symbols A, B, C: he does however deal with the *addition* of points, viz. if

$$P=xA+yB+zC, \quad P'=x'A+y'B+z'C,$$

then

$$P+\lambda P', \ =(x+\lambda x')A+(y+\lambda y')B+(z+\lambda z')C,$$

(where λ is indeterminate) denotes any point whatever in the line PP', (so also $P = xA + yB + zC$, where x, y, z are regarded as functions of a variable parameter θ, denotes any point whatever in the curve denoted by writing x, y, z proportional to such functions of θ). And he also considers (not quite generally) the question of the transformation of the fundamental points: this is really a treatment of the A, B, C as extraordinaries, the complete theory being as follows; assuming

$$A = \alpha A_1 + \beta B_1 + \gamma C_1, \quad B = \alpha' A_1 + \beta' B_1 + \gamma' C_1, \quad C = \alpha'' A_1 + \beta'' B_1 + \gamma'' C_1,$$

(if $\alpha + \beta + \gamma = 1$, $\alpha' + \beta' + \gamma' = 1$, $\alpha'' + \beta'' + \gamma'' = 1$, then these equations give conversely

$$A_1 = aA + a'B + a''C, \quad B_1 = bA + b'B + b''C, \quad C_1 = cA + c'B + c''C,$$

where $a + a' + a'' = 1$, $b + b' + b'' = 1$, $c + c' + c'' = 1$), then we have

$$P_1 = x\alpha + yB + zC_1 = (x\alpha + y\alpha' + z\alpha'')\, A_1 + (x\beta + y\beta' + z\beta'')\, B_1 + (x\gamma + y\gamma' + z\gamma'')\, C_1,$$

where the sum of the coefficients is

$$x(\alpha + \beta + \gamma) + y(\alpha' + \beta' + \gamma') + z(\alpha'' + \beta'' + \gamma''),$$

which is $= x + y + z$, $= 1$.

Equipollences: Bellavitis. Art. Nos. 32, 33.

32. There are earlier papers in the years 1832 and 1833, but the method is explained in the Memoir:—

Bellavitis. "Saggio di applicazioni di un nuovo metodo di geometria analitica (Calcolo delle Equipollenze)," *Ann. Lomb. Veneto*, t. V. (1835), pp. 244—259,

and more fully in Memoirs, t. VII. (1837), and t. VIII. (1838); the author notices that his results were obtained independently of those of Möbius but that, on studying the *Barycentrische Calcul*, he had recognised that the two methods started from the same principles, and could easily be reduced to identity, but that they speedily separated from each other as well in regard to object as to form. But, in fact, the principles are rather those of Warren than of Möbius. They may be thus stated: 1°. two lines, equal, parallel and in the same sense, are said to be *Equipollent*, $AB \equiv CD$ (where the sign $\equiv$ is used instead of a special sign employed by Bellavitis); this first assumption in effect reduces the whole theory to that of lines drawn through a fixed point; 2°. $AB + BC \equiv AC$ (addition); and 3°. $AB \equiv \dfrac{CD \,.\, EF}{GH}$, means not only that the lengths are connected by this relation, but that the inclination of AB to any fixed axis is $=$ inc. $CD +$ inc. $EF -$ inc. GH (proportion); or in particular, if GH be a line, length unity, in the direction of the axis, then $AB \equiv CD \,.\, EF$, viz. the length $AB =$ product of lengths CD and EF; and, further,

$$\text{inc. } AB = \text{inc. } CD + \text{inc. } EF \text{ (multiplication)};$$

the agreement with Warren is thus complete.

33. The method is applied very elegantly to the solution of problems, for instance, t. v. p. 248; given two lines AB and CD, to find a point H, such that the triangles ABH, CDH may be directly similar, that is, such that they can be, by a rotation of one of them, brought to be similarly situate. The condition for this (a figure is easily supplied) is

$$\frac{AH}{AB} \equiv \frac{CH}{CD}, \text{ that is, } AH.CD \equiv AB.CH, \ = AB(AH - AC),$$

whence $AB.AC = AH(AB - CD)$: or constructing a point E such that $BE \equiv DC$, viz. the line BE is drawn from B equal and parallel to and in the same sense with DC, then $AB - CD \equiv AE$, and the foregoing equation becomes

$$AH.AE \equiv AB.AC, \text{ or say } \frac{AH}{AC} \equiv \frac{AB}{AE},$$

viz. the required point H is such that the triangle ACH is directly similar to AEB. A solution of the like problem with the triangles ABH, CDH inversely similar is given, t. VIII., p. 27. I notice an expression t. VIII., p. 19, for the area of a triangle ABC,

$$= \frac{1}{4i}(BC.\text{cj}.AB - AB.\text{cj}.BC),$$

where cj. AB, conjugate of AB, is an equal line through A with an inclination $= -$ incl. AB; also a proof, t. VIII., p. 86, of the general theorem that an algebraical equation of any order has a root (this is, in a geometrical form, equivalent to the proofs given by Gauss and Cauchy); also t. VIII., p. 111, the notion of the "punti fittizj-conjugati," or real antipoints of a pair of conjugate imaginary points. The section "Delle figure a tre dimensioni," t. VIII., pp. 115—121, is confessedly quite incomplete, and without any anticipation of the quaternion or other three-dimensional imaginaries.

We have subsequently by the author

"Sposizione del Metodo delle Equipollenze," *Mem. Soc. Ital.*, t. XXV. (1855), pp. 225—309: published separately, Modena, 1854;

and in French under the title

Exposition de la Méthode des Equipollences, par G. Bellavitis, traduit de l'italien par C. A. Laisant: 8vo. Paris, 1874.

Quaternions: Sir W. R. Hamilton. Art. Nos. 34—42.

34. I give the following references to early papers and systematic works:

Sir W. R. Hamilton. "Abstract of a paper On Quaternions or on a new system of imaginaries in Algebra, with some geometrical illustrations." *Proc. R. I. Acad.*, vol. III., Nov. 11, 1844, pp. 1—16. A note added during the printing refers to Cayley *infra*.

—— "On Quaternions; or a new system of imaginaries in Algebra." *Phil. Mag.*, vol. XXV. (1844), pp. 489—495.

Cayley. "On certain results relating to Quaternions." *Phil. Mag.*, vol. XXVI. (1845), pp. 141—144, [20].

Sir W. R. Hamilton. *Lectures on Quaternions.* 8vo. Dublin, 1853.

—— *Elements of Quaternions.* 8vo. Dublin, 1866.

Tait, P. G. *An Elementary Treatise on Quaternions.* 8vo. Oxford, 1859.

Kelland, P. and **Tait, P. G.** *Introduction to Quaternions.* 8vo. London, 1873.

35. A quaternion is the sum of a scalar or mere magnitude w, plus a vector $ix+jy+kz$, which represents a line drawn from an origin O to the point whose rectangular coordinates are x, y, z; we have thus a quaternion $Q, =w+ix+jy+kz$, representing a line, and in connexion therewith a scalar w. Hamilton refers to the theory of Warren and Peacock as having in part suggested his investigations, but the contrast is very striking; neither the form $x+iy$, where the directed magnitude x is unaccompanied by an extraordinary, nor the form $kx+iy$ with any multiplication table for k, i, is in any wise analogous to the expression $ix+jy+kz$ of a vector. We have, for addition, the ordinary formula $Q+Q'=w+w'+i(x+x')+j(y+y')+k(z+z')$; and then we have the multiplication table

	1	i	j	k
1	1	i	j	k
i	i	-1	k	$-j$
j	j	$-k$	-1	i
k	k	j	$-i$	-1

,

or, what is the same thing, we have

$$i^2=j^2=k^2=-1,\quad i=jk=-kj,\quad j=ki=-ik,\quad k=ij=-ji,$$

and thence for the product of two quaternions

$$\begin{aligned}(w+ix+jy+kz)(w'+ix'+jy'+kz')\\ = \quad & ww'-xx'-yy'-zz'\\ & +i\,(wx'+xw'+yz'-y'z)\\ & +j\,(wy'+yw'+zx'-z'x)\\ & +k\,(wz'+zw'+xy'-x'y);\end{aligned}$$

these, of course, include the forms for the addition and multiplication of two vectors $ix+jy+kz$ and $ix'+jy'+kz'$.

But for the multiplication of two vectors, the form is simplified: viz. we have

$$\begin{aligned} VV' &= (ix+jy+kz)(ix'+jy'+kz') \\ &= -(xx'+yy'+zz') + i(yz'-y'z) + j(zx'-z'x) + k(xy'-x'y). \end{aligned}$$

36. We may consider unit-vectors,

$$U = ix+jy+kz, \text{ where } x^2+y^2+z^2=1,$$

and unit-quaternions

$$O = w+ix+jy+kz, \text{ where } w^2+x^2+y^2+z^2=1;$$

obviously the general form of a vector is $V=rU$, or, say the length is r, and the cosine-inclinations are x, y, z; similarly, the general form of a unit-quaternion is

$$O = \cos\delta + \sin\delta \,.\, U,$$

and that of a quaternion is

$$Q = TO = T(\cos\delta + \sin\delta \,.\, U),$$

T is the tensor, δ the amplitude, and x, y, z the cosine-inclinations of the unit-vector.

For the product of two unit-vectors, we have

$$UU' = -(xx'+yy'+zz') + i(yz'-y'z) + j(zx'-z'x) + k(xy'-x'y),$$

which is a unit-quaternion. Let δ, considered as a positive angle less than π, so that

$$\cos\delta = \pm, \quad \sin\delta = +,$$

be the inclination of the two vectors; we have $\cos\delta = xx'+yy'+zz'$. We can at right angles to the plane of U, U' draw in either of two opposite senses a unit-vector U''; and we may choose the sense in such wise that the rotation from U through angle δ to U' about U'' shall be in the sense Ox through angle $\frac{1}{2}\pi$ to Oy about Oz, say each of these is a right-handed rotation. This being so, we may write

$$\begin{aligned} &i(yz'-y'z) + j(zx'-z'x) + k(xy'-x'y), \\ &= \sin\delta\,(ix''+jy''+kz'') = \sin\delta \,.\, U'', \end{aligned}$$

and we have

$$UU' = -\cos\delta + \sin\delta \,.\, U'';$$

viz. the product is a unit-quaternion, the scalar part $=-\cos\delta$, and the vector part having a length $=\sin\delta$, and being at right angles to the plane of the two vectors U, U'; δ being, as above, the inclination of the two vectors. And similarly

$$U'U = -\cos\delta - \sin\delta \,.\, U''.$$

Fig. 3.

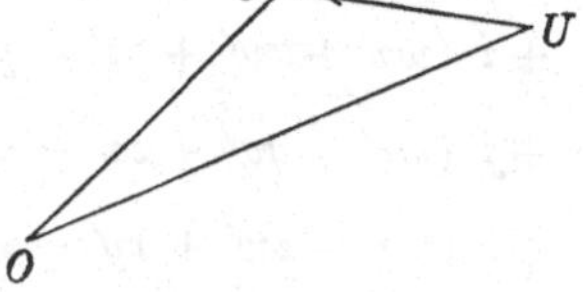

Representing the vectors as lines OU, OU', the product UU' will be represented, as in fig. 3, by means of an arrow drawn in the angle δ from U to U'; and of course, for the product $U'U$, there would be the same figure with only the arrow drawn in the opposite sense from U' to U. The figure serves also to represent the unit-quaternion $-\cos\delta + \sin\delta \,.\, U''$, which is the product of the two vectors. Observe that, in thus expressing a quaternion as a product of two vectors, only the plane of the vectors and the angle between them are determinate; we may rotate the angle UOU' in its own plane at pleasure.

In particular, if the two vectors are at right angles, then $\delta = 90°$, and the product is $UU' = U''$, viz. the product of the two unit-vectors at right angles to each other is a unit-vector at right angles to the plane of the two unit-vectors. If $\delta = 0$, then the two unit-vectors coincide, and we have $U^2 = -1$, viz. the square of a unit-vector is the scalar -1.

The product of any two vectors is given by the like formulæ; if V, $V' = rU$, $r'U'$ respectively, then $VV' = rr' \,.\, UU'$.

37. The product of a quaternion by a vector is given by the formula

$$QV' = (w + ix + jy + kz)(ix' + jy' + kz'),$$
$$= -(xx' + yy' + zz') + i(wx' + yz' - y'z) + j(wy' + zx' - z'x) + k(wz' + xy' - x'y).$$

In particular, if the quaternion and the vector are at right angles (that is, if the vector part of Q and V' are at right angles), $xx' + yy' + zz' = 0$, and the product is a vector. We may use this formula to express a given quaternion as the product of two vectors. Writing, for convenience,

$$U = i\xi + j\eta + k\zeta,$$

a unit-vector at right angles to Q, so that $\xi^2 + \eta^2 + \zeta^2 = 1$ and $\xi x + \eta y + \zeta z = 0$, then

$$QU = (w + ix + jy + kz)(i\xi + j\eta + k\zeta)$$
$$= i(w\xi + y\zeta - z\eta) + j(w\eta + z\xi - x\zeta) + k(w\zeta + x\eta - y\xi),$$

which is a vector, say it is

$$V = iX + jY + kZ;$$

and then

$$Q = -QU^2 = -VU = -(iX + jY + kZ)(i\xi + j\eta + k\zeta),$$

that is,

$$w + ix + jy + kz = \xi X + \eta Y + \zeta Z + i(\eta Z - \zeta Y) + j(\zeta X - \xi Z) + k(\xi Y - \eta X),$$

as is at once verified by substituting for X, Y, Z their values

$$w\xi + y\zeta - z\eta, \quad w\eta + z\xi - x\zeta, \quad w\zeta + x\eta - y\xi.$$

Observe that

$$Xx + Yy + Zz = w(\xi x + \eta y + \zeta z) = 0.$$

38. And we hence reduce the multiplication of quaternions to that of vectors. For, considering another quaternion Q', take $U = i\xi + j\eta + k\zeta$ as before, with the further condition that it is at right angles to Q', so that $\xi^2 + \eta^2 + \zeta^2 = 1$, $\xi x + \eta y + \zeta z = 0$, $\xi x' + \eta y' + \zeta z' = 0$; say U is the unit-vector at right angles to the vectors of Q, Q': then we have

$$-UQ' = -(i\xi + j\eta + k\zeta)(w' + ix' + jy' + kz')$$
$$= i(-w'\xi + y'\zeta - z'\eta) + j(-w'\eta + z'\xi - x'\zeta) + k(-w'\zeta + x'\eta - y'\xi),$$

which is a vector, say it is

$$V' = iX' + jY' + kZ';$$

and then

$$Q' = -U^2Q' = UV' = (i\xi + j\eta + k\zeta)(iX' + jY' + kZ'),$$

that is,

$$w' + ix' + jy' + kz' = -\xi X' - \eta Y' - \zeta Z' + i(\eta Z' - \zeta Y') + j(\zeta X' - \xi Z') + k(\xi Y' - \eta X'),$$

as may be verified by substituting for X', Y', Z' their values

$$-w'\xi + y'\zeta - z'\eta, \quad -w'\eta + z'\xi - x'\zeta, \quad -w'\zeta + x'\eta - y'\xi;$$

and it may be observed that

$$X'x' + Y'y' + Z'z', \quad = -w(\xi x' + \eta y' + \zeta z') = 0.$$

We hence find

$$\begin{aligned} QQ' = QU \,.\, -UQ' = VV' &= (iX + jY + kZ)(iX' + jY' + kZ'), \\ &= -XX' - YY' - ZZ' + i(YZ' - Y'Z) + j(ZX' - Z'X) + k(XY' - X'Y). \end{aligned}$$

39. This should, of course, be identically equal to

$$\begin{aligned} & ww' - xx' - yy' - zz' \\ & + i\,(wx' + xw' + yz' - y'z) \\ & + j\,(wy' + yw' + zx' - z'x) \\ & + k\,(wz' + zw' + xy' - x'y), \end{aligned}$$

on substituting for X, Y, Z, X', Y', Z' their values as above.

Thus, forming the value of $XX' + YY' + ZZ'$, and adding thereto the expression $(\xi x + \eta y + \zeta z)(\xi x' + \eta y' + \zeta z')$, which is identically $= 0$, we obtain

$$\begin{aligned} XX' + YY' + ZZ' &= (-ww' + xx' + yy' + zz')(\xi^2 + \eta^2 + \zeta^2), \\ &= -ww' + xx' + yy' + zz'; \end{aligned}$$

and similarly the other three relations may be verified.

40. The steps are, taking U the unit-vector at right angles to Q, Q', then the vectors V, V' are defined by $V = QU$, $V' = -UQ'$; and we then have $Q = -VU$, $Q' = UV'$, and thence $QQ' = VV'$.

We may look at the process as follows: taking first $Q = -VU$, we have V, U vectors in a determinate plane and at a determinate inclination to each other; taking then $Q' = U'V'$, we have U', V' vectors in a determinate plane and at a determinate inclination to each other; we then rotate the vector-pairs (V, U) and (U', V') each in its own plane and in the senses V to U and U' to V' respectively, in such wise as to bring the vectors U, U' into coincidence (along one or other of the opposite vectors which form the intersection of the two planes), and this being so, and U, U' being unit-vectors, we have $U = U'$, and consequently $QQ' = -VU \,.\, U'V' = VV'$ as above.

41. There is a kind of quaternion operator $Q^{-1}(\ )\,Q$, or, what is the same thing, if $\Lambda = i\lambda + j\mu + k\nu$, $(1-\Lambda)(\ )(1+\Lambda)$, which is the symbol of a rotation; viz. the operand, say the vector $ix + jy + kz$, is to be placed within the vacant ( ), say

$$Q^{-1}(\ )\,Q\,.\,(ix+jy+kz) \text{ means } Q^{-1}(ix+jy+kz)\,Q\,;$$

and this being so, the result is a vector $ix_1 + jy_1 + kz_1$, where the x_1, y_1, z_1 are what the x, y, z become by a rotation of the vector $ix + jy + kz$ about an axis and through an angle determined by the quaternion Q; viz. if as above

$$Q = 1 + i\lambda + j\mu + k\nu,$$

then writing

$$\lambda,\ \mu,\ \nu = \tan\tfrac{1}{2}\theta\cos f,\quad \tan\tfrac{1}{2}\theta\cos g,\quad \tan\tfrac{1}{2}\theta\cos h,$$

where $\cos^2 f + \cos^2 g + \cos^2 h = 1$, and therefore $\lambda^2 + \mu^2 + \nu^2 = \tan^2\tfrac{1}{2}\theta$, then f, g, h are the inclinations of the axis, and θ is the angle of rotation. The actual formula is

$$(1 - i\lambda - j\mu - k\nu)(ix + jy + kz)(1 + i\lambda + j\mu + k\nu) = ix_1 + jy_1 + kz_1,$$

where

$$ix_1 + jy_1 + kz_1 = \frac{1}{1+\lambda^2+\mu^2+\nu^2}\left\{\begin{aligned} & i\,[(1+\lambda^2-\mu^2-\nu^2)\,x + 2\,(\lambda\mu+\nu)\,y + 2\,(\lambda\nu-\mu)\,z] \\ + & j\,[2\,(\lambda\mu-\nu)\,x + (1-\lambda^2+\mu^2-\nu^2)\,y + 2\,(\mu\nu+\lambda)\,z] \\ + & k\,[2\,(\nu\lambda+\mu)\,x + 2\,(\mu\nu-\lambda)\,y + (1-\lambda^2-\mu^2+\nu^2)\,z] \end{aligned}\right\},$$

viz. the form on the right-hand side is

$$\begin{aligned} & i\,(\alpha x + \alpha' y + \alpha'' z) \\ + & j\,(\beta x + \beta' y + \beta'' z) \\ + & k\,(\gamma x + \gamma' y + \gamma'' z), \end{aligned}$$

where the coefficients are those of a rectangular transformation.

42. I call to mind that a binary matrix may be regarded as a quaternion; viz. writing

$$\begin{vmatrix} a, & b \\ c, & d \end{vmatrix} = (a+d) - \lambda(a-d)\,i + (b-c)\,j - \lambda(b+c)\,k,$$

where i, j, k are the quaternion imaginaries, and λ is written to denote the $i = \sqrt{(-1)}$ of ordinary analysis, then we at once deduce the equation

$$\begin{vmatrix} a, & b \\ c, & d \end{vmatrix} . \begin{vmatrix} a', & b' \\ c', & d' \end{vmatrix} = \begin{matrix} (a, & b) \\ (c, & d) \end{matrix} \left|\overline{\ \ \begin{matrix} (a', c'), (b', d') \\ \\ \end{matrix}\ \ }\right| = \begin{vmatrix} aa' + bc', & ab' + bd' \\ ca' + dc', & cb' + dd' \end{vmatrix}$$

for the product of two matrices.

The Ausdehnungslehre: Grassmann. Art. Nos. 43—63.

43. **Grassmann.** *Die lineale Ausdehnungslehre, ein neuer Zweig der Mathematik*, Leipzig, 1844; 2nd edition, with the title *Die Ausdehnungslehre von* 1844, Leipzig, 1878.

—— *Die Ausdehnungslehre, vollständig und in strenger Form bearbeitet*, Berlin, 1862, referred to as *Die Ausdehnungslehre von* 1862.

44. Plane Geometry. The representation of a point is that employed by Möbius in the *Barycentrische Calcul* (1827), viz. considering in the plane three fixed points A_1, A_2, A_3, and in regard to the triangle formed by these points, taking x_1, x_2, x_3 as the areal coordinates of a point x (x_1, x_2, x_3 are equal to the areas of the triangles xA_2A_3, xA_3A_1, xA_1A_2, each divided by the area of the triangle $A_1A_2A_3$; whence $x_1+x_2+x_3=1$), we write

$$x, \; = x_1\epsilon_1 + x_2\epsilon_2 + x_3\epsilon_3$$

as the representation of the point x. Here ϵ_1, ϵ_2, ϵ_3 are extraordinaries, or they may also be regarded as points, viz. as the points (1, 0, 0), (0, 1, 0), (0, 0, 1), or say the points A_1, A_2, A_3 respectively. If the sum $x_1+x_2+x_3$, instead of being $=1$ has any other value $=w$, then $x_1\epsilon_1+x_2\epsilon_2+x_3\epsilon_3$ represents not a point simpliciter, but a point of the weight w; a point simpliciter thus means a point of the weight 1.

45. Considering then

$$\lambda x, \; = \lambda(x_1\epsilon_1+x_2\epsilon_2+x_3\epsilon_3), \quad (x_1+x_2+x_3=1),$$

a point of the weight λ; and similarly

$$\mu y, \; = \mu(y_1\epsilon_1+y_2\epsilon_2+y_3\epsilon_3), \quad (y_1+y_2+y_3=1),$$

a point of the weight μ; we form herewith a sum $\lambda x+\mu y$,

$$= (\lambda x_1+\mu y_1)\,\epsilon_1 + (\lambda x_2+\mu y_2)\,\epsilon_2 + (\lambda x_3+\mu y_3)\,\epsilon_3,$$

which is a point of the weight $\lambda+\mu$, at the C. G. of the two given points, considered as being of the weights λ and μ respectively; if the weights λ and μ are regarded as indeterminate, then the ratio $\lambda:\mu$ is a variable parameter, and the sum will be any point in the line through the given points. It may be remarked that, if $\lambda+\mu$ be $=1$, then the sum will be a point simpliciter; in particular,

$$\tfrac{1}{2}(x+y), \; = \tfrac{1}{2}(x_1+y_1)\,\epsilon_1 + \tfrac{1}{2}(x_2+y_2)\,\epsilon_2 + \tfrac{1}{2}(x_3+y_3)\,\epsilon_3,$$

will be the point M, midway between the given points x and y.

Observe that $x-y$ is not properly a point; it may be regarded as a point, weight zero, at an infinite distance.

46. The extraordinaries ϵ_1, ϵ_2, ϵ_3 are assumed to be such that

$$\epsilon_1^2=0, \quad \epsilon_2^2=0, \quad \epsilon_3^2=0, \quad \epsilon_2\epsilon_1=-\epsilon_1\epsilon_2, \quad \epsilon_3\epsilon_2=-\epsilon_2\epsilon_3, \quad \epsilon_1\epsilon_3=-\epsilon_3\epsilon_1\,;$$

the product of the two points x, y is thus

$$x\,.\,y = (x_2y_3-x_3y_2)\,\epsilon_2\epsilon_3 + (x_3y_1-x_1y_3)\,\epsilon_3\epsilon_1 + (x_1y_2-x_2y_1)\,\epsilon_1\epsilon_2\,;$$

depending on the new extraordinaries $\epsilon_2\epsilon_3$, $\epsilon_3\epsilon_1$, $\epsilon_1\epsilon_2$. Grassmann further assumes $\epsilon_1\epsilon_2\epsilon_3 = 1$, and he calls the foregoing combinations the complements (Ergänzungen) of ϵ_1, ϵ_2, ϵ_3, denoting them by $/\epsilon_1$, $/\epsilon_2$, $/\epsilon_3$ respectively; it is better to use new letters, and I call them η_1, η_2, η_3 respectively; we thus have

$$\epsilon_1^2 = 0, \quad \epsilon_2^2 = 0, \quad \epsilon_3^2 = 0,$$

$$\epsilon_2\epsilon_3 = -\epsilon_3\epsilon_2 = \eta_1, \quad \epsilon_3\epsilon_1 = -\epsilon_1\epsilon_3 = \eta_2, \quad \epsilon_1\epsilon_2 = -\epsilon_2\epsilon_1 = \eta_3.$$

But we require further assumptions for the combinations of the η's *inter se* and with the ϵ's; the forms are

$$\eta_1^2 = 0, \quad \eta_2^2 = 0, \quad \eta_3^2 = 0,$$

$$-\eta_2\eta_3 = \eta_3\eta_2 = \epsilon_1; \quad -\eta_3\eta_1 = \eta_1\eta_3 = \epsilon_2, \quad -\eta_1\eta_2 = \eta_2\eta_1 = \epsilon_3;$$

$$\epsilon_1\eta_1 = \eta_1\epsilon_1 = 1, \quad \epsilon_2\eta_2 = \eta_2\epsilon_2 = 1, \quad \epsilon_3\eta_3 = \eta_3\epsilon_3 = 1, \quad \epsilon_1\eta_2 = \eta_2\epsilon_1 = 0, \text{ \&c.},$$

(viz. each such combination is $= 0$); or, what is the same thing, we have for 1, ϵ_1, ϵ_2, ϵ_3, η_1, η_2, η_3 the multiplication table

	1	ϵ_1	ϵ_2	ϵ_3	η_1	η_2	η_3
1	1	ϵ_1	ϵ_2	ϵ_3	η_1	η_2	η_3
ϵ_1	ϵ_1	0	η_3	$-\eta_2$	1	0	0
ϵ_2	ϵ_2	$-\eta_3$	0	η_1	0	1	0
ϵ_3	ϵ_3	η_2	$-\eta_1$	0	0	0	1
η_1	η_1	1	0	0	0	$-\epsilon_3$	ϵ_2
η_2	η_2	0	1	0	ϵ_3	0	$-\epsilon_1$
η_3	η_3	0	0	1	$-\epsilon_2$	ϵ_1	0

It is proper to remark that in Grassmann's theory there is no interpretation for the general linear symbol

$$\omega + x_1\epsilon_1 + x_2\epsilon_2 + x_3\epsilon_3 + y_1\eta_1 + y_2\eta_2 + y_3\eta_3,$$

which presents itself in connexion with the seven extraordinaries (1, ϵ_1, ϵ_2, ϵ_3, η_1, η_2, η_3).

47. It is to be noticed that the symbols are not associative; assuming them to be so, we should have $\epsilon_2\epsilon_3 \cdot \epsilon_3\epsilon_1 = \epsilon_2 \cdot \epsilon_3^2 \cdot \epsilon_1 = 0$, which is inconsistent with $\eta_1\eta_2 = -\epsilon_3$; and so in other cases. But any *three* ϵ's are associative, and hence a product

$$x \cdot y \cdot z, \; = (x_1\epsilon_1 + x_2\epsilon_2 + x_3\epsilon_3)(y_1\epsilon_1 + y_2\epsilon_2 + y_3\epsilon_3)(z_1\epsilon_1 + z_2\epsilon_2 + z_3\epsilon_3),$$

is associative; in fact, if this be regarded as standing for $(x \,.\, y)\, z$, the value is

$$= \{(x_2y_3 - x_3y_2)\,\eta_1 + (x_3y_1 - x_1y_3)\,\eta_2 + (x_1y_2 - x_2y_1)\,\eta_3\}\,(z_1\epsilon_1 + z_2\epsilon_2 + z_3\epsilon_3),$$

which is

$$= (x_2y_3 - x_3y_2)\,z_1 + (x_3y_1 - x_1y_3)\,z_2 + (x_1y_2 - x_2y_1)\,z_3, \;= \begin{vmatrix} x_1, & x_2, & x_3 \\ y_1, & y_2, & y_3 \\ z_1, & z_2, & z_3 \end{vmatrix};$$

and similarly, regarding it as standing for $x\,(y \,.\, z)$, the value is

$$= (x_1\epsilon_1 + x_2\epsilon_2 + x_3\epsilon_3)\,\{(y_2z_3 - y_3z_2)\,\eta_1 + (y_3z_1 - y_1z_3)\,\eta_2 + (y_1z_2 - y_2z_1)\,\eta_3\},$$

which is

$$= x_1\,(y_2z_3 - y_3z_2) + x_2\,(y_3z_1 - y_1z_3) + x_3\,(y_1z_2 - y_2z_1), \;= \begin{vmatrix} x_1, & x_2, & x_3 \\ y_1, & y_2, & y_3 \\ z_1, & z_2, & z_3 \end{vmatrix},$$

as before. The product of the three points is thus a scalar, viz. it is equal to the area of the triangle formed by the three points divided by that of the triangle $A_1A_2A_3$.

48. But a product $x \,.\, Y \,.\, z$,

$$= (x_1\epsilon_1 + x_2\epsilon_2 + x_3\epsilon_3)\,(Y_1\eta_1 + Y_2\eta_2 + Y_3\eta_3)\,(z_1\epsilon_1 + z_2\epsilon_2 + z_3\epsilon_3),$$

is not associative, and has thus no meaning until the grouping of the factors is determined. In fact, $(x \,.\, Y)\, z$ will be

$$= (x_1Y_1 + x_2Y_2 + x_3Y_3)\,(z_1\epsilon_1 + z_2\epsilon_2 + z_3\epsilon_3),$$

which denotes the point z_1 with a weight $= x_1Y_1 + x_2Y_2 + x_3Y_3$; whereas $x\,(Y \,.\, z)$ will be

$$= (x_1\epsilon_1 + x_2\epsilon_2 + x_3\epsilon_3)\,(y_1Z_1 + y_2Z_2 + y_3Z_3),$$

which denotes the point x_1 with a weight $= y_1Z_1 + y_2Z_2 + y_3Z_3$.

Hence also a product $x \,.\, y \,.\, z \,.\, w$,

$$= (x_1\epsilon_1 + x_2\epsilon_2 + x_3\epsilon_3)\,(y_1\epsilon_1 + y_2\epsilon_2 + y_3\epsilon_3)\,(z_1\epsilon_1 + z_2\epsilon_2 + z_3\epsilon_3)\,(w_1\epsilon_1 + w_2\epsilon_2 + w_3\epsilon_3),$$

of four factors is not associative, and has thus no meaning until the grouping of the factors is determined. Grassmann attributes to such a product the value $= 0$; but this is not a value corresponding to any grouping of the factors, and the equation $x \,.\, y \,.\, z \,.\, w = 0$ can only be regarded as an independent assumption.

For the product of a point into itself, or say the square of a point, we have

$$x^2 = (x_1\epsilon_1 + x_2\epsilon_2 + x_3\epsilon_3)^2, \;\; = (x_1\epsilon_1 + x_2\epsilon_2 + x_3\epsilon_3)\,(x_1\epsilon_1 + x_2\epsilon_2 + x_3\epsilon_3), \; = 0,$$

viz. this is $=$ zero.

49. The product of two points x and y is defined to be the "Strecke", the finite line or stroke (xy), considered as drawn from x to y, say we have $x \,.\, y = (xy)$; observe

that $y\,.\,x = (yx)$: by what precedes $y\,.\,x + x\,.\,y = 0$, whence also $(yx) + (xy) = 0$, which is an equation in the addition of strokes. Hence, if as before, the points x, y are $x = x_1\epsilon_1 + x_2\epsilon_2 + x_3\epsilon_3$, and $y = y_1\epsilon_1 + y_2\epsilon_2 + y_3\epsilon_3$, then for the stroke (xy) we have

$$(xy) = (x_1\epsilon_1 + x_2\epsilon_2 + x_3\epsilon_3)(y_1\epsilon_1 + y_2\epsilon_2 + y_3\epsilon_3),$$

$$= (x_2y_3 - x_3y_2)\,\eta_1 + (x_3y_1 - x_1y_3)\,\eta_2 + (x_1y_2 - x_2y_1)\,\eta_3,$$

which is, in fact,

$$= (A_1xy\,.\,\eta_1 + A_2xy\,.\,\eta_2 + A_3xy\,.\,\eta_3) \div A_1A_2A_3,$$

if A_1xy, A_2xy, A_3xy, $A_1A_2A_3$ denote the areas of the triangles formed by the points (A_1, x, y), (A_2, x, y), (A_3, x, y), (A_1, A_2, A_3) respectively. These triangles remain unaltered if the stroke (xy) is slid along the indefinite line joining the original points x, y, the absolute distance xy being unaltered; that is, strokes of equal length upon the same line are regarded as identical. Hence also strokes on the same line may be added together by adding their lengths.

50. Consider two strokes (xy), (xz), (fig. 4), having a common point x, then completing the parallelogram, we prove that the sum $(xy) + (xz)$ is equal to the diagonal (xw).

Fig. 4.

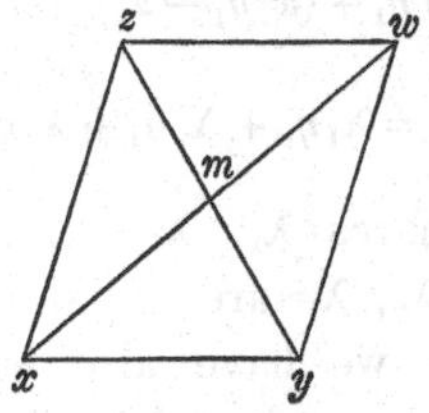

We have, in fact,

$$x + w = y + z,$$

and thence

$$x(x + w) = x(y + z),$$

that is,

$$x^2 + x\,.\,w = x\,.\,y + x\,.\,z,$$

or, since $x^2 = 0$ and $x\,.\,y = (xy)$, &c., this is

$$(xw) = (xy) + (xz),$$

the required formula for the addition of the strokes.

51. The same thing appears thus: writing

$$x = x_1\epsilon_1 + x_2\epsilon_2 + x_3\epsilon_3,$$

$$y = y_1\epsilon_1 + y_2\epsilon_2 + y_3\epsilon_3,$$

$$z = z_1\epsilon_1 + z_2\epsilon_2 + z_3\epsilon_3,$$

we have

$$(xy) = (x_2y_3 - x_3y_2)\,\eta_1 + (x_3y_1 - x_1y_3)\,\eta_2 + (x_1y_2 - x_2y_1)\,\eta_3,$$

$$(xz) = (x_2z_3 - x_3z_2)\,\eta_1 + (x_3z_1 - x_1z_3)\,\eta_2 + (x_1z_2 - x_2z_1)\,\eta_3.$$

Or, putting for a moment

$$m_1,\ m_2,\ m_3 = \tfrac{1}{2}(y_1+z_1),\quad \tfrac{1}{2}(y_2+z_2),\quad \tfrac{1}{2}(y_3+z_3),$$

we have for the point m, midway between y and z, the equation $m = m_1\epsilon_1 + m_2\epsilon_2 + m_3\epsilon_3$; and then

$$\begin{aligned}(xy)+(xz) &= 2\{(x_2m_3 - x_3m_2)\eta_1 + (x_3m_1 - x_1m_3)\eta_2 + (x_1m_2 - x_2m_1)\eta_3\}\\ &= 2(xm),\quad = xw,\end{aligned}$$

which is the theorem in question.

52. Recurring to the foregoing equation $x + w = y + z$, we deduce

$$xz - yw = zw - xy = (yzw) = (zwx) = -(wxy) = -(xyz),$$

if for shortness $(yzw) = yz + zw + wy$, and similarly for (zwx), (wxy), and (xyz). Observe here that $xz - yw$ and $zw - xy$ are each of them equal to the difference between two equal and parallel strokes; the strokes yz, zw, wy are the sides (in order) of a triangle; *quâ* forces, the expressions in question represent each of them a couple.

53. We have thus arrived at the representation of a stroke

$$(xy) = (x_2y_3 - x_3y_2)\eta_1 + (x_3y_1 - x_1y_3)\eta_2 + (x_1y_2 - x_2y_1)\eta_3,$$

or say

$$\lambda = \lambda_1\eta_1 + \lambda_2\eta_2 + \lambda_3\eta_3,$$

where the meaning of the coordinates λ_1, λ_2, λ_3 is such that, if x, y are the two extremities of the stroke, then λ_1, λ_2, λ_3 are equal to the areas A_1xy, A_2xy, A_3xy, each divided by the area $A_1A_2A_3$. And we have also established the rule for the addition of strokes: it will be noticed that this has been done without the aid of any expression for the length of the stroke.

54. *Metrical relations.* For the expression of the distance of two points or length of a stroke, and of the inclinations of a stroke to the sides of the fundamental triangle, &c., we require the values of the angles and sides of this triangle $A_1A_2A_3$, say the angles are A_1, A_2, A_3, and the sides are $= R\sin A_1$, $R\sin A_2$, $R\sin A_3$ respectively; R is thus equal to the diameter of the circumscribed circle. Moreover, considering a stroke xy, if through x we draw lines $x\theta_1$, $x\theta_2$, $x\theta_3$ in the senses A_2A_3, A_3A_1, A_1A_2 respectively, then taking the angles $yx\theta_1$, $yx\theta_2$, $yx\theta_3$ each of them in the same sense, or say each right-handedly, we call these angles θ_1, θ_2, θ_3 respectively, and take also ρ, $=(xy)$ for the length of the stroke. This being so, and if the two extremities are

$$x = x_1\epsilon_1 + x_2\epsilon_2 + x_3\epsilon_3,\ \text{ and }\ y = y_1\epsilon_1 + y_2\epsilon_2 + y_3\epsilon_3,$$

then we have, first,

$$\theta_2 - \theta_3 = \pi - A_1,\quad \theta_3 - \theta_1 = \pi - A_2,\quad \theta_1 - \theta_2 = -\pi - A_3,$$

giving

$$A_1 + A_2 + A_3 = \pi,$$

and

$$\sin A_1,\quad \sin A_2,\quad \sin A_3 = \sin(\theta_2 - \theta_3),\quad \sin(\theta_3 - \theta_1),\quad \sin(\theta_1 - \theta_2).$$

And then

$$y_1 = x_1 + \frac{\rho}{R}\frac{\sin\theta_1}{\sin A_2 \sin A_3},\quad y_2 = x_2 + \frac{\rho}{R}\frac{\sin\theta_2}{\sin A_3 \sin A_1},\quad y_3 = x_3 + \frac{\rho}{R}\frac{\sin\theta_3}{\sin A_1 \sin A_2}.$$

As regards these last equations, writing for a moment p_1, p_2, p_3 for the perpendicular distances of x, and q_1, q_2, q_3 for those of y from the sides of the triangle, then we have $q_1, q_2, q_3 = p_1 + \rho\sin\theta_1, p_2 + \rho\sin\theta_2, p_3 + \rho\sin\theta_3$, and we thence easily derive the equations in question.

55. *Expression for length of a stroke.* We hence, by a somewhat complicated analytical process, find an expression for ρ in terms of the coordinates (x_1, x_2, x_3), and (y_1, y_2, y_3), which enter into it through the combinations

$$x_2y_3 - x_3y_2,\quad x_3y_1 - x_1y_3,\quad x_1y_2 - x_2y_1.$$

In fact, writing for shortness

$$\frac{x_1}{\sin A_1},\ \frac{x_2}{\sin A_2},\ \frac{x_3}{\sin A_3} = K_1,\ K_2,\ K_3,$$

we have

$$(x_2y_3 - x_3y_2)\sin A_1 = \frac{\rho}{R}(K_2\sin\theta_3 - K_3\sin\theta_2),$$

$$(x_3y_1 - x_1y_3)\sin A_2 = \frac{\rho}{R}(K_3\sin\theta_1 - K_1\sin\theta_3),$$

$$(x_1y_2 - x_2y_1)\sin A_3 = \frac{\rho}{R}(K_1\sin\theta_2 - K_2\sin\theta_1);$$

and hence also, if $(*\between\xi, \eta, \zeta)^2$ denote the quadric function

$$(1, 1, 1, -\cos A_1, -\cos A_2, -\cos A_3\between\xi, \eta, \zeta)^2,$$

we have

$$(*\between(x_2y_3 - x_3y_2)\sin A_1, \ldots)^2$$
$$= \frac{\rho^2}{R^2}(*\between K_2\sin\theta_3 - K_3\sin\theta_2,\ K_3\sin\theta_1 - K_1\sin\theta_3,\ K_1\sin\theta_2 - K_2\sin\theta_1)^2,$$

where the quadric function on the right-hand side is in fact

$$= (K_1\sin A_1 + K_2\sin A_2 + K_3\sin A_3)^2,$$

that is,

$$= (x_1 + x_2 + x_3)^2,\ = 1;$$

and we thus finally obtain

$$\rho^2 = R^2(*\between(x_2y_3 - x_3y_2)\sin A_1,\ (x_3y_1 - x_1y_3)\sin A_2,\ (x_1y_2 - x_2y_1)\sin A_3)^2,$$

the required formula for the length of the stroke xy.

56. To effect the foregoing reduction of the quadric function

$$(*\between K_2\sin\theta_3 - K_3\sin\theta_2,\ K_3\sin\theta_1 - K_1\sin\theta_3,\ K_1\sin\theta_2 - K_2\sin\theta_1)^2,$$

observe that the coefficient herein of K_1^2 is

$$= \sin^2\theta_2 + \sin^2\theta_3 + 2\cos A_1 \sin\theta_2 \sin\theta_3,$$

where $\theta_2 - \theta_3 = \pi - A_1$, and thence

$$\cos\theta_2\cos\theta_3 + \sin\theta_2\sin\theta_3 = -\cos A_1;$$

that is, we have

$$\cos A_1 + \sin\theta_2\sin\theta_3 = -\cos\theta_2\cos\theta_3,$$

and thence

$$1 - \sin^2\theta_2 - \sin^2\theta_3 + \sin^2\theta_2\sin^2\theta_3 = \cos^2 A_1 + 2\cos A_1\sin\theta_2\sin\theta_3 + \sin^2\theta_2\sin^2\theta_3;$$

whence the coefficient in question is $= \sin^2 A_1$. And finding in like manner the coefficients of K_2^2, K_3^2, K_2K_3, &c., the whole function is, as already mentioned,

$$= (K_1\sin A_1 + K_2\sin A_2 + K_3\sin A_3)^2, \text{ that is, } = 1.$$

57. *New representation of a stroke.* Representing the stroke in the form

$$\lambda = \lambda_1\eta_1 + \lambda_2\eta_2 + \lambda_3\eta_3,$$

we have only to replace $x_2y_3 - x_3y_2$, $x_3y_1 - x_1y_3$, $x_1y_2 - x_2y_1$ by their values $= \lambda_1, \lambda_2, \lambda_3$ respectively, viz. for the length ρ of the stroke, we have

$$\rho^2 = R^2 (*\between \lambda_1\sin A_1, \lambda_2\sin A_2, \lambda_3\sin A_3)^2.$$

Say this equation is

$$\rho^2 = R^2 . \Omega, \text{ then } \rho = R\sqrt{(\Omega)}, \text{ or } 1 = \frac{\rho}{R\sqrt{(\Omega)}},$$

and the stroke may be represented in the form

$$\lambda = \frac{\rho}{R\sqrt{(\Omega)}}(\lambda_1\eta_1 + \lambda_2\eta_2 + \lambda_3\eta_3);$$

or, what is the same thing, if the absolute magnitudes of the coordinates $\lambda_1, \lambda_2, \lambda_3$ are such that

$$(*\between \lambda_1\sin A_1, \lambda_2\sin A_2, \lambda_3\sin A_3)^2 = 1,$$

then the expression is

$$\lambda = \frac{\rho}{R}(\lambda_1\eta_1 + \lambda_2\eta_2 + \lambda_3\eta_3),$$

where ρ is the length of the stroke.

58. It has been seen that the coefficients of η_1, η_2, η_3 have the values A_1xy, A_2xy, A_3xy, each divided by $A_1A_2A_3$; or, what is the same thing, if p_1, p_2, p_3 are the lengths of the perpendiculars on the stroke from the points A_1, A_2, A_3 respectively, then these coefficients are $(\tfrac{1}{2}\rho p_1, \tfrac{1}{2}\rho p_2, \tfrac{1}{2}\rho p_3)$, each divided by $A_1A_2A_3$; or, what is the same thing, the perpendicular distances p_1, p_2, p_3 are

$$= \frac{2A_1A_2A_3}{R\sqrt{\Omega}}(\lambda_1, \lambda_2, \lambda_3)$$

respectively, where $A_1A_2A_3$ is the area of the fundamental triangle.

59. *Expression for the mutual inclination of two strokes.* In connexion with the stroke $\lambda = \lambda_1\eta_1 + \lambda_2\eta_2 + \lambda_3\eta_3$, considering a stroke $\mu = \mu_1\eta_1 + \mu_2\eta_2 + \mu_3\eta_3$, suppose the inclinations hereof to the axes are ϕ_1, ϕ_2, ϕ_3, then

$$\phi_2 - \phi_3 = \pi - A_1, \quad \phi_3 - \phi_1 = \pi - A_2, \quad \phi_1 - \phi_2 = -\pi - A_3,$$

and we have

$$\theta_1 - \phi_1 = \theta_2 - \phi_2 = \theta_3 - \phi_3,$$

viz. each of these equal angles is the inclination of the two strokes to each other, say this angle is $= \delta$. The expression for this angle is

$$\cos\delta = (*)(\lambda_1 \sin A_1,\ \lambda_2 \sin A_2,\ \lambda_3 \sin A_3)(\mu_1 \sin A_1,\ \mu_2 \sin A_2,\ \mu_3 \sin A_3)$$
$$\div \sqrt{\{(1, \ldots)(\lambda_1 \sin A_1,\ \lambda_2 \sin A_2,\ \lambda_3 \sin A_3)^2\}} \,.\, \sqrt{\{(1, \ldots)(\mu_1 \sin A_1,\ \mu_2 \sin A_2,\ \mu_3 \sin A_3)^2\}}.$$

In fact, considering (as we may do) the two strokes as proceeding from a common point $x = x_1\epsilon_1 + x_2\epsilon_2 + x_3\epsilon_3$, then the function in question is

$$= (*)(K_2 \sin\theta_3 - K_3 \sin\theta_2,\ K_3 \sin\theta_1 - K_1 \sin\theta_3,\ K_1 \sin\theta_2 - K_2 \sin\theta_1)(K_2 \sin\phi_3 - K_3 \sin\phi_2, \ldots),$$

which is

$$= (K_1 \sin A_1 + K_2 \sin A_2 + K_3 \sin A_3)^2 \cos\delta, \quad = (x_1 + x_2 + x_3)^2 \cos\delta, \quad = \cos\delta.$$

60. In verification, observe that the whole coefficient of K_1^2 is

$$= \sin\theta_2 \sin\phi_2 + \sin\theta_3 \sin\phi_3 + \cos A_1 (\sin\theta_2 \sin\phi_3 + \sin\theta_3 \sin\phi_2),$$

where $\theta_2 - \theta_3 = \pi - A_1$, $\phi_2 - \phi_3 = \pi - A_1$. Hence

$$\sin\theta_2 = -\sin(\theta_3 - A_1), \quad \sin\phi_2 = -\sin(\phi_3 - A_1);$$

and the expression in question is

$$\begin{aligned} = \ & \sin(\theta_3 - A_1)\sin(\phi_3 - A_1) \\ & - \cos A_1 \sin(\theta_3 - A_1) \sin\phi_3, \\ & - \cos A_1 \sin(\phi_3 - A_1) \sin\theta_3, \\ & + \sin\theta_3 \sin\phi_3, \end{aligned}$$

which, expanding the sines, and writing the last term in the form

$$-\sin\theta_3 \sin\phi_3 (\cos^2 A_1 + \sin^2 A_1)$$

is

$$\begin{aligned} = \ & \cos^2 A_1 (\sin\theta_3 \sin\phi_3 - \sin\theta_3 \sin\phi_3 - \sin\theta_3 \sin\phi_3 + \sin\theta_3 \sin\phi_3) \\ & + \sin^2 A_1 (\cos\theta_3 \cos\phi_3 + \sin\theta_3 \sin\phi_3) \\ & + \cos A_1 \sin A_1 (-\sin\theta_3 \cos\phi_3 - \sin\phi_3 \cos\theta_3 + \sin\phi_3 \cos\theta_3 + \sin\theta_3 \cos\phi_3), \\ = \ & \sin^2 A_1 \cos(\theta_3 - \phi_3), \quad = \sin^2 A_1 \cos\delta, \end{aligned}$$

and reducing in like manner the coefficients of K_2^2, K_3^2, K_2K_3, &c., the whole expression becomes

$$= (K_1 \sin A_1 + K_2 \sin A_2 + K_3 \sin A_3)^2 \cos\delta, \quad = \cos\delta \text{ as above.}$$

61. The foregoing formulæ agree with the theorem, No. 50, that the sum of the strokes

$$\lambda, = \frac{\rho}{R\sqrt{(\Omega)}}(\lambda_1\eta_1 + \lambda_2\eta_2 + \lambda_3\eta_3),$$

and

$$\mu, = \frac{\sigma}{R\sqrt{(\Upsilon)}}(\mu_1\eta_1 + \mu_2\eta_2 + \mu_3\eta_3),$$

is a stroke

$$= \frac{\tau}{R\sqrt{(\Pi)}}(\nu_1\eta_1 + \nu_2\eta_2 + \nu_3\eta_3),$$

which is the diagonal of the parallelogram constructed with the given strokes λ and μ; the proof is a little simplified by assuming (as is allowable) $\Omega = 1$ and $\Upsilon = 1$, in which case we have

$$\lambda + \mu = \frac{1}{R}\{(\rho\lambda_1 + \sigma\mu_1)\eta_1 + (\rho\lambda_2 + \sigma\mu_2)\eta_2 + (\rho\lambda_3 + \sigma\mu_3)\eta_3\},$$

from which the length and inclination may be calculated.

As already appearing, the product x, y, z of any three points is the scalar

$$\begin{vmatrix} x_1, & x_2, & x_3 \\ y_1, & y_2, & y_3 \\ z_1, & z_2, & z_3 \end{vmatrix},$$

which is equal to the area of the triangle xyz, divided by that of the triangle $A_1A_2A_3$. We have in like manner the product of a point x into a stroke λ, viz. this is

$$= (x_1\epsilon_1 + x_2\epsilon_2 + x_3\epsilon_3)(\lambda_1\eta_1 + \lambda_2\eta_2 + \lambda_3\eta_3),$$

which is $= x_1\lambda_1 + x_2\lambda_2 + x_3\lambda_3$, the two factors being in this case commutative; the value is equal to the area of the triangle formed by the point and the stroke, divided by the area $A_1A_2A_3$. Of course, if in the one case the three points are in a line, or if in the other the point is in the line of the stroke, then the product is $= 0$.

62. We have yet to consider the product of two strokes; say these are

$$\lambda = \lambda_1\eta_1 + \lambda_2\eta_2 + \lambda_3\eta_3, \text{ and } \mu = \mu_1\eta_1 + \mu_2\eta_2 + \mu_3\eta_3,$$

then we have

$$\lambda\,.\,\mu = (\lambda_1\eta_1 + \lambda_2\eta_2 + \lambda_3\eta_3)(\mu_1\eta_1 + \mu_2\eta_2 + \mu_3\eta_3)$$
$$= -\{(\lambda_2\mu_3 - \lambda_3\mu_2)\epsilon_1 + (\lambda_3\mu_1 - \lambda_1\mu_3)\epsilon_2 + (\lambda_1\mu_2 - \lambda_2\mu_1)\epsilon_3\},$$

which is a point regarded as having weight. If we take the two strokes to be

$$\lambda = (xy) = (x_2y_3 - x_3y_2)\eta_1 + (x_3y_1 - x_1y_3)\eta_2 + (x_1y_2 - x_2y_1)\eta_3,$$

and

$$\mu = (xz) = (x_2z_3 - x_3z_2)\eta_1 + (x_3z_1 - x_1z_3)\eta_2 + (x_1z_2 - x_2z_1)\eta_3,$$

then the product is

$$\lambda\,.\,\mu = -(x_1\epsilon_1 + x_2\epsilon_2 + x_3\epsilon_3)\begin{vmatrix} x_1, & x_2, & x_3 \\ y_1, & y_2, & y_3 \\ z_1, & z_2, & z_3 \end{vmatrix},$$

viz. this is the common point x, with a weight = area of triangle $xyz \div$ area $A_1A_2A_3$, or say = the area of triangle formed by the two strokes $\div$ area $A_1A_2A_3$. The product is non-commutative: we, in fact, have $\lambda\,.\,\mu = -\mu\,.\,\lambda$.

63. The product of three strokes is associative; and it is a scalar. If the three strokes are the sides of a triangle LMN, then the value is $=(LMN)^2 \div (A_1A_2A_3)^2$, where LMN and $A_1A_2A_3$ denote the areas of the triangle LMN and the triangle $A_1A_2A_3$ respectively; and if any stroke instead of being a side of the triangle LMN be a part only of this side, then the value is diminished proportionally; hence the value is $=\dfrac{\rho\sigma\tau}{MN\,.\,NL\,.\,LM}\{(LMN)^2 \div (A_1A_2A_3)^2\}$, where ρ, σ, τ are the lengths of the three strokes, MN, NL, LM the lengths of the sides of the triangle LMN, and LMN, $A_1A_2A_3$ are the areas as above.

64. It is to be remarked that the form

$$(*\between \xi,\ \eta,\ \zeta)^2,\ =(1,\ 1,\ 1,\ -\cos A_1,\ -\cos A_2,\ -\cos A_3\between \xi,\ \eta,\ \zeta)^2$$

is the product of two linear factors; it is

$$=(\xi - \eta\cos A_3 - \zeta\cos A_2)^2 + (\eta\sin A_3 - \zeta\sin A_2)^2,$$

and thus

$$=(\xi - \eta e^{iA_3} - \zeta e^{-iA_2})(\xi - \eta e^{-iA_3} - \zeta e^{iA_2}).$$

This corresponds to the theorem that the distance of two points is = 0, when the line joining them passes through one of the circular points at infinity; the coordinates of one of these points may be written under the three equivalent forms

$$\begin{aligned} x_1 : x_2 : x_3 = &\ \sin A_1 &&: -\sin A_2 e^{-iA_3} &&: -\sin A_3 e^{-iA_2}, \\ = &-\sin A_1 e^{-iA_3} &&: \sin A_2 &&: -\sin A_3 e^{iA_1}, \\ = &-\sin A_1 e^{iA_2} &&: -\sin A_2 e^{-iA_1} &&: \sin A_3, \end{aligned}$$

and those of the other are of course obtained therefrom by the change of i into $-i$.

*(To be continued)**.

[* This paper remains incomplete; no continuation appears to have been prepared by Professor Cayley.]

866.

NOTE ON KIEPERT'S *L*-EQUATIONS, IN THE TRANSFORMATION OF ELLIPTIC FUNCTIONS.

[From the *Mathematische Annalen*, t. XXX. (1887), pp. 75—77.]

It appears, by comparison with Klein's paper "Ueber die Transformation u. s. w.," *Math. Annalen*, t. XIV. (1878), see p. 144, that Kiepert's L made use of in the Memoir "Ueber Theilung und Transformation der elliptischen Functionen," *Math. Annalen*, t. XXVI. (1886), pp. 369—454, is, in fact, the square of the multiplier, "für das durch $\sqrt[12]{\Delta}$ normirte Integral," viz. considering the general quartic function $(a, \ldots)(x, 1)^4 = (a, b, c, d, e)(x, 1)^4$, and the transformed function $(a_1, \ldots)(y, 1)^4$, then we have

$$\frac{L^2 \sqrt[12]{\Delta}\,dx}{\sqrt{(a, \ldots)(x, 1)^4}} = \frac{\sqrt[12]{\Delta_1}\,dy}{\sqrt{(a_1, \ldots)(y, 1)^4}},$$

where if

$$I = ae - 4bd + 3c^2,$$

$$J = ace - ad^2 - b^2e + 2bcd - c^3,$$

and similarly I_1, J_1, are the invariants of the two functions, then Δ, Δ_1 are the discriminants

$$\Delta = I^3 - 27J^2, \quad \Delta_1 = I_1^3 - 27J_1^2,$$

and the γ_2, γ_3 of Kiepert's equations are

$$\gamma_2 = I \div \sqrt[3]{\Delta}, \quad \gamma_3 = J \div \sqrt{\Delta},$$

whence

$$\gamma_2^3 - 27\gamma_3^2 = 1.$$

In particular, if the forms are

$$1 - x^2 . 1 - k^2x^2, \text{ and } 1 - y^2 . 1 - \lambda^2y^2,$$

and if as usual $k=u^4$, $\lambda=v^4$, and M is the multiplier for the form

$$\frac{dx}{\sqrt{1-x^2\,.\,1-k^2x^2}}=\frac{M\,dy}{\sqrt{1-y^2\,.\,1-\lambda^2y^2}}$$

then we have

$$I=\tfrac{1}{12}(1+14u^8+u^{16}),$$

$$J=\tfrac{1}{216}(1+u^8)(1-34u^8+u^{16}),$$

$$\Delta=\tfrac{1}{16}u^8(1-u^8)^4,\quad \Delta_1=\tfrac{1}{16}v^8(1-v^8)^4,$$

$$\gamma_2=\tfrac{1}{6}\sqrt[3]{2}\,\frac{1+14u^8+u^{16}}{u^{\frac{8}{3}}(1-u^8)^{\frac{4}{3}}},\quad \gamma_3=\tfrac{1}{54}\frac{(1+u^8)(1-34u^8+u^{16})}{u^4(1-u^8)^2},$$

and thence

$$L^2=\frac{v^{\frac{2}{3}}(1-v^8)^{\frac{1}{3}}}{u^{\frac{2}{3}}(1-u^8)^{\frac{1}{3}}}\,\frac{1}{M},$$

which last equation is the expression for L^2 in terms of the Jacobian symbols u, v, M.

As an easy verification in a particular case, suppose $n=5$. We have here

$$L^2=\frac{v^{\frac{2}{3}}(1-v^8)^{\frac{1}{3}}}{u^{\frac{2}{3}}(1-u^8)^{\frac{1}{3}}}\cdot\frac{1}{M},\quad M=\frac{v(1-uv^3)}{v-u^5},\quad \left(=\frac{v+u^5}{5u(1+u^3v)}\right),$$

$$u^6-v^6+5u^2v^2(u^2-v^2)+4uv(1-u^4v^4)=0,$$

$$\gamma_2=\tfrac{1}{6}\sqrt[3]{2}\,\frac{1+14u^8+u^{16}}{u^{\frac{8}{3}}(1-u^8)^{\frac{4}{3}}};$$

and it should be possible, by eliminating u, v, M, to deduce hence the L-equation

$$L^{12}+10L^6-12\gamma_2L^2+5=0.\quad \text{(Kiepert, p. 428.)}$$

It does not seem in any wise easy to do this in the case of an arbitrary modulus; but writing the modular equation in the form

$$(u^2-v^2)(u^4+6u^2v^2+v^4)+4uv(1-u^4v^4)=0,$$

we satisfy this by

$$uv-1=0,\quad u^4+6u^2v^2+v^4=0,$$

or say by

$$v=\frac{1}{u},\quad u^8+6u^4+1=0,$$

and the equation may be verified for this particular modulus.

We have

$$1+14u^8+u^{16}=48u^8,\quad (1-u^8)^2=32u^8,$$

and consequently

$$\gamma_2=\tfrac{1}{6}\sqrt[3]{2}\,.\,\frac{48u^8}{u^{\frac{8}{3}}(32u^8)^{\frac{2}{3}}},\quad =\frac{1}{2\,.\,3}\,2^{\frac{1}{3}}\,\frac{2^4\,.\,3u^8}{u^{\frac{8}{3}}\,.\,2^{\frac{10}{3}}\,.\,u^{\frac{16}{3}}}=1,\ \text{(whence also }\gamma_3=0).$$

Moreover

$$M = \frac{\frac{1}{u} - \frac{u}{u^4}}{\frac{1}{u} - u^5} = -\frac{1}{u^2}\frac{1-u^2}{1-u^6} = \frac{-1}{u^2(1+u^2+u^4)},$$

and thence

$$L^2 = \frac{\frac{1}{u^{\frac{2}{3}}}\left(1 - \frac{1}{u^8}\right)^{\frac{1}{3}}}{u^{\frac{2}{3}}(1-u^8)^{\frac{1}{3}}}\frac{1}{M} = \frac{(u^8-1)^{\frac{1}{3}}}{u^4(1-u^8)^{\frac{1}{3}}}\frac{1}{M} = -\frac{1}{u^4 M},$$

that is,

$$L^2 = 1 + u^2 + u^{-2}.$$

But we have

$$u^4 = -3 + 2\sqrt{2},$$

and thence

$$u^2 = i(1-\sqrt{2}),\quad u^{-2} = i(1+\sqrt{2}),$$

and

$$L^2 = \quad 1 + 2i,$$

whence

$$L^{12} + 10L^6 - 12\gamma_2 L^2 + 5 = (117 + 44i) + 10(-11 - 2i) - 12(1 + 2i) + 5 = 0,$$

or the L-equation is satisfied.

Cambridge, 14 *March* 1887.

867.

NOTE ON THE JACOBIAN SEXTIC EQUATION.

[From the *Mathematische Annalen*, t. XXX. (1887), pp. 78—84.]

In the Jacobian sextic equation

$$(z-a)^6 - 4a(z-a)^5 + 16b(z-a)^3 - 4c(z-a) + 5b^2 - 4ac = 0,$$

if A, B, C, D, E, F are the square roots (each with a determinate sign) of the roots $z_0, z_1, z_2, z_3, z_4, z_\infty$ of the equation, and if ϵ be an imaginary fifth root of unity, such that $\epsilon + \epsilon^4 - \epsilon^2 - \epsilon^3 = \sqrt{5}$, then we have between the square roots a system of three linear equations

$$\begin{aligned} 0 &= -A\sqrt{5} + B + C + D + E + F, \\ 0 &= B + \epsilon^3 C + \epsilon D + \epsilon^4 E + \epsilon^2 F, \\ 0 &= B + \epsilon^2 C + \epsilon^4 D + \epsilon E + \epsilon^3 F, \end{aligned}$$

see Brioschi's *Funzioni ellittiche* (Milan, 1880), third appendix, p. 402.

The sextic equation is irreducible, and there is thus nothing to distinguish any one square root from any other square root. But in the foregoing system of equations, A is distinguished from the other square roots; hence the system must be one of 6 systems, wherein the place occupied by A is occupied by A, B, C, D, E, F respectively. Observe however that the letters being square roots, it may very well happen, and (as will appear) it does in fact happen, that the passage from the first to another system implies a change of sign in certain of the letters A, B, C, D, E, F.

The selection of one of the square roots as $=A$, does not determine which of the other square roots shall be $=B, C, D, E, F$ respectively: and in fact each system of equations may be represented in 10 different forms: viz. by multiplying the second and third equations by $\epsilon, \epsilon^2, \epsilon^3, \epsilon^4$ respectively, we obtain four new forms: we have thus five forms: and then in each of them transposing the second and third equations, we have five other forms: we have thus in all 10 forms.

Write *ABCDEF* to denote the foregoing system of three equations

$$0=-A\sqrt{5}+B+\ C+\ D+\ E+\ F,$$
$$0=\qquad B+\epsilon^3C+\ \epsilon D+\epsilon^4E+\epsilon^2F,$$
$$0=\qquad B+\epsilon^2C+\epsilon^4D+\ \epsilon E+\epsilon^3F\,;$$

the 10 forms of the system are

$A\,B\,C\,D\,E\,F$	$A\,B\,F\,E\,D\,C$
$A\,C\,D\,E\,F\,B$	$A\,C\,B\,F\,E\,D$
$A\,D\,E\,F\,B\,C$	$A\,D\,C\,B\,F\,E$
$A\,E\,F\,B\,C\,D$	$A\,E\,D\,C\,B\,F$
$A\,F\,B\,C\,D\,E$	$A\,F\,E\,D\,C\,B,$

where observe that the five forms in the same column are connected by cyclical substitutions on the last five letters; and that the forms in each line are connected by two inversions of the last four letters.

It is hardly necessary to remark that *ABFEDC* means the system of three equations

$$0=-A\sqrt{5}+B+\ F+\ E+\ D+\ C,$$
$$0=\qquad B+\epsilon^3F+\ \epsilon E+\epsilon^4D+\epsilon^2C,$$
$$0=\qquad B+\epsilon^2F+\epsilon^4E+\ \epsilon D+\epsilon^3C,$$

which are the same with the equations of the original form; and similarly that *ACDEFB* means the system of three equations

$$0=-A\sqrt{5}+C+\ D+\ E+\ F+\ B,$$
$$0=\qquad C+\epsilon^3D+\ \epsilon E+\epsilon^4F+\epsilon^2B,$$
$$0=\qquad C+\epsilon^2D+\epsilon^4E+\ \epsilon F+\epsilon^3B,$$

where the first equation is the same with the first equation, and the second and third equations only differ by a factor from the second and third equations respectively of the original form. And similarly as regards the others of the 10 forms.

I say that we have a second system $BCAF\bar{D}\bar{E}$ (where $\bar{D}$, $\bar{E}$ mean $-D$, $-E$ respectively): in fact, this denotes the system of three equations

$$0=-B\sqrt{5}+C+\ A+\ F-\ D-\ E,$$
$$0=\qquad C+\epsilon^3A+\ \epsilon F-\epsilon^4D-\epsilon^2E,$$
$$0=\qquad C+\epsilon^2A+\epsilon^4F-\ \epsilon D-\epsilon^3E,$$

which are deducible from the original three equations.

We have, in fact, the identities

$$
\begin{aligned}
&-\sqrt{5}\,\{-B\sqrt{5} + C + A + F - D - E\} \\
&= 1\,(B + C - A\sqrt{5} + F + D + E) \\
&\quad + 2\,(B + \epsilon^3 C + \epsilon^2 F + \epsilon D + \epsilon^4 E) \\
&\quad + 2\,(B + \epsilon^2 C + \epsilon^3 F + \epsilon^4 D + \epsilon E),
\end{aligned}
$$

$$
\begin{aligned}
&(1 - \epsilon - \epsilon^3 + \epsilon^4)\,\{C + \epsilon^3 A + \epsilon F - \epsilon^4 D - \epsilon^2 E\} \\
&= 1\,(B + C - A\sqrt{5} + F + D + E) \\
&\quad + (\epsilon + \epsilon^4)\,(B + \epsilon^3 C + \epsilon^2 F + \epsilon D + \epsilon^4 E) \\
&\quad + (\epsilon^2 + \epsilon^3)\,(B + \epsilon^2 C + \epsilon^3 F + \epsilon^4 D + \epsilon E),
\end{aligned}
$$

and

$$
\begin{aligned}
&(1 + \epsilon - \epsilon^2 - \epsilon^4)\,\{C + \epsilon^2 A + \epsilon^4 F - \epsilon D - \epsilon^3 E\} \\
&= 1\,(B + C - A\sqrt{5} + F + D + E) \\
&\quad + (\epsilon^2 + \epsilon^3)\,(B + \epsilon^3 C + \epsilon^2 F + \epsilon D + \epsilon^4 E) \\
&\quad + (\epsilon + \epsilon^4)\,(B + \epsilon^2 C + \epsilon^3 F + \epsilon^4 D + \epsilon E),
\end{aligned}
$$

all which identities are at once verified on recollecting that

$$1 = -\epsilon - \epsilon^2 - \epsilon^3 - \epsilon^4 \text{ and } \sqrt{5} = \epsilon - \epsilon^2 - \epsilon^3 + \epsilon^4.$$

We can now write down the 6 systems, each of them under its 10 forms: these, in fact, are

$A\,B\,C\,D\,E\,F$	$A\,B\,F\,E\,D\,C$
$A\,C\,D\,E\,F\,B$	$A\,C\,B\,F\,E\,D$
$A\,D\,E\,F\,B\,C$	$A\,D\,C\,B\,F\,E$
$A\,E\,F\,B\,C\,D$	$A\,E\,D\,C\,B\,F$
$A\,F\,B\,C\,D\,E$	$A\,F\,E\,D\,C\,B$
$B\,C\,A\,F\,\bar{D}\,\bar{E}$	$B\,C\,\bar{E}\,\bar{D}\,F\,A$
$B\,A\,F\,\bar{D}\,\bar{E}\,C$	$B\,A\,C\,\bar{E}\,\bar{D}\,F$
$B\,F\,\bar{D}\,\bar{E}\,C\,A$	$B\,F\,A\,C\,\bar{E}\,\bar{D}$
$B\,\bar{D}\,\bar{E}\,C\,A\,F$	$B\,\bar{D}\,F\,A\,C\,\bar{E}$
$B\,\bar{E}\,C\,A\,F\,\bar{D}$	$B\,\bar{E}\,\bar{D}\,F\,A\,C$
$C\,D\,A\,B\,\bar{E}\,\bar{F}$	$C\,D\,\bar{F}\,\bar{E}\,B\,A$
$C\,A\,B\,\bar{E}\,\bar{F}\,D$	$C\,A\,D\,\bar{F}\,\bar{E}\,B$
$C\,B\,\bar{E}\,\bar{F}\,D\,A$	$C\,B\,A\,D\,\bar{F}\,\bar{E}$
$C\,\bar{E}\,\bar{F}\,D\,A\,B$	$C\,\bar{E}\,B\,A\,D\,\bar{F}$
$C\,\bar{F}\,D\,A\,B\,\bar{E}$	$C\,\bar{F}\,\bar{E}\,B\,A\,D$

$D E A C \bar{F} \bar{B}$	$D E \bar{B} \bar{F} C A$
$D A C \bar{F} \bar{B} E$	$D A E \bar{B} \bar{F} C$
$D C \bar{F} \bar{B} E A$	$D C A E \bar{B} \bar{F}$
$D \bar{F} \bar{B} E A C$	$D \bar{F} C A E \bar{B}$
$D \bar{B} E A C \bar{F}$	$D \bar{B} \bar{F} C A E$
$E F A D \bar{B} \bar{C}$	$E F \bar{C} \bar{B} D A$
$E A D \bar{B} \bar{C} F$	$E A F \bar{C} \bar{B} D$
$E D \bar{B} \bar{C} F A$	$E D A F \bar{C} \bar{B}$
$E \bar{B} \bar{C} F A D$	$E \bar{B} D A F \bar{C}$
$E \bar{C} F A D \bar{B}$	$E \bar{C} \bar{B} D A F$
$F A E \bar{C} \bar{D} B$	$F A B \bar{D} \bar{C} E$
$F E \bar{C} \bar{D} B A$	$F E A B \bar{D} \bar{C}$
$F \bar{C} \bar{D} B A E$	$F \bar{C} E A B \bar{D}$
$F \bar{D} B A E \bar{C}$	$F \bar{D} \bar{C} E A B$
$F B A E \bar{C} \bar{D}$	$F B \bar{D} \bar{C} E A$,

where, as before, the superscript line denotes a change of sign, $\bar{A} = -A$, &c.

As to the formation of this table, observe that we have $ABCDEF$ and $BCAF\bar{D}\bar{E}$; repeating the substitution, we have

$$\begin{array}{c} A\,B\,C\,D\,E\,F \\ B\,C\,A\,F\,\bar{D}\,\bar{E} \\ C\,A\,B\,\bar{E}\,\bar{F}\,D \\ \hline A\,B\,C\,D\,E\,F; \end{array}$$

viz. we have thus the $CAB\bar{E}\bar{F}D$ which belongs to the third system; but nothing further, for the substitution is periodic of the third order, and the three forms are merely repeated. But if, upon

$$\begin{array}{c} A\,B\,C\,D\,E\,F \\ B\,C\,A\,F\,\bar{D}\,\bar{E}, \end{array}$$

we operate successively with the substitutions which change the upper line into the three forms of the first system

$$A\,D\,E\,F\,B\,C,\quad A\,E\,F\,B\,C\,D,\quad A\,F\,B\,C\,D\,E,$$

then the lower line is changed into the three forms

$$D\,E\,A\,C\,\bar{F}\,\bar{B},\quad E\,F\,A\,D\,\bar{B}\,\bar{C},\quad F\,B\,A\,E\,\bar{C}\,\bar{D},$$

which are forms belonging to the fourth, fifth, and sixth systems respectively. By way of verification, observe that, for instance repeating upon the second line the substitution

$$A\,B\,C\,D\,E\,F,$$

in place of

$$D\ E\ A\ C\ \bar{F}\ \dot{B},$$

we obtain

$$C\ \bar{F}\ D\ A\ B\ \bar{E},$$

which is one of the forms of the third system, assumed to have been previously found; and so in other instances.

Reverting to the three equations belonging to the form $ABCDEF$, by subtracting the third from the second equation, we obtain a linear relation between the four square roots C, D, E, F, viz. this is

$$0 = (\epsilon^3 - \epsilon^2)(C - F) + (\epsilon - \epsilon^4)(D - E);$$

and the same equation is obtained by means of the form $ABFEDC$. We thus obtain from the thirty lines of either column of the preceding table, thirty such equations, but obviously the number of such equations should be fifteen, for there can be but one relation between any four square roots C, D, E, F; consequently each equation will be obtained twice; and, in fact, it is clear that the forms $ABCDEF$ and $BAF\bar{D}\bar{E}C$, and so in general any two forms which begin with the same pair of letters, give the same equation. But for greater symmetry, I write down the thirty equations in the order in which they are given by the left-hand column of the table: the equations are

$0 = (\epsilon^3 - \epsilon^2)$ multiplied by	$+ (\epsilon - \epsilon^4)$ multiplied by
$C - F$	$D - E$
$D - B$	$E - F$
$E - C$	$F - B$
$F - D$	$B - C$
$B - E$	$C - D$
$A + E$	$F + D$
$F - C$	$-D + E$
$-D - A$	$-E - C$
$-E - F$	$C - A$
$C + D$	$A - F$
$A + F$	$B + E$
$B - D$	$-E + F$
$-E - A$	$-F - D$
$-F - B$	$D - A$
$D + E$	$A - B$

$0 = (\epsilon^3 - \epsilon^2)$ multiplied by	$+ (\epsilon - \epsilon^4)$ multiplied by
$A + B$	$C + F$
$C - E$	$-F + B$
$-F - A$	$-B - E$
$-B - C$	$E - A$
$E + F$	$A - C$
$A + C$	$D + B$
$D - F$	$-B + C$
$-B - A$	$-C - F$
$-C - D$	$F - A$
$F + B$	$A - D$
$E - B$	$C + D$
$-C - A$	$-D - B$
$-D - E$	$B - A$
$B + C$	$A - E$
$A + D$	$E + C$

where it is to be observed that, adding together the five equations given by any one of the systems, we obtain the identical result $0 = 0$.

I write down the 15 equations in a different order, in some cases changing the sign of the whole equation, as follows

$0 = (\epsilon^3 - \epsilon^2)$ multiplied by	$+ (\epsilon - \epsilon^4)$ multiplied by
$A + C$	$B + D$
$B + C$	$A - E$
$A + B$	$C + F$
$D + E$	$A - B$
$B + F$	$A - D$
$A + F$	$B + E$
$A + D$	$C + E$
$C + D$	$A - F$
$E + F$	$A - C$
$A + E$	$D + F$
$B - E$	$C - D$
$F - D$	$B - C$
$E - C$	$F - B$
$D - B$	$E - F$
$C - F$	$D - E$

The first three of these equations, or writing $\lambda = \epsilon^3 - \epsilon^2$, $\mu = \epsilon - \epsilon^4$, say

$$\lambda (A + C) + \mu (B + D) = 0,$$
$$\lambda (B + C) + \mu (A - E) = 0,$$
$$\lambda (A + B) + \mu (C + F) = 0,$$

constitute the entire system of independent linear relations between the square roots A, B, C, D, E, F. The coefficients λ, μ are such that

$$\lambda^2 - \mu^2 = \lambda\mu \,(= \epsilon - \epsilon^2 - \epsilon^3 + \epsilon^4, \ = \sqrt{5}),$$

and it is hence easy to verify that the remaining twelve equations can be deduced from these by the elimination of one or two of the square roots A, B, C. For instance, to eliminate A from the first and second equations, multiplying by $-\mu$, λ and adding, we obtain

$$(-\lambda\mu + \lambda^2)\, C + (-\mu^2 + \lambda^2)\, B - \mu^2 D - \lambda\mu E = 0,$$

that is,

$$\mu^2 C + \qquad \lambda\mu B - \mu^2 D - \lambda\mu E = 0,$$

or finally

$$\lambda (B - E) + \mu (C - D) = 0,$$

which is one of the equations. And so again eliminating A from the first and third equations, we find

$$\lambda (B - C) + \mu (C + F - B - D) = 0,$$

that is,

$$(\lambda - \mu)(B - C) + \mu (F - D) = 0,$$

or multiplying by λ,

$$\mu^2 (B - C) + \lambda\mu (F - D) = 0,$$

that is, finally

$$\lambda\, (F - D) + \ \mu\, (B - C) = 0,$$

which is one of the equations.

Cambridge, 21 *March* 1887.

868.

ON THE INTERSECTION OF CURVES.

[From the *Mathematische Annalen*, t. XXX. (1887), pp. 85—90.]

It is only recently that I have studied Bacharach's paper "Ueber den Cayley'schen Schnittpunktsatz," *Math. Ann.*, t. XXVI. (1885), pp. 275—299: his theorem in regard to the case where the δ points lie on a curve of the order $\gamma-3$ is a very interesting and valuable one, but I consider it rather as an addition than as a correction to my original theorem; and I cannot by any means agree that the method by counting of constants is to be rejected as not trustworthy; on the contrary, it seems to me to be the proper foundation of the whole theory; it must of course be employed with due consideration of special cases. I reproduce the theorem in what appears to me the complete form.

Writing with Bacharach

$$r \geqq m, \quad r \geqq n, \quad r \leqq m+n-3, \quad \gamma = m+n-r, \quad \delta = \tfrac{1}{2}(\gamma-1)(\gamma-2),$$

and assuming $n \geqq m$, (these equations and inequalities are to be attended to throughout the present paper), I consider two curves of the orders m, n respectively meeting in δ points B, and in $(mn-\delta)$ points A; and I state the theorem as follows:

1°. The $mn-\delta$ points A are in general such that a curve of the order r, which passes through $mn-\delta-1$ of these points, does not of necessity pass through the remaining point; and in this case the general curve of the order r, which passes through the $mn-\delta$ points A, has for its form of equation

$$0 = L_{r-m}P_m + M_{r-n}Q_n,$$

where $P_m=0$, $Q_n=0$ are the equations of the given curves and L_{r-m}, M_{r-n} denote functions of the orders $r-m$, $r-n$ respectively; and it thus appears that the curve of the order r through the $mn-\delta$ points A passes also through the δ points B.

2°. If however the δ points B are on a curve of the order $\gamma-3$, then the $mn-\delta$ points A are a system such that every curve of the order r passing through $mn-\delta-1$ of these points passes through the remaining point; and in this case the general curve of the order r, which passes through the $mn-\delta$ points A, has for its form of equation

$$0=\Omega_r+L_{r-m}P_m+M_{r-n}Q_n,$$

$\Omega_r=0$ is a particular curve through the $mn-\delta$ points A, which does not go through any of the points B; and consequently the curve of the order r does not pass through any of the points B.

For the proof of the theorem I premise as follows:

A curve of the order r depends upon $\frac{1}{2}r(r+3)$ constants, or to use a convenient expression, its Postulandum is $=\frac{1}{2}r(r+3)$: if the curve has to pass through a given point, this imposes a single relation upon the constants, or say the Postulation is $=1$; similarly, if the curve has to pass through k given points, this imposes k relations, or say the Postulation is $=k$. The points may be however a special system, for instance, they may be such that every curve of the order r which passes through $k-1$ of the points, will pass through the remaining point; the Postulation is in this case $=k-1$; and so in other cases. Assuming the Postulation of the k points to be $=k$, then the Postulandum of a curve of the order r through the k points is $=\frac{1}{2}r(r+3)-k$. I stop to remark that the Postulation has reference to the particular curve or other entity in question; thus in the case of a curve passing through k points, the Postulation for a curve of a certain order may be $=k$, and for a curve of a different order it may be less than k.

Considering now, as above, two given curves of the orders m and n intersecting in the $mn-\delta$ points A and the δ points B, then assuming that the $mn-\delta$ points are not a special system, viz. that their Postulation in regard to a curve of the order r is $=mn-\delta$, the Postulandum of a curve of the order r through the $mn-\delta$ points is

$$=\tfrac{1}{2}r(r+3)-mn+\delta,$$

which is

$$=\tfrac{1}{2}(r-m+1)(r-m+2)+\tfrac{1}{2}(r-n+1)(r-n+2)-1,$$

viz. this is identically true when for δ we write its value

$$=\tfrac{1}{2}(m+n-r-1)(m+n-r-2).$$

But we have through the $mn-\delta$ points A, a curve

$$L_{r-m}P_m+M_{r-n}Q_n=0$$

with the proper Postulandum: viz. L_{r-m} contains $\frac{1}{2}(r-m+1)(r-m+2)$ constants, M_{r-n} contains $\frac{1}{2}(r-n+1)(r-n+2)$ constants, and there is a diminution -1 for the constant which divides out; hence this is the general equation of the curve of the order r through the $mn-\delta$ points A; and the curve passes through the remaining δ points B.

In the case where the δ points B are on a curve of the order $\gamma - 3$, (observe that this is a single condition imposed on the δ, $= \frac{1}{2}(\gamma - 1)(\gamma - 2)$ points, for a curve of the order $(\gamma - 3)$ can be drawn through $\frac{1}{2}\gamma(\gamma - 3)$ points), it is to be shown that the Postulation of the $mn - \delta$ points A is $= mn - \delta - 1$; for, this being so, the Postulandum of the curve of the order r through the $(mn - \delta)$ points A will be

$$= \tfrac{1}{2}r(r+3) - mn + \delta + 1,$$

and the equation of the curve will no longer be of the foregoing form, but it will be of the form

$$\Omega_r + L_{r-m}P_m + M_{r-n}Q_n = 0,$$

$\Omega_r = 0$ being a particular curve through the $mn - \delta$ points A, which does not pass through any of the points B. The proof depends on the theory of Residuation: which for the present purpose may be presented under the following form.

Let $A, B, \ldots$ denote systems of points upon a given Basis-curve, for instance the foregoing curve $P_m = 0$, of the order m. And let $A = ci$ denote that the points A are the complete intersection of the basis-curve by some other curve; (this implies that the number of the points is $= km$, a multiple of m, and the intersecting curve is then of course a curve of the order k). It is clear that, if $A = ci$, and $B = ci$, then also $A + B = ci$. But conversely we have the theorem that, if $A + B = ci$ and $A = ci$, then also $B = ci$. And we at once deduce the further theorem: if $A + B = ci$, $B + C = ci$, $C + D = ci$, then also $A + D = ci$. For the first and third relations give $A + B + C + D = ci$, and the second relation then gives $A + D = ci$.

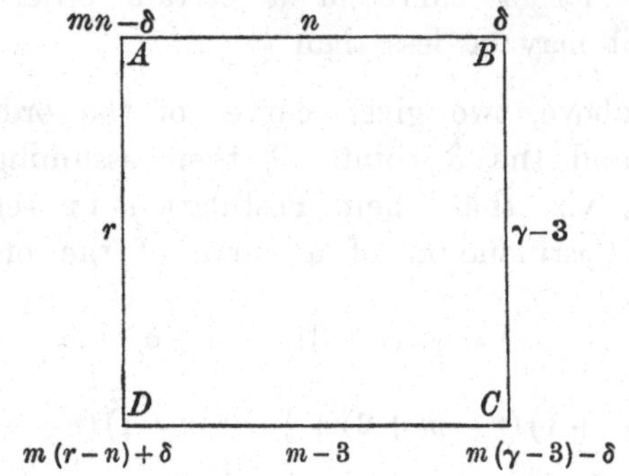

Starting now (see the diagram) with the δ points B on a curve of the order $\gamma - 3$, suppose that we have through these points the basis-curve $P_m = 0$ of the order m, and another given curve $Q_n = 0$, of the order n; and let these besides meet in the $mn - \delta$ points A. Let the curve of the order $\gamma - 3$ besides meet the basis-curve in the $m(\gamma - 3) - \delta$ points C; and through these let there be drawn a curve of the order $m - 3$, which besides meets the basis-curve in the $m(r - n) + \delta$ points D. We have here $A + B = ci$, $B + C = ci$, $C + D = ci$; consequently $A + D = ci$, that is, the $mn - \delta$ points A and the $m(r - n) + \delta$ points D lie on a curve of the order r. The curve of the order $m - 3$ passes through the $m(\gamma - 3) - \delta$ points C; its Postulandum is thus

$$= \tfrac{1}{2}m(m - 3) - m(\gamma - 3) + \delta,$$

which is

$$=\tfrac{1}{2}(r-n+1)(r-n+2).$$

In fact, substituting for γ, δ their values $=m+n-r$, and $\tfrac{1}{2}(m+n-r-1)(m+n-r-2)$ respectively, this equation is satisfied identically. The system of the $m(r-n)+\delta$ points D thus depends upon $\tfrac{1}{2}(r-n+1)(r-n+2)$ constants, or say the Postulandum of the points D is $=\tfrac{1}{2}(r-n+1)(r-n+2)$. It follows that the curve of the order r through the $(mn-\delta)$ points A and the $m(r-n)+\delta$ points D *cannot* have an equation of the form

$$L_{r-m}P_m+M_{r-n}Q_n=0;$$

for the intersections of this curve with the basis-curve $P_m=0$ are given by the equation $M_{r-n}Q_n=0$, which contains only the

$$\tfrac{1}{2}(r-n+1)(r-n+2)-1$$

constants of M_{r-n} (one constant of course divides out, giving the diminution -1). The equation must have the more general form

$$\Omega_r+L_{r-m}P_m+M_{r-n}Q_n=0;$$

and it thus appears that the Postulation of the $mn-\delta$ points A, instead of being $=mn-\delta$, must be $=mn-\delta-1$. This completes the proof.

I notice that, combining the last-mentioned identity

$$\tfrac{1}{2}m(m-3)-m(\gamma-3)+\delta=\tfrac{1}{2}(r-n+1)(r-n+2)$$

with the like identity

$$\tfrac{1}{2}n\,(n-3)-n(\gamma-3)+\delta=\tfrac{1}{2}(r-m+1)(r-m+2),$$

we obtain

$$\tfrac{1}{2}m(m-3)+\tfrac{1}{2}n(n-3)-(m+n)(\gamma-3)+2\delta$$
$$=\tfrac{1}{2}(r-n+1)(r-n+2)+\tfrac{1}{2}(r-m+1)(r-m+2),$$

and consequently, referring to a former result, the left-hand side should be

$$=\tfrac{1}{2}r(r+3)-mn+\delta+1;$$

substituting for γ, δ their values, this is at once verified.

As appears by what precedes, Bacharach's special case is that in which the δ, $=\tfrac{1}{2}(\gamma-1)(\gamma-2)$ points B satisfy the single condition of lying on a curve of the order $\gamma-3$. We may have between the points B more than a single relation; in particular, the points B may be such as to include among themselves the complete intersection of two curves of the orders a, b respectively ($ab \gtreqless \delta$): this will be the case, if the given curves are of the form

$$P_m=\lambda_a S_{m-a}-\mu_b R_{m-b},$$
$$Q_n=\lambda_a V_{n-a}-\mu_b U_{n-b},$$

it being understood, here and in what follows, that the values of a, b are such that the suffixes are none of them negative.

The two curves here intersect in the ab points ($\lambda_a = 0$, $\mu_b = 0$), and in the $mn - ab$ points

$$\left\| \begin{matrix} \lambda_a, & R_{m-b}, & U_{n-b} \\ \mu_b, & S_{m-a}, & V_{n-a} \end{matrix} \right\| = 0,$$

say the $(mn - \delta)$ points A are $mn - ab - \Theta$ of the last-mentioned points, and the δ points B are the remaining Θ points together with the ab points. Here the general form of the curve of the order r passing through the $mn - ab$ points, and therefore through the $mn - \delta$ points A, is

$$\left| \begin{matrix} L_{r-m-n+a+b}, & M_{r-n}, & N_{r-m} \\ \lambda_a \quad\quad , & R_{m-b}, & U_{n-b} \\ \mu_b \quad\quad , & S_{m-a}, & V_{n-a} \end{matrix} \right| = 0,$$

where $L_{r-m-n+a+b}$, M_{r-n}, N_{r-m} are arbitrary functions of the orders indicated by the respective suffixes. The theory in regard to the number of constants is of course altogether different from that which belongs to the case of the general functions P_m, Q_n; and it is probable that much interesting theory would present itself in the consideration of particular cases.

Cambridge, 22 March 1887.

869.

ON THE TRANSFORMATION OF ELLIPTIC FUNCTIONS.

[From the *American Journal of Mathematics*, vol. IX. (1887), pp. 193—224.]

THE algebraical theory of the Transformation of Elliptic Functions was established by Jacobi in a remarkably simple and elegant form, but it has not hitherto been developed with much completeness or success. The cases $n=3$ and $n=5$ are worked out very completely in the *Fundamenta Nova* (1829); viz. considering the equation

$$\frac{Mdy}{\sqrt{1-y^2 \,.\, 1-\lambda^2y^2}}=\frac{dx}{\sqrt{1-x^2 \,.\, 1-k^2x^2}},$$

($k=u^4$, $\lambda=v^4$; say this is the $Mk\lambda$- or Muv-form), Jacobi finds, in the two cases respectively, a modular equation between the fourth roots u, v, say the uv-modular equation, and, as rational functions of u, v, the value of M and the values of the coefficients of the several powers of x in the numerator and denominator of the fraction which gives the value of y; but there is no attempt at a like development of the general case. I shall have occasion to speak of other researches by Jacobi, Brioschi and myself; but I will first mention that my original idea in the present memoir was to develop the following mode of treatment of the theory:

In place of the $Mk\lambda$-form, using the $\rho\alpha\beta$-form

$$\frac{dy}{\sqrt{1-2\beta y^2+y^4}}=\frac{\rho dx}{\sqrt{1-2\alpha x^2+x^4}}$$

(I write for greater convenience 2α, 2β in place of the α of Jacobi and Brioschi and the β of Brioschi), we can, by expanding each side in a series, integrating, and reverting the resulting series for y, obtain y in the form

$$y=\rho x\,(1+\Pi_1x^2+\Pi_2x^4+\ldots),$$

where $\Pi_1, \Pi_2, \Pi_3, \ldots$ denote given functions of ρ, α, β. Taking n odd and $=2s+1$ we assume for y an expression

$$y = \frac{x(A_s + A_{s-1}x^2 + \ldots + A_1 x^{2s-2} + x^{2s})}{1 + A_1 x^2 + \ldots + A_{s-1}x^{2s-2} + A_s x^{2s}},$$

where the last coefficient A_s is at once seen to be $=\rho$. Comparing with the series-value $y = \rho x(1 + \Pi_1 x^2 + \Pi_2 x^4 + \ldots)$, we have an infinite series of equations. The first of these is, in fact, $A_s = \rho$; the next $(s-1)$ equations give linearly $A_1, A_2, \ldots, A_{s-1}$ in terms of the coefficients Π; that is, of ρ, α, β: the two which follow serve in effect to determine ρ, β as functions of α: and then, ρ and β having these values, all the remaining equations will be satisfied identically.

The process is an eminently practical one, so far as regards the determination of the coefficients $A_1, A_2, \ldots, A_{s-1}$ as functions of ρ, α, β; it is less so, and requires eliminations more or less complicated, as regards the determination of the relations between ρ, α, β. As to this, it may be remarked that the problem is not so much the determination of the equation between ρ and α (or say the $\rho\alpha$-multiplier equation, or simply the $\rho\alpha$-equation), and of the equation between β, α (or say the $\alpha\beta$-modular equation, or simply the $\alpha\beta$-equation), as it is to determine the complete system of relations between ρ, α, β; treating these as coordinates, we have what may be called the multiplier-modular-curve, or say the MM-curve, and the relations in question are those which determine this curve.

In the absence of special exceptions, it follows from general principles that the coefficients $A_1, A_2, \ldots, A_{s-1}$, *quà* rational functions of ρ, α, β, must also be rational functions of α, β or of α, ρ; and I think it may be assumed that this is the case; the method, however, affords but little assistance towards thus expressing them.

In connexion with the foregoing theory, I consider the solutions of the problem of transformation given by Jacobi's partial differential equation ("Suite de Notices sur les Fonctions elliptiques," *Crelle*, t. IV. (1829), pp. 185—193), and by what I call the Jacobi-Brioschi differential equations. The first and third of these were obtained by Jacobi in the memoir*, "De functionibus ellipticis Commentatio," *Crelle*, t. IV. (1829), pp. 371—390 (see p. 376); but the second equation, which completes the system, was, I believe, first given by Brioschi in the second appendix to his translation of my *Elliptic Functions: Trattato elementare delle Funzioni ellittiche:* Milan, 1880. I had, strangely enough, overlooked the great importance of these equations. I shall have occasion also to refer to results, and further develop the theory contained in my memoir, "On the Transformation of Elliptic Functions," *Phil. Trans.*, t. CLXIV. (1874), pp. 397—456, [578], and the addition thereto, *Phil. Trans.*, t. CLXXXIX. (1878), pp. 419—424, [692].

I remark that, while I have only worked out the formulæ for the cases $n=3$ and $n=5$, and a few formulæ for the case $n=7$, the memoir is intended to be a contribution to the general theory of the $\rho\alpha\beta$-transformation; I hope to be able to complete the theory for the case $n=7$.

[* *Ges. Werke*, bd. I., pp. 295—318; in particular, p. 303.]

Comparison of the $Mk\lambda$- and $\rho\alpha\beta$-Forms. The Modular and Multiplier Equations.

Art. Nos. 1 to 12.

1. The equation

$$\frac{M\mathrm{dy}}{\sqrt{1-\mathrm{y}^2 \,.\, 1-\lambda^2\mathrm{y}^2}} = \frac{\mathrm{dx}}{\sqrt{1-\mathrm{x}^2 \,.\, 1-k^2\mathrm{x}^2}},$$

if we write therein

$$\mathrm{x} = \frac{x}{\sqrt{k}}, \;= \frac{x}{u^2}; \quad \mathrm{y} = \frac{y}{\sqrt{\lambda}}, \;= \frac{y}{v^2},$$

becomes

$$\frac{Mdy}{v^2\sqrt{1-(v^4+v^{-4})\,y^2+y^4}} = \frac{dx}{u^2\sqrt{1-(u^4+u^{-4})\,x^2+x^4}};$$

viz. this is

$$\frac{dy}{\sqrt{1-2\beta y^2+y^4}} = \frac{\rho dx}{\sqrt{1-2\alpha x^2+x^4}},$$

if only

$$2\alpha = u^4 + \frac{1}{u^4}, \quad 2\beta = v^4 + \frac{1}{v^4}, \quad \rho = \frac{v^2}{u^2 M}.$$

2. We have a uv-modular equation, and, as shown in my Transformation Memoir*, p. 450, this may be converted into a u^4v^4-modular equation; in particular, $n=3$, the equation is

$$y^4 + 6x^2y^2 + x^4 - 4xy\,(4x^2y^2 - 3x^2 - 3y^2 + 4) = 0,$$

where x, y denote u^4, v^4 respectively; say the equation is

$$F(x,\, y), \;= x^4 + x^3(-16y^3+12y) + x^2(6y^2) + x(12y^3-16y) + y^4, \;= 0.$$

From the equation $F(x, y) = 0$, we derive

$$x^{-2}F(x,\, y)\,.\,x^{-2}F(x,\, y^{-1}) = 0;$$

say this is

$$(Ax^2 + Bx + C + Dx^{-1} + Ex^{-2})\,(A'x^2 + B'x + C' + D'x^{-1} + E'x^{-2}) = 0,$$

viz. the equation is

$$AA'x^4 + (AB' + A'B)\,x^3 + \ldots + EE'x^{-4} = 0,$$

where, by reason of the symmetry of $F(x, y)$, the coefficients AA', EE' of x^4, x^{-4}, those of x^3, x^{-3}, &c., have equal values; the form thus is

$$\mathfrak{A}\,(x^4+x^{-4}) + \mathfrak{B}\,(x^3+x^{-3}) + \mathfrak{C}\,(x^2+x^{-2}) + \mathfrak{D}\,(x+x^{-1}) + \mathfrak{E} = 0,$$

where x^4+x^{-4}, x^3+x^{-3}, x^2+x^{-2}, are given functions of $x+x^{-1}$, $=2\alpha$; viz. we have

$$\begin{aligned} x + x^{-1} &= 2\alpha, \\ x^2 + x^{-2} &= 4\alpha^2 - 2, \\ x^3 + x^{-3} &= 8\alpha^3 - 6\alpha, \\ x^4 + x^{-4} &= 16\alpha^4 - 16\alpha^2 + 2. \end{aligned}$$

[* This Collection, vol. IX., p. 170.]

The coefficients $\mathfrak{A}, \mathfrak{B}, \ldots$ are in like manner expressible as functions of $y+y^{-1}, =2\beta$; thus we have $\mathfrak{A}=1$,

$$\begin{aligned}\mathfrak{B} &= AB' + A'B \\ &= -16(y^3+y^{-3}) + 12(y+y^{-1}), \ = -16(8\beta^3 - 6\beta) + 12 \, . \, 2\beta;\end{aligned}$$

or, finally, $\mathfrak{B} = -128\beta^3 + 120\beta$; and so for the other coefficients. The numerical coefficients contain, all of them, the factor 16; and, throwing this out, we obtain, for $n=3$, the $\alpha\beta$-modular equation in the form

	α^4	α^3	α^2	α	1	
β^4					+ 1	
β^3		− 64		+ 60		
β^2			− 186		+ 192	= 0,
β		+ 60		− 64		
1	+ 1		+ 192		− 192	
	+ 1	− 4	+ 6	− 4	+ 1	

where observe that the form is symmetrical as regards α, β; and, further, that the sums of the numerical coefficients in the lines or columns are the binomial coefficients $1, -4, +6, -4, +1$. Observe, further, that the sums in the direction of the sinister diagonal are $-64, -64, +320, -192$; viz. dividing by -64, it thus appears that, writing $\beta=\alpha$, the equation becomes

$$\alpha^6 + \alpha^4 - 5\alpha^2 + 3 = 0;$$

that is, $(\alpha^2-1)^2(\alpha^2+3)=0$.

Again, writing $\beta=-\alpha$, then dividing by 16, the equation becomes

$$4\alpha^6 - 19\alpha^4 + 28\alpha^2 - 12 = 0;$$

that is,

$$(4\alpha^2-3)(\alpha^2-2)^2 = 0.$$

3. So also, for $n=5$, we have the u^4v^4-modular equation in the form

$$\left.\begin{aligned} & x^6 + 655x^4y^2 + 655x^2y^4 + y^6 - 640x^2y^2 - 640x^4y^4 \\ & \quad + xy\,(-256 + 320x^2 + 320y^2 - 70x^4 - 660x^2y^2 - 70y^4 \\ & \qquad + 320x^4y^2 + 320x^2y^4 - 256x^4y^4) \end{aligned}\right\} = 0;$$

and in precisely the same manner, we obtain the $\alpha\beta$-modular equation; viz. casting out a factor 64, this is

	β^6	β^5	β^4	β^3	β^2	β	1
α^6							+ 1
α^5		− 4096		+ 6400		− 2310	
α^4			+ 69120		− 172785		+ 103680
α^3		+ 6400		− 133140		+ 126720	
α^2			− 172785		+ 276480		− 103680
α		− 2310		+ 126720		− 124416	
1	+ 1		+ 103680		− 103680		
	+ 1	− 6	+ 15	− 20	+ 15	− 6	+ 1

= 0,

where the form is symmetrical as regards α, β; the sums of the numerical coefficients in the lines or columns are 1, − 6, + 15, − 20, + 15, − 6, + 1. The sums in the direction of the sinister diagonal all divide by − 4096; viz. throwing out this factor, we have, for $\beta = \alpha$, the equation

$$\alpha^{10} - 20\alpha^8 + 118\alpha^6 - 180\alpha^4 + 81\alpha^2 = 0;$$

that is,

$$\alpha^2 (\alpha^2 - 1)^2 (\alpha^2 - 9)^2 = 0.$$

If $\beta = -\alpha$, the coefficients divide by 64; and throwing out this factor, the equation is

$$64\alpha^{10} + 880\alpha^8 - 3247\alpha^6 + 3600\alpha^4 - 1296\alpha^2 = 0;$$

that is,

$$\alpha^2 (\alpha^4 + 16\alpha^2 - 16)(8\alpha^2 - 9)^2 = 0.$$

4. We have a Mu-multiplier equation of the form $F\left(\frac{1}{M},\ 2u^8 - 1\right) = 0$ (see Memoir*, pp. 420—422), but we cannot, by the preceding formulæ, deduce thence a $\rho\alpha$-multiplier equation; in fact, writing therein $\frac{1}{M} = \frac{u^2\rho}{v^2}$, the resulting equation is $F\left(\frac{u^2\rho}{v^2},\ 1 - 2u^8\right) = 0$, which is a $\rho\alpha$-multiplier equation only on the assumption that $1 - 2u^8$, u^2 and v^2 are therein regarded as given functions of α. But it is very

[* This Collection, vol. x., pp. 334, 335.]

remarkable that the $\rho\alpha$-equation, in fact, is $F(\rho, \alpha)=0$. To prove this, assume that the equation

$$\frac{dy}{\sqrt{1-2\beta y^2+y^4}}=\frac{\rho dx}{\sqrt{1-2\alpha x^2+x^4}}$$

has a $\rho\alpha$-multiplier equation $F(\rho, \alpha)=0$. Starting from the equation

$$\frac{M d\mathrm{y}}{\sqrt{1-\mathrm{y}^2 \,.\, 1-\lambda^2\mathrm{y}^2}}=\frac{d\mathrm{x}}{\sqrt{1-\mathrm{x}^2 \,.\, 1-k^2\mathrm{x}^2}},$$

we may, by effecting on each side a quadric transformation, convert this into

$$\frac{dy}{\sqrt{1-2\,(2v^8-1)\,y^2+y^4}}=\frac{M^{-1}dx}{\sqrt{1-2\,(2u^8-1)\,x^2+x^4}};$$

and this being so, we have, between M^{-1} and $2u^8-1$, the relation

$$F\left(\frac{1}{M},\ 2u^8-1\right)=0;$$

or, conversely, if this be the form of the Mu-multiplier equation, then the $\rho\alpha$-multiplier equation is $F(\rho, \alpha)=0$.

5. The quadric transformations are

$$x=\frac{\sqrt{1-\mathrm{x}^2}}{\mathrm{x}\sqrt{1-k^2\mathrm{x}^2}},\quad y=\frac{\sqrt{1-\mathrm{y}^2}}{\mathrm{y}\sqrt{1-\lambda^2\mathrm{y}^2}}.$$

We have then only to show that

$$\frac{dx}{\sqrt{1-2\,(2u^8-1)\,x^2+x^4}}=\frac{d\mathrm{x}}{\sqrt{1-\mathrm{x}^2 \,.\, 1-k^2\mathrm{x}^2}};$$

for then, in like manner,

$$\frac{dy}{\sqrt{1-2\,(2v^8-1)\,y^2+y^4}}=\frac{d\mathrm{y}}{\sqrt{1-\mathrm{y}^2 \,.\, 1-\lambda^2\mathrm{y}^2}},$$

and we pass from the assumed differential relation between x, y to the above-mentioned differential equation between x, y.

6. For the quadric transformation between x, x, write

$$\theta^{\frac{1}{2}}=k-ik',\quad \theta^{-\frac{1}{2}}=k+ik',$$

$\left(\text{whence also } \theta=\dfrac{k-ik'}{k+ik'}\right)$, and therefore

$$\theta^{\frac{1}{2}}+\theta^{-\frac{1}{2}}=2k,\quad \theta+\theta^{-1}=2k^2-2k'^2=2\,(2k^2-1),\quad =2\,(2u^8-1);$$

we have

$$1-\theta x^2=1-\theta\frac{1-\mathrm{x}^2}{\mathrm{x}^2\,(1-k^2\mathrm{x}^2)},\quad =\frac{1}{\mathrm{x}^2\,(1-k^2\mathrm{x}^2)}\{-\theta+(\theta+1)\,\mathrm{x}^2-k^2\mathrm{x}^4\},$$

$$=\frac{-\theta}{\mathrm{x}^2\,(1-k^2\mathrm{x}^2)}\,(1-\theta^{-\frac{1}{2}}k\mathrm{x}^2)^2;$$

and similarly,

$$1-\theta^{-1}x^2=\frac{-\theta^{-1}}{\mathrm{x}^2(1-k^2\mathrm{x}^2)}(1-\theta^{\frac{1}{2}}k\mathrm{x}^2)^2.$$

Consequently,

$$(1-\theta x^2)(1-\theta^{-1}x^2)=1-2(2u^8-1)x^2+x^4=\frac{1}{\mathrm{x}^4(1-k^2\mathrm{x}^2)^2}(1-2k^2\mathrm{x}^2+k^2\mathrm{x}^4)^2;$$

or say

$$\sqrt{1-2(2u^8-1)x^2+x^4}=\frac{1}{\mathrm{x}^2(1-k^2\mathrm{x}^2)}(1-2k^2\mathrm{x}^2+k^2\mathrm{x}^4).$$

Moreover,

$$dx=\frac{d\mathrm{x}}{\mathrm{x}^2(1-\mathrm{x}^2)^{\frac{1}{2}}(1-k^2\mathrm{x}^2)^{\frac{3}{2}}}(1-2k^2\mathrm{x}^2+k^2\mathrm{x}^4),$$

and thence the required equation

$$\frac{dx}{\sqrt{1-2(1-2u^8)x^2+x^4}}=\frac{d\mathrm{x}}{\sqrt{1-\mathrm{x}^2\,.\,1-k^2\mathrm{x}^2}};$$

this completes the proof.

7. Thus, referring to the *Mu*-equations given in the place referred to, we obtain the following $\rho\alpha$-multiplier equations. When $n=3$, we have

$$\rho^4-6\rho^2-8\alpha\rho-3=0.$$

This may be written in the forms

$$\begin{aligned}8\alpha\rho &= \rho^4-6\rho^2-3,\\ 8(\alpha+1)\rho &= (\rho-1)^3(\rho+3),\\ 8(\alpha-1)\rho &= (\rho+1)^3(\rho-3).\end{aligned}$$

Next, for $n=5$, we have $\rho^6-10\rho^5+35\rho^4-60\rho^3+55\rho^2+(38-64\alpha^2)\rho+5=0$.

This may be written in the two forms

$$64\alpha^2\rho=(\rho^2-4\rho-1)^2(\rho^2-2\rho+5)$$

and

$$64(\alpha^2-1)\rho=(\rho-1)^5(\rho-5).$$

And, for $n=7$, we have

$$\rho^8-28\rho^6-112\alpha\rho^5-210\rho^4-224\alpha\rho^3+(-1484+1344\alpha^2)\rho^2+(-560\alpha+512\alpha^3)\rho+7=0.$$

8. The relation between ρ and β, or say the $\rho\beta$-multiplier equation, may be obtained by a known property of elliptic functions; viz. writing $\rho\sigma=\pm n$ (the sign is $-$ for $n=3$, $n=7$, or generally for any prime value $4p+3$: and it is $+$ for $n=5$ and generally for any prime value $=4p+1$), then we have between σ, β the same relation as between ρ, α. Thus, if $n=3$, $\sigma=-\dfrac{3}{\rho}$, for ρ, α writing σ, β, the equation is $\sigma^4-6\sigma^2-8\beta\sigma-3=0$; or, as this may be written,

$$\rho^4+8\beta\rho^3+18\rho^2-27=0;$$

and so for the other cases; but it is perhaps more convenient to retain the σ; thus, if $n=5$, $\sigma=\dfrac{5}{\rho}$, we have

$$\sigma^6-10\sigma^5+35\sigma^4-60\sigma^3+55\sigma^2+(38-64\beta^2)\sigma+5=0.$$

9. We are hence able to express β as a rational function of ρ, α. We, in fact, have

$$8\alpha = \frac{1}{\sqrt{\rho}}(\rho^2 - 4\rho - 1)\sqrt{\rho^2 - 2\rho + 5}, \quad 8\beta = -\frac{1}{\sqrt{\sigma}}(\sigma^2 - 4\sigma - 1)\sqrt{\sigma^2 - 2\sigma + 5},$$

(the signs must be opposite), and then for σ, substituting its value $= \frac{1}{5}\rho$, and observing that $\sigma^2 - 2\sigma + 5$ is thus $= \frac{5}{\rho^2}(\rho^2 - 2\rho + 5)$, we find

$$\frac{\beta}{\alpha} = \frac{\rho^2 + 20\rho - 25}{\rho^2(\rho^2 - 4\rho - 1)},$$

which is the required formula.

Observe that, for $\rho = \sigma = \sqrt{5}$, the formulæ with the sign $-$, as above, give $\beta = -\alpha$, whereas with the sign $+$ they would have given $\beta = \alpha$. For the value in question, $\rho = \sqrt{5}$, the equation

$$64\alpha^2 = \frac{1}{\rho}(\rho^2 - 4\rho - 1)^2 (\rho^2 - 2\rho + 5),$$

gives

$$64\alpha^2 = \frac{1}{\sqrt{5}} 16 (1 - \sqrt{5})^2 (10 - 2\sqrt{5});$$

that is,

$$\alpha^2 = \frac{1}{\sqrt{5}}(3 - \sqrt{5})(5 - \sqrt{5}), \quad = (3 - \sqrt{5})(\sqrt{5} - 1);$$

that is, $\alpha^2 = -8 + 4\sqrt{5}$, or $\alpha^4 + 16\alpha^2 - 16 = 0$; it appears, *ante* No. 3, that this value belongs to the case $\beta = -\alpha$ and not to $\beta = \alpha$.

10. But there is another way of arriving at a formula containing β. Starting from Jacobi's equation

$$nM^2 = \frac{\lambda\lambda'^2}{kk'^2} \cdot \frac{dk}{d\lambda},$$

and introducing for λ, λ', k, k', M their values in terms of u, v, we have

$$\frac{nv^4}{u^4\rho^2} = \frac{v^4(1 - v^8)\, u^3 du}{u^4(1 - u^8)\, v^3 dv};$$

that is,

$$\frac{dv}{du} = \frac{\rho^2}{n} \frac{u^3(1 - v^8)}{v^3(1 - u^8)};$$

but, from the values of α, β, we find

$$\frac{dv}{du} = \frac{v^5(1 - u^8)}{u^5(1 - v^8)} \frac{d\beta}{d\alpha},$$

and, combining these results,

$$\frac{d\beta}{d\alpha} = \frac{\rho^2}{n} \frac{u^8}{v^8} \cdot \frac{(1 - v^8)^2}{(1 - u^8)^2}, \quad = \frac{\rho^2}{n} \frac{(v^4 - v^{-4})^2}{(u^4 - u^{-4})^2};$$

that is,

$$\frac{d\beta}{d\alpha} = \frac{\rho^2}{n}\frac{\beta^2-1}{\alpha^2-1}.$$

We have, consequently,

$$\frac{d\beta}{\beta^2-1} = \frac{\rho^2 d\alpha}{n(\alpha^2-1)},$$

and therefore

$$\tfrac{1}{2}\log\frac{\beta-1}{\beta+1} = \frac{1}{n}\int\frac{\rho^2 d\alpha}{\alpha^2-1},$$

where ρ^2 must be regarded as a function of α, or α of ρ; and from the form of the equation, it appears that the integral must be expressible as the logarithm of an algebraic function of ρ, α.

11. Thus, when $n=3$, we have

$$8\alpha = \rho^3 - 6\rho - \frac{3}{\rho};$$

whence

$$8\frac{d\alpha}{d\rho} = 3\rho^2 - 6 + \frac{3}{\rho^2}, \quad = \frac{3}{\rho^2}(\rho^2-1)^2,$$

and thence easily

$$\tfrac{1}{2}\log\frac{\beta-1}{\beta+1} = \int\frac{8\rho^2 d\rho}{\rho^2-1\,.\,\rho^2-9}, \quad = -\int\frac{d\rho}{\rho^2-1} + 9\int\frac{d\rho}{\rho^2-9},$$

$$= -\tfrac{1}{2}\log\frac{\rho-1}{\rho+1} + \tfrac{3}{2}\log\frac{\rho-3}{\rho+3};$$

that is,

$$\frac{\beta-1}{\beta+1} = \frac{(\rho-3)^3(\rho+1)}{(\rho+3)^3(\rho-1)},$$

as may be at once verified.

12. In the case $n=5$, I verify the equation under the form

$$\frac{d\beta}{\beta^2-1} = \frac{\rho^2}{5}\cdot\frac{d\alpha}{\alpha^2-1}.$$

From the equations

$$64(\alpha^2-1) = \frac{1}{\rho}(\rho-1)^5(\rho-5), \text{ and } 8\alpha = \frac{1}{\sqrt{\rho}}(\rho^2-4\rho-1)\sqrt{\rho^2+2\rho-5},$$

we have

$$\frac{128\alpha\, d\alpha}{\alpha^2-1} = \frac{5(\rho^2-4\rho-1)\,d\rho}{\rho(\rho-1)(\rho-5)},$$

and thence

$$\frac{16 d\alpha}{\alpha^2-1} = \frac{5 d\rho}{(\rho-1)(\rho-5)\sqrt{\rho(\rho^2-2\rho+5)}}.$$

Similarly, observing the $-$ sign of 8β, we have

$$\frac{16 d\beta}{\beta^2-1} = \frac{-5 d\sigma}{(\sigma-1)(\sigma-5)\sqrt{\sigma(\sigma^2-2\sigma+5)}},$$

whence, substituting for σ its value $=\dfrac{5}{\rho}$, we have

$$\frac{16d\beta}{\beta^2-1}=\frac{\rho^2\,d\rho}{(\rho-1)(\rho-5)\sqrt{\rho(\rho^2-2\rho+5)}},\quad =\frac{\rho^2}{5}\cdot\frac{16d\alpha}{\alpha^2-1},$$

which is right.

Connexion of the Mkλ- and ραβ-Theories. Order of Modular Equation.
Art. Nos. 13 to 18.

13. In the Transformation Memoir [578], starting from the equation

$$\frac{1-\mathrm{y}}{1+\mathrm{y}}=\frac{1-\mathrm{x}}{1+\mathrm{x}}\left(\frac{P-Q\mathrm{x}}{P+Q\mathrm{x}}\right)^2,$$

I sought to determine the coefficients of P, Q by the consideration that the relation between x, y remains unaltered when x, y are changed into $\dfrac{1}{k\mathrm{x}}$, $\dfrac{1}{\lambda\mathrm{y}}$ respectively. This comes to saying that, when for x, y we write $\dfrac{x}{u^2}$, $\dfrac{y}{v^2}$ respectively, the relation between x and y presents itself in the form

$$y=\frac{x(A_s+A_{s-1}x^2+\ldots+A_0x^{2s})}{A_0+A_1x^2+\ldots+A_sx^{2s}},$$

where $s=\frac{1}{2}(n-1)$, as before. For instance, when $n=7$, $P=\alpha+\gamma\mathrm{x}^2$, $Q=\beta+\delta\mathrm{x}^2$.

If, solving for y, we then for x, y write $\dfrac{x}{u^2}$, $\dfrac{y}{v^2}$, we find

$$y=\frac{v^2u^{-2}x\{(\alpha^2+2\alpha\beta)+(2\alpha\gamma+\beta^2+2\alpha\delta+2\beta\gamma)x^2u^{-4}+(\gamma^2+2\beta\delta+2\gamma\delta)x^4u^{-8}+\delta^2x^6u^{-12}\}}{\alpha^2+(2\alpha\gamma+\beta^2+2\alpha\beta)x^2u^{-4}+(\gamma^2+2\beta\delta+2\alpha\delta+2\beta\gamma)x^4u^{-8}+(\delta^2+2\gamma\delta)x^6u^{-12}};$$

and comparing this with

$$y=\frac{x(A_3+A_2x^2+A_1x^4+A_0x^6)}{A_0+A_1x^2+A_2x^4+A_3x^6},$$

we have for each of the coefficients A two different expressions. Equating these and making a slight change of form, we obtain the relations between u, v, α, β, γ, δ used in the Memoir: thus,

$$A_0=\alpha^2=v^2u^{-14}\delta^2,\ A_1=v^2u^{-10}(\gamma^2+2\beta\delta+2\gamma\delta)=u^{-4}(2\alpha\gamma+\beta^2+2\alpha\beta),\ \&c.;$$

in the Memoir, $k\,(=u^4)$ is used instead of u, and $\Omega\,(=v^2u^{-2})$ instead of v, and the equations thus are

$$k^3\alpha^2=\Omega\delta^2,$$
$$k(2\alpha\gamma+2\alpha\beta+\beta^2)=\Omega(\gamma^2+2\gamma\delta+2\beta\delta),$$
$$\gamma^2+2\beta\gamma+2\alpha\delta+2\beta\delta=\Omega k(2\alpha\gamma+2\beta\gamma+2\alpha\delta+\beta^2),$$
$$\delta^2+2\gamma\delta=\Omega k^3(\alpha^2+2\alpha\beta);$$

viz. these are the equations [*Coll. Math. Papers*, vol. IX., p. 119]. The idea in the present Memoir is that of considering the coefficients A in the stead of α, β, ...

14. We have here, and in general for any odd value of n, equations of the form

$$(\Omega =)\frac{U}{U'}=\frac{V}{V'}=\ldots,$$

where $U, V, \ldots, U', V', \ldots$ are quadric functions of the coefficients $\alpha, \beta, \gamma, \ldots$; and these equations serve to establish between Ω and k a relation called the Ωk-modular equation, and which in regard to Ω is of the same degree as the uv-modular equation is in regard to v. Leaving out the equation $(\Omega =)$, we have

$$\begin{vmatrix} U, & V, & W, \ldots \\ U', & V', & W', \ldots \end{vmatrix}=0;$$

and to each system of values of $\alpha, \beta, \gamma, \delta, \ldots$ (or say of their ratios) given by these equations, there corresponds a single value of Ω; the number of values of Ω, or the degree in Ω, of the Ωk-equation is thus found as $=(n+1)\,2^{\frac{1}{2}(n-3)}$. This is far too high; for $n=3, 5, 7, \ldots$, the degrees are 4, 12, 32, ...; those of the proper Ωk-equations are 4, 6, 8, ...

15. I showed, or endeavoured to show, that in the case $n=5$, the extraneous factor was $(\Omega-1)^6$, $(\Omega-1=0$, the Ωk-modular equation belonging to $n=1$, for which the transformation is the trivial one $y=x)$, and that in the case $n=7$, the extraneous factor was $\{(\Omega,\ 1)^4\}^6$, $((\Omega,\ 1)^4=0$, the Ωk-modular equation for the case $n=3)$; generally the extraneous factors seem to depend on the Ωk-functions for the values $n-4$, $n-8$, &c. The ground for this is that, in the assumed formula for any given value n, we may take P, Q to contain a common factor $1\pm k\mathrm{x}^2$ $\Big($observe that, to a factor près, this is unaltered by the change x into $\frac{1}{k\mathrm{x}}$, viz. it becomes $\frac{1}{k\mathrm{x}^2}(1\pm k\mathrm{x}^2)$, a condition which is necessary$\Big)$, and we thereby reduce the equation to

$$\frac{1-\mathrm{y}}{1+\mathrm{y}}=\frac{1-\mathrm{x}}{1+\mathrm{x}}\left(\frac{P'-Q'\mathrm{x}}{P'+Q'\mathrm{x}}\right)^2,$$

in which equation the degrees of the numerator and the denominator are each diminished by 4, and the equation thus belongs to the value $n-4$.

16. I remark here that, in the case of n an odd prime, the degree of the modular equation is $=n+1$; but for any other odd value, the degree is

$$\sigma'(n),\ =n\left(1+\frac{1}{a}\right)\left(1+\frac{1}{b}\right)\ldots,$$

where $a, b, \ldots$ are all the unequal prime factors of n; thus, if $n=a^\alpha$, the degree is

$$a^\alpha\left(1+\frac{1}{a}\right),\ =a^{\alpha-1}(a+1).$$

In the case of a number $n=abc\ldots$, without any squared factor, the degree is

$$abc\ldots\left(1+\frac{1}{a}\right)\left(1+\frac{1}{b}\right)\left(1+\frac{1}{c}\right)\ldots,\ =(a+1)(b+1)(c+1)\ldots,$$

the sum of the factors of n. We have

$$\sigma'(n) = \text{coeff. } x^n \text{ in } \Sigma\phi(x^N),$$

where

$$\phi x = x + 3x^3 + 5x^5 + \ldots, = \frac{x(1+x^2)}{(1-x^2)^2},$$

and the summation extends to all odd values of N having no squared factor; thus,

$$\begin{array}{llllllllll}
\phi(x) & = x & + 3x^3 & + 5x^5 & + 7x^7 & + 9x^9 & + 11x^{11} & + 13x^{13} & + 15x^{15} \ldots \\
\phi(x^3) = & & 1x^3 & & & + 3x^9 & & & + 5x^{15}, \\
\phi(x^5) = & & & 1x^5 & & & & & + 3x^{15}, \\
\phi(x^7) = & & & & 1x^7 & & & & \\
\phi(x^{11}) = & & & & & & 1x^{11} & & \\
\phi(x^{13}) = & & & & & & & 1x^{13} & \\
\phi(x^{15}) = & & & & & & & & 1x^{15}, \\
\ldots & & & & & & & & \\
\hline
\Sigma\phi(x^N) & = x & + 4x^3 & + 6x^5 & + 8x^7 & + 12x^9 & + 12x^{11} & + 14x^{13} & + 24x^{15} \ldots
\end{array}$$

17. Supposing that the reduction is completely accounted for as above, then, to obtain the numerical relations, the numbers 1, 4, 12, 32, ..., $(n+1)\,2^{\frac{1}{2}(n-3)}$ have to be expressed linearly in terms of 1, 4, 6, 8, ..., $\sigma'(n)$, viz. $(n+1)\,2^{\frac{1}{2}(n-3)}$ as a linear function of $\sigma'(n)$, $\sigma'(n-4)$, $\sigma'(n-8)$, ..., and we have

$$\begin{aligned}
1 &= 1, \\
4 &= 4, \\
12 &= 6 + 6\,.\,1, \\
32 &= 8 + 6\,.\,4, \\
80 &= 12 + 6\,.\,6 + 32\,.\,1, \\
192 &= 12 + 6\,.\,8 + 33\,.\,4, \\
448 &= 14 + 6\,.\,12 + 33\,.\,6 + 164\,.\,1, \\
1024 &= 24 + 6\,.\,12 + 33\,.\,8 + 166\,.\,4, \\
2304 &= 18 + 6\,.\,14 + 33\,.\,12 + 166\,.\,6 + 810\,.\,1, \\
5120 &= 20 + 6\,.\,24 + 33\,.\,12 + 166\,.\,8 + 817\,.\,4, \\
11264 &= 32 + 6\,.\,18 + 33\,.\,14 + 166\,.\,12 + 817\,.\,6 + 3768\,.\,1, \\
24576 &= 24 + 6\,.\,20 + 33\,.\,24 + 166\,.\,12 + 817\,.\,8 + 3778\,.\,4, \\
&\ldots
\end{aligned}$$

but it is of course very doubtful whether these relations have any value in regard to the present theory.

18. In the same way that, by assuming a common factor, $1+kx^2$, in the values of P and Q, we pass from the case n to the case $n-4$, so, by assuming a common factor, $1+x^2$, in the numerator and denominator of the expression for y in terms of x and the coefficients B, we pass from the case n to the case $n-2$. Contrariwise, in the solutions given by the Jacobi-Brioschi differential equations and by the Jacobi partial differential equation, the solution for a given value of n does not thus contain the solution for an inferior value of n; see *post* Nos. 36 and 43.

I pass now to the theory before referred to.

The Development $y=\rho x(1+\Pi_1 x^2+\Pi_2 x^4+\ldots)$. Art. Nos. 19 and 20.

19. Starting from the equation

$$\frac{dy}{\sqrt{1-2\beta y^2+y^4}}=\frac{\rho\, dx}{\sqrt{1-2\alpha x^2+x^4}},$$

and writing for shortness

$$\begin{aligned}
R_1&=\tfrac{1}{3}\,\alpha, & S_1&=\tfrac{1}{3}\,\beta,\\
R_2&=\tfrac{1}{5}\left(\tfrac{3}{2}\,\alpha^2-\tfrac{1}{2}\right), & S_2&=\tfrac{1}{5}\left(\tfrac{3}{2}\,\beta^2-\tfrac{1}{2}\right),\\
R_3&=\tfrac{1}{7}\left(\tfrac{5}{2}\,\alpha^3-\tfrac{3}{2}\,\alpha\right), & S_3&=\tfrac{1}{7}\left(\tfrac{5}{2}\,\beta^3-\tfrac{3}{2}\,\beta\right),\\
R_4&=\tfrac{1}{9}\left(\tfrac{35}{8}\,\alpha^4-\tfrac{15}{4}\,\alpha^2+\tfrac{3}{8}\right), & S_4&=\tfrac{1}{9}\left(\tfrac{35}{8}\,\beta^4-\tfrac{15}{4}\,\beta^2+\tfrac{3}{8}\right),\\
&\ldots\ldots & &\ldots\ldots
\end{aligned}$$

(viz. save as to the exterior factors $\tfrac{1}{3}$, $\tfrac{1}{5}$, ..., $3R_1$, $5R_2$, ... are the Legendrian functions of α, and $3S_1$, $5S_2$, ... the Legendrian functions of β), we have

$$dy\,(1+3S_1y^2+5S_2y^4+\ldots)=\rho\, dx\,(1+3R_1x^2+5R_2x^4+\ldots),$$

whence integrating, so that y may vanish with x, we have

$$y+S_1y^3+S_2y^5+\ldots=\rho\,(x+R_1x^3+R_2x^5+\ldots),$$

say this is

$$=u.$$

20. We then have $y=u-fy$, where $fy=S_1y^3+S_2y^5+\ldots$; and thence, expanding by Lagrange's theorem,

$$y=u-fu+\frac{1}{2}(f^2u)'-\frac{1}{2\,.\,3}(f^3u)''+\frac{1}{2\,.\,3\,.\,4}(f^4u)'''-\ldots,$$

we have

$$fu\;=S_1u^3+S_2u^5+S_3u^7+S_4u^9+\ldots,$$

and thence

$$\begin{aligned}
f^2u&=S_1^2u^6+2S_1S_2u^8+(2S_1S_3+S_2^2)\,u^{10}+\ldots,\\
f^3u&=S_1^3u^9+3S_1^2S_2u^{11}+\ldots,\\
f^4u&=S_1^4u^{12}+\ldots;
\end{aligned}$$

consequently,

$$\begin{aligned} y = \; & u, \\ & + u^3 (-S_1), \\ & + u^5 (-S_2 + 3S_1^2), \\ & + u^7 (-S_3 + 8S_1S_2 - 12S_1^3), \\ & + u^9 (-S_4 + 10S_1S_3 + 5S_2^2 - 55S_1^2S_2 + 55S_1^4) \\ & + \ldots\ldots \\ & + \ldots\ldots, \end{aligned}$$

and writing herein

$$\begin{aligned} u &= \rho \; \{x + R_1x^3 + R_2x^5 + R_3x^7 + R_4x^9 + \ldots\}, \\ u^3 &= \rho^3 \{x^3 + 3R_1x^5 + (3R_2 + 3R_1^2)\, x^7 + (3R_3 + 6R_1R_2 + R_1^3)\, x^9 + \ldots\}, \\ u^5 &= \rho^5 \{x^5 + 5R_1x^7 + (5R_2 + 10R_1^2)\, x^9 + \ldots\}, \\ u^7 &= \rho^7 \{x^7 + 7R_1x^9 + \ldots\}, \\ u^9 &= \rho^9 \{x^9 + \ldots\}, \end{aligned}$$

we have the required series

$$y = \rho x \{1 + \Pi_1x^2 + \Pi_2x^4 + \Pi_3x^6 + \ldots\},$$

where the values of the coefficients are

$$\begin{aligned} \Pi_1 &= R_1 + (-S_1)\, \rho^2, \\ \Pi_2 &= R_2 + (-S_1)\, 3R_1\rho^2 + (-S_2 + 3S_1^2)\, \rho^4, \\ \Pi_3 &= R_3 + (-S_1)(3R_2 + 3R_1^2)\, \rho^2 + (-S_2 + 3S_1^2)\, 5R_1\rho^4 + (-S_3 + 8S_1S_2 - 12S_1^3)\, \rho^6, \\ \Pi_4 &= R_4 + (-S_1)(3R_3 + 6R_1R_2 + R_1^3)\, \rho^2 + (-S_2 + 3S_1^2)(5R_2 + 10R_1^2)\, \rho^4 \\ & \quad + (-S_3 + 8S_1S_2 - 12S_1^3)\, 7R_1\rho^6 + (-S_4 + 10S_1S_3 + 5S_2^2 - 55S_1^2S_2 + 55S_1^4)\, \rho^8, \\ & \ldots\ldots, \\ & \ldots\ldots, \end{aligned}$$

and so on, as far as we please.

The Cubic Transformation, $n = 3$. Art. Nos. 21 to 28.

21. We have here

$$\frac{\rho + x^2}{1 + \rho x^2} = \rho\, (1 + \Pi_1x^2 + \Pi_2x^4 + \ldots);$$

whence, developing the left-hand side and equating coefficients,

$$\rho\Pi_1 = -\rho^2 + 1, \quad \rho\Pi_2 = \rho^3 - \rho, \quad \rho\Pi_3 = -\rho^4 + \rho^2, \ldots$$

It will be convenient to write

$$\begin{aligned}
\Theta_1 &= \rho\Pi_1 + \rho^2 - 1, &&= -S_1\rho^3 + \rho^2 + R_1\rho - 1,\\
\Theta_2 &= \ \Pi_2 - \rho^2 + 1, &&= (-S_2 \ + 3S_1^2)\,\rho^4 - (3R_1S_1 + 1)\,\rho^2 + R_2 + 1,\\
\Theta_3 &= \ \Pi_3 + \rho^3 - \rho, &&= (-S_3 \ + 8S_1S_2 - 12S_1^3)\,\rho^6\\
& && \quad + (-5R_1S_2 + 15R_1S_1^2)\,\rho^4\\
& && \quad + \rho^3\\
& && \quad + (-3R_2S_1 - \ 3R_1^2S_1)\,\rho^2\\
& && \quad - \rho\\
& && \quad + R_3,
\end{aligned}$$

....

....

where observe the difference of form in the function Θ_1, and in the subsequent functions Θ_2, Θ_3, ... In these last, a factor ρ is thrown out.

22. The two equations $\Theta_1 = 0$ and $\Theta_2 = 0$ serve to determine ρ, β in terms of α; the subsequent equations $\Theta_3 = 0$, $\Theta_4 = 0$, ... will then be, all of them, satisfied identically. This implies that Θ_3, Θ_4, ... are each of them a linear function of Θ_1, Θ_2. The *à posteriori* verification and determination of the factors is by no means easy; I have effected it only for Θ_3; we have

$$7\Theta_3 = (\rho^3 - 3S_1\rho^2 - 2\rho + 27R_1)\,\Theta_1 + (-S_1\rho^2 - 10\rho + 25R_1)\,\Theta_2,$$

or, at full length,

$$7\left|\begin{array}{r}
(-S_3 + 8S_1S_2 - 12S_1^3)\,\rho^6\\
+ (-5R_1S_2 \ + 15R_1S_1^2)\,\rho^4\\
+\rho^3\\
+ (-3R_2S_1 \ - \ 3R_1^2S_1)\,\rho^2\\
-\rho\\
+R_3
\end{array}\right|$$

$$\begin{aligned}
&= (\rho^3 - 3S_1\rho^2 - \ 2\rho + 27R_1)\,(-S_1\rho^3 + \rho^2 + R_1\rho - 1)\\
&\quad + (- \ S_1\rho^2 - 10\rho + 25R_1)\,((-S_2 \ + 3S_1^2)\,\rho^4 - (3R_1S_1 + 1)\,\rho^2 + R_2 + 1);
\end{aligned}$$

in verifying which we must, of course, take account of the relations between the expressions R and those between the expressions S; we have

$$\alpha = 3R_1 \text{ and thence } 10R_2 = 27R_1^2 - 1, \quad 14R_3 = 135R_1^3 - 9R_1;$$

similarly,

$$10S_2 = 27S_1^2 - 1, \quad 14S_3 = 135S_1^3 - 9S_1.$$

Equating the coefficients of ρ^6, we have

$$-7S_3 + 56S_1S_2 - 84S_1^3 = -S_1 + S_1S_2 - 3S_1^2;$$

viz. multiplying by 2, this is

$$-14S_3+110S_1S_2-162S_1^3+2S_1=0;$$

or, finally, it is

$$(-135S_1^3+9S_1)+(297S_1^3-11S_1)-162S_1^3+2S_1=0,$$

an identity, as it should be. The identity of the coefficients of ρ^5, ρ^4, ρ^3, ρ^2, ρ, 1 may be verified in like manner.

23. Considering α as known, the values of ρ and β are determined by the foregoing equations $\Theta_1=0$, $\Theta_2=0$; that is,

$$-S_1\rho^3+\rho^2+R_1\rho-1=0,$$

$$(-S_2+3S_1^2)\rho^4-(3R_1S_1+1)\rho^2+R_2+1=0,$$

where, of course, the R's and S's are regarded as given functions of α and β respectively.

It is to be observed that the equations are satisfied by $\rho^2=1$, $\alpha=\beta$; viz. we have the transformation $y=\dfrac{x(\pm 1+x^2)}{1\pm x^2}$; that is, $y=\pm x$, which is the transformation of the first order, $n=1$. The two equations represent surfaces of the orders 4 and 6 respectively, and they have thus a complete intersection of the order 24. As part of this, we have, as just shown, each of the two lines ($\rho=1$, $\alpha=\beta$) and ($\rho=-1$, $\alpha=\beta$); but there is a more considerable reduction of order to be accounted for, the proper MM-curve being, as will appear, a unicursal curve of the order $=6$.

24. Multiplying the second equation by $10\rho^2$, and for $10R_2$ and $10S_2$ writing their values $27R_1^2-1$ and $27S_1^2-1$ respectively, we have

$$(3S_1^2+1)\rho^6-(30R_1S_1+10)\rho^4+(27R_1^2+9)\rho^2=0;$$

and if herein we substitute for $S_1\rho^3$ its value from the first equation, $=\rho^2+R_1\rho-1$, we have

$$3(\rho^2+R_1\rho-1)^2+\rho^6-30R_1\rho(\rho^2+R_1\rho-1)-10\rho^4+(27R_1^2+9)\rho^2=0;$$

that is,

$$\rho^6-7\rho^4-24R_1\rho^3+3\rho^2+24R_1\rho+3=0;$$

viz. this is

$$(\rho^2-1)(\rho^4-6\rho^2-24R_1\rho-3)=0,$$

containing, as it ought to do, the factor ρ^2-1. Throwing this out, and repeating the first equation, we have

$$\begin{aligned}-S_1\rho^3+\rho^2+R_1\rho-1&=0,\\ \rho^4\qquad-6\rho^2-24R_1\rho-3&=0,\end{aligned}$$

which two equations may be replaced by

$$\begin{aligned}\rho^4-24S_1\rho^3+18\rho^2\qquad\qquad-27&=0,\\ \rho^4\qquad\qquad-6\rho^2-24R_1\rho-\ 3&=0,\end{aligned}$$

which are the $\rho\beta$- and $\rho\alpha$-equations respectively. Recollecting that R_1 and S_1 denote $\frac{1}{3}\alpha$ and $\frac{1}{3}\beta$, they agree with the results obtained in No. 7. The $\alpha\beta$-modular equation is obtained by the elimination of ρ from these two equations, and may be at once written down in the form, Det. $=0$, where the determinant is of the order 8, but contains S_1 and R_1, that is, β and α, each of them, in the fourth order only: the form is thus the same with that of the $\alpha\beta$-equation obtained in No. 2; but the identification would be a work of some labour.

25. The equations may be written

$$24S_1\rho^3 = \rho^4 + 18\rho^2 - 27,$$

$$24R_1\rho = \rho^4 - 6\rho^2 - 3,$$

and, treating R_1, S_1, ρ as coordinates, it hence appears that the MM-curve is (as mentioned above) a unicursal curve of the order 6; in fact, we have R_1, S_1, each of them given as a rational function of ρ; and cutting the curve by an arbitrary plane $AR_1 + BS_1 + C\rho + D = 0$, the substitution of the values of R_1, S_1 in this equation gives for ρ an equation of the order 6.

26. The same conclusion may be obtained from the foregoing system of a cubic and a quartic equation in ρ. Considering R_1, S_1, ρ as coordinates, they represent, each of them, a surface of the order 4, and the complete intersection is of the order 16; but this is made up of a line in the plane infinity counting 10 times, and of the MM-curve, which is thus of the order $16-10$, $=6$. In fact, introducing, for homogeneity, a fourth coordinate θ, the two equations are

$$-S_1\rho^3 + \rho^2\theta^2 + R_1\rho\theta^2 - \theta^4 = 0,$$

$$\rho^4 - 6\rho^2\theta^2 - 24R_1\rho\theta^2 - 3\theta^4 = 0,$$

and the line $\rho=0$, $\theta=0$ is thus a triple line on each of these surfaces; viz. cutting them by an arbitrary plane, we have for the first surface an ordinary triple point, as shown by the continuous lines of the annexed figure, and for the second surface a triple point = cusp + two nodes, as shown by the dotted lines of the figure. There is, moreover, as shown in the figure, a contact of two branches, and the number of intersections is thus $=10$.

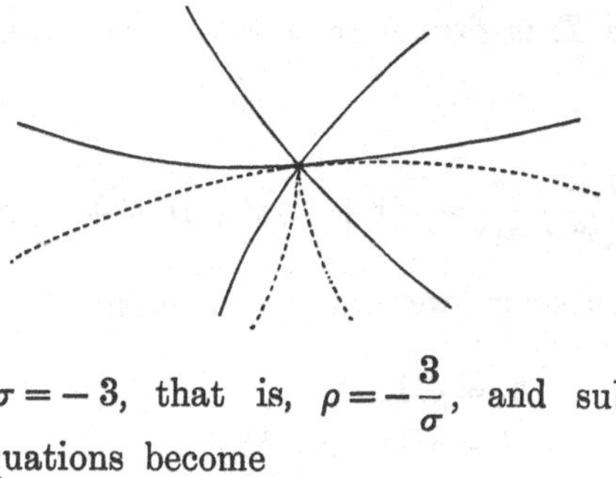

27. If we assume $\rho\sigma = -3$, that is, $\rho = -\dfrac{3}{\sigma}$, and substitute this value in the equation for S_1, the two equations become

$$24S_1\sigma = \sigma^4 - 6\sigma^2 - 3,$$

$$24R_1\rho = \rho^4 - 6\rho^2 - 3;$$

viz. β is the same function of $\sigma\left(=-\frac{3}{\rho}\right)$ that α is of ρ. This accords with the theorem in Elliptic Functions that a combination of two transformations leads to a multiplication.

28. We have

$$24(R_1+\tfrac{1}{3})\rho=\rho^4-6\rho^2+8\rho-3,\ =(\rho-1)^3(\rho+3),$$

or, what is the same thing,

$$24(R_1+\tfrac{1}{3})\quad=(\rho-1)^2(\rho+\sigma+2);$$

and, in like manner,

$$24(R_1-\tfrac{1}{3})\rho=\rho^4-6\rho^2-8\rho-3,\ =(\rho+1)^3(\rho-3),$$

and, consequently,

$$24(R_1-\tfrac{1}{3})\quad=(\rho+1)^2.(\rho+\sigma-2);$$

with the like equations between S_1, σ, ρ. It will be recollected that

$$R_1=\tfrac{1}{3}\alpha,\ =\tfrac{1}{6}\left(u^4+\frac{1}{u^4}\right);$$

hence

$$24(R_1\pm\tfrac{1}{3})=4\left(u^4+\frac{1}{u^4}\pm2\right),\ =4\left(u^2\pm\frac{1}{u^2}\right)^2.$$

The formulæ just obtained are useful for obtaining the *uv*-modular equation from the foregoing equations; or say

$$4\left(v^4+\frac{1}{v^4}\right)\sigma=\sigma^4-6\sigma^2-3,$$

$$4\left(u^4+\frac{1}{u^4}\right)\rho=\rho^4-6\rho^2-3,$$

where $\rho\sigma=-3$, and we have to eliminate ρ and σ; the elimination gives

$$\frac{v^2}{u^2}-\frac{u^2}{v^2}+2vu-\frac{2}{uv}=0,$$

that is,

$$v^4+2v^3u^3-2vu-u^4=0.$$

The Quintic Transformation, $n=5$. Art. Nos. 29 to 32.

29. We have here

$$\frac{\rho+A_1x^2+x^4}{1+A_1x^2+\rho x^4}=\rho(1+\Pi_1x^2+\Pi_2x^4+\Pi_3x^6+\ldots),$$

and multiplying by $1+A_1x^2+\rho x^4$, we obtain an infinite series of equations, the first three of which are

$$A_1=\rho\Pi_1+A_1\rho,$$
$$1=\rho\Pi_2+A_1\rho\Pi_1,$$
$$0=\rho\Pi_3+A_1\rho\Pi_2+\rho^2\Pi_1,$$

The first of these gives

$$A_1 = \frac{\rho\Pi_1}{-\rho+1},\ = \frac{\Theta_1-\rho^2+1}{-\rho+1};$$

and the other two equations then determine the MM-curve. These being satisfied, the remaining equations will be satisfied identically. It is proper to introduce Θ_1, Θ_2, Θ_3 into the equations instead of Π_1, Π_2, Π_3. We have first

$$1 = \rho\Pi_2 + \rho\Pi_1\frac{\Theta_1-\rho^2+1}{-\rho+1} + \rho^2,$$

that is,

$$0 = \rho\,(\Theta_2+\rho^2-1) - \frac{(\Theta_1-\rho^2+1)^2}{\rho-1} + \rho^2 - 1;$$

viz. this is

$$\rho\,(\rho-1)(\Theta_2+\rho^2-1) - (\Theta_1-\rho^2+1)^2 + (\rho^2-1)(\rho-1) = 0;$$

or, finally, this is

$$\rho\,(\rho-1)\,\Theta_2 - \Theta_1^2 + 2\Theta_1(\rho^2-1) = 0.$$

Secondly, we have

$$0 = \Pi_3 + \frac{\rho\Pi_1\Pi_2}{-\rho+1} + \rho\Pi_1 = 0;$$

that is,

$$\Theta_3 - \rho^3 + \rho + \frac{(\Theta_1-\rho^2+1)(\Theta_2+\rho^2-1)}{-\rho+1} + \Theta_1 - \rho^2 + 1 = 0;$$

viz. this is

$$(\Theta_3+\Theta_1-\rho^3-\rho^2+\rho+1)(-\rho+1) + (\Theta_1-\rho^2+1)(\Theta_2+\rho^2-1) = 0;$$

or, finally, it is

$$\Theta_3(-\rho+1) + \Theta_1\Theta_2 + \Theta_1(\rho^2-\rho) - \Theta_2(\rho^2-1) = 0.$$

30. We have thus the two equations

$$(\rho^2-\rho)\,\Theta_2 - \Theta_1^2 + 2\Theta_1(\rho^2-1) = 0,$$
$$\Theta_3(-\rho+1) + \Theta_1\Theta_2 + \Theta_1(\rho^2-\rho) - \Theta_2(\rho^2-1) = 0;$$

and recollecting that Θ_3 is of the form $L\Theta_1 + M\Theta_2$, we see that each of these equations is satisfied if only $\Theta_1 = 0$, $\Theta_2 = 0$ (the formulæ belonging to the cubic transformation). This ought to be the case, for we can, by writing $A_1 = \rho + 1$, reduce the expression $\dfrac{x(\rho + A_1x^2 + x^4)}{1 + A_1x^2 + \rho x^4}$ to the form $\dfrac{x(\rho+x^2)}{1+\rho x^2}$, which belongs to the cubic transformation (see *ante* No. 17). The equations may be written

$$\rho\Theta_2 = -(2\rho+2)\,\Theta_1 + \frac{\Theta_1^2}{\rho-1},$$

$$\rho\Theta_3 = \quad (3\rho^2+4\rho+2)\,\Theta_1 - (3\rho+3)\frac{\Theta_1^2}{\rho-1} + \frac{\Theta_1^3}{(\rho-1)^2}.$$

31. The investigation may be presented in a slightly different form by introducing the functions Θ at an earlier stage; viz. writing

$$\rho\Pi_1 = \Theta_1 - \rho^2 + 1,\ \rho\Pi_2 = \rho\Theta_2 + \rho^3 - \rho, \ldots,$$

we have

$$\frac{\rho + A_1 x^2 + x^4}{1 + A_1 x^2 + \rho x^4} = \rho + (\Theta_1 - \rho^2 + 1)\, x^2 + (\rho\Theta_2 + \rho^3 - \rho)\, x^4 + \ldots$$

$$= \frac{\rho + x^2}{1 + \rho x^2} + \Theta_1 x^2 + \rho\Theta_2 x^4 + \rho\Theta_3 x^6 + \ldots$$

Transposing, reducing, and dividing by x^2, we have

$$\frac{(1 - x^2)\,[\rho^2 - 1 + A_1(-\rho + 1)]}{(1 + \rho x^2)(1 + A_1 x^2 + \rho x^4)} = \Theta_1 + \rho\Theta_2 x^2 + \rho\Theta_3 x^4 + \ldots,$$

whence clearly $\rho^2 - 1 + A_1(-\rho + 1) = \Theta_1$, giving for A_1 the before-mentioned value; and we then have

$$1 + A_1 x^2 + \rho x^4 = 1 + (\rho + 1)\, x^2 + \rho x^4 - \frac{\Theta_1 x^2}{\rho - 1},\ = (1 + x^2)(1 + \rho x^2) - \frac{\Theta_1 x^2}{\rho - 1}.$$

The equation thus becomes

$$\frac{(1 - x^2)\,\Theta_1}{(1 + x^2)(1 + \rho x^2)^2 \left(1 - \dfrac{\Theta_1 x^2}{\rho - 1 \,.\, 1 + x^2 \,.\, 1 + \rho x^2}\right)} = \Theta_1 + \rho\Theta_2 x^2 + \rho\Theta_3 x^4 + \ldots,$$

and expanding the left-hand side, first in the form

$$\frac{(1 - x^2)\,\Theta_1}{(1 + x^2)(1 + \rho x^2)^2} + \frac{(1 - x^2)\, x^2 \Theta_1^2}{(\rho - 1)(1 + x^2)^2 (1 + \rho x^2)^3} + \frac{(1 - x^2)\, x^4 \Theta_1^3}{(\rho - 1)(1 + x^2)^3 (1 + \rho x^3)^4} + \ldots$$

and then each of these terms separately in powers of x^2, and comparing with $\Theta_1 + \rho\Theta_2 x^2 + \rho\Theta_3 x^4 + \ldots$, we have the two equations in the last-mentioned form, and an infinite series of other equations, which will be satisfied identically.

32. The successive coefficients might be called $\Phi_2, \Phi_3, \ldots$; say

$$\Phi_2 = (\rho^2 - \rho)\,\Theta_2 - \Theta_1^2 + 2\,(\rho^2 - 1)\,\Theta_1,$$

$$\Phi_3 = (-\rho + 1)\,\Theta_3 + \Theta_1\Theta_2 + (\rho^2 - \rho)\,\Theta_1 - (\rho^2 - 1)\,\Theta_2,$$

and similarly for $\Phi_4, \ldots$; and it would then be proper to show *à posteriori* that each of the equations $\Phi_4 = 0$, $\Phi_5 = 0, \ldots$ is satisfied identically in virtue of the two equations $\Phi_2 = 0$, $\Phi_3 = 0$, or, what is the same thing, that the functions $\Phi_4, \Phi_5, \ldots$ are each of them a linear function (with coefficients which are functions of ρ) of the two functions Φ_2 and Φ_3. I do not attempt to do this, nor even to discuss the MM-curve by means of the equations $\Phi_2 = 0$, $\Phi_3 = 0$; but I will obtain equivalent results, and complete the solution by means of the Jacobi-Brioschi equations, in effect reproducing the investigation contained in the third appendix of the *Funzioni ellittiche.*

The General Transformation, $n=2s+1$. Art. No. 33.

33. The equation here is

$$\frac{\rho+A_{s-1}x^2+\dots}{1+A_1x^2+\dots}=\rho(1+\Pi_1x^2+\dots).$$

The general theory is sufficiently illustrated by the preceding particular cases, and I wish at present only to notice the equation obtained by comparing the coefficients of x^2; viz. this is $A_{s-1}-\rho A_1=\rho\Pi_1$, or, substituting for Π_1 its value,

$$A_{s-1}-\rho A_1=\tfrac{1}{3}(\alpha\rho-\beta\rho^3).$$

The Jacobi-Brioschi Equations. Art. Nos. 34 to 42.

34. These were obtained for the differential equation

$$\frac{dx}{\sqrt{a'x^4+b'x^3+c'x^2+d'x+e'}}=\frac{dy}{\sqrt{ay^4+by^3+cy^2+dy+e}};$$

viz. if this be satisfied by $y=U\div V$, where U, V are rational and integral functions of x of the degrees n and $n-1$ respectively, then, writing for shortness

$$\phi=a'x^4+b'x^3+c'x^2+d'x+e',$$

and using accents to denote differentiation in regard to x, the numerator and denominator U, V satisfy the equations

$$\begin{aligned}(VV''-V'^2)\phi+\tfrac{1}{2}VV'.\phi' \quad &+\ aU^2+\tfrac{1}{2}bUV+\ pV^2=0,\\ -(VU''+V''U-2V'U')\phi-\tfrac{1}{2}(VU'+V'U)\phi+\tfrac{1}{2}bU^2&+(c-2p)UV+\tfrac{1}{2}dV^2=0,\\ (UU''-U'^2)\phi+\tfrac{1}{2}UU'.\phi' \quad &+\ pU^2+\tfrac{1}{2}dUV+\ eV^2=0,\end{aligned}$$

where p is a function $=\mathrm{a}x^2+\mathrm{b}x+\mathrm{c}$, with coefficients a, b, c, the values of which have to be determined. By way of verification, observe that, multiplying by U^2, UV, V^2, and adding, we obtain

$$-(VU'-V'U)^2\phi+aU^4+bU^3V+cU^2V^2+dUV^3+eV^4=0;$$

that is,

$$-\frac{1}{V^2}(VU'-V'U)^2(a'x^4+b'x^3+c'x^2+d'x+e')+ay^4+by^3+cy^2+dy+e=0,$$

the result obtained by substituting for y its value, $=U\div V$, in the differential equation.

35. Considering the foregoing special form

$$\frac{dx}{\sqrt{1-2\alpha x^2+x^4}}=\frac{dy}{\rho\sqrt{1-2\beta y^2+y^4}},$$

so that a, b, c, d, e have the values ρ^2, 0, $-2\beta\rho^2$, 0, ρ^2 and ϕ is $=1-2\alpha x^2+x^4$, the equations are

$$\begin{aligned}(VV''-V'^2)\phi+\quad&\tfrac{1}{2}VV'.\phi'+\rho^2U^2+\quad pV^2=0,\\ -(VU''+V''U-2V'U')\phi+&\tfrac{1}{2}(VU'+V'U)\phi'-(2\beta\rho^2+2p)UV\quad=0,\\ (UU''-U'^2)\phi+\quad&\tfrac{1}{2}UU'.\phi'+pU^2+\quad\rho^2V^2=0,\end{aligned}$$

where, writing as before, $n=2s+1$, and assuming that the last coefficient, $A_{\frac{1}{2}(n-1)}$ or A_s, is $=\rho$, we have

$$U=x(\rho+A_{s-1}x^2+A_{s-2}x^4+\ldots+\quad A_1x^{2s-2}+\quad x^{2s}),$$
$$V=\quad 1+\quad A_1x^2+\quad A_2x^4+\ldots+A_{s-1}\,x^{2s-2}+\rho x^{2s},$$

and where, as is easily shown, p has the value $=-\{2A_1+(2s+1)\,x^2\}$. In comparing with Brioschi, it will be recollected that 2α, 2β are written in place of his α, β.

36. The equations contain n, and they are not satisfied by values of U, V belonging to any inferior value of n; U, V may each of them be multiplied by any common constant factor at pleasure, but not by a common variable factor P; viz. it is assumed that the fraction $U\div V$ is in its least terms, and consequently that (save as to a constant factor) U, V are determinate functions. It is easy to verify that the equations (being verified by U, V) are not verified by PU, PV, but it is interesting to show *à priori* why this is so. The equations are obtained as follows. Consider the differential equation in the form $\dfrac{dx}{\sqrt{X}}=\dfrac{dy}{\sqrt{Y}}$, and suppose that an integral equation is given in the form $F=0$ (F a rational and integral function of x, y); we thence deduce a relation $Ldx+Mdy=0$ between the differentials, and this must agree with the given differential equation; that is, we have $L\sqrt{X}+M\sqrt{Y}=0$, or, rationalizing, $L^2X-M^2Y=0$; viz. this last equation must agree with the equation $F=0$, or, what is the same thing, L^2X-M^2Y must contain F as a factor; say we have

$$L^2X-M^2Y=F\,.\,G,$$

where G is a function of x, y. In the particular case where the integral is of the form

$$y=U\div V,$$

we have

$$F=Vy-U,$$

and we have therefore

$$L^2X-M^2Y=G\,(Vy-U);$$

and it is by means of this identity that the equations are obtained. But suppose that there is a common factor P, or that we have $y=PU\div PV$; then, if we write $F=PVy-PVU$, $=P\,(Vy-U)$, there is no necessity that L^2X-M^2Y should contain as a factor this expression of F, and it will not in fact contain it; all that is necessary is that L^2X-M^2Y shall contain the factor $Vy-U$; and thus the equations obtained for U, V do not apply to PU, PV. We might, of course, introduce an arbitrary *constant* factor Θ; contrast herewith the solution by means of the Jacobi partial differential equation, *post* No. 43, where Θ is not arbitrary but has a determinate value.

37. In virtue of the assumed forms of U, V, the first and the third equations give each of them the same relations between the coefficients A; and only one of these equations, say the first, need be attended to. It will be observed that this equation

does not contain β; it consequently serves to determine the coefficients A in terms of ρ, α, and to establish a relation between ρ, α, that is, the multiplier equation. We can from this, as will be explained, deduce the equation between ρ, β; the theory thus depends entirely upon the first equation; say this is

$$(VV''-V'^2)(1-2\alpha x^2+x^4)+VV'(-2\alpha x+2x^3)+\rho^2U^2-\{2A_1+(2s+1)x^2\}V^2=0.$$

38. We have $V=1+A_1x^2+A_2x^4+\ldots$, but the equation contains the quadric functions $VV''-V'^2$, VV', and V^2; it is convenient to write

$$VV''-V'^2=K_1+K_2x^2+K_3x^4+\ldots,$$
$$V^2=L_0+L_1x^2+L_2x^4+\ldots;$$

whence of course

$$VV'=2L_1x+4L_2x^3+\ldots,$$

and we have

$K_1=$	$K_2=$	$K_3=$	$K_4=$	$K_5=$	$K_6=$	$K_7=$	$K_8=$	...
$2A_1$	$12A_2$	$30A_3$	$56A_4$	$90A_5$	$132A_6$	$182A_7$	$240A_8$	,
	$-2A_1^2$	$-2A_1A_2$	$+8A_1A_3$	$+26A_1A_4$	$+52A_1A_5$	$+86A_1A_6$	$+128A_1A_7$	
			$-4A_2^2$	$-6A_2A_3$	$+4A_2A_4$	$+22A_2A_5$	$+48A_2A_6$	
					$-6A_3^2$	$-10A_3A_4$	$0A_3A_5$	
							$-8A_4^2$	

$L_0=$	$L_1=$	$L_2=$	$L_3=$	$L_4=$	...
1	$2A_1$	$2A_2$	$2A_3$	$2A_4$	.
		$+A_1^2$	$+2A_1A_2$	$+2A_1A_3$	
				$+A_2^2$	

The coefficients of U^2 are at once obtained; say we have $U^2=\Lambda_0x^2+\Lambda_1x^4+\Lambda_2x^6+\ldots$,

$\Lambda_0=$	$\Lambda_1=$	$\Lambda_2=$	$\Lambda_3=$	$\Lambda_4=$	...
ρ^2	$2\rho A_{s-1}$	$2\rho A_{s-2}$	$2\rho A_{s-3}$	$2\rho A_{s-4}$	.
		$+A^2_{s-1}$	$+2A_{s-1}A_{s-2}$	$+2A_{s-1}A_{s-3}$	
				$+A^2_{s-2}$	

Substituting in the equation and equating to zero the coefficients of the several powers of x^2, we find

$$\begin{aligned}
&K_1 \qquad - 2A_1L_0 = 0,\\
&K_2 \qquad - 2A_1L_1 + (-2s-1)L_0 - 2\alpha(K_1 + L_1) + \rho^2\Lambda_0 = 0,\\
&K_3 + K_1 - 2A_1L_2 + (-2s+1)L_1 - 2\alpha(K_2 + 2L_2) + \rho^2\Lambda_1 = 0,\\
&K_4 + K_2 - 2A_1L_3 + (-2s+3)L_2 - 2\alpha(K_3 + 3L_3) + \rho^2\Lambda_2 = 0,\\
&K_5 + K_3 - 2A_1L_4 + (-2s+5)L_3 - 2\alpha(K_4 + 4L_4) + \rho^2\Lambda_3 = 0,\\
&\ldots\ldots\ldots\ldots\ldots\ldots\ldots\ldots\\
&\ldots\ldots\ldots\ldots\ldots\ldots\ldots\ldots
\end{aligned}$$

The number of equations is $=2(s+1)$, for the equation contains terms in $x^0, x^2, x^4, \ldots, x^{4s+2}$; but the first equation, and also the last and last but one equations, are in fact identities; there remain thus $2(s+1)-3$, $=2s-1$ equations; but these are equivalent to s independent equations, serving to determine the $s-1$ coefficients $A_1, A_2, \ldots, A_{s-1}$, and to determine the relation between ρ and α. In writing down the equations for a determinate value of s, the coefficients A_0, A_s must be taken to be $=0$ and ρ respectively; and coefficients with a negative suffix or a suffix greater than s, must be taken to be each $=0$.

39. Thus, $(n=3)$ $s=1$, we have the $2(s+1)$, $=4$ equations:

$$\begin{aligned}
&2\rho \qquad - 2\rho\,.\,1 = 0,\\
&-2\rho^2 \qquad - 2\rho\,.\,2\rho + (-3)\,1 - 2\alpha(\ 2\rho + 2\rho) + \rho^2\,.\,\rho^2 = 0,\\
&0 + 2\rho - 2\rho\,.\,\rho^2 + (-1)\,2\rho - 2\alpha(-2\rho^2 + 2\rho^2) + \rho^2\,.\,2\rho = 0,\\
&0 - 2\rho^2 - 2\rho\,.\,0 + (+1)\,\rho^2 - 2\alpha(\ 0 + 3\,.\,0) + \rho^2\,.\,1 = 0,
\end{aligned}$$

where the first, third and fourth equations are each of them an identity; the second equation is $-2\rho^2 - 4\rho^2 - 3 - 8\alpha\rho + \rho^4 = 0$; viz. in accordance with what precedes, writing $\alpha = 3R_1$, this is the foregoing equation

$$\rho^4 - 6\rho^2 - 24R_1\rho - 3 = 0.$$

To complete the solution, we use the theorem in elliptic functions referred to *ante* (No. 8); viz. writing $\rho\sigma = -3$, then we have β the same function of σ that α is of ρ; whence

$$\sigma^4 - 6\sigma^2 - 24S_1\sigma - 3 = 0,$$

and we thus have two equations giving the MM-curve.

40. In the case, $(n=5)$ $s=2$, we have the $2(s+1)$, $=6$ equations:

$$\begin{aligned}
&2A_1 \qquad - 2A_1\,.\,1 = 0,\\
&12\rho - 2A_1^2 - 2A_1(2A_1) \quad - 5\,.\,1 \quad - 2\alpha\,.\,2A_1 \quad + \rho^2\,.\,\rho^2 = 0,\\
&-2\rho A_1 + 2A_1 - 2A_1(2\rho + A_1^2) \quad - 3\,.\,2A_1 \quad - 2\alpha\{12\rho - 2A_1^2 + 2(2\rho + A_1^2)\} + \rho^2\,.\,2\rho A_1 = 0,\\
&-4\rho^2 + 12\rho \quad - 2A_1^2 - 2A_1\,.\,2A_1\rho - 1\,(2\rho + A_1^2) - 2\alpha\{-2\rho A_1 + 3\,.\,2A_1\rho\} + \rho^2(2\rho + A_1^2) = 0,\\
&0 \quad - 2\rho A_1 \quad - 2A_1\,.\,\rho^2 \quad + 1\,.\,2A_1\rho \quad - 2\alpha\{-4\rho^2 \quad + 4\,.\,\rho^2\} \quad + \rho^2\,.\,2A_1 = 0,\\
&0 \quad - 4\rho^2 \quad - 2A_1\,.\,0 \quad + 3\,.\,\rho^2 \quad - 2\alpha\{\ 0 \quad + 5\,.\,0\} \quad + \rho^2\,.\,1 = 0,
\end{aligned}$$

where the first, fifth and sixth equations are each of them an identity. The remaining equations are

$$(\rho^2-2\rho+5)(\rho^2+2\rho-1)-6A_1^2-8A_1\alpha=0,$$
$$2\rho^3A_1-6\rho A_1-32\rho\alpha-2A_1^3-4A_1=0,$$
$$2\rho(\rho^2-2\rho+5)+10\rho-4A_1^2\rho-8\alpha A_1\rho+3A_1^2=0.$$

41. Writing the first and third of these in the forms

$$-6A_1^2 \qquad -8A_1\alpha \ +(\rho^2-2\rho+5)(\rho^2+2\rho-1)=0,$$
$$A_1^2(\rho^2-4\rho+3)-8A_1\alpha\rho+(\rho^2-2\rho+5)\,2\rho \qquad =0,$$

they determine A_1^2, $8A_1\alpha$ in terms of ρ; viz. we find

$$A_1^2=(\rho^2-2\rho+5)\ \rho,$$
$$8A_1\alpha=(\rho^2-2\rho+5)(\rho^2-4\rho-1);$$

and then, writing the second equation in the form

$$(\rho^3-3\rho-2)A_1^2-16\rho\alpha A_1-A_1^4=0,$$

and substituting these values of A_1^2 and $8A_1\alpha$, and omitting the factor $\rho^2-2\rho-5$, we have the identity

$$\rho(\rho^3-3\rho-2)-2\rho(\rho^2-4\rho-1)-\rho^2(\rho^2-2\rho+5)=0;$$

viz. the second equation is then also satisfied.

Forming the square of $8A_1\alpha$, and for A_1^2 substituting its value, then omitting a factor $\rho^2-2\rho+5$, we find

$$64\rho\alpha^2=(\rho^2-2\rho+5)(\rho^2-4\rho-1)^2,$$
$$=\rho^6-10\rho^5+35\rho^4-60\rho^3+55\rho^2+38\rho+5;$$

or, as this may also be written,

$$64\rho\ (\alpha^2-1)=(\rho-1)^5(\rho-5),$$

and we then have also, as before,

$$64\sigma(\beta^2-1)=(\sigma-1)^5(\sigma-5),$$

which two equations determine the MM-curve.

The coefficient A_1 is given by the foregoing equation for $8A_1\alpha$, say the value is

$$A_1=\frac{1}{8\alpha}(\rho^2-2\rho+5)(\rho^2-4\rho-1).$$

The value $A_1=\dfrac{\rho\Pi_1}{-\rho+1}$, obtained in No. 29, on substituting for Π_1 its value, is

$$A_1=\frac{\frac{1}{3}(\beta\rho^3-\alpha\rho)}{\rho-1},$$

and these two values are, in fact, equivalent in virtue of the value of β obtained in No. 9.

42. I consider the case $n=7$, in order to show the form of the equations which have to be solved; these equations are

$$2A_1 - 2A_1 . 1 = 0,$$

$$12A_2 - 2A_1^2 - 2A_1 . 2A_1 - 7 . 1 - 2\alpha (2A_1 + 1 . 2A_1) + \rho^2 . \rho^2 = 0,$$

$$30\rho - 2A_1A_2 + 2A_1 - 2A_1(2A_2 + A_1^2) - 5 . 2A_1 - 2\alpha (12A_2 - 2A_1^2 + 2(2A_2 + A_1^2)) + \rho^2 . 2\rho A_2 = 0,$$

$$8A_1\rho - 4A_2^2 + 12A_2 - 2A_1^2 - 2A_1(2\rho + 2A_1A_2) - 3(2A_2 + A_1^2) - 2\alpha (30\rho - 2A_1A_2 + 3(2\rho + 2A_1A_2)) + \rho^2 (2\rho A_1 + A_2^2) = 0,$$

$$-6A_2\rho + 30\rho - 2A_1A_2 - 2A_1(2A_1\rho + A_2^2) - 1(2\rho + 2A_1A_2) - 2\alpha (8A_1\rho - 4A_2^2 + 4(2A_1\rho + A_2^2)) + \rho^2(2\rho + 2A_1A_2) = 0,$$

$$-6\rho^2 \quad + 8A_1\rho - 4A_2^2 - 2A_1(2A_2\rho) + 1(2A_1\rho + A_2^2) - 2\alpha(-6A_2\rho + 5(2A_2\rho)) + \rho^2(2A_2 + A_1^2) = 0,$$

$$0 \quad -6A_2\rho - 2A_1 . \rho^2 + 3(2A_2\rho) - 2\alpha(-6\rho^2 + 6 . \rho^2) + \rho^2 . 2A_1 = 0,$$

$$0 \quad -6\rho^2 + 5 . \rho^2 - 2\alpha(0 + 7 . 0) + \rho^2 . 1 = 0;$$

viz. the first, seventh and eighth equations are satisfied identically, and there remain five equations connecting ρ, α, A_1, A_2.

These equations* should lead to the before-mentioned $\alpha\beta$-modular equation

$$\rho^8 - 28\rho^6 - 112\alpha\rho^5 - 210\rho^4 - 224\alpha\rho^3 + (-1484 + 1344\alpha^2)\rho^2 + (-560\alpha + 512\alpha^3)\rho + 7 = 0,$$

and to expressions for A_1, A_2 as rational functions of α, ρ: and they should be, all five of them, satisfied by these results; but I do not see how the results are to be worked out; there is, so far as appears, no clue to the discovery of the rational functions of α, ρ.

The Jacobi Partial Differential Equation. Art. Nos. 43 to 48.

43. Writing, as above, 2α in place of Jacobi's α, this is

$$(1 - 2\alpha x^2 + x^4)\frac{d^2z}{dx^2} + (n-1)(2\alpha x - 2x^3)\frac{dz}{dx} + n(n-1)x^2z - 4n(\alpha^2 - 1)\frac{dz}{dx} = 0,$$

satisfied by the numerator and denominator U, V, each of them taken with the same proper value of the coefficient A_0, or, what is the same thing, by the values

$$U = \Theta x(A_s + A_{s-1}x^2 + A_{s-2}x^4 + \ldots + A_1x^{2s-2} + x^{2s}),$$

$$V = \Theta\ (1 + A_1x^2 + A_2x^4 + \ldots + A_{s-1}x^{2s-2} + A_sx^{2s}),$$

* [See this volume, p. 535.]

where now $A_s = \rho$ as before: Θ has its proper value; viz. disregarding an arbitrary merely numerical factor which might of course be introduced, the value is

$$\Theta = \sqrt{\frac{1}{M}\frac{\lambda'}{k'}}, \quad = \sqrt{\frac{u^2\rho}{v^2}\frac{\sqrt{1-v^8}}{\sqrt{1-u^8}}}, \quad = \sqrt{\rho}\,\frac{\sqrt[4]{v^4-v^{-4}}}{\sqrt[4]{u^4-u^{-4}}},$$

or, what is the same thing,

$$\Theta = \sqrt{\rho}\sqrt[8]{\frac{\beta^2-1}{\alpha^2-1}}.$$

If for z we write $\Theta\zeta$, then the equation becomes

$$(1-2\alpha x^2+x^4)\frac{d^2\zeta}{dx^2}+(n-1)(2\alpha x-2x^3)\frac{d\zeta}{dx}+n(n-1)x^2\zeta-4n(\alpha^2-1)\left(\frac{d\zeta}{d\alpha}+\frac{1}{\Theta}\frac{d\Theta}{d\alpha}\zeta\right)=0,$$

satisfied by the foregoing values without the factor Θ or, attending only to the denominator, say by the value

$$V = 1 + A_1x^2 + A_2x^4 + \ldots + A_{s-1}x^{2s-2} + \rho x^{2s}.$$

44. To calculate the value of $\frac{1}{\Theta}\frac{d\Theta}{d\alpha}$, we have

$$\frac{1}{\Theta}\frac{d\Theta}{d\alpha} = \frac{\frac{1}{2}}{\rho}\frac{d\rho}{d\alpha} + \frac{\frac{1}{4}\beta}{\beta^2-1}\frac{d\beta}{d\alpha} - \frac{\frac{1}{4}\alpha}{\alpha^2-1};$$

but it has been seen (No. 10) that we have

$$\frac{d\beta}{d\alpha} = \frac{\rho^2}{n}\frac{\beta^2-1}{\alpha^2-1},$$

and the formula thus becomes

$$\frac{1}{\Theta}\frac{d\Theta}{d\alpha} = \frac{\frac{1}{2}}{\rho}\frac{d\rho}{d\alpha} + \frac{\frac{1}{4n}(\beta\rho^2-n\alpha)}{\alpha^2-1}.$$

We have, as the first of the equations obtained by substituting in the partial differential equation,

$$2A_1 - 4n(\alpha^2-1)\frac{1}{\Theta}\frac{d\Theta}{d\alpha} = 0,$$

and we have hence the value of the first coefficient,

$$A_1 = n(\alpha^2-1)\frac{1}{\rho}\frac{d\rho}{d\alpha} + \tfrac{1}{2}(\beta\rho^2-n\alpha);$$

or we may, by means of this result, get rid of the term $\frac{1}{\Theta}\frac{d\Theta}{d\alpha}$ from the partial differential equation; viz. the equation may be written

$$(1-2\alpha x^2+x^4)\frac{d^2\zeta}{dx^2}+(n-1)(2\alpha x-2x^3)\frac{d\zeta}{dx}+\{n(n-1)x^2-2A_1\}\zeta-4n(\alpha^2-1)\frac{d\zeta}{d\alpha}=0.$$

Before going further, I remark that the last of the equations obtained by the substitution gives the coefficient A_{s-1}; but this is also given in terms of A_1 by the formula No. 33, $A_{s-1} - \rho A_1 = \frac{1}{3}(\alpha\rho - \beta\rho^3)$; combining the two formulæ, we have

$$A_1 = \frac{1}{\rho} n(\alpha^2-1)\frac{d\rho}{d\alpha} - \tfrac{1}{2}n\alpha + \tfrac{1}{2}\beta\rho^2,$$

$$A_{s-1} = n(\alpha^2-1)\frac{d\rho}{d\alpha} + (-\tfrac{1}{2}n+\tfrac{1}{3})\,\alpha\rho + \tfrac{1}{6}\beta\rho^3.$$

45. In the case $n=3$, we have $A_{s-1} = A_0 = 1$, $A_1 = \rho$; the two equations become

$$3(\alpha^2-1)\frac{d\rho}{d\alpha} - \tfrac{3}{2}\alpha\rho - \rho^2 + \tfrac{1}{2}\beta\rho^3 = 0,$$

$$3(\alpha^2-1)\frac{d\rho}{d\alpha} - 1 - \tfrac{7}{6}\alpha\rho + \tfrac{1}{6}\beta\rho^3 = 0,$$

each of which is easily verified.

I remark also that, in the same case, $(n=3)$, we have

$$\rho^4\frac{\beta^2-1}{\alpha^2-1} = \left(\frac{\rho^2-9}{\rho^2-1}\right)^2, \text{ and thence } \Theta = \sqrt{\rho}\sqrt[4]{\frac{\beta^2-1}{\alpha^2-1}} = \sqrt[4]{\frac{\rho^2-9}{\rho^2-1}};$$

hence

$$\frac{1}{\Theta}\frac{d\Theta}{d\rho} = \frac{4\rho}{(\rho^2-1)(\rho^2-9)};$$

and writing the equation $A_1 - 2n(\alpha^2-1)\dfrac{1}{\Theta}\dfrac{d\Theta}{d\alpha} = 0$ in the form

$$\rho - 2n(\alpha^2-1)\frac{1}{\Theta}\frac{d\Theta}{d\rho}\frac{d\rho}{d\alpha} = 0,$$

we can verify this equation.

46. In the case $n=5$, we have for A_1 two equations, each ultimately giving the foregoing value

$$A_1 = \frac{1}{8\alpha}(\rho^2-2\rho+5)(\rho^2-4\rho-1).$$

Moreover, the equation $\Theta = \sqrt{\rho}\sqrt[4]{\dfrac{\beta^2-1}{\alpha^2-1}}$ gives, without difficulty, $\Theta = \dfrac{1}{\sqrt{\rho}}\dfrac{\rho-5}{\rho-1}$.

47. In the case $n=7$, the formulæ give the two coefficients A_1, A_2; viz. we have

$$A_1 = \frac{1}{\rho}7(\alpha^2-1)\frac{d\rho}{d\alpha} - \tfrac{7}{2}\alpha + \tfrac{1}{2}\beta\rho^2,$$

$$A_2 = 7(\alpha^2-1)\frac{d\rho}{d\alpha} - \tfrac{19}{6}\alpha\rho + \tfrac{1}{6}\beta\rho^3,$$

where the value of $\dfrac{d\rho}{d\alpha}$ must of course be obtained from the before-mentioned $\rho\alpha$-equation (given in No. 7). I have not considered these results nor endeavoured to compare them with the results for this case, obtained in the Transformation Memoir and the addition thereto, [578, 692].

48. Substituting the value $1+A_1x^2+A_2x^4+\ldots+A_{s-1}x^{2s-2}+\rho x^{2s}$ in the last-mentioned form of the partial differential equation, we obtain

$$\begin{aligned}
2A_1 &= \quad 2A_1,\\
12A_2 &= -\ \ 4(n-2)\,\alpha A_1+2A_1^2 \quad - n(n-1) \qquad\qquad + 4n(\alpha^2-1)\frac{dA_1}{d\alpha},\\
30A_3 &= -\ \ 8(n-4)\,\alpha A_2+2A_1A_2-(n-2)(n-3)\,A_1+4n(\alpha^2-1)\frac{dA_2}{d\alpha},\\
56A_4 &= -12(n-6)\,\alpha A_3+2A_1A_3-(n-4)(n-5)\,A_2+4n(\alpha^2-1)\frac{dA_3}{d\alpha},\\
&\ldots\ldots\ldots\ldots\ldots\ldots\ldots\ldots\ldots\ldots\\
&\ldots\ldots\ldots\ldots\ldots\ldots\ldots\ldots\ldots\ldots
\end{aligned}$$

The number of equations is of course finite and $=s+2$, but the last equation is an identity. To obtain the last equation but one, it is convenient to write down the general equation; viz. this is

$$\begin{aligned}
(2r+1)(2r+2)\,A_{r+1} = &-4r(n-2r)\,\alpha A_r+2A_1A_r\\
&-(n-2r+1)(n-2r+2)\,A_{r-1}+4n(\alpha^2-1)\frac{dA_r}{d\alpha};
\end{aligned}$$

and then, writing herein $r=s$, we have

$$\begin{aligned}
0 = &-4s(n-2s)\,\alpha\rho+2A_1\rho\\
&-(n-2s+1)(n-2s+2)\,A_{s-1}+4n(\alpha^2-1)\frac{d\rho}{d\alpha};
\end{aligned}$$

viz. for n substituting its value $2s+1$, the equation is

$$0=-2(n-1)\,\alpha\rho+2A_1\rho-6A_{s-1}+4n(\alpha^2-1)\frac{d\rho}{d\alpha}.$$

Recapitulation of Formulæ for the Cases $n=3$ and $n=5$. Art. Nos. 49 and 50.

49. In conclusion, it will be convenient to collect the formulæ as follows:

$$n=3, \qquad y=\frac{x(\rho+x^2)}{1+\rho x^2}, \quad \Theta=\sqrt{\frac{\rho^2-9}{\rho^2-1}},$$

$$8\alpha\rho=\rho^4-6\rho^2-3,$$

$$8(\alpha+1)\,\rho=(\rho-1)^3(\rho+3),\ \ 8(\alpha-1)\,\rho=(\rho+1)^3(\rho-3),$$

$$\sigma=-\frac{3}{\rho}, \qquad 8\beta\sigma=\sigma^4-6\sigma^2-3,$$

$$8(\beta+1)\,\sigma=(\sigma-1)^3(\sigma+3),\ \ 8(\beta-1)\,\sigma=(\sigma-1)^3(\sigma+3);$$

$\alpha\beta$-equation, see No. 2.

50. $n=5,\quad y=\dfrac{x(\rho+A_1x^2+x^4)}{1+A_1x^2+\rho x^4},\quad \Theta=\dfrac{1}{\sqrt{\rho}}\dfrac{\rho-5}{\rho-1},\quad A_1=\dfrac{1}{8\alpha}(\rho^2-2\rho+5)(\rho^2-4\rho-1),$

$$64\alpha^2\rho=(\rho^2-4\rho-1)^2(\rho^2-2\rho+5),$$

or say

$$8\alpha\sqrt{\rho}=(\rho^2-4\rho-1)\sqrt{\rho^2-2\rho+5}:$$

$$64(\alpha^2-1)\rho=(\rho-1)^5(\rho-5),$$

$$\sigma=\frac{5}{\rho},\qquad 64\beta^2\sigma=(\sigma^2-4\sigma-1)^2(\sigma^2-2\sigma+5),$$

$$-8\beta\sqrt{\sigma}=(\sigma^2-4\sigma-1)\sqrt{\sigma^2-2\sigma+5},$$

$$64(\beta^2-1)\sigma=(\sigma-1)^5(\sigma-5),$$

$$\frac{\beta}{\alpha}=-\frac{\rho^2+20\rho-5}{\rho^2(\rho^2-4\rho-1)};$$

$\alpha\beta$-equation, see No. 3.

The $\rho\alpha$-equations for the cases in question, $n=3$ and $n=5$, are the so-called Jacobian equations of the fourth and the sixth degrees, studied by Brioschi (in the third appendix above referred to) and by others: the foregoing $\alpha\beta$-equations have not (so far as I am aware) been previously obtained; as rationally connected with the $\rho\alpha$-equations, they must belong to the same class of equations.

Cambridge, England, December 18, 1886.

870.

ON THE TRANSFORMATION OF ELLIPTIC FUNCTIONS (SEQUEL).

[From the *American Journal of Mathematics*, vol. x. (1888), pp. 71—93.]

THE chief object of the present paper is the further development of the $\rho\alpha\beta$-theory in the case $n=7$. I recall that the forms are

$$\frac{dy}{\sqrt{1-2\beta y^2+y^4}}=\frac{\rho dx}{\sqrt{1-2\alpha x^2+x^4}},$$

where

$$y=\frac{x(\rho+A_2x^2+A_1x^4+x^6)}{1+A_1x^2+A_2x^4+\rho x^6}.$$

The paragraphs are numbered consecutively with those of the former paper "On the Transformation of Elliptic Functions," vol. IX., pp. 193—224, [869].

The Seventhic Transformation: the ρα-Equation. Art. Nos. 51 to 57.

51. The equation is given incorrectly Nos. 7 and 42; there was an error of sign in a term $512\alpha^3\rho$, which affected also the coefficient of $\alpha\rho$, and an error of sign in the absolute term 7. The correct form is

$$\rho^8-28\rho^6-112\alpha\rho^5-210\rho^4-224\alpha\rho^3+(-1484+1344\alpha^2)\rho^2+(464\alpha-512\alpha^3)\rho-7=0;$$

or, arranging in powers of α, this is

$$\begin{aligned}&\alpha^3\,.\,512\rho\\ &+\alpha^2\,.\,-1344\rho^2\\ &+\alpha\,.\,112\rho^5+224\rho^3-464\rho\\ &\qquad-(\rho^8-28\rho^6-210\rho^4-1484\rho^2-7)=0.\end{aligned}$$

This may also be written in the forms

$$(\alpha-1)\{\alpha^2 . 512\rho+\alpha(-1344\rho^2+512\rho)+112\rho^5+224\rho^3-1344\rho^2+48\rho\}-(\rho+1)^7(\rho-7)=0,$$

and

$$(\alpha+1)\{\alpha^2 . 512\rho+\alpha(-1344\rho^2-512\rho)+112\rho^5+224\rho^3+1344\rho^2+48\rho\}-(\rho-1)^7(\rho+7)=0.$$

To simplify the $\rho\alpha$-equation, we assume $A=8\rho\alpha-7\rho^2$; then the $A\rho$-equation is

$$\begin{aligned}&A^3\\ &+A\rho^2(14\rho^4-119\rho^2-58)\\ &-\ \ \rho^2(\rho^8-126\rho^6+280\rho^4-1078\rho^2-7)=0;\end{aligned}$$

viz. this is a cubic equation wanting its second term, and so at once solvable by Cardan's formula: say the equation is

$$A^3+A\rho^2q_1-\rho^2r_1=0,$$

where

$$\begin{aligned}q_1&=14\rho^4-119\rho^2-58,\\ r_1&=\ \ \rho^8-126\rho^6+280\rho^4-1078\rho^2-7.\end{aligned}$$

It is convenient to recall here that, writing $\sigma=-\frac{7}{\rho}$, and $B=8\sigma\beta-7\sigma^2$, we have between σ, β, B precisely the same equations as between ρ, α, A; $\rho=1$ gives $\sigma=-7$, and we have as corresponding values $\alpha=-1$, $A=-15$, $\beta=-1$, $B=-287$: these are very convenient for verification of the formulæ. Similarly, $\rho=-7$ gives $\sigma=1$, and then $\alpha=-1$, $A=-287$, $\beta=-1$, $B=-15$; but I have, in general, used the former values only.

52. We have

$$A=f+g,$$

where

$$\begin{aligned}3fg&=-\rho^2q_1,\\ f^3+g^3&=\ \ \rho^2r_1,\end{aligned}$$

and thence

$$f^3-g^3=\rho^2\sqrt{r_1^2+\frac{4\rho^2q_1^3}{27}}.$$

We have identically

$$\begin{aligned}27(\rho^8-126\rho^6+280\rho^4-1078\rho^2-7)^2+4\rho^2(14\rho^4-119\rho^2-58)^3&\\ =(\rho^6+75\rho^4-141\rho^2+1)^2(27\rho^4+122\rho^2+1323)&:\end{aligned}$$

if $\rho=1$, this is $27 . 930^2+4(-163)^3=64^2 . 1472$; that is, $23352300-17322988=6029312$, which is right; but it is convenient to divide by 27, so as instead of $27\rho^4+122\rho^2+1323$ to have in the formulæ $\rho^4+\frac{122}{27}\rho^2+49$, or say

$$\rho^4+K\rho^2+49\ \ (K=\tfrac{122}{27}).$$

Hence writing

$$t_1 = \rho^6 + 75\rho^4 - 141\rho^2 + 1,$$

$$\delta = \rho^4 + K\rho^2 + 49,$$

we have

$$r_1^2 + \tfrac{4}{27}\rho^2 q_1 = t_1^2\delta,$$

and consequently

$$2f^3 = \rho^2(r_1 + t_1\sqrt{\delta}),$$

$$2g^3 = \rho^2(r_1 - t_1\sqrt{\delta}).$$

53. It was easy to foresee that the cube root of $r_1 \pm t_1\sqrt{\delta}$ would break up into the form $(U \pm \sqrt{\delta})\sqrt[3]{W \pm \sqrt{\delta}}$, and I was led to the actual expressions by the identities

$$20(14\rho^4 - 119\rho^2 - 58) = (19\rho^2 - 53)^2 - 3(27\rho^4 + 122\rho^2 + 1323);$$

that is,

$$20q_1 = (19\rho^2 - 53)^2 - 81\delta,$$

and

$$27(\rho^2 - 7)^2 - (27\rho^4 + 122\rho^2 + 1323) = -500\rho^2,$$

$$27(\rho^2 + 7)^2 - (27\rho^4 + 122\rho^2 + 1323) = 256\rho^2;$$

or, as these may be written,

$$(\rho^2 - 7)^2 - \delta = -\tfrac{500}{27}\rho^2, \quad (\rho^2 + 7)^2 - \delta = \tfrac{256}{27}\rho^2.$$

We, in fact, have further the two identities

$$\begin{aligned} &1000(\rho^6 + 75\rho^4 - 141\rho^2 + 1) \\ &\quad = \{(19\rho^2 - 53)^3 + 243(19\rho^2 - 53)(\rho^4 + K\rho^2 + 49)\} \\ &\qquad + \{27(19\rho^2 - 53)^2 + 729(\rho^4 + K\rho^2 + 49)\}(-\rho^2 + 7), \end{aligned}$$

$$\begin{aligned} &-1000(\rho^8 - 126\rho^6 + 280\rho^4 - 1078\rho^2 - 7) \\ &\quad = \{(19\rho^2 - 53)^3 + 243(19\rho^2 - 53)(\rho^4 + K\rho^2 + 49)\}(-\rho^2 + 7) \\ &\qquad + \{27(19\rho^2 - 53)^2 + 729(\rho^4 + K\rho^2 + 49)\}(\rho^4 + K\rho^2 + 49), \end{aligned}$$

viz. writing

$$19\rho^2 - 53 = 9U, \quad -\rho^2 + 7 = W,$$

these equations become

$$\tfrac{1000}{729}t_1 = U^3 + 3U\delta + (3U^2 + \delta)W,$$

$$-\tfrac{1000}{729}r_1 = (U^3 + 3U\delta)W + (3U^2 + \delta)\delta,$$

and we have thus

$$-\tfrac{1000}{729}(r_1 - t_1\sqrt{\delta}) = (U + \sqrt{\delta})^3(W + \sqrt{\delta}),$$

and the like equation with $-\sqrt{\delta}$ in place of $\sqrt{\delta}$.

54. In part verification of the last-mentioned identities, observe that, in the first of them, putting $\rho = 1$, and comparing first the coefficients of ρ^6 and then the coefficients of ρ^0, we ought to have

$$1000 = 19^3 + 243.19 - (27.19^2 + 729), \; = 11476 - 10476,$$

$$1000 = (-53^3 - 243.53.49) + (27.53^2 + 729.49)7, \; = -779948 + 780948,$$

which are right; and similarly in the second equation, comparing first the coefficients of ρ^8 and next those of ρ^0, we have

$$\begin{aligned}-1000 &= (19^3+243.19)(-1)+(27.19^2+729), = -11476+10476,\\ +7000 &= (-53^3-243.53.49)(7)+(27.53^2+729.49)49,\\ &= -5459636+5466636,\end{aligned}$$

which are right.

55. We have now $A=f+g$, where

$$f=-\tfrac{9}{10}(U-\sqrt{\delta})\sqrt[3]{\tfrac{1}{2}\rho^2(W-\sqrt{\delta})},$$
$$g=-\tfrac{9}{10}(U+\sqrt{\delta})\sqrt[3]{\tfrac{1}{2}\rho^2(W+\sqrt{\delta})}:$$

observe that, multiplying these two values, we have

$$fg=\tfrac{81}{100}(U^2-\delta)\sqrt[3]{\tfrac{1}{4}\rho^4.(W^2-\delta)}, =\tfrac{81}{100}(U^2-\delta)\sqrt[3]{\tfrac{1}{4}\rho^4.\frac{-500}{27}\rho^2},$$
$$=\tfrac{81}{100}(U^2-\delta)(-\tfrac{5}{3}\rho^2);$$

that is,

$$fg=-\tfrac{27}{20}\rho^2(U^2-\delta), =-\tfrac{27}{20}\rho^2.\frac{20q_1}{81}=-\tfrac{1}{3}\rho^2q_1,$$

which is right. Or, finally, substituting for U, W, δ their values, we have, for the solution of the $A\rho$-equation, $A=f+g$, where

$$f=-\tfrac{9}{10}(19\rho^2-53-\sqrt{\rho^4+K\rho^2+49})\sqrt[3]{\tfrac{1}{2}\rho^2\{-\rho^2+7-\sqrt{\rho^4+K\rho^2+49}\}},\ (K=\tfrac{122}{27}),$$
$$g=-\tfrac{9}{10}(19\rho^2-53+\sqrt{\rho^4+K\rho^2+49})\sqrt[3]{\tfrac{1}{2}\rho^2\{-\rho^2+7+\sqrt{\rho^4+K\rho^2+49}\}}.$$

56. In the case $\rho=1$, α has a value $=-1$, giving for A, $=8\rho\alpha-7\rho^2$, the value -15; and, in fact, here $\rho^2=1$, and the $A\rho$-equation becomes

$$A^3-163A+930=0,$$

that is,

$$(A+15)(A^2-15A+62)=0,$$

the roots thus being

$$A=-15,\quad A=\tfrac{1}{2}(15\pm i\sqrt{23}).$$

To verify in this case the values given by the solution of the cubic equation, observe that, for $\rho^2=1$, we have $\delta=50+\tfrac{122}{27}, =\tfrac{1472}{27}$, and therefore $\sqrt{\delta}=\dfrac{8\sqrt{23}}{3\sqrt{3}}, =\dfrac{8\sqrt{69}}{9}$; also, $U=\dfrac{19\rho^2-53}{9}, =\dfrac{-34}{9}$, and $W=-\rho^2+7$, $=6$. Hence $U+\sqrt{\delta}=\dfrac{-34+8\sqrt{69}}{9}$, and

$$\sqrt[3]{\tfrac{1}{2}\rho^2(W+\sqrt{\delta})}=\sqrt[3]{3+\frac{8\sqrt{69}}{9}}, =\frac{\sqrt[3]{81+12\sqrt{69}}}{3};$$

hence

$$g=-\tfrac{9}{10}\frac{2(-17+4\sqrt{69})}{9}\tfrac{1}{3}\sqrt[3]{81+12\sqrt{69}}, =\tfrac{1}{15}(17-4\sqrt{69})\sqrt[3]{81+12\sqrt{69}};$$

but the cube root is $=\frac{1}{2}(3+\sqrt{69})$, and we have $(17-4\sqrt{69})(3+\sqrt{69})=-225+5\sqrt{69}$, $=5(-45+\sqrt{69})$; that is, $g=\frac{1}{6}(-45+\sqrt{69})$. Similarly, $f=\frac{1}{6}(-45-\sqrt{69})$. We have thus the real root $f+g=-15$, and the imaginary roots

$$f\omega+g\omega^2 \text{ or } f\omega^2+g\omega, \ =-\tfrac{15}{2}(\omega+\omega^2)+\tfrac{1}{6}\sqrt{69}\,(\omega-\omega^2),$$

viz. the first term is $=\frac{15}{2}$ and the second is $\pm\frac{1}{6}\sqrt{69}\,.\,i\sqrt{3}$, $=\pm\frac{1}{2}i\sqrt{23}$; thus the roots are $\frac{1}{2}(15\pm i\sqrt{23})$, as they should be.

57. I found, by considerations arising out of the new theory Nos. 72 *et seq.*, that writing for shortness $m=i\sqrt{3}$, then, for $\rho=m-2$, the $\rho\alpha$-equation has a root $\alpha=m$; the corresponding values of A, ρ^2 thus are $A=12m-31$, $\rho^2=-4m+1$, viz. substituting this value for ρ^2 in the $A\rho$-equation, there should be a root $A=12m-31$. The equation becomes

$$A^3+A\,(3704m-7653)+148306m+206162=0,$$

or, as this may be written,

$$(A-12m+31)\,\{A^2+A\,(12m-31)+2960m+4062\}=0,$$

and the roots thus are

$$A=\ 12m-31,$$
$$A=-6m+\tfrac{31}{2}\pm\tfrac{1}{2}\sqrt{-12584m-16777},$$

where the square root is not expressible as a rational function of m.

Expression of β as a Rational Function of α, ρ. Art. Nos. 58 to 66.

58. Writing $\sigma=-\dfrac{7}{\rho}$, we have β the same function of σ that ρ is of α; hence if $B=8\sigma\beta-7\sigma^2$, the $B\sigma$-equation is

$$\begin{aligned}&B^3\\&+B\sigma^2(14\sigma^4-119\sigma^2-58)\\&-\sigma^2(\sigma^8-126\sigma^6+280\sigma^4-1078\sigma^2-7)=0;\end{aligned}$$

and the expression for B in terms of σ is obtained from that of A by the mere change of ρ into σ. Say we have $B=f'+g'$, where

$$f'=-\tfrac{9}{10}(U'-\sqrt{\delta'})\sqrt[3]{\tfrac{1}{2}\sigma^2(W'-\sqrt{\delta'})},$$
$$g'=-\tfrac{9}{10}(U'+\sqrt{\delta'})\sqrt[3]{\tfrac{1}{2}\sigma^2(W'+\sqrt{\delta'})};$$

then we have

$$\tfrac{1}{2}\sigma^2(W'+\sqrt{\delta'})=\tfrac{1}{2}\frac{49}{\rho^2}\left(-\frac{49}{\rho^2}+7+\sqrt{\frac{2401}{\rho^4}+\frac{49K}{\rho^2}+49}\right)$$

$$=-\tfrac{1}{2}\,.\,\frac{343}{\rho^4}(-\rho^2+7-\sqrt{\rho^4+K\rho^2+49})$$

$$=-\frac{343}{\rho^6}\,.\,\tfrac{1}{2}\rho^2(W-\sqrt{\delta}),$$

or, say

$$\sqrt[3]{\tfrac{1}{2}\sigma^2(W'-\sqrt{\delta'})} = -\frac{7}{\rho^2}\sqrt[3]{\tfrac{1}{2}\rho^2(W-\sqrt{\delta})};$$

and similarly

$$\sqrt[3]{\tfrac{1}{2}\sigma^2(W'+\sqrt{\delta'})} = -\frac{7}{\rho^2}\sqrt[3]{\tfrac{1}{2}\rho^2(W+\sqrt{\delta})}.$$

The cube roots which enter into the expression of B are thus identical with those in the expression of A, and it hence appears that B can be expressed rationally in terms of A, ρ; or, what is the same thing, β can be expressed rationally in terms of α, ρ.

59. The *à priori* reason is obvious: the $\rho\alpha$-equation is a cubic in α, but of the order 8 in ρ; hence, to a given value of α, there correspond 8 values of ρ. Similarly, the $\sigma\beta$-equation is a cubic in β, but it is of the order 8 in σ; or if for σ we substitute its value $= -\frac{7}{\rho}$, then we have a $\rho\beta$-equation which is a cubic in β, but it is of the order 8 in ρ. In the absence of any special relation between this $\rho\beta$-equation and the $\rho\alpha$-equation, there would correspond, to each of the 8 values of ρ, 3 values of β; that is, to a given value of α there would correspond 8×3, $=24$ values of β. But, in fact, to a given value of α, there correspond only 8 values of β, and the two cubic equations are related to each other in such wise that this is so; viz. the relation between them is such that it is possible by means of them to express β as a rational function of ρ, α.

60. Returning to the investigation, we have

$$9U' = 19\sigma^2 - 53, \ = \frac{19 \cdot 49}{\rho^2} - 53;$$

or, writing

$$63\overline{U} = 53\rho^2 - 931,$$

this is

$$U' = -\frac{7}{\rho^2}\overline{U}, \text{ whence } U' \pm \sqrt{\delta'} = -\frac{7}{\rho^2}(\overline{U} \mp \sqrt{\delta}).$$

Hence writing

$$\theta = \sqrt[3]{\tfrac{1}{2}\rho^2(W-\sqrt{\delta})}, \quad \phi = \sqrt[3]{\tfrac{1}{2}\rho^2(W+\sqrt{\delta})},$$

we have

$$f = -\tfrac{9}{10}(U-\sqrt{\delta})\,\theta, \quad f' = -\tfrac{9}{10}\frac{49}{\rho^4}(\overline{U}+\sqrt{\delta})\,\phi,$$

$$g = -\tfrac{9}{10}(U+\sqrt{\delta})\,\phi, \quad g' = -\tfrac{9}{10}\frac{49}{\rho^4}(\overline{U}-\sqrt{\delta})\,\theta,$$

so that, putting for shortness

$$L = -\tfrac{9}{10}(U-\sqrt{\delta}) \ , \ \overline{L} = -\tfrac{9}{10}\frac{49}{\rho^4}(\overline{U}-\sqrt{\delta}),$$

$$M = -\tfrac{9}{10}(U+\sqrt{\delta}) \ , \ \overline{M} = -\tfrac{9}{10}\frac{49}{\rho^4}(\overline{U}+\sqrt{\delta}),$$

we have

$$A = L\theta + M\phi,\quad B = \bar{L}\theta + \bar{M}\phi,$$

where θ^3, ϕ^3 and $\theta\phi$ are each of them free from any cube root; we have, in fact,

$$\theta\phi = \sqrt[3]{\tfrac{1}{4}\rho^4 (W^2 - \delta)},\quad = \sqrt[3]{\tfrac{1}{4}\rho^4 . \frac{-500}{27}\rho^2},\quad = -\tfrac{5}{3}\rho^2,$$

and it may be added that

$$3LM\theta\phi = -\rho^2 q_1, \text{ whence } LM = \tfrac{1}{5}q_1,$$

$$L^3\theta^3 + M^3\phi^3 = \rho^2 r_1,$$

$$L^3\theta^3 - M^3\phi^3 = \rho^2 t_1\sqrt{\delta};$$

these are, in fact, only the equations obtained by writing $L\theta$, $M\phi$ in place of f, g respectively.

61. In the case $\rho = 1$, we have $\sigma = -7$; the equation for B becomes

$$B^3 + 1358525B + 413536578 = 0;$$

that is,

$$(B + 287)(B^2 - 287B + 1440894) = 0,$$

and the roots are

$$-287 \text{ and } \tfrac{1}{2}(287 \pm 497i\sqrt{23}), \text{ or, say } -7.41 \text{ and } \tfrac{7}{2}(41 \pm 71i\sqrt{23}).$$

We have as before, $\sqrt{\delta} = \dfrac{8\sqrt{69}}{9}$, and $\sqrt[3]{\tfrac{1}{2}\rho^2(W + \sqrt{\delta})} = \tfrac{1}{3}\sqrt[3]{81 + 12\sqrt{69}} = \theta$; also $\bar{U} = \dfrac{-878}{63}$, whence $\bar{U} + \sqrt{\delta} = \dfrac{2(-439 + 28\sqrt{69})}{63}$. We thus have

$$f' = -\tfrac{9}{10}49 . \frac{2(-439 + 28\sqrt{69})}{63}\tfrac{1}{3}\sqrt[3]{81 + 12\sqrt{69}},$$

$$= -\tfrac{7}{15}(-439 + 28\sqrt{69})\sqrt[3]{81 + 12\sqrt{69}},$$

or, putting for the cube root its value $= \tfrac{1}{2}(3 + \sqrt{69})$, this is

$$f' = -\tfrac{7}{30}(-439 + 28\sqrt{69})(3 + \sqrt{69}),\quad = -\tfrac{287}{2} + \tfrac{497}{6}\sqrt{69}.$$

Similarly, $g' = -\tfrac{287}{2} - \tfrac{497}{6}\sqrt{69}$; and forming the values $f' + g'$, $\omega f' + \omega^2 g'$, $\omega^2 f' + \omega g'$, we have the real root -287 and the imaginary roots $\tfrac{1}{2}(287 \pm 497i\sqrt{23})$, as above.

62. We have the equations

$$B = \bar{L}\theta \quad + \bar{M}\phi,$$

$$A = L\theta \quad + M\phi,$$

$$A^2 - 2LM\theta\phi = \frac{M^2\phi^3}{\theta\phi}\theta + \frac{L^2\theta^3}{\theta\phi}\phi,$$

from which, eliminating θ and ϕ so far as they present themselves linearly on the right-hand side, and in the resulting equation replacing $\theta\phi$ and $LM\theta\phi$ by their values, we have

$$\begin{vmatrix} B, & \overline{L}, & \overline{M} \\ A, & L, & M \\ -\tfrac{5}{3}\rho^2(A^2+\tfrac{2}{3}\rho^2 q_1), & M^2\phi^3, & L^2\theta^3 \end{vmatrix} = 0;$$

that is,

$$B(L^3\theta^3 - M^3\phi^3) = A(L^2\overline{L}\theta^3 - M^2\overline{M}\phi^3) - \tfrac{5}{3}\rho^2(A^2+\tfrac{2}{3}\rho^2 q_1)(L\overline{M} - \overline{L}M).$$

This may be written

$$B\rho^2 t_1\sqrt{\delta} = A\left\{-\tfrac{729}{1000}\frac{49}{\rho^4}[(U-\sqrt{\delta})^2(\overline{U}-\sqrt{\delta})\tfrac{1}{2}\rho^2(W-\sqrt{\delta}) - (U+\sqrt{\delta})^2(\overline{U}+\sqrt{\delta})\tfrac{1}{2}\rho^2(W+\sqrt{\delta})]\right\}$$
$$-\tfrac{5}{3}\rho^2(A^2+\tfrac{2}{3}\rho^2 q_1)\tfrac{81}{100}\frac{49}{\rho^4}[(U-\sqrt{\delta})(\overline{U}+\sqrt{\delta}) - (U+\sqrt{\delta})(\overline{U}-\sqrt{\delta})],$$

where the terms in [ ] contain each of them the factor $\sqrt{\delta}$. Omitting this factor from the equation, and multiplying by ρ^2, we have

$$B\rho^4 t_1 = \tfrac{81}{100}49\{\tfrac{9}{10}A[(U^2+2U\overline{U}+\delta)W + U^2\overline{U} + (2U+\overline{U})\delta] - \tfrac{10}{3}(A^2+\tfrac{2}{3}\rho^2 q_1)(U-\overline{U})\},$$

which I verify at this stage by writing, as before, $\rho=1$. We have $B=-287$, $A=-15$, $t_1=-64$, $q_1=-163$, $W=6$, $U=-\tfrac{34}{9}$, $\overline{U}=\dfrac{-878}{63}$; and, omitting intermediate steps, the equation becomes

$$287\,.\,64 = \frac{81\,.\,49}{100}\left(\tfrac{2496000}{567} - \tfrac{2233600}{567}\right),\ = \frac{81\,.\,49}{100\,.\,567}262400,\ = 18368,$$

which is right.

63. We require the values of $(U^2+2U\overline{U}+\delta)W + U^2\overline{U} + (2U+\overline{U})\delta$, and of $U-\overline{U}$: I insert some of the steps of the calculation. We have

$$U^2+2U\overline{U}+\delta = \frac{1}{63^2}\{(133\rho^2-371)(239\rho^2-2233) + 63^2(\rho^4+49) + 3\,.\,49\,.\,122\rho^2\}$$
$$= \frac{1}{63^2}\{35756\rho^4 - 367724\rho^2 + 1022924\}$$
$$= \frac{4}{567}\{\ 1277\rho^4 - \ 13133\rho^2 + \ \ 36533\}.$$

Multiplying by W, $=-\rho^2+7$, we have

$$(U^2+2U\overline{U}+\delta)W = \frac{4}{567}\{-\ 1277\rho^6 + \ 22072\rho^4 - \ 128464\rho^2 + \ 255731\}$$
$$= \frac{4}{5103}\{-11493\rho^6 + 198648\rho^4 - 1156176\rho^2 + 2301579\},$$

$$U^2\overline{U} = \frac{1}{81\,.\,63}(19\rho^2-53)^2(53\rho^2-931) = \frac{1}{5103}\{19133\rho^6 - 442833\rho^4 + 2023911\rho^2 - 2615179\},$$

$$(2U+\overline{U})\,\delta=\frac{1}{63\,.\,27}(319\rho^2-1673)(27\rho^4+122\rho^2+1323)$$

$$=\frac{1}{1701}\{8613\rho^6-6253\rho^4+217931\rho^2-221379\}$$

$$=\frac{1}{5103}\{25839\rho^6-\ 18759\rho^4+\ 653793\rho^2-6640137\},$$

whence

$$U^2\overline{U}+(2U+\overline{U})\,\delta=\frac{1}{5103}\{44972\rho^6-461592\rho^4+2677704\rho^2-9255316\}$$

$$=\frac{4}{5103}\{11243\rho^6-115398\rho^4+\ 669426\rho^2-2313829\}.$$

Hence, adding, we obtain

$$(U^2+2U\overline{U}+\delta)\,W+U^2\overline{U}+(2U+\overline{U})\,\delta$$

$$=\frac{4}{5103}\{-250\rho^6+83250\rho^4-486750\rho^2-12250\}$$

$$=\frac{-1000}{5103}\{\quad \rho^6-\ 333\rho^4+\ 1947\rho^2+\ \ 49\}:$$

and we have at once

$$U-\overline{U}=\frac{1}{63}(80\rho^2+560)=\frac{80}{63}(\rho^2+7).$$

64. We now find

$$B\rho^4t_1=-7A\,(\rho^6-333\rho^4+1947\rho^2+49)$$
$$-56\,(3A^2+2\rho^2q_1)(\rho^2+7),$$

viz. substituting for t_1, q_1 their values, this is

$$B\rho^4(\rho^6+75\rho^4-141\rho^2+1)=-7A\,(\rho^6-333\rho^4+1947\rho^2+49)$$
$$-56\,(3A^2+2\rho^2(14\rho^4-119\rho^2+1))(\rho^2+7),$$

which is the value of B, expressed rationally in terms of ρ, A; it will be observed that B is obtained as a quadric function of A, which is the proper form.

Writing $\rho=-1$, we have $A=-15$, $B=-287$, $t_1=-64$, $q_1=-163$, and the equation is

$$287\,.\,64=105\,.\,1664-56\,.\,349\,.\,8,\ =174720-156352,\ =18368,$$

which is right.

65. Writing for B, A their values $=-\dfrac{56}{\rho}\beta-\dfrac{343}{\rho^2}$, and $8\rho\alpha-7\rho^2$, we have

$$\rho^4\left(-\frac{56}{\rho}\beta-\frac{343}{\rho^2}\right)t_1=(-56\rho\alpha+49\rho^2)(\rho^6-333\rho^4+1947\rho^2+49)$$
$$-56\,(192\rho^2\alpha^2-336\rho^3\alpha+147\rho^4+2\rho^2q_1)(\rho^2+7);$$

that is,

$$\begin{aligned}-56\rho^3\beta t_1 = &-56\,.\,192\rho^2(\rho^2+7)\alpha^2\\ &-56\rho\alpha(\rho^6-333\rho^4+1947\rho^2+49)\\ &+56\,.\,336\rho^3\alpha(\rho^2+7)\\ &+49\rho^2(\rho^6-333\rho^4+1947\rho^2+49)\\ &-56(147\rho^4+2\rho^2(14\rho^4-119\rho^2-58))(\rho^2+7)\\ &+343\rho^2(\rho^6+75\rho^4-141\rho^2+1),\end{aligned}$$

where the fourth and sixth lines unite into a term divisible by 56, viz. omitting in the first instance a factor 49, the lines are

$$\rho^8-333\rho^6+1947\rho^4+49\rho^2,$$

and

$$7\rho^8+525\rho^6-\ 987\rho^4+\ 7\rho^2,$$

which together are

$$=8\rho^8+192\rho^6+960\rho^4+56\rho^2,$$

and hence, restoring the factor 49, the lines are

$$=392(\rho^8+24\rho^6+120\rho^4+7\rho^2),$$

and the formula now easily becomes

$$\begin{aligned}\rho^2\beta t_1 = &\ 192\rho(\rho^2+7)\alpha^2\\ &+(\rho^6-669\rho^4-405\rho^2+49)\,\alpha\\ &+\rho(21\rho^6-63\rho^4-1593\rho^2-861),\end{aligned}$$

where the last line is

$$=\rho(\rho^2+7)(21\rho^4-210\rho^2-123).$$

66. Hence, finally, substituting for t_1 its value, we have

$$\beta\rho^2(\rho^6+75\rho^4-141\rho^2+1)=3\rho(\rho^2+7)(64\alpha^2+7\rho^4-70\rho^2-41)+\alpha(\rho^6-669\rho^4-405\rho^2+49),$$

which is the expression for β as a rational function of ρ, α.

Here $\rho=1$, $\alpha=-1$, $\beta=-1$ give $64=-960+1024$, which is right; and again, $\rho=-7$, $\alpha=-1$, $\beta=-1$ give

$$\begin{aligned}-49(117649+180075-6909+1)=&-21\,.\,56(64+16807-3430-41)\\ &-(117649-1606269-19845+49);\end{aligned}$$

that is,

$$-49\,.\,290816=-1176\,.\,13400+1508416,$$

or,

$$-14249984\ =-15758400\ \ +1508416,$$

which is right.

The $\alpha\beta$-Differential Equation. Art. No. 67.

67. We have, No. 10,

$$\frac{d\beta}{\beta^2-1}=\frac{\rho^2}{7}\frac{d\alpha}{\alpha^2-1};$$

it should, of course, be possible to verify this equation by means of the $\rho\alpha$-equation and the value just obtained for β. But the expression for $\frac{d\rho}{d\alpha}$ given by the $\rho\alpha$-equation is of so complicated a form that I do not see in what way the verification will come out, and I have not attempted to effect it.

The Coefficients A_1 and A_2. Art. Nos. 68 to 71.

68. These are given by the formulæ No. 47, viz. we have

$$A_1=\frac{1}{\rho}7(\alpha^2-1)\frac{d\rho}{d\alpha}-\tfrac{7}{2}\alpha\ +\tfrac{1}{2}\beta\rho^2,$$

$$A_2=\ 7(\alpha^2-1)\frac{d\rho}{d\alpha}-\tfrac{19}{6}\alpha\rho+\tfrac{1}{6}\beta\rho^3,$$

where $\frac{d\rho}{d\alpha}$ and β have each of them to be expressed in terms of ρ, α; we have thus A_1 and A_2, each of them expressible rationally in terms of ρ, α; but I have not attempted to effect the substitutions.

69. The five equations of No. 42, merely collecting the terms, are

$$12A_2-6A_1^2-8\alpha A_1+\rho^4-7=0,$$

$$(-6A_1-32\alpha+2\rho^3)A_2-2A_1^3-8A_1+30\rho=0,$$

$$(\rho^2-4)A_2^2+(-4A_1^2-8\alpha A_1+6)A_2-5A_1^2+(2\rho^3+4\rho)A_1-72\alpha\rho=0,$$

$$-2A_1A_2^2+\{(2\rho^2-4)A_1-6\rho\}A_2-4\rho A_1^2-32\rho\alpha A_1+2\rho^3+28\rho=0,$$

$$-3A_2^2+(-4\rho A_1+2\rho^2-8\alpha\rho)A_2+\rho^2A_1^2+10\rho A_1-6\rho^2=0,$$

which would, of course, be all of them satisfied by the values of A_1, A_2 as rational functions of ρ, α, viz. the substitution of these values in any one of the equations would give a function of ρ and α, containing as a factor the expression on the left-hand side of the $\rho\alpha$-equation.

70. Or again, the equations should determine A_1 and A_2 as rational functions of ρ, α, but there is no obvious way of finding such values in a simple form. We, of course, have

$$12A_2=6A_1^2+8\alpha A_1-\rho^4+7,$$

and using this value to eliminate A_2 from the remaining equations, we find the following four equations:

$$A_1^3 . 30 + A_1^2 (120\alpha - 6\rho^3) + A_1 \{128\alpha^2 - 8\rho^3\alpha - 3\rho^4 + 69\} + \alpha(-16\rho^4 + 112) + \rho^7 - 7\rho^3 - 180\rho = 0,$$

$$\begin{aligned} A_1^4 (36\rho^2 - 432) &+ A_1^3 \alpha (96\rho^2 - 1344) \\ &+ A_1^2 \{\alpha^2 (64\rho^2 - 1024) - 12\rho^6 + 48\rho^4 + 84\rho^2 - 624\} \\ &+ A_1 \{\alpha(-16\rho^6 + 160\rho^4 + 112\rho^2 - 544) + 288\rho^3 + 576\rho\} \\ &+ \{-10368\alpha\rho + \rho^{10} - 4\rho^8 - 14\rho^6 - 16\rho^4 + 49\rho^2 - 308\} = 0, \end{aligned}$$

$$\begin{aligned} A_1^5 . 36 + A_1^4 . 96\alpha &+ A_1^3 \{64\alpha^2 - 12\rho^4 - 72\rho^2 + 208\} \\ &+ A_1^2 \{\alpha(-16\rho^4 - 96\rho^2 + 304) + 504\rho\} \\ &+ A_1 \{\alpha . 2592\rho + \rho^8 - 12\rho^6 + 10\rho^4 + 84\rho^2 - 119\} \\ &+ 36\rho^5 - 144\rho^3 - 2268\rho = 0, \end{aligned}$$

$$\begin{aligned} A_1^4 . 36 + A_1^3 (96\alpha + 96\rho) &+ A_1^2 \{64\alpha^2 + 320\alpha\rho - 12\rho^4 - 96\rho^2 + 84\} \\ &+ A_1 \{256\alpha^2\rho + \alpha(-80\rho^4 + 112) - 16\rho^5 - 368\rho\} \\ &+ \{\alpha(-32\rho^5 + 224\rho) + \rho^8 + 8\rho^6 - 14\rho^4 + 232\rho^2 + 49\} = 0, \end{aligned}$$

and we could from these equations obtain various rational expressions for A_1 and its powers; but these would apparently be of degrees far too high in ρ and α.

71. It is to be remarked that, for $\rho = 1$, $\alpha = -1$, the values of A_1, A_2 are $A_1 = A_2 = 3$, viz. these belong to the solution

$$y = \frac{x(1 + 3x^2 + 3x^4 + x^6)}{1 + 3x^2 + 3x^4 + x^6}, \ = x, \text{ of } \frac{dy}{1 + y^2} = \frac{dx}{1 + x^2};$$

and that for $\rho = -7$, $\alpha = -1$, the values are $A_1 = -21$, $A_2 = 35$, viz. these belong to the solution

$$y = \frac{-7x + 35x^3 - 21x^5 + x^7}{1 - 21x^2 + 35x^4 - 7x^6} \text{ of } \frac{dy}{1 + y^2} = \frac{-7dx}{1 + x^2}.$$

For example, the equation $12A_2 = 6A_1^2 + 8\alpha A_1 - \rho^4 + 7$ becomes, for the first set of values, $36 = 54 - 24 - 1 + 7$, and for the second set of values, $420 = 2646 + 168 - 2401 + 7$, which are each of them right.

New Form of the Seventhic Transformation. Art. Nos. 72 to 83.

72. For the quartic function $1 - 2\alpha x^2 + x^4$, the coefficients a, b, c, d, e are $= 1, 0, -\frac{1}{3}\alpha, 0, 1$, and hence the invariants I, J and the discriminant Δ are

$$I = 1 + \tfrac{1}{3}\alpha^2, \ = \tfrac{1}{3}(\alpha^2 + 3),$$

$$J = -\tfrac{1}{3}\alpha + \tfrac{1}{27}\alpha^3, \ = \tfrac{1}{27}(\alpha^3 - 9\alpha),$$

$$\Delta = I^3 - 27J^2, \ = \tfrac{1}{27}\{(\alpha^2 + 3)^3 - (\alpha^3 - 9\alpha)^2\}, \ = (\alpha^2 - 1)^2, \text{ whence } \sqrt[12]{\Delta} = \sqrt[6]{\alpha^2 - 1}.$$

This being so, then assuming

$$\rho = p\frac{\sqrt[6]{\alpha^2-1}}{\sqrt[6]{\beta^2-1}},$$

the differential equation

$$\frac{dy}{\sqrt{1-2\beta y^2+y^4}} = \frac{\rho\, dx}{\sqrt{1-2\alpha x^2+x^4}}$$

becomes

$$\frac{\sqrt[6]{\beta^2-1}\,dy}{\sqrt{1-2\beta y^2+y^4}} = \frac{p\sqrt[6]{\alpha^2-1}\,dx}{\sqrt{1-2\alpha x^2+x^4}},$$

viz. this is, for the radicals $\sqrt{1-2\alpha x^2+x^4}$ and $\sqrt{1-2\beta y^2+y^4}$, the form considered by Klein in the paper "Ueber die Transformation der elliptischen Functionen und die Auflösung der Gleichungen fünften Grades," *Math. Ann.*, t. XIV. (1879), pp. 111—172. I notice that there is some error as to a factor 7, and that p is equal to the z of p. 148, not, as might appear, equal to $\frac{1}{7}z$.

73. The modular equation presents itself in the form given, *l.c.*, p. 143, viz. this is

$$\mathbf{J} : \mathbf{J}-1 : 1 = (\tau^2+13\tau+49)(\tau^2+5\tau+1)^3 : (\tau^4+14\tau^3+63\tau^2+70\tau-7)^2 : 1728\tau,$$

with the like relation in $\mathbf{J}'$, τ'; and then $\tau\tau'=49$. We have thus $\mathbf{J}$, $\mathbf{J}'$ each given as a function of τ; and thence by elimination of τ, we have the modular equation as a relation between the absolute invariants $\mathbf{J}$, $\mathbf{J}'$. But $\tau = p^2$, and for the form $1-2\alpha x^2+x^4$, as appears above, we have

$$\mathbf{J}-1, \;=\frac{27J^2}{\Delta};\; = \frac{\frac{1}{27}(\alpha^3-9\alpha)^2}{(\alpha^2-1)^2};$$

hence Klein's equation

$$\mathbf{J}-1 = \frac{(\tau^4+14\tau^3+63\tau^2+70\tau-7)^2}{1728\tau}$$

becomes

$$\frac{\alpha^3-9\alpha}{\alpha^2-1} = \frac{p^8+14p^6+63p^4+70p^2-7}{8p};$$

or, say

$$p^8+14p^6+63p^4+70p^2-8\left(\frac{\alpha^3-9\alpha}{\alpha^2-1}\right)p-7=0,$$

which is the equation, *l.c.*, p. 148 with p for z; viz. this is the $p\alpha$-equation connecting α with the new multiplier p. It will be observed that it is of the degree 8 in p, and the degree 3 in α, viz. it resembles herein the foregoing $\rho\alpha$-equation, but the form is very much more simple, inasmuch as the α enters into a single coefficient only. The equation may also be written

$$(p^4+5p^2+1)^3(p^4+13p^2+49)-64\frac{(\alpha^2+3)^3}{(\alpha^2-1)^2}p^2=0.$$

74. Using for shortness a single letter m to denote the value $i\sqrt{3}$, we have

$$\frac{\alpha^3-9\alpha+3m(\alpha^2-1)}{\alpha^3-9\alpha-3m(\alpha^2-1)} = \frac{p^8+14p^6+63p^4+70p^2+24mp-7}{p^8+14p^6+63p^4+70p^2-24mp-7};$$

that is,

$$\left(\frac{\alpha+m}{\alpha-m}\right)=\frac{(p^2-mp+1)^3(p^2+3mp-7)}{(p^2+mp+1)^3(p^2-3mp-7)},$$

or, say

$$\frac{\alpha+m}{\alpha-m}=\frac{p^2-mp+1}{p^2+mp+1}\sqrt[3]{\frac{p^2+3mp-7}{p^2-3mp-7}},$$

which is another form of the $p\alpha$-equation.

75. We had $\tau=p^2$; and similarly, writing $\tau'=q^2$, then $\tau\tau'=49=p^2q^2$; it must be assumed that $pq=-7$; β is then the same function of q which α is of p, viz. we have

$$\frac{\beta+m}{\beta-m}=\frac{q^2-mq+1}{q^2+mq+1}\sqrt[3]{\frac{q^2+3mq-7}{q^2-3mq-7}}.$$

These equations in α and β contain the same cubic radical, viz. we have

$$q^2+3mq-7, \;=\frac{49}{p^2}-\frac{21m}{p}-7, \;=-\frac{7}{p^2}(p^2+3mp-7),$$

and similarly

$$q^2-3mq-7 \qquad =-\frac{7}{p^2}(p^2-3mp-7).$$

Moreover

$$q^2-\;mq+1, \;=\frac{49}{p^2}+\frac{7m}{p}+1, \;=\;\frac{1}{p^2}(p^2+7mp+49),$$

and similarly

$$q^2+\;mq+1 \qquad =\;\frac{1}{p^2}(p^2-7mp+49),$$

and we thus obtain

$$\frac{\beta+m}{\beta-m}=\frac{p^2+7mp+49}{p^2-7mp+49}\sqrt[3]{\frac{p^2+3mp-7}{p^2-3mp-7}},$$

whence, eliminating the cubic radical,

$$\frac{\beta+m}{\beta-m}=\frac{p^2+7mp+49}{p^2-7mp+49}\,\frac{p^2+mp+1}{p^2-mp+1}\,\frac{\alpha+m}{\alpha-m},$$

viz. this gives β as a rational function of α, p. We in fact have

$$\beta=\frac{\alpha(p^4+29p^2+49)-24p(p^2+7)}{8\alpha p(p^2+7)+(p^4+29p^2+49)}.$$

76. The differential relation

$$\frac{d\beta}{\beta^2-1}=\frac{\rho^2}{7}\,\frac{d\alpha}{\alpha^2-1},$$

on substituting therein for ρ its value, becomes

$$\frac{d\beta}{(\beta^2-1)^{\frac{2}{3}}}=\frac{p^2}{7}\,\frac{d\alpha}{(\alpha^2-1)^{\frac{2}{3}}}.$$

But, from the expression for $\dfrac{\alpha+m}{\alpha-m}$, we obtain

$$d\alpha\left(\frac{1}{\alpha+m}-\frac{1}{\alpha-m}\right)=dp\left\{\left(\frac{2p-m}{p^2-mp+1}-\frac{2p+m}{p^2+mp+1}\right)+\tfrac{1}{3}\left(\frac{2p+3m}{p^2+3mp-7}-\frac{2p-3m}{p^2-3mp-7}\right)\right\},$$

or, omitting from each side a factor $-2m$,

$$\frac{d\alpha}{\alpha^2+3}=dp\left(\frac{-p^2+1}{p^4+5p^2+1}+\frac{p^2+7}{p^4+13p^2+49}\right)=\frac{56dp}{(p^4+5p^2+1)(p^4+13p^2+49)}.$$

But we have, No. 73,

$$\frac{\alpha^2+3}{(\alpha^2-1)^{\frac{2}{3}}}=\frac{(p^4+5p^2+1)(p^4+13p^2+49)^{\frac{1}{3}}}{4p^{\frac{2}{3}}},$$

and thence

$$\frac{d\alpha}{(\alpha^2-1)^{\frac{2}{3}}}=\frac{14dp}{p^{\frac{2}{3}}(p^4+13p^2+49)^{\frac{2}{3}}},$$

and similarly

$$\frac{d\beta}{(\beta^2-1)^{\frac{2}{3}}}=\frac{14dq}{q^{\frac{2}{3}}(q^4+13q^2+49)^{\frac{2}{3}}}.$$

The equation $q=-\dfrac{7}{p}$ gives

$$dq=\frac{7dp}{p^2},\quad q^{\frac{2}{3}}(q^4+13q^2+49)^{\frac{2}{3}}=49p^{-\frac{10}{3}}(p^4+13p^2+49)^{\frac{2}{3}},$$

and we thence have

$$\frac{d\beta}{(\beta^2-1)^{\frac{2}{3}}}=\frac{2p^{\frac{4}{3}}\,dp}{(p^4+13p^2+49)^{\frac{2}{3}}},\quad =\frac{p^2}{7}\,\frac{d\alpha}{(\alpha^2-1)^{\frac{2}{3}}},$$

the required relation.

77. From the value of ρ, we have

$$\frac{d\rho}{\rho}=\frac{dp}{p}+\frac{\frac{1}{3}\alpha d\alpha}{\alpha^2-1}-\frac{\frac{1}{3}\beta d\beta}{\beta^2-1},$$

which, substituting for $d\beta$ its value, becomes

$$=\frac{dp}{p}+\frac{\frac{1}{3}d\alpha}{(\alpha^2-1)^{\frac{2}{3}}}\left\{\frac{\alpha}{(\alpha^2-1)^{\frac{1}{3}}}-\frac{\beta}{(\beta^2-1)^{\frac{1}{3}}}\,\frac{p^2}{7}\right\},$$

or, say

$$\frac{1}{\rho}\frac{d\rho}{d\alpha}=\frac{1}{p}\frac{dp}{d\alpha}+\frac{\frac{1}{3}}{(\alpha^2-1)^{\frac{2}{3}}}\left\{\frac{\alpha}{(\alpha^2-1)^{\frac{1}{3}}}-\frac{\beta}{(\beta^2-1)^{\frac{1}{3}}}\,\frac{p^2}{7}\right\},$$

which, however, is more conveniently written

$$\frac{1}{\rho}\frac{d\rho}{d\alpha}=\frac{1}{p}\frac{dp}{d\alpha}+\frac{\frac{1}{3}}{\alpha^2-1}(\alpha-\beta\rho^2);$$

and then, substituting in the formulæ for A_1, A_2, we find

$$A_1 = 7(\alpha^2-1)\frac{1}{p}\frac{dp}{d\alpha} - \tfrac{7}{6}\alpha + \tfrac{1}{6}\beta\rho^2,$$

$$\frac{1}{\rho}A_2 = 7(\alpha^2-1)\frac{1}{p}\frac{dp}{d\alpha} - \tfrac{5}{6}\alpha - \tfrac{1}{6}\beta\rho^2,$$

expressions which give, as they should do, $A_2 - \rho A_1 = \tfrac{1}{3}(\alpha\rho - \beta\rho^3)$. In these last formulæ, ρ is to be regarded as standing for its value, $= p\dfrac{\sqrt[6]{\alpha^2-1}}{\sqrt[6]{\beta^2-1}}$.

78. To further reduce these values, consider the expression of β given No. 75. If for a moment we represent this by

$$\beta = \frac{F\alpha - 3G}{G\alpha + F}, \text{ where } F = p^4 + 29p^2 + 49,\ G = 8p(p^2+7),$$

then we have

$$\beta^2 - 1 = \frac{(F^2-G^2)\alpha^2 - 8FG\alpha + 9G^2 - F^2}{(G\alpha+F)^2},$$

or, multiplying the numerator and denominator each by $G\alpha + F$, so as to make the denominator a perfect cube, the numerator becomes

$$G(F^2-G^2)(\alpha^3-9\alpha) + F(F^2-9G^2)(\alpha^2-1);$$

and putting for the factor G of the first term its value $= 8p(p^2+7)$, we thus obtain

$$\frac{\beta^2-1}{\alpha^2-1} = \frac{(F^2-G^2)(p^2+7)\,8p\left(\dfrac{\alpha^3-9\alpha}{\alpha^2-1}\right) + F(F^2-9G^2)}{(G\alpha+F)^3},$$

viz. in virtue of the $p\alpha$-equation, this is

$$\frac{\beta^2-1}{\alpha^2-1} = \frac{(F^2-G^2)(p^2+7)(p^8+14p^6+63p^4+70p^2-7) + F(F^2-9G^2)}{(G\alpha+F)^3}.$$

This numerator is $= (p^4+5p^2+1)^3p^6$; in fact, we have

$$\begin{aligned}(F^2-G^2)(p^2+7) &= p^{10} + p^8 + p^6 + 7p^4 + 343p^2 + 16807,\\ F^2 - 9G^2 &= \qquad p^8 - 518p^6 - 7125p^4 - 25382p^2 + 2401,\end{aligned}$$

and thence forming the two terms of the numerator and adding them together—for shortness I write down only the coefficients—we have

1	15	78	154	567	22113	257390	1082802	1174089	− 117649
			1	− 489	− 22098	− 257389	− 1082802	− 1174089	117649
= 1	15	78	155	78	15	1	0	0	0,

viz. these are the coefficients of $(p^4+5p^2+1)^3p^6$. Hence

$$\frac{\beta^2-1}{\alpha^2-1} = \frac{(p^4+5p^2+1)^3p^6}{(G\alpha+F)^3};$$

or, extracting the cube root, and for G, F substituting their values,

$$\frac{\sqrt[3]{\beta^2-1}}{\sqrt[3]{\alpha^2-1}}=\frac{(p^4+5p^2+1)\,p^2}{8p\,(p^2+7)\,\alpha+p^4+29p^2+49},$$

and thence also

$$\rho^2=\frac{8p\,(p^2+7)\,\alpha+p^4+29p^2+49}{p^4+5p^2+1},$$

viz. we have thus ρ^2 expressed as a rational function of p, α.

79. It will presently appear that ρ is, in fact, expressible as a rational function of p, α: but I am unable to obtain this expression in a simple form. Admitting that ρ is thus expressible, a direct process for obtaining the expression is as follows. Writing

$$\xi=\frac{8p\,(p^2+7)\,\alpha+p^4+29p^2+49}{p^4+5p^2+1}\;(=\rho^2),$$

and by means hereof introducing ξ in place of α into the equation

$$p^8+14p^6+63p^4+70p^2-8p\,\frac{\alpha^3-9\alpha}{\alpha^2-1}-7=0,$$

we have for ξ a cubic equation,

$$a\xi^3+b\xi^2+c\xi+d=0,$$

where the coefficients a, b, c, d are given rational functions of p. This equation may be written

$$a\xi\,(\xi+\vartheta)^2+b'\xi^2+c'\xi+d=0,$$

where $b'=b-2a\vartheta$, $c'=c-a\vartheta^2$; and the last three terms will be a square if only $c'^2-4b'd=0$; that is, if

$$(a\vartheta^2-c)^2+4d\,(2a\vartheta-b)=0,$$

a biquadratic equation in ϑ which (ρ being expressible as above) must have one of its roots = a rational function of p. Calling this ϑ, we then have

$$a\xi\,(\xi+\vartheta)^2+\frac{1}{b'}(b'\xi+\tfrac{1}{2}c')^2=0,\text{ or say }a\rho^2\,(\xi+\vartheta)^2+\frac{1}{b'}(b'\xi+\tfrac{1}{2}c')^2=0,$$

hence

$$\rho=\sqrt{\frac{-1}{ab'}}\cdot\frac{b'\xi+\frac{1}{2}c'}{\xi+\vartheta},$$

where ξ denotes a linear function of α as above; the quadric radical will have a rational value, and the form of the equation thus is

$$\rho=\frac{A\alpha+B}{C\alpha+D},$$

where A, B, C, D are rational and integral functions of p. But I am not able to carry out the process.

80. As shown, No. 78, we have

$$\rho^2 = \frac{8p(p^2+7)\alpha + p^4 + 29p^2 + 49}{p^4 + 5p^2 + 1}.$$

Multiplying by the value of β, *ante* No. 75, we find

$$\beta\rho^2 = \frac{(p^4 + 29p^2 + 49)\alpha - 24p(p^2+7)}{p^4 + 5p^2 + 1};$$

and we can hence find A_1 and A_2 by the formulæ

$$A_1 = 7(\alpha^2 - 1)\frac{1}{p}\frac{dp}{d\alpha} - \tfrac{7}{6}\alpha + \tfrac{1}{6}\beta\rho^2,$$

$$\frac{1}{\rho}A_2 = 7(\alpha^2 - 1)\frac{1}{p}\frac{dp}{d\alpha} - \tfrac{5}{6}\alpha - \tfrac{1}{6}\beta\rho^2,$$

or, for the second of these we may write

$$\frac{1}{\rho}A_2 = A_1 + \tfrac{1}{3}(\alpha - \beta\rho^2).$$

But in a different point of view, regarding only ρ^2, but not ρ, as a given function of p, α, we must to these equations join the equation $12A_2 = 6A_1^2 + 8\alpha A_1 - \rho^4 + 7$, *ante* No. 69: and we have thus equations for the determination of A_1, A_2, and ρ.

81. We have

$$A_1 = \frac{(p^4 + 5p^2 + 1)(p^4 + 13p^2 + 49)}{8p}\,\frac{\alpha^2 - 1}{\alpha^2 + 3}$$
$$- \tfrac{7}{6}\alpha + \frac{\alpha(p^4 + 29p^2 + 49) - 24p(p^2+7)}{6(p^4 + 5p^2 + 1)},$$

where the second line is

$$= \frac{\alpha(-p^4 - p^2 + 7) - 4p(p^2+7)}{p^4 + 5p^2 + 1}.$$

Uniting the two terms, we have a denominator $8p(p^4 + 5p^2 + 1)$, and in the numerator a term $8p\alpha^3$ which may be got rid of by means of the $p\alpha$-equation; the numerator thus becomes

$$= 96p(-p^4 - p^2 + 7) - 128p^2(p^2+7)\alpha$$
$$+ (\alpha^2 - 1)\{(-p^4 - p^2 + 7)(p^8 + 14p^6 + 63p^4 + 70p^2 - 7)\}$$
$$+ (p^4 + 5p^2 + 1)^2(p^4 + 13p^2 + 49) - 32p^2(p^2+7),$$

where the whole divides by $8p$; and we finally obtain

$$A_1 = \frac{12(-p^4 - p^2 + 7) - 16p(p^2+7)\alpha + (\alpha^2 - 1)p(p^8 + 17p^6 + 102p^4 + 225p^2 + 97)}{(\alpha^2 + 3)(p^4 + 5p^2 + 1)}.$$

Proceeding to calculate the value of $A_1 + \tfrac{1}{3}(\alpha - \beta\rho^2)$, we then have

$$\tfrac{1}{3}(\alpha - \beta\rho^2) = \frac{-8(p^2+2)\alpha + 8p(p^2+7)}{p^4 + 5p^2 + 1}.$$

Multiplying the numerator and denominator by α^2+3, we have in the numerator a term in $8\alpha^3$, which may be got rid of by means of the $p\alpha$-equation; the numerator for $\frac{1}{\rho}A_2$ thus becomes

$$12(-p^4-9p^2-9)+16p(p^2+7)\alpha$$
$$+(\alpha^2-1)p\left[\{p^8+17p^6+102p^4+225p^2+97\}-\frac{p^2+2}{p^2}(p^8+14p^6+63p^4+70p^2-7)+8(p^2+7)\right],$$

and we finally obtain

$$\frac{1}{\rho}A_2=\frac{12(-p^4-9p^2-9)+16p(p^2+7)\alpha+(\alpha^2-1)p^{-1}(p^8+11p^6+37p^4+20p^2+2)}{(\alpha^2+3)(p^4+5p^2+1)}.$$

82. The expressions obtained above for ρ^2, A_1, A_2 are of the form

$$\rho^2=\frac{M+N\alpha}{S},\quad A_1=\frac{P_1+Q_1\alpha+R_1\alpha^2}{S(\alpha^2+3)},\quad \frac{1}{\rho}A_2=\frac{P_2+Q_2\alpha+R_2\alpha^2}{S(\alpha^2+3)},$$

where

$$M=p^4+29p^2+49;\qquad N=8p(p^2+7);\qquad S=p^4+5p^2+1,$$
$$P_1=12(-p^4-p^2+7)-p(p^8+17p^6+102p^4+225p^2+97),\quad Q_1=-16p(p^2+7),$$
$$R_1=p(p^8+17p^6+102p^4+225p^2+97);$$
$$P_2=12(-p^4-9p^2-9)-p^{-1}(p^8+11p^6+37p^4+20p^2+2),\quad Q_2=16p(p^2+7),$$
$$R_2=p^{-1}(p^8+11p^6+37p^4+20p^2+2);$$

substituting these values in the foregoing equation

$$12A_2=6A_1^2+8\alpha A_1-\rho^4+7,$$

we obtain

$$12\rho\left\{\frac{P_2+Q_2\alpha+R_2\alpha^2}{S(\alpha^2+3)}\right\}=\left\{\frac{6(P_1+Q_1\alpha+R_1\alpha^2)^2}{S^2(\alpha^2+3)^2}+8\alpha\frac{P_1+Q_1\alpha+R_1\alpha^2}{S(\alpha^2+3)}-\frac{(M+N\alpha)^2}{S^2}+7\right\};$$

that is,

$$\rho=\frac{1}{12(P_2+Q_2\alpha+R_2\alpha^2)^2(3+\alpha^2)S}\{6(P_1+Q_1\alpha+R_1\alpha^2)^2+8\alpha S(3+\alpha^2)(P_1+Q_1\alpha+R_1\alpha^2)$$
$$-(M+N\alpha)^2(3+\alpha^2)^2+7S^2(3+\alpha^2)^2\},$$

which, by means of the $p\alpha$-equation

$$p^8+14p^6+63p^4+70p^2-\left(\frac{\alpha^3-9\alpha}{\alpha^2-1}\right)8p-7=0,$$

should be reducible to the form

$$\rho=A\alpha^2+B\alpha+C,\text{ or }\rho=\frac{A\alpha+B}{C\alpha+D};$$

but I have not been able to obtain, in either of these forms, a simple expression of ρ as a function of p, α. Supposing it obtained, the $\rho\alpha$-equation, *ante* No. 51, would of course be thereby transformable into the foregoing $p\alpha$-equation. And considering p as an auxiliary parameter thus introduced into the formulæ in place of ρ, then β and the coefficients A_1, A_2 are, by what precedes, expressed in terms of p, α, that is, in effect in terms of ρ, α; and we thus have the formulæ of transformation for the $\rho\alpha\beta$-form.

83. There exists a remarkably simple particular case. Write for convenience $\theta = \sqrt{7}$, the $p\alpha$-equation is satisfied by the values $p = -\theta$, $\alpha = -\frac{3}{8}\theta$. In fact, these values give $8p\alpha = 3\theta^2$, $= 21$, $\frac{\alpha^2 - 9}{\alpha^2 - 1} = (\frac{63}{64} - 9) \div (\frac{63}{64} - 1)$, $= 513$; the term in α is thus $21 . 513$, $= 10773$; but, assuming $p^2 = 7$, we have

$$p^8 + 14p^6 + 63p^4 + 70p^2 - 7 = 2401 + 4802 + 3087 + 490 - 7, = 10773,$$

and the equation is thus satisfied. And these values, $p = -\theta$, $\alpha = -\frac{3}{8}\theta$, give $\rho^2 = 7$, $\beta = \frac{3}{8}\theta$, $A_1 = 2\theta$, $A_2 = \rho\theta$; the equation $12A_2 = 6A_1{}^2 + 8\alpha A_1 - \rho^4 + 7$ thus becomes $12\rho\theta = 168 - 42 - 49 + 7$, $= 84$; that is, $\rho\theta = 7$, $= \theta^2$, or $\rho = \theta\,(= -p)$. We have $\alpha^2 - 1$, $= \beta^2 - 1$, $= -\frac{1}{64}$; but from the equation $\rho = p\frac{\sqrt[6]{\alpha^2 - 1}}{\sqrt[6]{\beta^2 - 1}}$, it appears that the sixth roots must be equal with opposite signs, say $\sqrt[6]{\alpha^2 - 1} = \frac{i}{2}$, $\sqrt[6]{\beta^2 - 1} = \frac{-i}{2}$. Retaining θ to stand for its value $= \sqrt{7}$, the differential equation is

$$\frac{dy}{\sqrt{1 - \frac{3}{4}\theta y^2 + y^4}} = \frac{\theta dx}{\sqrt{1 + \frac{3}{4}\theta x^2 + x^4}},$$

and it is satisfied by

$$y = \frac{x(\theta + 7x^2 + 2\theta x^4 + x^6)}{1 + 2\theta x^2 + 7x^4 + \theta x^6}.$$

It may be remarked that the quartic functions of y and x resolved into their linear factors are

$$\left\{y + \frac{3i + \theta}{2\sqrt{2}(1+i)}\right\}\left\{y + \frac{3i - \theta}{2\sqrt{2}(1+i)}\right\}\left\{y + \frac{-3i + \theta}{2\sqrt{2}(1-i)}\right\}\left\{y + \frac{-3i - \theta}{2\sqrt{2}(1-i)}\right\},$$

and

$$\left\{x + \frac{3 - i\theta}{2\sqrt{2}(1+i)}\right\}\left\{x + \frac{3 + i\theta}{2\sqrt{2}(1+i)}\right\}\left\{x + \frac{3 - i\theta}{2\sqrt{2}(1-i)}\right\}\left\{x + \frac{3 + i\theta}{2\sqrt{2}(1-i)}\right\}$$

and that for the first of the y-factors, substituting for y its value, we have

$$x^7 + 2\theta x^5 + 7x^3 + \theta x + \frac{3i + \theta}{2\sqrt{2}(1+i)}(\theta x^6 + 7x^4 + 2\theta x^2 + 1)$$

$$= \left\{x + \frac{3 - i\theta}{2\sqrt{2}(1+i)}\right\}\left\{x^3 + \frac{1 + i\theta}{\sqrt{2}(1+i)}x^2 + \tfrac{1}{2}(i + \theta)x + \frac{1+i}{\sqrt{2}}\right\}^2,$$

with like expressions for the other y-factors respectively.

Brioschi's Transformation Theory. Art. No. 84.

84. M. Brioschi has kindly referred me to two papers by him, "Sur une Formule de Transformation des Fonctions Elliptiques," *Comptes Rendus*, t. LXXIX. (1874), pp. 1065—1069, and *ibid.* t. LXXX. (1875), pp. 261—264. They relate to the form

$$\frac{dx}{\sqrt{4x^3 - g_2 x - g_3}} = \frac{dy}{\sqrt{4y^3 - G_2 y - G_3}},$$

with a formula of transformation

$$y=\frac{U}{T^2},\quad T=x^\nu+\alpha_1x^{\nu-1}+\alpha_2x^{\nu-2}+\ldots+\alpha_\nu,\text{ where }\nu=\tfrac{1}{2}(n-1),$$

$$U=x^n+\alpha_1x^{n-2}+\alpha_2x^{n-3}+\ldots+\alpha_\nu.$$

The general theory for any value of n is developed to a considerable extent, and it would without doubt give very interesting results for the case $n=7$; but the formulæ are only completely worked out for the preceding two cases $n=3$ and $n=5$. For these cases the formulæ are as follows:

Cubic transformation: $n=3$,

$$y=\frac{x^3+\alpha_1x^2+\alpha_2x+\alpha_3}{(x+a_1)^2}.$$

Corresponding to the modular equation, we have

$$a_1^4-\tfrac{1}{2}g_2a_1^2+g_3a_1-\tfrac{1}{48}g_2^2=0,$$

and then

$$G_2-9g_2=6\,(20a_1^2-3g_2),\quad G_3+27g_3=-14\,(20a_1^2-3g_2)\,a_1,$$

whence also

$$a_1=-\tfrac{3}{7}\,\frac{G_3+27g_3}{G_2-9g_2};$$

and by the general theory α_1, α_2, α_3 are given rationally in terms of a_1, g_2, g_3.

Quintic transformation: $n=5$,

$$y=\frac{x^5+\alpha_1x^4+\alpha_2x^3+\alpha_3x^2+\alpha_4x+\alpha_5}{(x^2+a_1x+a_2)^2}.$$

We have

$$a_1X-2Y=0,\quad (12a_1^2+g_2)\,X-30a_1Y=0,$$

where

$$X=a_1^3-6a_1^2a_2+\tfrac{1}{2}g_2a_1-g_3,$$

$$Y=5a_2^2-a_1^2a_2+\tfrac{1}{2}g_2a_2-g_3a_1+\tfrac{1}{16}g_2^2.$$

The first of these gives

$$a_2=\frac{1}{6a_1}(a_1^3+\tfrac{1}{2}g_2a_1-g_3);$$

then eliminating a_2, we have, corresponding to the modular equation,

$$a_1^6-5g_2a_1^4+40g_3a_1^3-5g_2^2a_1^2+8g_2g_3a_1-5g_3^2=0.$$

We then have

$$G_2-25g_2=\frac{8}{a_1}(10a_1^3-8g_2a_1+5g_3),\quad G_3+125g_3=-14\,(10a_1^3-8g_2a_1+5g_3);$$

whence also

$$a_1=-\tfrac{4}{7}\,\frac{G_3+125g_3}{G_2-25g_2};$$

and by the general theory α_1, α_2, α_3, α_4, α_5 are given rationally in terms of a_1, g_2, g_3.

These results are contained in the former of the papers above referred to; the latter contains some properties of these modular equations.

871.

A CASE OF COMPLEX MULTIPLICATION WITH IMAGINARY MODULUS ARISING OUT OF THE CUBIC TRANSFORMATION IN ELLIPTIC FUNCTIONS.

[From the *Proceedings of the London Mathematical Society*, vol. XIX. (1888), pp. 300, 301.]

THE case in question is referred to in my "Note on the Theory of Elliptic Integrals," *Math. Ann.*, t. XII. (1877), pp. 143—146, [657]; but I here work it out directly.

In the cubic transformation, the modular equation is

$$u^4 - v^4 + 2uv(1 - u^2v^2) = 0;$$

and we have

$$y = \frac{\left(1 + \frac{2u^3}{v}\right)x + \frac{u^6}{v^2}x^3}{1 + vu^2(v + 2u^3)x^2},$$

giving

$$\frac{dy}{\sqrt{1 - y^2 \,.\, 1 - v^8y^2}} = \frac{\left(1 + \frac{2u^3}{v}\right)dx}{\sqrt{1 - x^2 \,.\, 1 - u^8x^2}}.$$

We thus have a case of complex multiplication if $v^8 = u^8$, or say $v = \gamma u$, where $\gamma^8 = 1$, or γ denotes an eighth root of unity. Substituting in the modular equation, this becomes

$$u^4(1 - \gamma^4) + 2\gamma u^2(1 - \gamma^2u^4) = 0,$$

or, throwing out the factor u^2 and reducing,

$$u^4 - \tfrac{1}{2}u^2(\gamma^5 - \gamma) - \gamma^6 = 0,$$

that is,

$$\frac{u^2}{\gamma} = \tfrac{1}{4}(\gamma^4 - 1 \pm \sqrt{\gamma^8 + 14\gamma^4 + 1}),$$

or, what is the same thing,

$$=\tfrac{1}{4}\{\gamma^4-1\pm\sqrt{14\gamma^4+2}\}.$$

We have $\gamma^8=1$, that is, $\gamma^4=\pm 1$. Considering first the case $\gamma^4=1$, here

$$\frac{u^2}{\gamma}=\pm 1,$$

and thence

$$1+\frac{2u^3}{v}=1+\frac{2u^2}{\gamma}, \ =1\pm 2, \ =3 \text{ or } -1;$$

moreover, $u^8=v^8=1$. We have thus only the non-elliptic formulæ

$$\frac{dy}{1-y^2}=\frac{-dx}{1-x^2}, \text{ satisfied by } y=-x,$$

and

$$\frac{dy}{1-y^2}=\frac{3dx}{1-x^2}, \qquad \text{by } y=\frac{3x+x^3}{1+3x^2}.$$

If however, $\gamma^4=-1$, then

$$\frac{u^2}{\gamma}=\tfrac{1}{4}(-2\pm\sqrt{-12}),$$

viz. this is

$$\frac{u^2}{\gamma}=\tfrac{1}{2}(-1\pm i\sqrt{3})=\omega,$$

if ω be an imaginary cube root of unity $(\omega^2+\omega+1=0)$; hence

$$u^8=(\gamma\omega)^4=-\omega.$$

Moreover,

$$1+\frac{2u^3}{v}=1+\frac{2u^2}{\gamma}, \ =1+2\omega,$$

or say,

$$=\omega-\omega^2, \quad [=\sqrt{-3}, \text{ if } \omega=\tfrac{1}{2}(-1+i\sqrt{3})];$$

and we thus have, as in the above-mentioned Note,

$$y=\frac{(\omega-\omega^2)\,x+\omega^2x^3}{1-\omega^2(\omega-\omega^2)\,x^2},$$

giving

$$\frac{dy}{\sqrt{1-y^2\,.\,1+\omega y^2}}=\frac{(\omega-\omega^2)\,dx}{\sqrt{1-x^2\,.\,1+\omega x^2}};$$

or, what is the same thing, for the modulus $k^2=-\omega$, we have

$$\text{sn}\,(\omega-\omega^2)\,\theta=\frac{(\omega-\omega^2)\,\text{sn}\,\theta+\omega^2\,\text{sn}^3\,\theta}{1-\omega^2(\omega-\omega^2)\,\text{sn}^2\,\theta};$$

the values of $\text{cn}\,(\omega-\omega^2)\,\theta$ and $\text{dn}\,(\omega-\omega^2)\,\theta$ are thence found to be

$$\text{cn}\,(\omega-\omega^2)\,\theta=\frac{\text{cn}\,\theta\,(1-\omega^2\,\text{sn}^2\,\theta)}{1-\omega^2(\omega-\omega^2)\,\text{sn}^2\,\theta};$$

and

$$\text{dn}\,(\omega-\omega^2)\,\theta=\frac{\text{dn}\,\theta\,(1+\omega^2\,\text{sn}^2\,\theta)}{1-\omega^2(\omega-\omega^2)\,\text{sn}^2\,\theta};$$

which are the formulæ of transformation for the elliptic functions.

872.

ON THE FINITE NUMBER OF THE COVARIANTS OF A BINARY QUANTIC.

[From the *Mathematische Annalen*, t. XXXIV. (1889), pp. 319, 320.]

The proof of Gordan's theorem given in this paper is insufficient; the paper accordingly is not reprinted.

873.

SYSTEM OF EQUATIONS FOR THREE CIRCLES WHICH CUT EACH OTHER AT GIVEN ANGLES.

[From the *Messenger of Mathematics*, vol. XVII. (1888), pp. 18—21.]

CONSIDER a triangle ABC, angles A, B, C, $(A+B+C=\pi)$: to fix the absolute magnitude, assume that the radius of the circumscribed circle is $=1$; the lengths of the sides are therefore $=2\sin A$, $2\sin B$, $2\sin C$ respectively. On the three sides as bases, outside of each, describe isosceles triangles aBC, bCA, cAB, the base angles whereof are $=\alpha$, β, γ respectively. If we draw a circle touching aB, aC at the points B, C respectively; a circle touching bC, bA at the points C, A respectively; and a circle touching cA, cB at the points A, B respectively; then these circles form a curvilinear triangle ABC, the angles whereof are $A+\beta+\gamma$, $B+\gamma+\alpha$, $C+\alpha+\beta$ respectively. Taking as origin the centre of the circumscribed circle, and through this point, for axis of x, an arbitrary line, its position determined by the angle θ, I write for convenience

$$\begin{array}{lll} F=\theta+2B, & F'=\theta-A, & A'=A+\beta+\gamma, \\ G=\theta+2B+2C, & G'=\theta+B, & B'=B+\gamma+\alpha, \\ H=\theta, & H'=\theta+2B+C, & C'=C+\alpha+\beta; \end{array}$$

then the coordinates of the angular points A, B, C are $(\cos F, \sin F)$, $(\cos G, \sin G)$, $(\cos H, \sin H)$ respectively; and the equations of the three circles are

$$\left\{x+\frac{\sin(A-\alpha)}{\sin\alpha}\cos F'\right\}^2+\left\{y+\frac{\sin(A-\alpha)}{\sin\alpha}\sin F'\right\}^2=\frac{\sin^2 A}{\sin^2\alpha},$$

$$\left\{x+\frac{\sin(B-\beta)}{\sin\beta}\cos G'\right\}^2+\left\{y+\frac{\sin(B-\beta)}{\sin\beta}\sin G'\right\}^2=\frac{\sin^2 B}{\sin^2\beta},$$

$$\left\{x+\frac{\sin(C-\gamma)}{\sin\gamma}\cos H'\right\}^2+\left\{y+\frac{\sin(C-\gamma)}{\sin\gamma}\sin H'\right\}^2=\frac{\sin^2 C}{\sin^2\gamma},$$

respectively.

In verification, observe that we have

$$\begin{aligned} G-H &= 2\pi-2A, & G'-H' &= -\pi+A, & G-F' &= 2\pi-A, & H-F' &= A,\\ H-F &= \quad -2B, & H'-F' &= \pi+B, & H-G' &= \quad -B, & F-G' &= B,\\ F-G &= \quad -2C, & F'-G' &= -\pi+C, & F-H' &= \quad -C, & G-H' &= C; \end{aligned}$$

hence

$$(\cos G-\cos H)^2+(\sin G-\sin H)^2=2-2\cos(G-H),\ =2(1-\cos 2A),\ =4\sin^2 A,$$

and we thus see that the sides are $=2\sin A,\ 2\sin B,\ 2\sin C$ respectively.

The first circle should pass through the points $(\cos G,\ \sin G)$, $(\cos H,\ \sin H)$; we ought therefore to have, for the first of these points,

$$1+2\frac{\sin(A-\alpha)}{\sin\alpha}\cos(G-F)+\frac{\sin^2(A-\alpha)}{\sin^2\alpha}=\frac{\sin^2 A}{\sin^2\alpha},$$

that is,

$$1+2\frac{\sin(A-\alpha)}{\sin\alpha}\cos A+\frac{\sin^2(A-\alpha)}{\sin^2\alpha}=\frac{\sin^2 A}{\sin^2\alpha}:$$

and, for the second of the points, the same equation. Write for a moment

$$X=\frac{\sin A}{\sin\alpha},\ \text{ then }\ \frac{\sin(A-\alpha)}{\sin\alpha}=X\cos\alpha-\cos A;$$

then the equation is

$$1+2(X\cos\alpha-\cos A)\cos A+(X\cos\alpha-\cos A)^2=X^2,$$

that is,

$$1-\cos^2 A=X^2\sin^2\alpha,$$

which is right.

The second and third circles should intersect at the angle A', that is, we ought to have

$$\left\{\frac{\sin(B-\beta)}{\sin\beta}\cos G'-\frac{\sin(C-\gamma)}{\sin\gamma}\cos H'\right\}^2+\left\{\frac{\sin(B-\beta)}{\sin\beta}\sin G'-\frac{\sin(C-\gamma)}{\sin\gamma}\sin H'\right\}$$
$$=\frac{\sin^2 B}{\sin^2\beta}+\frac{\sin^2 C}{\sin^2\gamma}+2\frac{\sin B\sin C}{\sin\beta\sin\gamma}\cos A',$$

or, reducing and for $\cos(G'-H')$ substituting its value, $=-\cos A$, the equation is

$$\frac{\sin^2(B-\beta)}{\sin^2\beta}+\frac{\sin^2(C-\gamma)}{\sin^2\gamma}+2\frac{\sin(B-\beta)\sin(C-\gamma)}{\sin\beta\sin\gamma}\cos A=\frac{\sin^2 B}{\sin^2\beta}+\frac{\sin^2 C}{\sin^2\gamma}+2\frac{\sin B\sin C}{\sin\beta\sin\gamma}\cos A'.$$

Writing here

$$\frac{\sin B}{\sin\beta}=Y,\ \frac{\sin C}{\sin\gamma}=Z,$$

the equation is

$$(Y\cos\beta-\cos B)^2+(Z\cos\gamma-\cos C)^2+2(Y\cos\beta-\cos B)(Z\cos\gamma-\cos C)\cos A$$
$$=Y^2+Z^2+2YZ\cos A',$$

viz. this is

$$\begin{aligned}&Y^2\cos^2\beta + Z^2\cos^2\gamma + 2YZ\cos\beta\cos\gamma\cos A\\ &- 2Y\cos\beta\,(\cos B + \cos C\cos A) - 2Z\cos\gamma\,(\cos C + \cos A\cos B)\\ &+ \cos^2 B + \cos^2 C + 2\cos A\cos B\cos C = Y^2 + Z^2 + 2YZ\cos A'.\end{aligned}$$

Reducing by the relation $A + B + C = \pi$, this becomes

$$\begin{aligned}&-2Y\cos\beta\sin A\sin C - 2Z\cos\gamma\sin A\sin B + 1 - \cos^2 A\\ &\qquad = Y^2\sin^2\beta + Z^2\sin^2\gamma + 2YZ\,(\cos A' - \cos\beta\cos\gamma\cos A).\end{aligned}$$

Here $A' = A + \beta + \gamma$, and thence

$$\cos A' = \cos A\,(\cos\beta\cos\gamma - \sin\beta\sin\gamma) - \sin A\,(\sin\gamma\cos\beta + \sin\beta\cos\gamma);$$

hence the right-hand is

$$= Y^2\sin^2\beta + Z^2\sin^2\gamma - 2YZ\,(\cos A\sin\beta\sin\gamma + \sin A\sin\gamma\cos\beta + \sin A\sin\beta\cos\gamma),$$

or, reducing by

$$Y\sin\beta = \sin B,\quad Z\sin\gamma = \sin C,$$

this is

$$= \sin^2 B + \sin^2 C - 2\sin B\sin C\cos A - 2Y\cos\beta\sin A\sin C - 2Z\cos\gamma\sin A\sin B,$$

and the terms in Y, Z are equal to the like terms on the left-hand; the whole equation thus becomes

$$-1 + \cos^2 A + \sin^2 B + \sin^2 C - 2\cos A\sin B\sin C = 0,$$

where the last term is

$$\begin{aligned}&= 2\cos A\,\{\cos(B + C) - \cos B\cos C\},\\ &= -2\cos^2 A - 2\cos A\cos B\cos C,\\ &= -2\cos^2 A + (\cos^2 A + \cos^2 B + \cos^2 C - 1),\\ &= -\cos^2 A + \cos^2 B + \cos^2 C - 1;\end{aligned}$$

the equation is thus

$$-1 + \cos^2 A + \sin^2 B + \sin^2 C - \cos^2 A + \cos^2 B + \cos^2 C - 1 = 0,$$

or, finally, it is $-1 + 1 + 1 - 1 = 0$, which is an identity. The formulæ for the intersection of the third and first circles, and for that of the first and second circles, are of course precisely similar to the above formula for the intersection of the second and third circles; and the verifications are thus completed.

Cambridge, April 7, 1887.

874.

NOTE ON THE LEGENDRIAN COEFFICIENTS OF THE SECOND KIND.

[From the *Messenger of Mathematics*, vol. XVII. (1888), pp. 21—23.]

As regards the integration of the equation

$$(1-x^2)\frac{d^2y}{dx^2}-2x\frac{dy}{dx}+n(n+1)y=0$$

(n a positive integer), it seems to me that sufficient prominence is not given to the solution

$$y=\tfrac{1}{2}P_n\log\frac{x+1}{x-1}-Z_n\,(=Q_n),$$

where P_n is the Legendrian integral of the first kind, a rational and integral function of x of the degree n, and Z_n is a rational and integral function of the degree $n-1$; viz. we have here a solution containing no transcendental function other than the logarithm, and which should thus be adopted as a second particular integral in preference to the form $y=Q_n$, in which we have the infinite series Q_n which is an unknown transcendental function.

Moreover, the expression usually given for Z_n, viz.

$$Z_n=\frac{2n-1}{1\,.\,n}P_{n-1}+\frac{2n-5}{3(n-1)}P_{n-3}+\frac{2n-9}{5(n-2)}P_{n-5}+\ldots\text{ (to term in } P_1 \text{ or } P_0),$$

is a very simple and elegant one; but the more natural definition (and that by which Z_n is most readily calculated) is that Z_n is the integral part of $\tfrac{1}{2}P_n\log\frac{x+1}{x-1}$, when the logarithm is expanded in descending powers of x, viz. it is the integral part of

$$P_n\left(\frac{1}{x}+\frac{1}{3}\frac{1}{x^3}+\frac{1}{5}\frac{1}{x^5}+\ldots\right),$$

whence also Q_n is the portion containing negative powers only of this same series.

The expressions for $P_0, P_1, \ldots, P_{10}$ are given in Ferrers' *Elementary Treatise on Spherical Harmonics, &c.*, London, 1877, pp. 23—25. Reproducing these, and joining to them the values of $Z_0, Z_1, \ldots, Z_{10}$ we have as follows: read $P_2 = \frac{3}{2}x^2 - \frac{1}{2}$, and so in other cases.

$$
\begin{array}{ll}
P_0 = & 1\ , \\
P_1 = x & 1\ , \\
P_2 = (x^2,\ \ 1) & \frac{3}{2} - \frac{1}{2}\ , \\
P_3 = (x^3,\ \ x) & \frac{5}{2} - \frac{3}{2}\ , \\
P_4 = (x^4 \ldots 1) & \frac{35}{8} - \frac{15}{4} + \frac{3}{8}\ , \\
P_5 = (x^5 \ldots x) & \frac{63}{8} - \frac{35}{4} + \frac{15}{8}\ , \\
P_6 = (x^6 \ldots 1) & \frac{231}{16} - \frac{315}{16} + \frac{105}{16} - \frac{5}{16}\ , \\
P_7 = (x^7 \ldots x) & \frac{429}{16} - \frac{693}{16} + \frac{315}{16} - \frac{35}{16}\ , \\
P_8 = (x^8 \ldots 1) & \frac{6435}{128} - \frac{3003}{32} + \frac{3465}{64} - \frac{315}{32} + \frac{35}{128}\ , \\
P_9 = (x^9 \ldots x) & \frac{12155}{128} - \frac{6435}{32} + \frac{9009}{64} - \frac{1155}{32} + \frac{315}{128}\ , \\
P_{10} = (x^{10} \ldots 1) & \frac{46189}{256} - \frac{109395}{256} + \frac{45045}{128} - \frac{15015}{128} + \frac{3465}{256} - \frac{63}{256}. \\
Z_0 = & 0\ , \\
Z_1 = & 1\ , \\
Z_2 = x & \frac{3}{2}\ , \\
Z_3 = (x^2,\ \ 1) & \frac{5}{2} - \frac{2}{3}\ , \\
Z_4 = (x^3,\ \ x) & \frac{35}{8} - \frac{55}{24}\ , \\
Z_5 = (x^4 \ldots 1) & \frac{63}{8} - \frac{49}{8} + \frac{8}{15}\ , \\
Z_6 = (x^5 \ldots x) & \frac{231}{16} - \frac{119}{8} + \frac{231}{80}\ , \\
Z_7 = (x^6 \ldots 1) & \frac{429}{16} - \frac{275}{8} + \frac{849}{80} - \frac{16}{35}\ , \\
Z_8 = (x^7 \ldots x) & \frac{6435}{128} - \frac{9867}{128} + \frac{4213}{128} - \frac{11659}{4480}\ , \\
Z_9 = (x^8 \ldots 1) & \frac{12155}{128} - \frac{65065}{384} + \frac{11869}{128} - \frac{14179}{896} + \frac{128}{315}\ , \\
Z_{10} = (x^9 \ldots x) & \frac{46189}{256} - \frac{281996}{768} + \frac{157157}{640} - \frac{26741}{448} + \frac{61567}{16128}.
\end{array}
$$

I notice that the numerical values of $P_1, P_2, \ldots, P_7$, for $x = 0{\cdot}00$, $0{\cdot}01$, ..., $1{\cdot}00$ are given (*Report of the British Association for* 1879, "Report on Mathematical Tables"); as the functions contain only powers of 2 in their denominators, the decimal values terminate, and the complete values are given. The functions Z have not been tabulated; the denominators contain other prime factors, and the decimal values would not terminate.

Cambridge, March 29, 1887.

875.

ON THE SYSTEM OF THREE CIRCLES WHICH CUT EACH OTHER AT GIVEN ANGLES AND HAVE THEIR CENTRES IN A LINE.

[From the *Messenger of Mathematics*, vol. XVII. (1888), pp. 60—69.]

IN the system considered in the paper "System of equations for three circles which cut each other at given angles," *Messenger*, t. XVII. pp. 18—21, [873], we may consider the particular case where the centres of the circles are in a line. The condition in order that this may be so is obviously

$$\begin{vmatrix} \sin(A-\alpha)\cos F', & \sin(A-\alpha)\sin F', & \sin\alpha \\ \sin(B-\beta)\cos G', & \sin(B-\beta)\sin G', & \sin\beta \\ \sin(C-\gamma)\cos H', & \sin(C-\gamma)\sin H', & \sin\gamma \end{vmatrix} = 0,$$

that is,

$$\sin(B-\beta)\sin(C-\gamma)\sin\alpha\sin(G'-H') + \ldots = 0\,;$$

or since $\sin(G'-H')$, $\sin(H'-F')$, $\sin(F'-G')$ are $=\sin A$, $\sin B$, $\sin C$ respectively, this is

$$\sin(B-\beta)\sin(C-\gamma)\sin A\sin\alpha + \ldots = 0,$$

viz. this is

$$\frac{\sin A\sin\alpha}{\sin(A-\alpha)} + \frac{\sin B\sin\beta}{\sin(B-\beta)} + \frac{\sin C\sin\gamma}{\sin(C-\gamma)} = 0,$$

or, as this may also be written,

$$\frac{1}{\cot A - \cot\alpha} + \frac{1}{\cot B - \cot\beta} + \frac{1}{\cot C - \cot\gamma} = 0.$$

But assuming this equation to be satisfied, it does not appear that there is any simple expression for the equation of the line through the three centres; nor would it be easy to transform the equations so as to have this line for one of the axes.

The case in question (which is a very important one from its connexion with Poincaré's theory of the Fuchsian functions) may be considered independently.

Taking the line of centres for the axis of x, and writing α, β, γ for the abscissæ of the centres, and P, Q, R for the radii, then the equations of the circles are

$$(x-\alpha)^2+y^2=P^2,$$
$$(x-\beta)^2+y^2=Q^2,$$
$$(x-\gamma)^2+y^2=R^2;$$

and then, if the pairs of circles cut at the angles A, B, C respectively, we have

$$Q^2+2QR\cos A+R^2=(\beta-\gamma)^2,$$
$$R^2+2RP\cos B+P^2=(\gamma-\alpha)^2,$$
$$P^2+2PQ\cos C+Q^2=(\alpha-\beta)^2,$$

which are the equations connecting $\alpha, \beta, \gamma, P, Q, R$.

It is to be remarked in regard hereto that, if A, B, C are used to denote the interior angles of the curvilinear triangle ABC, then the angles $\gamma A\beta$, $\alpha B\gamma$, $\beta C\alpha$ are $=\pi-A$, B, C respectively; whence, if P, Q, R were used to denote the three radii taken positively, the first equation would be

$$Q^2+2QR\cos A+R^2=(\beta-\gamma)^2,$$

as above; but the other two equations would be

$$R^2-2RP\cos B+P^2=(\gamma-\alpha)^2,$$
$$P^2-2PQ\cos C+Q^2=(\alpha-\beta)^2;$$

hence, in order that the equations may be as above, it is necessary that P denote the radius of the circle, centre α, taken *negatively*; and it in fact appears that, in a limiting case afterwards considered, the value of P comes out negative. Similarly as regards the curvilinear triangle $AB'C'$; here A, $B\,(=B')$ and $C\,(=C')$ are the interior angles of the triangle; and the radius of the circle, centre α', must be regarded as negative.

Considering A, B, C as given, we have an equation between the radii P, Q, R. In fact, this is at once obtained in the irrational form $\sqrt{(X)}+\sqrt{(Y)}+\sqrt{(Z)}=0$; and proceeding to rationalise this, we obtain

$$-2\sqrt{(YZ)}=Y+Z-X,$$

that is,

$$-\sqrt{\{(P^2+2PR\cos B+R^2)(P^2+2PQ\cos C+Q^2)\}}=P^2+P(Q\cos C+R\cos B)-QR\cos A.$$

Hence, squaring and reducing, we find without difficulty

$$\begin{aligned}0=\ & Q^2R^2\sin^2 A+R^2P^2\sin^2 B+P^2Q^2\sin^2 C\\ &+2P^2QR(\cos A+\cos B\cos C)+2PQ^2R(\cos B+\cos C\cos A)\\ &+2PQR^2(\cos C+\cos A\cos B),\end{aligned}$$

or putting herein $P, Q, R = \frac{\sin A}{\xi}, \frac{\sin B}{\eta}, \frac{\sin C}{\zeta}$, this is

$$\left(1, 1, 1, \frac{\cos A + \cos B \cos C}{\sin B \sin C}, \frac{\cos B + \cos C \cos A}{\sin C \sin A}, \frac{\cos C + \cos A \cos B}{\sin A \sin B}\right)(\xi, \eta, \zeta)^2 = 0;$$

and it may be remarked that, in this quadric form, the three coefficients are each less than 1, or each greater than 1, according as $A + B + C > \pi$, or $A + B + C < \pi$.

Suppose 1°, $A + B + C > \pi$; the coefficients are here $= \cos\lambda, \cos\mu, \cos\nu$, the form is

$$(1, 1, 1, \cos\lambda, \cos\mu, \cos\nu \between \xi, \eta, \zeta)^2,$$

that is,

$$(\xi + \eta\cos\nu + \zeta\cos\mu)^2 + (\eta^2\sin^2\nu + 2\eta\zeta\cos\lambda + \zeta^2\sin^2\mu),$$

namely, this is

$$(\xi + \eta\cos\nu + \zeta\cos\mu)^2 + \left\{\eta\sin\nu + \zeta\frac{\cos\lambda - \cos\mu\cos\nu}{\sin\nu}\right\}^2 + \zeta^2\left\{\sin^2\mu - \left(\frac{\cos\lambda - \cos\mu\cos\nu}{\sin\nu}\right)^2\right\};$$

where the last term is

$$= \frac{\zeta^2}{\sin^2\nu}\{\sin^2\mu\sin^2\nu - (\cos\lambda\cos\mu\cos\nu)^2\};$$

the coefficient in { } is

$$= 1 - \cos^2\lambda - \cos^2\mu - \cos^2\nu + 2\cos\lambda\cos\mu\cos\nu,$$

namely, substituting for $\cos\lambda, \cos\mu, \cos\nu$ their values, this is

$$= \frac{1}{\sin^2 A\sin^2 B\sin^2 C}(1 - \cos^2 A - \cos^2 B - \cos^2 C - 2\cos A\cos B\cos C)^2.$$

It thus appears that the form is the sum of three squares, and is thus constantly positive: it therefore only vanishes for imaginary values of the radii; or the case does not arise for any real figure.

Hence, 2°, if the figure be real, $A + B + C < \pi$, that is, the sum of the angles of the curvilinear triangle is less than two right angles: the radii are connected as above by the equation

$$\left(1, 1, 1, \frac{\cos A + \cos B\cos C}{\sin B\sin C}, \frac{\cos B + \cos C\cos A}{\sin C\sin A}, \frac{\cos C + \cos A\cos B}{\sin A\sin B}\right)\left(\frac{\sin A}{P}, \frac{\sin B}{Q}, \frac{\sin C}{R}\right)^2 = 0,$$

in which form the three coefficients are each greater than 1. Restoring therein ξ, η, ζ, and regarding these as rectangular coordinates, the equation represents a real cone which might be constructed without difficulty; and then taking ξ, η, ζ as the coordinates of any particular point on the conical surface, we have

$$P, Q, R = \frac{\sin A}{\xi}, \frac{\sin B}{\eta}, \frac{\sin C}{\zeta}.$$

Obviously, points on the same generating line of the cone give values of P, Q, R which differ in their absolute magnitudes only, their ratios being the same: the original equations, in fact, remain unaltered when P, Q, R, α, β, γ are each affected with any common factor.

Supposing P, Q, R taken so as to satisfy the equation in question, then taking the radicals

$$\sqrt{(Q^2 + 2QR\cos A + R^2)},\quad \sqrt{(R^2 + 2RP\cos B + P^2)},\quad \sqrt{(P^2 + 2PQ\cos C + Q^2)},$$

with the proper signs, we have a sum $= 0$, and these give the values of $\beta - \gamma$, $\gamma - \alpha$, $\alpha - \beta$, respectively; and the construction of the figure would be thus completed.

I look at the question from a different point of view; taking Q, R, $\beta - \gamma$ such as to satisfy the first equation

$$Q^2 + 2QR\cos A + R^2 = (\beta - \gamma)^2,$$

that is, starting from the two circles $(x - \beta)^2 + y^2 = Q^2$, $(x - \gamma)^2 + y^2 = R^2$ which cut each other at a given angle A: then the problem is to find a circle $(x - \alpha)^2 + y^2 = P^2$, cutting these at given angles C, B respectively. To determine the coordinate of the centre α, and the radius P, we have the remaining two equations

$$R^2 + 2RP\cos B + P^2 = (\gamma - \alpha)^2,$$
$$P^2 + 2PQ\cos C + Q^2 = (\alpha - \beta)^2,$$

namely, considering α, P as the coordinates of a point (in reference to the foregoing origin and axes), and for greater clearness writing $\alpha = \mathrm{x}$, $P = \mathrm{y}$, we have

$$\mathrm{y}^2 + 2\mathrm{y}R\cos B + R^2 - (\mathrm{x} - \gamma)^2 = 0,$$
$$\mathrm{y}^2 + 2\mathrm{y}Q\cos C + Q^2 - (\mathrm{x} - \beta)^2 = 0,$$

or, as these may be written,

$$(\mathrm{y} + R\cos B)^2 - (\mathrm{x} - \gamma)^2 = -R^2\sin^2 B,$$
$$(\mathrm{y} + Q\cos C)^2 - (\mathrm{x} - \beta)^2 = -Q^2\sin^2 C,$$

namely, the first of these equations denotes a rectangular hyperbola, coordinates of centre ($\mathrm{x} = \gamma$, $\mathrm{y} = -R\cos B$), transverse semi-axes $= R\sin B$; and the second of them a rectangular hyperbola, coordinates of centre ($\mathrm{x} = \beta$, $\mathrm{y} = -Q\cos C$), transverse semi-axes $= Q\sin C$: as similar and similarly situate hyperbolas, these intersect in two points only; namely, the points are the intersections of either of them with the common chord

$$2\mathrm{y}(R\cos B - Q\cos C) + 2(\gamma - \beta)\{\mathrm{x} - \tfrac{1}{2}(\gamma + \beta)\} + R^2 - Q^2 = 0.$$

It is possible to construct a circle through the two points of intersection, and so to obtain these points as the intersections of a line and circle; but the construction by the two rectangular hyperbolas is practically by no means an inconvenient one. I remark in passing that, for a rectangular hyperbola, the radius of curvature at the vertex is equal to the transverse semi-axis, and thus by drawing a small circular arc and by means of the asymptotes, we lay down a rectangular hyperbola graphically, without difficulty and with a fair amount of accuracy.

But the analytical solution may be carried somewhat further: we may without loss of generality write $\gamma = -\beta$, for this comes only to taking the origin midway between the centres of the circles β and γ: doing this, and for greater simplicity writing also for the moment $Q \cos C = M$, $R \cos B = N$, the equations become

$$y^2 + 2yN + R^2 - (x + \beta)^2 = 0,$$

$$y^2 + 2yM + Q^2 - (x - \beta)^2 = 0,$$

where x is now the abscissa of the centre of the circle α (measured from the last-mentioned midway point) and y is the radius of this circle. We deduce

$$2(N - M) y + R^2 - Q^2 - 4\beta x = 0,$$

or say

$$4\beta (x - \beta) = 2(N - M) y + R^2 - Q^2 + 4\beta^2,$$

and thence, from the first equation multiplied by $16\beta^2$, we have

$$16\beta^2 (y^2 + 2yN + R^2) - \{2(N - M) y + R^2 - Q^2 + 4\beta^2\}^2 = 0,$$

that is,

$$\begin{aligned} &4y^2 \{4\beta^2 - (N - M)^2\} \\ &+ 4y \{4\beta^2 (N + M) - (N - M)(R^2 - Q^2)\} \\ &+ \{8\beta^2 (R^2 + Q^2) - 16\beta^4 - (R^2 - Q^2)^2\} = 0, \end{aligned}$$

say this is $4y^2\mathfrak{A} + 4y\mathfrak{B} + \mathfrak{C} = 0$.

This gives

$$(2\mathfrak{A}y + \mathfrak{B})^2 = \mathfrak{B}^2 - \mathfrak{A}\mathfrak{C},$$

and we find without difficulty, restoring for M, N their values $Q \cos C$ and $R \cos B$,

$$\begin{aligned} \mathfrak{B}^2 - \mathfrak{A}\mathfrak{C} = 4\beta^2 \{&16\beta^4 - 8\beta^2 (Q^2 + R^2 - 2QR \cos B \cos C) \\ &+ (P^4 - 4P^3Q \cos B \cos C + 2P^2Q^2 (2 \cos^2 B + 2 \cos^2 C - 1) - 4PQ^3 \cos B \cos C + Q^4)\}, \end{aligned}$$

which is

$$= 4\beta^2 \{[4\beta^2 - (Q^2 - 2QR \cos B \cos C + R^2)]^2 - 4Q^2R^2 \sin^2 B \sin^2 C\}.$$

But we have

$$Q^2 + 2QR \cos A + R^2 = 4\beta^2,$$

and this equation thus becomes

$$\begin{aligned} \mathfrak{B}^2 - \mathfrak{A}\mathfrak{C} &= 16\beta^2Q^2R^2 \{(\cos A + \cos B \cos C)^2 - \sin^2 B \sin^2 C\} \\ &= 16\beta^2Q^2R^2 (-1 + \cos^2 A + \cos^2 B + \cos^2 C + 2 \cos A \cos B \cos C). \end{aligned}$$

We have therefore

$$\begin{aligned} 2\{4\beta^2 - (R \cos B - Q \cos C)^2\} y + \{4\beta^2 (R \cos B + Q \cos C) - (R \cos B - Q \cos C)(R^2 - Q^2)\} \\ = \pm 4\beta QR \sqrt{\{-(1 - \cos^2 A - \cos^2 B - \cos^2 C - 2 \cos A \cos B \cos C)\}}, \end{aligned}$$

$$4(R \cos B - Q \cos C) y + R^2 - Q^2 = 4\beta x,$$

or, completing the reduction by the substitution of the value of $4\beta^2$, this is

$$\begin{aligned}
&y\{(Q^2 \sin^2 C + R^2 \sin^2 B) + 2QR(\cos A + \cos B \cos C)\} \\
&\quad + QR\{Q(\cos B + \cos C \cos A) + R(\cos C + \cos A \cos B)\} \\
&\quad = \pm 4\beta QR \sqrt{\{-(1-\cos^2 A - \cos^2 B - \cos^2 C - 2\cos A \cos B \cos C)\}},
\end{aligned}$$

viz. we have thus two values of the radius $y\,(= P)$; and to each of these there corresponds a single value of the abscissa x, given by

$$4\beta x = R^2 - Q^2 + 2(R\cos B - Q\cos C)\,y.$$

The two values become equal, if $A + B \pm C = \pi$; in this case the three circles meet in a pair of points (x_1, y_1), $(x_1, -y_1)$. In fact, writing $A + B + C = \pi$, and thence

$$\cos A = -\cos(B + C), \quad = -\cos B \cos C + \sin B \sin C, \text{ \&c.},$$

we find

$$\begin{aligned}
&\{Q^2 \sin^2 C + 2QR(\cos A + \cos B \cos C) + R^2 \sin^2 B\}\,y \\
&\qquad + QR\{Q(\cos B + \cos C \cos A) + R(\cos C + \cos A \cos B)\} = 0,
\end{aligned}$$

that is,

$$(Q\sin C + R\sin B)^2\, y + QR(Q\sin C + R\sin B)\sin A = 0,$$

or, throwing out the factor $Q\sin C + R\sin B$, this is

$$(Q\sin C + R\sin B)\,y + QR\sin A = 0,$$

and we then have

$$\begin{aligned}
4\beta x &= R^2 - Q^2 - 2(R\cos B - Q\cos C)\frac{QR\sin A}{R\sin B + Q\sin C} \\
&= \frac{1}{R\sin B + Q\sin C}\{(R\sin B + Q\sin C)(R^2 - Q^2) - 2(R\cos B - Q\cos C)\,QR\sin A\}.
\end{aligned}$$

The term in { } is here

$$\begin{aligned}
&R^3(\ \sin B) \\
&+ R^2Q(\ \sin C - 2\sin A\cos B) \\
&+ RQ^2(-\sin B + 2\sin A\cos C) \\
&+ \ Q^3(-\sin C),
\end{aligned}$$

which is

$$\begin{aligned}
= \ &R^3(\ \sin B) \\
&+ R^2Q(-\sin C + 2\sin B\cos A) \\
&+ RQ^2(\ \sin B - 2\sin C\cos A) \\
&+ \ Q^3(-\sin C) \\
= \ &(R^2 + Q^2 + 2RQ\cos A)(R\sin B - Q\sin C), \\
= \ &4\beta^2(R\sin B - Q\sin C),
\end{aligned}$$

or, finally

$$y = \frac{-QR\sin A}{R\sin B + Q\sin C},$$

$$x = \frac{\beta(R\sin B - Q\sin C)}{R\sin B + Q\sin C}.$$

In these equations, y, x should be replaced by P, α respectively; and in obtaining them, it was assumed that $\gamma = -\beta$; restoring the general values of β, γ, the equations become

$$P = \frac{-QR\sin A}{R\sin B + Q\sin C},$$

$$\alpha - \tfrac{1}{2}(\beta+\gamma) = \frac{\tfrac{1}{2}(\beta-\gamma)(R\sin B - Q\sin C)}{R\sin B + Q\sin C},$$

viz. this last equation becomes

$$\alpha = \frac{\beta R\sin B + \gamma Q\sin C}{R\sin B + Q\sin C},$$

or, say

$$\alpha(R\sin B + Q\sin C) - \beta R\sin B - \gamma Q\sin C = 0,$$

which by means of the first equation becomes

$$\alpha\frac{QR}{P}\sin A + \beta R\sin B + \gamma Q\sin C = 0.$$

It thus appears that the two equations are

$$\frac{\sin A}{P} + \frac{\sin B}{Q} + \frac{\sin C}{R} = 0,$$

$$\frac{\alpha\sin A}{P} + \frac{\beta\sin B}{Q} + \frac{\gamma\sin C}{R} = 0,$$

viz. these equations, wherein $A+B+C=\pi$, belong to the case where the three circles intersect in the same pair of points; hence, if the coordinates x, y refer to the points of intersection of the three circles, we have simultaneously the equations of the three circles, and the three equations which determine the angles at which they intersect, viz. we have the six equations

$$\begin{aligned} (x-\alpha)^2 + y^2 &= P^2, & Q^2 + R^2 + 2QR\cos A &= (\beta-\gamma)^2, \\ (x-\beta)^2 + y^2 &= Q^2, & R^2 + P^2 + 2RP\cos B &= (\gamma-\alpha)^2, \\ (x-\gamma)^2 + y^2 &= R^2, & P^2 + Q^2 + 2PQ\cos C &= (\alpha-\beta)^2, \end{aligned}$$

viz. from these six equations, with the condition $A+B+C=\pi$, it must be possible to deduce the last-mentioned pair of equations.

In the general case, where $A+B+C<\pi$, and the three circles do not meet in a point, then taking the circles $(x-\beta)^2+y^2=Q^2$, $(x-\gamma)^2+y^2=R^2$ to be circles cutting each other at the angle A, or, what is the same thing, the values Q, R, β, γ to be such as to satisfy the relation

$$Q^2 + R^2 + 2QR\cos A = (\beta-\gamma)^2;$$

the two equations for the determination of the abscissa of the centre α, and the radius P of the remaining circle give, by what precedes,

$$\begin{aligned} &2\{(\beta-\gamma)^2 - (R\cos B - Q\cos C)^2\}P \\ &\quad + \{(\beta-\gamma)^2(R\cos B + Q\cos C) - (R\cos B - Q\cos C)(R^2-Q^2)\} \\ &\quad = \pm 2(\beta-\gamma)QR\sqrt{\{-(1-\cos^2 A - \cos^2 B - \cos^2 C - 2\cos A\cos B\cos C)\}}, \\ &4(R\cos B - Q\cos C)P + (R^2-Q^2) = (\beta-\gamma)(2\alpha-\beta-\gamma), \end{aligned}$$

viz. we have thus the two circles $(x-\alpha)^2+y^2=P^2$, each of them cutting the circles $(x-\beta)^2+y^2=Q^2$, and $(x-\gamma)^2+y^2=R^2$ at the angles C, B respectively.

876.

ON SYSTEMS OF RAYS.

[From the *Messenger of Mathematics*, vol. XVII. (1888), pp. 73—78.]

SIR W. R. HAMILTON'S Memoir, "Theory of Systems of Rays" (I do not at present consider the three Supplements), dated June, 1827, is printed *Trans. R. I. Acad.*, vol. XV. (1828), pp. 69—174. There is one page of Introduction, and pp. 70 to 80, an elaborate Contents. Part First: On ordinary systems of Reflected Rays. Part Second: On ordinary systems of Refracted Rays. Part Third: On extraordinary systems and systems of Rays in general. But only the First Part was published. This is considered under the headings: (I) Analytic Expressions of the Law of Ordinary Reflexion. (II) Theory of Focal Mirrors. (III) Surfaces of Constant Action. (IV) Classification of Systems of Rays. (V) On the Pencils of a Reflected System. (VI) On the Developable Pencils, the Two Foci of a Ray and the Caustic Curves and Surfaces. (VII) Lines of Reflexion on a Mirror. (VIII) On Osculating Focal Mirrors. (IX) On Thin and Undevelopable Pencils. (X) On the Axes of a Reflected System. (XI) Images Formed by Mirrors. (XII) Aberrations. (XIII) Density. And we have, p. 174, a "Conclusion to the First Part," wherein this first part is described as "an attempt to establish general principles respecting the system of rays produced by the ordinary reflexion of light at any mirror or combination of mirrors shaped and placed in any manner whatever; and to show that the mathematical properties of such a system may all be deduced by analytic methods from the form of ONE CHARACTERISTIC FUNCTION: as in the application of Analysis to Geometry, the properties of a plane curve or of a curve surface may all be deduced by uniform methods from the form of the function which characterises its equation."

The foregoing headings (I) to (V) may be regarded as containing the general theory, and the remaining ones (VI) to (XIII) as containing applications and developments.

I remark on the theory as follows:

Considering a congruence or doubly infinite system of lines (or say of rays), suppose that for any particular ray the cosine-inclinations are α, β, γ $(\alpha^2+\beta^2+\gamma^2=1)$, and that the coordinates of a point on the ray are (x, y, z). We may look at the system in two ways:

1°. The rays are considered as emanating from the points of a surface: here, if (x, y, z) are considered as belonging to a point on the surface, then z is a given function of (x, y) (or more generally x, y, z are given functions of two arbitrary parameters u, v); and to determine the congruence, we must have α, β, γ each of them a given function of (x, y) or of (u, v), such that identically $\alpha^2+\beta^2+\gamma^2=1$, but with no other condition as to the form of the functions.

2°. The rays may be considered irrespectively of any surface: here (α, β, γ) are each of them a given function of (x, y, z), such that always $\alpha^2+\beta^2+\gamma^2=1$; but there are further conditions, viz. it is assumed that we have one and the same ray, whatever is the point (x, y, z) on the ray; or, what is the same thing (using ρ to denote an arbitrary distance), that α, β, γ regarded as functions of x, y, z remain unaltered by the change of x, y, z into $x+\rho\alpha$, $y+\rho\beta$, $z+\rho\gamma$ respectively; this implies that we have

$$(\alpha\delta_x+\beta\delta_y+\gamma\delta_z)\,\alpha,\quad (\alpha\delta_x+\beta\delta_y+\gamma\delta_z)\,\beta,\quad (\alpha\delta_x+\beta\delta_y+\gamma\delta_z)\,\gamma,$$

each $=0$; and, conversely, it may be shown that when these relations are satisfied then α, β, γ remain unaltered by the change in question.

The equation $\alpha^2+\beta^2+\gamma^2=1$, gives

$$\alpha\delta_x\alpha+\beta\delta_x\beta+\gamma\delta_x\gamma,\quad \alpha\delta_y\alpha+\beta\delta_y\beta+\gamma\delta_y\gamma,\quad \alpha\delta_z\alpha+\beta\delta_z\beta+\gamma\delta_z\gamma,$$

each $=0$; and combining with the last-mentioned system of equations, it follows that

$$\delta_z\beta-\delta_y\gamma,\quad \delta_x\gamma-\delta_z\alpha,\quad \delta_y\alpha-\delta_x\beta,$$

are proportional to α, β, γ; or say $=k\alpha, k\beta, k\gamma$ respectively.

Hamilton considers whether the rays are such that they are cut at right angles by a surface; supposing this is so (say in this case the rays are a rectangular, or orthotomic system, or that they are the normals of a surface), then if x, y, z refer to the point on the surface, we must have

$$\alpha\,dx+\beta\,dy+\gamma\,dz=0;$$

this implies

$$\alpha\,(\delta_z\beta-\delta_y\gamma)+\beta\,(\delta_x\gamma-\delta_z\alpha)+\gamma\,(\delta_y\alpha-\delta_x\beta)=0,$$

a condition which, by what precedes, is $k\,(\alpha^2+\beta^2+\gamma^2)=0$; viz. we must have $k=0$, and therefore

$$\delta_z\beta-\delta_y\gamma,\quad \delta_x\gamma-\delta_z\alpha,\quad \delta_y\alpha-\delta_x\beta$$

each $=0$; that is, $\alpha\,dx+\beta\,dy+\gamma\,dz$ must be a complete differential, say it is $=dV$. And we then have $V=c$, the equation of a system of parallel surfaces each of them cutting the rays at right angles. Evidently $\alpha, \beta, \gamma=\delta_x V, \delta_y V, \delta_z V$ respectively, and the function V satisfies the partial differential equation

$$(\delta_x V)^2+(\delta_y V)^2+(\delta_z V)^2=1.$$

Hamilton in effect considers only systems of rays of the form in question, viz. those which are the normals of a surface (or, what is the same thing, the normals of a system of parallel surfaces), and it is such a system which is said to have the characteristic function V. It is shown that a system of rays originally of this kind remains a system of this kind after any number of reflexions (or ordinary refractions); in particular, if the rays originally emanate from a point, then, after any number of reflexions at mirrors of any form whatever, they are a system of rays cut at right angles by a surface. And moreover, there is given for the surface a simple construction, viz. starting from any surface which cuts the rays at right angles, and measuring off on the path of each ray (as reflected at the mirror or succession of mirrors) one and the same arbitrary distance, we have a set of points forming a surface which cuts at right angles the system of rays as reflected at the mirror or last of the mirrors.

The ray-systems considered by Hamilton are thus the normals of a surface $V-c=0$, and a large part of the properties of the system are thus included under the known theory of the normals of a surface; it may be remarked that the analytical formulæ are somewhat simplified by the circumstance that V instead of being (as usual) an arbitrary function of (x, y, z) is a function satisfying the partial differential equation $(\delta_x V)^2+(\delta_y V)^2+(\delta_z V)^2=1$. In particular, we have the theorem that each ray is intersected by two consecutive rays in foci which are the centres of curvature of the normal surface; the intersecting rays are rays proceeding from the curves of curvature of the normal surface, &c. There are other properties easily deducible from, but not actually included in, the theory of the normals; for instance, the intersecting rays aforesaid are rays proceeding from certain curves on the mirror, analogous to, but which obviously are not, the curves of curvature of the mirror. The natural mode of treatment of this part of the theory is to regard the rays as proceeding not from the normal surface, but from the mirror, and (by an investigation perfectly analogous to that for the normals of a surface) to enquire as to the intersection of the ray by rays proceeding from consecutive points of the mirror; it would thus appear that there are on the mirror two directions, such that proceeding along either of them to a consecutive point, the ray from the original point is intersected by the ray from the consecutive point, but that these directions are *not* in general at right angles, &c.

But in regard to such an investigation, the restriction introduced by the Hamiltonian theory is altogether unnecessary; there is no occasion to consider the rays which proceed from the several points of the mirror as being rays which are the normals of a surface, and the question is considered from the more general point of view as well by Malus in his *Théorie de la Double Refraction, &c.*, Paris, 1810, as more recently by Kummer in the Memoir "Allgemeine Theorie der gradlinigen Strahlensysteme," *Crelle*, t. LVII. (1860), pp. 189—230, viz. we have in Kummer a surface of any form whatever (defined according to the Gaussian theory, x, y, z given functions of the arbitrary parameters u, v), and from the several points thereof rays proceeding according to any law whatever, viz. the cosine-inclinations α, β, γ (or as Kummer writes them ξ, η, ζ) being given functions (such of course that $\alpha^2+\beta^2+\gamma^2=1$) of the same parameters u, v. It may be remarked: 1° that Kummer, while considering the simplifications of the general theory presenting themselves in the case where the rays are normals of the surface, does not specifically consider the case where, not being such normals, they are (as in the Hamiltonian theory) normals of a surface: 2° that some interesting investigations in regard to the shortest distances between consecutive rays, while naturally connecting themselves with the normals of the surface, or with that of the rays which are normals of another surface, do not properly belong to the "Allgemeine Theorie" of a congruence, which is independent of the notion of rectangularity.

It has been already remarked that the system may be looked at in the two ways 1° and 2°, and it is in the former of these that the question is considered by Kummer; it is interesting to work out part of the theory in the latter of the two ways. Taking X, Y, Z as current coordinates, we have, for a line through the point (x, y, z), the equations

$$X, Y, Z = x + \alpha\rho, \quad y + \beta\rho, \quad z + \gamma\rho;$$

α, β, γ are functions of (x, y, z), satisfying identically the equation $\alpha^2 + \beta^2 + \gamma^2 = 1$ (and therefore the derived equations in regard to x, y, z respectively); and also satisfying the equations

$$(\alpha\delta_x + \beta\delta_y + \gamma\delta_z)\,\alpha = 0, \quad (\alpha\delta_x + \beta\delta_y + \gamma\delta_z)\,\beta = 0, \quad (\alpha\delta_x + \beta\delta_y + \gamma\delta_z)\,\gamma = 0.$$

It should be remarked that, if these equations were not satisfied, then instead of a congruence there would be a complex, or triply infinite system of lines, viz. to the several points of space (x, y, z) there would correspond lines $X, Y, Z = x + \alpha\rho,\ y + \beta\rho,\ z + \gamma\rho$ as above, which lines would not reduce themselves to a doubly infinite system.

Suppose that the line through the point x, y, z is met by the line through a consecutive point $(x + dx,\ y + dy,\ z + dz)$; then, if X, Y, Z refer to the point of intersection of the two lines, we have

$$dx + \alpha d\rho + \rho d\alpha, \quad dy + \beta d\rho + \rho d\beta, \quad dz + \gamma d\rho + \rho d\gamma = 0;$$

and consequently

$$\begin{vmatrix} dx, & d\alpha, & \alpha \\ dy, & d\beta, & \beta \\ dz, & d\gamma, & \gamma \end{vmatrix} = 0$$

as a relation connecting the increments dx, dy, dz, in order that the lines may intersect, viz. this is a quadric relation $(*\between dx, dy, dz)^2 = 0$ between the increments. In the case of a complex, this equation represents a cone (passing evidently through the line $dx : dy : dz = \alpha : \beta : \gamma$), but in the case of a congruence the cone must break up into a pair of planes intersecting in the line in question $dx : dy : dz = \alpha : \beta : \gamma$. To verify *à posteriori* that this is so, observe that the differential equations satisfied by α, β, γ give, as above,

$$\delta_y\gamma - \delta_z\beta, \quad \delta_z\alpha - \delta_x\gamma, \quad \delta_x\beta - \delta_y\alpha,$$

proportional to α, β, γ, or say $= 2k\alpha,\ 2k\beta,\ 2k\gamma$; and it hence follows that the differentials $\delta\alpha, \delta\beta, \delta\gamma$ can be expressed in the forms

$$\begin{aligned} d\alpha &= a\,dx + h\,dy + g\,dz + k\,(\beta\,dz - \gamma\,dy), \\ d\beta &= h\,dx + b\,dy + f\,dz + k\,(\gamma\,dx - \alpha\,dz), \\ d\gamma &= g\,dx + f\,dy + c\,dz + k\,(\alpha\,dy - \beta\,dx), \end{aligned}$$

where

$$0 = a\alpha + h\beta + g\gamma,$$
$$0 = h\alpha + b\beta + f\gamma,$$
$$0 = g\alpha + f\beta + c\gamma.$$

The equation

$$\begin{vmatrix} dx, & d\alpha, & \alpha \\ dy, & d\beta, & \beta \\ dz, & d\gamma, & \gamma \end{vmatrix} = 0$$

thus assumes the form

$$(a, b, c, f, g, h \between dx, dy, dz \between \gamma\, dy - \beta\, dz, \alpha\, dz - \gamma\, dx, \beta\, dx - \alpha\, dy)$$
$$+ k\{(\gamma\, dy - \beta\, dz)^2 + (\alpha\, dz - \gamma\, dx)^2 + (\beta\, dx - \alpha\, dy)^2\} = 0.$$

Write for shortness

$$\gamma\, dy - \beta\, dz, \quad \alpha\, dz - \gamma\, dx, \quad \beta\, dx - \alpha\, dy = \xi, \ \eta, \ \zeta;$$

then putting for a moment $a\, dx + h\, dy + g\, dz = \Theta$, from this equation and $a\alpha + h\beta + g\gamma = 0$ we deduce

$$\alpha\Theta = -h(\beta\, dx - \alpha\, dy) + g(\alpha\, dz - \gamma\, dx),$$

that is, $= -h\zeta + g\eta$; or, putting for Θ its value and forming the analogous equations, we have

$$\alpha\,(a\, dx + h\, dy + g\, dz) = -h\zeta + g\eta,$$
$$\beta\,(h\, dx + b\, dy + f\, dz) = -f\xi + h\zeta,$$
$$\gamma\,(g\, dx + f\, dy + c\, dz) = -g\eta + f\xi,$$

and the quadric equation in (dx, dy, dz) thus becomes

$$\frac{\xi}{\alpha}(-h\zeta + g\eta) + \frac{\eta}{\beta}(-f\xi + h\zeta) + \frac{\zeta}{\gamma}(-g\eta + f\xi) + k(\xi^2 + \eta^2 + \zeta^2) = 0,$$

which, in virtue of the linear relation $\alpha\xi + \beta\eta + \gamma\zeta = 0$ connecting the (ξ, η, ζ), breaks up into a pair of planes, each passing through the line $\xi = 0$, $\eta = 0$, $\zeta = 0$, (that is, $dx : dy : dz = \alpha : \beta : \gamma$).

We obtain at once, as the condition that the two planes may be at right angles to each other, $k = 0$; that is,

$$\delta_y\gamma - \delta_z\beta, \quad \delta_z\alpha - \delta_x\gamma, \quad \delta_x\beta - \delta_y\alpha \text{ each } = 0;$$

or, what is the same thing, $\alpha\, dx + \beta\, dy + \gamma\, dz = dV$, a complete differential; and, as was seen, this is the condition in order that the lines may be the normals of a surface.

It thus appears that in a congruence each line is intersected by two consecutive lines, which determine respectively two planes through the line; and further, that if for every line of the congruence, the two planes are at right angles to each other, then the consecutive lines are the normals of a surface.

877.

NOTE ON THE TWO RELATIONS CONNECTING THE DISTANCES OF FOUR POINTS ON A CIRCLE.

[From the *Messenger of Mathematics*, vol. XVII. (1888), pp. 94, 95.]

CONSIDER a quadrilateral $BACD$ inscribed in a circle; and let the sides BA, AC, CD, DB and diagonals BC and AD be $=c$, b, h, g, a, $-f$ respectively; f is for convenience taken negative, so that the equation connecting the sides and diagonals may be

$$\Delta, = af + bg + ch, = 0.$$

We have between the sides and diagonals another relation

$$V, = abc + agh + bhf + cfg, = 0,$$

as is easily proved geometrically; in fact, recollecting that the opposite angles are supplementary to each other, the double area of the quadrilateral is $=(bc+gh)\sin A$, and it is also $=(bh+cg)\sin B$; that is, we have

$$(bc+gh)\sin A - (bh+cg)\sin B = 0.$$

But from the triangles BAD and BAC, in which the angles D, C are equal to each other, we have

$$\frac{c}{\sin D} = -\frac{f}{\sin B}, \quad \frac{c}{\sin C} = \frac{a}{\sin A};$$

that is,

$$f\sin A + a\sin B = 0;$$

and thence the required relation

$$a(bc+gh) + f(bh+cg) = 0.$$

The distances of the four points on the circle are thus connected by the two equations $\Delta=0$, $V=0$. Considering a, b, c, f, g, h as the distances from each other of any four points in the plane, we have between them the relation

$$\begin{aligned}\Omega, = \quad & a^2f^2(-a^2-f^2+b^2+g^2+c^2+h^2)\\ &+b^2g^2(\ \ a^2+f^2-b^2-g^2+c^2+h^2)\\ &+c^2h^2(\ \ a^2+f^2+b^2+g^2-c^2-h^2)\\ &-a^2b^2c^2-a^2g^2h^2-b^2h^2f^2-c^2f^2g^2, =0;\end{aligned}$$

and it is clear that this equation should be a consequence of the equations $\Delta=0$, $V=0$. To verify this, forming the sum $\Omega+V^2$, we have

$$\begin{aligned}\Omega+V^2= \quad & (a^2+f^2)(-a^2f^2+b^2g^2+c^2h^2+2bgch)\\ &+(b^2+g^2)(-b^2g^2+c^2h^2+a^2f^2+2chaf)\\ &+(c^2+h^2)(-c^2h^2+a^2f^2+b^2g^2+2afbg);\end{aligned}$$

viz. this is

$$\begin{aligned}= \quad & (a^2+f^2)\{-a^2f^2+(\Delta-af)^2\}\\ &+(b^2+g^2)\{-b^2g^2+(\Delta-bg)^2\}\\ &+(c^2+h^2)\{-c^2h^2+(\Delta-ch)^2\};\end{aligned}$$

or, since

$$-a^2f^2+(\Delta-af)^2=\Delta(\Delta-2af)=\Delta(-af+bg+ch), \text{ \&c.},$$

this is

$$\begin{aligned}\Omega+V^2=\Delta[\ & (a^2+f^2)(-af+bg+ch)\\ &+(b^2+g^2)(\ \ af-bg+ch)\\ &+(c^2+h^2)(\ \ af+bg-ch)],\end{aligned}$$

which proves the theorem.

It may be remarked that the equation $V=0$ may be written

$$a(bc+gh)+f(bh+cg)=0;$$

viz. multiplying by a, and for af writing its value, $=-(bg+ch)$ from the equation $\Delta=0$, this gives

$$-a^2(bc+gh)+(bg+ch)(bh+cg)=0,$$

that is,

$$bc(g^2+h^2-a^2)+gh(b^2+c^2-a^2)=0,$$

which expresses that the angles A, D are supplementary to each other; and, similarly, by the elimination of any other of the six quantities from the equations $\Delta=0$, $V=0$, we have five other like equations.

878.

NOTE ON THE ANHARMONIC RATIO EQUATION.

[From the *Messenger of Mathematics*, vol. XVII. (1888), pp. 95, 96.]

GIVEN any four quantities $\alpha, \beta, \gamma, \delta$, if θ be one of the values of the anharmonic ratio, the other values are

$$\frac{1}{\theta},\quad -(1+\theta),\quad -\frac{1}{1+\theta},\quad -\frac{\theta}{1+\theta},\quad -\frac{1+\theta}{\theta};$$

and hence the equation having these six roots is

$$(x-\theta)\left(x-\frac{1}{\theta}\right)(x+1+\theta)\left(x+\frac{1}{1+\theta}\right)\left(x+\frac{\theta}{1+\theta}\right)\left(x+\frac{1+\theta}{\theta}\right)=0;$$

or, multiplying out, the equation, as is well known, takes the form

$$(x^2+x+1)^3-\frac{(\theta^2+\theta+1)^3}{\theta^2(\theta+1)^2}x^2(x+1)^2=0.$$

But to effect the multiplication in the easiest manner we may proceed as follows: writing

$$a,\ b,\ c=(\alpha-\delta)(\beta-\gamma),\quad (\beta-\delta)(\gamma-\alpha),\quad (\gamma-\delta)(\alpha-\beta),$$

so that $a+b+c=0$, the equation is

$$\left(x-\frac{b}{c}\right)\left(x-\frac{c}{b}\right)\left(x-\frac{c}{a}\right)\left(x-\frac{a}{c}\right)\left(x-\frac{a}{b}\right)\left(x-\frac{b}{a}\right)=0.$$

The product of the first pair of factors is

$$x^2+1-\left(\frac{b}{c}+\frac{c}{b}\right)x,\ =(x+1)^2-\frac{a^2}{bc}x;$$

thus the equation is

$$\left\{(x+1)^2-\frac{a^2}{bc}x\right\}\left\{(x+1)^2-\frac{b^2}{ca}x\right\}\left\{(x+1)^2-\frac{c^2}{ab}x\right\}=0\,;$$

that is,

$$(x+1)^6-\left(\frac{a^2}{bc}+\frac{b^2}{ca}+\frac{c^2}{ab}\right)x(x+1)^4+\left(\frac{bc}{a^2}+\frac{ca}{b^2}+\frac{ab}{c^2}\right)x^2(x+1)^2-x^3=0\,;$$

and recollecting that $a+b+c=0$, and writing $q=bc+ca+ab$, $r=abc$, the equation becomes

$$(x+1)^6-3(x+1)^4x+\left(3+\frac{q^3}{r^2}\right)(x+1)^2x^2-x^3=0\,;$$

that is,

$$(x^2+x+1)^3+\frac{q^3}{r^2}(x+1)^2x^2=0.$$

But, writing $\theta=\dfrac{b}{a}$, we have

$$(\theta^2+\theta+1)^3+\frac{q^3}{r^2}(\theta+1)^2\theta^2=0\,;$$

or finally,

$$(x^2+x+1)^3-\frac{(\theta^2+\theta+1)^3}{\theta^2(\theta+1)^2}x^2(x+1)^2=0,$$

the required result.

879.

NOTE ON THE DIFFERENTIAL EQUATION

$$\frac{dx}{\surd(1-x^2)}+\frac{dy}{\surd(1-y^2)}=0.$$

[From the *Messenger of Mathematics*, vol. XVIII. (1889), p. 90.]

We have

$$\frac{\sin u-\sin v}{\cos u-\cos v}=-\cot\tfrac{1}{2}(u+v), \;=-\sqrt{\left\{\frac{1+\cos(u+v)}{1-\cos(u+v)}\right\}},$$

and thence, writing $\cos u=x$, $\sin u=\surd(1-x^2)=\surd(X)$, and similarly

$$\cos v=y,\; \sin v=\surd(1-y^2)=\surd(Y),$$

we have

$$\frac{\surd(X)-\surd(Y)}{x-y}=-\sqrt{\left\{\frac{1+xy-\surd(XY)}{1-xy+\surd(XY)}\right\}},$$

an identical equation which, in the form

$$\frac{2-x^2-y^2-2\surd(XY)}{(x-y)^2}=\frac{1+xy-\surd(XY)}{1-xy+\surd(XY)},$$

may be verified directly without any difficulty. The integral of the proposed differential equation can of course be taken to be $c=xy-\surd(XY)$; and we have thus another form of integral

$$\frac{\surd(X)-\surd(Y)}{x-y}=-\sqrt{\left(\frac{1+c}{1-c}\right)}, \text{ say } =\surd(C),$$

viz. we have the integral

$$\left\{\frac{\surd(X)-\surd(Y)}{x-y}\right\}^2=C,$$

which is what Lagrange's integral of the differential equation

$$\frac{dx}{\surd(X)}+\frac{dy}{\surd(Y)}=0$$

becomes when the quartic functions X, Y reduce themselves to the quadric functions $1-x^2$ and $1-y^2$ respectively.

880.

NOTE ON THE RELATION BETWEEN THE DISTANCE OF FIVE POINTS IN SPACE.

[From the *Messenger of Mathematics*, vol. XVIII. (1889), pp. 100—102.]

IN Lagrange's paper "Solutions analytiques de quelques problèmes sur les pyramides triangulaires" (*Berlin Memoirs*, 1773; *Œuvres*, t. III., see p. 677), there is contained a formula for the relation between the distances from each other of five points in space; viz. this is

$$\begin{aligned}4\Delta^2 f = \;& \alpha(a+f-g)^2+\alpha'(a'+f-g')^2+\alpha''(a''+f-g'')^2\\ &+2\beta(a'+f-g')(a''+f-g'')+2\beta'(a+f-g)(a''+f-g'')\\ &+2\beta''(a+f-g)(a'+f-g'),\end{aligned}$$

or, in a slightly altered notation, say

$$\Pi = -4\Delta^2 f + (\alpha_1, \alpha_2, \alpha_3, \beta_1, \beta_2, \beta_3 \between a_1+f-g_1, a_2+f-g_2, a_3+f-g_3)^2 = 0,$$

where, if the points are called 1, 2, 3, 4, 5, then

c_1, c_2, c_3	are the squared distances	23, 31, 12,
a_1, a_2, a_3	"	41, 42, 43,
g_1, g_2, g_3	"	51, 52, 53,

and

f	is the squared distance	45.

The values of $\alpha_1, \alpha_2, \alpha_3, \beta_1, \beta_2, \beta_3$, in terms of these squared distances, are

$$\begin{aligned}\alpha_1 &= -(a_2-a_3)^2+2c_1(a_2+a_3)-c_1^2,\\ \alpha_2 &= -(a_3-a_1)^2+2c_2(a_3+a_1)-c_2^2,\\ \alpha_3 &= -(a_1-a_2)^2+2c_3(a_1+a_2)-c_3^2,\\ \beta_1 &= -(a_3-a_1)(a_1-a_2)+2c_1a_1-c_2(a_1+a_2)-c_3(a_3+a_1)+c_2c_3,\\ \beta_2 &= -(a_1-a_2)(a_2-a_3)+2c_2a_2-c_3(a_2+a_3)-c_1(a_1+a_2)+c_3c_1,\\ \beta_3 &= -(a_2-a_3)(a_3-a_1)+2c_3a_3-c_1(a_3+a_1)-c_2(a_2+a_3)+c_1c_2.\end{aligned}$$

After some reductions, the value of $4\Delta^2$, in terms of these squared distances, is found to be

$$\begin{aligned}4\Delta^2 = \ & c_1(a_3-a_1)(a_1-a_2)+a_1(-c_1^2+c_1c_2+c_1c_3)-c_1c_2c_3\\ &+c_2(a_1-a_2)(a_2-a_3)+a_2(-c_2^2+c_2c_3+c_2c_1)\\ &+c_3(a_2-a_3)(a_3-a_1)+a_3(-c_3^2+c_3c_1+c_3c_2),\end{aligned}$$

which is, in fact,

$$8\Delta^2 = \begin{vmatrix} . & 1, & 1, & 1, & 1 \\ 1, & . & c_3, & c_2, & a_1 \\ 1, & c_3, & . & c_1, & a_2 \\ 1, & c_2, & c_1, & . & a_3 \\ 1, & a_1, & a_2, & a_3, & . \end{vmatrix},$$

where observe that each term of the determinant contains the factor 2.

By the formula in my paper "On a theorem in the Geometry of Position," *Camb. Math. Jour.*, t. II. (1841), pp. 267—271, [1], (introducing into it the present notation), the relation between the distances of the five points is given in the form

$$\Omega = \begin{vmatrix} . & 1, & 1, & 1, & 1, & 1 \\ 1, & . & c_3, & c_2, & a_1, & g_1 \\ 1, & c_3, & . & c_1, & a_2, & g_2 \\ 1, & c_2, & c_1, & . & a_3, & g_3 \\ 1, & a_1, & a_2, & a_3, & . & f \\ 1, & g_1, & g_2, & g_3, & f, & . \end{vmatrix} = 0.$$

The equations $\Pi=0$, $\Omega=0$ should therefore agree with each other; we have in Ω the term

$$-f^2 \begin{vmatrix} 1, & c_3, & . & c_1 \\ 1, & c_2, & c_1, & . \\ . & 1, & 1, & 1 \\ 1, & . & c_3, & c_2 \end{vmatrix},$$

which is

$$=-f^2(c_1^2+c_2^2+c_3^2-2c_2c_3-2c_3c_1-2c_1c_2);$$

and similarly in Π we have the term

$$f^2(\alpha_1+\alpha_2+\alpha_3+2\beta_1+2\beta_2+2\beta_3),$$

which is easily shown to be

$$=-f^2(c_1^2+c_2^2+c_3^2-2c_2c_3-2c_3c_1-2c_1c_2);$$

and it thus appears that we have identically $\Pi=\Omega$.

It is to be remarked that, in Lagrange's form, the points 4 and 5 are regarded as determined each of them by means of its squared distances from the vertices of the triangle 123, and that the formula gives (by a quadratic equation) the squared distance 45; but that nevertheless the two points 4 and 5 do not present themselves symmetrically in the formula; in fact, Δ^2 and the coefficients $(\alpha_1, \alpha_2, \alpha_3, \beta_1, \beta_2, \beta_3)$ of the formula relate all of them to the tetrahedron 4123; as noticed in the paper, $\Delta = 6 \times$ volume of tetrahedron; $\sqrt{(\alpha_1)}$, $\sqrt{(\alpha_2)}$, $\sqrt{(\alpha_3)}$ are the doubled areas of the faces 423, 431, 412 respectively, and the doubled area of the face 123 is

$$= \sqrt{(\alpha_1 + \alpha_2 + \alpha_3 + 2\beta_1 + 2\beta_2 + 2\beta_3)};$$

it may be added that

$$\beta_1 \div \sqrt{(\alpha_2\alpha_3)},\ \ \beta_2 \div \sqrt{(\alpha_3\alpha_1)},\ \ \beta_3 \div \sqrt{(\alpha_1\alpha_2)},$$

are equal to the cosines of the dihedral angles at the edges 41, 42, 43 respectively.

881.

ON HERMITE'S H-PRODUCT THEOREM.

[From the *Messenger of Mathematics*, vol. XVIII. (1889), pp. 104—107.]

I GIVE this name to a theorem relating to the product of an even number of Eta-functions, established by M. Hermite in his "Note sur le calcul différential et le calcul integral," forming an appendix to the sixth edition of Lacroix's *Differential and Integral Calculus*, and separately printed, 8vo. Paris, 1862. It is the theorem stated p. 65, in the form

$$\phi(x) = F(z^2) + \frac{dz}{dx} z F_1(z^2),$$

where

$$\phi(x) = \frac{AH(x-\alpha_1)\,H(x-\alpha_2)\ldots H(x-\alpha_{2n})}{\Theta^{2n}(x)},$$

where $\alpha_1+\alpha_2+\ldots+\alpha_{2n}=0$, and $z = \operatorname{sn} x$, $\operatorname{cn} x$ or $\operatorname{dn} x$ at pleasure; $F(z^2)$, $F_1(z^2)$ denote rational and integral functions of z^2 of the degrees n and $n-2$ respectively; A is a constant, which we may if we please so determine that in $F(z^2)$ the coefficient of the highest power z^{2n} shall be $=1$.

If, for shortness, we write s, c, d for $\operatorname{sn} x$, $\operatorname{cn} x$, $\operatorname{dn} x$ respectively; and to fix the ideas, assume $z = \operatorname{sn} x$, $=s$, then the theorem is

$$\frac{AH(x-\alpha_1)\,H(x-\alpha_2)\ldots H(x-\alpha_{2n})}{\Theta^{2n}(x)} = F(s^2) + scdF_1(s^2);$$

viz. the theorem is that the product of the $2n$ H-functions ($\alpha_1+\alpha_2+\ldots+\alpha_{2n}=0$ as above), divided by $\Theta^{2n}(x)$, is a function of the elliptic functions sn, cn, dn, of the form in question.

Hermite uses the theorem for the demonstration of Abel's theorem, as applied to the elliptic functions; or as I would rather express it, he uses the theorem for the determination of the sn, cn, and dn of $\alpha_1+\alpha_2+\ldots+\alpha_{2n-1}$.

To show how this is, observe that $F(s^2)$, *quà* rational and integral function of s^2 of the degree n, with its first coefficient $=1$, contains n arbitrary coefficients; and $F_1(s^2)$, *quà* rational and integral function of s^2 of the degree $n-2$, contains $n-1$ arbitrary coefficients: hence $F(s^2)+scdF_1(s^2)$ contains $2n-1$ arbitrary coefficients; and considering $\alpha_1, \alpha_2, \ldots, \alpha_{2n-1}$ as given, the function in question must vanish for each of the values $x=\alpha_1, \alpha_2, \ldots, \alpha_{2n-1}$; and we have therefore $2n-1$ equations for obtaining the $2n-1$ coefficients, which are thus completely determined: in particular, the constant term, say L, of $F_1(s^2)$ is a given function of $\alpha_1, \alpha_2, \ldots, \alpha_{2n-1}$, that is, of the sn, cn, and dn of these quantities; and the theorem shows that the function thus determined vanishes also for $x=\alpha_{2n}$, that is, $=-(\alpha_1+\alpha_2+\ldots+\alpha_{2n-1})$.

Now writing $-x$ for x in the formula, and recollecting that H is an odd function, Θ an even function, we find

$$\frac{AH(x+\alpha_1)\,H(x+\alpha_2)\ldots H(x+\alpha_{2n})}{\Theta^{2n}(x)}=F(s^2)-scdF_1(s^2);$$

and multiplying together the two sides of these equations respectively,

$$A^2\frac{H(x-\alpha_1)\,H(x+\alpha_1)}{\Theta^2(x)}\ldots\frac{H(x-\alpha_{2n})\,H(x+\alpha_{2n})}{\Theta^2(x)}=\{F(s^2)\}^2-s^2c^2d^2\{F_1(s^2)\}^2,$$

where the right-hand side is a rational and integral function of s^2 of the degree $2n$, and the coefficient of the highest term s^{4n} is $=1$; in fact, this term arises only from the square of $F(s^2)$, which has its highest term $=s^{2n}$.

Now $\dfrac{H(x-\alpha_1)\,H(x+\alpha_1)}{\Theta^2(x)}$ is a mere constant multiple of $\text{sn}^2 x-\text{sn}^2\alpha_1$, or say of $s^2-\text{sn}^2\alpha_1$; (this well-known theorem is, in fact, the particular case $n=2$ of Hermite's theorem); and similarly for the other terms: we must clearly have A^2, multiplied by the product of the factors thus introduced, $=1$; and thus the theorem becomes

$$(s^2-\text{sn}^2\alpha_1)(s^2-\text{sn}^2\alpha_2)\ldots(s^2-\text{sn}^2\alpha_{2n})=\{F(s^2)\}^2-s^2c^2d^2\{F_1(s^2)\}^2.$$

And putting herein $s=0$, and writing as before L for the constant term of $F(s^2)$, we have

$$\text{sn}^2\alpha_1\,\text{sn}^2\alpha_2\ldots\text{sn}^2\alpha_{2n}=L^2,$$

or, the sign $\pm$ being properly determined, say

$$\text{sn}\,\alpha_1\,\text{sn}\,\alpha_2\ldots\text{sn}\,\alpha_{2n}=\pm L,$$

where, by what precedes, L is a given function of the sn, cn, and dn of $\alpha_1, \alpha_2, \ldots, \alpha_{2n-1}$. Hence we have $\text{sn}\,\alpha_{2n}$, that is, $-\text{sn}(\alpha_1+\alpha_2+\ldots+\alpha_{2n-1})$ as a given function of the sn, cn, and dn of $\alpha_1, \alpha_2, \ldots, \alpha_{2n-1}$.

Similarly writing $z=\text{cn}\,x$, $=c$, and $z=\text{dn}\,x$, $=d$, we have $\text{cn}(\alpha_1+\alpha_2+\ldots+\alpha_{2n-1})$ and $\text{dn}(\alpha_1+\alpha_2+\ldots+\alpha_{2n-1})$ each of them as a given function of the sn, cn, and dn of $\alpha_1, \alpha_2, \ldots, \alpha_{2n-1}$.

It is hardly necessary to remark that $F(z^2)+z\dfrac{dz}{dx}F_1(z^2)$ is a function of the same form, whether we have $z=s$, c or d; in fact, the functions F and F_1 are rational in s^2, c^2, or d^2, and we have $z\dfrac{dz}{dx}=scd$, $-scd$, and $-k^2scd$ for the three values respectively.

The number of terms $\alpha_1, \alpha_2, \ldots, \alpha_{2n-1}$ has been odd, but by taking one of them $=0$, the formulæ give the values of the sn, cn, and dn for the sum of an even number of terms.

It has been seen that Hermite's H-product theorem gives, say Abel's theorem, in the form

$$\Pi\,(s^2-\operatorname{sn}^2\alpha)=\{F(s^2)\}^2-s^2c^2d^2\{F_1(s^2)\}^2,$$

each side of this relation being the product of two factors, viz. for the left-hand side the factors are

$$A\Pi\frac{H(x-\alpha)}{\Theta(x)},\quad A\Pi\frac{H(x+\alpha)}{\Theta(x)},$$

and for the right-hand side they are the rational functions of s^2,

$$F(s^2)+scdF_1(s^2),\quad F(s^2)-scdF_1(s^2);$$

these factors are by Hermite's theorem equal each to each; viz. this is the relation in which Hermite's stands to Abel's theorem.

The H-product theorem is given as one out of a group of four theorems; the other three may be called the H-product, H_1-product and Θ_1-product, odd theorems respectively,

$$\{\Theta_1(x)=\Theta(x+K),\quad H_1(x)=H(x+K)\},$$

viz. these are

$$\left\{\begin{aligned}&\frac{A_1H(x-\alpha_1)\,H(x-\alpha_2)\ldots H(x-\alpha_{2n+1})}{\Theta^{2n+1}(x)}=sF(s^2)+\quad cd\,\phi(s^2),\\&\frac{A_2H_1(x-\alpha_1)\,H_1(x-\alpha_2)\ldots H_1(x-\alpha_{2n+1})}{\Theta^{2n+1}(x)}=cF(c^2)-\quad sd\,\phi(c^2),\\&\frac{A_3\Theta_1(x-\alpha_1)\,\Theta_1(x-\alpha_2)\ldots\Theta_1(x-\alpha_{2n+1})}{\Theta^{2n+1}(x)}=dF(d^2)-k^2sc\,\phi(d^2),\end{aligned}\right.$$

where F, ϕ are rational and integral functions of the degrees n and $n-1$, having their proper values in the three equations respectively, and in each case

$$\alpha_1+\alpha_2+\ldots+\alpha_{2n+1}=0.$$

It was seen above that, for $n=1$, the H-product theorem became

$$\frac{AH(x-\alpha)\,H(x+\alpha)}{\Theta^2(x)}=\operatorname{sn}^2x-\operatorname{sn}^2\alpha,$$

which is the most simple case; for the odd theorems, the most simple case is $n=0$, viz. we then have

$$\frac{A_1H(x)}{\Theta(x)}=\operatorname{sn}x,\quad\frac{A_2H_1(x)}{\Theta(x)}=\operatorname{cn}x,\quad\frac{A_3\Theta_1(x)}{\Theta(x)}=\operatorname{dn}x;$$

to complete the formulæ observe that the values of the constants are

$$A=\frac{2k'K}{k\pi\Theta^2(\alpha)},\quad A_1=\frac{1}{\sqrt{(k)}},\quad A_2=\sqrt{\left(\frac{k'}{k}\right)},\quad A_3=\sqrt{(k')}.$$

The three theorems may be used, in like manner with the H-product theorem, to give the values of the sn, cn, and dn respectively of the sum $\alpha_1+\alpha_2+\ldots+\alpha_{2n}$.

882.

A CORRESPONDENCE OF CONFOCAL CARTESIANS WITH THE RIGHT LINES OF A HYPERBOLOID.

[From the *Messenger of Mathematics*, vol. XVIII. (1889), pp. 128—130.]

TAKE α, β, γ arbitrary, $A, B, C = \beta - \gamma, \gamma - \alpha, \alpha - \beta$ (so that $A+B+C=0$), and writing ρ, σ, τ for rectangular coordinates, consider the hyperboloid

$$A\rho^2 + B\sigma^2 + C\tau^2 + ABC = 0.$$

Let ρ_0, σ_0, τ_0 be the coordinates of a point on the surface $(A\rho_0^2 + B\sigma_0^2 + C\tau_0^2 + ABC = 0)$. The equations of a line through this point are $\rho, \sigma, \tau = \rho_0 + f\Omega, \sigma_0 + g\Omega, \tau_0 + h\Omega$ (Ω indeterminate); and if this lies on the surface, we have

$$A\rho_0 f + B\sigma_0 g + C\tau_0 h = 0,$$
$$Af^2 + Bg^2 + Ch^2 = 0,$$

which equations determine the ratios $f : g : h$; the equations give

$$(A\rho_0 f + B\sigma_0 g)^2 = C\tau_0^2 . Ch^2, \quad = -C\tau_0^2 (Af^2 + Bg^2);$$

that is,

$$(A^2\rho_0^2 + AC\tau_0^2) f^2 + 2AB\rho_0\sigma_0 fg + (B^2\sigma_0^2 + BC\tau_0^2) g^2 = 0,$$

whence

$$\begin{aligned}&\{(B^2\sigma_0^2 + BC\tau_0^2) g + AB\rho_0\sigma_0 f\}^2\\ &= \{A^2B^2\rho_0^2\sigma_0^2 - (A^2\rho_0^2 + AC\tau_0^2)(B^2\sigma_0^2 + BC\tau_0^2)\} f^2,\\ &= -ABC(A\rho_0^2 + B\sigma_0^2 + C\tau_0^2)\tau_0^2 f^2,\\ &= A^2B^2C^2\tau_0^2 f^2;\end{aligned}$$

that is,

$$\{(B\sigma_0^2 + C\tau_0^2) g + A\rho_0\sigma_0 f\}^2 = A^2C^2\tau_0^2 f^2,$$

or say

$$(B\sigma_0^2 + C\tau_0^2) g + A(\rho_0\sigma_0 \pm C\tau_0) f = 0,$$

which equation, together with $A\rho_0 f + B\sigma_0 g + C\tau_0 h = 0$, determines the ratios $f : g : h$. We have thus the two lines through the point $(\rho_0, \sigma_0, \tau_0)$.

But the equations of the line may be conveniently represented in a different form; writing the equation first obtained in the form

$$\sigma_0(B\sigma_0 g + A\rho_0 f) + C\tau_0^2 g \pm AC\tau_0 f = 0,$$

this is

$$-\sigma_0 C\tau_0 h + C\tau_0^2 g \pm AC\tau_0 f = 0,$$

viz.

$$-h\sigma_0 + g\tau_0 \pm Af = 0;$$

and we have the like equations

$$-f\tau_0 + h\rho_0 \pm Bg = 0,$$
$$-g\rho_0 + f\sigma_0 \pm Ch = 0,$$

where the sign is the same in each of the three equations.

The equations of the line on the surface may be written

$$\begin{array}{llll} . & h\sigma & -g\tau & -h\sigma_0 + g\tau_0 = 0, \\ -h\rho & . & +f\tau & -f\tau_0 + h\rho_0 = 0, \\ g\rho & -f\sigma & . & -g\rho_0 + f\sigma_0 = 0, \\ (h\sigma_0 - g\tau_0)\rho & +(f\tau_0 - h\rho_0)\sigma & +(g\rho_0 - f\sigma_0)\tau & . = 0; \end{array}$$

and hence from the foregoing three equations, taking the sign $-$, we have

$$\begin{array}{lllll} . & h\sigma & -g\tau & +Af & = 0, \\ -h\rho & . & +f\tau & +Bg & = 0, \\ g\rho & -f\sigma & . & +Ch & = 0, \\ -Af\rho & -Bg\sigma & -Ch\tau & . & = 0, \end{array}$$

where $Af^2 + Bg^2 + Ch^2 = 0$, for the equations of a line on the surface.

In like manner, taking the sign $+$, and for f, g, h writing new values f', g', h', we have

$$\begin{array}{lllll} . & h'\sigma & -g'\tau & -Af' & = 0, \\ -h'\rho & . & +f'\tau & -Bg' & = 0, \\ g'\rho & -f'\sigma & . & -Ch' & = 0, \\ Af'\rho & +Bg'\sigma & +Ch'\tau & . & = 0, \end{array}$$

where $Af'^2 + Bg'^2 + Ch'^2 = 0$, for the equations of a line on the surface.

The two systems of equations evidently belong to the lines of the two different kinds respectively. Writing for shortness $P, Q, R = gh' + g'h, hf' + h'f, fg' + f'g$, the two lines in fact intersect in a point, the coordinates say $(\rho_0, \sigma_0, \tau_0)$ whereof are $= \Theta QR, \Theta RP, \Theta PQ$, where

$$\Theta = \frac{A}{g^2h'^2 - g'^2h^2} = \frac{B}{h^2f'^2 - h'^2f^2} = \frac{C}{f^2g'^2 - f'^2g^2},$$

the three expressions for Θ being equal to each other in virtue of the equations

$$Af^2 + Bg^2 + Ch^2 = 0, \quad Af'^2 + Bg'^2 + Ch'^2 = 0.$$

Take now, in a plane, P, Q, R points on any line, say the axis of x, at distances α, β, γ from the origin, then for a point of the plane, coordinates (x, y), if ρ, σ, τ be the distances of the point from these three points, or say foci, we have

$$\begin{aligned} \rho^2 &= (x-\alpha)^2 + y^2, \\ \sigma^2 &= (x-\beta)^2 + y^2, \\ \tau^2 &= (x-\gamma)^2 + y^2; \end{aligned}$$

and if as before $A, B, C = \beta - \gamma,\ \gamma - \alpha,\ \alpha - \beta$, we thence have

$$A\rho^2 + B\sigma^2 + C\tau^2 + ABC = 0.$$

A point, coordinates (ρ, σ, τ), of the hyperboloid thus corresponds to a point in the plane, distances ρ, σ, τ from the three foci R, S, T respectively; and to any line

$$\begin{array}{rrrrl} . & h\sigma - & g\tau + & Af & = 0, \\ -\ h\rho & . & +\ f\tau + & Bg & = 0, \\ g\rho - & f\sigma & . & +\ Ch & = 0, \\ -\ Af\rho & -\ Bg\sigma & -\ Ch\tau & . & = 0, \end{array}$$

corresponds the Cartesian represented by these linear equations. Similarly, to the line represented by the other system of equations

$$\begin{array}{rrrrl} . & h'\sigma - & g'\tau - & Af' & = 0, \\ -\ h'\rho & . & +\ f'\tau - & Bg' & = 0, \\ g'\rho - & f'\sigma & . & -\ Ch' & = 0, \\ Af'\rho + & Bg'\sigma + & Ch'\tau & . & = 0, \end{array}$$

corresponds the Cartesian represented by these equations; the two curves intersect in the point $\rho_0, \sigma_0, \tau_0 = \Theta QR, \Theta RP, \Theta PQ$, corresponding to the intersection of the lines on the hyperboloid; and moreover, *quà* confocal Cartesians, they intersect at right angles.

883.

ANALYTICAL FORMULÆ IN REGARD TO AN OCTAD OF POINTS.

[From the *Messenger of Mathematics*, vol. XVIII. (1889), pp. 149—152.]

THE term "tetrad" is used in two distinct but not inconsistent senses, viz. a tetrad denotes any four points; and it also denotes the four vertices of a self-conjugate tetrahedron in regard to a quadric surface. In fact, given any four points, there exists a triply infinite series of quadric surfaces such that, in regard to any one of them, the four points form a self-conjugate tetrahedron.

Two or more tetrads, in regard to one and the same quadric surface, are called similar tetrads.

The eight points of intersection of any three quadric surfaces are an octad; and we have the theorem that any seven points determine the octad, viz. the quadric surfaces which pass through any seven given points pass also through a uniquely determinate eighth point.

We have the further theorem that any two similar tetrads form an octad.

In particular, the vertices of the tetrahedron

$$(x=0,\ y=0,\ z=0,\ w=0),$$

or, say the points 1000, 0100, 0010, 0001 are a tetrad in regard to the quadric surface $x^2+y^2+z^2+w^2=0$. The points (x_1, y_1, z_1, w_1), (x_2, y_2, z_2, w_2), (x_3, y_3, z_3, w_3), (x_4, y_4, z_4, w_4), or say the points 1, 2, 3, 4, will be a tetrad in regard to the same quadric surface, if only

$$x_1x_2+y_1y_2+z_1z_2+w_1w_2=0, \text{ \&c., (six equations).}$$

Hence, these equations being satisfied, the two tetrads form an octad; the equation of a general quadric surface through the points of the first tetrad is

$$fyz+gzx+hxy+lxw+myw+nzw=0,$$

and we can from three such equations eliminate any two terms; the two terms eliminated may be such as xy, xz, (12 forms) or such as yz, xw, (3 forms); and the equation may thus be presented either in the form

$$fyz + (lx + my + nz)\,w = 0,$$

or in the form

$$gzx + hxy + myw + nzw = 0.$$

The absolute magnitudes of the coordinates

$$(x_1,\ y_1,\ z_1,\ w_1),\ \&c.,$$

of these points are properly indeterminate; but we may if we please fix the absolute magnitudes by assuming

$$x_1^2 + y_1^2 + z_1^2 + w_1^2 = M,\ \&c.,$$

(four equations, the same quantity M in each of them), and we have as before

$$x_1x_2 + y_1y_2 + z_1z_2 + w_1w_2 = 0,\ \&c.,\ \text{(six equations)},$$

viz. the coordinates are taken to satisfy these 10 equations; and this being so, it is to be shown that there exist quadric surfaces as above. The 10 equations may be expressed in a correlative form in like manner with the ordinary six equations for the transformation of the rectangular coordinates (x, y, z), viz. the new form is

$$x_1^2\ + x_2^2\ + x_3^2\ + x_4^2\ = M,\ \&c.,\ \text{(four equations)},$$

and

$$x_1y_1 + x_2y_2 + x_3y_3 + x_4y_4 = 0,\ \&c.,\ \text{(six equations)},$$

and the coordinates thus also satisfy these 10 equations.

Thus the equation $fyz + (lx + my + nz)\,w = 0$, if we determine the coefficients in such wise that the surface passes through the points 1, 2, 3, becomes

$$\begin{vmatrix} yz\ , & xw\ , & yw\ , & zw \\ y_1z_1, & x_1w_1, & y_1w_1, & z_1w_1 \\ y_2z_2, & x_2w_2, & y_2w_2, & z_2w_2 \\ y_3z_3, & x_3w_3, & y_3w_3, & z_3w_3 \end{vmatrix} = 0,$$

and this equation must therefore be satisfied on writing therein $(x_4,\ y_4,\ z_4,\ w_4)$ for (x, y, z, w); viz. the equation thus obtained, det. (1, 2, 3, 4) = 0, must be satisfied in virtue of the equations which connect the coordinates of the four points. But, instead of effecting this verification, it is better to exhibit the equation in (x, y, z, w) in a form wherein the coordinates of the four points enter symmetrically; and the new equation, *quâ* transformation of the original form, is satisfied for $(x, y, z, w) = (x_1, y_1, z_1, w_1)$: and then of course, by reason of the symmetry, it is also satisfied for

$$(x,\ y,\ z,\ w) = (x_4,\ y_4,\ z_4,\ w_4).$$

The transformation may be effected directly; but I prefer to write down the final result, and afterwards verify it.

The form is

$$\begin{aligned} yz\,.\,Mw_1w_2w_3w_4 + xw\,[&- x_1y_1z_1\,.\,w_2w_3w_4 - x_2y_2z_2\,.\,w_3w_4w_1 - x_3y_3z_3\,.\,w_4w_1w_2 - x_4y_4z_4\,.\,w_1w_2w_3] \\ + yw\,[&\;\;\, x_1^2w_1\,.\,z_2z_3z_4 + x_2^2w_2\,.\,z_3z_4z_1 + x_3^2w_3\,.\,z_4z_1z_2 + x_4^2w_4\,.\,z_1z_2z_3] \\ + zw\,[&\;\;\, x_1^2w_1\,.\,y_2y_3y_4 + x_2^2w_2\,.\,y_3y_4y_1 + x_3^2w_3\,.\,y_4y_1y_2 + x_4^2w_4\,.\,y_1y_2y_3] = 0. \end{aligned}$$

We have to show that the equation is satisfied by writing therein

$$(x,\ y,\ z,\ w) = (x_1,\ y_1,\ z_1,\ w_1).$$

Calling the resulting value Ω_1, we have

$$\begin{aligned} \Omega_1 \div w_1 = My_1z_1w_2w_3w_4 &- x_1^2y_1z_1w_2w_3w_4 - x_1\,(x_2y_2z_2w_3w_4w_1 + x_3y_3z_3w_4w_1w_2 + x_4y_4z_4w_1w_2w_3) \\ &+ x_1^2w_1\,(y_1z_2z_3z_4 + z_1y_2y_3y_4) + y_1z_1\,[x_2^2w_2\,(z_3z_4 + y_3y_4) + x_3^2w_3\,(z_4z_2 + y_4y_2) + x_4^2w_4\,(z_2z_3 + y_2y_3)], \end{aligned}$$

which is

$$\begin{aligned} &= y_1z_1w_2w_3w_4\,(x_2^2 + x_3^2 + x_4^2) - x_1\,(x_2y_2z_2w_3w_4w_1 + x_3y_3z_3w_4w_1w_2 + x_4y_4z_4w_1w_2w_3) \\ &\quad + x_1^2w_1\,(y_1z_2z_3z_4 + z_1y_2y_3y_4) - y_1z_1\,[x_2^2w_2\,(x_3x_4 + w_3w_4) + x_3^2w_3\,(x_2x_4 + w_2w_4) + x_4^2w_4\,(x_2x_3 + w_2w_3)]. \end{aligned}$$

The expression contains terms in $w_2w_3w_4$ which destroy each other; omitting them, we have

$$\begin{aligned} \Omega_1 \div w_1 = &- x_1\,(x_2y_2z_2w_3w_4w_1 + x_3y_3z_3w_4w_1w_2 + x_4y_4z_4w_1w_2w_3) \\ &+ x_1^2w_1\,(y_1z_2z_3z_4 + z_1y_2y_3y_4) \\ &- x_2x_3x_4y_1z_1\,(x_2w_2 + x_3w_3 + x_4w_4), \end{aligned}$$

where the last line is

$$= x_2x_3x_4y_1z_1\,.\,x_1w_1.$$

Hence the whole divides by x_1w_1, or we have

$$\begin{aligned} \Omega_1 \div x_1w_1^2 = &- x_2y_2z_2\,.\,w_3w_4 - x_3y_3z_3\,.\,w_4w_2 - x_4y_4z_4\,.\,w_2w_3 \\ &+ x_1y_1z_2z_3z_4 + x_1z_1y_2y_3y_4 + y_1z_1x_2x_3x_4, \end{aligned}$$

viz. this is

$$\begin{aligned} = \quad & x_2x_3x_4y_1z_1 + y_2y_3y_4z_1x_1 + z_2z_3z_4x_1y_1 \\ &+ x_2y_2z_2\,(x_3x_4 + y_3y_4 + z_3z_4) \\ &+ x_3y_3z_3\,(x_2x_4 + y_2y_4 + z_2z_4) \\ &+ x_4y_4z_4\,(x_2x_3 + y_2y_3 + z_2z_3)\,; \end{aligned}$$

or, finally, it is

$$\begin{aligned} = \quad & x_2x_3x_4\,(y_1z_1 + y_2z_2 + y_3z_3 + y_4z_4) \\ &+ y_2y_3y_4\,(z_1x_1 + z_2x_2 + z_3x_3 + z_4x_4) \\ &+ z_2z_3z_4\,(x_1y_1 + x_2y_2 + x_3y_3 + x_4y_4), \end{aligned}$$

which is $=0$; that is, we have $\Omega_1 = 0$, the required result.

The foregoing equation of the quadric surface can be by mere interchanges of the letters exhibited in 12 different forms.

We may, in like manner, first establish and then verify the equation

$$\begin{aligned} & zx\,(-z_1^2w_1y_2y_3y_4 - z_2^2w_2y_3y_4y_1 - z_3^2w_3y_4y_1y_2 - z_4^2w_4y_1y_2y_3) \\ & + xy\,(\ \ y_1^2w_1z_2z_3z_4 + y_2^2w_2z_3z_4z_1 + y_3^2w_3z_4z_1z_2 + y_4^2w_4z_1z_2z_3) \\ & + yw\,(-y_1^2x_1z_2z_3z_4 - y_2^2x_2z_3z_4z_1 - y_3^2x_3z_4z_1z_2 - y_4^2x_4z_1z_2z_3) \\ & + zw\,(\ \ z_1^2x_1y_2y_3y_4 + z_2^2x_2y_3y_4y_1 + z_3^2x_3y_4y_1y_2 + z_4^2x_4y_1y_2y_3) = 0\,; \end{aligned}$$

this can be by a cyclical interchange of the letters (x, y, z) exhibited in 3 different forms.

Of course, the fifteen equations belong each of them to a quadric surface through the 8 points. Any three of the fifteen equations, say $U=0$, $V=0$, $W=0$, may be used for determining the octad; the equation of any other quadric through the octad is then $\alpha U+\beta V+\gamma W=0$.

884.

NOTE SUR LES SURFACES MINIMA ET LE THÉORÈME DE JOACHIMSTHAL.

[From the *Comptes Rendus*, t. CVI. (1888), pp. 995, 996.]

I. La seule surface minima réglée est l'hélicoïde à plan directeur: ce beau théorème de M. Catalan se démontre par des considérations géométriques très simples.

Soit un système de droites ..., P, Q, R, ..., tel que, pour trois droites successives P, Q, R quelconques, tout plan Π' perpendiculaire à Q rencontre les trois droites en des points p', q', r', situés sur une droite. En particulier, si qr est la distance la plus petite des droites Q et R, le plan perpendiculaire Π, qui passe par le point q, passera par le point r et rencontrera la droite P en un point p, tel que les points p, q, r seront sur une même droite Λ, perpendiculaire à chacune des droites P, Q, R; c'est-à-dire, chacune de ces droites rencontre perpendiculairement une seule et même droite Λ. Cela étant, pour que les points de rencontre p', q', r' avec un autre plan Π' soient sur la même droite, il faut encore une condition; et, en supposant (ce que l'on peut faire sans perte de généralité) que les distances pq, qr soient égales, cette condition sera: que l'inclinaison des plans ($P\Lambda$, $Q\Lambda$) est égale à l'inclinaison des plans ($Q\Lambda$, $R\Lambda$).

En considérant de même les droites Q, R, S, on démontre d'abord que la droite S rencontre perpendiculairement la droite Λ; puis, en supposant que les distances qr, rs soient égales, on obtient la condition que l'inclinaison des plans ($Q\Lambda$, $R\Lambda$) est égale à celle des plans ($R\Lambda$, $S\Lambda$), et ainsi de suite: savoir, on obtient une série de droites ..., P, Q, R, ..., qui rencontrent perpendiculairement la droite Λ et sont telles qu'en supposant que les distances ..., pq, qr, rs, ... soient égales, les inclinaisons ..., ($P\Lambda$, $Q\Lambda$), ($Q\Lambda$, $R\Lambda$), ($R\Lambda$, $S\Lambda$), ... sont égales: c'est-à-dire, en considérant des droites ..., P, Q, R, S, ... consécutives, on a les génératrices d'une hélicoïde à plan directeur; et l'on voit ainsi que cette surface est la seule surface réglée, qui est telle que tout plan perpendiculaire à une génératrice quelconque rencontre la surface selon une courbe qui a, au point de rencontre avec la génératrice, une inflexion (ou, ce qui est la même chose, un rayon infini de courbure). Mais, comme l'avait remarqué M. Catalan, c'est là la condition pour que les deux rayons principaux de courbure soient égaux et opposés, ou enfin pour que la surface soit une surface minima.

II. Le théorème de Joachimsthal et aussi le théorème plus général de Bonnet et Serret, par rapport aux lignes de courbure planes ou sphériques, se déduisent immédiatement de ce théorème élémentaire de Géométrie: En considérant, dans des plans différents, deux triangles isoscèles $PP'O$ et $PP'N$ avec une base commune PP' ($OP=OP'$ et $NP=NP'$), les angles OPN et $OP'N$ seront égaux. En effet, si, pour une surface quelconque, PP' est l'élément d'une courbe de courbure sphérique, les normales à la surface aux points P et P' se rencontrent dans un point N, et les rayons de la sphère aux mêmes points se rencontrent dans un point O; on a ainsi les deux triangles isoscèles $PP'O$ et $PP'N$, et de là les angles égaux OPN et $OP'N$, c'est-à-dire que, pour deux points consécutifs P et P' de la ligne de courbure sphérique, l'inclinaison de la normale de la surface au rayon de la sphère a la même valeur, et cette inclinaison a ainsi la même valeur pour tous les points de la ligne de courbure. En prenant le point O à l'infini, on obtient le théorème pour une ligne de courbure plane, ou, si l'on veut, le théorème pour ce cas se déduit directement de celui-ci: Une droite quelconque PO, perpendiculaire à la base PP' d'un triangle isoscèle $PP'N$, est également inclinée sur les deux droites PN et $P'N$.

885.

ON THE DIOPHANTINE RELATION, $y^2+y'^2$=SQUARE.

[From the *Proceedings of the London Mathematical Society*, vol. XX. (1889), pp. 122—127.]

THE diophantine relation $y^2+y'^2=$ Square, where y is a function of x, and y' denotes $\frac{dy}{dx}$, is considered by Prof. Sylvester in his paper "On Reducible Cyclodes," *Proc. Lond. Math. Soc.*, t. I. (1865—66), pp. 137—160. It is at once seen that there exists a solution

$$y=(x+a)^\alpha (x+b)^\beta (x+c)^\gamma (x+d)^\delta \ldots,$$

where the roots $a, b, c, d, \ldots$ are essentially unequal, and the number of simple factors $x+a$, $x+b$, $x+c$, $x+d, \ldots$ is even; the exponents $\alpha, \beta, \gamma, \delta, \ldots$ are taken to be positive integer numbers. Sylvester assumes, and it will be shown, that the factors must separate themselves into two sets, or, as he calls them, diptychs, each containing the same number of simple factors and such that the sum of the exponents for the one diptych is equal to the sum of the exponents for the other diptych; viz. the form is $y=UU_1$, where

$$U=(x+a)^\alpha (x+b)^\beta \ldots,\quad U_1=(x+a_1)^{\alpha_1}(x+b_1)^{\beta_1}\ldots,$$

with the same number of simple factors, and with the relation $\alpha+\beta+\ldots=\alpha_1+\beta_1+\ldots$ between the exponents. Hence, if the number of simple factors be called the class and the sum of the exponents be called the order, the class and the order are each of them even; or, what is the same thing, the semi-class (say μ) and the semi-order (say ν) are each of them integral.

The separation of the factors into two diptychs is a remarkable theorem. I consider the analytical theory; for greater simplicity, first in the case, class = 2, and secondly in the case, class = 4; but it is easy to see that the like process is applicable to the case of any even value whatever of the class.

I write as usual $i=\sqrt{-1}$; the equation $y^2+y'^2=$ square, implies $y+iy'=$ square, and $y-iy'=$ square; at least this is so, save as to a common denominator, as will appear.

First, if the class is $=2$; we have

$$y=(x+a)^\alpha (x+b)^\beta;$$

hence

$$y+iy'=y\left(1+\frac{i\alpha}{x+a}+\frac{i\beta}{x+b}\right),$$

$$y-iy'=y\left(1-\frac{i\alpha}{x+a}-\frac{i\beta}{x+b}\right),$$

say these are $=\dfrac{(x+l)^2\,y}{x+a\,.\,x+b}$ and $\dfrac{(x+m)^2\,y}{x+a\,.\,x+b}$ respectively; and, this being so, we have

$$y^2+y'^2=\frac{y^2(x+l)^2(x+m)^2}{(x+a)^2(x+b)^2},\ =(x+a)^{2\alpha-2}(x+b)^{2\beta-2}(x+l)^2(x+m)^2.$$

It is to be shown that the assumed relations lead to $\alpha=\beta$. Resolving the last-mentioned expressions for $y+iy'$, $y-iy'$ each into simple fractions, we have

$$i\alpha\,(b-a)=(l-b)^2,\quad -i\alpha\,(b-a)=(m-b)^2,$$

$$i\beta\,(a-b)=(l-a)^2,\quad -i\beta\,(a-b)=(m-a)^2.$$

Hence

$$(l-b)^2+(m-b)^2=0,\quad (l-a)^2+(m-a)^2=0;$$

these cannot give

$$(l-b)+i\,(m-b)=0,\quad (l-a)+i\,(m-a)=0,$$

with the same sign for i in the two equations; for we should then have

$$(1+i)\,(b-a)=0,$$

but $1+i$ is not $=0$, and a, b are essentially unequal. Hence, taking (as we may do) $+i$ in the first equation, we must have $-i$ in the second equation, and the two equations are

$$l-b+i\,(m-b)=0,\ \text{that is,}\ l+im=(1+i)\,b,$$

$$l-a-i\,(m-a)=0,\quad\text{,,}\quad l-im=(1-i)\,a,$$

and thence

$$2l\ =(1+i)\,(b-ia),$$

$$2m=(1+i)\,(b+ia).$$

Hence also

$$2\,(l-b)=\ \ (1-i)\,(a-b),\quad 2i\,(m-b)=(1-i)\,(b-a),$$

$$2\,(l-a)=-(1+i)\,(a-b),\quad 2i\,(m-a)=(1+i)\,(b-a);$$

consequently

$$2\,(l-b)^2=-i\,(a-b)^2,\ =-2i\alpha\,(a-b),$$

$$2\,(l-a)^2=\ \ i\,(a-b)^2,\ =\ \ 2i\beta\,(a-b).$$

Hence $\alpha=\beta=\frac{1}{2}(a-b)$, and the solution thus is

$$y=(x+a)^{\alpha}(x+b)^{\alpha},\quad \alpha=\tfrac{1}{2}(a-b),$$
$$l=\tfrac{1}{2}\{a+b+i(a-b)\},$$
$$m=\tfrac{1}{2}\{a+b-i(a-b)\},$$
$$y^2+y'^2=(x+a)^{2\alpha-2}(x+b)^{2\alpha-2}(x+l)^2(x+m)^2.$$

The class is here $=2$, and the order is $=2\alpha$; considering the order as given, say it is $=2\nu$, we have $\alpha=\nu$, and the equation $\nu=\frac{1}{2}(a-b)$ then shows that one of the roots a, b is arbitrary. Taking it to be a, we have $b=a-2\nu$, or the solution, class 2 and order 2ν, is

$$y=(x+a)^{\nu}(x+a-2\nu)^{\nu},$$
$$l=a-\nu+i\nu,\quad m=a-\nu-i\nu,$$
$$y^2+y'^2=(x+a)^{2\nu-2}(x+a-2\nu)^{2\nu-2}(x+l)^2(x+m)^2.$$

Considering next for the case, class $=4$, the solution

$$y=(x+a)^{\alpha}(x+b)^{\beta}(x+c)^{\gamma}(x+d)^{\delta},$$

we have

$$y+iy'=y\left(1+\frac{i\alpha}{x+a}+\frac{i\beta}{x+b}+\frac{i\gamma}{x+c}+\frac{i\delta}{x+d}\right),$$

$$y-iy'=y\left(1-\frac{i\alpha}{x+a}-\frac{i\beta}{x+b}-\frac{i\gamma}{x+c}-\frac{i\delta}{x+d}\right);$$

or, putting these

$$=\frac{(x+l)^2(x+p)^2\,y}{x+a\,.\,x+b\,.\,x+c\,.\,x+d}\text{ and }\frac{(x+m)^2(x+q)^2\,y}{x+a\,.\,x+b\,.\,x+c\,.\,x+d}$$

respectively, we have

$$y^2+y'^2=\frac{y^2(x+l)^2(x+p)^2(x+m)^2(x+q)^2}{(x+a)^2(x+b)^2(x+c)^2(x+d)^2},$$

$$=(x+a)^{2\alpha-2}(x+b)^{2\beta-2}(x+c)^{2\gamma-2}(x+d)^{2\delta-2}(x+l)^2(x+p)^2(x+m)^2(x+q)^2.$$

Also, by decomposing the expressions for $y+iy'$, $y-iy'$ into simple fractions and comparing with the original values, we find

$$i\alpha\,(b-a)(c-a)(d-a)=(a-l)^2(a-p)^2,$$
$$i\beta\,(a-b)(c-b)(d-b)=(b-l)^2(b-p)^2,$$
$$i\gamma\,(a-c)(b-c)(d-c)=(c-l)^2(c-p)^2,$$
$$i\delta\,(a-d)(b-d)(c-d)=(d-l)^2(d-p)^2,$$

$$-i\alpha\,(b-a)(c-a)(d-a)=(a-m)^2(a-q)^2,$$
$$-i\beta\,(a-b)(c-b)(d-b)=(b-m)^2(b-q)^2,$$
$$-i\gamma\,(a-c)(b-c)(d-c)=(c-m)^2(c-q)^2,$$
$$-i\delta\,(a-d)(b-d)(c-d)=(d-m)^2(d-q)^2.$$

Hence

$$(a-l)^2(a-p)^2+(a-m)^2(a-q)^2=0,$$
$$(b-l)^2(b-p)^2+(b-m)^2(b-q)^2=0,$$
$$(c-l)^2(c-p)^2+(c-m)^2(c-q)^2=0,$$
$$(d-l)^2(d-p)^2+(d-m)^2(d-q)^2=0;$$

we cannot from these obtain *three* equations

$$(a-m)(a-q)-i(a-l)(a-p)=0,$$
$$(b-m)(b-q)-i(b-l)(b-p)=0,$$
$$(c-m)(c-q)-i(c-l)(c-p)=0,$$

with the same sign for i; in fact these would give

$$(1+i)(b-c)(c-a)(a-b)=0,$$

but $1+i$ is not $=0$, and the a, b, c are essentially unequal. Hence we must have equations such as

$$(a-m)(a-q)-i(a-l)(a-p)=0;\ (c-m)(c-q)+i(c-l)(c-p)=0,$$
$$(b-m)(b-q)-i(b-l)(b-p)=0;\ (d-m)(d-q)+i(d-l)(d-p)=0,$$

two of them with $-i$, and two of them with $+i$; viz. the a, b, c, d divide themselves into pairs which are taken to be a, b and c, d.

We hence easily obtain

$$a+b-m-q-i(a+b-l-p)=0,\quad ab-mq-i(ab-lp)=0,$$
$$c+d-m-q-i(c+d-l-p)=0,\quad cd-mq-i(cd-lp)=0,$$

and thence

$$a+b-c-d=i(a+b+c+d)-2i(l+p),$$
$$ab-cd=i(ab+cd)\qquad -2ilp.$$

Forming from these values of $l+p$, lp the expression for $2i(a-l)(a-p)$, we find $2i(a-l)(a-p)=(i+1)(a-c)(a-d)$; and we have thus the set of equations

$$2i(a-l)(a-p)=(i+1)(a-c)(a-d),$$
$$2i(b-l)(b-p)=(i+1)(b-c)(b-d),$$
$$2i(c-l)(c-p)=(i-1)(c-a)(c-b),$$
$$2i(d-l)(d-p)=(i-1)(d-a)(d-b).$$

Hence also

$$2(a-l)^2(a-p)^2=-i(a-c)^2(a-d)^2,$$
$$2(b-l)^2(b-p)^2=-i(b-c)^2(b-d)^2,$$
$$2(c-l)^2(c-p)^2=\quad i(c-a)^2(c-b)^2,$$
$$2(d-l)^2(d-p)^2=\quad i(d-a)^2(d-b)^2;$$

and, substituting these values in a former set of equations, we obtain

$$2\alpha\,(b-a)=-(a-c)(a-d),$$
$$2\beta\,(a-b)=-(b-c)(b-d),$$
$$2\gamma\,(d-c)=\quad(c-a)(c-b),$$
$$2\delta\,(c-d)=\quad(d-a)(d-b);$$

and thence

$$2(\alpha+\beta)= a+b-c-d,$$
$$2(\gamma+\delta)=-(c+d-a-b),$$

that is, $\alpha+\beta=\gamma+\delta$; viz. there are, in this case also, two diptychs.

If, as before, the order is taken to be $=2\nu$, then $\alpha+\beta=\nu$, $\gamma+\delta=\nu$; supposing that ν is a given positive integer, and that α, β, γ, δ are positive integers satisfying these equations $\alpha+\beta=\nu$, $\gamma+\delta=\nu$, then the last-mentioned four equations between α, β, γ, δ and a, b, c, d are equivalent to three relations serving to determine the differences of a, b, c, d (say $a-d$, $b-d$, $c-d$) in terms of α, β, γ, δ. And we then further have

$$\begin{aligned}
(a-l)(a-p)&=-(1-i)\,\alpha\,(b-a), & (a-m)(a-q)&=-(1+i)\,\alpha\,(b-a),\\
(b-l)(b-p)&=-(1-i)\,\beta\,(a-b), & (b-m)(b-q)&=-(1+i)\,\beta\,(a-b),\\
(c-l)(c-p)&=\ \ (1+i)\,\gamma\,(d-c), & (c-m)(c-q)&=\ \ (1-i)\,\gamma\,(c-d),\\
(d-l)(d-p)&=\ \ (1+i)\,\delta\,(c-d), & (d-m)(d-q)&=\ \ (1-i)\,\delta\,(d-c),
\end{aligned}$$

each set equivalent to two equations; or, as these may be written,

$$\begin{aligned}
2(l+p)&=a+b+c+d+i(a+b-c-d),\\
2lp&=ab+cd\quad +i(ab-cd),\\
2(m+q)&=a+b+c+d-i(a+b-c-d),\\
2mq&=ab+cd\quad -i(ab-cd),
\end{aligned}$$

serving to determine l, p, m, q in terms of a, b, c, d.

Observe also that, u being arbitrary, we have

$$2(u-l)(u-p)=(1+i)(u-a)(u-b)+(1-i)(u-c)(u-d),$$
$$2(u-m)(u-q)=(1-i)(u-a)(u-b)+(1+i)(u-c)(u-d),$$

(which equations, writing therein $u=a$, b, c, or d, in fact reproduce the two systems of four equations).

We have also

$$\begin{aligned}
l+p+m+q&=a+b+c+d, & l+p-m-q&=i(a+b-c-d),\\
lp+mq&=ab+cd, & lp-mq&=i(ab-cd);
\end{aligned}$$

and moreover

$$4(l-p)^2=\ \ 2i\{(a-b)^2-(c-d)^2\}+4(a+b)(c+d)-8(ab+cd),$$
$$4(m-q)^2=-2i\{(a-b)^2-(c-d)^2\}+4(a+b)(c+d)-8(ab+cd),$$

which equations, combined with the foregoing values of $2(l+p)$ and $2(m+q)$, give the values of l, p, m, q. We have thus the complete solution for the case class $=4$, order $=20$; say

$$y=(x+a)^\alpha(x+b)^\beta.(x+c)^\gamma(x+d)^\delta;\ \alpha+\beta=\gamma+\delta=\nu,$$
$$y^2+y'^2=(x+a)^{2\alpha-2}(x+b)^{2\beta-2}(x+c)^{2\gamma-2}(x+d)^{2\delta-2}(x+l)^2(x+p)^2(x+m)^2(x+q)^2,$$

with the foregoing relations between the constants.

886.

ON THE SURFACES WITH PLANE OR SPHERICAL CURVES OF CURVATURE.

[From the *American Journal of Mathematics*, vol. XI. (1889), pp. 71—98; pp. 293—306.]

THE theory is considered in two nearly cotemporaneous papers—Bonnet, "Mémoire sur les surfaces dont les lignes de courbure sont planes ou sphériques," *Jour. de l'École Polyt.*, t. XX. (1853), pp. 117—306, and Serret, "Mémoire sur les surfaces dont toutes les lignes de courbure sont planes ou sphériques," *Liouville*, t. XVIII. (1853), pp. 113—162. I desire to reproduce in a more compact form, and with some additional developments, the chief results obtained in these elaborate memoirs.

The basis of the theory is a theorem by Lancret, 1806. In any curve described upon a surface, the angle between the osculating planes at consecutive points is equal to the difference of the angles between the osculating planes and the corresponding tangent planes of the surface.

This includes as a particular case Joachimsthal's theorem, *Crelle*, t. XXX. (1846): If a surface have a plane curve of curvature, then at any point thereof the angle between the plane of the curve and the tangent plane of the surface has a constant value.

Bonnet and Serret each deduce the like theorem for a spherical curve of curvature, viz.: If a surface have a spherical curve of curvature, then at any point thereof the angle between the tangent plane of the sphere and the tangent plane of the surface has a constant value. Bonnet (Mémoire, p. 235) says that this follows from Lancret's theorem. Serret (Mémoire, p. 128) obtains it, by the transformation by reciprocal radius vectors, from Joachimsthal's theorem.

I remark that the theorem for a spherical curve of curvature, and (as a particular case thereof) that for a plane curve of curvature, are obtained at once from the most elementary geometrical considerations, viz. if we have (in the same plane or in different planes) the two isosceles triangles NPP', OPP' on a common base PP', then the angle OPN is equal to the angle $OP'N$. For take P, P' consecutive points on a spherical curve of curvature; then at P, P' the normals of the surface meet in a point N, and the normals (or radii) of the sphere meet in the centre O, and we have angle $OPN =$ angle $OP'N$, that is, at each of these points the inclination of the normal of the surface to the normal of the sphere has the same value; and this value being thus the same for any two consecutive points, must be the same for all points of the curve of curvature. The proof applies to the plane curve of curvature; but in this case, the fundamental theorem may be taken to be, a line at right angles to the base PP' of the isosceles triangle NPP' is equally inclined to the two equal sides NP, NP'.

A surface may have one set of its curves of curvature plane or spherical. To include the two cases in a common formula, the equation may be written

$$k(x^2+y^2+z^2)-2ax-2by-2cz-2u=0;$$

$k=1$ in the case of a sphere, $=0$ in that of a plane; and the expression a sphere may be understood to include a plane. I write in general A, B, C to denote the cosines of the inclinations of the normal of the surface at the point (x, y, z) to the axes of coordinates (consequently $A^2+B^2+C^2=1$). Hence considering a surface, and writing down the equations

$$k(x^2+y^2+z^2)-2ax-2by-2cz-2u=0,$$

$$(kx-a)A+(ky-b)B+(kz-c)C=l,$$

where (a, b, c, u, l) are regarded as functions of a parameter t. The first of these equations is that of a variable sphere; and the second equation expresses that at a point of intersection of the surface with the sphere, the inclination of the tangent plane of the surface to the tangent plane of the sphere has a constant value l, viz. this is a value depending only on the parameter t, and therefore constant for all points of the curve of intersection of the sphere and surface: by what precedes, the curve of intersection is a curve of curvature of the surface, and the surface will thus have a set of spherical curves of curvature.

Supposing the surface defined by means of expressions of its coordinates (x, y, z) as functions of two variable parameters, we may for one of these take the parameter t which enters into the equation of the sphere; and if the other parameter be called θ, then the expressions of the coordinates are of the form $x, y, z = x(t, \theta), y(t, \theta), z(t, \theta)$ respectively; these give equations $dx, dy, dz = adt + a'd\theta, bdt + b'd\theta, cdt + c'd\theta$, where of course (a, b, c, a', b', c') are in general functions of t, θ; and we have A, B, C proportional to $bc'-b'c$, $ca'-c'a$, $ab'-a'b$, viz. the values are equal to these expressions each divided by the square root of the sum of their squares. In order that the surface may have a set of spherical curves of curvature, the above three equations must be satisfied identically by means of the values of

$$a, b, c, u, l, A, B, C, x, y, z,$$

as functions of (t, θ); and it may be seen without difficulty that we are thereby led to a partial differential equation of the first order for the determination of the surface. But I do not at present further consider this question of the determination of a surface having one set of its curves of curvature (plane or) spherical.

Suppose now that there is a second set of (plane or) spherical curves of curvature. We have in like manner

$$\kappa(x^2+y^2+z^2)-2\alpha x-2\beta y-2\gamma z-2\upsilon \;=0,$$

$$(\kappa x-\alpha)A+(\kappa y-\beta)B+(\kappa z-\gamma)C-\lambda=0,$$

where κ is $=1$ or $=0$ according as the curves are spherical or plane, and $(\alpha, \beta, \gamma, \upsilon, \lambda)$ are functions of a variable parameter θ. We take the t of the former set of equations and the θ of these equations as the two parameters in terms of which the coordinates (x, y, z) are expressed. This being so (the former equations being satisfied as before), if these equations are satisfied identically by the values of $\alpha, \beta, \gamma, \upsilon, \lambda, A, B, C, x, y, z$ as functions of (t, θ), then the surface will have its other set of curves of curvature also spherical. It will be recollected that by hypothesis a, b, c, u, l are functions of the parameter t only, and that $\alpha, \beta, \gamma, \upsilon, \lambda$ functions of the parameter θ only. The foregoing equations, together with the assumed relations

$$A^2+B^2+C^2=1,$$

$$Adx+Bdy+Cdz=0,$$

are the "six equations" for the determination of a surface having its two sets of curves of curvature each of them (plane or) spherical.

Assuming now the values of a, b, c, l, u as functions of t, and $\alpha, \beta, \gamma, \lambda, \upsilon$ as functions of θ, the question at once arises whether we can then satisfy the six equations. These equations other than $Adx+Bdy+Cdz=0$, or say the five equations, in effect determine any five of the eight quantities $A, B, C, x, y, z, t, \theta$, in terms of the remaining three, say they determine A, B, C, t, θ as functions of x, y, z: we thus have a differential equation $Adx+Bdy+Cdz=0$, wherein A, B, C are to be regarded as given functions of (x, y, z). An equation of this form is not in general integrable; and if the equation in question be not integrable, then clearly the system of equations cannot be satisfied by any value of z as a function of (x, y), or, what is the same thing, by any values of (x, y, z) as functions of (t, θ). We thus arrive at the condition that the equation may be integrable, viz. the condition is

$$\nabla, =A\left(\frac{dB}{dz}-\frac{dC}{dy}\right)+B\left(\frac{dC}{dx}-\frac{dA}{dz}\right)+C\left(\frac{dA}{dy}-\frac{dB}{dx}\right), =0.$$

If this be satisfied, then we have an integral equation $I=0$ (containing a constant of integration which is an absolute constant) and which is, in fact, the equation of the required surface. But it is proper to look at the question somewhat differently. Supposing that the condition $\nabla=0$ is satisfied, then we have the integral equation $I=0$, and this equation, together with the five equations, in effect determine any six of the quantities $A, B, C, x, y, z, t, \theta$ in terms of the remaining two of them, or, what is the same thing, they determine a relation between any three of these quantities. We can, from the five equations and their differentials, and from the equation $Adx + Bdy + Cdz = 0$, obtain a differential equation between any three of the eight quantities: and it has just been seen that corresponding hereto we have an integral relation between the same three quantities; that is, the condition $\nabla = 0$ being satisfied, we can from the six equations obtain between any three of the quantities $A, B, C, x, y, z, t, \theta$ a linear differential equation of the foregoing form (for instance $Zdz + Tdt + \Theta d\theta = 0$, where Z, T, Θ are given functions of z, t, θ) which will *ipso facto* be integrable, furnishing between z, t, θ an integral equation which may be used instead of the before-mentioned integral equation $I = 0$. And we thus have (without any further integration) in all six equations which serve to determine any six of the quantities $A, B, C, x, y, z, t, \theta$ in terms of the remaining two. It is often convenient to seek in this way for the expressions of (A, B, C and) x, y, z as functions of t, θ, in preference to seeking for the integral equation $I = 0$ between the coordinates x, y, z.

The condition $\nabla = 0$ is in fact the condition which expresses that at any point of the surface the two curves of curvature intersect at right angles. Serret (and after him Bonnet) in effect obtain the condition by the assumption of this geometrical relation, without showing that the geometrical relation is in fact the necessary condition for the coexistence of the six equations. They give the condition in the form $dx\delta x + dy\delta y + dz\delta z = 0$, where dx, dy, dz are the increments of (x, y, z) along one of the curves of curvature, and δx, δy, δz the increments along the other curve of curvature. The equations give

$$\begin{aligned}(kx - a)\,dx + (ky - b)\,dy + (kz - c)\,dz &= 0,\\ Adx + \qquad\quad Bdy + \qquad\quad Cdz &= 0,\end{aligned}$$

and similarly

$$\begin{aligned}(\kappa x - \alpha)\,\delta x + (\kappa y - \beta)\,\delta y + (\kappa z - \gamma)\,\delta z &= 0,\\ A\delta x + \qquad\quad B\delta y + \qquad\quad C\delta z &= 0.\end{aligned}$$

We thence have

$$dx : dy : dz = B(kz - c) - C(ky - b) : C(kx - a) - A(kz - c) : A(ky - b) - B(kx - a),$$

and

$$\delta x : \delta y : \delta z = B(\kappa z - \gamma) - C(\kappa y - \beta) : C(\kappa x - \alpha) - A(\kappa z - \gamma) : A(\kappa y - \beta) - B(\kappa x - \alpha).$$

We have thus the required condition, in a form which is readily changed into

$$\begin{aligned}(A^2 + B^2 + C^2)\,\{(kx - a)(\kappa x - \alpha) + (ky - b)(\kappa y - \beta) + (kz - c)(\kappa z - \gamma)\}\\ - \{A(kx - a) + B(ky - b) + C(kz - c)\}\,\{A(\kappa x - \alpha) + B(\kappa y - \beta) + C(\kappa z - \gamma)\} = 0,\end{aligned}$$

and writing herein $A^2 + B^2 + C^2 = 1$, this becomes

$$\begin{aligned}&\tfrac{1}{2}\kappa\,\{k(x^2 + y^2 + z^2) - 2ax - 2by - 2cz\}\\ &+ \tfrac{1}{2}k\,\{\kappa(x^2 + y^2 + z^2) - 2\alpha x - 2\beta y - 2\gamma z\}\\ &+ (a\alpha + b\beta + c\gamma) - l\lambda = 0,\end{aligned}$$

that is,

$$a\alpha + b\beta + c\gamma - l\lambda + \kappa u + kv = 0.$$

I proceed to show that this is the condition $\nabla = 0$ for the integrability of the differential equation $Adx + Bdy + Cdz = 0$. Writing as before

$$\nabla = A\left(\frac{dB}{dz} - \frac{dC}{dy}\right) + B\left(\frac{dC}{dx} - \frac{dA}{dz}\right) + C\left(\frac{dA}{dy} - \frac{dB}{dx}\right),$$

we have from the six equations

$$\begin{aligned}
&AdA + BdB + CdC = 0,\\
&(kx - a)\,dA + (ky - b)\,dB + (kz - c)\,dC = -k\,(Adx + Bdy + Cdz) + (Aa_1 + Bb_1 + Cc_1 + l_1)\,dt,\\
&(\kappa x - \alpha)\,dA + (\kappa y - \beta)\,dB + (\kappa z - \gamma)\,dC = -\kappa\,(Adx + Bdy + Cdz) + (A\alpha' + B\beta' + C\gamma' + \lambda')\,d\theta,\\
&(kx - a)\,dx + (ky - b)\,dy + (kz - c)\,dz = (a_1x + b_1y + c_1z + u_1)\,dt,\\
&(\kappa x - \alpha)\,dx + (\kappa y - \beta)\,dy + (\kappa z - \gamma)\,dz = (\alpha'x + \beta'y + \gamma'z + \upsilon')\,d\theta,
\end{aligned}$$

where a_1, b_1, c_1, l_1, u_1 denote derived functions in regard to t, and α', β', γ', λ', υ' derived functions in regard to θ. Putting for shortness

$$\Omega = \begin{vmatrix} A, & B, & C \\ kx - a, & ky - b, & kz - c \\ \kappa x - \alpha, & \kappa y - \beta, & \kappa z - \gamma \end{vmatrix},$$

we readily obtain

$$\begin{aligned}
\Omega dA = {}& [(\kappa y - \beta)\,C - (\kappa z - \gamma)\,B]\Big\{-k\,(Adx + Bdy + Cdz)\\
&\qquad + \frac{Aa_1 + Bb_1 + Cc_1 + l_1}{a_1x + b_1y + c_1z + u_1}\{(kx - a)\,dx + (ky - b)\,dy + (kz - c)\,dz\}\Big\}\\
&- [(ky - b)\,C - (kz - c)\,B]\Big\{-\kappa\,(Adx + Bdy + Cdz)\\
&\qquad + \frac{A\alpha' + B\beta' + C\gamma' + \lambda'}{\alpha'x + \beta'y + \gamma'z + \upsilon'}\{(\kappa x - \alpha)\,dx + (\kappa y - \beta)\,dy + (\kappa z - \gamma)\,dz\}\Big\};
\end{aligned}$$

say this is

$$\begin{aligned}
\Omega dA = {}& [(\kappa y - \beta)\,C - (\kappa z - \gamma)\,B]\Big\{-k\,(Adx + Bdy + Cdz)\\
&\qquad + \frac{L}{P}\{(kx - a)\,dx + (ky - b)\,dy + (kz - c)\,dz\}\Big\}\\
&- [(ky - b)\,C - (kz - c)\,B]\Big\{-\kappa\,(Adx + Bdy + Cdz)\\
&\qquad + \frac{\Lambda}{\Pi}\{(\kappa x - \alpha)\,dx + (\kappa y - \beta)\,dy + (\kappa z - \gamma)\,dz\}\Big\},
\end{aligned}$$

or, introducing further abbreviations, and writing down the analogous values of ΩdB and ΩdC, we have

$$\begin{aligned}
\Omega dA &= [(\kappa y - \beta)\,C - (\kappa z - \gamma)\,B]\,U - [(ky - b)\,C - (kz - c)\,B]\,\Upsilon,\\
\Omega dB &= [(\kappa z - \gamma)\,A - (\kappa x - \alpha)\,C]\,U - [(kz - c)\,A - (kx - a)\,C]\,\Upsilon,\\
\Omega dC &= [(\kappa x - \alpha)\,B - (\kappa y - \beta)\,A]\,U - [(kx - a)\,B - (ky - b)\,A]\,\Upsilon.
\end{aligned}$$

We hence find

$$\Omega\frac{dB}{dz} = \quad [(\kappa z-\gamma)A-(\kappa x-\alpha)C]\left\{-kC+\frac{L}{P}(kz-c)\right\}$$

$$-[(kz-c)A-(kx-a)C]\left\{-\kappa C+\frac{\Lambda}{\Pi}(\kappa z-\gamma)\right\},$$

$$-\Omega\frac{dC}{dy}=-[(\kappa x-\alpha)B-(\kappa y-\beta)A]\left\{-kB+\frac{L}{P}(ky-b)\right\}$$

$$+[(kx-a)B-(ky-b)A]\left\{-\kappa B+\frac{\Lambda}{\Pi}(\kappa y-\beta)\right\}.$$

Combining these two terms, in the resulting value of $\Omega\left(\frac{dB}{dz}-\frac{dC}{dy}\right)$, first, the term without L or Λ is found to be

$$=-kA\{A(\kappa x-\alpha)+B(\kappa y-\beta)+C(\kappa z-\gamma)\}$$

$$-k(\kappa x-\alpha)(A^2+B^2+C^2)$$

$$+\kappa(kx-a)(A^2+B^2+C^2)$$

$$+\kappa A\{A(kx-a)+B(ky-b)+C(kz-c)\},$$

which is

$$=-kA\lambda+k(\kappa x-\alpha)-\kappa(kx-a)+\kappa Al,$$

$$=\quad A(\kappa l-k\lambda)-k\alpha+\kappa a.$$

Next, the coefficient of $\frac{L}{P}$ is

$$A(kz-c)(\kappa z-\gamma)-C(\kappa x-\alpha)(kz-c)$$

$$+A(ky-b)(\kappa y-\beta)-B(\kappa x-\alpha)(ky-b),$$

which is

$$=A[(kx-a)(\kappa x-\alpha)+(ky-b)(\kappa y-\beta)+(kz-c)(\kappa z-\gamma)]$$

$$-(\kappa x-\alpha)[A(kx-a)+B(ky-b)+C(kz-c)]$$

$$=AM+(\kappa x-\alpha)l,$$

if for shortness

$$M=(kx-a)(\kappa x-\alpha)+(ky-b)(\kappa y-\beta)+(kz-c)(\kappa z-\gamma);$$

and similarly, the coefficient of $\frac{\Lambda}{\Pi}$ is

$$-A(kz-c)(\kappa z-\gamma)+C(kx-a)(\kappa x-\alpha)$$

$$-A(ky-b)(\kappa y-\beta)+B(kx-b)(\kappa y-\beta),$$

which is

$$=-A[(kx-a)(\kappa x-\alpha)+(ky-b)(\kappa y-\beta)+(kz-c)(\kappa z-\gamma)]$$

$$-(kx-a)[A(\kappa x-\alpha)+B(\kappa y-\beta)+C(\kappa z-\gamma)]$$

$$=-AM-(kx-a)\lambda.$$

We thus obtain

$$\Omega\left(\frac{dB}{dz}-\frac{dC}{dy}\right)=A(\kappa l-k\lambda)-k\alpha+\kappa a+\frac{L}{P}\{AM+(\kappa x-\alpha)l\}-\frac{\Lambda}{\Pi}\{AM+(kx-a)\lambda\},$$

and similarly

$$\Omega\left(\frac{dC}{dx}-\frac{dA}{dz}\right)=B(\kappa l-k\lambda)-k\beta+\kappa b+\frac{L}{P}\{BM+(\kappa y-\beta)l\}-\frac{\Lambda}{\Pi}\{BM+(ky-b)\lambda\},$$

$$\Omega\left(\frac{dA}{dy}-\frac{dB}{dx}\right)=C(\kappa l-k\lambda)-k\gamma+\kappa c+\frac{L}{P}\{CM+(\kappa z-\gamma)l\}-\frac{\Lambda}{\Pi}\{CM+(kz-c)\lambda\};$$

hence multiplying by A, B, C and adding, we obtain

$$\Omega\nabla=\kappa l-k\lambda-k(A\alpha+B\beta+C\gamma)+\kappa(Aa+Bb+Cc)+\frac{L}{P}(M-l\lambda)-\frac{\Lambda}{\Pi}(M-l\lambda),$$

where the first four terms are together

$$=\kappa l-k\lambda+k\{\kappa(Ax+By+Cz)-\lambda\}-\kappa\{k(Ax+By+Cz)-l\},$$

viz. these destroy each other, and the equation becomes

$$\Omega\nabla=\left(\frac{L}{P}-\frac{\Lambda}{\Pi}\right)(M-l\lambda).$$

But we have

$$\begin{aligned}M-l\lambda=\ &\tfrac{1}{2}\kappa\{k(x^2+y^2+z^2)-2ax-2by-2cz\}\\&+\tfrac{1}{2}k\{\kappa(x^2+y^2+z^2)-2\alpha x-2\beta y-2\gamma z\}+(a\alpha+b\beta+c\gamma)-l\lambda,\end{aligned}$$

which is

$$=a\alpha+b\beta+c\gamma-l\lambda+\kappa u+kv,$$

or we find

$$\Omega\nabla=\left(\frac{L}{P}-\frac{\Lambda}{\Pi}\right)(a\alpha+b\beta+c\gamma-l\lambda+\kappa u+kv),$$

viz. the condition $\nabla=0$ is

$$a\alpha+b\beta+c\gamma-l\lambda+\kappa u+kv=0,$$

the result which was to be proved.

If we consider separately the cases where the two sets of curves of curvature are each plane, the first plane and the second spherical, and each of them spherical; or say the cases PP, PS and SS, then in these cases respectively the condition is

$$\begin{aligned}&a\alpha+b\beta+c\gamma-l\lambda=0,\\&a\alpha+b\beta+c\gamma-l\lambda+u=0,\\&a\alpha+b\beta+c\gamma-l\lambda+u+v=0:\end{aligned}$$

we have, in each case, to take the italic letters functions of t and the greek letters functions of θ, satisfying identically the appropriate equation, but otherwise arbitrary; and then, in each case, the six equations lead to a differential equation $Adx+Bdy+Cdz=0$ (or say $Zdz+Tdt+\Theta d\theta=0$) between three variables, which equation is *ipso facto* integrable; and we thus obtain a new integral equation which, with the original five integral equations, gives the solution of the problem. The condition is, in each case, of the form $\Sigma a\alpha = 0$, the number of terms $a\alpha$ being 4, 5 or 6. Considering for instance the form

$$a\alpha + b\beta + c\gamma + d\delta + e\epsilon + f\phi = 0$$

with 6 terms, it is easy to see how such an equation is to be satisfied by values of a, b, c, d, e, f which are functions of t, and values of $\alpha, \beta, \gamma, \delta, \epsilon, \phi$ which are functions of θ. Suppose that $t_1, t_2, \ldots$ are particular values of t, and $a_1, b_1, \ldots, f_1$; $a_2, b_2, \ldots, f_2$, &c., the corresponding values of $a, b, \ldots, f$, these values being of course absolute constants; we have $\alpha, \beta, \ldots, \phi$, functions of θ, satisfying all the equations

$$(a_1, b_1, c_1, d_1, e_1, f_1 \between \alpha, \beta, \gamma, \delta, \epsilon, \phi) = 0,$$
$$(a_2, b_2, c_2, d_2, e_2, f_2 \between \quad \text{,,} \quad) = 0,$$
&c.,

and if 6 or more of these equations were independent, the equations could, it is clear, be satisfied only by the values $\alpha = \beta = \gamma = \delta = \epsilon = \phi = 0$. To obtain a proper solution, only some number less than 6 of these equations can be independent. Suppose, for instance, that only two of the equations are independent; we then have $\alpha, \beta, \gamma, \delta, \epsilon, \phi$ functions of θ satisfying these equations, but otherwise arbitrary; or, what is the same thing, we may take $\alpha, \beta, \gamma, \delta, \epsilon, \phi$ linear functions of $6-2, = 4$ arbitrary functions, say P, Q, R, S of θ; say we have

$$\alpha = (\alpha_0, \alpha_1, \alpha_2, \alpha_3 \between P, Q, R, S),$$
$$\beta = (\beta_0, \ldots\ldots\ldots\ldots \between \quad \text{,,} \quad),$$
$$\ldots\ldots\ldots\ldots\ldots\ldots\ldots\ldots$$
$$\ldots\ldots\ldots\ldots\ldots\ldots\ldots\ldots$$
$$\ldots\ldots\ldots\ldots\ldots\ldots\ldots\ldots$$
$$\phi = (\phi_0, \ldots\ldots\ldots\ldots \between \quad \text{,,} \quad),$$

where the suffixed greek letters denote absolute constants; and this being so, in order to satisfy the proposed equation $a\alpha + b\beta + c\gamma + d\delta + e\epsilon + f\phi = 0$, we must have

$$(\alpha_0, \beta_0, \gamma_0, \delta_0, \epsilon_0, \phi_0 \between a, b, c, d, e, f) = 0,$$
$$(\alpha_1, \ldots\ldots\ldots\ldots\ldots\ldots \between \quad \text{,,} \quad) = 0,$$
$$(\alpha_2, \ldots\ldots\ldots\ldots\ldots\ldots \between \quad \text{,,} \quad) = 0,$$
$$(\alpha_3, \ldots\ldots\ldots\ldots\ldots\ldots \between \quad \text{,,} \quad) = 0,$$

viz. a, b, c, d, e, f will then be functions of t satisfying these four equations, but otherwise arbitrary. The above is a solution for the partition $2+4$ of the number 6. We have in like manner a solution for any other partition of 6; or if we disregard the extreme cases $a=b=c=d=e=f=0$ and $\alpha=\beta=\gamma=\delta=\epsilon=\phi=0$, then we have in this manner solutions for the several partitions 15, 24, 33, 42 and 51 of the number 6.

But applying this theory to the actual problem, there is a good deal of difficulty as regards the enumeration of the really distinct cases. I use the letters P, S to denote that a set of curves of curvature is plane or spherical as the case may be, the surfaces to be considered are thus *PP*, *PS*, and *SS*. First, for the *PP* problem where the equation is $a\alpha + b\beta + c\gamma - l\lambda = 0$: the two systems (a, b, c, l) and $(\alpha, \beta, \gamma, \lambda)$ are symmetrically related to each other, and instead of the solutions 13, 22 and 31, it is sufficient to consider the solutions 13 and 22. But here (a, b, c, l) are not a system of four symmetrically related functions, (a, b, c) are a symmetrical system, and l is a distinct term: and the like for the system $(\alpha, \beta, \gamma, \lambda)$. In the *PS* problem, where the equation is $a\alpha + b\beta + c\gamma - l\lambda + u = 0$, and thus the systems (a, b, c, l, u), $(\alpha, \beta, \gamma, \lambda, 1)$ are of different forms, we should consider the solutions 14, 23, 32 and 41: but here again, in each of the systems separately, the terms are not symmetrically related to each other. Lastly, in the *SS* problem where the equation is $a\alpha + b\beta + c\gamma - l\lambda + u + v = 0$: the systems $(a, b, c, l, u, 1)$ and $(\alpha, \beta, \gamma, \lambda, 1, v)$ are of the same form, it is enough to consider the solutions 15, 24 and 33; but in this case also, in each of the systems separately, the terms are not symmetrically related to each other. I do not at present further consider the question, but simply adopt Serret's enumeration.

It is to be remarked that for a developable (but not for a skew surface) the generating lines may be curves of curvature, and regarding the generating lines as plane curves we might have developables *PP* or *PS*; but a straight line is not a curve in a determinate plane, and it is better to consider the case apart from the general theory. Again, the curves of curvature of one set or those of each set may be circles; and a circle may be regarded either as a plane or a spherical curve; regarding it, however, as a spherical curve, it is a curve not in a determinate sphere. The cases in question, of the curves of curvature of the one set or of those of each set being circles, are therefore also to be considered apart from the general theory. The surfaces referred to present themselves for consideration among Serret's cases *PP*, 1°, 2°, 3°; *PS*, 1°, 2°, 3°, 4°, 5°, 6°, 7°; and *SS*, 1°, 2°, 3°, 4°; but they are excluded from his enumeration, and he in fact reckons in his "Conclusion," pp. 161, 162, two kinds of surfaces *PP*, three kinds *PS*, and two kinds *SS*.

It is very easily seen that, if a surface has a plane or a spherical curve of curvature, then on any parallel surface the corresponding curve is a plane or a spherical curve of curvature: and thus if a surface be *PP*, *PS*, or *SS*, then the parallel surfaces are respectively *PP*, *PS*, or *SS*. The solutions obtained include for the most part all the parallel surfaces, and thus there is no occasion to make use of this theorem; but see in the continuation of the present paper the case considered under the subheading *post*, *PS*, 4° = Serret's third case of *PS*.

If a surface have a plane or a spherical curve of curvature, then transforming the surface by reciprocal radius vectors (or inverting in regard to an arbitrary point), then in the transformed surface the corresponding curve is a spherical curve of curvature. Hence if a surface be *PP*, *PS* or *SS*, the transformed surface is *SS*. Conversely, as shown by Bonnet and Serret, and as will appear, every surface *SS* is in fact an inversion of a surface *PP* or *PS*.

I proceed to the enumeration, developing the theory only in regard to the two, three, and two, cases *PP*, *PS* and *SS* respectively.

PP, The Two Sets of Curves of Curvature each Plane.

The six equations are

$$\begin{aligned} A^2+B^2 \quad &+C^2 \quad = 1,\\ ax+by \quad &+cz \quad +u=0,\\ Aa+Bb \quad &+Cc \quad +l=0,\\ \alpha x+\beta y \quad &+\gamma z \quad +v=0,\\ A\alpha+B\beta \quad &+C\gamma \quad +\lambda=0,\\ Adx+Bdy+Cdz &=0; \end{aligned}$$

the condition is

$$a\alpha+b\beta+c\gamma-l\lambda=0,$$

not containing u or v, so that these remain arbitrary functions of t, θ respectively. The cases are

	a	b	c	l	α	β	γ	λ
PP, 1°	1	0	c	0	0	1	0	λ
PP, 2°	0	1	0	$-m$	α	$-m\lambda$	γ	λ
PP, 3°	1	0	c	mc	0	1	$m\lambda$	λ;

m is an arbitrary constant; and in the body of the table, c is an arbitrary function of t, and α, γ, λ arbitrary functions of θ.

PP, 1° is Serret's first case of PP, included in his second case.

PP, 2° gives a developable.

PP, 3° is Serret's second case of PP.

I consider the case

PP, 3° = Serret's Second Case of PP.

Writing for greater symmetry $m=g$, $\dfrac{1}{m}=f$, so that $fg=1$; also $m\lambda=\gamma$, and consequently $\lambda=f\gamma$, we take c and γ for the two parameters respectively, or write $c=t$, $\gamma=\theta$; also changing the letters u, v, we write

a	b	c	l	u	α	β	γ	λ	v
$=1$,	0,	t	gt	P	0	1	θ	$f\theta$	Π,

and the six equations thus are

$$\begin{aligned} A^2+B^2 \quad &+C^2 &&=1,\\ x+tz \quad &-P &&=0,\\ A+tC \quad &-gt &&=0,\\ y+\theta z \quad &-\Pi &&=0,\\ B+\theta C \quad &-f\theta &&=0,\\ Adx+Bdy+Cdz &&&=0. \end{aligned}$$

We seek for the differential equation in z, t, θ. We have

$$A^2+B^2+C^2=1,\quad A=t(g-C),\quad B=\theta(f-C),$$

and thence

$$t^2(g-C)^2+\theta^2(f-C)^2+C^2=1,$$

that is,

$$C^2(1+t^2+\theta^2)+2C(gt^2+f\theta^2)=1-g^2t^2-f^2\theta^2,$$

or multiplying by $1+t^2+\theta^2$ and completing the square,

$$\begin{aligned}\{(1+t^2+\theta^2)C-gt^2-f\theta^2\}^2&=(1-g^2t^2-f^2\theta^2)(1+t^2+\theta^2)-(gt^2+f\theta^2)^2\\&=\{f+(f-g)t^2\}\{g+(g-f)\theta^2\}\\&=\frac{1}{T^2\Theta^2},\end{aligned}$$

if

$$\frac{1}{T^2}=f+(f-g)t^2,$$

$$\frac{1}{\Theta^2}=g+(g-f)\theta^2;$$

and thence, giving a determinate sign to the square root, say

$$(1+t^2+\theta^2)C=gt^2+f\theta^2-\frac{1}{T\Theta},$$

an equation which may also be written

$$C=\frac{fT-g\Theta}{f-g}.$$

In fact, observing that $\frac{1}{T^2}-\frac{1}{\Theta^2}=(f-g)(1+t^2+\theta^2)$, we deduce from the original form

$$\begin{aligned}\left(\frac{1}{T^2}-\frac{1}{\Theta^2}\right)C&=(f-g)(gt^2+f\theta^2)-\frac{f-g}{T\Theta},\\&=g\left(\frac{1}{T^2}-f\right)-f\left(\frac{1}{\Theta^2}-g\right)-\frac{f-g}{T\Theta}\\&=\left(\frac{g}{T}-\frac{f}{\Theta}\right)\left(\frac{1}{T}+\frac{1}{\Theta}\right),\end{aligned}$$

or throwing out the factor $\frac{1}{T}+\frac{1}{\Theta}$ and reducing, we have the required value; and thence forming the values of A and B, we have

$$A=-tT\frac{f-g}{T-\Theta},\quad B=-\theta\Theta\frac{f-g}{T-\Theta},\quad C=\frac{fT-g\Theta}{T-\Theta};$$

we have, moreover,

$$x+tz=P,\quad y+\theta z=\Pi,$$

or differentiating, and writing P_1 and Π' for the derived functions in regard to t and θ respectively,

$$dx = -tdz - zdt + P_1dt, \quad dy = -\theta dz - zd\theta - \Pi' d\theta.$$

The equation $Adx + Bdy + Cdz = 0$ thus becomes

$$-Ttdx - \Theta\theta dy + \frac{fT - g\Theta}{f-g}dz = 0,$$

viz. this is

$$[-tT(-tdz - zdt + P_1dt) - \theta\Theta(-\theta dz - zd\theta + \Pi'dt)](f-g) + (fT - g\Theta)dz = 0,$$

or collecting,

$$[\{f + (f-g)t^2\}T - \{g + (g - ft^2)\}\Theta]dz + (tTzdt + \theta\Theta zd\theta)(f-g)$$
$$-(tTP_1dt + \theta\Theta\Pi'd\theta)(f-g) = 0,$$

that is,

$$\left(\frac{1}{\Theta} - \frac{1}{T}\right)dz + (f-g)z(tTdt + \theta\Theta d\theta) - (f-g)(tTP_1dt + \theta\Theta\Pi'd\theta) = 0,$$

which is an integrable form as it should be; viz. the equation is

$$d\left(\frac{1}{T} - \frac{1}{\Theta}\right)z - (f-g)(tTP_1dt + \theta\Theta\Pi'd\theta) = 0,$$

and we obtain

$$\left(\frac{1}{T} - \frac{1}{\Theta}\right)z - (f-g)\int(tTP_1dt + \theta\Theta\Pi'd\theta) = 0,$$

the constant of integration being considered as included in the integral. But it is proper to alter the form of the second term. Take F, Φ arbitrary functions of t, θ respectively: and writing F_1, Φ' for the derived functions: assume $P = \frac{gF_1}{T^3}$, $\Pi = \frac{f\Phi'}{\Theta^3}$; we have

$$\int(tTP_1dt + \theta\Theta\Pi'd\theta) = \int\left(gtT\left(\frac{F_1}{T^3}\right)_1 dt + f\theta\Theta\left(\frac{\Phi'}{\Theta^3}\right)' d\theta\right)$$
$$= -F + \frac{gtF_1}{T^2} - \Phi + \frac{f\theta\Phi'}{\Theta^2}.$$

In fact, this will be true if only

$$\left(-F + \frac{gtF_1}{T^2}\right)_1 = gtT\left(\frac{F_1}{T^3}\right)_1, \quad \left(-\Phi + \frac{f\theta\Phi'}{\Theta^2}\right)' = f\theta\Theta\left(\frac{\Phi'}{\Theta^3}\right)',$$

which are equations of like form in t, θ respectively; it will be sufficient to verify the first of them. Effecting the differentiation, the terms in F_{11} destroy each other, and there remain only terms containing the factor F_1; throwing this out, we obtain

$$-1 + \frac{g}{T^2} + \frac{gtT_1}{T^3} = 0,$$

viz. this is

$$-1+g\{f+(f-g)t^2\}-gt^2(f-g)=0,$$

which is identically true, and the equation is thus verified.

The foregoing result is

$$\left(\frac{1}{T}-\frac{1}{\Theta}\right)z+(f-g)\left\{F+\Phi-\frac{gtF_1}{T^2}-\frac{f\theta\Phi'}{\Theta^2}\right\}=0;$$

we then have

$$x+tz-\frac{gF_1}{T^3}=0,\quad y+\theta z-\frac{f\Phi'}{\Theta^3}=0,$$

and hence, repeating also the equation for z,

$$\left(\frac{1}{T}-\frac{1}{\Theta}\right)x+(f-g)\left\{-t(F+\Phi)+\frac{ft\theta\Phi'}{\Theta^2}\right\}+\left(-1+\frac{g}{\Theta T}\right)\frac{F_1}{T^2}=0,$$

$$\left(\frac{1}{T}-\frac{1}{\Theta}\right)y+(f-g)\left\{-\theta(F+\Phi)+\frac{gt\theta F_1}{T^2}\right\}+\left(1-\frac{f}{\Theta T}\right)\frac{\Phi'}{\Theta^2}=0,$$

$$\left(\frac{1}{T}-\frac{1}{\Theta}\right)z+(f-g)\left\{F+\Phi-\frac{gtF_1}{T^2}-\frac{f\theta\Phi'}{\Theta^2}\right\}=0,$$

equations which give the values of the coordinates x, y, z in terms of the parameters t, θ. It will be recollected that $fg=1$ (f or g being arbitrary), then the values of T, Θ are

$$\frac{1}{T^2}=f+(f-g)t^2,\quad \frac{1}{\Theta^2}=g+(g-f)\theta^2,$$

and that F, Φ denote arbitrary functions of t, θ respectively. I repeat also the foregoing equations

$$A,\ B,\ C=-tT\frac{f-g}{T-\Theta},\ -\theta\Theta\frac{f-g}{T-\Theta},\ \frac{fT-g\Theta}{T-\Theta}.$$

The equations may be presented under a different form; we have

$$-tTx-\theta\Theta y+\frac{fT-g\Theta}{f-g}z+F+\Phi=0,$$

$$-fT^3(x+tz)+F_1=0,$$

$$-g\Theta^3(y+\theta z)+\Phi'=0,$$

where it will be observed that the second and third equations are the derivatives of the first equation in regard to t and θ respectively. We thus have the required surface as the envelope of the plane represented by the first equation, regarding therein t, θ as variable parameters. Moreover, the second equation (which contains only the parameter t) represents the planes of the curves of curvature of the one set; and the third equation (which contains only the parameter θ) represents the planes of the curves of curvature of the other set. It is to be observed that, from the equations for l, λ, viz. $A+tC=gt$ and $B+\theta C=f\theta$, then for any plane of the first set the inclination to a tangent plane of the surface is $=\cos^{-1}\dfrac{gt}{\sqrt{1+t^2}}$, and that for any plane of the second set the inclination is $=\cos^{-1}\dfrac{f\theta}{\sqrt{1+\theta^2}}$.

It may be remarked that the last-mentioned results may be arrived at by the consideration of an equation $Ax+By+Cz+D=0$, where the coefficients are functions of t and θ (A a function of t only, and B a function of θ only), such that the derived equations $A_1x+C_1z+D_1=0$ and $B'x+C'z+D'=0$ depend the former of them upon t only, and the latter of them upon θ only.

A very simple case of the equation is when $f=g=1$; here $T=\Theta=1$, and the surface is the envelope of the plane $z-tx-\theta y+F+\Phi=0$.

Returning to the general form

$$-tTx-\theta\Theta y+\frac{fT-g\Theta}{f-g}z+F+\Phi=0,$$

I transform this, by introducing therein in place of t, θ two variable parameters α, β which are such that $k\alpha=-tT$, $k\beta=\theta\Theta$ $\left(k \text{ a constant which is presently put } =\dfrac{1}{\sqrt{f-g}}\right)$; we find

$$t^2=\frac{fk^2\alpha^2}{1-(f-g)k^2\alpha^2},\quad \theta^2=\frac{gk^2\beta^2}{1-(g-f)k^2\beta^2},$$

and thence

$$T=\frac{1}{\sqrt{f}}\sqrt{1-(f-g)k^2\alpha^2},\quad \Theta=\frac{1}{\sqrt{g}}\sqrt{1-(g-f)k^2\beta^2},$$

or putting $k=\dfrac{1}{\sqrt{f-g}}$, these last values are

$$T=\frac{1}{\sqrt{f}}\sqrt{1-\alpha^2},\quad \Theta=\frac{1}{\sqrt{g}}\sqrt{1+\beta^2},$$

and we hence obtain

$$\frac{fT-g\Theta}{f-g}=\frac{\sqrt{f}}{f-g}\sqrt{1-\alpha^2}-\frac{\sqrt{g}}{f-g}\sqrt{1+\beta^2},$$

$$=\frac{1}{\sqrt{f-g}}\left\{\frac{\sqrt{f}}{\sqrt{f-g}}\sqrt{1-\alpha^2}-\frac{\sqrt{g}}{\sqrt{f-g}}\sqrt{1+\beta^2}\right\},$$

say this is

$$=k\{\lambda\sqrt{1-\alpha^2}-\mu\sqrt{1+\beta^2}\},$$

where $\lambda=\dfrac{\sqrt{f}}{\sqrt{f-g}}$, $\mu=\dfrac{\sqrt{g}}{\sqrt{f-g}}$, and therefore $\lambda^2-\mu^2=1$ or $\mu=\sqrt{\lambda^2-1}$.

Hence writing $F+\Phi=k(A+B)$, k times the sum of two arbitrary functions of α and β respectively, the equation becomes

$$\alpha x-\beta y+z\{\lambda\sqrt{1-\alpha^2}-\sqrt{\lambda^2-1}\sqrt{1+\beta^2}\}+A+B=0,$$

viz. the surface is given as the envelope of this plane considering α, β as two variable parameters. This is the solution given by Darboux, *Leçons sur la théorie générale des surfaces, &c.*, t. I., Paris, 1887, pp. 128—131. He obtains it in a very elegant manner, starting from the following theorem: Take A, A_1, &c., functions of the parameter α, and B, B_1, &c., functions of the parameter β; then, if we have identically

$$(A_1-B_1)^2+(A_2-B_2)^2+(A_3-B_3)^2=(A_4-B_4)^2,$$

the required surface will be obtained as the envelope of the plane

$$(A_1-B_1)\,x+(A_2-B_2)\,y+(A_3-B_3)\,z=A-B,$$

where A, B are two new functions of α, β respectively.

The foregoing identity is the condition in order that each sphere of the one series $(x-A_1)^2+(y-A_2)^2+(z-A_3)^2=A_4^2$ may touch each sphere of the other series $(x-B_1)^2+(y-B_2)^2+(z-B_3)^2=B_4^2$; the two series of spheres thus envelope one and the same surface which will have its curves of curvature of each set circles: viz. this will be the surface of the fourth order called Dupin's Cyclide, the normals whereof pass through an ellipse and hyperbola which are focal curves one of the other, and which contain the centres of all the spheres touching the surface along its curves of curvature. The equations of the ellipse and the hyperbola may be taken to be

$$x^2+\frac{z^2}{\lambda^2}=1,\ y=0, \text{ and } y^2-\frac{z^2}{\lambda^2-1}=-1,\ x=0,$$

respectively, and we thence obtain the required PP surface as the envelope of the plane

$$\alpha x-\beta y+(\lambda\sqrt{1-\alpha^2}-\sqrt{\lambda^2-1}\sqrt{1+\beta^2})\,z+A+B=0.$$

The Case PP, 1° = Serret's First Case of PP.

We deduce this from the second case by writing therein $m=0$, that is, $g=0$, $f=\infty$; but it is necessary to make also a transformation upon the parameter θ, viz. in place thereof we introduce the new parameter ϕ, where $\theta^2=\dfrac{g\phi^2}{f-g\phi^2}$. This gives

$$\frac{1}{\Theta^2}=g+(g-f)\,\theta^2=g\left\{1+\frac{(g-f)\,\phi^2}{f-g\phi^2}\right\}=\frac{gf(1-\phi^2)}{f-g\phi^2},\quad \theta^2=\frac{g\phi^2}{f-g\phi^2},$$

and thence

$$\theta\Theta=\frac{\theta}{\sqrt{g+(g-f)\,\theta^2}}=\frac{\phi}{\sqrt{f}\sqrt{1-\phi^2}};\quad \frac{fT-g\Theta}{f-g} \text{ for } g=0 \text{ is } =T.$$

We have also $T=\dfrac{1}{\sqrt{f+(f-g)\,t^2}}, =\dfrac{1}{\sqrt{f}\sqrt{1+t^2}}$ when $g=0$, and substituting these values, considering Φ as a function of ϕ, and for $F+\Phi$ writing as we may do $\dfrac{F+\Phi}{\sqrt{f}}$, the equation becomes

$$\frac{-t}{\sqrt{f}\sqrt{1+t^2}}\,x-\frac{\phi y}{\sqrt{f}\sqrt{1-\phi^2}}+\frac{z}{\sqrt{f}\sqrt{1+t^2}}+\frac{F+\Phi}{\sqrt{f}}=0,$$

where the divisor $\sqrt{f}$ is to be omitted. Hence finally, instead of ϕ restoring the original letter θ, and again considering Φ as a function of θ, the equation is

$$\frac{z-tx}{\sqrt{1+t^2}}-\frac{\theta y}{\sqrt{1-\theta^2}}+F+\Phi=0,$$

viz. here F, Φ are arbitrary functions of t, θ respectively, and the surface is the envelope of this plane considering t, θ as variable.

We obtain an imaginary special form of PP, 1°, by writing in this equation $k\theta$ for θ and then putting $k=\infty$; the Φ remains an arbitrary function of the new θ, and the equation is

$$\frac{z-tx}{\sqrt{1+t^2}}+iy+F+\Phi=0,$$

($i=\sqrt{-1}$ as usual). This is, in fact, the equation which is obtained from PP, 3° by simply writing therein $g=0$ without the transformation upon θ.

PS, The Sets of Curves of Curvature, the First Plane, the Second Spherical.

The six equations are

$$\begin{aligned}
&A^2+B^2+C^2 &&=1,\\
&ax+by+cz+u &&=0,\\
&Aa+Bb+Cc+l &&=0,\\
&x^2+y^2+z^2-2\alpha x-2\beta y-2\gamma z-2v &&=0,\\
&A(x-\alpha)+B(y-\beta)+C(z-\gamma)-\lambda &&=0,\\
&Adx+Bdy+Cdz &&=0.
\end{aligned}$$

The condition is

$$a\alpha+b\beta+c\gamma-l\lambda+u=0,$$

not containing v, so that this remains an arbitrary function of θ. The cases are

	a	b	c	l	u	α	β	γ	λ
PS, 1°	a	b	c	0	0	0	0	0	λ
PS, 2°	a	b	c	l	ml	0	0	0	m
PS, 3°	a	b	c	$-mc$	0	0	0	γ	$\frac{1}{m}\gamma$
PS, 4°	a	b	0	l	ml	0	0	γ	m
PS, 5°	a	b	0	0	0	0	0	γ	λ
PS, 6°	0	b	0	l	ml	α	0	γ	m
PS, 7°	a	b	0	ma	0	α	0	γ	$-\frac{1}{m}\alpha$,

where m is an arbitrary constant; in the body of the table, the other italic letters are arbitrary functions of t, and the greek letters arbitrary functions of θ.

PS, 1° is Serret's first case of *PS*, included in his second case.

PS, 2° gives developable.

PS, 3° is Serret's second case of *PS*.

PS, 4° is Serret's third case of *PS*.

PS, 5° gives circular sections (surfaces of revolution).

PS, 6° gives circular sections (tubular surfaces).

PS, 7° gives circular sections.

I consider

PS, 3° = *Serret's Second Case of PS.*

The six equations are

$$\begin{aligned} A^2 &+ B^2 + C^2 = 1, \\ ax &+ by + cz = 0, \\ Aa &+ Bb + Cc = -cm, \\ x^2 &+ y^2 + (z - m\phi)^2 = \theta + m^2\phi^2, \\ Ax &+ By + C(z - m\phi) = \phi, \\ Adx &+ Bdy + Cdz = 0, \end{aligned}$$

where a, b, c are assumed such that $a^2 + b^2 + c^2 = 1$. We easily obtain

$$(1 - c^2) A = - ac(C + m) - b\sqrt{\Omega},$$
$$(1 - c^2) B = - bc(C + m) + a\sqrt{\Omega},$$

and thence

$$aB - bA = \sqrt{\Omega},$$

where

$$\Omega = (1 - c^2)(1 - C^2) - c^2(C + m)^2, \ = 1 - c^2 - C^2 - 2c^2Cm - c^2m^2;$$

also

$$\begin{aligned} x\sqrt{1 - c^2m^2} &= \qquad\qquad A\phi\sqrt{1 - c^2m^2} + (bC - cB)\sqrt{\theta + (m^2 - 1)\phi^2}, \\ y\sqrt{1 - c^2m^2} &= \qquad\qquad B\phi\sqrt{1 - c^2m^2} + (cA - aC)\sqrt{\theta + (m^2 - 1)\phi^2}, \\ z\sqrt{1 - c^2m^2} &= (C + m)\phi\sqrt{1 - c^2m^2} + (aB - bA)\sqrt{\theta + (m^2 - 1)\phi^2}. \end{aligned}$$

We seek for the differential equation in C, t, θ. From the equation

$$Ax + By + (C - m\phi)z = \phi,$$

and attending to

$$Adx + Bdy + Cdz = 0,$$

we deduce

$$xdA + ydB + (z - m\phi)dC - (1 + Cm)\phi' d\theta = 0,$$

and we have herein to substitute for dA, dB their values in terms of dC, dt, $d\theta$. We have

$$AdA + BdB = - CdC,$$
$$adA + bdB = - cdC - Q,$$

if for shortness

$$Q = Ada + Bdb + (C + m)\, dc.$$

Hence

$$\sqrt{\Omega}\, dA = (-cB + bC)\, dC - BQ,$$
$$\sqrt{\Omega}\, dB = (-aC + cA)\, dC + AQ.$$

We find without difficulty,

$$(1 - c^2)\, Q = (C + m)\, dc + (adb - bda)\sqrt{\Omega},$$

and consequently,

$$(1 - c^2)\sqrt{\Omega}\, dA = \{\quad b\,(C + c^2m) - ac\sqrt{\Omega}\}\, dC - B\,\{(C + m)\, dc + (adb - bda)\sqrt{\Omega}\},$$
$$(1 - c^2)\sqrt{\Omega}\, dB = \{-a\,(C + c^2m) - bc\sqrt{\Omega}\}\, dC + A\,\{(C + m)\, dc + (adb - bda)\sqrt{\Omega}\}.$$

Substituting these values, we have

$$\begin{aligned}&\{(bx - ay)(C + c^2m) - (ax + by)\, c\sqrt{\Omega}\}\, dC \\ &- (Bx - Ay)\,\{(C + m)\, dc + (adb - bda)\sqrt{\Omega}\} \\ &+ (1 - c^2)\sqrt{\Omega}\,\{(z - m\phi)\, dC - (1 + Cm)\,\phi' d\theta\} = 0,\end{aligned}$$

viz. this is

$$\begin{aligned}&\{(bx - ay)(C + c^2m) - (ax + by)\, c\sqrt{\Omega} + (1 - c^2)(z - m\phi)\sqrt{\Omega}\}\, dC \\ &- (Bx - Ay)\,\{(C + m)\, dc + (adb - bda)\sqrt{\Omega}\} \\ &- (1 - c^2)(1 + Cm)\sqrt{\Omega}\,\phi' d\theta = 0.\end{aligned}$$

The coefficient of dC contains a term $-(ax + by + cz)\, c\sqrt{\Omega}$ which is $= 0$. Moreover, we have

$$bx - ay = -\phi\sqrt{\Omega} + \frac{C + c^2m}{\sqrt{1 - c^2m^2}}\sqrt{\theta + (m^2 - 1)\,\phi^2},$$

and then

$$\begin{aligned}(1 - c^2)(Bx - Ay) &= -c\,(C + m)(bx - ay) - cz\sqrt{\Omega} \\ &= -c\,(C + m)\left\{-\phi\sqrt{\Omega} + \frac{C + c^2m}{\sqrt{1 - c^2m^2}}\sqrt{\theta + (m^2 - 1)\,\phi^2}\right\} \\ &\quad - c\sqrt{\Omega}\left\{(C + m)\,\phi + \frac{\sqrt{\Omega}\sqrt{\theta + (m^2 - 1)\,\phi^2}}{\sqrt{1 - c^2m^2}}\right\},\end{aligned}$$

which, observing that the terms in $C\phi\sqrt{\Omega}$ destroy each other, and that we have

$$(C + m)(C + c^2m) + \Omega = (1 - c^2)(1 + Cm),$$

gives

$$Bx - Ay = \frac{-c\,(1 + Cm)}{\sqrt{1 - c^2m^2}}\sqrt{\theta + (m^2 - 1)\,\phi^2},$$

and the equation becomes

$$\left\{\left(-\phi\sqrt{\Omega}+\frac{C+c^2m}{\sqrt{1-c^2m^2}}\sqrt{\theta+(m^2-1)\phi^2}\right)(C+c^2m)+z\sqrt{\Omega}-(1-c^2)\,m\phi\sqrt{\Omega}\right\}dC$$

$$-c(1+Cm)\frac{\sqrt{\theta+(m^2-1)\phi^2}}{\sqrt{1-c^2m^2}}\left\{(C+m)\,dc+(adb-bda)\sqrt{\Omega}\right\}$$

$$-(1-c^2)\sqrt{\Omega}\,(1+Cm)\,\phi' d\theta=0.$$

Here the coefficient of dC is

$$=[z-(C+m)\,\phi]\sqrt{\Omega}+\frac{(C+c^2m)^2}{\sqrt{1-c^2m^2}}\sqrt{\theta+(m^2-1)\,\phi^2},$$

or, substituting for $z-(C+m)\sqrt{\Omega}$ its value $=\dfrac{\sqrt{\Omega}\sqrt{\theta+(m^2-1)\,\phi^2}}{\sqrt{1-c^2m^2}}$, and observing that $\Omega+(C+c^2m)^2=(1-c^2m^2)(1-c^2)$, this coefficient is found to be

$$=\sqrt{1-c^2m^2}\,(1-c^2)\sqrt{\theta+(m^2-1)\,\phi^2},$$

and we have

$$\sqrt{1-c^2m^2}\,(1-c^2)\sqrt{\theta+(m^2-1)\,\phi^2}\,dC$$

$$-c(1+Cm)\frac{\sqrt{\theta+(m^2-1)\,\phi^2}}{\sqrt{1-c^2m^2}}\left\{(C+m)\,dc+(adb-bda)\sqrt{\Omega}\right\}$$

$$-(1-c^2)(1+Cm)\sqrt{\Omega}\,\phi' d\theta=0,$$

or, as this may be written,

$$\frac{1}{\sqrt{\Omega}}\left\{\frac{\sqrt{1-c^2m^2}\,dC}{1+Cm}-\frac{(C+m)\,cdc}{(1-c^2)\sqrt{1-c^2m^2}}\right\}-\frac{c\,(adb-bda)}{(1-c^2)\sqrt{1-c^2m^2}}-\frac{\phi' d\theta}{\sqrt{\theta+(m^2-1)\,\phi^2}}=0,$$

where from the foregoing value of Ω we have identically

$$\Omega\,(1-m^2)=(1-c^2)(1+Cm)^2-(1-c^2m^2)(C+m)^2.$$

Here a, b, c are functions of t; and we have thus the required differential equation in C, t, θ.

It is convenient to multiply by the constant factor $\sqrt{1-m^2}$. The first term is an exact differential, viz. writing

$$\sin\zeta=\frac{\sqrt{1-c^2m^2}}{\sqrt{1-c^2}}\,\frac{C+m}{1+Cm},\ \text{and therefore}\ \cos\zeta=\frac{\sqrt{1-m^2}\sqrt{\Omega}}{\sqrt{1-c^2}\,(1+Cm)},$$

we have

$$d\zeta=\frac{\sqrt{1-m^2}}{\sqrt{\Omega}}\left\{\frac{\sqrt{1-c^2m^2}\,dC}{1+Cm}+\frac{(C+m)\,cdc}{(1-c^2)\sqrt{1-c^2m^2}}\right\},$$

as may easily be verified. And the second and third terms are obviously the differentials of a function of t and a function of θ respectively. But to obtain the integral functions, a transformation of each term is required.

First, for the term $\dfrac{\sqrt{1-m^2}c\,(adb-bda)}{(1-c^2)\sqrt{1-c^2m^2}}$; we take a, b, c functions of t which are such that $a^2+b^2+c^2=1$; and then writing a_1, b_1, c_1 for the derived functions so that $aa_1+bb_1+cc_1=0$, we assume $a', b', c' = Va_1, Vb_1, Vc_1$ where $\dfrac{1}{V^2}=a_1^2+b_1^2+c_1^2$; we have therefore $aa'+bb'+cc'=0$, and $a'^2+b'^2+c'^2=1$; and then writing $a'', b'', c'' = bc'-b'c$, $ca'-c'a$, $ab'-a'b$ respectively, we have

$$aa''+bb''+cc''=0,\quad a'a''+b'b''+c'c''=0,\quad a''^2+b''^2+c''^2=1;$$

thus a, b, c, a′, b′, c′, a″, b″, c″ are a set of rectangular coefficients. We then write

$$a,\ b,\ c=\frac{1}{\rho}(a'+mb''),\quad \frac{1}{\rho}(b'-ma''),\quad \frac{1}{\rho}c',$$

determining ρ so that $a^2+b^2+c^2=1$ as above, viz. we thus have

$$\rho^2=(1+cm)^2+c'^2m^2.$$

Observe that we thus have $\rho^2(1-c^2)=\rho^2-c'^2$ and $\rho^2(1-c^2m^2)=(1+cm)^2$.

Writing now

$$T=\tan^{-1}\frac{c+m}{c''\sqrt{1-m^2}},\ \text{and therefore}\ \sin T=\frac{c+m}{\sqrt{\rho^2-c'^2}},\quad \cos T=\frac{c''\sqrt{1-m^2}}{\sqrt{\rho^2-c'^2}},$$

we find that

$$dT=\frac{\sqrt{1-m^2}c\,(adb-bda)}{(1-c^2)\sqrt{1-c^2m^2}}.$$

The verification is somewhat long, but it is very interesting. We have

$$dT=\frac{\sqrt{1-m^2}\,\{c''dc-(c+m)\,dc''\}}{\rho^2-c'^2},$$

or observing that $c''=ab'-a'b$, $=V(ab_1-a_1b)$, $dc=c_1dt$, this is

$$dT=\frac{\sqrt{1-m^2}}{\rho^2-c'^2}\{V(ab_1-a_1b)\,c_1-(c+m)\,[V_1(ab_1-a_1b)+V(ab_{11}-a_{11}b)]\,dt\},$$

where we have

$$\frac{1}{V^2}=a_1^2+b_1^2+c_1^2,\ \text{and therefore}\ -\frac{V_1}{V^3}=a_1a_{11}+b_1b_{11}+c_1c_{11};$$

also from $aa_1+bb_1+cc_1=0$, we have $a_1^2+b_1^2+c_1^2+aa_{11}+bb_{11}+cc_{11}=0$, and we thence obtain

$$\begin{aligned}dT=\frac{\sqrt{1-m^2}\,V^3dt}{\rho^2-c'^2}\{&-(ab_1-a_1b)\,c_1\,(aa_{11}+bb_{11}+cc_{11})\\&-(c+m)\,[-(a_1a_{11}+b_1b_{11}+c_1c_{11})\,(ab_1-a_1b)+(a_1^2+b_1^2+c_1^2)\,(ab_{11}-ba_{11})]\},\end{aligned}$$

the term in [ ] is found to be $= -c_1\{a_{11}(bc_1 - b_1c) + b_{11}(ca_1 - c_1a) + c_{11}(ab_1 - a_1b)\}$, hence c_1 appears as a factor of the whole expression; and reducing the part independent of m, we find

$$dT = \frac{\sqrt{1-m^2}\,V^3c_1\,dt}{\rho^2 - c'^2}\{(a_1b_{11} - a_{11}b_1) + m\,[a_{11}(bc_1 - b_1c) + b_{11}(ca_1 - c_1a) + c_{11}(ab_1 - a_1b)]\}.$$

Next, calculating the value of $adb - bda$, we have

$$a = \frac{V}{\rho}\{a_1 + m\,(ca_1 - c_1a)\},\quad b = \frac{V}{\rho}\{b_1 - m\,(bc_1 - b_1c)\},$$

or, as these may be written,

$$a = \frac{V}{\rho}\{a_1(1+cm) - ac_1m\},\quad b = \frac{V}{\rho}\{b_1(1+cm) - bc_1m\},$$

and we thence easily obtain

$$adb - bda = \frac{V^2dt}{\rho^2}(1+cm)\{(a_1b_{11} - a_{11}b_1) + m\,[a_{11}(bc_1 - b_1c) + b_{11}(ca_1 - c_1a) + c_{11}(ab_1 - a_1b)]\},$$

viz. the factor in { } has the same value as in the expression for dT, and we thus have

$$\frac{dT}{adb - bda} = \frac{\sqrt{1-m^2}\,Vc_1\rho^2}{(1+cm)(\rho^2 - c'^2)} = \frac{c\sqrt{1-m^2}}{\sqrt{1-c^2m^2}\,(1-c^2)},$$

that is,

$$dT = \frac{\sqrt{1-m^2}\,c\,(adb - bda)}{(1-c^2)\sqrt{1-c^2m^2}},$$

the required equation.

Secondly, for the term $\dfrac{\sqrt{1-m^2}\,\phi' d\theta}{\sqrt{\theta + (m^2-1)\phi^2}}$, we introduce Φ a function of θ, such that writing Φ' for the derived function we have

$$\phi = \frac{\Phi - 2\theta\Phi'}{\sqrt{1 - 4(1-m^2)\Phi\Phi' + 4(1-m^2)\theta\Phi'^2}},\ = \frac{\Phi - 2\theta\Phi'}{\sqrt{M}}\ \text{suppose},$$

whence also

$$\sqrt{\theta + (m^2-1)\phi^2} = \frac{\sqrt{\theta + (m^2-1)\Phi^2}}{\sqrt{M}},\quad \frac{\phi}{\sqrt{\theta + (m^2-1)\phi^2}} = \frac{\Phi - 2\theta\Phi'}{\sqrt{\theta + (m^2-1)\Phi^2}}.$$

Then writing

$$\sin\Theta\ = \frac{\Phi\sqrt{1-m^2}}{\sqrt{\theta}},\quad \cos\Theta\ = \frac{\sqrt{\theta + (m^2-1)\Phi^2}}{\sqrt{\theta}},$$

$$\sin\Theta_0 = \frac{\phi\sqrt{1-m^2}}{\sqrt{\theta}},\quad \cos\Theta_0 = \frac{\sqrt{\theta + (m^2-1)\phi^2}}{\sqrt{\theta}},$$

we find

$$\cos\Theta\,d\Theta = -\frac{\frac{1}{2}\sqrt{1-m^2}\,(\Phi - 2\theta\Phi')\,d\theta}{\theta\sqrt{\theta}},$$

that is,

$$d\Theta = \frac{-\tfrac{1}{2}\sqrt{1-m^2}\,(\Phi - 2\theta\Phi')\,d\theta}{\theta\sqrt{\theta+(m^2-1)\,\Phi^2}}\,;$$

and similarly

$$\cos\Theta_0 d\Theta_0 = \frac{-\tfrac{1}{2}\sqrt{1-m^2}\,(\phi - 2\theta\phi')\,d\theta}{\theta\sqrt{\theta}}\,,$$

that is,

$$d\Theta_0 = \frac{-\tfrac{1}{2}\sqrt{1-m^2}\,(\phi - 2\theta\phi')\,d\theta}{\theta\sqrt{\theta+(m^2-1)\,\phi^2}}\,.$$

Hence

$$-d\Theta + d\Theta_0 = \frac{\sqrt{1-m^2}}{2\theta}\left\{\frac{\Phi-2\theta\Phi'}{\sqrt{\theta+(m^2-1)\,\Phi^2}} - \frac{\phi-2\theta\phi'}{\sqrt{\theta+(m^2-1)\,\phi^2}}\right\}d\theta$$

$$= \frac{\sqrt{1-m^2}}{2\theta}\left\{\frac{\phi-(\phi-2\theta\phi')}{\sqrt{\theta+(m^2-1)\,\phi^2}}\right\}d\theta,\ = \frac{\sqrt{1-m^2}\,\phi' d\theta}{\sqrt{\theta+(m^2-1)\,\phi^2}}\,,$$

the required equation.

We find, moreover,

$$\sin(\Theta-\Theta_0) = \frac{2\Phi'\sqrt{1-m^2}\sqrt{\theta+(m^2-1)\,\Phi^2}}{\sqrt{M}},\quad \cos(\Theta-\Theta_0) = \frac{1-2\,(1-m^2)\,\Phi\Phi'}{\sqrt{M}}\,,$$

which will be presently useful.

The differential equation now is $d\zeta - dT + d\Theta - d\Theta_0 = 0$, hence the integral equation (taking the constant of integration $=0$) is $\zeta = T - \Theta + \Theta_0$, or say

$$\sin\zeta = \sin(T - \Theta + \Theta_0),$$

viz. substituting for $\sin T$ and $\cos T$ their values, and observing that

$$\sin\zeta = \frac{\sqrt{1-c^2m^2}}{\sqrt{1-c^2}}\,\frac{C+m}{1+Cm},\ = \frac{1+cm}{\sqrt{\rho^2-c'^2}}\,\frac{C+m}{1+Cm}\,,$$

the factor $\dfrac{1}{\sqrt{\rho^2-c'^2}}$ multiplies out, and we have

$$(1+cm)\frac{C+m}{1+Cm} = (c+m)\cos(\Theta-\Theta_0) - c''\sqrt{1-m^2}\sin(\Theta-\Theta_0).$$

And I further remark here that a former equation is

$$\Omega\,(1-m^2) = (1-c^2)\,(1+Cm)^2 - (1-c^2m^2)\,(C+m)^2,$$

that is,

$$\Omega\,\frac{1-m^2}{(1+Cm)^2} = (1-c^2)\left\{1 - \frac{(1-c^2m^2)\,(C+m)^2}{(1-c^2)\,(1+Cm)^2}\right\} = (1-c^2)\cos^2\zeta.$$

We thus have

$$\sqrt{\Omega} = \frac{1+Cm}{\sqrt{1-m^2}}\,\frac{\sqrt{\rho^2-c'^2}}{\rho}\cos\zeta,$$

$$= \frac{1+Cm}{\rho\sqrt{1-m^2}}\,\{c''\sqrt{1-m^2}\cos(\Omega-\Omega_0) + (c+m)\sin(\Omega-\Omega_0)\}.$$

We have thus C, and consequently also A, B, x, y, z, all of them given as functions of t, θ; but the formulæ admit of further development.

Write

$$\varsigma = \frac{C+m}{1+Cm}, \text{ whence also } C = \frac{\varsigma - m}{1 - m\varsigma}.$$

We have $C(1-m\varsigma)+m=\varsigma$, and hence $(1+cm)\{C(1-m\varsigma)+m\}$, $=(1+cm)\varsigma$, $=(c+m)\cos(\Theta-\Theta_0)-c''\sqrt{1-m^2}\sin(\Theta-\Theta_0)$. Using the value of C given by this equation, and calculating from it those of A, B; then writing for shortness

$$\begin{aligned} X &= a\sqrt{1-m^2}\cos(\Theta-\Theta_0)-(a''-mb')\sin(\Theta-\Theta_0),\\ Y &= b\sqrt{1-m^2}\cos(\Theta-\Theta_0)-(b''+ma')\sin(\Theta-\Theta_0),\\ Z &= \quad (c+m)\cos(\Theta-\Theta_0)-c''\sqrt{1-m^2}\sin(\Theta-\Theta_0), \end{aligned}$$

we have

$$\begin{aligned} A(1-m\varsigma)(1+cm) &= \sqrt{1-m^2}X,\\ B(1-m\varsigma)(1+cm) &= \sqrt{1-m^2}Y,\\ C(1-m\varsigma)(1+cm) &= \qquad Z-m(1+cm), \end{aligned}$$

to which I join

$$\varsigma(1+cm) = \qquad Z.$$

By way of verification, observe that $A^2+B^2+C^2=1$, and that the equations give

$$(1-m\varsigma)^2(1+cm)^2=(1-m^2)(X^2+Y^2+Z^2)+m^2Z^2-2mZ(1+cm)+m^2(1+cm)^2;$$

we have

$$X^2+Y^2+Z^2=(1+cm)^2,\quad Z=\varsigma(1+cm),$$

and hence the identity

$$(1-m\varsigma)^2(1+cm)^2=(1-m^2+m^2\varsigma^2-2m\varsigma+m^2)(1+cm)^2.$$

Proceeding to calculate the values of x, y, z, recollecting that

$$\sqrt{1-c^2m^2}=\frac{1}{\rho}(1+cm),$$

we have

$$\begin{aligned} x(1+cm) &= A\phi(1+cm)+\rho(bC-cB)\sqrt{\theta+(m^2-1)\phi^2},\\ &= A\phi(1+cm)+\{(b'-ma'')C-c'B\}\sqrt{\theta+(m^2-1)\phi^2}, \end{aligned}$$

that is,

$$\begin{aligned} x(1+cm)(1-m\varsigma) &= \phi\sqrt{1-m^2}X+\frac{1}{1+cm}\{(b'-ma'')(Z-m(1+cm))\\ &\qquad -c'\sqrt{1-m^2}Y\}\sqrt{\theta+(m^2-1)\phi^2}\\ &= \phi\sqrt{1-m^2}X+\frac{1}{1+cm}\{(b'-ma'')Z-c'\sqrt{1-m^2}Y\}\sqrt{\theta+(m^2-1)\phi^2}\\ &\qquad -m(b'-ma'')\sqrt{\theta+(m^2-1)\phi^2}, \end{aligned}$$

where the term $(b' - ma'')Z - c'\sqrt{1-m^2}Y$ contains the factor $1 + cm$; in fact, this is

$$= (b' - ma'')\{\quad (c+m)\cos(\Theta - \Theta_0) - c''\sqrt{1-m^2}\sin(\Theta - \Theta_0)\}$$
$$- c'\sqrt{1-m^2}\{b\sqrt{1-m^2}\cos(\Theta - \Theta_0) - (b'' + ma')\sin(\Theta - \Theta_0)\}.$$

The coefficient of the cosine is $(b' - ma'')(c+m) - bc'(1-m^2)$, which is

$$= b'c - bc' + m(b' - ca'') + m^2(-a'' + bc'),\ = -a'' + m(b' - ca'') + m^2(-b'c),$$
$$= (1 + cm)(-a'' + mb'),$$

and similarly the coefficient of $\sqrt{1-m^2}$, multiplied by the sine, is

$$-c''(b' - ma'') + c'(b'' + ma'),\ = -b'c'' + b''c' + m(a'c' + a''c''),$$
$$= -a + m(-ac),\ = (1 + cm)(-a).$$

Calculating in like manner the values of y and z, and putting for shortness

$$X_1 = (-a'' + mb')\quad \cos(\Theta - \Theta_0) - a\sqrt{1-m^2}\sin(\Theta - \Theta_0),$$
$$Y_1 = (-b'' - ma')\quad \cos(\Theta - \Theta_0) - b\sqrt{1-m^2}\sin(\Theta - \Theta_0),$$
$$Z_1 = (-c''\sqrt{1-m^2})\cos(\Theta - \Theta_0) + \quad (c+m)\sin(\Theta - \Theta_0),$$

we have

$$x = \qquad \phi\sqrt{1-m^2}X + X_1\sqrt{\theta + (m^2-1)\phi^2} - m(b' - ma'')\sqrt{\theta + (m^2-1)\phi^2},$$
$$y = \qquad \phi\sqrt{1-m^2}Y + Y_1\sqrt{\theta + (m^2-1)\phi^2} + m(a' + mb'')\sqrt{\theta + (m^2-1)\phi^2},$$
$$z = \sqrt{1-m^2}\{\phi\sqrt{1-m^2}Z + Z_1\sqrt{\theta + (m^2-1)\phi^2}\},$$

which are the required expressions of x, y, z in terms of t and θ. It will be noticed that X, X_1, Y, Y_1, Z, Z_1, each contain a term with $\cos(\Theta - \Theta_0)$ and one with $\sin(\Theta - \Theta_0)$; but as the terms in X_1, Y_1, Z_1 are each multiplied by $\sqrt{\theta + (m^2-1)\phi^2}$, the cosine and sine terms of X, X_1, of Y, Y_1 and of Z, Z_1 do not in any case unite into a single term.

I remark that we have identically

$$aX + bY + c\sqrt{1-m^2}Z = 0,$$
$$aX_1 + bY_1 + c\sqrt{1-m^2}Z_1 = 0.$$

The foregoing values of x, y, z thus satisfy $ax + by + cz = 0$, which is one of the six equations. The others of them might be verified without difficulty. I recall that we have $a, b, c = \frac{1}{\rho}(a' + mb''), \frac{1}{\rho}(b' - ma''), \frac{1}{\rho}c'$; the six equations might therefore be written

$$\begin{aligned}
&A^2 + B^2 + C^2 &&= 1,\\
&(a' + mb'')\ x + (b' - ma'')\ y + c'z &&= 0,\\
&(a' + mb'')\ A + (b' - ma'')\ B + c'C &&= -c'm,\\
&x^2 + y^2 + (z - m\phi)^2 &&= \theta + m^2\phi^2,\\
&Ax + By + C(z - m\phi) &&= \phi,\\
&Adx + Bdy + Cdz &&= 0.
\end{aligned}$$

The Case PS, 1° = Serret's First Case of PS.

This is at once deduced from *PS*, 3° by writing therein $m=0$; the formulæ are a good deal more simple. We introduce, as before, the rectangular coefficients a, b, c, a′, b′, c′, a″, b″, c″; and the values of a, b, c then are a′, b′, c′. The six equations, using therein these values for a, b, c, are

$$\begin{aligned} A^2+B^2\ +C^2\ &=1,\\ \mathrm{a}'x+\mathrm{b}'y\ +\mathrm{c}'z\ &=0,\\ \mathrm{a}'A+\mathrm{b}'B\ +\mathrm{c}'C\ &=0,\\ x^2+y^2\ +z^2\ &=\theta,\\ Ax+By\ +Cz\ &=\phi,\\ Adx+Bdy+Cdz&=0.\end{aligned}$$

The function Φ is such that

$$\phi=\frac{\Phi-2\theta\Phi'}{\sqrt{1-4\Phi\Phi'+4\theta\Phi'^2}}=\frac{\Phi-2\theta\Phi'}{\sqrt{M}}.$$

We have

$$\sin\Theta=\frac{\Phi}{\sqrt{\theta}}\cos\Theta=\frac{\sqrt{\theta-\Phi^2}}{\sqrt{\theta}},$$

$$\sin\Theta_0=\frac{\phi}{\sqrt{\theta}}\cos\Theta_0=\frac{\sqrt{\theta-\phi^2}}{\sqrt{\theta}},$$

and thence

$$\sin(\Theta-\Theta_0)=\frac{2\Phi'\sqrt{\theta-\Phi^2}}{\sqrt{M}};\quad \cos(\Theta-\Theta_0)=\frac{1-2\Phi\Phi'}{\sqrt{M}}.$$

Also

$$\sin\zeta=\frac{C}{\sqrt{1-\mathrm{c}^2}},\quad \cos\zeta=\frac{\sqrt{1-\mathrm{c}^2-C^2}}{\sqrt{1-\mathrm{c}^2}};\quad \sin T=\frac{\mathrm{c}}{\sqrt{1-\mathrm{c}'^2}},\quad \cos T=\frac{\mathrm{c}''}{\sqrt{1-\mathrm{c}^2}},$$

$$\zeta=T-\Theta+\Theta_0,\quad C=\mathrm{c}\cos(\Theta-\Theta_0)-\mathrm{c}''\sin(\Theta-\Theta_0),$$

$$\sqrt{1-\mathrm{c}^2-C^2}=\mathrm{c}''\cos(\Theta-\Theta_0)+\mathrm{c}\sin(\Theta-\Theta_0).$$

We have

$$\begin{aligned} A&=X=\mathrm{a}\cos(\Theta-\Theta_0)-\mathrm{a}''\sin(\Theta-\Theta_0);\quad X_1=\mathrm{a}''\cos(\Theta-\Theta_0)+\mathrm{a}\sin(\Theta-\Theta_0),\\ B&=Y=\mathrm{b}\cos(\Theta-\Theta_0)-\mathrm{b}''\sin(\Theta-\Theta_0);\quad Y_1=\mathrm{b}''\cos(\Theta-\Theta_0)+\mathrm{b}\sin(\Theta-\Theta_0),\\ C&=Z=\mathrm{c}\cos(\Theta-\Theta_0)-\mathrm{c}''\sin(\Theta-\Theta_0);\quad Z_1=\mathrm{c}''\cos(\Theta-\Theta_0)+\mathrm{c}\sin(\Theta-\Theta_0),\end{aligned}$$

and then

$$\begin{aligned} x&=X\phi+X_1\sqrt{\theta-\phi^2},\\ y&=Y\phi+Y_1\sqrt{\theta-\phi^2},\\ z&=Z\phi+Z_1\sqrt{\theta-\phi^2},\end{aligned}$$

which are the expressions of the coordinates in terms of the parameters t and θ.

I consider next the case

PS, 4° = *Serret's Third Case of PS.*

The six equations are

$$\begin{aligned} A^2 &+ B^2 + C^2 &&= 1\,, \\ ax &+ by &&= lm, \\ Aa &+ Bb &&= l\,, \\ x^2 &+ y^2 + z^2 - 2\theta z &&= 2v, \\ Ax &+ By + C(z-\theta) &&= m\,, \\ Adx &+ Bdy + Cdz &&= 0\,; \end{aligned}$$

where θ has been written in place of γ: m is a given constant; a, b, l are functions of t; v is a function of θ. The equation $ax + by = lm$ evidently denotes that the planes of the plane curves of curvature are all of them parallel to the axis of z, or, what is the same thing, they envelope a cylinder; in the particular case $m = 0$, they all of them pass through the axis of z. In the general case, the required surface is the parallel surface, at the normal distance m, to the surface which belongs to the particular case $m = 0$. This is not assumed in the investigation which follows; but it will be readily perceived how the theorem is involved in, and in fact proved by, the investigation.

I obtain the solution synthetically as follows:

Taking T, a, b functions of t, $\mathrm{a}^2 + \mathrm{b}^2 = 1$; T_1, a_1, b_1 their derived functions, $\mathrm{a}\mathrm{a}_1 + \mathrm{b}\mathrm{b}_1 = 0$; $\Omega = \dfrac{T_1^2}{4T^2} + \mathrm{a}_1^2 + \mathrm{b}_1^2$; Θ a function of θ, Θ' its derived function,

$$M = \frac{2\Theta}{\Theta'}\,;\quad P = \frac{2\sqrt{T\Theta}}{T+\Theta}\,,\quad Q = \frac{T-\Theta}{T+\Theta}\,,$$

and therefore $P^2 + Q^2 = 1$; then writing

$$A_0 = \frac{1}{\sqrt{\Omega}}\left(-\,\mathrm{b}\,\frac{T_1}{2T} + \mathrm{b}_1 Q\right),$$

$$B_0 = \frac{1}{\sqrt{\Omega}}\left(\ \ \mathrm{a}\,\frac{T_1}{2T} - \mathrm{a}_1 Q\right),$$

$$C_0 = \frac{-1}{\sqrt{\Omega}}\,(\mathrm{a}\mathrm{b}_1 - \mathrm{a}_1\mathrm{b})\,P,$$

where $A_0^2 + B_0^2 + C_0^2 = 1$, we assume

$$\begin{aligned} x &= mA_0 &&+ \mathrm{a}MP, \\ y &= mB_0 &&+ \mathrm{b}MP, \\ z &= mC_0 + \theta &&+ MQ, \end{aligned}$$

equations which determine x, y, z as functions of the parameters t and θ. As will presently be shown, we have $A_0dx+B_0dy+C_0dz=0$; and we have thus $A, B, C=A_0, B_0, C_0$; and this being so, we easily verify the six equations

$$\begin{aligned}
&A^2 \quad +B^2+C^2 &&=1,\\
&\mathrm{b}x \quad -\mathrm{a}y &&=m\left(-\frac{T_1}{2T}\frac{1}{\sqrt{\Omega}}\right),\\
&\mathrm{b}A \quad -\mathrm{a}B &&=\left(-\frac{T_1}{2T}\frac{1}{\sqrt{\Omega}}\right),\\
&x^2 \quad +y^2+(z-\theta)^2 &&=m^2+M^2+\theta^2,\\
&Ax \quad +By+C(z-\theta) &&=m,\\
&Adx+Bdy+Cdz &&=0,
\end{aligned}$$

which are the six equations of the problem with the values $a=\mathrm{b}$, $b=-\mathrm{a}$, $l=-\dfrac{2T_1}{T}\dfrac{1}{\sqrt{\Omega}}$, $2v=m^2+M^2$, for a, b, l and $2v$.

We in fact at once obtain the third equation $\mathrm{b}A_0-\mathrm{a}B_0=-\dfrac{T_1}{2T}\dfrac{1}{\sqrt{\Omega}}$, and thence the second equation $\mathrm{b}x-\mathrm{a}y=m(\mathrm{b}A_0-\mathrm{a}B_0)$, $=m\left(-\dfrac{T_1}{2T}\dfrac{1}{\sqrt{\Omega}}\right)$; then for the fifth equation, we have

$$A_0x+B_0y+C_0(z-\theta)=m+M\{(A_0\mathrm{a}+B_0\mathrm{b})P+C_0Q\},\ =m,$$

since $(A_0\mathrm{a}+B_0\mathrm{b})P+C_0Q=0$; and for the fourth equation, we have

$$x^2+y^2+(z-\theta)^2=m^2+2m\{M(A_0\mathrm{a}+B_0\mathrm{b})P+C_0Q\}+M^2,\ =m^2+M^2.$$

It remains only to prove the assumed equation $A_0dx+B_0dy+C_0dz=0$. Writing for a moment $X, Y, Z=\mathrm{a}MP, \mathrm{b}MP, \theta+MQ$, we have

$$\begin{aligned}
A_0dx+B_0dy+C_0dz&=A_0(mdA_0+dX)+B_0(mdB_0+dY)+C_0(mdC_0+dZ),\\
&=A_0dX+B_0dY+C_0dZ,
\end{aligned}$$

since $A_0dA_0+B_0dB_0+C_0dC_0=0$ in virtue of $A_0^2+B_0^2+C_0^2=1$.

We have thus to show that, if $X, Y, Z=\mathrm{a}MP, \mathrm{b}MP, \theta+MQ$, then

$$A_0dX+B_0dY+C_0dZ=0;$$

say we have

$$\begin{aligned}
dX&=pdt+p'd\theta,\\
dY&=qdt+q'd\theta,\\
dZ&=rdt+r'd\theta,
\end{aligned}$$

then the required values of A_0, B_0, C_0 are proportional to $qr'-q'r$, $rp'-r'p$, $pq'-p'q$, and the sum of their squares is $=1$. Writing for shortness $MP=R$, $MQ=S$, we have

$$\begin{aligned}
p&=\mathrm{a}_1R+\mathrm{a}R_1, & p'&=\mathrm{a}R',\\
q&=\mathrm{b}_1R+\mathrm{b}R_1, & q'&=\mathrm{b}R',\\
r&=\qquad\quad S_1, & r'&=1+S';
\end{aligned}$$

hence

$$qr' - q'r = \mathrm{b}\,[R_1(1+S') - R'S_1] + \mathrm{b}_1 R(1+S'),$$
$$rp' - r'p = -\mathrm{a}\,[R_1(1+S') - R'S_1] - \mathrm{a}_1 R(1+S'),$$
$$pq' - p'q = -(\mathrm{ab}_1 - \mathrm{a}_1\mathrm{b})\,RR'.$$

Here

$$R' = MP' + M'P, \quad R_1 = MP_1,$$
$$S' = MQ' + M'Q, \quad S_1 = MQ_1,$$

and hence

$$RS' - R'S = M^2(PQ' - P'Q),$$
$$R_1S' - R'S_1 = M^2(P_1Q' - P'Q_1) + MM'(P_1Q - PQ_1);$$

moreover, from the values of P and Q, we have

$$P' = \frac{\Theta'}{2\Theta}PQ, = \frac{PQ}{M}, \quad P_1 = -\frac{T_1}{2T}PQ,$$
$$Q' = -\frac{\Theta'}{2\Theta}P^2 \quad = -\frac{P^2}{M}, \quad Q_1 = \frac{T_1}{2T}P^2,$$

and thence

$$P_1Q - PQ_1 = -\frac{T_1P}{2T}; \quad PQ' - P'Q = -\frac{P}{M}; \quad P_1Q' - P'Q_1 = 0;$$

also

$$R' = PQ + M'P, = P(Q+M'), \qquad 1 + S = 1 - P^2 + M'Q, = Q(Q+M');$$
$$R_1 = -\frac{T_1}{2T}MPQ, \quad RS' - R'S = -PM, \qquad R_1S' - R'S_1 = -\frac{T_1}{2T}MPM',$$

and consequently

$$R_1(1+S') - R'S_1 = -\frac{T_1}{2T}MP\ (Q+M'),$$
$$R\ (1+S') \quad = \quad MPQ\ (Q+M'),$$
$$RR' \quad = \quad MP^2\ (Q+M').$$

Hence the foregoing expressions for $qr' - q'r$, $rp' - r'p$, $pq' - p'q$ each contain the factor $MP(Q+M')$; omitting this factor, the expressions are

$$\left\{-\mathrm{b}\frac{T_1}{2T} + \mathrm{b}_1Q\right\}, \quad \left\{\mathrm{a}\frac{T_1}{2T} - \mathrm{a}_1Q\right\}, \quad -(\mathrm{ab}_1 - \mathrm{a}_1\mathrm{b})\,P;$$

the sum of the squares of these values is $= \frac{T_1^2}{4T^2} + \mathrm{a}_1^2 + \mathrm{b}_1^2$, $= \Omega$, and we have thus the required values

$$A_0 = \frac{1}{\sqrt{\Omega}}\left\{-\mathrm{b}\frac{T_1}{2T} + \mathrm{b}_1Q\right\},$$
$$B_0 = \frac{1}{\sqrt{\Omega}}\left\{\ \mathrm{a}\frac{T_1}{2T} - \mathrm{a}_1Q\right\},$$
$$C_0 = \frac{-1}{\sqrt{\Omega}}(\mathrm{ab}_1 - \mathrm{a}_1\mathrm{b})\,P,$$

which completes the proof.

In the case $m=0$, the solution is

$$x,\ y,\ z = \frac{2\Theta}{\Theta'}\,\mathrm{a}\,\frac{2\sqrt{T\Theta}}{T+\Theta},\quad \frac{2\Theta}{\Theta'}\,\mathrm{b}\,\frac{2\sqrt{T\Theta}}{T+\Theta},\quad \theta+\frac{2\Theta}{\Theta'}\,\frac{T-\Theta}{T+\Theta}.$$

Bonnet, in the paper (*Jour. Ecole Polyt.* t. XX.) referred to at the beginning of this memoir, gives for this case (see p. 199) a solution which he says is equivalent to that obtained by Joachimsthal in the paper "Demonstrationes theorematum ad superficies curvas spectantium," *Crelle*, t. XXX. (1846), pp. 347—350; viz. Joachimsthal's form is

$$x = \frac{\mu \sin L \sin \lambda}{1+\cos L \cos M},$$

$$y = \frac{\mu \sin L \cos \lambda}{1+\cos L \cos M},$$

$$z = \frac{\mu \cos L \sin M}{1+\cos L \cos M} + \int \cot M\, d\mu,$$

where L, M denote arbitrary functions of the parameters λ, μ respectively. To identify these with the foregoing form, I write

$$\sin\lambda = -\mathrm{a},\quad \cos L = -\frac{T-1}{T+1},\quad \cos M = \frac{\Theta-1}{\Theta+1},\quad \mu = \frac{4\Theta\sqrt{\Theta}}{\Theta'(\Theta+1)};$$

$$\cos\lambda = -\mathrm{b},\quad \sin L = \frac{-2\sqrt{T}}{T+1},\quad \sin M = \frac{-2\sqrt{\Theta}}{\Theta+1};$$

we thus have

$$\frac{\sin L \sin\lambda}{1+\cos L\cos M} = \mathrm{a}\,\frac{2\sqrt{T}}{T+1} \div 1 - \frac{(T-1)(\Theta-1)}{(T+1)(\Theta+1)},\quad = \frac{\mathrm{a}\sqrt{T}(\Theta+1)}{T+\Theta},$$

and thence

$$x = \frac{2\Theta}{\Theta'}\,\mathrm{a}\,\frac{2\sqrt{T\Theta}}{T+\Theta},\quad y = \frac{2\Theta}{\Theta'}\,\mathrm{b}\,\frac{2\sqrt{T\Theta}}{T+\Theta}.$$

Moreover,

$$\frac{\cos L \sin M}{1+\cos L \cos M} = \frac{\sqrt{\Theta}(T-1)}{T+\Theta},$$

and thence the first term of z is $=\dfrac{2\Theta}{\Theta'}\,\dfrac{2\Theta(T-1)}{(\Theta+1)(T+\Theta)}$; or observing that

$$2\Theta(T-1) = (\Theta+1)(T-\Theta) + (\Theta-1)(T+\Theta),$$

this is

$$= \frac{2\Theta}{\Theta'}\,\frac{T-\Theta}{T+\Theta} + \frac{2\Theta(\Theta-1)}{\Theta'(\Theta+1)},$$

or we have

$$z = \frac{2\Theta}{\Theta'}\,\frac{T-\Theta}{T+\Theta} + \frac{2\Theta(\Theta-1)}{\Theta'(\Theta+1)} + \int \cot M\, d\mu.$$

Here

$$\cot M\,d\mu = -\frac{\Theta-1}{2\sqrt{\Theta}}\,\frac{4\Theta\sqrt{\Theta}}{\Theta(\Theta+1)}\left\{\frac{\tfrac{3}{2}\Theta'}{\Theta}-\frac{\Theta'}{\Theta+1}-\frac{\Theta''}{\Theta'}\right\}d\theta$$

$$=\frac{\Theta(\Theta-1)}{\Theta+1}\left\{-\frac{3}{\Theta}+\frac{2}{\Theta+1}+\frac{2\Theta''}{\Theta'^2}\right\}d\theta.$$

But writing

$$\xi=\frac{2\Theta(\Theta-1)}{\Theta'(\Theta+1)},$$

we have

$$d\xi=\frac{2\Theta(\Theta-1)}{\Theta'(\Theta+1)}\left\{\frac{\Theta'}{\Theta}+\frac{\Theta'}{\Theta-1}-\frac{\Theta'}{\Theta+1}-\frac{\Theta''}{\Theta'}\right\}d\theta$$

$$=\frac{\Theta(\Theta-1)}{\Theta+1}\left\{\frac{2}{\Theta}+\frac{2}{\Theta-1}-\frac{2}{\Theta+1}-\frac{2\Theta''}{\Theta'^2}\right\}d\theta,$$

and thence

$$d\xi+\cot M\,d\mu=\frac{\Theta(\Theta-1)}{\Theta+1}\left\{-\frac{1}{\Theta}+\frac{2}{\Theta-1}\right\}d\theta,\ =d\theta,$$

and consequently

$$\xi+\int\cot M\,d\mu=\theta,$$

and the value of z thus is

$$z=\theta+\frac{2\Theta}{\Theta'}\,\frac{T-\Theta}{T+\Theta},$$

which completes the identification.

Bonnet's formulæ just referred to, making a slight change of notation and correcting a sign, are

$$x=\frac{\Gamma'\sin\theta}{\cos i(c+\Theta)},$$

$$y=\frac{\Gamma'\cos\theta}{\cos i(c+\Theta)},$$

$$z=\Gamma+i\Gamma'\tan i(c+\Theta),$$

where Γ, Θ are arbitrary functions of the parameters c, θ respectively. To identify these with Joachimsthal's, write

$$\sin\lambda=\sin\theta,\quad \cos M=i\cot ic,\quad \cos L=i\cot i\Theta,\quad \mu=-\Gamma'\operatorname{cosec} ic,$$

$$\cos\lambda=\cos\theta,\quad \sin M=\operatorname{cosec} ic,\quad \sin L=\operatorname{cosec} i\Theta,$$

$$\cot M=i\cos ic,\quad \cot L=i\cos i\Theta;$$

we have

$$x=\frac{\mu\operatorname{cosec} i\Theta\sin\theta}{1-\cot ic\cot i\Theta}=\frac{-\mu\sin ic\sin\theta}{\cos i(c+\Theta)},\ =\frac{\Gamma'\sin\theta}{\cos i(c+\Theta)};$$

and similarly

$$y=\frac{\Gamma'\cos\theta}{\cos i(c+\Theta)}.$$

Moreover, the first term of z is

$$\frac{\mu i \cot i\Theta \operatorname{cosec} ic}{1-\cot ic \cot i\Theta} = \frac{-\mu i \cos i\Theta}{\cos i\,(c+\Theta)} = \frac{i\Gamma' \cos i\Theta}{\sin ic \cos i\,(c+\Theta)};$$

or since $i\Theta = i\,(c+\Theta) - ic$, and thence

$$\cos i\Theta = \cos ic \cos i\,(c+\Theta) + \sin ic \sin i\,(c+\Theta),$$

this is

$$= i\Gamma' \,\{\cot ic + \tan i\,(c+\Theta)\},$$

and we have

$$z = i\Gamma' \tan i\,(c+\Theta) + i\Gamma' \cot ic + \int \cot M\, d\mu.$$

But from the equation $\mu = -\Gamma' \operatorname{cosec} ic$, we obtain

$$d\mu = (-\Gamma'' \operatorname{cosec} ic + i\Gamma' \operatorname{cosec} ic \cot ic)\, dc\,;$$

whence

$$\cot M\, d\mu = (-i\Gamma'' \cot ic - \Gamma' \cot^2 ic)\, dc,$$

and thence

$$d\left(i\Gamma' \cot ic + \int \cot M\, d\mu\right) = (i\Gamma'' \cot ic + \Gamma' \operatorname{cosec}^2 ic)\, dc$$
$$+ (-i\Gamma'' \cot ic - \Gamma' \cot^2 ic)\, dc, \; = \Gamma' dc\,;$$

that is,

$$i\Gamma' \cot ic + \int \cot M\, d\mu = \Gamma,$$

and consequently

$$z = \Gamma + i\Gamma' \tan i\,(c+\Theta),$$

which completes the identification of Bonnet's formula with Joachimsthal's.

SS, The Sets of Curves of Curvature each Spherical.

The six equations are

$$\begin{aligned}
&A^2 + B^2 + C^2 &&= 1,\\
&x^2 + y^2 + z^2 - 2ax - 2by - 2cz - 2u &&= 0,\\
&A\,(x-a) + B\,(y-b) + C\,(z-c) - l &&= 0,\\
&x^2 + y^2 + z^2 - 2\alpha x - 2\beta y - 2\gamma z - 2v &&= 0,\\
&A\,(x-\alpha) + B\,(y-\beta) + C\,(z-\gamma) - \lambda &&= 0,\\
&A dx + B dy + C dz &&= 0\,;
\end{aligned}$$

the condition being

$$a\alpha + b\beta + c\gamma - l\lambda + u + v = 0.$$

The cases are

	a	b	c	l	u	α	β	γ	λ	v
SS, 1°	0	0	0	l	$\frac{1}{2}(ml+m')$	α	β	γ	$\frac{1}{2}m$	$-\frac{1}{2}m'$
SS, 2°	0	0	c	l	$\frac{1}{2}(ml+m')$	α	β	0	$\frac{1}{2}m$	$-\frac{1}{2}m'$
SS, 3°	0	0	c	$mc+\frac{1}{2}m'$	$-\frac{1}{2}m''c-m'''$	α	β	$m\lambda+\frac{1}{2}m''$	λ	$\frac{1}{2}m'\lambda+m'''$
SS, 4°	0	b	c	$mc+m'$	$mm''c+m'm''-m'''$	α	0	γ	$\frac{1}{m}\gamma+m''$	$\frac{m'}{m}\gamma+m'''$,

where m, m', m'', m''' are constants; b, c, l functions of t; $\alpha, \beta, \gamma, \lambda$ functions of θ.

SS, 1° gives circles (i.e. the curves of curvature of one set are circles).

SS, 2° is Serret's first case of SS.

SS, 3° gives circles.

SS, 4° is Serret's second case of SS.

SS, 2° = *Serret's First Case of SS.*

Writing for convenience $m'=-f^2$, the six equations are

$$\begin{aligned} &A^2+B^2+C^2 &&=1,\\ &x^2+y^2+z^2-2cz-ml+f^2 &&=0,\\ &Ax+By+C(z-c)-l &&=0,\\ &x^2+y^2+z^2-2\alpha x-2\beta y-f^2 &&=0,\\ &A(x-\alpha)+B(y-\beta)+C(z-\lambda) &&=0,\\ &Adx+Bdy+Cdz &&=0,\end{aligned}$$

where m, f are constants; c, l are functions of t; α, β, λ functions of θ. The first set of spheres have no points in common, but the second set have in common the two points $x=0$, $y=0$, $z=\pm f$. Hence inverting (by reciprocal radius vectors) with one of these points, say $(0, 0, f)$ as centre, the spheres of the first set will continue spheres, but the spheres of the second set will be changed into planes, and the required surface is thus the inversion of a surface PS, which is in fact PS, 3: say this surface PS is the "Inversion" of SS. We invert by the formulæ

$$x=\frac{K^2X}{\Omega},\quad y=\frac{K^2Y}{\Omega},\quad z-f=\frac{K^2(Z-f)}{\Omega},$$

where $\Omega=X^2+Y^2+(Z-f)^2$.

Writing the equation for the second set of spheres in the form

$$x^2+y^2+(z-f)^2-2\alpha x-2\beta y+2f(z-f)=0,$$

the transformed equation is at once found to be

$$-2\alpha X-2\beta Y+2f(Z-f)+K^2=0,$$

or say

$$\alpha X+\beta Y-fZ+f^2-\tfrac{1}{2}K^2=0;$$

viz. this gives the planes of the Inversion.

Similarly for the first set of spheres, writing the equation in the form

$$x^2+y^2+(z-f)^2+2(f-c)(z-f)+2f(f-c)-ml=0,$$

the transformed equation is found to be

$$\{2f(f-c)-ml\}\{X^2+Y^2+(Z-f)^2\}+2(f-c)K^2(Z-f)+K^4=0;$$

viz. this is

$$\{2f(f-c)-ml\}(X^2+Y^2+Z^2)+2Z\{(f-c)(-2f^2+K^2)+fml\}$$
$$+\{2f(f-c)(f^2-K^2)-f^2ml+K^4\}=0,$$

which gives the spheres of the Inversion. The two equations take a more simple form if we write therein $K^2=2f^2$; viz. they then become

$$\alpha X+\beta Y-fZ=0,$$
$$(2f^2-2fc-ml)(X^2+Y^2+Z^2)+2Zfml+f^2(2f^2+2cf-ml)=0;$$

or, say these are

$$\begin{cases}\dfrac{\alpha X+\beta Y-fZ}{\sqrt{\alpha^2+\beta^2+f^2}}=0,\\[2mm] X^2+Y^2+Z^2+\dfrac{2fml}{2f^2-2fc-ml}Z+\dfrac{f^2(2f^2+2cf-ml)}{2f^2-2fc-ml}=0.\end{cases}$$

Interchanging the parameters so as to have t in the first equation and θ in the second equation, these are of the form

$$aX+bY+cZ=0,$$
$$X^2+Y^2+Z^2-2\gamma Z-2v=0,$$

where $a^2+b^2+c^2=1$; and the Inversion is thus a surface PS, 3^0.

SS, 4^0 = *Serret's Second Case of SS.*

Writing for convenience $m'=-mf$, $m'''=\tfrac{1}{2}(e^2+f^2)$, $mm''=-g$, and therefore $m'm''=fg$, the six equations are

$$\begin{aligned}
&A^2+B^2+C^2 &&=1,\\
&x^2+y^2+z^2-2by-2c(z-g)-2fg+e^2+f^2 &&=0,\\
&Ax+B(y-b)+C(z-c)+m(f-c) &&=0,\\
&x^2+y^2+z^2-2\alpha x-2\gamma(z-f)-e^2-f^2 &&=0,\\
&A(x-\alpha)+By+C(z-\gamma)-\frac{1}{m}(g+\gamma) &&=0,\\
&A\,dx+B\,dy+C\,dz &&=0,
\end{aligned}$$

where e, f, g, m are constants; b, c are functions of t; α, γ functions of θ.

The spheres of the first set pass all of them through the two points

$$x = \pm\sqrt{2fg - e^2 - f^2 - g^2},\quad y = 0,\quad z = g,$$

and those of the second set pass all of them through the two points

$$x' = 0,\quad y' = \pm e,\quad z' = f,$$

where observe that these are such that

$$(x - x')^2 + (y - y')^2 + (z - z')^2 = 0;$$

viz. the distance of each point of the first pair from each point of the second pair is $= 0$. The pairs of points are one real, the other imaginary: but this is quite consistent with the reality of the spheres.

The first pair of points lies in a line parallel to the axis of x, meeting the axis of z at the point $z = g$; and the second pair in a line parallel to the axis of y, cutting the axis of z at the point $z = f$. It is clear that we can, without loss of generality, by moving the origin along the axis of z, in effect make g to be $= -f$; the equations of the two sets of spheres thus become

$$x^2 + y^2 + z^2 - 2by - 2c\,(z + f) + e^2 + 3f^2 = 0,$$
$$x^2 + y^2 + z^2 - 2\alpha x - 2\gamma\,(z - f) - e^2 - f^2 = 0,$$

or, if in these equations for e^2 we write $e^2 - 2f^2$, the equations become

$$x^2 + y^2 + z^2 - 2by - 2c\,(z + f) + e^2 + f^2 = 0,$$
$$x^2 + y^2 + z^2 - 2\alpha x - 2\gamma\,(z - f) - e^2 + f^2 = 0,$$

which are very symmetrical forms.

The spheres of the first set pass through the two points

$$\pm\sqrt{-e^2 - 2f^2},\quad 0\ ,\quad -f,$$

and those of the second set through the two points

$$0\ ,\quad \pm\sqrt{e^2 - 2f^2},\quad f,$$

where, of course, the two pairs of points are related as is mentioned above.

By taking as centre of inversion a point of the first pair, we invert the first set of spheres into planes and the second set into spheres; and similarly, by taking a point of the second pair, we invert the first set of spheres into spheres and the second set into planes. By reason of the symmetry of the system, it is quite indifferent which point is chosen; and taking it to be a point of the second pair, and writing for convenience $n = \sqrt{e^2 - 2f^2}$ (n is, in fact, the quantity originally denoted by e), then the points of the first pair are

$$\pm\sqrt{-n^2 - 4f^2},\quad 0,\quad -f,$$
$$0\ ,\quad \pm n,\quad f,$$

and I take for centre of inversion the point $(0, n, f)$.

Observe that, if $e=0$, $f=0$, then the four points coincide at the origin, and taking this as centre of inversion, the two sets of spheres are each changed into planes, and the Inversion of the surface *SS* is thus a surface *PP*; this particular case will be considered further on, but I first consider the general case.

The formulæ of inversion are

$$x=\frac{K^2X}{\Omega},\quad y-n=\frac{K^2(Y-n)}{\Omega},\quad z-f=\frac{K^2(Z-f)}{\Omega},$$

where

$$\Omega=X^2+(Y-n)^2+(Z-f)^2.$$

Writing the equation of the second set of spheres in the form

$$x^2+(y-n)^2+(z-f)^2-2\alpha x+2n(y-n)+2(f-\gamma)(z-f)=0,$$

the transformed equation is

$$-\alpha X+n(Y-n)+(f-\gamma)(Z-f)+\tfrac{1}{2}K^2=0,$$

which gives the planes of the Inversion.

Similarly, writing the equation of the first set of spheres in the form

$$x^2+(y-n)^2+(z-f)^2+2(n-b)(y-n)+2(f-c)(z-f)+2n^2-2bn+4f(f-c)=0,$$

the transformed equation is

$$\{n^2-bn+2f(f-c)\}\{X^2+(Y-n)^2+(Z-f)^2\}$$
$$+K^2\{(n-b)(Y-n)+(f-c)(Z-f)\}+\tfrac{1}{2}K^4=0,$$

which gives the spheres of the Inversion.

Changing the origin, the two equations may be written

$$-\alpha X+nY+(f-\gamma)Z+\tfrac{1}{2}K^2=0,$$
$$\{n^2-bn+2f(f-c)\}(X^2+Y^2+Z^2)+K^2\{(n-b)Y+(f-c)Z\}+\tfrac{1}{2}K^4=0.$$

I stop to consider a particular case. Suppose $n=0$; the equations are

$$-\alpha X+(f-\gamma)Z+\tfrac{1}{2}K^2=0,$$
$$X^2+Y^2+Z^2-\frac{bK^2}{2f(f-c)}Y+\frac{K^2}{2f}Z+\frac{K^4}{4f(f-c)}=0,$$

or, interchanging herein Y and Z, they are

$$-\alpha X+(f-\gamma)Y+\tfrac{1}{2}K^2=0,$$
$$X^2+Y^2+Z^2+\frac{K^2}{2f}Y-\frac{bK^2}{2f(f-c)}Z+\frac{K^4}{4f(f-c)}=0;$$

and if for Y we write $Y - \frac{K^2}{4f}$, then the equations become

$$-\alpha X + (f-\gamma) Y + K^2 \frac{f+\gamma}{4f} = 0,$$

$$X^2 + Y^2 + Z^2 - \frac{bK^2}{2f(f-c)} Z + \frac{K^4 (5f-4c)}{f^2 (f-c)} = 0,$$

viz. interchanging the parameters so as to have t in the first equation and θ in the second equation, these are of the form

$$aX + bY = lm,$$

$$X^2 + Y^2 + Z^2 - 2\theta Z = 2v,$$

which belong to PS, 4°. Hence, in this particular case, $n = 0$; the Inversion is PS, 4°.

Reverting to the general case, and to the two equations obtained above, observe that, in the second of the two equations, the terms in Y, Z have the variable coefficients $n - b$ and $f - c$, so that it does not at first sight seem as if these terms could by a transformation of coordinates be reduced to a single term.

But if, again changing the origin, we write $Y - \frac{\frac{1}{2}K^2}{n}$ for Y, the two equations become

$$-\alpha X + nY + (f-\gamma) Z = 0,$$

$$\{n^2 - bn + 2f(f-c)\} (X^2 + Y^2 + Z^2) + \frac{K^2}{n} (f-c)(-2fY + nZ)$$
$$+ \frac{K^4}{4n^2} \{n^2 + bn + 2f(f-c)\} = 0,$$

where, in the second equation, the terms in Y, Z present themselves in the combination $-2fY + nZ$ with the constant coefficients $-2f$ and n. Hence writing

$$\sqrt{n^2 + 4f^2}\, Y = nY' - 2fZ',$$

$$\sqrt{n^2 + 4f^2}\, Z = 2fY' + nZ',$$

and consequently $-2fY + nZ = \sqrt{n^2 + 4f^2}\, Z'$, and (after the transformation) removing the accents, the equations become

$$-\alpha X + \frac{1}{\sqrt{n^2 + 4f^2}} [\{n^2 + 2f(f-\gamma)\} Y - n(f+\gamma) Z] = 0,$$

$$X^2 + Y^2 + Z^2 + \frac{K^2 (f-c) \sqrt{n^2 + 4f^2}}{n \{n^2 - bn + 2f(f-c)\}} Z + \frac{K^4 \{n^2 + bn + 2f(f-c)\}}{n^2 \{n^2 - bn + 2f(f-c)\}} = 0,$$

viz. interchanging the parameters so as to have t in the first equation and θ in the second equation, these are of the form

$$aX + bY + cZ = 0,$$

$$X^2 + Y^2 + Z^2 - 2m\phi Z = \theta,$$

which belong to the case PS, 3°. Hence, in this general case, the Inversion is a surface PS, 3°.

I have spoken above of the particular case $e=0$, $f=0$: here the equations of the two sets of spheres are

$$x^2+y^2+z^2-2by-2cz=0,$$
$$x^2+y^2+z^2-2\alpha x-2\gamma z=0,$$

which have the origin as a common point. Taking this as the centre of inversion, or writing

$$x=\frac{K^2X}{\Omega},\quad y=\frac{K^2Y}{\Omega},\quad z=\frac{K^2Z}{\Omega},\ \text{where } \Omega=X^2+Y^2+Z^2,$$

the transformed equations are

$$bY+cZ-\tfrac{1}{2}K^2=0,$$
$$\alpha X\quad+\gamma Z-\tfrac{1}{2}K^2=0,$$

or, interchanging X and Y, say

$$bX\quad+cZ-\tfrac{1}{2}K^2=0,$$
$$\alpha Y+\gamma Z-\tfrac{1}{2}K^2=0,$$

which are of the form

$$X\quad+tZ-P\quad=0,$$
$$Y+\theta Z-\Pi\quad=0,$$

belonging to a surface PP, 3°. Hence, in this case, the Inversion is a surface PP, 3°.

It thus appears that the surface SS, 4° has an Inversion which is either PS, 3°, PS, 4° or PP, 3°. The inversion has in some cases to be performed in regard to an imaginary centre of inversion.

It was previously shown that the surface SS, 3° had an Inversion PS, 3°, and we thus arrive at the conclusion that a surface SS, with its two sets of curves of curvature each spherical, is in every case the Inversion of a surface PS with one set plane and the other spherical, or else of a surface PP with each set plane. Serret notices that the centre of inversion may be imaginary: this (he says) presents no difficulty, but he adds that it is easy to see that the centres of inversion may be taken to be real, provided that we join to the surfaces thus obtained all the parallel surfaces.

It seems to me that there is room for further investigation as to the surfaces SS: first, without employing the theory of inversion, it would be desirable to obtain the several forms by direct integration, as was done in regard to the surfaces PP and PS; secondly, starting from the several surfaces PP and PS considered as known forms, it would be desirable to obtain from these, by inversion in regard to an arbitrary centre, or with regard to a centre in any special position, the several forms of the surfaces SS. But I do not at present propose to consider either of these questions.

In conclusion, I remark that I have throughout assumed Serret's *negative* conclusions, viz. that the several cases, other than those considered in the present memoir, give only developable surfaces, or else surfaces having circles for one set of their curves of curvature. These being excluded from consideration, there remain

PP, Serret's two cases PP, 1^0, PP, 3^0;

PS, his three cases PS, 1^0, PS, 3^0, PS, 4^0;

SS, his two cases SS, 2^0 and SS, 4^0;

but PP, 1^0 is a particular case of, and so may be included in, PP, 3^0; and similarly PS, 1^0 is a particular case of, and may be included in, PS, 3^0; the cases considered thus are

PP, 3^0; PS, 3^0, PS, 4^0; SS, 2^0 and SS, 4^0.

It would however appear by what precedes that the case SS, 4^0 includes several cases which it is possible might properly be regarded as distinct; and the classification of the surfaces SS can hardly be considered satisfactory; it would seem that there should be at any rate 3 cases, viz. the surfaces which are the Inversions of PP, 3^0, PS, 3^0 and PS, 4^0 respectively.

I regard the present memoir as a development of the analytical theory of the surfaces PP, 3^0, PS, 3^0 and PS, 4^0.

887.

ON THE THEORY OF GROUPS.

[From the *American Journal of Mathematics*, vol. XI. (1889), pp. 139—157.]

I REFER to my papers on the theory of groups as depending on the symbolic equation $\theta^n = 1$, *Phil. Mag.*, vol. VII. (1854), pp. 40—47 and 408, 409, [125, 126]; also vol. XVIII. (1859), pp. 34—37, [243]; and "On the Theory of Groups," *American Journal of Mathematics*, vol. I. (1878), pp. 50—52, and "The Theory of Groups: Graphical Representation," *ibid.*, pp. 174—176, [694]; also to Mr Kempe's "Memoir on the Theory of Mathematical Form," *Phil. Trans.*, vol. CLXXVII. (1886), pp. 1—70, see the section "Groups containing from one to twelve units," pp. 37—43, with the diagrams given therein. Mr Kempe's paper has recalled my attention to the method of graphical representation explained in the second of the two papers of 1878, and has led me to consider, in place of a diagram as there given for the independent substitutions, a diagram such as those of his paper, for all the substitutions. I call this a colourgroup; viz. for the representation of a substitution-group of s substitutions upon the same number of letters, or say of the order s, we employ a figure of s points (in space or in a plane) connected together by coloured lines, and called a colourgroup.

I remark that up to $s = 11$, the first case of any difficulty is that of $s = 8$, and that the 5 groups of this order were determined in my papers of 1854 and 1859. For the order 12, Mr Kempe has five groups, but one of these is non-existent, and there is a group omitted; the number is thus $= 5$.

The colourgroup consists of s points joined in pairs by $\frac{1}{2}s(s-1)$ coloured lines under prescribed conditions. A line joining two points is in general regarded as a vector drawn *from* one *to* the other of the two points; the currency is shown by an arrow, and in speaking of a line ab we mean the line from a to b. But we may have a line regarded as a double line, drawn from each to the other of the two points; the arrow is then omitted, and in speaking of such a line ab we mean the line from b to a and from a to b. A fresh condition is that for a given colour there shall be one and only one line *from* each of the points, and one and only one line *to* each of the points. We may have through two points a, b only the line ab of the given colour; this is then a double line regarded as drawn from a to b and from b to a; and there is thus one and only one line of the colour from each of these points and to each of these points. The condition implies that the lines of a given colour form either a single polygon or a set of polygons, with a continuous currency round each polygon; for instance, there may be a pentagon *abcde*, meaning thereby the pentagon formed by the lines drawn from a to b, from b to c, from c to d, from d to e, and from e to a. An arrow on one of the sides is sufficient to indicate the currency. In the case of a double line we have a polygon of two points, or say a digon.

There is a further condition which, after the necessary explanation of the meaning of the terms, may be concisely expressed as follows: Each route must be of independent effect, and (as will readily be seen) this implies that the lines of a given colour must form either a single polygon or else two or more polygons each of the same number of points: thus if $\theta = k\theta_1$, they may form k θ_1-gons; in particular, if θ be even, they may form $\frac{1}{2}\theta$ digons.

To explain the foregoing statement, first as to the term "route." I denote the several colours by capital letters, R = red, G = green, B = blue, &c. Any capital or combination of capitals determines a route; R means go along a red line; $RRBG$, go along a red line, a red line, a blue line, a green line, and so in other cases. Given the starting point, or initial, the route determines the several points passed through, and the point arrived at, or terminal: thus $aRRBG = abefk, = k$, means that the route $RRBG$ leads from a through b, e, f to k, viz. that the red line from a leads to b, the red line from b leads to e, the blue line from e leads to f, and the green line from f leads to k. We may give in this way the Itinerary, or write simply $aRRBG = k$, meaning that the route leads from a to k. We may of course write R^2 for RR, and so in other cases. A single capital, as already mentioned, is a route, but it may for distinction be called a stage. A stage, and thence also a route, may be *reversed*; R^{-1} means go along the red line drawn to the point; if $aR = b$, then $bR^{-1} = a$; and so if $aRRBG = abefk, = k$, then $kG^{-1}B^{-1}R^{-1}R^{-1} = kfeba, = a$; $R^{-1}R^{-1} = R^{-2}$, and so in other cases.

The effect of a route depends in general on the initial point: thus, a route may lead from a point a to itself, or say it may be a circuit from a; and it may not be a circuit from another point b. And similarly, two different routes each leading from a point a, to one and the same point x, or say two routes equivalent for the initial point a, may not be equivalent for a different initial point b. Thus we cannot in general say simpliciter that a route is a circuit, or that two different routes are equivalent. But the figure may be such as to render either of these locutions, and if either, then each of them, admissible. For it is easy to see that if every route which is a circuit from any one initial point is also a circuit from every other initial point, then two routes which are equivalent for any one initial point will be equivalent for every other initial point. And conversely, if in every case where two different routes are equivalent for any one initial point, they are equivalent for every other initial point, then every route which is a circuit from any one initial point is a circuit from every other initial point; and we express this by saying that every route is of independent effect: this explains the meaning of the foregoing statement of the condition which is to be satisfied by a colourgroup.

It is at once evident that a colourgroup, *quâ* figure where each route is of independent effect, furnishes a graphical representation of the substitution-group and gives the square by which we define such group. For, in the colourgroup of s points, we have the route from a point to itself and the routes to each of the other $(s-1)$ points, in all s non-equivalent routes; and if starting from a given arrangement, say *abcd* ..., of the s points, we go by one of these routes from the several points *a*, *b*, *c*, *d*, ... successively, we obtain a different arrangement of these points. Observe that this is so; the same point cannot occur twice, for if it did, there would be a route leading from two different points *b*, *f* to one and the same point *x*, or the reverse route from *x* would lead to two different points *b*, *f*. The route from a point to itself which leaves each point unaltered, and thus gives the primitive arrangement *abcd* ..., may be called the route 1. Taking this route and the other $(s-1)$ routes successively, we obtain s different arrangements of the points, or say a square, each line of which is a different arrangement of the points. And not only are the arrangements different, but we cannot have the same point twice in any column, for this would mean that there were two different routes leading from a point to one and the same point *x*; hence each column of the square will be an arrangement of the s points. We have thus the substitution-group of the s points or letters; the s routes, or say the route 1 and the other $(s-1)$ routes, are the substitutions of the group.

The complete figure is called the colourgroup. As already mentioned, the lines of any colour form either a single polygon or two or more polygons each of the same number of points. The number of lines of a given colour is thus $= s$, or when the polygons are digons (which implies s even), the number is $=\frac{1}{2}s$. The number of colours is thus $=\frac{1}{2}(s-1)$ at least, and $=(s-1)$ at most. A general description of the figure may be given as in the annexed Table. Thus, for the group $6B$, we have

$$\begin{array}{rr} R.\ \ 2\ \text{3gons} = & 6 \\ B,\ G,\ Y.\ \ (3\ \text{2gons})^3 = & 9 \\ \hline & 15\,; \end{array}$$

we have the red lines forming two trigons, 6 lines, and the blue, green and yellow lines each forming three digons, together 3×3, $=9$ lines, in all 15, $=\frac{1}{2}6.5$ lines. Such description, however, does not indicate the currencies, and it is thus insufficient for the determination of the figure. But the figure is completely determined by means of the substitutions as given in the outside column of the square; thus $R=(abc)(dfe)$ shows that the red lines form the two triangles *abc*, *dfe* with these currencies, $G=(ad)(be)(cf)$, that the green lines form the three digons *ad*, *be*, *cf*, and so for the other two colours B and Y.

The line of a colour may be spoken of as a colour, and the lines of a colour or of two or more colours as a colourset. The colourset either does not connect together all the points, and it is then a broken set; or it does connect together all the points, and it is then a bondset. A bondset not containing any superfluous colour is termed a bond, viz. a bond is a colourset which connects together all the points, but which is moreover such that if any one of the colours be omitted it becomes a broken set. The word colour is used as a prefix, colourset as above, colourbond, &c., and so also with a numeral, a twocolourbond is a bond with two colours, and so in other cases. Observe that we may very well have for instance a threecolourbond, and also a twocolour or a onecolourbond, only the colours or colour hereof must not be included among those of the threecolourbond, for this would then contain a superfluous colour or colours and would not be a bond.

A colourgroup may contain a onecolourbond, viz. this is the case when all the points form a single polygon; it is then said to be unibasic. If it contains no onecolourbond but contains a twocolourbond, it is bibasic; if it contains no onecolourbond or twocolourbond but contains a threecolourbond, it is tribasic, and so on. In all cases, the number of bonds (onecolour-, twocolour-, &c.) may very well be and in general is greater than one; thus a unibasic colourgroup will in general contain several onecolourbonds, a bibasic colourgroup several twocolourbonds, and so on.

The bond of the proper number of colours completely determines the colourgroup; in fact, the colourbond gives the route from any one point to each of the other $(s-1)$ points; that is, it determines all the s routes, and consequently the colourgroup. The only type of onecolourbond is the polygon of the s points; we have thus, for any value whatever of s, a unibasic colourgroup which may be called sA. The theory is well known. If s be a prime number, the number of colours is $=\frac{1}{2}(s-1)$, each colour gives a polygon through the s points, so that we have here only onecolourbonds; but in other cases we have broken sets, and there will be in general (but not for all such values of s) twocolourbonds. Observe, moreover, that, for s a prime number, the only colourgroup is the foregoing unibasic group sA. I have just employed, and shall again do so, the word type; the sense in which it is used does not, I think, require explanation.

Passing next to the bibasic colourgroups sB: there will be in general, for a given composite value of s, several of these, and in the absence of a more complete classification they may be called $sB1$, $sB2$, &c. In regard hereto, observe that, supposing for a given value of s that we know all the different types of twocolourbond, each one of these gives rise to a group; but this is not in every case a group sB: any twocolourbond contained in the corresponding group sA would give rise to the group sA which contained it, and not to a group sB. We have thus, in the first instance, to reject those twocolourbonds which are contained in the group sA. But attending only to the remaining twocolourbonds, these give rise each of them to a group sB, but the groups thus obtained are not in every case distinct groups. For looking at the converse question, suppose that, for a given value of s, we know the group sA and also the several groups sB. In any one of these groups, combining in pairs the several colours hereof RG, RY, GY, &c., we ascertain how many of these combinations are distinct types of twocolourbond, and in this manner reproduce the whole series of types of twocolourbond, not in general singly, but in sets, those which arise from sA, those which arise from $sB1$, those which arise from $sB2$, &c.; and we thus have (it may be) several types of two-colourbond each leading to the unibasic group sA, several types each leading to the bibasic group $sB1$, several each leading to $sB2$, and so on.

The like considerations would apply to the tribasic colourgroups sC. Supposing that we had, for a given value of s, the several distinct types of threecolourbond, it would be necessary first to exclude from consideration those which give rise to a unibasic group sA or a bibasic group sB, and then to consider what sets out of the remaining types give rise to distinct tribasic groups sC. But in the table we have only one case $8C$ of a tribasic group.

I give now a table of the several groups $s=2$ to 12, viz. these are as above: A, unibasic; B, bibasic; C, tribasic; the several groups being

$2A$,	$3A$,	$4A$,	$5A$,	$6A$,	$7A$,	$8A$,	$9A$,	$10A$,	$11A$,	$12A$,
		$4B$,		$6B$,		$8B1$,	$9B$,	$10B$,		$12B1$,
						$8B2$,				$12B2$,
						$8B3$,				$12B3$,
						$8C$,				$12B4$,

in all 23 groups.

Table of the Groups 2 to 12.

$2A$ — 1 colour.

a	b	1	$=1$	
b	a	$R=(ab)=R$		

R. 1 digon $\frac{1}{\overline{1}}$

$3A$ — 1 colour.

a	b	c	$1=1$	$=1$
b	c	a	$R=(abc)=R$	
c	a	b	$R^2=(acb)=R^{-1}$	

R. 1 3gon $\frac{3}{\overline{3}}$

$4A$ — 2 colours.

a	b	c	d	$1=1$	$=1$
b	c	d	a	$R=(abcd)$	$=R$
c	d	a	b	$R^2=(ac)(bd)$	$=G$
d	a	b	c	$R^3=(adcb)$	$=R^{-1}$

R. 1 4gon 4
G. 2 digons 2
$\overline{6}$

4B

3 colours.

a	b	c	d
b	a	d	c
c	d	a	b
d	c	b	a

$$1 = 1 = 1$$
$$R = (ab)(cd) = R$$
$$G = (ac)(bd) = G$$
$$RG = (ad)(bc) = Y$$

$R, G, Y.$ (2 digons)³ $\dfrac{6}{6}$

5A

2 colours.

a	b	c	d	e
b	c	d	e	a
c	d	e	a	b
d	e	a	b	c
e	a	b	c	d

$$1 = 1 = 1$$
$$R = (abcde) = R$$
$$R^2 = (acebd) = G$$
$$R^3 = (adbec) = G^{-1}$$
$$R^4 = (aedcb) = R^{-1}$$

$R, G.$ (1 5gon)² $\dfrac{10}{10}$

6A

3 colours.

a	b	c	d	e	f
b	c	d	e	f	a
c	d	e	f	a	b
d	e	f	a	b	c
e	f	a	b	c	d
f	a	b	c	d	e

$$1 = 1 = 1$$
$$R = (abcdef) = R$$
$$R^2 = (ace)(bdf) = G$$
$$R^3 = (ad)(be)(cf) = Y$$
$$R^4 = (aec)(bfd) = G^{-1}$$
$$R^5 = (afedcb) = Y^{-1}$$

R.   1 6gon   6
G.   2 3gons   6
Y.   3 digons   3
                15

6B

4 colours.

a	b	c	d	e	f
b	c	a	f	d	e
c	a	b	e	f	d
d	e	f	a	b	c
e	f	d	c	a	b
f	d	e	b	c	a

$$1 = 1 = 1$$
$$R = (abc)(dfe) = R$$
$$R^2 = (acb)(def) = R^{-1}$$
$$G = (ad)(be)(cf) = G$$
$$RG = (ae)(bf)(cd) = Y$$
$$R^2G = (af)(bd)(ce) = B$$

R.   2 3gons   6
G, Y, B.   (3 digons)³   9

7A

a	*b*	*c*	*d*	*e*	*f*	*g*
b	*c*	*d*	*e*	*f*	*g*	*a*
c	*d*	*e*	*f*	*g*	*a*	*b*
d	*e*	*f*	*g*	*a*	*b*	*c*
e	*f*	*g*	*a*	*b*	*c*	*d*
f	*g*	*a*	*b*	*c*	*d*	*e*
g	*a*	*b*	*c*	*d*	*e*	*f*

3 colours.

$1 = 1 \quad = 1$

$R = (abcdefg) = R$

$R^2 = (acegbdf) = G$

$R^3 = (adgcfbe) = Y$

$R^4 = (aebfcgd) = Y^{-1}$

$R^5 = (afdbgec) = G^{-1}$

$R^6 = (agfedcb) = R^{-1}$

R, G, Y. $(1\ \text{7gon})^3$ 21

21

8A

a	*b*	*c*	*d*	*e*	*f*	*g*	*h*
b	*c*	*d*	*e*	*f*	*g*	*h*	*a*
c	*d*	*e*	*f*	*g*	*h*	*a*	*b*
d	*e*	*f*	*g*	*h*	*a*	*b*	*c*
e	*f*	*g*	*h*	*a*	*b*	*c*	*d*
f	*g*	*h*	*a*	*b*	*c*	*d*	*e*
g	*h*	*a*	*b*	*c*	*d*	*e*	*f*
h	*a*	*b*	*c*	*d*	*e*	*f*	*g*

4 colours.

$1 = 1 \quad = 1$

$R = (abcdefgh) \quad = R$

$R^2 = (aceg)(bdhf) \quad = Y$

$R^3 = (adgbehcf) \quad = G$

$R^4 = (ae)(bf)(cg)(dh) = B$

$R^5 = (afchebgd) \quad = G^{-1}$

$R^6 = (agec)(bhfd) \quad = Y^{-1}$

$R^7 = (ahgfedcb) \quad = R^{-1}$

R, G. $(1\ \text{8gon})^2$ 16
Y. 2 4gons 8
B. 4 digons 4

28

8B1

a	b	c	d	e	f	g	h
b	c	d	a	f	g	h	e
c	d	a	b	g	h	e	f
d	a	b	c	h	e	f	g
e	f	g	h	a	b	c	d
f	g	h	e	b	c	d	a
g	h	e	f	c	d	a	b
h	e	f	g	d	a	b	c

5 colours.

$1\ \ =1\qquad\qquad =1$

$R\ \ =(abcd)(efgh)\qquad =R$

$R^2\ =(ac)(bd)(eg)(fh)=Y$

$R^3\ =(adcb)(ehgf)\quad =R^{-1}$

$G\ \ =(ae)(bf)(cg)(dh)=G$

$RG\ =(afch)(bgde)\quad =I$

$R^2G=(ag)(bh)(ce)(df)=B$

$R^3G=(ahcf)(bedg)\quad =I^{-1}$

R, I.	(2 4gons)²	16
Y, G, B.	(4 digons)³	12
		28

8B2

a	b	c	d	e	f	g	h
b	c	d	a	h	e	f	g
c	d	a	b	g	h	e	f
d	a	b	c	f	g	h	e
e	f	g	h	a	b	c	d
f	g	h	e	d	a	b	c
g	h	e	f	c	d	a	b
h	e	f	g	b	c	d	a

6 colours.

$1\ \ =1\qquad\qquad =1$

$R\ \ =(abcd)(ehgf)\qquad =R$

$R^2\ =(ac)(bd)(eg)(fh)=Y$

$R^3\ =(adcb)(efgh)\quad =R^{-1}$

$G\ \ =(ae)(bf)(cg)(dh)=G$

$RG\ =(af)(bg)(ch)(de)=I$

$R^2G=(ag)(bh)(ce)(df)=B$

$R^3G=(ah)(be)(cf)(dg)=O$

R.	2 4gons	8
Y, G, I, B, O.	(4 digons)⁵	20
		28

8*B*3

4 colours.

a	*b*	*c*	*d*	*e*	*f*	*g*	*h*	$1 \quad =1$	$=1$
b	*c*	*d*	*a*	*h*	*e*	*f*	*g*	$R \quad =(abcd)\,(efgh)$	$=R$
c	*d*	*a*	*b*	*g*	*h*	*e*	*f*	$R^2 \; =(ac)\,(bd)\,(ef)\,(gh)$	$=Y$
d	*a*	*b*	*c*	*f*	*g*	*h*	*e*	$R^3 \; =(adcb)\,(ehgf)$	$=R^{-1}$
e	*f*	*g*	*h*	*a*	*b*	*c*	*d*	$G \quad =(aecg)\,(bhdf)$	$=G$
f	*g*	*h*	*e*	*d*	*a*	*b*	*c*	$R^3G=(afch)\,(bedg)$	$=B$
g	*h*	*e*	*f*	*c*	*d*	*a*	*b*	$R^2G=(agce)\,(bfdh)$	$=G^{-1}$
h	*e*	*f*	*g*	*b*	*c*	*d*	*a*	$RG \; =(ahcf)\,(bgde)$	$=B^{-1}$

R, *G*, *B*. (2 4gons)^3 24
Y. 4 digons 4
28

8*C*

7 colours.

a	*b*	*c*	*d*	*e*	*f*	*g*	*h*	$1 \quad =1$	$=1$
b	*a*	*d*	*c*	*f*	*e*	*h*	*g*	$R \quad =(ab)\,(cd)\,(ef)\,(gh)$	$=R$
c	*d*	*a*	*b*	*g*	*h*	*e*	*f*	$G \quad =(ac)\,(bd)\,(eg)\,(fh)$	$=G$
d	*c*	*b*	*a*	*h*	*g*	*f*	*e*	$RG \; =(ad)\,(bc)\,(eh)\,(fg)$	$=B$
e	*f*	*g*	*h*	*a*	*b*	*c*	*d*	$Y \quad =(ae)\,(bf)\,(cg)\,(dh)$	$=Y$
f	*e*	*h*	*g*	*b*	*a*	*d*	*c*	$RY \; =(af)\,(be)\,(ch)\,(dg)$	$=I$
g	*h*	*e*	*f*	*c*	*d*	*a*	*b*	$GY \; =(ag)\,(bh)\,(ce)\,(df)$	$=O$
h	*g*	*f*	*e*	*d*	*c*	*b*	*a*	$RGY=(ah)\,(bg)\,(cf)\,(de)$	$=V$

R, *G*, *B*, *Y*, *I*, *O*, *V*. (4 digons)^7 28
28

9A

a	*b*	*c*	*d*	*e*	*f*	*g*	*h*	*i*	$1=1$	$=1$
b	*c*	*d*	*e*	*f*	*g*	*h*	*i*	*a*	$R=(abcdefghi)$	$=R$
c	*d*	*e*	*f*	*g*	*h*	*i*	*a*	*b*	$R^2=(acegibdfh)$	$=G$
d	*e*	*f*	*g*	*h*	*i*	*a*	*b*	*c*	$R^3=(adg)\,(beh)\,(cfi)$	$=Y$
e	*f*	*g*	*h*	*i*	*a*	*b*	*c*	*d*	$R^4=(aeidhcgbf)$	$=B$
f	*g*	*h*	*i*	*a*	*b*	*c*	*d*	*e*	$R^5=(afbgchdie)$	$=B^{-1}$
g	*h*	*i*	*a*	*b*	*c*	*d*	*e*	*f*	$R^6=(agd)\,(bhe)\,(cif)$	$=Y^{-1}$
h	*i*	*a*	*b*	*c*	*d*	*e*	*f*	*g*	$R^7=(ahfdbigec)$	$=G^{-1}$
i	*a*	*b*	*c*	*d*	*e*	*f*	*g*	*h*	$R^8=(aihgfedcb)$	$=R^{-1}$

4 colours.

R, G, B. $(1\ \text{9gon})^3$ 27
Y. 3 3gons 9
36

9B

a	*b*	*c*	*d*	*e*	*f*	*g*	*h*	*i*	1	$=1$
b	*c*	*a*	*e*	*f*	*d*	*h*	*i*	*g*	R	$=(abc)\,(def)\,(ghi)=R$
c	*a*	*b*	*f*	*d*	*e*	*i*	*g*	*h*	R^2	$=(acb)\,(dfe)\,(gih)=R^{-1}$
d	*e*	*f*	*g*	*h*	*i*	*a*	*b*	*c*	G	$=(adg)\,(beh)\,(cfi)=G$
e	*f*	*d*	*h*	*i*	*g*	*b*	*c*	*a*	RG	$=(aei)\,(bfg)\,(cdh)=B$
f	*d*	*e*	*i*	*g*	*h*	*c*	*a*	*b*	R^2G	$=(afh)\,(bdi)\,(ceg)=Y$
g	*h*	*i*	*a*	*b*	*c*	*d*	*e*	*f*	G^2	$=(agd)\,(bhe)\,(cif)=G^{-1}$
h	*i*	*g*	*b*	*c*	*a*	*e*	*f*	*d*	RG^2	$=(ahf)\,(bid)\,(cge)=Y^{-1}$
i	*g*	*h*	*c*	*a*	*b*	*f*	*d*	*e*	R^2G^2	$=(aie)\,(bgf)\,(chd)=B^{-1}$

4 colours.

R, G, B, Y. $(3\ \text{3gons})^4$ 36
36

10A

5 colours.

a	*b*	*c*	*d*	*e*	*f*	*g*	*h*	*i*	*j*	$1 = 1$	$= 1$
b	*c*	*d*	*e*	*f*	*g*	*h*	*i*	*j*	*a*	$R = (abcdefghij)$	$= R$
c	*d*	*e*	*f*	*g*	*h*	*i*	*j*	*a*	*b*	$R^2 = (acegi)\ (bdfhj)$	$= G$
d	*e*	*f*	*g*	*h*	*i*	*j*	*a*	*b*	*c*	$R^3 = (adgjcfibeh)$	$= Y$
e	*f*	*g*	*h*	*i*	*j*	*a*	*b*	*c*	*d*	$R^4 = (aeicg)\ (bfdjh)$	$= B$
f	*g*	*h*	*i*	*j*	*a*	*b*	*c*	*d*	*e*	$R^5 = (af)\ (bg)\ (ch)\ (di)\ (ej) = O$	
g	*h*	*i*	*j*	*a*	*b*	*c*	*d*	*e*	*f*	$R^6 = (agcie)\ (bhjdf)$	$= B^{-1}$
h	*i*	*j*	*a*	*b*	*c*	*d*	*e*	*f*	*g*	$R^7 = (ahebifcjgd)$	$= Y^{-1}$
i	*j*	*a*	*b*	*c*	*d*	*e*	*f*	*g*	*h*	$R^8 = (aigec)\ (bjhfd)$	$= G^{-1}$
j	*a*	*b*	*c*	*d*	*e*	*f*	*g*	*h*	*i*	$R^9 = (ajihgfedcb)$	$= R^{-1}$

R, *G*, *Y*, *B*. (1 10gon)[4] 40
O. 5 digons 5
45

10B

7 colours.

a	*b*	*c*	*d*	*e*	*f*	*g*	*h*	*i*	*j*	$1 = 1$	$= 1$
b	*c*	*d*	*e*	*a*	*j*	*f*	*g*	*h*	*i*	$R = (abcde)\ (fjihg)$	$= R$
c	*d*	*e*	*a*	*b*	*i*	*j*	*f*	*g*	*h*	$R^2 = (acebd)\ (figjh)$	$= Y$
d	*e*	*a*	*b*	*c*	*h*	*i*	*j*	*f*	*g*	$R^3 = (adbec)\ (fhjgi)$	$= Y^{-1}$
e	*a*	*b*	*c*	*d*	*g*	*h*	*i*	*j*	*f*	$R^4 = (aedcb)\ (fghij)$	$= R^{-1}$
f	*g*	*h*	*i*	*j*	*a*	*b*	*c*	*d*	*e*	$G = (af)\ (bg)\ (ch)\ (di)\ (ej) = G$	
g	*h*	*i*	*j*	*f*	*e*	*a*	*b*	*c*	*d*	$RG = (ag)\ (bh)\ (ci)\ (dj)\ (ef) = B$	
h	*i*	*j*	*f*	*g*	*d*	*e*	*a*	*b*	*c*	$R^2G = (ah)\ (bi)\ (cj)\ (df)\ (eg) = O$	
i	*j*	*f*	*g*	*h*	*c*	*d*	*e*	*a*	*b*	$R^3G = (ai)\ (bj)\ (cf)\ (dg)\ (eh) = V$	
j	*f*	*g*	*h*	*i*	*b*	*c*	*d*	*e*	*a*	$R^4G = (aj)\ (bf)\ (cg)\ (dh)\ (ei) = I$	

R, *Y*. (2 5gons)[2] 20
G, *B*, *O*, *V*, *I*. (5 digons)[5] 25
45

11A

5 colours.

a	*b*	*c*	*d*	*e*	*f*	*g*	*h*	*i*	*j*	*k*	$1 = 1$	$= 1$
b	*c*	*d*	*e*	*f*	*g*	*h*	*i*	*j*	*k*	*a*	$R = (abcdefghijk) = R$	
c	*d*	*e*	*f*	*g*	*h*	*i*	*j*	*k*	*a*	*b*	$R^2 = (acegikbdfhj) = G$	
d	*e*	*f*	*g*	*h*	*i*	*j*	*k*	*a*	*b*	*c*	$R^3 = (adgjbehkcfi) = Y$	
e	*f*	*g*	*h*	*i*	*j*	*k*	*a*	*b*	*c*	*d*	$R^4 = (aeibfjcgkdh) = B$	
f	*g*	*h*	*i*	*j*	*k*	*a*	*b*	*c*	*d*	*e*	$R^5 = (afkejdichbg) = O$	
g	*h*	*i*	*j*	*k*	*a*	*b*	*c*	*d*	*e*	*f*	$R^6 = (agbhcidjekf) = O^{-1}$	
h	*i*	*j*	*k*	*a*	*b*	*c*	*d*	*e*	*f*	*g*	$R^7 = (ahdkgcjfbie) = B^{-1}$	
i	*j*	*k*	*a*	*b*	*c*	*d*	*e*	*f*	*g*	*h*	$R^8 = (aifckhebjgd) = Y^{-1}$	
j	*k*	*a*	*b*	*c*	*d*	*e*	*f*	*g*	*h*	*i*	$R^9 = (ajhfdbkigec) = G^{-1}$	
k	*a*	*b*	*c*	*d*	*e*	*f*	*g*	*h*	*i*	*j*	$R^{10} = (akjihgfedcb) = R^{-1}$	

R, G, Y, B, O. $(1\ 11\text{gon})^5$ 55

55

12A

6 colours.

a	*b*	*c*	*d*	*e*	*f*	*g*	*h*	*i*	*j*	*k*	*l*	$1 = 1$	$= 1$
b	*c*	*d*	*e*	*f*	*g*	*h*	*i*	*j*	*k*	*l*	*a*	$R = (abcdefghijkl)$	$= R$
c	*d*	*e*	*f*	*g*	*h*	*i*	*j*	*k*	*l*	*a*	*b*	$R^2 = (acegik)\ (bdfhjl)$	$= G$
d	*e*	*f*	*g*	*h*	*i*	*j*	*k*	*l*	*a*	*b*	*c*	$R^3 = (adgj)\ (behk)\ (cfil)$	$= Y$
e	*f*	*g*	*h*	*i*	*j*	*k*	*l*	*a*	*b*	*c*	*d*	$R^4 = (aei)\ (bfj)\ (cgk)\ (dhl)$	$= B$
f	*g*	*h*	*i*	*j*	*k*	*l*	*a*	*b*	*c*	*d*	*e*	$R^5 = (afkdibglejch)$	$= O$
g	*h*	*i*	*j*	*k*	*l*	*a*	*b*	*c*	*d*	*e*	*f*	$R^6 = (ag)\ (bh)\ (ci)\ (dj)\ (ek)\ (fl) = V$	
h	*i*	*j*	*k*	*l*	*a*	*b*	*c*	*d*	*e*	*f*	*g*	$R^7 = (ahcjelgbidkf)$	$= O^{-1}$
i	*j*	*k*	*l*	*a*	*b*	*c*	*d*	*e*	*f*	*g*	*h*	$R^8 = (aie)\ (bjf)\ (ckg)\ (dlh)$	$= B^{-1}$
j	*k*	*l*	*a*	*b*	*c*	*d*	*e*	*f*	*g*	*h*	*i*	$R^9 = (ajgd)\ (bkhe)\ (clif)$	$= Y^{-1}$
k	*l*	*a*	*b*	*c*	*d*	*e*	*f*	*g*	*h*	*i*	*j*	$R^{10} = (akigec)\ (bljhfd)$	$= G^{-1}$
l	*a*	*b*	*c*	*d*	*e*	*f*	*g*	*h*	*i*	*j*	*k*	$R^{11} = (alkjihgfedcb)$	$= R^{-1}$

R, O.	$(1\ 12\text{gon})^2$		24
G.	2	6gons	12
Y.	3	4gons	12
B.	4	3gons	12
V.	6	digons	6
			66

12B1

7 colours.

a	*b*	*c*	*d*	*e*	*f*	*g*	*h*	*i*	*j*	*k*	*l*	$1 = 1$	$= 1$
b	*c*	*d*	*e*	*f*	*a*	*h*	*i*	*j*	*k*	*l*	*g*	$R = (abcdef)(ghijkl)$	$= R$
c	*d*	*e*	*f*	*a*	*b*	*i*	*j*	*k*	*l*	*g*	*h*	$R^2 = (ace)(bdf)(gik)(hjl)$	$= Y$
d	*e*	*f*	*a*	*b*	*c*	*j*	*k*	*l*	*g*	*h*	*i*	$R^3 = (ad)(be)(cf)(gj)(hk)(il)$	$= B$
e	*f*	*a*	*b*	*c*	*d*	*k*	*l*	*g*	*h*	*i*	*j*	$R^4 = (aec)(bfd)(gki)(hlj)$	$= Y^{-1}$
f	*a*	*b*	*c*	*d*	*e*	·*l*	*g*	*h*	*i*	*j*	*k*	$R^5 = (afedcb)(glkjih)$	$= R^{-1}$
g	*h*	*i*	*j*	*k*	*l*	*a*	*b*	*c*	*d*	*e*	*f*	$G = (ag)(bh)(ci)(dj)(ek)(fl)$	$= G$
h	*i*	*j*	*k*	*l*	*g*	*b*	*c*	*d*	*e*	*f*	*a*	$RG = (ahcjel)(bidkfg)$	$= P$
i	*j*	*k*	*l*	*g*	*h*	*c*	*d*	*e*	*f*	*a*	*b*	$R^2G = (aiegck)(bjfhdl)$	$= O$
j	*k*	*l*	*g*	*h*	*i*	*d*	*e*	*f*	*a*	*b*	*c*	$R^3G = (aj)(bk)(cl)(dg)(eh)(fi)$	$= V$
k	*l*	*g*	*h*	*i*	*j*	*e*	*f*	*a*	*b*	*c*	*d*	$R^4G = (akcgei)(bdlhfj)$	$= O^{-1}$
l	*g*	*h*	*i*	*j*	*k*	*f*	*a*	*b*	*c*	*d*	*e*	$R^5G = (alejch)(bgfkdi)$	$= P^{-1}$

R, P, O.	$(2\ 6\text{gons})^3$	36
Y.	4 3gons	12
B, G, V.	$(6\ \text{digons})^3$	18
		66

12B2

9 colours.

a	*b*	*c*	*d*	*e*	*f*	*g*	*h*	*i*	*j*	*k*	*l*	$1 = 1$	$= 1$
b	*c*	*d*	*e*	*f*	*a*	*l*	*g*	*h*	*i*	*j*	*k*	$R = (abcdef)(glkjih)$	$= R$
c	*d*	*e*	*f*	*a*	*b*	*k*	*l*	*g*	*h*	*i*	*j*	$R^2 = (ace)(bdf)(gki)(hlj)$	$= Y$
d	*e*	*f*	*a*	*b*	*c*	*j*	*k*	*l*	*g*	*h*	*i*	$R^3 = (ad)(be)(cf)(gj)(hk)(il)$	$= B$
e	*f*	*a*	*b*	*c*	*d*	*i*	*j*	*k*	*l*	*g*	*h*	$R^4 = (aec)(bfd)(gik)(hjl)$	$= Y^{-1}$
f	*a*	*b*	*c*	*d*	*e*	*h*	*i*	*j*	*k*	*l*	*g*	$R^5 = (afedcb)(ghijkl)$	$= R^{-1}$
g	*h*	*i*	*j*	*k*	*l*	*a*	*b*	*c*	*d*	*e*	*f*	$G = (ag)(bh)(ci)(dj)(ek)(fl)$	$= G$
h	*i*	*j*	*k*	*l*	*g*	*f*	*a*	*b*	*c*	*d*	*e*	$RG = (ah)(bi)(cj)(dk)(el)(fg)$	$= P$
i	*j*	*k*	*l*	*g*	*h*	*e*	*f*	*a*	*b*	*c*	*d*	$R^2G = (ai)(bj)(ck)(dl)(eg)(fh)$	$= O$
j	*k*	*l*	*g*	*h*	*i*	*d*	*e*	*f*	*a*	*b*	*c*	$R^3G = (aj)(bk)(cl)(dg)(eh)(fi)$	$= V$
k	*l*	*g*	*h*	*i*	*j*	*c*	*d*	*e*	*f*	*a*	*b*	$R^4G = (ak)(bl)(cg)(dh)(ei)(fj)$	$= I$
l	*g*	*h*	*i*	*j*	*k*	*b*	*c*	*d*	*e*	*f*	*a*	$R^5G = (al)(bg)(ch)(di)(ej)(fk)$	$= S$

R.	2 6gons	12
Y.	4 3gons	12
B, G, P, O, V, I, S.	$(6\ \text{digons})^7$	42
		66

12B3 — 6 colours.

a	*b*	*c*	*d*	*e*	*f*	*g*	*h*	*i*	*j*	*k*	*l*	$1 = 1$	$= 1$
b	*c*	*d*	*e*	*f*	*a*	*l*	*g*	*h*	*i*	*j*	*k*	$R = (abcdef)(glkjih)$	$= R$
c	*d*	*e*	*f*	*a*	*b*	*k*	*l*	*g*	*h*	*i*	*j*	$R^2 = (ace)(bdf)(gki)(hlj)$	$= B$
d	*e*	*f*	*a*	*b*	*c*	*j*	*k*	*l*	*g*	*h*	*i*	$R^3 = (ad)(be)(cf)(gj)(hk)(il)$	$= Y$
e	*f*	*a*	*b*	*c*	*d*	*i*	*j*	*k*	*l*	*g*	*h*	$R^4 = (aec)(bfd)(gik)(hjl)$	$= B^{-1}$
f	*a*	*b*	*c*	*d*	*e*	*h*	*i*	*j*	*k*	*l*	*g*	$R^5 = (afedcb)(ghijkl)$	$= R^{-1}$
g	*h*	*i*	*j*	*k*	*l*	*d*	*e*	*f*	*a*	*b*	*c*	$G = (agdj)(bhek)(cifl)$	$= G$
h	*i*	*j*	*k*	*l*	*g*	*c*	*d*	*e*	*f*	*a*	*b*	$RG = (ahdk)(biel)(cjfg)$	$= O$
i	*j*	*k*	*l*	*g*	*h*	*b*	*c*	*d*	*e*	*f*	*a*	$R^2G = (aidl)(bjeg)(ckfh)$	$= P$
j	*k*	*l*	*g*	*h*	*i*	*a*	*b*	*c*	*d*	*e*	*f*	$R^3G = (ajdg)(bkeh)(clfi)$	$= G^{-1}$
k	*l*	*g*	*h*	*i*	*j*	*f*	*a*	*b*	*c*	*d*	*e*	$R^4G = (akdh)(blei)(cgfj)$	$= O^{-1}$
l	*g*	*h*	*i*	*j*	*k*	*e*	*f*	*a*	*b*	*c*	*d*	$R^5G = (aldi)(bgej)(chfk)$	$= P^{-1}$

R.	2 6gons	12
G, O, P.	(3 4gons)^3	36
B.	4 3gons	12
Y.	6 digons	6
		66

12B4 — 7 colours.

a	*b*	*c*	*d*	*e*	*f*	*g*	*h*	*i*	*j*	*k*	*l*	$1 = 1$	$= 1$
b	*c*	*a*	*e*	*f*	*d*	*h*	*i*	*g*	*k*	*l*	*j*	$R = (abc)(def)(ghi)(jkl)$	$= R$
c	*a*	*b*	*f*	*d*	*e*	*i*	*g*	*h*	*l*	*j*	*k*	$R^2 = (acb)(dfe)(gih)(jlk)$	$= R^{-1}$
d	*l*	*g*	*a*	*i*	*j*	*c*	*k*	*e*	*f*	*h*	*b*	$RGR^2 = (ad)(bl)(cg)(ei)(fj)(hk)$	$= Y$
e	*j*	*h*	*b*	*g*	*k*	*a*	*l*	*f*	*d*	*i*	*c*	$RG = (aeg)(bjd)(chl)(fki)$	$= B$
f	*k*	*i*	*c*	*h*	*l*	*b*	*j*	*d*	*e*	*g*	*a*	$RGR = (afl)(bkg)(cid)(ehj)$	$= O$
g	*d*	*l*	*j*	*a*	*i*	*e*	*c*	*k*	*b*	*f*	*h*	$GR^2 = (age)(bdj)(clh)(fik)$	$= B^{-1}$
h	*e*	*j*	*k*	*b*	*g*	*f*	*a*	*l*	*c*	*d*	*i*	$G = (ah)(be)(cj)(dk)(fg)(il)$	$= G$
i	*f*	*k*	*l*	*c*	*h*	*d*	*b*	*j*	*a*	*e*	*g*	$GR = (aij)(bfh)(cke)(dlg)$	$= P$
j	*h*	*e*	*g*	*k*	*b*	*l*	*f*	*a*	*i*	*c*	*d*	$R^2G = (aji)(bhf)(cek)(dgl)$	$= P^{-1}$
k	*i*	*f*	*h*	*l*	*c*	*j*	*d*	*b*	*g*	*a*	*e*	$R^2GR = (ak)(bi)(cf)(dh)(el)(gj)$	$= V$
l	*g*	*d*	*i*	*j*	*a*	*k*	*e*	*c*	*h*	*b*	*f*	$R^2GR^2 = (alf)(bgk)(cdi)(ejh)$	$= O^{-1}$

R, B, P, O.	(4 3gons)^4	48
Y, G, V.	(6 digons)^3	18
		66

Extracting from these colourgroups the twocolourbonds contained in them respectively, we have the twocolourbonds shown in the annexed series of figures. I have in each case given the number $4B$, $6A$, &c., of the colourgroup in which the bond is contained, and which colourgroup is given conversely by the twocolourbond. The several points may have letters a, b, c, d, &c., attached to them at pleasure; but as the particular letters are quite immaterial, it seemed to me better to give the several figures without any letters.

$4B$

$6A$

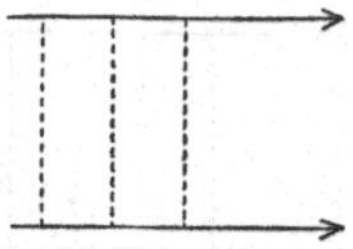

$6B$

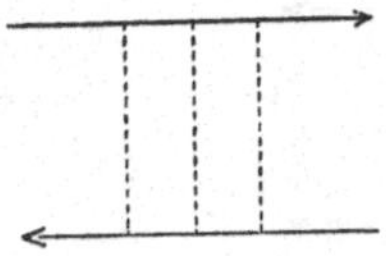

$6B$

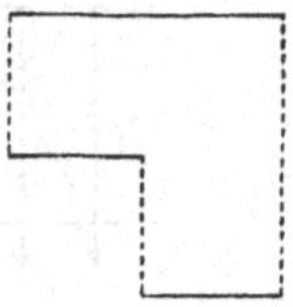

$8B1$

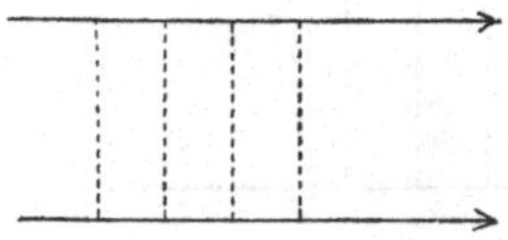

$8B1$

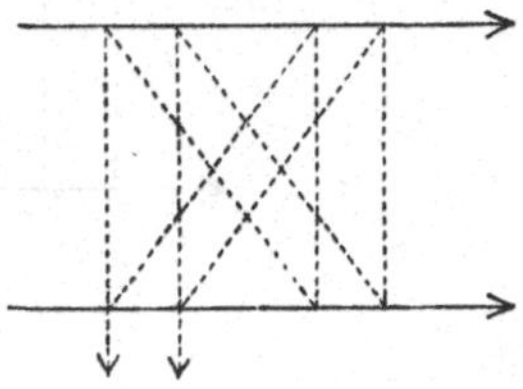

$8B2$

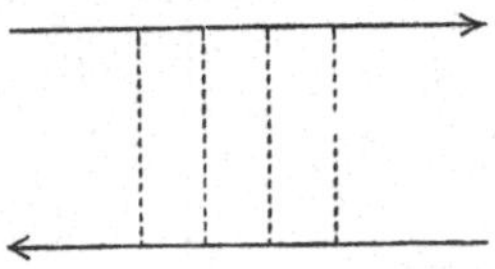

$8B2$

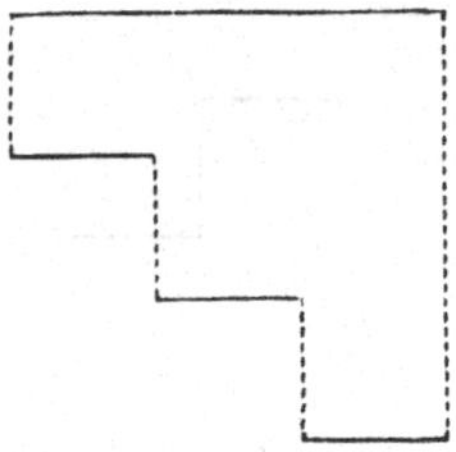

$8B3$

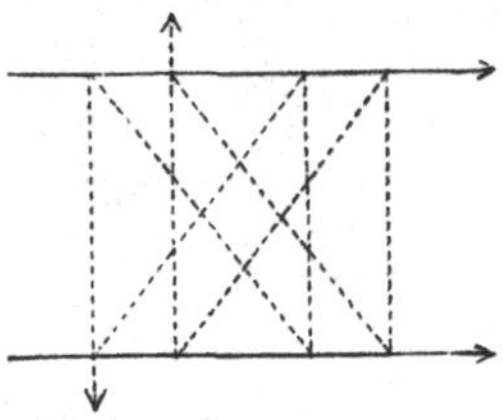

$9B$

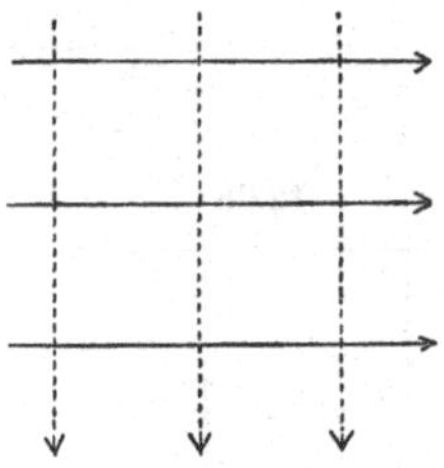

$10A$

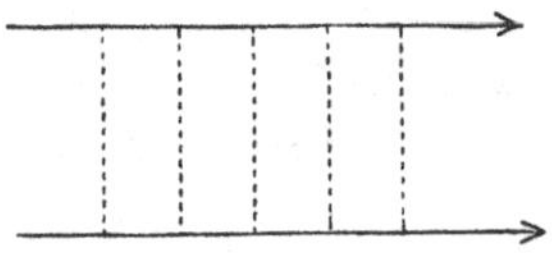

$10B$

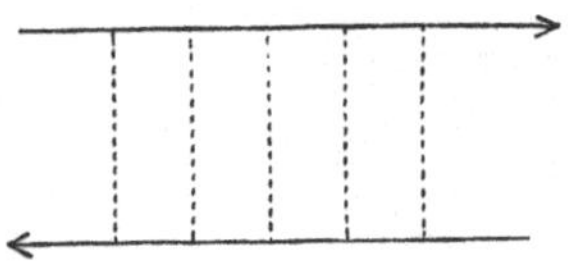

$10B$

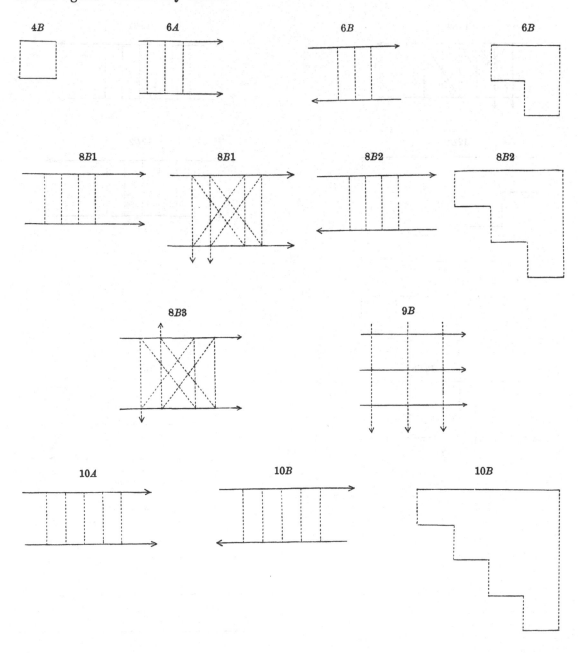

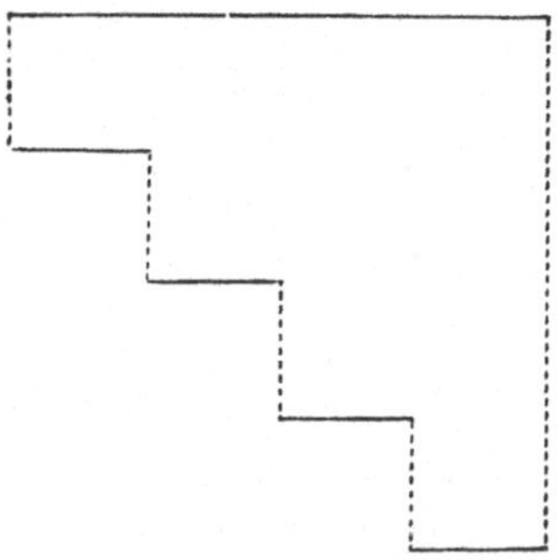

12A

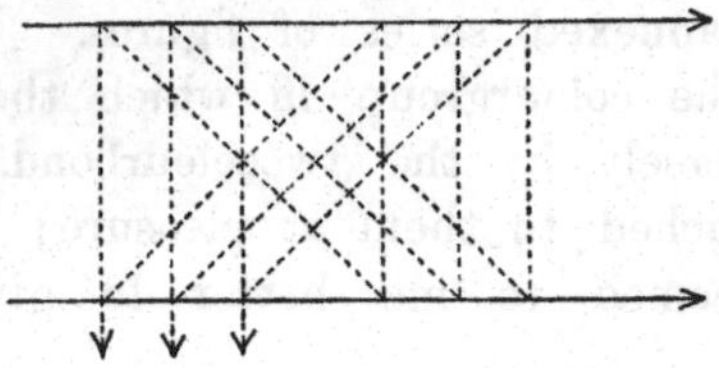

12A

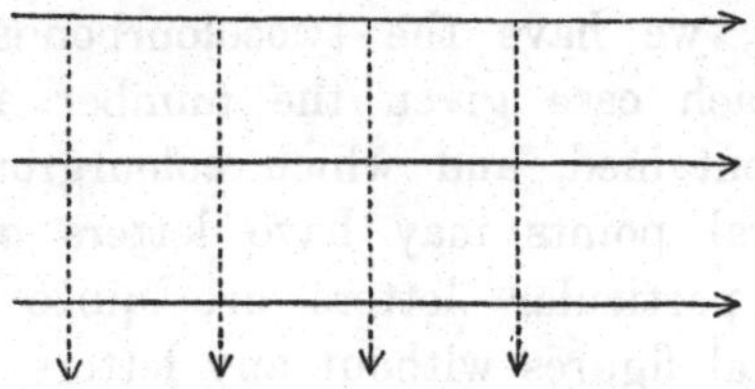

12B1

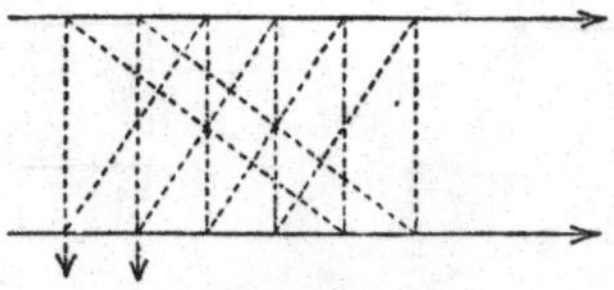

12B1

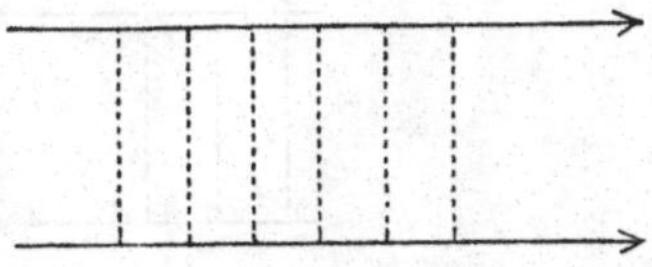

12B2

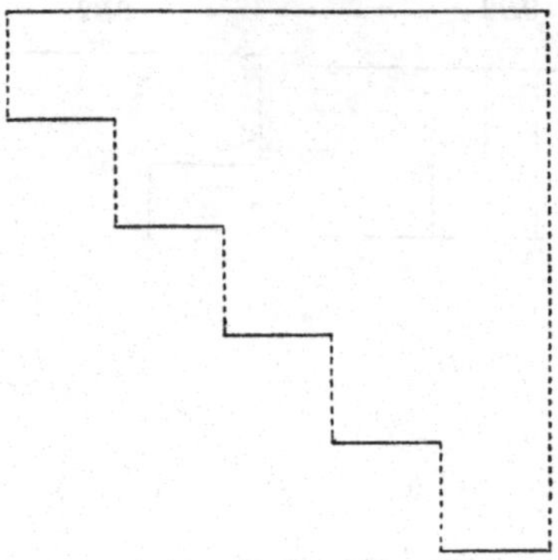

12B2

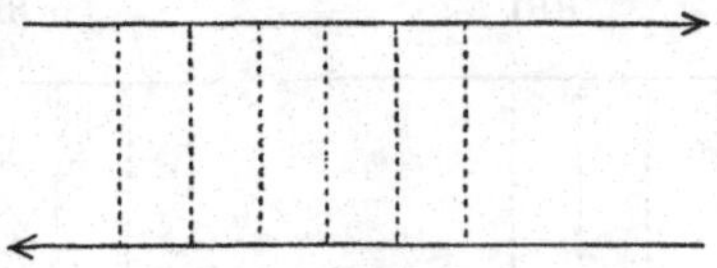

12B3

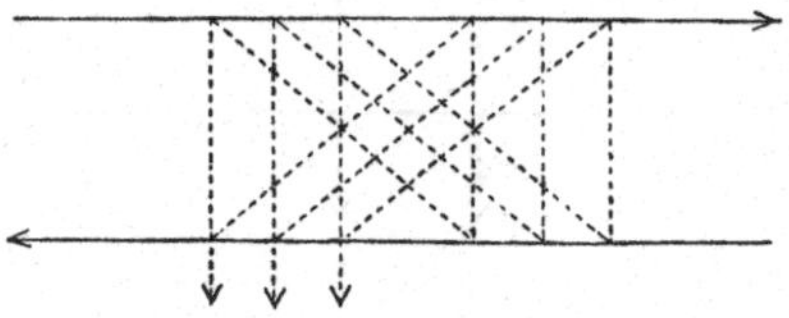

12B3

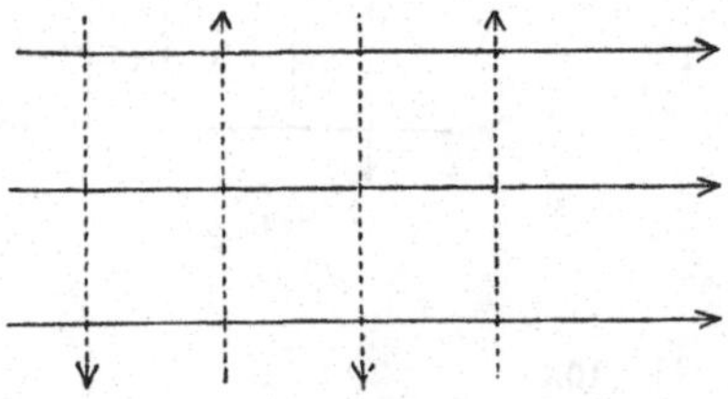

12B4

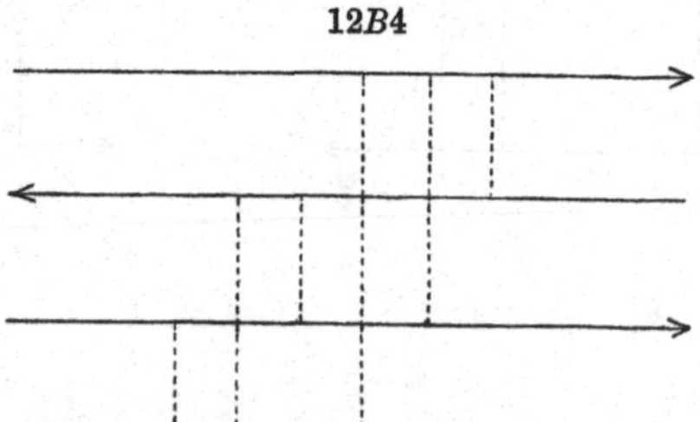

In any one of the foregoing forms of twocolourbond, each point, in its relations to the other points, is indistinguishable from each of the other points. This would seem to be a relation of symmetry equivalent to the before-mentioned condition that each route is of independent effect; and it would moreover seem as if the relation of symmetry were satisfied for each of the following forms:

12 (wrong form).

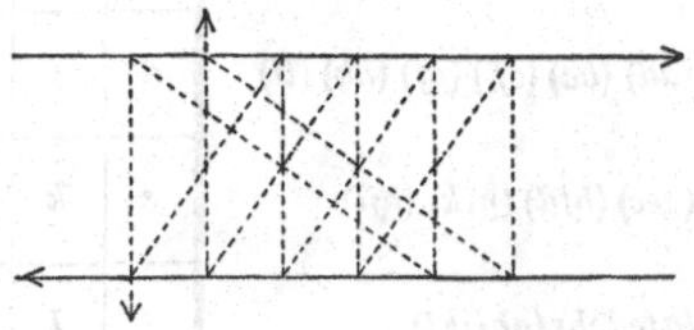

12 (wrong form).

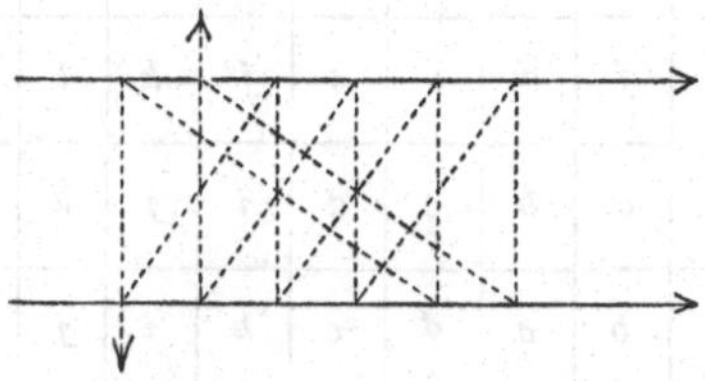

12 (wrong form).

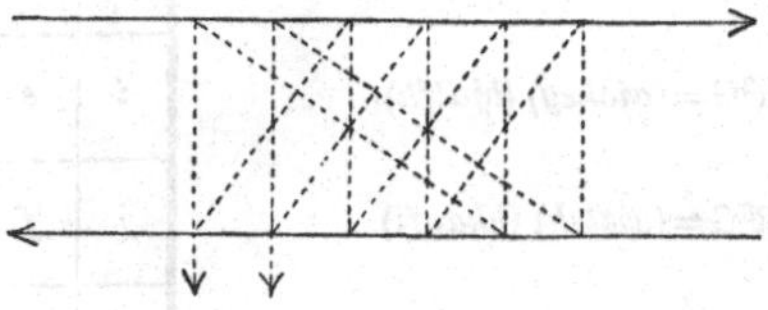

12 (wrong form).

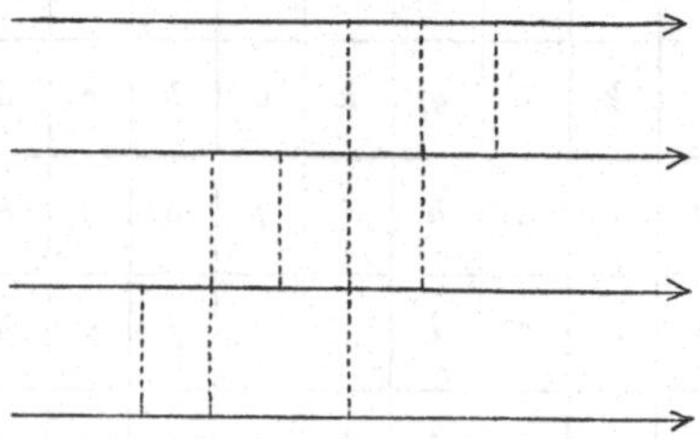

Each of these is, however, a wrong form, not satisfying the condition that each route is of independent effect. As to this, observe that, when the condition is satisfied, there are in all ($\vartheta =$) 12 non-equivalent routes, and there is thus a completely determinate square. When the condition is not satisfied, there are more than this number of non-equivalent routes, and there may very well be ϑ routes giving rise to a latin square, viz. a square each line of which, and also each column of which, contains all the letters, and which thus seems at first sight to represent a substitution-group; but the substitutions, by which each line of the square is derived from itself and the other lines of the square, are not the same as those by which each line is derived from the top line, and thus the square does not represent a group. Thus in one of the above wrong forms, starting from the routes $R = (abcdef)\,(glkjih)$ and $G = (agciek)\,(bhdjfl)$, we have

12 (wrong form).

a	*b*	*c*	*d*	*e*	*f*	*g*	*h*	*i*	*j*	*k*	*l*	$1 = 1$
b	*c*	*d*	*e*	*f*	*a*	*l*	*g*	*h*	*i*	*j*	*k*	$R = (abcdef)(glkjih)$
c	*d*	*e*	*f*	*a*	*b*	*k*	*l*	*g*	*h*	*i*	*j*	$R^2 = (ace)(bdf)(gki)(hlj)$
d	*e*	*f*	*a*	*b*	*c*	*j*	*k*	*l*	*g*	*h*	*i*	$R^3 = (ad)(be)(cf)(gj)(hk)(il)$
e	*f*	*a*	*b*	*c*	*d*	*i*	*j*	*k*	*l*	*g*	*h*	$R^4 = (aec)(bfd)(gik)(hjl)$
f	*a*	*b*	*c*	*d*	*e*	*h*	*i*	*j*	*k*	*l*	*g*	$R^5 = (afedcb)(ghijkl)$
g	*h*	*i*	*j*	*k*	*l*	*c*	*d*	*e*	*f*	*a*	*b*	$G = (agciek)(bhdjfl)$
h	*i*	*j*	*k*	*l*	*g*	*b*	*c*	*d*	*e*	*f*	*a*	$RG = (ahcjel)(bidkfg)$
i	*j*	*k*	*l*	*g*	*h*	*a*	*b*	*c*	*d*	*e*	*f*	$R^2G = (aickeg)(bjdlfh)$
j	*k*	*l*	*g*	*h*	*i*	*f*	*a*	*b*	*c*	*d*	*e*	$R^3G = (ajcleh)(bkdgfi)$
k	*l*	*g*	*h*	*i*	*j*	*e*	*f*	*a*	*b*	*c*	*d*	$R^4G = (akcgei)(bldhfj)$
l	*g*	*h*	*i*	*j*	*k*	*d*	*e*	*f*	*a*	*b*	*c*	$R^5G = (alchej)(bgdifk)$

	G	R^2	G
a	*g*	*k*	*a*
b	*h*	*l*	*b*
c	*i*	*g*	*c*
d	*j*	*h*	*d*
e	*k*	*i*	*e*
f	*l*	*j*	*f*
g	*c*	*e*	*k*
h	*d*	*f*	*l*
i	*e*	*a*	*g*
j	*f*	*b*	*h*
k	*a*	*c*	*i*
l	*b*	*d*	*j*

which is not a group; there is no substitution $G^{-1} = (akeicg)(blfjdh)$. And we see that, in fact, each route is not of independent effect; the route GR^2G leads as shown from the primitive arrangement *abcdefghijkl* to *abcdefklghij*, viz. it is a circuit from each of the points *a*, *b*, *c*, *d*, *e*, *f*, but not from any one of the remaining points *g*, *h*, *i*, *j*, *k*, *l*.

END OF VOL. XII.

CAMBRIDGE: PRINTED BY J. AND C. F. CLAY, AT THE UNIVERSITY PRESS.

Printed in the United States